Environmental Science
Managing Biological & Physical Resources

Environmental Science
Managing Biological & Physical Resources

Michael D. Morgan
University of Wisconsin-Green Bay

Joseph M. Moran
University of Wisconsin-Green Bay

James H. Wiersma
University of Wisconsin-Green Bay

WCB **Wm. C. Brown Publishers**
Dubuque, Iowa • Melbourne, Australia • Oxford, England

Book Team

Editor *Kevin Kane*
Developmental Editor *Margaret J. Kemp*
Production Editor *Reneé A. Menne*
Visuals/Design Consultant *Donna Slade*
Art Editor *Carla Goldhammer*
Photo Editor *Robin Storm*
Permissions Editor *Gail I. Wheatley*

Wm. C. Brown Publishers
A Division of Wm. C. Brown Communications, Inc.

Vice President and General Manager *Beverly Kolz*
National Sales Manager *Vincent R. Di Blasi*
Director of Marketing *John W. Calhoun*
Marketing Manager *Carol J. Mills*
Advertising Manager *Amy Schmitz*
Director of Production *Colleen A. Yonda*
Manager of Visuals and Design *Faye M. Schilling*

Design Manager *Jac Tilton*
Art Manager *Janice Roerig*
Publishing Services Manager *Karen J. Slaght*
Permissions/Records Manager *Connie Allendorf*

Wm. C. Brown Communications, Inc.

President and Chief Executive Officer *G. Franklin Lewis*
Corporate Vice President, President of WCB Manufacturing *Roger Meyer*
Vice President and Chief Financial Officer *Robert Chesterman*

Cover Photos:

Environmental Science: Managing Biological and Physical Resources:
© Wm. C. Brown Communications, Inc./Doug Sherman
A roaring mountain stream, trees bursting forth in blossom, and plants emerging on the forest floor attest to the dynamism of the environment.

Volume I: Environmental Science: Economics and Ecology:
© Dale Jorgenson/Tom Stack and Associates
The combination of abundant rainfall, ceaseless spray from the falls, and the shade of an overhanging cliff promotes the luxuriant growth of this community of ferns.

Volume II: Environmental Science: Managing Biological Resources:
© Terry Donnelly/Tom Stack and Associates
Around the world, humans exploit a variety of biological resources. In these Montana foothills, upper slopes are managed for livestock grazing while hay for winter forage is grown on the valley floor.

Volume III: Environmental Science: Managing Physical Resources:
© Wm. C. Brown Communications, Inc./Doug Sherman
In view of the adverse environmental impacts of human exploitation of nonrenewable physical resources, there is increasing interest in less disruptive and sustainable alternatives such as this wind farm.

Copyedited by Nick Murray

The credits section for this book begins on page C1 and is considered an extension of the copyright page.

Library of Congress Catalog Card Number: **Life:** 91–77992

ISBN **Environmental Science** Casebound, recycled interior stock: 0–697–10801–5
ISBN **Environmental Science** Paper binding, recycled interior stock: 0–697–16305–9
ISBN **Volume I: Economics and Ecology** Paper binding, recycled interior stock: 0–697–16307–5
ISBN **Volume II: Managing Biological Resources** Paper binding, recycled interior stock: 0–697–16308–3
ISBN **Volume III: Managing Physical Resources** Paper binding, recycled interior stock: 0–697–16309–1
ISBN **Environmental Science** Boxed set, recycled interior stock: 0–697–16306–7

Printed in the United States of America by Wm. C. Brown Communications, Inc., 2460 Kerper Boulevard, Dubuque, IA 52001

10 9 8 7 6 5 4 3 2 1

Dedicated to our patient wives: Gloria, Jennifer, and Ruth

"A personal library is a lifelong source of enrichment and distinction. Consider this book an investment in your future and add it to your personal library."

Publisher's Note

Binding Option	Description	ISBN
Environmental Science, casebound	Complete Text	10801
Environmental Science, paperbound	Complete Text	16305
Volume 1, paperbound *Principles of Ecology and Human Population*	Volume 1 features parts I–III, or the first seven chapters of the text, covering the scientific method, basic ecological concepts, and human population dynamics and regulation. It is available at a fraction of the price of the full-length text.	16307
Volume 2, paperbound *Biological Resources*	Volume 2 features Part IV, or chapters 8–12, which emphasize biological aspects of environmental resources and quality. The topics represent the most current coverage of food production, pest management strategies, public land management, and biodiversity management. The text is complete, with the inclusion of the Epilogue, Glossary, Appendix, and Index. It is available for a fraction of the price of the full-length text.	16308
Volume 3, paperbound *Physical Resources*	Volume 3 features Part V, or chapters 13–21, which places a greater emphasis on environmental studies relevant to the physical sciences. The topics represent the most current coverage of air, water, and mineral resources as well as management and attainment of future energy needs. The text is complete, with Epilogue, Glossary, Appendix and Index.	16309
Boxed set, paperbound	The boxed set features a complete set of volumes 1, 2, and 3.	16306

All text options, ancillaries and advertisement pieces for this title are printed exclusively on recycled paper.

TO THE STUDENT: A study guide for this textbook is available through your college bookstore under the title *Student Study Guide to accompany Environmental Science: Managing Biological and Physical Resources* by Michael Morgan and Joseph Moran. This study guide will help you master course material by acting as a tutorial, review, and guide to testing proficiency. If you don't see a copy of it on your bookstore's shelves, ask the bookstore manager to order a copy for you.

Brief Contents

Contents

Chapter

11

Managing for Biodiversity 215

Chapter

12

Environmental Toxicology 237

Part

V

Managing Physical Resources and Environmental Quality 259

Chapter

13

The Water Cycle and Freshwater Resources 260

Chapter

14

Water Quality 282

Chapter 15

The Atmosphere: Weather and Climate 314

Chapter 16

Air Quality 338

Chapter 17

Rock and Mineral Resources 366

Chapter 18

Current Energy Supply and Demand 389

Chapter 19

Meeting Future Energy Demands 412

Chapter 20

Waste Management and Resource Recovery 435

Chapter 21

Land-Use Conflicts 458

Epilogue

Toward a Sustainable Environment E1

List of Special Topics

List of Issues

Guided Tour through the *Environmental Science* Learning System

Chapter Outlines

Chapter outlines offer the student a quick overview of key topics and how they are organized within the chapter.

Chapter 3

Ecosystems
Energy Flow and the Cycling of Materials

This scene of a feeding polar bear illustrates one step in the flow of energy within an ecosystem.
© Jack Stein Grove/Tom Stack & Associates

30

Boldfaced Terms

Important terms are boldfaced within the narrative when they are initially introduced. They are also defined within the context of the sentence and are again listed at the end of the chapter and in the glossary.

Figure 10.6   A close clustering of large, clearcut plots can lead to a loss of species diversity and accelerated soil erosion.
© Jack Swenson/Tom Stack & Associates

Figure 10.7   Partial cutting of a forest stand leaves behind trees that serve as seed sources.
© Milton H. Tierney, Jr./Visuals Unlimited

The Forest Service is now officially planning to make the transition from a timber/grazing/watershed/mining orientation to one with more emphasis on fisheries, wildlife, recreation, and environmentally friendly resource use. However, transferring such strategic planning to actual forest management will be most challenging. These new initiatives require significant additional funding, which will be difficult to obtain from Congress. The forest-products industry remains a powerful economic and political force, and many people within the Forest Service remain strongly committed to timber production. Furthermore, if timber harvests are to be reduced, retraining should be made available to those in the timber industry who will lose their jobs. There is a consensus that forestry practices must be more ecologically based and that the Forest Service must achieve a better balance in its management of the diverse resources on NFS lands. However, progress will be slow as research and management evolve to enhance the compatibility of multiple resource uses and to ensure a long-term, sustained supply of resources from National Forest Service lands.

National Rangelands

Rangelands are essentially unsuited for rain-fed crop cultivation or forestry. Grasses or shrubs such as sagebrush (fig. 10.8) dominate these arid and semiarid tracts. Essentially all of the rangeland in the United States is west of the Mississippi River. About one-third of it is publicly owned, most of that land is managed by the Forest Service or the Bureau of Land Management (BLM).

Federal rangelands are primarily managed for livestock production. In 1989, ranchers leased nearly 108 million hectares (267 million acres) of federal land for grazing their livestock. Although grazing has long been the traditional use of federal rangelands, management of public rangelands primarily for this single purpose is coming under increasing fire from conservation groups.

closely clustered clear-cut plots on steep slopes are subject to severe soil erosion and stream sedimentation.

After clearing, a clear-cut plot may reseed naturally. Increasingly, however, the plot is planted with seedlings selected for high growth rates. Clear-cutting thus eventually leads to establishment of an even-aged stand, often composed of only one or two species of trees. Such plots are more vulnerable to pest outbreaks, fires, and perhaps even to climate change.

Partial cutting covers a continuum of activities that range from selectively cutting only a few trees (by desired size and species, for example) to harvesting most but not all (typically 85% to 90%) of the trees on a site (fig. 10.7). In the latter situation, the remaining trees serve as seed sources and as refugia from which species can invade the surrounding developing forest as it becomes established and matures. Aggregating partially cut sites reduces habitat fragmentation and thus helps maintain species diversity. Furthermore, soil erosion is usually less of a problem than on clear-cut sites.

196   **Managing Biological Resources and Environmental Quality**

Special Topic 5.1

Mutualism: Species Helping Each Other

Mutualism—an interaction between species that is beneficial to each—occurs widely and is important to many populations. Such an interaction can benefit a species in several ways, including (1) nutrition, either through the digestion of food for the partner or the synthesis of vitamins or proteins for the partner; (2) protection, either from enemies or from environmental change, or (3) transport, such as the dispersal of pollen or the movement of materials from unsuitable to suitable environments.

Interactions between ruminants (e.g., cows and sheep) and the microorganisms that inhabit their stomachs are good examples of nutritional mutualism. A ruminant's diet consists mainly of cellulose, but like most animals, ruminants do not produce the necessary enzymes to break down cellulose. Unlike most animals, however, they have microorganisms in their stomachs that can digest cellulose. Furthermore, these microorganisms can also synthesize proteins from ammonia and urea that are present in the stomach. Some of these microorganisms are subsequently digested, thus providing the ruminant with protein. They can also synthesize all of the B-complex vitamins and vitamin K. Hence microbial activity enables ruminants to gain additional energy and nutrients. The microorganisms, in turn, benefit from free food and good housing—the warm, wet interior of the stomach provides a favorable environment for their growth and reproduction.

In their mutualistic relationship, these red ants obtain nourishment from aphids and protect the aphids from predators. © Hans Pfletschinger/Peter Arnold, Inc.

The head of this bronzy hermit hummingbird is about to be dusted by pollen as it feeds upon nectar from a passion flower. © Michael Fogden/Animals Animals

We, too, have a complex assemblage of mutualistic microorganisms residing in our digestive systems. Unlike those of ruminants, our microbes cannot digest cellulose, but they do synthesize the B-complex vitamins and vitamin K.

Many organisms provide protection from enemies and in turn may be provided with food, protection, housing, or all of these. In East Africa, for instance, certain ants live in the swollen thorns of acacia trees. The thorns provide an ideal shelter and a balanced and almost complete diet for all stages of ant development. The ants, in turn, serve as protectors. As soon as a branch is touched by a browsing animal or some other organism, the ants feel the vibrations and respond by pouring out of their holes in the thorns and racing toward the ends of the branches. Along the way, they release a repulsive odor that usually drives the enemy away.

Ants are also involved in protective mutualisms with sap-feeding insects, such as aphids. Plant sap often contains too much sugar for the aphids to utilize. Consequently, aphids extrude a sugary solution, called *honeydew*, which serves the ants as a major source of food. In return, ants protect the aphids by attacking and driving away predators (left photo).

Flowering plants and their pollinators exemplify transport mutualism. A wide variety of animals (e.g., butterflies, beetles, hummingbirds, and bats) visit flowers to obtain food (right photo). Flowers produce pollen, which is rich in protein, and nectar, a nutritious sugary fluid. As the visitors feed, they brush against the male structures (which produce pollen), and their bodies are dusted with pollen. When they visit another flower, the pollen from the first flower is transferred inadvertently to the female structures of the second flower. This pollination process may then lead to fertilization and the production of seeds and fruits. Several advantages accrue to flowering plants from pollination by animals. Because the animals move directly from one flower to another, less pollen is wasted than in pollination that depends on the random movements of the wind. Furthermore, when plants are relatively few and widely scattered, the food-seeking activities of pollinators make pollination more likely than it is with wind pollination.

How do mutualistic interactions form? Are the mutualistic microbes that live in us the descendants of microbes that were once intestinal parasites in our ancestors? Are mutualistic interactions stable, or do they frequently break down? For example, instead of protecting their partners, do aphid-tending ants sometimes turn on the aphids and devour them? Ecologists still have much to learn about these fascinating population interactions.

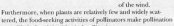

Special Topics

Throughout the text, special topic boxes highlight additional topics at a more extended level. Students will find them a lively and interesting feature as they investigate the process of science.

Fire Management Policies in National Parks

The year 1988 brought national attention to the National Park Service in general and to Yellowstone National Park in particular. The spectacular Yellowstone fires that summer produced such headlines as "A Legacy in Ashes," "Yellowstone Destroyed," "Valuable Timber Lost," and "Survival of Yellowstone in Jeopardy." These headlines along with sensational video footage of wildfires roaring through Yellowstone gave the public the erroneous impression that the entire park had been reduced to cinders. Such media exposure also reinforced the popular notion that all fires are destructive. Because Yellowstone National Park had a fire management policy that under certain conditions allowed lightning-started fires to burn out on their own, the Park Service was blamed for the conflagration. In fact, many of the major fires were human-caused, and some originated on adjacent U.S. Forest Service lands (fig. 10.18). Nevertheless, federal, state, and local politicians, business people of the gateway towns on the park's perimeter, and some members of the media severely criticized the Park Service for its fire management policy. Let us evaluate the ecological basis for this policy.

The legislative mandate of the National Park Service is to maintain, as near as possible, a primitive ecological situation. Ecologists have known for some time that naturally caused fires are among the factors that have shaped the evolution of vegetation and wildlife for millennia. The aftermath of the Yellowstone fires of 1988 supports this perspective.

Contrary to popular opinion, less than 1% of the park experienced the high-intensity fires that left little more than ashes. In fact, as the fires progressed, they produced in their wake a mosaic of burned, partially burned, and nonburned areas. As a result, new habitats were created that will follow various pathways of ecological succession. The fires increased the heterogeneity (patchiness) of the landscape, which sustains greater species diversity.

Plant growth began almost immediately after the fires. Less than 1% of the soil was subjected to heat extreme enough to penetrate more than 2.5 centimeters (1 inch). Thus, the fires left most seeds unharmed, and many germinated when soil moisture conditions became favorable. Furthermore, the fires released billions of seeds of the lodgepole pine, a species that provides nearly 80% of the forest cover in Yellowstone. Lodgepole pines produce two types of cones. Some shed seeds when they reach maturity. Others are sealed with a resin coating and will not open until they are exposed to high temperatures from a fire. Hence, they may remain unopened for years, but they release millions of seeds within a few days after a fire. Following the 1988 fires, densities of lodgepole pine seeds in some areas of the park ranged from 50,000 to 100,000 seeds per acre. While many of these seeds nourished birds, mice,

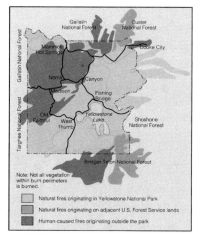

Figure 10.18 Extent of natural and human-caused fires in the Greater Yellowstone Area in 1988. Not all vegetation within burn parameters was burned.
Source: *Yellowstone Fire, 1988*, National Park Service.

and squirrels, some escaped predation, germinated, and will eventually produce forests that are much like the ones that burned. Other plants, such as willows and grasses, quickly resprouted from underground structures that were not harmed by the fires. Thus, within only one or two growing seasons, the charred forest floor was replaced with herbaceous plants, shrubs, and the seedlings of tomorrow's forests (fig. 10.19).

Rather than destroying the vegetation of the park forever, the fires actually spurred a renewal of life in Yellowstone. Prior to the fires, plant growth had slowed considerably on some sites where the soils are inherently infertile. The relatively dry and cool summer weather in Yellowstone, which slows decomposer activity, also contributes to low soil fertility. Without fire, most of the nutrients are tied up in living and dead vegetation. The fires of 1988 triggered a resurgence of plant growth because they released plant nutrients in the form of ash and because they exposed the previously shaded forest floor to the sun. Herbivores flourished after the

(a)   (b)

Figure 5.17 Differences in the structural adaptations of (*a*) the hairy woodpecker and (*b*) the downy woodpecker enable them to exploit somewhat different food sources. This reduces competition between the two species.
(*a*) © John Gerlach/Animals Animals. (*b*) © Rod Planck/Photo Researchers, Inc.

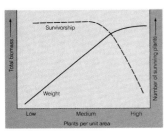

Figure 5.18 As plant density increases, total biomass production increases until a maximum production level is reached. Increasing plant density also eventually results in lowered survivorship.

that were previously occupied by bluebunch wheatgrass. Although ranchers prefer to graze cattle on wheatgrass, it has been almost impossible to reestablish wheatgrass, even where grazing has been eliminated.

Field and laboratory experiments indicate that cheatgrass can exclude wheatgrass for several reasons. The seeds of both species germinate at approximately the same time, during the fall, when the soil is moist. Then during the winter, wheatgrass becomes dormant, while cheatgrass continues to grow. As a result, the extensive root system of cheatgrass grows deeper and gains control of the soil before the wheatgrass seedlings can become better established. Also, cheatgrass matures four to six weeks earlier than wheatgrass, so it withdraws most of the water from the soil before the wheatgrass has a chance to tap it. During the typically hot and dry summers, wheatgrass seedlings usually succumb to severe soil moisture deficits. Hence in places where summers are relatively dry, cheatgrass excludes wheatgrass by out-competing it for soil moisture. In areas where plentiful rain falls in the summer,

A wide array of beautiful photos and informative illustrations are featured throughout the text, including numerous graphs that convey analytical information in an easy-to-read format.

Table 8.1	Sources and deficiency symptoms of some important vitamins	
Vitamin	**Major Sources**	**Deficiency Symptoms**
A	egg yolk, green or yellow vegetables, fruits, liver, butter	night blindness; dry, flaky skin
B_1 (thiamine)	pork, whole grains	beriberi: fatigue, heart failure, indigestion, nerve damage
B_2 (riboflavin)	milk, eggs, liver, whole grains	inflammation and cracked skin, including lips; swollen tongue
B_6 (niacin)	whole grains, liver, and other meats	pellagra: cracked skin, digestive disorders, mental changes
Folic acid	liver, leafy vegetables	anemia, digestive disorders
C	citrus fruits, tomatoes, green leafy vegetables, potatoes	scurvy: bleeding of gums and skin, slow healing of wounds
D	fish oils, liver, fortified milk and other dairy products, sunlight on skin	rickets: defective bone formation
K	synthesis by intestinal bacteria, leafy vegetables	bleeding, internal hemorrhaging

Figure 8.6 Children in Nigeria suffering from a form of protein-calorie malnutrition called *kwashiorkor*.
© Dourdin/Photo Researchers, Inc.

Protein malnutrition is perhaps the greatest single cause of human disease in the world. **Kwashiorkor** (a form of protein malnutrition) occurs in many less-developed nations where young children receive insufficient protein to support their rapidly growing bodies. Symptoms of this devastating illness include lethargy, depressed mental abilities, failure to grow, hair loss, and swelling caused by accumulation of fluids beneath the skin (fig. 8.6). Kwashiorkor commonly develops when a child is switched from highly nutritious mother's milk to a diet of mostly cereals or roots, such as sweet potato and cassava—foods rich in carbohydrates, but poor in essential amino acids. Fortunately, permanent retardation can be avoided if a child is put on a balanced diet soon enough.

To maintain good health, humans also require about twenty vitamins, most of which the body's cells cannot manufacture. Deficiencies of each vitamin are associated with specific metabolic effects. Table 8.1 lists some of the essential vitamins, their major food sources, and associated deficiency symptoms. Vitamin A deficiency, for example, results in dry, flaky skin that is susceptible to invasion by microorganisms. Vitamin A deficiency also reduces the ability of the eye to respond to light. Mild deficiencies usually lead to an inability to see at night (night blindness). In extreme cases, blindness may occur. Vitamin A is also known as an anti-infection vitamin. People with insufficient Vitamin A are more prone to infections in the eyes, kidneys, or respiratory tract. Severe deficiencies in children dramatically increase deaths due to diarrhea, measles, and other maladies.

Recall from chapter 3 that we also require a variety of minerals for proper nutrition. Among these are iron, a component of hemoglobin, which carries oxygen in the blood; calcium, a component of bones; and potassium, required for transmission of nerve impulses and for muscle contraction. Table 8.2 lists these and other minerals and their sources. Iodine is an example of an essential mineral that is lacking in the diet of some regions. Iodine is a component of the hormone thyroxin, which is produced by the thyroid gland. Without sufficient iodine, thyroxin production declines, and the thyroid gland, located in the neck, swells to form what is known as a *goiter*. Because thyroxin stimulates metabolism, other symptoms of low thyroxin production include mental and muscular sluggishness and extreme lethargy. A victim often sleeps 14-16 hours a day. In extreme cases, people develop deafness and arteriosclerosis (hardening of the arteries). Goiters and associated symptoms are common in mountainous regions of Asia, Africa, and Latin America, where

152   **Managing Biological Resources and Environmental Quality**

Molds that grow on seeds and fruits of food plants synthesize a variety of natural toxins, including carcinogens, that apparently help the molds to survive, and some of these carcinogens find their way into food. Aflatoxin, one of the most potent carcinogens in laboratory rats, is formed by molds that grow on corn, wheat, and nuts and thus enters the human food chain. For example, peanut butter marketed in the United States contains on average about 2.0 ppb aflatoxin. Because of their potency, mold toxins may be one of the most important sources of carcinogens in our food supply. The concern is even greater in less-developed nations, where poor storage facilities and warm, humid climates greatly enhance mold growth. In more-developed nations, modern storage facilities that use fungicides greatly reduce mold contamination.

From our discussion of risk assessment, it is clear that we still have much to learn about exposure to chemicals, especially carcinogens, mutagens, and teratogens. Research suggests that for society as a whole, pollution only contributes minimally to the cancer hazard compared to background levels. This does not mean that we should be complacent about pollution, but it does indicate that we need a more equitable balance between the fear of exposure and the actual risk. Currently, the possibility of exposure to even minute quantities of almost any chemical is stringently controlled by regulations that make compliance costly for industry. In many cases, these costs provide society with no tangible benefits. On the other hand, there still remain many occupations in the United States where lifetime exposure to the chemicals of the trade can result in permanent illness and a shortened life span.

Conclusions

We have examined human exposure to chemicals in an effort to understand the risks we face from continual exposure to synthetic and naturally occurring toxins. Some species are significantly more sensitive to toxins than we are, while other are more tolerant. For all organisms, we know that the severity of symptoms increases as the dose of noncarcinogenic toxins increases. The sensitivity of organisms varies because species differ in their ability to detoxify toxins.

When risks from activities such as driving a car, jaywalking, or sunbathing are compared to risks from chemical exposure, we see that today's society perceives risks from many toxins as being much higher that they actually are. The result has been strict regulation of activities that involve exposure to toxins. We will apply the fundamental concepts of toxicology again when we examine the impact of water and air pollutants (chapters 14 and 16) and the management of hazardous waste (chapter 20).

Key Terms

toxicology	oncogene
toxins	promoter
heavy metals	suppressor gene
bioaccumulation	mutagen
acute exposure	mutation
chronic exposure	teratogen
toxicity	detoxification mechanisms
enzymes	metabolites
no-observed-effect level (NOEL)	risk
LD_{50}	individual risk
LC_{50}	societal risk
carcinogen	risk assessment
initiator	

Summary Statements

Toxins enter organisms via four possible routes: inhalation, absorption, ingestion, and injection. Inhalation is the most frequent route of human exposure to toxins.

Normally, we ingest only minor amounts of toxins. The amounts we ingest from animals and plants contaminated with toxins through bioaccumulation are exceptions.

Short-term, or acute, exposure to toxins involves exposure to high levels of toxins with almost immediate symptoms. Long-term, or chronic, exposure to toxins involves exposure to low levels of toxins over lengthy periods, usually with long-delayed symptoms and illnesses.

The response to most toxins is dose-dependent; higher doses produce more severe symptoms.

The LD_{50} value, the dose required to kill 50% of a test population, is one way to express the toxicity of a substance.

A carcinogen is a substance or radiation that causes cancer. Carcinogens include initiators and promoters that act on DNA.

Mutagens are substances that cause heritable changes in genes.

Teratogens cause defects between conception and birth.

The body defends against toxins by excreting or metabolizing them. The liver is the primary organ for metabolizing toxins.

Risk is a measure of the likelihood of something going wrong. People vary in their response to risk; typically, they underestimate the risk associated with familiar activities and overestimate the risk associated with unfamiliar activities or highly publicized risks.

Toxicological data are used to determine the allowable levels (risk) of exposure of humans to toxins.

Key Terms

A list of key terms for each chapter will help students identify and review important terms.

Summary Statements

Summary statements offer students a synopsis of each chapter's key topics for quick study reference and for emphasis of important concepts.

Conclusions

Every chapter ends with statements that reiterate key concepts and present corollaries that are relevant to students' lives.

Expanded use of biofuels as an energy source is possible, but large-scale expansion would require tremendous amounts of land, which, in turn, would aggravate environmental problems.

Wind generation of electricity by moderate-sized windmills is economically competitive with conventional electric generation methods but is unreliable because winds are highly variable.

The United States has used approximately 42% of its favorable hydroelectric dam construction sites. Construction of new dams is highly controversial because dams cause social and environmental disruptions and are expensive to build.

In some regions it is possible to extract energy from geothermal resources, particularly for generating electricity. Some favorable sites cannot or should not be used because extraction of energy might jeopardize geothermal features in public parks. Geothermal sites produce pollutants that must be dealt with.

Nuclear fusion yields enormous quantities of energy. Development of nuclear fusion technology that will permit sustained generation of electricity is technologically challenging, and thus, extremely expensive. Its ultimate development is still highly questionable.

The mix of fossil fuels, nuclear fuels, and renewable resources will change somewhat by the year 2010, with coal and renewable sources picking up most of the increase (25 quads) in demand.

Questions for Review

1. List the top three energy-consuming activities in the average household.
2. What measures are some utilities taking to reduce the demand for electricity in their service areas?
3. What passive features can be incorporated into a new home to cut its energy consumption? Which of these features could be incorporated into existing buildings?
4. How much more efficient are the new screw-in-fluorescent bulbs than ordinary incandescent bulbs?
5. List six ways to conserve energy in your home or apartment. Rank them in order of pay-back time and the likelihood of implementation.
6. Why is fuel efficiency in automobiles of primary concern? What are some of the factors that keep fuel efficiency from reaching its full potential?
7. How can a community increase the efficiency of its transportation network? How could you cut the amount of energy you use for personal transportation?
8. Cite reasons why car and van pools are not used more widely. Suggest some ways to overcome some of the barriers to their use.
9. Give two examples of passive and active solar collection systems.
10. Use figure 19.6 to determine the seasonal variation in the amount of solar radiation in your region. What are some of the more common methods used in your region to take advantage of solar energy?
11. What are photovoltaic cells? Why are they considered to be an ideal solar energy collection device. What are some of their limitations?

12. How are biofuels used to supplement traditional energy sources? Can the use of biofuels be expanded?
13. Identify the advantages and disadvantages of harvesting the sun's energy using the following methods: solar collectors, solar photovoltaic cells, windmills, hydroelectric dams, and biofuels. Which systems work best for homes? Which systems work best for large-scale applications?
14. Summarize the arguments of those who favor and those who oppose expansion of hydroelectric power in the United States.
15. Describe why geothermal energy should be considered a nonrenewable resource.
16. Peak electrical demand in most communities occurs at about 4 P.M. each day. On an annual basis, peak demand occurs during the summer. What are the advantages of reducing peak demand? Suggest some policies that would lower peak demand.
17. More efficient energy use during the last decade is the result of substantially higher energy prices. If energy costs continue to rise, will it be possible to conserve as much energy as was conserved in the 1980s?
18. Distinguish between nuclear fission and nuclear fusion.
19. What efforts are underway to make nuclear fusion a viable technology?
20. What changes in the energy delivery system are possible for the United States over the next two decades? Are these changes likely to be controversial?

Projects

1. Locate a solar home in your community, and obtain some data on its performance. Report on changes that might make it even more energy-efficient.
2. Do you think mandatory performance standards imposed by federal or local governments for vehicles, appliances, and new buildings are a good idea? Explain why you agree or disagree with them.
3. Contact your local electric utility and determine which energy conservation measures they are attempting to implement in your community. What types of incentives are they using to reach their energy conservation goals?

Selected Readings

Abelson, P. H. "Energy Futures." *American Scientist* 75 (1987):584-93. A discussion focusing on why energy policies must change if the United States is to avoid severe shortages of liquid fuels.

Conn, R. W., V. A. Chuyanov, N. Inoue, and D. R. Sweetman. "The International Thermonuclear Experimental Reactor." *Scientific American* 266 (April 1992):102-10. A discussion of efforts underway to build a single large experimental nuclear fusion reactor.

Dostrovsky, I. "Chemical Fuels from the Sun" *Scientific American* 265 (Dec. 1991):102-7. A description of fuels that could be produced using sunlight and then transported to the point of use.

Echeverria, J. et al. *Rivers at Risk: The Concerned Citizen's Guide to Hydropower.* Covelo, Calif.: Island Press, 1989. Discusses how to become involved when a hydroelectric generating dam is proposed for a scenic area.

Projects

To encourage active participation in environmental awareness, students can choose one or more of the projects that appear at the end of most chapters.

Selected Readings

For further reference on specific topics, students can use the list of selected readings from current journals and texts.

Questions for Review

Each chapter ends with Questions for Review to help students test their mastery of terms and concepts and to promote critical thinking and writing skills.

Preface

This book introduces the college nonscience major to the principles and issues of environmental science. It was written for a one-semester course, but is comprehensive enough for most two-semester courses. In it, we demonstrate how environmental issues relate to the reader's everyday life, illustrate scientific principles by building on worldwide and familiar examples, and encourage students to become personally involved with solving environmental problems. So while we cover large-scale or global issues, we encourage students to act locally as well.

Over the past twenty years, we have coauthored five textbooks on the environment. Each of us brings to this work the experience of teaching undergraduates for more than twenty-two years and the complementary perspectives of different scientific backgrounds. One of us is an ecologist, one a meteorologist, and the other an environmental chemist. Through team-teaching and collaboration, we have learned much from each other, giving our writing a genuine sense of interdisciplinary instruction, a writing style nonscience students with diverse backgrounds should find appealing.

The study of environmental science is dynamic, and we have attempted to keep pace with rapid advances in the field. This text reflects the continued maturation of environmental science, which is characterized by an improved scientific understanding of how environmental systems work and a better ability to predict the consequences of human and nonhuman disruptions of the environment. In addition, this book responds to the renewed public interest in environmental quality by providing a solid scientific basis for understanding and demonstrating how various environmental issues are linked.

Improper disposal of toxic and hazardous wastes, deterioration of groundwater quality, the buildup of greenhouse gases and the resulting possibility of global climatic change, and the thinning of the stratospheric ozone layer are among the major environmental concerns we discuss. The linkage between human population growth and environmental deterioration is appropriately emphasized as well. Increased demand for resources, which threatens the quality and availability of adequate fresh water, food, energy, and living space, is also fully discussed. On the positive side, significant progress is frequently cited in air quality issues, the reclamation of surface-mined land, recycling, the preservation of wilderness, protection of endangered species, and greater energy efficiency.

We have tried to meet the challenge of writing not just a *timely* environmental science textbook but a *timeless* one by emphasizing scientific principles and the natural functioning of the environment. Although the importance of certain environmental issues may wax and wane with time, the underlying principles that govern physical and biological systems do not change. Hence, the readers' grounding in basic ecological concepts provides them with the interpretive skills and flexibility needed to analyze new environmental issues as they arise.

We have made every effort to describe and analyze environmental problems as objectively as possible. While we convey the seriousness of those problems, we do not editorialize. Rather, we encourage readers to become personally involved in solving environmental problems and to form their own opinions on controversial issues.

Important Features of the Text

Highlights of this text include the following:

A unique, three volume organization based on the three basic elements of most courses: economics and ecology, biological resources, and physical resources. The latter two should allow instructors with backgrounds in either biological or physical science to move through the text more efficiently. Because each part may be adopted individually in paperback "separates," instructors can customize the book to their particular course needs.

Extensive treatment of ecological principles and application to environmental issues.

Separate chapters on the nature of environmental science (chapter 1) and economic and public policy aspects of environmental issues (chapter 2).

Two chapters on human population growth (chapters 6 and 7) precede discussion of environmental problems.

An analysis of various strategies to enhance food production and of the future role of sustainable agriculture (chapter 8).

Chapter-length treatment of pest management (chapter 9).

Chapter-length coverage of the management of public lands (chapter 10).

Extensive application of ecological principles to managing biodiversity (chapter 11).

An entire chapter on environmental toxicology (chapter 12).

Concise treatment of contemporary concerns in water quality (chapters 13 and 14) and air quality (chapters 15 and 16) that is based upon physical principles.

An extensive treatment of mining, land reclamation, and management of rock and mineral resources (chapter 17).

Thorough analyses of present energy sources and future energy options (chapters 18 and 19).

A chapter on waste management that combines coverage of nonhazardous, hazardous, and radioactive wastes (chapter 20).

A unique look at land-use conflicts in the context of coastal zone management, geological hazards, and urban growth (chapter 21).

Color photographs and illustrations. Ninety - five percent of the book's visuals are in full color.

Inclusion of two types of readings called "Issues" or "Special Topics."

Metric and British units of measure are used throughout text.

Many case studies and practical examples.

Numerous suggestions for individual action.

Includes key pedagogical aids such as key terms, summary statements, questions for review, suggestions for projects, and an annotated bibliography.

An epilogue entitled "Toward a Sustainable Environment" that summarizes numerous themes developed throughout the text. It appears at the end of the full text or at the end of Parts II and III, if adopted as "separates."

A complete glossary.

An appendix addressing some of the special environmental concerns of Canada, Mexico, and Hawaii.

Organization and Content

Environmental Science: Managing Biological and Physical Resources remains both a basic science text and a text on environmental issues. This approach is evident throughout the book and is reflected in the basic organization of its twenty-one chapters. The organization is based upon suggestions of reviewers and an in-depth survey of potential adopters.

Volume I: Economics and Ecology

Part I consists of two introductory chapters. Chapter 1 introduces the reader to environmental science, how scientists do science, and the successes and limitations of scientific inquiry. Many of the ideas presented in chapter 1 (such as scientific models) are woven through the entire text. Chapter 2, authored by Richard J. Tobin, covers the economic and policy-making dimensions of environmental issues.

Basic principles of ecology are treated in **Part II** (chapters 1 – 5). We consider the flow of energy and cycling of materials within the environment, the nature of environmental disturbance and pollution, how organisms and ecosystems respond to environmental change, and the various biological and physical factors that govern population growth. These fundamental ecological principles are applied to a wide variety of environmental issues throughout the text.

Part III (chapters 6 and 7) focuses on human population growth. Exponential growth of global human population and the consequent soaring demand for resources are at the very core of many environmental problems. Hence, we treat this topic extensively and introduce it prior to our discussion of environmental issues.

Volume II: Managing Biological Resources

Part IV (chapters 8 – 12) deals with problems of environmental quality and management that are primarily biological in nature. Thus, we describe the challenge of meeting the food requirements of a soaring human population, pest management strategies, management of public lands in the face of pressures from visitors and resource developers, and threats to biodiversity. This section closes with a chapter on environmental toxicology, reflecting the importance of assessing the risks and response of humans and other organisms to toxins.

Volume III: Managing Physical Resources

Part V (chapters 13 – 21) covers environmental issues that are chiefly, but not exclusively, physical. Thus, we consider water quality, air quality, exploitation of mineral resources, energy supply and demand, waste management, and land-use conflicts and natural hazards. For each problem area, we describe the natural functioning of the environment, the nature of the problem, and how that problem is managed.

The text closes with an epilogue that underscores the need for sustainable development and the role of the individual in environmental problem solving and with an appendix on the special environmental concerns of Canada, Mexico, and Hawaii.

The three-volume organization of the text provides the instructor with considerable flexibility in course planning and can provide students with significant price savings. For example, a course that emphasizes biological problems may opt for a packaging of Chapters 1 – 7 plus 8 – 12, or volumes I and II, while a course with a greater emphasis on physical science might utilize a packaging of Chapters 1 – 5 plus 13 – 21, or volumes I and III. Chapters in volumes II and III may be covered in any order, and any topic may be dropped without loss of continuity.

Ancillaries

An *Instructor's Manual* and *Test Item File,* prepared by Barry Wulff, is available free to adopters as additional support material. The instructor's manual provides a list of key terms/concepts, a chapter outline, learning objectives, and ideas/activities for each chapter. The test item file contains 30 – 40 objective questions for each chapter that can be used for exam purposes. (ISBN 0 – 697 – 10833 – 3)

A *Student Study Guide,* prepared by Michael Morgan and Joseph Moran, is available to students as a resource for extensive self-testing of basic scientific concepts and environmental issues. (ISBN 0 – 697 – 10834 – 1)

Also available to instructors is *TestPak,* a computerized testing service that includes a database of objective questions and a grade-recording program. Disks are available in IBM, Apple, and Macintosh formats. If a computer is not available, the instructor can choose questions from the *Test Item File* and phone or FAX in their request for a printed exam, which will be returned within 48 hours.

There are 100 two- and four-color overhead *Transparencies* available free to adopters. These acetates feature key illustrations from the text that will enhance your lecture or course outlines. (ISBN 0 – 697 – 10835 – X)

Transparency Masters are also available in the instructor's manual. These can be used to prepare handouts or to project as overhead transparencies, depending on the instructor's needs. (ISBN 0 – 697 – 16358 – X)

You Can Make a Difference, by Judith Getis, is a short, inexpensive supplement that offers students practical guidelines for recycling, conserving energy, disposing of hazardous wastes, and other environmentally sound practices. It can be shrink-wrapped with the text, at minimal additional cost. (ISBN 0 – 697 – 13923 – 9)

Acknowledgments

In preparing *Environmental Science: Managing Biological and Physical Resources,* we profited greatly from our interactions with the talented and dedicated professionals at Wm. C. Brown Communications, Inc. We especially thank Kevin Kane for his extraordinary faith in our ideas and Marge Kemp for her incredible patience and guidance over the past several years. We are indebted to Nick Murray for his skill and diligence in making our manuscript more engaging. In addition, we thank the following people for their dedication to excellence: Renee Menne, Gail Wheatley, Robin Storm, and Carla Goldhammer.

Our very special thanks go to Joy Phillips and Pattie Dimmer for typing portions of the manuscript. To the students of Environmental Science 102—your curiosity and concerns about the environment have provided the motivation for writing this book. To our families and friends—your years of encouragement and understanding are much appreciated. And without the support, patience, and sacrifices of our wives, Gloria, Jennifer, and Ruth, there would be no book.

In revising the previous edition of this text, we were fortunate in having the constructive criticism of many outstanding reviewers. We are especially grateful to the following people who reviewed this edition.

Laura Tamber
Nassau Community College

Dorothea Sager
University of Wisconsin-Green Bay

Kenneth J. Van Dellen
Macomb Community College

Douglas N. Reynolds
Eastern Kentucky University

J. Ernest Bernice
Edinboro University of Pennsylvania

Keith L. Bildstein
Winthrop College

George Fister
McNeese State University

Jack S. Wood
Western Michigan University

Penelope M. Koines
University of Maryland

William Rogers
Winthrop University

Peggy J. Guthrie
University of Central Oklahoma

Barry L. Wulff
Eastern Connecticut State University

Clayton H. Reitan
Northern Illinois University

Lynda Swander
Johnson County Community College

Robert W. Howe
University of Wisconsin-Green Bay

James K. Lein
Ohio University

David J. O'Neill
Dundalk Community College

Mary G. Kohrell
University of Wisconsin-Green Bay

Diane E. Anderson
University of Wisconsin-Green Bay

Richard G. Bjorklund
Bradley University

H. Fred Siewert
Ball State University

B. Thomas Lowe
Ball State University

Edward J. Hopkins
University of Wisconsin-Madison

Brenda J. Boleyn
Cape Cod Community College

Richard W. Wahl
Shippenburg University of Pennsylvania

Nat B. Frazer
Mercer University

Finally, we are grateful to our colleagues at the University of Wisconsin-Green Bay, particularly to Thomas McIntosh, Charles Rhyner, Paul Sager, Ronald Starkey, and Ronald Stieglitz for their valuable contributions.

If we may be of assistance to you in utilizing this book, or if you have suggestions or comments, please feel free to contact us at the University of Wisconsin—Green Bay, College of Environmental Sciences, Green Bay, WI 54311. FAX: 414 – 465 – 2376; VOICE: 414 – 465 – 2371.

Michael D. Morgan
Joseph M. Moran
James H. Wiersma

Volume I

Environmental Science

Economics and Ecology

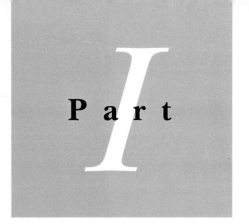

Science, Public Policy, and Economics

An Introduction to Environmental Science

*S*cientists, public policymakers, and economists perceive and analyze environmental issues in different ways. In part 1 we summarize those contrasting perspectives and introduce many concepts that we weave through the entire text. Chapter 1 deals with science and how scientists do science. Scientists view the environment as made up of numerous interacting systems, and they strive to understand the principles that govern those systems. We describe the scientific method, how and why scientists conduct laboratory and field experiments, and the value and limitations of scientific models.

Chapter 2 traces the key steps involved in the evolution of public policy regarding the environment. We describe the types of environmental issues that are most likely to attract the attention of policymakers, and we explain the absence of a single comprehensive environmental law. We also demonstrate how a combination of government regulations and economic forces shapes the nation's environmental policy. Furthermore, we describe various analytic tools utilized by economists in environmental decision making.

The environment encompasses all the living and nonliving things that surround us. The well-being of the environment is essential to our own personal well-being.
© Sharon Gerig/Tom Stack & Associates

Science and the Environment

Scenes such as this one in the Pacific Northwest
illustrate the diverse components of the
environment.
© Wolfgang Kaehler

*D*estruction of tropical rain forests, threats to the ozone shield, potential global warming, nuclear waste disposal, groundwater pollution, and world hunger are some of the major environmental problems that are likely to be with us for many years to come (fig. 1.1). Newspapers and television specials sound the warning that something must be done about these and other problems that threaten to erode the quality of life on earth. Yet many of us seem unconcerned about these problems. Perhaps this is because we are not particularly aware that these problems affect us personally or that our own survival depends upon our environment and its well-being.

Few of us spend much time outdoors communing with nature. For example, rather than walk, many of us prefer to (or must) drive to work or school in air-conditioned autos, buses, or subways. We not only work and reside in climate-controlled buildings, but also many of our leisure activities take place inside, in domed stadiums, malls, or indoor fitness centers, for example. Few of us may know the ultimate source of our food and water. Food is trucked to the supermarket, and garbage is hauled away; water simply flows from a faucet, and wastewater is flushed somewhere.

Even when we do venture outdoors, we usually find ourselves in a highly modified environment that is home to only a few species of plants and animals. Most city parks, for example, feature mowed lawns and formal flower gardens laid out in pleasing (but unnatural) symmetrical patterns. The only animals may be some squirrels and gophers scavenging for scraps of food, and the only nesting birds are starlings, sparrows, and pigeons. Certainly, there is very little of nature's wilderness in such places.

Because of our perception that we are set apart from nature, we often find it difficult to relate to environmental issues unless, of course, they take place in our own backyards.

(a)

(b)

(c)

(d)

Figure 1.1 Our utilization and disposal of resources have disturbed the environment in ways that will create a legacy of environmental problems for generations to come: (*a*) Smog continues to plague Los Angeles and many other metropolitan areas. (*b*) Waste dumped at sea washes up on shorelines and forces the closing of recreational beaches. (*c*) Strip mining for coal for energy disrupts the landscape. (*d*) Abandoned toxic dump sites threaten groundwater quality.

(*a*) © Tom McHugh/Photo Researchers (*b*) © Hank Morgan/Photo Researchers (*c*) © Kenneth Murray/Photo Researchers (*d*) © Ray Pfortner/Peter Arnold, Inc.

Even then, perhaps more often than we care to admit, our foremost concern is likely to be the expense of correcting an environmental problem. For example, our first reaction to a referendum on newspaper recycling, may be "How much more will I have to pay for waste disposal if the city decides to go ahead with such a program?"

While we may have heard of such problems as the thinning of the ozone layer, acid rain, soil erosion, or potential global warming, they may appear to have little direct relevance to our daily lives. We are usually much more concerned about doing well on the next exam, finishing that overdue business report, taking the car in for repairs, paying for the children's education, or wondering if our favorite football team will make the playoffs. Too often, we are so busy attending to day-to-day matters and trying to get ahead that we fail to appreciate that our survival depends upon an environment that is conducive to life. We *are* a part of nature, and our well-being depends upon a healthy environment.

To better appreciate our dependence on the environment, we must realize that we, like all living organisms, are extremely complex, and our survival depends upon the availability of vital resources and the proper functioning of many intricate internal processes, for example, breathing, ingesting water, and digesting food. We cannot live more than five minutes without oxygen, seven days without water, or five to six weeks without food.

Food is processed, or *metabolized,* to provide for repair, growth, and reproduction. For example, each of us produces 2.5 million new red blood cells (carriers of oxygen in the blood) per second to replace those that die. Metabolism produces waste by-products that must be excreted into the environment. The hundreds of thousands of people in the United States who suffer from kidney disorders attest to the essential function of this waste disposal organ. In addition, all organisms must be able to receive and respond to stimuli from their surroundings. Our environment is continually changing, and our survival hinges upon our ability to recognize change and to respond appropriately.

For any organism (including humans) to remain alive, each of these vital and complex processes must function properly. But what if not enough food is available? The consequence may be malnutrition or starvation. What if the drinking water is contaminated? The consequence may be hepatitis, dysentery, cholera, or other diseases. What if the air we breathe is contaminated with too much carbon monoxide? The consequence may be asphyxiation. What if we are exposed to too much solar ultraviolet radiation? The consequence may be skin cancer.

From the discussion thus far, we can draw two very important conclusions: (1) To live, all organisms must be able to perform certain vital functions, and (2) the environment controls the destiny of all organisms (including ourselves). The same fundamental rules of life apply to all organisms, and the environment is the arena in which the rules are played out.

In this book, we explore principles that govern the workings of the environment. In so doing, we come to better understand how human activities can adversely affect the environment. We will also be better able to evaluate current environmental management strategies and develop more effective means of coping with environmental change. As the global human population continues its rapid growth and makes ever-increasing demands on the planet's finite resources, a thorough understanding of environmental principles becomes more important to our well-being, and perhaps to our very survival.

The Environment and Environmental Problems

Living organisms depend on the environment for their survival. But what do we mean by the *environment*? By definition, the **environment** encompasses the entire assemblage of external factors or conditions that influence living organisms in any way. Thus, the environment includes all living organisms (plants, animals, and microorganisms) and nonliving things (such as air, rock, and water) that influence organisms.

Determining exactly what constitutes an environmental problem, however, is not always straightforward. For example, what people consider to be a problem, and what they are willing to do about it, largely depends on their personal values and interests. Hence, what some people consider to be a serious problem, others may see as a welcome solution to a quite different problem. For example, consider how different people might react to an announcement that operations at a local industrial plant will expand. The plant manager and employees greet the news with elation. For them, it is a signal that the local economy is improving and that their jobs are secure. The Chamber of Commerce proclaims that plant expansion will create new jobs and pour money into the local economy. Thus many people view the plant expansion positively—as a sign of prosperity. However, others may see it as yet another affront to the quality of the environment. They argue that the costs of added air and water pollution that would accompany plant expansion outweigh any benefits to the local economy. These people clearly consider the plant expansion as an environmental problem.

Even when most of us agree that a particular situation constitutes a serious environmental problem that calls for immediate remedial action, economic constraints may discourage us from doing much about it. The private sector (industry, for example) assumes much of the cost of controlling air and water quality. Typically, these costs are passed on to consumers in the form of higher prices for goods and services. In some cases, the added costs of environmental quality control along with other factors (such as higher labor costs) put domestic industries at a competitive disadvantage compared to foreign manufacturers, who may not have to contend with such strict environmental regulations. This competitive disadvantage may eventually translate into closed industrial plants and more unemployment.

Economic constraints also arise in the public sector. We are well aware that public tax coffers are not bottomless, and funding for environmental quality control must compete with funding for education, construction projects (such as highways and public housing), national defense, social welfare programs, and all the other projects and programs of our federal, state, and local governments. Because funds for environmental concerns are limited, the expenditure of $10 million on flood control, for example, means that less money is available for new sewage treatment facilities or garbage recycling. Hence, people must choose which problems are to have highes priority. Multiple demands on limited financial resources require us to set priorities and make informed decisions. Thus, we cannot regard environmental issues simply as scientific or technical problems; they almost always have political and economic dimensions as well.

We begin our study of environmental science by examining some of the scientific tools that enable us to conceptualize

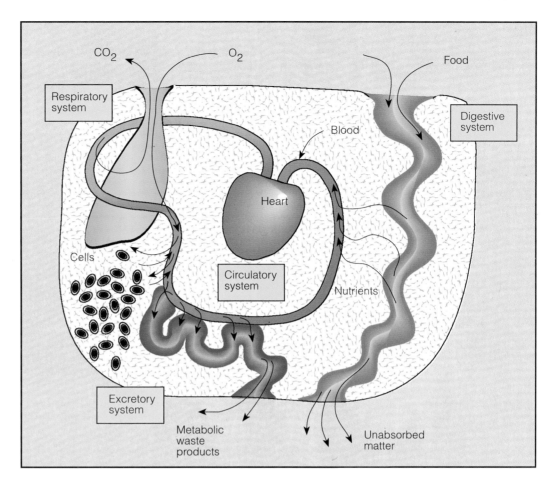

Figure 1.2   Schematic representation of the various internal systems that must function properly for life to be maintained.

and understand the workings of the environment. Then, in chapter 2, we discuss political and economic aspects of environmental issues; for example, we consider the political factors that shape public environmental policy and how people determine the costs of dealing with environmental problems. While the principal focus of this book is on scientific aspects, we also discuss political and economic dimensions where doing so helps us to appreciate the full ramifications of environmental problems and their potential solutions.

The Scientific Perspective

The various components of the environment are not isolated from one another; rather, they are strongly interdependent. For example, green plants acquire essential resources from the physical realm (water and minerals from the soil, carbon dioxide from the atmosphere, and light-energy from the sun). Animals depend on plants or other animals as sources of food. We harvest food from the land and the sea, and mine minerals and fuels from rock in the earth's crust.

Linkages among components of the environment are actually pathways along which energy flows and materials cycle. For example, plants absorb solar energy and convert it to food energy, which is harvested by animals who eat the plants.

Then, when an organism dies or is consumed by another organism, its chemical components are cycled into other parts of the environment, including the soil. Heavy rains wash soil particles into nearby streams, or strong winds dry the soil and carry soil particles into other regions. We will learn more about the cycling of materials and the flow of energy within the environment in chapter 3.

Interactions among components of the environment are not arbitrary or capricious acts of nature; they are systematic. Behavior that is systematic is orderly and organized because it is governed by certain principles or laws of nature. A **system** is made up of components that behave (function and interact with one another) in some regular and predictable manner. The environment consists of numerous systems, and each system is an assemblage of interacting living and nonliving things.

Most of us are familiar with the term *system* in reference to the human body, in which the digestive, circulatory, respiratory, and excretory systems must function separately and together in an orderly manner in order for us to stay alive (fig. 1.2). For example, the digestive system breaks down and absorbs nutrients from the food that we eat, and the respiratory system takes in oxygen from the atmosphere. The circulatory system performs the essential function of transporting oxygen from the lungs and nutrients from the small intestine

(a)

(b)

(c)

Figure 1.3   Three ecosystems. (*a*) An agricultural field is a relatively simple ecosystem composed of one dominant plant species (in this case, corn), several species of weeds, and numerous insect species and microorganisms. (*b*) A rain forest is a considerably more complex ecosystem containing hundreds of plant species and thousands of species of animals (mostly insects) and microorganisms. (*c*) A city also qualifies as an ecosystem because it is made of living and nonliving things that are interdependent. Would you describe an urban ecosystem as simple or complex?

(*a*) U.S. Department of Agriculture (*b*) © Brian Parker/Tom Stack and Associates (*c*) © James H. Karalas/Peter Arnold, Inc.

to all cells of the body. Within these cells, oxygen and nutrients are used in metabolism, which in turn produces metabolic wastes. If allowed to accumulate, these wastes would eventually poison the cells. Thus, the excretory system (the kidneys, for example) rids the body of these potentially lethal substances. Furthermore, these four systems function properly only if our coordinating systems (the nervous and hormonal systems) also operate properly.

Analogous to the human body, planet Earth comprises numerous interacting systems. Our understanding of the planet is aided by examining its component systems, just as our understanding of the human body benefits from study of its component systems. Earth is a mosaic of different climates, soil types, and plant and animal communities. Hence, it is useful to view the environment as being subdivided into a number of ecosystems (fig. 1.3). An **ecosystem** consists of communities of organisms and associated physical and chemical characteristics within a specific geographical area. Deserts, tropical rain forests, tundra, estuaries, lakes, and streams are examples of ecosystems.

The immediate cause of an environmental problem is usually some disturbance of the orderly functioning of an ecosystem, that is, a disruption of the usual operations (or interactions) of physical and biological processes. Consider some examples. A rural lake that is noted for its clear water and fine trout fishing becomes choked with aquatic weeds and devoid of fish several years after vacation homes are built along the lakeshore. A tropical rain forest is cleared to provide land for agriculture, but within a few years, soil fertility declines to the point that further cultivation is futile. Huge earth-moving equipment strips away one of the few remnants of native prairie to get at subsurface deposits of coal.

Ultimately, environmental problems stem from more and more of us making increasing demands on the planet's finite resources. Global human population is soaring, and most people strive for a life-style that is more resource-intensive than in the past. More people are living in urban areas where they depend on inefficient and wasteful systems to supply their food, water, energy, and other needs. In addition, we continue to rely upon economic systems in which the price of goods and services fails to account for the environmental cost of producing them.

Environmental science is the study of the fundamental interactions between living and nonliving things. Its goal is to understand the natural functioning of the environment as well as contribute to solving the problems that arise when the environment is disturbed. Because the environment is composed of many interacting systems, environmental science is a broad endeavor that must draw on the contributions of many sciences; that is, it is **interdisciplinary.**

Each of the traditional scientific disciplines contributes to our analysis of the environment and, consequently, to our understanding of environmental problems. Physics, perhaps the most fundamental of the sciences, is essentially the study of matter and energy, so it is concerned with laws that govern the cycling of materials and the flow of energy within and between ecosystems. Chemistry, which focuses on the composition of materials and their reactions, enables us to determine the origins and fate of air and water pollutants. Biology is concerned with the functioning of organisms and their interactions; it enables us to understand how organisms respond to contaminants. Meteorology, the study of atmospheric circulation, provides insight on how air pollutants are transported from one place to another. Geology deals with the earth's structure and composition (rocks, minerals, and fuel); hence, it is concerned with the supply and geographic distribution of important resources. Together, these sciences facilitate our understanding of the environment and our impact on it.

Consider an illustration of how the unique perspectives offered by different scientific disciplines are brought to bear on a specific environmental problem. One day, almost all of the residents of a tiny village fell ill, developing debilitating symptoms of nausea and persistent diarrhea. After ruling out other potential causes, local health officials blamed the illness on the well that supplied drinking water to everyone in the village. Pending results of chemical and biological analysis of the well water, authorities closed down the well. Sure enough, laboratory tests detected large numbers of coliform bacteria, an indication that the well water was contaminated by raw sewage. To help determine the origin of the contaminants, geologists and hydrologists were called in to study the flow of groundwater into the well. Eventually, the contaminants were traced to a ruptured sewage holding tank. Thus, resolving this problem required the expertise of several scientists (biologist, chemist, hydrologist, geologist, and medical personnel).

The Scientific Method

Scientists gather information and draw conclusions about the workings of the environment by applying the **scientific method,** a systematic form of inquiry that involves observation, speculation, and reasoning. The ultimate objective of the scientific method is to understand physical and biological phenomena.

The investigation of a problem that developed in some lakes in the Adirondack Mountains of New York State provides an example of applying the scientific method. Those lakes had long been noted for their abundance of game fish and other aquatic life, and good fishing had helped to make the lakes a popular attraction with sportsmen and other vaca-

tioners. Beginning in the early 1970s, however, populations of fish in some Adirondack lakes began to decline. Local businesses became concerned that poor fishing would hurt the region's recreational industry, and conservationists worried that the lakes had become yet another casualty of pollution.

Biologists initially proposed that toxic materials (poisons) were entering the lakes and killing aquatic life, but they did not know the nature of the toxins or their origins. Subsequent chemical testing of water samples failed to find hazardous concentrations of toxins. The tests did reveal, however, that the lake waters were abnormally acidic, and scientists hypothesized that acidic rainwater (and snowmelt) was causing the lake waters to become excessively acidic. Laboratory studies have shown that excessively acidic waters are lethal to young fish.

Why was the rainwater (and snowmelt) so acidic? Rainwater is normally slightly acidic, because it dissolves some of the carbon dioxide in the air, which forms the same weak and harmless acid found in carbonated beverages; however, rainwater samples from the vicinity of the Adirondack lakes were at least one hundred times more acidic than expected. Additional laboratory studies identified sulfuric and nitric acids in the rainwater, which the investigators knew were formed as a result of oxides of sulfur and nitrogen—common industrial air pollutants—dissolving in rainwater.

The next question concerned the source of the sulfur and nitrogen oxides. Because the Adirondack Mountains are downwind of some major industrial sources of those air pollutants, such as coal-fired power plants, scientists reasoned that the loss of fish in Adirondack lakes was linked, at least circumstantially, to industrial air pollution. The hypothesis that excessively acidic rainwater led to the fish kills is generally accepted because it is consistent with our knowledge of the limited tolerance of fish to acidic waters, and of the chemical reactions that involve rainwater and air pollutants.

Ideally, we may consider the scientific method as a sequence of steps in which scientists (1) identify specific questions related to the problem at hand; (2) propose an answer to one of these questions in the form of an educated guess; (3) state the educated guess in such a way that it can be tested, that is, formulate a **hypothesis;** (4) predict what the outcome of the test would be if the hypothesis were correct; (5) test the hypothesis by checking to see if the prediction is correct; and (6) revise or restate the hypothesis if the prediction is wrong.

In actual practice, scientists often do not follow these steps exactly or in the order given. Furthermore, some steps (e.g., steps 3 and 4) are often combined. Nor is the scientific method a formula for creativity, since it does not provide the key idea, the hunch, or the educated guess that spontaneously springs to mind and forms the original basis of the hypothesis. Rather, it is a method or technique that can be used to assess the validity or worth of a creative key idea, however and wherever it originates.

As in our acid rain example, a hypothesis is a tool that suggests new experiments or observations, or opens new avenues of inquiry. Thus, even an erroneous hypothesis may be fruitful. Above all, scientists must bear in mind that a hypothesis is merely a working assumption that may be accepted, modified, or rejected. They must be objective in evaluating a hypothesis and not allow personal biases or expectations to cloud that evaluation. In fact, scientists actively search for observations or information that could disprove their beliefs. If they find such evidence, they reorganize and redefine their beliefs to incorporate the disparate observations or information. Inquiry, creative thinking, and imagination are stifled when hypotheses are considered to be immutable.

A new hypothesis (or an old, resurrected one) may be hotly debated within the scientific community. History warns us of a natural human resistance to new ideas that threaten to displace long-held notions. Also, disagreement among scientists on a particularly controversial issue sometimes receives wide publicity, which may confuse the general public. As of this writing, for example, there is considerable public debate among scientists as to whether burning coal and oil for power is leading to global warming.

In some cases, the prevailing public reaction may be, "Well, if the so-called experts can't agree among themselves, who am I to believe? Is there really a problem after all?" However, debate and disagreement are essential steps in the process of reaching scientific understanding; they generate useful suggestions, stimulate new thinking, and uncover errors. In fact, such debate and skepticism buffer the scientific community from a too-hasty acceptance of new ideas. If a hypothesis survives the scrutiny and skepticism of scientists and the public, it is probably accurate.

Experimentation

In our acid rain example, scientific understanding was gained primarily through observation and monitoring. Alternatively, the scientific method may be applied in controlled field or laboratory experiments. An **experiment** is a procedure designed to study some phenomenon under known conditions with the goal of discovering or illustrating some scientific principle or identifying the cause of some problem. Experiments are conducted either in the laboratory or in the field. In fact, long-term field observations and experiments are responsible for much of what we understand about the relationships between organisms and their environment. Experimentation is not infallible, however. The investigator conducting the experiment must separate objective observation from subjective interpretation, pay close attention to detail, and carefully record all procedures and observations. Above all, the experiment must be reproducible so that other investigators may repeat it and independently confirm the findings.

A type of experiment that may be useful in environmental science is the **controlled experiment,** which entails comparing two very similar systems. One, the so-called **test system,** is perturbed, while the other, the **control system,** is not perturbed and is used as a standard for comparison. For example, scientists might assess the potential impact of acid rain on fish and other aquatic organisms by artificially acidifying a lake (the test system). A nearby lake with similar properties is

not disturbed and functions as the control system. In this way scientists are able to isolate changes in the lake ecosystem caused by acidification alone.

While field and laboratory experiments have certainly advanced scientific knowledge, many significant discoveries have originated from unexpected observations or chance occurrences. This underscores the need for scientists to be diligent observers. A classic illustration of the role of chance in scientific research is Sir Alexander Fleming's 1928 discovery of penicillin.

By today's rigorous standards, Fleming's laboratory was far from sterile; there was considerable dust in the air, and the risk of contamination was high. That is precisely what happened. While studying cultures of the bacterium *Staphylococcus* (a disease-causing microorganism), Fleming by chance exposed the cultures to the dusty air, and they were contaminated. Fleming later noticed that a mold had formed on petri dishes containing the cultures, and bacterium colonies surrounded by the mold had died. The mold turned out to be *Penicillium notatum,* the source of one of the world's most powerful antibiotics, penicillin. The development of therapeutic uses for penicillin is attributed to the later work of two other scientists, Sir Howard Florey and Ernst Chain. In 1945, Fleming, Florey, and Chain won a Nobel Prize for their work on penicillin.

Scientific Models

A road map is a useful model that is familiar to all of us. Such a map readily conveys information on the location of streets and highways, towns and cities, and other landmarks that aid our travels. A road map omits irrelevant information about the region, such as types of soil, bedrock, and climate. Other kinds of models facilitate the application of the scientific method. Thus, a **scientific model** is defined as an approximate representation or simulation of a real system; it includes only the essential variables or characteristics of that system. Such models help us to better understand complex systems or situations by reducing them to their essential components. Let us consider several examples of scientific models.

To learn how to improve the fuel efficiency of automobiles, we can observe a model automobile in a wind tunnel. An automobile can be designed to reduce its frictional (air) resistance, which can increase its fuel efficiency. Thus, the shape of the model automobile is the critical variable and becomes the focus of study. Other variables, such as color or whitewall tires, are irrelevant to the experiment and can be ignored.

Depending on their particular functions, scientific models may be classified as conceptual, graphic, physical, or numerical. A **conceptual model** describes the general relationships among components of a system. For example, a food web (chapter 3) is a conceptual model that portrays the dependency of one group of organisms on another group for food. A **graphic model** compiles and displays data in a form or pattern that readily conveys meaning. An example is a weather map (fig. 1.4), which integrates simultaneous weather

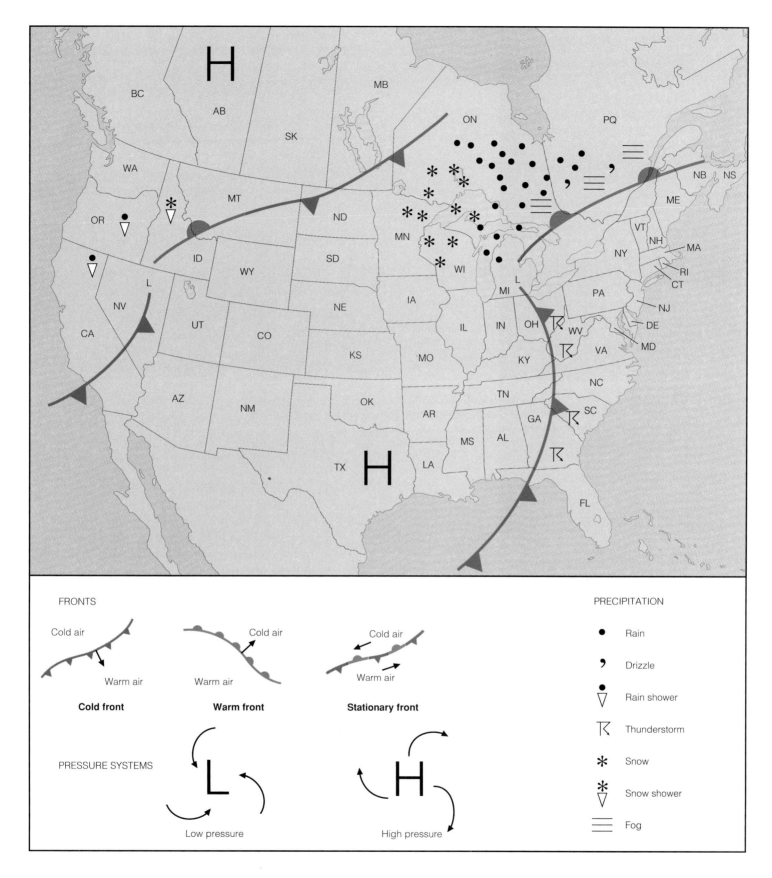

Figure 1.4 A weather map is a graphic model that represents the state of the atmosphere at a specific time over some geographical area.

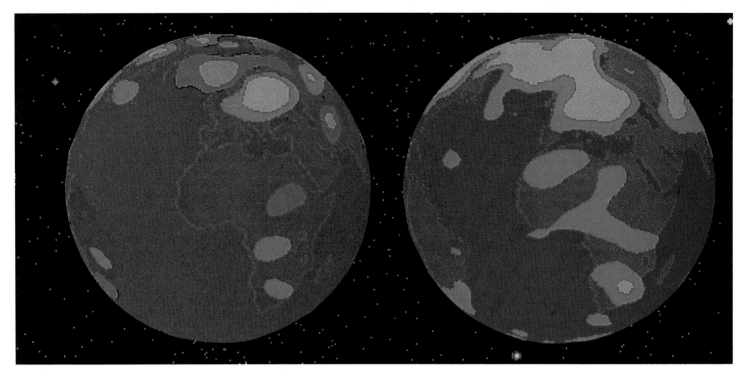

Figure 1.5 A computer model simulation of changes in the earth's surface temperature variation resulting from a doubling of atmospheric carbon dioxide concentration. Temperature fluctuations of 5–10F ° above normal (orange) or below normal (blue) under the planet's present condition (left) and in a projection of the planet with its CO_2 doubled (right).

observations at hundreds of locations into a coherent representation of the state of the atmosphere. (Some conceptual models, such as a food web, can also be presented graphically.) A **physical model** is a miniature version of a large system. For example, the U.S. Army Corps of Engineers operated a physical model of the Mississippi River on 89 hectares (220 acres) of land near Clinton, Mississippi. That model enabled engineers to forecast changes in the characteristics of the river's flow, including flooding, that occurs in response to changes within the river's watershed.

Today, modeling is greatly aided by electronic computers, which accommodate enormous quantities of data and perform many calculations very rapidly. Typically, a computer is programmed with a **numerical model** that consists of one or more mathematical equations that describe, for example, the behavior of a particular physical or biological system. The variables in the numerical model can be manipulated, individually or in groups, to assess their impact on the system.

Computerized numerical models of the atmosphere have been used to forecast the weather since the 1950s. More recently, they have been used to predict the climatic effects of rising levels of carbon dioxide in the atmosphere. As discussed in greater detail in chapters 3 and 15, the concentration of carbon dioxide in the atmosphere has been rising for some time, mostly because of the burning of coal and oil as fuel. As a result, we might expect warmer air temperatures, because carbon dioxide slows the loss of the earth's heat to space.

To predict how much warming might occur at the earth's surface, scientists follow three steps. First, a computerized numerical model of the atmosphere is designed to predict the present worldwide air-temperature pattern, given the current level of atmospheric carbon dioxide. Then, holding all other variables in the model constant, the carbon dioxide concentration is elevated (usually doubled), and the numerical model computes a new worldwide temperature pattern. Finally, the initial temperature distribution is subtracted from the final temperature distribution. Presumably, the net warming can be attributed to the elevated carbon dioxide concentration (fig. 1.5).

Thus scientific models are valuable tools in environmental science. Because they are not cluttered with extraneous and distracting detail, such models may provide important insights concerning how things interact, or they may trigger creative thinking about complex phenomena. Furthermore, models help us to predict the impact of environmental change.

It is important to remember that models are only approximations of physical and biological systems. As such, they are subject to error. For example, one potential difficulty with numerical models is the accuracy of their component equations. Usually, the equations are only approximations of the way that things really work in nature, and they may not include all the relevant variables. For example, a computerized numerical model can be used to predict the change in the concentration of a water pollutant in a river as it is transported and diluted downstream. However, the model's prediction may be unrealistic if, for example, it fails to account for chemical reactions that remove the pollutant from the water and thereby reduce its concentration.

Conclusions

In this introductory chapter, we have seen that all living things—including ourselves—depend on the environment for survival. The environment can be subdivided into ecosystems, assemblages of interdependent living and nonliving things that function in a systematic manner. In an effort to understand the workings of the environment, scientists employ the scientific method, engage in field and laboratory observation and experimentation, and call upon models to help them understand how the environment functions and how it responds to disturbances. However, we must keep in mind that environmental problems have political and economic as well as scientific dimensions. In the next chapter, we examine how public policymakers and economists view environmental issues.

Key Terms

environment
system
ecosystem
environmental science
interdisciplinary
scientific method
hypothesis
experiment

controlled experiment
test system
control system
scientific model
conceptual model
graphic model
physical model
numerical model

Summary Statements

All living organisms must perform the same essential functions, and the environment governs the destiny of all species, including our own.

The environment is the assemblage of external factors or conditions that influence living organisms in any way. The living and nonliving components of the environment are strongly interdependent, and these linkages are pathways along which energy flows and materials cycle.

Interactions among the various components of the environment are systematic because they are governed by certain principles or laws of nature. For convenience of study, we subdivide the environment into ecosystems, each of which is composed of all the associated organisms and physical features within a specific geographic area.

Environmental science is an interdisciplinary endeavor that is concerned with both the natural functioning of the environment and the problems that arise from disturbances of physical and biological processes.

Our understanding of the environment is aided by the scientific method, a systematic form of inquiry that involves observation, speculation, and reasoning.

As a scientific tool, a hypothesis is a working assumption that may suggest new experiments or observations or open new avenues of inquiry.

Scientists conduct laboratory or field experiments to test the validity of hypotheses. Above all, an experiment must be reproducible, so that other investigators can repeat it and independently confirm its findings.

Scientific models reduce a complex system to its essential components or variables. Depending on their particular functions, scientific models are classified as conceptual, graphic, physical, or numerical. All models are only approximations of systems, and as such they are subject to error.

Questions for Review

1. Provide some illustrations of the ways in which humans depend on nature for survival.
2. Environmental issues cannot be viewed as simply scientific or technical problems. Explain this statement.
3. In what way are interactions among components of the environment systematic?
4. Provide a definition for the term *ecosystem*. Also, give several examples of natural ecosystems and ecosystems that have been greatly modified by human activity.
5. Explain why environmental science must be an interdisciplinary endeavor.
6. Describe the scientific method in your own words. What role is played by hypotheses in the scientific method?
7. How do scientists test the validity of hypotheses?
8. Provide some examples of controlled experiments. What is the basic objective of a controlled experiment?
9. How do scientific models assist the scientific method?
10. Distinguish among conceptual, graphic, physical, and numerical models. What do all models have in common?

Selected Readings

Beveridge, W. I. B. *The Art of Scientific Investigation*. 3d ed. New York: Vintage Books, 1957. 239 pp. An excellent description of how scientists do science.

Leopold, A. *A Sand County Almanac*. New York: Ballantine Books, 1970. 295 pp. A classic description (first published in 1949) of the balance of nature and of human disturbance of the environment.

Nash, R. F. *American Environmentalism: Readings in Conservation History*. 3d ed. New York: McGraw-Hill, 1990. 364 pp. A collection of fifty-one essays that trace the historical roots of the environmental movement.

Chapter 2

Politics, Economics, and the Environment

by
Richard Tobin
State University of New York at Buffalo

A wilderness such as this is threatened with
development that scientists warn is likely to
destroy plant and animal habitats. Makers of
public policy can enact laws that protect and
preserve such areas while economists analyze the
costs and benefits of development. Thus,
environmental issues often involve political and
economic aspects.
© Tom Kitchin/Tom Stack and Associates

Despite our vast and growing scientific knowledge of how the environment works, environmental problems plague much of the world. If some scientists are correct, many of these problems, such as the thinning of the ozone layer, will only get worse, and we risk becoming victims of various environmental catastrophes. How can we explain our unparalleled scientific competence and our simultaneous exposure to such significant environmental problems?

There is no easy answer to this question, but our political and economic systems provide at least part of the answer. No matter how well we understand the environment and our impact on it, we must remember that management of a nation's environment is largely left to governmental institutions whose major goal is economic growth and prosperity. In short, if we want to understand environmental problems, we must also know something about the politics and economics of environmental management. This is the objective of this chapter.

Environmental Problems and Public Policies: Deciding What to Do

We can all contribute to improved environmental quality. We can lower our thermostats in winter; we can recycle aluminum cans and newspapers; we can ride bicycles rather than drive to the neighborhood store. Even with our knowledge (and our good intentions), however, few people voluntarily change their life-styles to benefit the environment unless they realize some gain. Others find that they can save money by not doing what is environmentally desirable. A factory owner can install an effective wastewater treatment system, or dump contaminated water into a nearby river at considerably less expense.

When we call for improved environmental quality, we normally expect governments to respond and to require or persuade people to act in an environmentally acceptable manner. Our calls for action reflect a belief that only governments have the political authority and financial resources to develop and implement successful public policies to protect the environment. For our purposes, we consider **public policies** to be those things that governments do or decide not to do. Furthermore, people who request more government programs believe that the government's role in solving environmental problems is preeminent; that is, if our policies and agencies were somehow more effective, then we would not have the serious problems that we have today.

Environmental quality often depends on what governments do, so it is important to understand how public policymakers typically perceive environmental problems and make decisions about them. Before discussing these topics, however, we should consider, from an environmental scientist's perspective, how we might expect governments to act when they address the problems of pollution or allocate resources such as water, fuels, or minerals.

First, if we want to reverse environmental degradation, public policies must address the root causes of pollution. How policymakers view these root causes will affect the public policies they choose. As an example, if policymakers believe that people who litter or factories that grossly pollute the air and water are the primary causes of pollution, then only moderate changes in public policies are probably needed. We can strengthen laws against littering or mandate the use of new pollution-control equipment. Another explanation might suggest that the root causes of pollution and resource exploitation can be traced to widely accepted patterns of consumption and the faulty use of finite natural resources. This explanation places the blame for pollution on a society that expects high and ever-increasing standards of living. Policymakers who accept this explanation realize that fundamental shifts in life-styles and existing public policies are necessary.

Second, we know that all parts of the environment are interrelated, so when one part of the environment is disrupted, other parts are also affected. Some of these effects are noticed immediately, while others become apparent only gradually, perhaps after several years. Therefore, given this scientific principle, public policies should be formulated that reflect a **holistic perspective,** that is, one that considers a problem in its entirety.

Third, the need for a holistic perspective also suggests that different government agencies should thoroughly integrate their activities concerning resources and the environment to avoid contradictory policies.

These three expectations may seem reasonable, but it is extremely difficult for any government to accommodate them. Policymakers are subject to constant and competing pressures from groups and individual citizens with different values and goals. The continuing debate between pro-choice and pro-life advocates offers a good example. It is impossible for the government to satisfy all individuals, so public policies inevitably favor some groups at the expense of others. When the government allocates scarce water supplies in arid regions of the western United States, for example, some groups do not receive as much as they want or believe they should get. Groups that have certain advantages want to maintain their privileged position and resist efforts to alter the status quo. This means that groups that want to change the status quo must persuade policymakers not only that existing policies are unsatisfactory but also that change is necessary. First, however, policymakers must be convinced that a problem exists and that government is responsible for providing a solution.

Selecting Environmental Problems

Policymakers rarely employ a rational, comprehensive approach to problem solving. Even when most scientists agree that an environmental problem exists and should be remedied, that consensus does not necessarily extend to policymakers, who include judges, elected officials, and senior bureaucrats in the government's administrative agencies. Unless these people view a situation as a public problem, it will not receive their attention.

Figure 2.1 One of today's most pressing environmental problems is the location and cleanup of abandoned hazardous-waste disposal sites.
© J. Cancalosi/Tom Stack & Associates

An environmental issue that passes the first hurdle and is recognized as a public problem must still compete for a place on the government's agenda. Thousands of public problems are potential candidates for attention, but policymakers can focus only on a small number of them. Indeed, the policymakers' agenda is already overloaded with problems that require their full attention. Moreover, their agenda reflects a preference for old, familiar, and recurring issues. Every year, for example, we can expect members of Congress to discuss taxes, the budget, national defense, health care, economic growth, and foreign policy.

The consequences of having such a long list of recurring agenda items is that any new problem that gains a position on the government's overloaded agenda must replace a problem that another group has already justified as deserving a place on that agenda. If the agenda is changed, some other group must lose its advantageous position, which explains why some

groups oppose any attempts to change agendas and why changes are so slow to occur.

Despite such opposition, changes in agendas do take place. When hundreds of abandoned hazardous-waste sites were discovered throughout the United States in the late 1970s, the issue was quickly placed before state and federal officials, who responded with new environmental laws and programs designed to require the cleanup of the sites (fig. 2.1).

The chances of adding new items to an agenda are increased when the new problems are easy to understand, can be decided by a yes or no response, or are the result of crisis or catastrophe. New problems are also likely to gain agenda status when their solution is not likely to be expensive, and when they attract widespread attention and produce forceful demands for change. Problems that do not share at least several of these attributes are unlikely to receive serious consideration. This means that policymakers tend to delay or neglect consideration of new problems that are complex, not easily understood, have uncertain short-term consequences, and that are potentially costly to solve. These features are exactly the ones that characterize many environmental problems. Global warming provides an excellent example. During the first years of the Bush administration, the president acknowledged that global warming might be a problem, but he said it was too complex and too little understood to justify immediate restrictions on the factories and power plants that might contribute to the problem by their emissions of carbon dioxide. President Bush was also concerned about the potential costs of an effective remedial program. His Council of Economic Advisers reported that the costs of achieving just a 20% reduction in emissions of carbon dioxide in the United States would range from $800 billion to $3.6 trillion, or several thousand dollars per person.

Even when an issue gains the status of a legitimate public problem, policymakers do not necessarily agree that a response is required. As an illustration, prevention of cancer has been on the government's agenda for many years, and many of the causes of the disease are well known. Tobacco is one of those causes, but the government does not prohibit smoking. Indeed, the national government actually encourages the growth of tobacco. Every year the government provides millions of dollars in agricultural subsidies for tobacco farmers and also encourages the export of tobacco products. Well-organized groups, such as cigarette companies, and their powerful political allies in the Congress, almost always have an advantage in the political process over individuals who share common concerns but are unorganized or are unable to sustain a strong lobbying effort over the many years that may be necessary to enact the needed legislation.

The Political Response

After a policymaker decides that a problem requires a response, what usually happens? Earlier, we suggested that policymakers should adopt a holistic perspective when responding to problems that are related to resources and the environment. Such an approach might be desirable, but policymakers rarely employ it. Instead, policymakers (especially elected officials such as governors, presidents, and state and national legislators) often make decisions on the basis of factors that have little to do with the issue at hand. Many legislators vote for policies that reinforce their political ties (considering whether their vote will embarrass their opponents or be to their party's partisan advantage), not according to scientific information or economic considerations.

Other factors besides partisan advantage similarly affect policymakers' responses to environmental issues. Of these factors, two of the most important are elections and the basis of representation in the United States. Most state legislators and all members of the U.S. House of Representatives face reelection campaigns every other year. Presidents have four-year terms, as do most governors. Only U.S. senators have six-year terms. If these politicians are to be reelected, their short terms mean that they must continually demonstrate their accomplishments in a manner that quickly reflects favor on them. This requirement encourages many legislators to support proposed laws that promise quick results. Furthermore, as a general rule, politicians favor programs that distribute benefits to the current generation of voters. In contrast, programs that might take many years to produce results or that promise benefits for future generations are viewed as being much less desirable.

The basis for political representation also influences the kinds of policies that legislators support. All state and federal legislators are elected to represent specific geographic areas, some of which are no more than a few square kilometers in size. To be reelected, a legislator only needs to satisfy a majority of voters in his or her electoral district. The result is that legislators can safely ignore the concerns of all other districts—or even the concerns of the country as a whole. Therefore, in responding to public problems, a legislator often makes choices only in terms of the potential implications for his or her constituents, such as whether a proposed law will increase the amount of money spent in his or her district, or whether the proposal will create new jobs for some constituents or cause others to lose them.

The consequences of narrowly based geographical representation should not be underestimated. It encourages a narrow view of environmental problems at the expense of a regional, national, or international perspective. Indeed, many legislators are likely to support programs that will help their constituents and oppose programs that will not. In discussing changes in the federal clean air legislation in 1990, for example, many senators from states with large deposits of high-sulfur coal consistently opposed provisions that would reduce the use of that fuel. One senator refused to support the proposed legislative changes unless they contained a provision to provide up to $500 million in compensation for coal miners who might lose their jobs. When the Senate voted on this specific proposal, it lost by only one vote.

Presidents also want to be reelected, and so they too must demonstrate quick successes, perhaps by introducing new ways to deal with existing problems. Moreover, presidents are traditionally assumed to be responsible for the nation's economic health; they can gain much favorable attention by manipulating the economy successfully. Thus, instead of promoting conservation and deferred consumption, presidential policies invariably favor growth and a booming economy.

Incremental Decision Making

Now that we have some understanding of the variables that affect policymakers, we can turn to the process that probably best describes the formulation of public policies, **incremental decision making,** that is, making small changes in existing policies or programs. The major assumption of incremental decision making is that existing approaches to public problems are satisfactory and preferred. According to this approach, when problems arise in the implementation of existing policies, only small, or incremental, changes are needed. As a result, although many laws and public policies are claimed to be new or innovative, most are actually little more than small revisions of existing practices.

For busy policymakers, incremental decision making has many appealing attributes. First, it assumes that the best guide to the future can be found in what is already being done. Second, the incremental approach is easy to use because it makes few demands on policymakers. As a rule, alternative solutions to problems are rarely examined, and those that are considered are often limited to slight variations in the status quo. Third, when new problems arise, public officials usually conclude that existing responses to similar problems can be applied. Such policymakers reject the belief that new problems require solutions that are different from existing ones.

Fourth, in contrast to decision-making processes that produce the most effective policy regardless of the public's reaction to it, incremental decision making allows elected officials to be responsive to public opinion and to engage in bargaining and compromise. By compromising, legislators can respond to competing demands and develop policies that are acceptable to as many powerful and well-organized groups as possible. Thus, many legislators satisfy groups who voice competing demands by voting for vague or ambiguous laws. For example, one goal of U.S. clean air legislation is the protection and enhancement of the nation's air resources in order "to promote the public health and welfare and the productive capacity of its population." No one would disagree that these are desirable goals, but we must also appreciate that the wording of the statement attempts to placate those who favor improved environmental quality as well as those who favor economic growth. In other words, every group is satisfied, at least in terms of how the goal is defined.

As a consequence of the reluctance or inability of legislators to be more specific about their ultimate goals, problems are frequently created for officials in executive-branch agencies who must interpret and implement the laws on a daily basis. These officials are expected to develop precise regulations and guidelines without knowing exactly what goals they are trying to achieve or the relative priorities of each goal before them. Should these officials interpret some environmental laws literally and possibly shut down entire industries?

Should the officials seek penalties from all those who violate environmental laws, or should they focus their limited resources on the worst violators?

Congressional consideration of the Endangered Species Act shows how incremental decision making dominates the policymaking process. This 1973 act is designed to protect endangered species of plants and animals. Subsequent amendments to the act, passed in 1977, 1978, 1979, 1980, 1982, and 1988, have modified the original law. Thus, instead of a series of well-coordinated environmental laws that reflect an integrated perspective, we have dozens of relatively uncoordinated (and occasionally conflicting) federal environmental laws and hundreds of equally uncoordinated state laws. Control and management of hazardous chemicals provides one of many examples. Relevant laws include the Clean Water Act, the Clean Air Act, the Safe Drinking Water Act, the Toxic Substances Control Act, the Resources Conservation and Recovery Act, and the Federal Insecticide, Fungicide, and Rodenticide Act (for which the U.S. Environmental Protection Agency has major responsibility), the Hazardous Materials Transportation Act (Department of Transportation), and the Surface Mining Control and Reclamation Act (Department of the Interior).

The fact that we have so many laws means that each resource or environmental problem is usually treated as if it had little or no relationship to other environmental problems. From a holistic perspective, these laws violate the scientific principle that each part of the environment is interrelated and cannot be effectively considered in isolation.

Why Environmental Policymaking Is Fragmented

Some understanding of how the Congress is organized helps to explain why we lack an integrated set of environmental laws. Briefly, whenever a proposed law is introduced to the U.S. Senate or House of Representatives, it is referred to a committee for consideration. No single committee in either the House or the Senate has sole responsibility for environmental issues. In fact, many committees in each chamber handle these issues. In the Senate, proposals that concern oceans can be sent to the Committee on Environment and Public Works, the Committee on Energy and Natural Resources, or the Committee on Commerce, Science, and Transportation. Moreover, because each chamber establishes its own committee structure, different committees in each chamber often address the same environmental issues.

The committee structure in the Congress (and, likewise, in most state legislatures) obviously fragments decision making and encourages a narrow view of environmental problems. Improved coordination among committees is desirable, but each committee jealously guards its own narrow jurisdictional responsibilities. Although the potential for a holistic perspective exists when a bill reaches the full Senate and House, they rarely overturn the substantive recommendations of their committees. Furthermore, both the Senate and the House have delegated the task of writing environmental laws to certain committees, but they allow entirely separate appropriations committees to decide how much money should be spent to implement the laws. Thus, much environmental legislation fails to receive sufficient funding to produce adequate results.

Given this situation, we might find it remarkable that any environmental laws emerge from the Congress. When they do, however, they consistently share several of the following features. We have already noted that lawmakers prefer incremental laws. In addition, most federal laws reflect a strong emphasis on regulatory as opposed to economic approaches and on engineering and technological solutions. As a result, our laws favor the use of such technologies as catalytic converters for motor vehicles, high-temperature incinerators for toxic substances, and secondary treatment plants for sewage. However well these controls may work, laws that require technological solutions to environmental problems shift much of the responsibility for environmental management and pollution control from individuals who ultimately cause pollution to manufacturers and industrial concerns who are responding to consumer demands for goods and services. In other words, reliance on technological "fixes," or solutions, reflects an assumption that changes in individual or group behavior are largely unnecessary (or, perhaps, impossible) and that technology is capable of solving whatever environmental problems we create.

Environmental laws also reveal a mix of science, politics, and economics, but not necessarily in equal doses. Science can show what is technically possible, but politics and economics temper what the laws require polluters to do. Scientists can provide information about the harmful effects of different levels of pollution, but government policymakers must decide which of these levels (and which levels of pollution control) are acceptable.

Furthermore, our environmental laws are subject to the same processes as other laws. Opportunities are provided for dissatisfied groups or individuals to challenge laws in state or federal courts. These opponents can argue that the law is unconstitutional, that the agency responsible for administering the law has misinterpreted it, or that the agency has exceeded its legal authority in implementing the law. To compound the problem, judges have become increasingly responsive to such complaints. For many years judges deferred to the expertise of government agencies and rarely overturned their decisions. More recently, however, many judges have adopted an activist approach, and they have not hesitated to intervene in environmental decision making. Although the activism is well meant, not everyone agrees that it should be encouraged. In many cases, judicial decisions have increased the effectiveness of environmental laws and have facilitated citizens' access to the courts. At the same time, however, critics of judicial activism complain that some judges consistently ignore the political and economic implications of their decisions. In addition, although judges are expected to reach unbiased decisions, some evidence reveals that many judges are sympathetic to local economic and industrial concerns and are often reluctant or unwilling to impose penalties on companies that are large employers albeit persistent polluters.

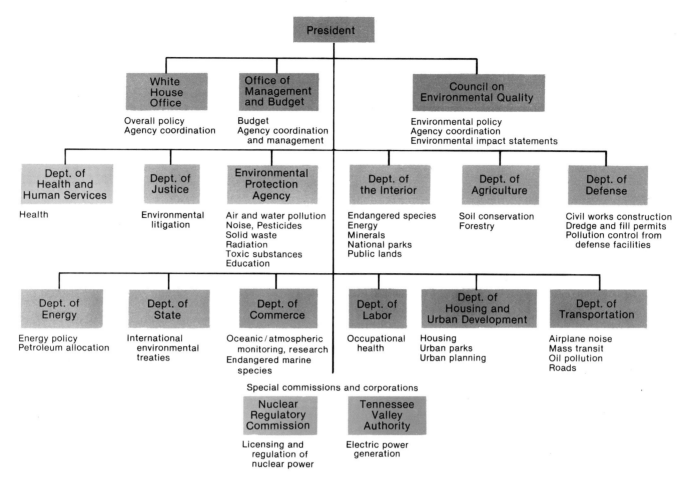

Figure 2.2 The major agencies of the federal government's executive branch that have environmental responsibilities.
Source: *Environmental Protection: Agenda for the 1980s,* U.S. General Accounting Office, Washington, D.C., 1982.

Polluters are not the only groups that encourage increased judicial activism. Frustrated with what they consider to be foot-dragging, environmental groups frequently ask courts to force administrative agencies to issue regulations or pollution standards that a law requires. For example, the Endangered Species Act requires the Department of Commerce to identify, list, and protect marine species in danger of extinction. Despite this requirement, through the 1980s the only time the Commerce Department acted in regard to such species was after it had been threatened with a lawsuit or presented with a petition requesting action.

Why Implementation of Environmental Policy Is Fragmented

If the U.S. government used a holistic perspective, it would coordinate its programs and policies. We can readily agree that such coordination would be desirable, but the distribution of responsibility among federal agencies discourages that type of coordination; many agencies share responsibility for environmental management. As figure 2.2 suggests, within the federal government's executive branch, the job of managing the environment is widely scattered and fragmented.

Although the Environmental Protection Agency (EPA) has programs for pesticides, drinking water, solid wastes, toxic substances, and air and water pollution, the agency must share responsibility for many of these programs with other state and federal agencies, whose concern for the environment is often less important than their primary goals. For example, while the EPA is attempting to reduce sulfur dioxide emissions from coal burning, the Department of Energy would like to increase the use of coal to reduce our dependence on imported petroleum.

Although other agencies may not be pursuing environmentally harmful policies, these agencies are expected to balance the allocation or development of resources with environmental protection. How effectively they achieve this balance is open to debate. During the Reagan administration, many people claimed that the Department of the Interior favored rapid exploitation of mineral and petroleum resources on federal lands at the expense of adequate safeguards for the environment.

Still other agencies are expected to promote urban growth, dredging of rivers and harbors, and the construction of highways and power plants and, simultaneously, to minimize the environmental disruptions associated with these activities. To help ensure that environmental goals are reached, the federal government and many states require the preparation and publication of **environmental impact statements.** At the national level, these statements must be prepared for all major federal actions "significantly affecting the quality of the human environment." Statements must discuss the environmental effects of the proposed action, the environmental impacts that cannot be avoided if the project is completed, and alternatives to the proposed action or project.

The requirement that environmental impact statements be prepared has reduced the adverse environmental impacts of many federal projects, but the process is not without problems. Once a statement is issued in draft form, interested groups and individuals can submit comments to the agency responsible for the project. The agency that prepared the statement must respond to all comments that are made, but it is not obligated to alter the project because of the comments. In addition, the EPA is required to review many environmental impact statements, but this task can be overwhelming. In a typical year, federal agencies prepare hundreds of statements, far more than the EPA can review adequately.

For other activities that do not require impact statements, there is little the EPA can do to ensure that its sister agencies are sensitive to environmental values. The EPA can express concern about practices that are detrimental to the environment, and it can attempt to negotiate changes. If these efforts fail, the EPA can theoretically initiate legal action against an uncooperative agency. In reality, that rarely occurs. The Department of Justice handles environmental litigation for the federal government, so the EPA must first convince that department that legal action is necessary. Furthermore, presidents actively discourage such legal action because they want to avoid the open and embarrassing conflict that would result if the Justice Department took another federal agency to court. Thus the lack of effective sanctions surely discourages any coordinating role that the EPA might wish to play.

The EPA encounters problems with coordination and cooperation not only with other federal departments but with the states as well. Most laws that the EPA administers at the federal level declare that the authority for actual implementation should be delegated to the states. The EPA develops guidelines and regulations for laws concerning pesticides, solid waste, drinking water, and air and water pollution, but the states apply and enforce those regulations. If all states shared a high degree of concern for the environment, few problems would arise with this arrangement. However, the states' commitment to environmental protection has varied. Some state governments sharply disagree with the intent of many federal laws and openly favor industrial growth at the expense of environmental protection. Other states compete with one another for new industrial development or are unable or unwilling to bear the costs of effective control programs. Still other states are reluctant to penalize polluters for fear that industrial plants will close, thus increasing unemployment.

Finally, we should note that the EPA is frequently criticized for its activities and, occasionally, for its leadership. During the petroleum-related energy shortages of the 1970s, for example, many people claimed that the EPA's regulations contributed to the problem. President Ronald Reagan complained that the EPA was not sufficiently concerned with the economic plight of American industry, and his initial choice

of a head for the agency provoked strong denunciations from environmental organizations. During the Bush administration, many environmental groups similarly complained that decisions on key environmental issues, like global warming, were made by officials in the White House rather than by the EPA. This concern seemed justified when one of the president's closest advisors complained that "Americans did not fight and win the wars of the twentieth century to make the world safe for green vegetables."

The consequences of the adversities that the EPA must face are not entirely clear. Nonetheless, we can suggest that any agency whose authority is in doubt and which experiences a constant stream of criticism will be much less assertive than it might otherwise be.

Environmental Problems and Economics

We all favor a clean and healthy environment, but we cannot have such an environment unless we are willing to change our life-styles dramatically or pay for the measures necessary to remedy environmental problems. Most of us are willing to pay for at least some environmental protection, but how many of us would be willing to pay twice as much as we do now for gasoline (fig. 2.3) to cover the environmental cost of burning that fuel in our autos? We must also decide how much "protection" we actually need. The more protection we demand (or the less risk we are willing to tolerate), the higher the cost. In short, decisions about costs and appropriate levels of protection are necessary every time new environmental programs are debated. In 1990, for example, the U.S. Congress considered revisions to the nation's clean air laws. Depending on who was making the estimates, the projected costs of the revisions to consumers and taxpayers ranged upwards of $50 billion per year.

Figure 2.3 While we might all favor a cleaner environment, the economic cost of achieving environmental quality is often a stumbling block.

TOLES copyright 1990 *The Buffalo News.* Reprinted with permission of UNIVERSAL PRESS SYNDICATE. All rights reserved.

Not only was this too expensive, but it would also have provided far more protection than was necessary or desirable, at least according to those who opposed the legislation.

Each of us should be interested in the cost of pollution control because it affects nearly everything we buy, from pizzas and soft drinks to automobiles and aircraft carriers. In 1990, the U.S. spent about $115 billion on pollution control and most of this was borne by consumers and private industry. Perhaps surprisingly, this represented about 40% of what the nation spent on defense during that same year. The EPA expects the cost of pollution control to top $170 billion annually by the end of the century.

Economists too are interested in the costs of pollution control, but their concern is with how a society uses its resources. Economists assume that resources such as money and raw materials are scarce, that each of us acts to maximize our self-interest, and that we want to obtain the greatest possible satisfaction for ourselves. Thus, one goal of economists is to determine how society's resources can be allocated most efficiently among all competing demands. An **efficient economy** is operating when resources are used in the best possible way to satisfy the largest number of consumer demands. An **inefficient economy** is operating when resources are not used to their greatest advantage, so that if they were more productively employed, they could be used to satisfy more demands. Unemployment suggests an inefficient economy because labor (a resource) is not being used effectively; that is, people who could produce something are not producing anything. Similarly, a manufacturer who uses 100 units of energy to make a product is considered to be inefficient if a change in process or technology would make it possible to produce the same product with only 80 units of energy. Improved efficiency is always desirable, but no country's economy is entirely efficient. Consequently, all societies have opportunities to improve their economic efficiency and the use of their scarce resources.

Economists often examine environmental issues to determine how they affect economic efficiency. One consequence is that environmental scientists and economists might disagree about what constitutes an environmental problem. For the scientist, environmental problems usually result from disturbances of the environment. In contrast, economists might not view some disturbances as problems requiring solutions unless they have an adverse effect on the economy. For example, although scientists have largely confirmed their hypothesis that acid rain damages aquatic resources, there is uncertainty about how best to address the problem without jeopardizing several hundred thousand jobs and the ready availability of electricity in the midwestern United States. An economist might thus argue that until more information is available about the costs of various remedies, it is premature and therefore economically inefficient to spend large amounts of money on them.

In contrast to possible disagreements about acid rain, scientists and economists are likely to reach quick agreement about the existence of problems in other areas. Both groups are likely to oppose policies that favor the rapid consumption of expensive virgin natural resources, which, in turn, discourages the use of recyclable materials, such as bottles, newsprint, and aluminum.

If there is an environmental problem and an opportunity to improve economic efficiency, economists will want to know how the persons, corporations, or industries that are responsible for the problem can be induced to change their behavior to improve efficiency. The range of possible approaches that can be used to bring about change is broad, but the initial choice usually involves questions about the relative role of market forces versus government regulation.

Free Markets

Free, competitive markets, as many economists have asserted, are inherently efficient. In **free markets,** resources are allocated and prices are established on the basis of individual, voluntary exchanges among producers and consumers. In a free market system, we get what we are willing and able to pay for, and markets adjust to accommodate changing preferences. If only a few of us want widgets, for example, few will be produced. Manufacturers that continue to produce unwanted widgets will soon go out of business, and resources will be transferred from less efficient to more efficient uses. When particular goods are popular, producers will have incentives to provide more of them. According to this view, supply, demand, and price tend to reach equilibrium, and the production of goods and services reflects the demand for them.

Critics of this laissez-faire, or market, approach correctly note that many economists idealize the role that markets play in the efficient allocation of a country's scarce resources. For markets to operate efficiently, a large number of wise buyers and sellers must have complete information about the price, quality, and availability of all products. No such market exists anywhere. It surely would be desirable for all of us to know how much environmental damage is associated with the manufacture, use, and disposal of the products we use. With this information, we could avoid those products that are particularly harmful to the environment. Unfortunately, however conscientious we may be about protecting the environment, we are not likely to have this information.

In addition, critics of the market approach also contend that it fails in at least two important respects with regard to resource management and environmental protection. The first problem is that not all of the consequences of economic activities are reflected in market transactions. Such unreflected consequences are called externalities, or spillovers. When we are advantaged by what someone else does, we reap external benefits and enjoy **positive externalities.** When we landscape our lawns, paint our houses, and repave our driveways, we increase not only the value of our own property but the overall quality of the neighborhood as well. Our neighbors thus benefit, even though they have not done anything to their own homes.

We have no reason to complain about positive externalities, but the opposite is true for **negative externalities.** These externalities, such as noise, smog, or littered beaches, are the

Figure 2.4 Two examples of public goods. Once fire protection and national parks are available to one user, they are available to all users at no additional cost to those users. Because public goods are essentially free to those who use them, private producers have no incentive to supply them. Thus, in most cases, public goods must be provided by government and financed by taxation.
© Cecile Brunswick/Peter Arnold, Inc.; © J. Lotter/Tom Stack & Associates

undesirable side effects of activities that do not involve us directly. When we experience negative externalities, we face external costs; we bear the negative consequences of someone else's activities. Steel mills produce air and water pollution, two externalities that can be detrimental to the health and well-being of thousands of people. Communities that surround these mills, and not the mills' owners, bear the costs and consequences of the pollution. Therefore, damages that are caused by the pollution are not reflected in the selling price of steel, and the owners have no economic reason to stop polluting the environment. In this case, the free market actually encourages pollution. In other instances, free markets might encourage extensive clear-cutting of forests, overgrazing by livestock on fragile ecosystems, or the destruction of habitats of endangered species.

From an environmental perspective, another criticism of a free market approach focuses on the difference between private goods and public goods. Individuals or companies own **private goods.** Their owners can control the product, prevent others from using or consuming it, and sell the goods to willing buyers at the prices determined by market forces. Cars, stereos, and factories are all examples of private goods. **Public goods,** in contrast, are considerably different; they include, for example, national defense, fire and police services, and environmental protection (fig. 2.4). When public goods are provided to one person, everyone else has access to them at no extra cost.

The distinction between private and public goods is an important one because free markets typically favor the former. On the one hand, a major factor that motivates an economic exchange is profit, so a producer must believe that he or she will realize a profit by providing something to a consumer. If you build a house, you can make a profit from someone who is willing to buy or rent it. Once the house is sold or rented, one less house is available on the market, and as long as more

demand exists, you have reason to build another house, which is a private good. Free markets, then, encourage the production of private goods, which someone can use to make a profit.

On the other hand, free markets discourage the production of public goods. By definition, public goods are available to everyone. Consequently, we have no reason to pay for them, and private producers in a market system have no economic incentive to provide them. Although we have many public goods, governments invariably rely on compulsory taxation to pay for them.

Market forces offer both advantages and disadvantages in terms of solving problems, but few economists are in favor of relying on them exclusively. In all countries, governments affect market forces through such things as welfare policies, the use of progressive taxes, and controls on the supply of money. Thus, we frequently use the term **mixed-market economies** to describe systems, such as the one used in the United States, that combine private, competitive enterprise with some government involvement.

Government Regulation

An alternative to a completely free market system is government regulation—either partial or total. Instead of relying exclusively on prices and market forces to affect behavior, regulation imposes command-and-control techniques and specifies exactly what the private sector must do. Thus, if government relies on regulation to protect the environment, laws could be passed to prohibit the discharge of pollutants into the air, require operators of surface mines to restore mined lands to their original contour and vegetation, or require that our cars and motorcycles have their emission-control devices inspected every six months.

Government regulation is by far the most prevalent method that is used to change the behavior of those who adversely affect environmental quality and our personal well-being. Despite widespread reliance on regulation, however, many people complain about regulation, including many economists. In brief, critics argue that government regulation, because it does not rely on the profit motive, discourages efficiency, innovation, and improved productivity. According to many who are subject to it, government regulation too often ignores the costs of compliance and the variations that exist among different industries and regions of the country. For example, economists ask whether it makes economic sense to require the same pollution-control devices on automobiles in rural areas and on those in Los Angeles or New York City. From the perspective of administrative efficiency, the answer may be yes; in terms of economic efficiency, however, the answer is a definite no. Vehicle owners in rural areas are forced to pay far more than is necessary to compensate for the environmental damage they cause, at least in the areas in which they live. In this case, economists would argue that government regulation causes a misallocation of scarce resources and contributes to economic inefficiency.

This example suggests that neither regulation nor a free market is a completely satisfactory way to affect behavior. Economists recognize this, so they often recommend approaches that combine government intervention with reliance on economic incentives. Such approaches can include preferential tax treatment for companies that install pollution-control equipment and subsidies for users of recycled products. Industries may be required to pay emission or effluent fees, which are based on either the amount of pollution emitted or the estimated value of the damage the pollution causes. Examples of other fee-related systems are widespread. Italians, as an illustration, pay a special tax every time they purchase a plastic bag, while industrial polluters in Finland are taxed for each metric ton of carbon emissions. In the United States, many of us likewise pay a small deposit on returnable beverage containers. In each case, the fees or taxes are included in the prices consumers pay, but fees have the advantage of internalizing, or accurately reflecting, the actual environmental costs of the product or its resulting pollution.

Costs, Risks, and Benefits—Some Analytic Tools

Economists are interested in more than just affecting behavior. One of their goals is to improve the allocation of a country's resources, so economists also want to know how much of a change in behavior is necessary—or how large an expenditure is desirable to control pollution—xin terms of economic efficiency. To answer these questions, economists use several analytic tools.

One of the most popular of these tools is **cost-benefit analysis,** in which theoretically all the gains, or benefits, of a project are compared with all the corresponding losses, or costs, of that project (fig. 2.5). If the benefits exceed the costs by a sufficient margin, a project or activity is usually deemed desirable and worthwhile to pursue. In spite of its apparent simplicity, however, cost-benefit analysis is a controversial approach. It assumes that everything, such as a human life, a clean river, or an endangered plant species, has a monetary value that we can measure.

Cost-effectiveness is a concept that is closely related to cost-benefit analysis. A **cost-effectiveness analysis** makes no judgments about the desirability of a project. Instead, given a particular goal, an economist tries to determine how the goal can be achieved for the lowest cost. If our goal is to reduce the incidence of cancer, for example, we need to decide whether we should ban the use of all known cancer-producing substances (carcinogens) through regulation, spend more money on research to prevent cancer, develop anticancer immunizations, or cut our exposure to ingested carcinogens by consuming more fiber. In assessing the cost-effectiveness of these alternatives, we consider social, economic, and administrative costs.

Another approach, **risk-benefit analysis,** weighs the risks of a product or activity against its benefits to determine whether the activity should be tolerated or allowed. We frequently apply risk-benefit analysis without realizing it. When we choose to live or work in a notoriously smoggy city, we make a judgment that the economic or other benefits of doing so outweigh the health hazards of breathing air that is badly polluted. When we eat cured meats, such as bacon, we make a judgment that the benefits of eating bacon vastly exceed the potential risks of exposure to the preservative that is in it, which is usually sodium nitrite. When sodium nitrite interacts with certain chemical compounds, small amounts of nitrosamines result, and these compounds are potent carcinogens. In contrast, if we eat cured meats that contain no preservatives, we increase the risk of contracting botulism, a potentially fatal type of poisoning that is caused by the ingestion of toxins (poisons) that are produced by certain bacteria.

Risk-benefit analysis has a certain appeal to people who are involved in environmental decision making. Most pollution-causing activities produce benefits that we enjoy on a daily basis, and we must decide whether those benefits outweigh the associated risks. Risk-benefit analysis can help us to make those decisions.

Marginal costs provide another useful analytic concept that is especially important in the control of pollution. In economists' terms, **marginal costs** reflect the additional expense of producing an extra unit of output or one less unit of pollution. Just as when we sweep a floor, the first increments of dust or pollution are the cheapest and easiest to remove or control. When a polluter seeks to reduce further increments of pollution, however, the costs of control rise faster than the amount of pollution that is reduced. For example, a company

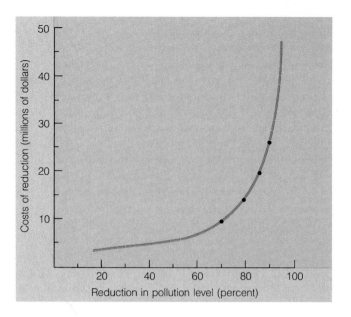

Figure 2.5 Cost-benefit analysis makes the assumption that a monetary value can be assigned to everything in the environment.
Used by permission of Chris Riddell.

Figure 2.6 The marginal cost of pollution control rises ever more steeply with each additional increment of control. In this example, a control level of 70% would cost $10 million, or approximately $143,000 for each percent of polluting substance that is removed. Increasing the control level to 80% would cost another $5 million, or $500,000 for each additional percent removed. And removing just 5% more would cost still another $5 million, or $1 million for each additional percent removed. The shape of the curve varies with actual conditions, but a steeply rising slope is typical of curves that plot the marginal costs of pollution control.

might find that 70% of its effluents can be reduced at a cost of $10 million (fig. 2.6). To reduce another 10% would cost another $5 million; to reduce another 5% would cost another $5 million. In other words, the company has high marginal costs—the costs of reducing the last increments of pollution are significantly greater than the costs associated with the first increments.

From an economist's perspective, marginal costs should be an important consideration in deciding how much pollution should be controlled. Does it make economic sense, for example, for a government to require one company to spend $5 million to reduce 5% of its emissions when another company could spend the same amount of money to control 40% of its pollution? Consideration of marginal costs and the use of cost-benefit or risk-benefit analysis can help to provide an answer to this question.

Although economists employ many other concepts and analytic tools, the ones we have described provide an initial framework for understanding how economists analyze environmental issues. If one theme characterizes this view, it is that decisions about the environment should be made rationally, with the goal of making the best use of our resources. How likely we are to reach this goal depends on our choice of decision-making processes and how responsive the political system is to both scientific and economic concerns about environmental problems.

A Global Perspective on Environmental Problems

However serious environmental problems may be in the United States, we must recognize that much of the world's environment is in even worse condition (fig. 2.7). Hundreds of millions of Asians, Africans, and South and Central Americans find themselves enmeshed in abysmal poverty, unable to feed, clothe, or shelter themselves properly, and without access to essential medical care or proper sanitary facilities. Most of these people are condemned to a life of illiteracy. Millions of children in these less-developed nations will die within a few years of birth because of starvation or illnesses such as measles, tetanus, or diarrhea that are easily prevented and readily cured, at least in the United States and other more-developed countries.*

If poverty is the criminal, then the environment is one of the victims. Due to the pressures of poverty, many people find that the only way to survive is to assault the environment. Tropical forests, which recycle much of the world's carbon dioxide and contain much of the world's biological diversity, are being destroyed at unprecedented rates in order to provide agricultural land or revenues from timber exports. In areas around Africa's Sahara Desert, the constant search for fuelwood denudes the landscape and leads to expansion of the desert.

*Less-developed nations are characterized by low levels of personal income, high infant mortality, illiteracy, and high population growth rates. These nations include most of the countries of Africa, Asia, and Central and South America. The *more-developed* nations are characterized by high levels of personal income, low infant mortality, high literacy, and low rates of population growth. Sometimes less-developed nations are called *developing nations,* but this term is not always satisfactory because many so-called developing nations are either not developing or are actually slipping backwards economically. More-developed nations are sometimes called *developed nations,* but some people argue that they are really overdeveloped.

Unfortunately, rapid population growth in many less-developed countries compounds the problems of poverty, and many countries find it increasingly difficult to grow sufficient food to maintain their populations. As long ago as the mid-1970s, for example, Kenya had the agricultural potential to support less than 30% of its population. If current rates of population growth continue, Kenya will see its population rise from about 25 million people in 1990 to 38 million by 2000. The consequences of such explosive population growth in Kenya and many other less-developed countries are dire. Such nations soon will exceed their **carrying capacity,** the maximum population that a given area's resources can sustain indefinitely.

Many environmental problems thus have global dimensions and require global cooperation for their solution. Although we may be primarily concerned with environmental problems that afflict our own communities, we cannot neglect what happens in the villages of Africa or the cities of Asia. If the world's environmental problems become even more severe, countries will increasingly turn to the United States and other more-developed nations for advice and leadership, and for money as well. Only a few less-developed countries can afford the technologies or pollution controls necessary to halt much of their environmental degradation. As a result, many political leaders in less-developed countries say that they will not be able to stem pollution, deforestation, or the loss of biological diversity—all of which ultimately affect each of us—without technical and financial assistance from the rich countries. Whether the rich countries provide this assistance depends, once again, on politics and economics. Are we responsible not only for solving our own environmental problems, but also those of our global neighbors? So far, our government's response has been uncertain.

What Does All This Mean?

The American government has not fared well in our evaluation of its response to existing environmental problems. We might be disappointed, but we should not be surprised. In their efforts to be responsive to groups that have different values and goals, all democratic governments attempt to achieve diverse and frequently competing goals. Concern for environmental quality and protection of our rivers, national parks, and endangered species represent only a few of these goals, and they must compete with efforts to ensure a strong economy, an adequate national defense, sufficient energy supplies, a healthy and adequately fed population, appropriate care for children, the elderly, and the disabled, and so on. Most people would agree

Figure 2.7 Scavengers pick through garbage at Manila's "Smokey Mountain" dump. In this less-developed nation (the Philippines), the poor sustain themselves on what others discard.
© Fred McConnaughey/Photo Researchers, Inc.

that these goals are reasonable, and few people would argue that governments should abandon such goals in favor of exclusive concern for the environment. Moreover, in a democratic system of government, a close (but not exact) relationship usually exists between what people want and what governments provide. Consequently, a government's relative attention to the environment is, to a large extent, a reflection of public preferences and priorities. Thus, the problem (and the solution) lies with individual citizens both in their choice of life-styles and in what they demand of local, state, and national governments.

At least two other considerations should also temper our disappointment with the government's handling of environmental issues. First, we should understand that governments are expected to resolve society's most complex and difficult problems. This means that governments find themselves saddled with issues that are highly controversial and do not have clear solutions. The debate over acid rain provides a premier example. Except for an extraordinary program of energy conservation, all proposed solutions will cost billions of dollars, raise electricity bills, discourage the use of readily available high-sulfur coal, and increase U.S. dependence on imported petroleum. Unfortunately for policymakers, in the absence of indisputable data that indicate exactly what should be done about acid rain, they will be criticized for whatever they do or do not accomplish.

Second, whatever their shortcomings may be, our environmental laws and agencies have produced measurable improvements in the quality of our environment. As the chapters that follow demonstrate, there is considerable cause for concern about the quality of our environment, but there is also reason for optimism. Many of our lakes, rivers, and streams are less polluted than in the past; it is easier to breathe in many metropolitan areas; we have begun to clean up abandoned hazardous-waste sites. Our endangered species program is one of the best in the world. On a per capita basis, few other nations spend as much as we do to protect the environment. (On the other hand, few nations pollute as much as we do.) Moreover, public opinion surveys consistently demonstrate a growing concern about pollution as well as an increasing desire to protect the environment. In one national survey conducted in 1981, as an example, 45% of the respondents agreed that improvements to the environment should be made regardless of the costs of doing so. By 1990, however, nearly 75% of the respondents in another national survey agreed with the same statement.

Our accomplishments may be impressive, but merely listing them obscures some important questions. Therefore, as you learn about the science underlying environmental functions in this book, ask yourself whether public policies in the United States and elsewhere are sufficiently responsive to our environmental and resource-management problems. Also, try to determine how governments should respond if these problems get worse in the future. Think about whether we need more government, less government, or perhaps the same level of government but with different policies and priorities. Think as well about how we are all affected by what happens to the world's environment—regardless of where we live.

Unfortunately, ready answers to these issues do not exist. Some people fear that our future environmental problems will be so severe that we will not be able to survive without vast increases in government authority. If these people are correct, our civil liberties will be diminished, and our standard of living will decline.

In contrast to those who anticipate the need for increased government authority, others argue with equal vigor that we already have too much government, that it creates unnecessary intrusions in our lives, and that many government programs cause, rather than solve, environmental problems. Advocates of this view also believe that without continued economic growth, we cannot sustain or improve the world's standard of living or continue to pay for environmental protection.

Although questions remain concerning what the appropriate role of government should be, we can be certain that our personal actions—today and tomorrow—will determine the answers. If our life-styles and those of our parents contributed to government as we know it today, then just as surely, what we and our children do in the future will ultimately determine what governments can, must, or should do in response.

Conclusions

Environmental science is a problem-oriented, interdisciplinary field of study. Although environmental science primarily involves physical, chemical, and biological processes that operate in the environment, we must recognize that environmental issues have important political and economic dimensions as well. We have learned that scientists, policymakers, and economists differ in important ways in their approaches to environmental issues. Scientists from many disciplines apply the scientific method in an attempt to understand the natural functioning of the environment and the causes of and solutions to environmental problems. Elected officials, agency bureaucrats, and other public officials respond to pressures from the public and special-interest groups in shaping policies for the environment. Finally, economists view environmental problems in the context of economic efficiency and use a variety of analytic tools to assess the costs of responding to environmental problems.

Key Terms

public policies	negative externalities
holistic perspective	private goods
incremental decision making	public goods
environmental impact statements	mixed-market economies
	cost-benefit analysis
efficient economy	cost-effectiveness analysis
inefficient economy	risk-benefit analysis
free markets	marginal costs
positive externalities	carrying capacity

Summary Statements

The management of a nation's environment is primarily the responsibility of governmental institutions whose major goal is economic growth and prosperity.

For politicians and key policymakers, concern for environmental protection is only one of many issues that deserve attention. The amount of attention that can be devoted to a single issue is not only limited but also subject to pressures from individuals and groups advocating a wide variety of competing goals.

Public policymaking for environmental protection does not reflect a holistic perspective; rather, it reveals considerable fragmentation at all levels of government. Most governmental policymaking tends to be incremental and to rely on existing solutions or approaches.

All decisions about environmental protection or management ultimately include some economic considerations; each decision involves who will bear the costs and consequences of those decisions.

There are contrasting views about the best way to achieve a healthy environment. Advocates of free, competitive markets assert that individuals, acting to maximize their self-interests, will produce an acceptable level of environmental quality. Other people favor government regulation because they believe that free markets discourage the production of public goods, including environmental protection.

The use of such economic tools as cost-benefit analysis, cost-effectiveness analysis, and risk-benefit analysis improves our understanding of the consequences of various environmental policies.

Many of the world's less-developed countries endure problems of widespread poverty and severe environmental degradation. These problems may not be immediately apparent to us, but how they are handled affects everyone.

Despite our ability to identify many shortcomings in the government's handling of environmental problems, we can point to many successes and increasing public support for efforts to remedy our environmental problems.

Politics and economics help us to understand our current environmental situation, and each of us has opportunities and responsibilities with regard to the environment.

Questions for Review

1. What are some criticisms of government regulation as an approach to environmental protection?
2. What are the basic characteristics of issues that are most likely to be considered by policymakers? Do these characteristics favor environmental issues?
3. Should environmental standards be established to protect some, most, or all of the population from the adverse consequences of pollution? Who should make these decisions, and what information should they consider?
4. What is incremental decision making, and why is it so popular among policymakers?
5. In seeking to satisfy the demands of competing groups, why do legislators often rely on vague or ambiguous laws? What are some possible consequences of such laws?
6. Explain why policymaking for environmental protection does not adopt a holistic perspective.
7. What are the advantages and disadvantages of relying on free, competitive markets to provide acceptable levels of environmental quality?
8. What are the differences between positive and negative externalities, and why are these differences important?
9. What might be some of the consequences of making environmental protection a private rather than a public good?
10. What assumptions must economists make when they employ cost-benefit and cost-effectiveness analysis? Are these assumptions valid?
11. Explain how environmental problems in less-developed countries potentially affect everyone. If these poor countries cannot afford to solve these problems, how can we expect the problems to be resolved?

Selected Readings

Conservation Foundation. *State of the Environment: A View Toward the Nineties.* Washington, D.C.: Conservation Foundation, 1987. A comprehensive assessment of U.S. programs to improve the condition of the U.S. environment and the management of its natural resources.

"The Environment: The Politics of Posterity." *The Economist,* 2 September 1989. This special supplement provides an excellent introduction to economics and its relationship to the environment, including that in less-developed countries.

Hines, L. G. *The Market, Energy, and the Environment.* Boston: Allyn and Bacon, 1988. An economist applies the tools of the trade, including cost-benefit analysis, to various environmental issues.

Hynes, H. P. *Earth Right, Every Citizen's Guide.* Rocklin, Calif.: Prima Publishing & Communications, 1990. Describes major contemporary environmental issues and identifies what each of us can do about them.

Milbrath, L. W. *Envisioning a Sustainable Society: Learning Our Way Out.* Albany: State University of New York Press, 1989. An eminent social scientist considers how society can cope with environmental change to avoid ecological catastrophe.

Tobin, R. *The Expendable Future: U.S. Politics and the Protection of Biodiversity.* Durham, N.C.: Duke Univ. Press, 1990. A valuable critical examination of political behavior in the context of preserving the world's diversity of plants and animals.

Vig, N. J., and M. E. Kraft. *Environmental Policy in the 1990s: Toward a New Agenda.* Washington, D.C.: CQ Press, 1990. A collection of seventeen perceptive essays by economists, philosophers, and political scientists.

Part II

The Science of Ecology

*T*o develop practical solutions to environmental problems, we must first understand how the environment works. In part 2 we consider some of the fundamental principles that govern nature's activities. In chapter 3, we examine the flow of energy through food webs and the cycling of materials within the environment. The environment is constantly changing, either naturally or as a consequence of human activities. Hence, in chapter 4, we consider the responses of ecosystems and organisms to such environmental stresses as water and air pollution, fire, and agriculture. In chapter 5, we describe the patterns of population growth as well as biological and physical controls on that growth. Only by understanding the dynamics of population growth can we control the pests that attack our crops and livestock, preserve biodiversity, and identify options for limiting human population growth. Applying these ecological principles to what we learn in subsequent sections of this book enables us to better understand the environmental consequences of human activities. As our increasing numbers continue to impact the planet, a better understanding of ecological principles becomes increasingly important to our personal well-being.

Ecology is the study of the interactions of living and nonliving components of the environment. This unique ecosystem on Kauai, Hawaii, is the rainiest spot on earth.
© Brian Parker/Tom Stack & Associates

Chapter 3

Ecosystems
Energy Flow and the Cycling of Materials

This scene of a feeding polar bear illustrates one step in the flow of energy within an ecosystem.
© Jack Stein Grove/Tom Stack & Associates

When scientists landed on Christmas Island in November of 1982, they could tell immediately that something was wrong. Expecting to find tens of thousands of seabirds and their nestlings on this coral atoll in the central Pacific Ocean, they discovered instead that virtually all the adult seabirds had disappeared from the island. They found many dead nestlings, and the few live nestlings were starving. Eighteen species of seabirds, including petrels, shearwaters, terns, and frigate birds experienced total reproductive failure. Subsequent reports from other scientists told of similar breeding failures among seabirds as well as marine mammals, including sea lions and fur seals, throughout the eastern Pacific equatorial zone. In addition, many areas reported high death rates even among adult seabirds and marine mammals. Although many adults had attempted to leave the afflicted areas, mortality among these migrants was also high. Why were so many animals dying?

The cause of the problems was an extreme occurrence of **El Niño.*** El Niño is the Spanish name given to a large-scale change in wind systems and ocean currents that can have a profound effect on the survival and reproduction of organisms that live in the eastern Pacific equatorial zone. Although occurrences of El Niño are rather frequent (twenty have been documented since 1900), they are highly variable in intensity, geographical extent, and duration.[†]

One of the first signs of El Niño is a wind shift in the equatorial Pacific Ocean. The normal east-to-west winds reverse direction and drag a large mass of warm water eastward toward the South American coast (fig. 3.1). This top layer of warm water prevents the upwelling of cold, nutrient-rich bottom water to the surface. As a result, the growth of microscopic algae that normally flourish in these nutrient-rich areas plummets. In 1982-83, El Niño was so intense that it triggered a twenty-fold reduction in the quantity of algae along the South American west coast. This decline reduced the number of anchovy, which feed heavily on the algae, to a record low. Other fish that depended on the algae, such as jack mackerel, also declined in number. With the reduction of fish populations, marine birds (e.g., frigate birds and terns) and marine mammals (e.g., fur seals and sea lions) also suffered major population declines as the loss of their food sources led to breeding failures. The impact of weak and moderate occurrences of El Niño is usually confined to the coastal regions of Ecuador and Peru. In 1982-83, however, El Niño was so severe that its influence extended far westward from the coast of South America to remote Christmas Island, located near the equator and south of Hawaii.

El Niño brings changes in the atmosphere as well as in ocean currents. For example, torrential rains between mid-November 1982 and late January 1983 caused the worst flooding of the century in Ecuador. In fact, much of the eastern

*El Niño usually first appears during the Christmas season, and El Niño is the Spanish term for the Christ child; hence the name.

†Changes in wind direction and ocean currents occur almost every year around Christmas, but they frequently cover a relatively small region and last only a few months. Scientists reserve the term El Niño to describe large-scale changes that may persist for over a year.

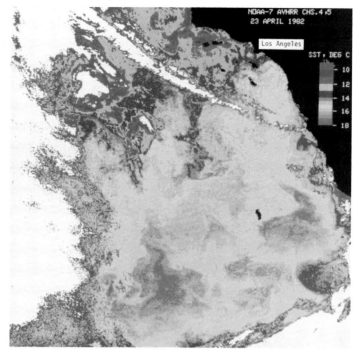

(a)

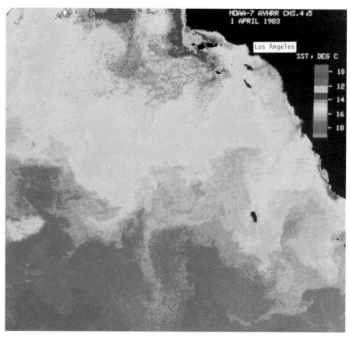

(b)

Figure 3.1   Sea-surface temperature patterns associated with the 1982–1983 El Niño illustrate the expansion of warm waters (red areas denote the warmest temperatures) from April 1982 (*a*) to April 1983 (*b*).
NOAA

equatorial Pacific zone experienced unusually heavy rainfall. Thus, in addition to reduced food resources, heavy rains and flooding probably contributed to the reproductive failure of birds on Christmas Island and elsewhere.

By July 1983, the return of more normal oceanic and atmospheric conditions triggered an increase in algal productivity, which in turn spurred renewed breeding activities among

(a)   (b)   (c)

Figure 3.2   Examples of terrestrial ecosystems: (*a*) a desert in Arizona, (*b*) pronghorns on grassland, (*c*) a shallow pond.
(*a*) © Charlie Ott/Photo Researchers, Inc. (*b*) © G. C. Kelley/Photo Researchers, Inc. (*c*) © Tom Stack/Tom Stack & Associates

marine mammals and seabirds. Many species have recovered to their former population levels, but others have shown little recovery. Because fishers have overexploited the anchovy fishery along the South American coast, food for seabirds and marine animals in this region remains quite limited despite the return of favorable growing conditions. Hence, many populations of seabirds and marine mammals in this region remain near record lows.

El Niño illustrates the interdependence of organisms, their vulnerability to environmental change, and some of the interactions that link organisms, atmosphere, land, and oceans. To better understand and predict the outcome of these relationships, we need to investigate the ecological principles that govern how plants, animals, and microorganisms interact with each other and with the chemical substances and physical conditions in their environment. In this and the following two chapters, our goal is to understand these principles and their major ramifications. Throughout the remainder of this book, we will apply these principles in order to understand the ecological basis of environmental problems and to evaluate alternative solutions to environmental problems.

We begin our exploration of ecological principles by considering two essential processes: the flow of energy and the cycling of materials. We can learn to understand these processes by examining how ecosystems function. Thus, we first consider two questions: What is an ecosystem, and what are its components?

Ecosystems: Types and Components

An **ecosystem** consists of communities of organisms that interact with one another and with the physical conditions and chemical substances that characterize their environment. To some extent, we are already familiar with terrestrial ecosystems such as forests, grasslands, or deserts (fig. 3.2). Most of us live in highly-modified terrestrial ecosystems such as large metropolitan areas, villages, and farms. We also interact with aquatic ecosystems, including lakes, streams, oceans, and wetlands such as marshes (treeless aquatic ecosystems dominated by grasses) and swamps (aquatic ecosystems dominated by

shrubs and trees). We will consider the interactions of humans and ecosystems frequently throughout this book.

All ecosystems have both living (**biotic**) and nonliving (**abiotic**) components. The biotic community consists of plants, animals, and microorganisms.* Green plants use sunlight as a source of energy to produce sugars from water obtained from the soil and carbon dioxide obtained from the atmosphere. Plants then convert these sugars to other chemicals, such as cellulose and starch. Plants, therefore, are often called **producers.** In contrast, because animals and most microorganisms cannot utilize solar energy directly, they must consume other organisms to obtain energy. Animals and most microorganisms are therefore called **consumers.**

Consumers fall into one of several categories based on their food source. Some consumers generally feed upon living organisms, whereas others feed upon dead remains of plants and animals. Those that feed upon living organisms fall into one of three major categories. A consumer that eats only plants is called a **herbivore** (plant-eater), whereas a consumer that eats only animals is called a **carnivore** (flesh-eater). A consumer that eats both plants and animals is called an **omnivore** (everything-eater). Familiar herbivores include rabbits, squirrels, chipmunks, deer, sheep, and cattle. Common carnivores include birds of prey (eagles and hawks) and members of the cat family (lions, cheetahs, bobcats, and the house cat) and the dog family (wolves and the domestic dog). Most humans are omnivores. When you eat a bacon, lettuce, and tomato sandwich, for example, you are consuming both plant and animal products. Bears and pigs are other examples of omnivores.

Have you ever wondered what happens to leaves after they fall to the forest floor in autumn? When plants (or their components, such as leaves) and animals die, they become resources for other organisms. Dead plant and animal remains are termed **detritus.** If the remains are not eaten immediately, they are usually colonized quickly by **decomposers,**

---

*Some instructors prefer to expand these three categories to the five kingdoms: bacteria, protists (including green algae and amoeba), fungi, plants, and animals. For our purposes, the three categories are sufficient.

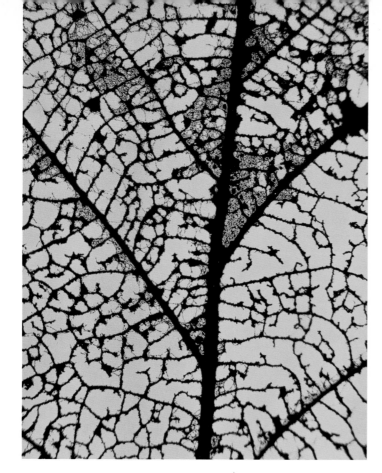

Figure 3.3   Skeleton of a leaf, the result of bacterial and fungal decomposition of the leaf's softer tissues.
© L. West/Photo Researchers, Inc.

Table 3.1   Chemical composition of a 70-kilogram (154-pound) person

	Grams	Percent
Inorganic chemicals		
water	41,400	59.1
calcium	1,160	1.7
phosphorus	670	1.0
potassium	150	0.2
sulfur	112	0.2
chlorine	85	0.1
sodium	63	0.1
magnesium	21	0.03
iron	3	0.004
Organic chemicals		
fat	12,600	18.0
protein	12,600	18.0
carbohydrates	300	0.4

Source: Data from A. C. Guyton, *Textbook of Medical Physiology,* 7th ed.
Copyright © 1986 W. B. Saunders College.

primarily bacteria and fungi. These organisms utilize detritus as a source of energy and nutrients. During the process of decomposition, complex chemicals are broken down into simpler products that return to the soil, air, and water where they can again be taken up by green plants and recycled through an ecosystem (fig. 3.3). The green mold (fungus) found on rotting (decomposing) fruit left too long in a refrigerator is a familiar example of a decomposer at work.

Frequently detritus and its associated decomposers are consumed by **detritivores** before decomposition is completed. Examples of detritivores include earthworms, snails, slugs, and the larvae (maggots) of certain flies. Although many people abhor even the sight of detritivores, these organisms, particularly earthworms, play essential roles in maintaining soil fertility.

The abiotic (nonliving) components of an ecosystem include chemical substances and physical conditions. Chemicals are either inorganic or organic. **Inorganic chemicals** include relatively simple substances such as water (H_2O), oxygen (O_2), and carbon dioxide (CO_2), and minerals such as calcium (Ca) and phosphorus (P). **Organic chemicals** usually have more complex structures than inorganic substances. Organic substances have carbon and hydrogen as primary elemental components and include carbohydrates, fats, proteins, and vitamins. (Although the gases carbon monoxide (CO) and carbon dioxide (CO_2) contain carbon, they are

usually classified as inorganic chemicals.) The survival of organisms requires the interaction of many inorganic and organic chemicals. For example, we are composed of about 60% water (an inorganic substance), 5% minerals (inorganic substances), and 35% organic compounds. Table 3.1 provides a more complete description of the chemical composition of humans.

Abiotic components of an ecosystem also include such physical factors as sunlight, temperature, wind, and fire. These components play major roles in determining the types and numbers of organisms that reside in an ecosystem.

The Flow of Energy

When we walk in a woods in the springtime, we sense activity all around us. Birds are busy building nests and raising their young. Bees fly from one flower to another gathering nectar to make honey. Chipmunks scurry across the forest floor in search of food. Trees and shrubs are leafing out, and wildflowers are springing from the soil. A breeze carries fresh scents to us. In the distance we may hear the faint sounds of an approaching thunderstorm. All of this activity expresses the basic dynamism of nature.

All of the activity within an ecosystem is made possible by the flow, or transfer, of energy. **Energy** is defined as the ability to do work or to produce change. An example from the physical world is the energy that drives the wind, which can uproot trees; an example from the biological world is the food energy that enables us to carry out our daily activities, such as walking and thinking. In this section, we focus primarily on how energy flows through and sustains organisms. In later chapters, we explore how energy drives the water cycle, the weather, and our industrial society.

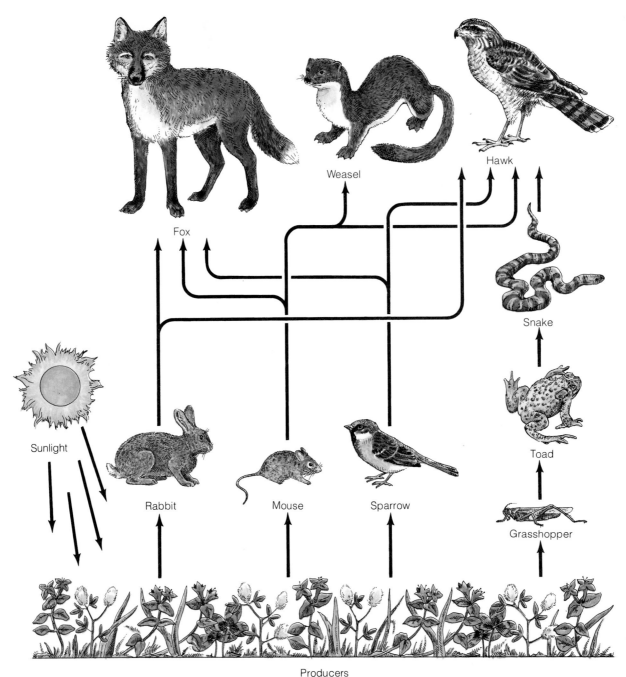

Figure 3.4 A simplified model of a food web, composed of a network of interconnected food chains.

Food Webs

Most of us eat without thinking much about how eating keeps us alive. Hunger pangs, however, are frequently our body's way of telling us that we need more energy so that we can continue to do the things that we need and like to do. We require a continual supply of energy for the replacement or repair of damaged or aging body parts. Recall, for example, in chapter 1 the discussion of our body's continual replacement of aging red blood cells. In addition, every time we eat, we form a link with other organisms, thereby playing a role in the continuous flow of energy through ecosystems.

All organisms, dead or alive, are potential sources of food energy for other organisms. That energy always travels in one direction only—from producers to consumers. In a clover field, for example, a rabbit may eat a clover plant and then fall prey to a fox. This simple energy pathway is called a **food chain.** In natural ecosystems energy flows through numerous interconnected food chains that form networks called **food webs.** Figure 3.4 illustrates a model food web; the arrows indicate the flow of energy, in the form of food, from one organism to another. Although it may appear complicated, this model, like most models, is overly simplistic. We would have to add hundreds more species to portray the actual complexity of almost any naturally occurring food web.

Each group of organisms along an energy pathway occupies a **trophic level,** that is, a feeding level. All green plants (producers) in an ecosystem belong to the first trophic level. Herbivores compose the second trophic level. Carnivores that eat herbivores make up the third trophic level. Carnivores that consume other carnivores are on the fourth trophic level, and so on. Omnivores function at more than one level. For example, a bear functions at a lower trophic level when eating berries than when eating a fish.

The sun is the ultimate source of the energy that sustains all organisms. On an annual basis, only about 1% of the solar energy that strikes plant leaves actually enters food webs. By the process of **photosynthesis,** clover plants, for example, use light energy from the sun to combine low-energy substances from their environment (carbon dioxide from the air and water from the soil) and produce sugar, which is a form of carbohydrate that contains a relatively large amount of energy. The process of photosynthesis is summarized as follows:

Carbon dioxide + water + light energy → sugar + oxygen

$$6CO_2 + 6H_2O + \text{light energy} \rightarrow C_6H_{12}O_6 + 6O_2$$

Plants convert sugars into other types of organic chemicals: proteins, fats, and starch (another form of carbohydrate). These chemicals play two roles: (1) they are the basic building blocks in the construction and repair of living **cells,** which are the basic structural and functional units of plants and all other organisms; (2) they serve as the energy source for performing the work of maintenance, growth, and reproduction.

Photosynthesis also releases oxygen. Plants use some of the oxygen they produce in various metabolic processes; the excess escapes into the atmosphere. Essentially all of the oxygen in the atmosphere is the product of photosynthesis over the past hundreds of millions of years. Figure 3.5 illustrates the sources of raw materials and the distribution of the products of photosynthesis within a plant.

Within an ecosystem, there are two types of food webs: grazing food webs and detritus food webs. The former are more familiar to us. **Grazing food webs** are based upon living plants. In a clover field, for example, rabbits, mice, and other herbivores feed on plants to obtain needed energy and nutrients. In turn, weasels, hawks, and other carnivores feed on the herbivores, gaining from them the energy and raw materials to sustain their metabolism and to grow and reproduce.

Not all of the plants growing within an ecosystem are harvested by herbivores. Eventually, the unharvested plants die, and their remains along with those of dead animals form the foundation of **detritus food webs.** Although detritus food webs can be quite important, they are generally less conspicuous than grazing food webs. We can get a sense of how a detritus food web works by examining the energy flow at the headwaters of a stream, as shown in figure 3.6. Because such streams are commonly shaded by a canopy of trees, there is too little light to support much plant growth. Most of the food, therefore, consists of detritus (dead insects and fallen leaves, for example) washed into the stream from adjacent

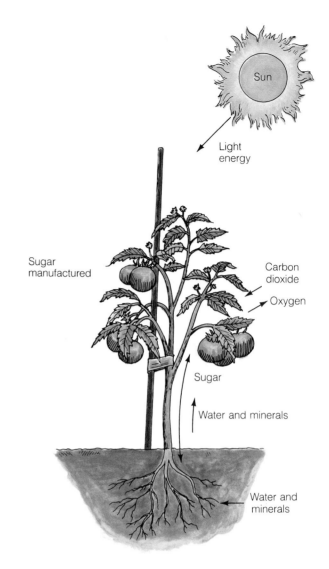

Figure 3.5   Sources of raw materials and the distribution of the products of photosynthesis in a tomato plant. Sugars can move upward through the stem to be used for growth or stored in seeds or fruits, or they can move downward to the roots to be used for growth or to be stored.

land. Such remains have either already been colonized by decomposers (bacteria and fungi), or they soon will be. These species initiate decomposition. In addition, certain detritivores, termed *shredders,* begin to break the larger fragments of detritus into smaller pieces. In doing so, shredders consume some of the detritus, while part of the remainder becomes food for another group of detritivores called *collectors.* Many of the collectors are filter-feeders; that is, they sift small particles of organic material from the flowing current of water. Larval forms of aquatic insects function as shredders and collectors. These species subsequently fall prey to immature forms of other insects, such as dragonflies, which in turn are eaten by predatory fish such as trout.

The relative importance of grazing and detritus food webs within an ecosystem depends upon the nature of the ecosystem. Detritus pathways usually dominate in such aquatic ecosystems as streams, rivers, and marshes, and in

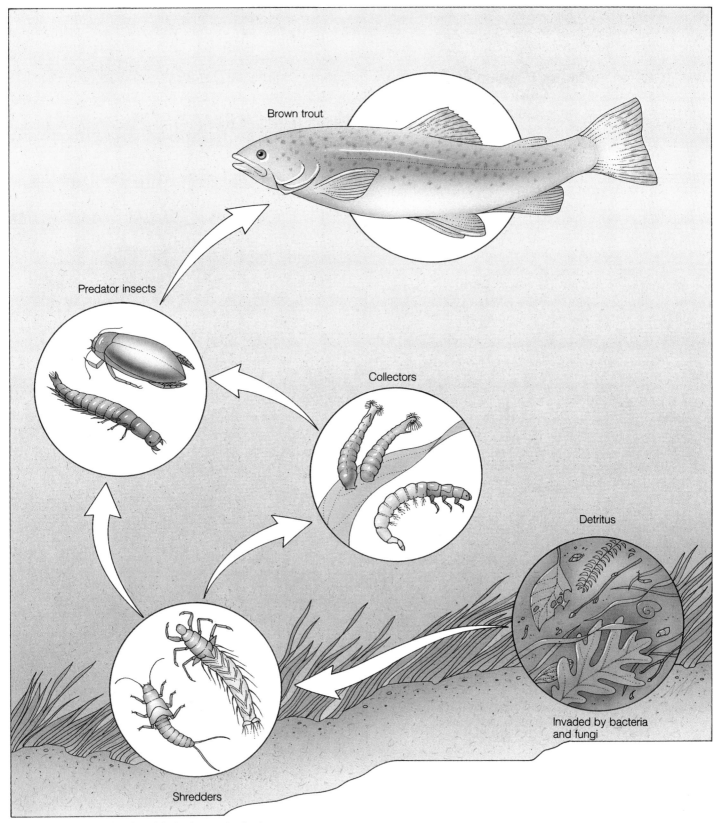

Figure 3.6   A simplified model of a detritus food web.

many terrestrial ecosystems such as forests. In fact, herbivores eat as little as 10% of the leaves in a forest; the remainder die and are funneled through detritus pathways. In some types of ecosystems, however, energy flows mostly through grazing food webs rather than detritus food webs. At sea, for example, **zooplankton** (microscopic free-floating animals), consume 90% of the chief producer organisms, **phytoplankton** (microscopic, free-floating plants); only 10% of the phytoplankton

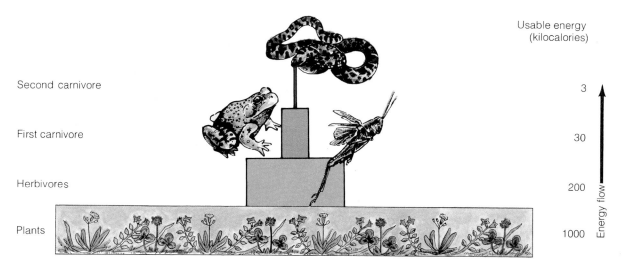

Second carnivore

First carnivore

Herbivores

Plants

Usable energy
(kilocalories)

3

30

200

1000

Energy flow

Figure 3.7 A pyramid of energy content, illustrating the low efficiency of energy transfer between trophic levels.

enter detritus pathways. Thus, the relative partitioning of the energy flow between grazing and detritus food webs varies with the type of ecosystem.

It is important to realize that no organism can create its own energy supply. Consumers at each trophic level depend on those at lower trophic levels for energy. If we stand in the sun for days without eating, we may get a terrific tan, but we will still be quite hungry. Our bodies cannot use solar energy directly to produce sugars; we depend totally upon green plants to make this transformation for us. Green plants, in turn, cannot create their own energy; they depend upon light energy to power their activities.

Energy flow into and within food webs is governed by the **law of energy conservation** (also known as the **first law of thermodynamics**), which states that energy cannot be created or destroyed but can change from one form to another. This law holds for all systems and governs every aspect of our lives. For example, because energy cannot be created out of nothing, we cannot drive our cars without gasoline, nor can we warm a meal in a microwave oven without electricity. The performance of any task requires a source of energy. Furthermore, because energy cannot be destroyed, all forms of energy are ultimately transformed into heat. Thus, an internal combustion engine requires a cooling system to dissipate the waste heat produced in the engine when gasoline is burned. Likewise, when we vigorously exercise, a portion of the waste heat produced by our body's metabolism is dissipated through evaporation of perspiration. Because the law of energy conservation governs all systems, natural and human-made, we must deal with its implications throughout our study of environmental issues.

The Efficiency of Food Webs

We can visualize the energy distribution within a grazing food web as a pyramid (fig. 3.7). In this model, producers form a broad base that contains most of the ecosystem's energy. Each successive trophic level contains considerably less energy than the one below it. Obviously, not all the energy that is available

at one trophic level is transferred to the next higher trophic level. That is, energy transfer within food webs is less than 100% efficient. **Energy efficiency** is defined as the fraction of the total energy input of a system that is transformed into work or some other usable form of energy. Before we look further at energy efficiencies in food webs, let us first consider several examples of energy efficiency from personal experience.

An incandescent light bulb transforms electrical energy into light energy. But how efficient is it? That is, how much of the electrical energy that enters the light bulb is actually transformed into light energy that leaves the light bulb? Our senses tell us that not only do we see light, but we feel that the bulb is hot. What is the source of that heat energy? In fact, only 5% of the electrical energy that enters the light bulb is converted to light energy, the desired form of energy. We say, therefore, that the light bulb is 5% efficient. The remaining 95% of the electrical energy is transformed directly into heat. In contrast, a fluorescent light, which has a higher efficiency (22%), is much cooler to the touch.

Energy efficiency also applies to humans. On a winter morning as we start out on a brisk walk, we initially feel uncomfortably cold. Within a few minutes, however, we begin to feel more comfortable. Why? To provide energy for walking, our muscles metabolize sugar (chemical energy), but less than 50% of the energy in sugar goes into muscle movement. The remainder is converted to heat, which our bodies use to help maintain an internal temperature of 37° C (98.6° F). The faster we walk, the more heat we produce, and the warmer we feel.

Energy inefficiency, as seen in these two examples, is an expected consequence of the **second law of thermodynamics.** This law, which applies to all systems, states that in every energy transformation, some energy is converted to heat that is thereafter unavailable to do further useful work. Whenever we attempt to increase the fuel efficiency (mileage) of an automobile, the electrical output of a power plant, or the yield of crop plants, we are faced with the limitations of this inviolable law of nature. Thus, we will be dealing with the ramifications of the second law of thermodynamics throughout our study of environmental issues.

Table 3.2	Estimates of net primary production for various ecosystems (in grams of biomass per square meter per year)

Type of Ecosystem	Net Primary Production
desert scrub	71
tundra	144
temperate grassland	500
savanna	700
boreal forest	800
temperate deciduous forest	1,200
tropical rain forest	2,000

Source: Adapted from R. Whittaker and G. Likens, "Carbon in the Biota" in G. Woodwell and E. Pecan, eds., *Carbon and the Biosphere*, Technical Information Center, USAEC, 1973, Washington, D.C.

Let us now return to food webs and consider the efficiency of energy transfer from one trophic level to the next higher trophic level. We begin with green plants, the producers. The rate at which plants accumulate energy during a specified period of time by means of growth and reproduction is termed **net primary production.** Because the energy content of plants (and also animals) is difficult to measure directly, production is usually determined by measuring a change in **biomass,** which is the total weight or mass of plants (or any specified group of organisms) in an area. For example, we can approximate the annual productivity of plants in a grassland by harvesting, drying, and then weighing the plants to estimate the biomass that was produced during the growing season. Tropical rain forests are among the most productive ecosystems, but productivity can vary greatly from one type of ecosystem to another (table 3.2).

Net primary production is the energy that is available to the consumers. What then is the efficiency of energy transfer from producers to higher trophic levels? Although energy efficiencies vary depending on the types of organisms and ecosystems, they are never very high. For many natural ecosystems, less than 1% of the energy at one trophic level becomes incorporated in the tissues of organisms that occupy the next higher trophic level. In contrast, efficiencies of up to 20% have been achieved by some domestic livestock, which are raised in human-controlled environments such as feed lots and poultry houses.

For ease of illustration, we assume an energy efficiency of 10% to estimate the amount of energy that is transferred through a food chain from one trophic level to the next higher level. For example, if we apply this so-called *10% rule* to the grain/beef cattle/people food chain, we predict that 100 kilograms (220 pounds) of grain will produce 10 kilograms (22 pounds) of beef cattle, which in turn will produce only 1 kilogram (2.2 pounds) of people.*

Since energy efficiency in food webs is low and each organism requires a certain amount of energy to survive, only a limited number of organisms can survive at a particular trophic level. This is true for all organisms, including people:

*See Appendix A for conversion factors for the English and metric systems.

The earth cannot support an unlimited number of consumers. Today the global human population totals over 5.4 billion, and it already consumes (often wastefully) nearly 40% of the planet's terrestrial plant production. Furthermore, population experts predict that the human population will nearly double over the next fifty years. Will there be enough primary production to feed a growing human population and still maintain a diversity of plant and animal life?

Causes of low efficiency

We have seen that the energy efficiency in natural food webs is often less than 1%. At first glance, these food webs appear to be extremely inefficient, but even the efficiency of agricultural food webs rarely exceeds 20%. Why then are food webs so inefficient? There are several reasons: (1) not all biomass at each trophic level is harvested, (2) not all the harvested biomass is ingested, (3) not all the ingested biomass is digested (assimilated), and (4) not all the assimilated biomass is converted to usable energy. This sequence is illustrated in figure 3.8.

Many organisms have characteristics that reduce their chances of being harvested. Even plants have ways to ward off herbivores; for example, needles and thorns on thistles and cacti discourage hungry herbivores (fig. 3.9). Also, many plants, including daisies, black-eyed susans, oaks, and citrus trees produce chemicals that repel insects or other animals that attack them. Animals, too, have characteristics that help them escape predators. For example, prairie dogs find refuge in their burrows, skunks emit a pungent odor, fleet-footed antelope outrun their predators, and, as figure 3.10 shows, a frog may blend unnoticed into its surroundings.

Predators, on the other hand, have many traits that enable them to capture their prey. For example, hawks and owls have an unparalleled system of color vision and can resolve details at three times the distance humans can. But even the hunting skills of these well-adapted predators are limited. In fact, carnivores usually succeed in taking down a prey in only one out of every ten attempts. Hence, typically no more than 10%-20% of the biomass at one trophic level is actually harvested by the organisms at the next higher trophic level. Organisms that escape their enemies live to produce the next generation, thereby preserving the species. Some of these offspring will fall prey to the next generation of predators, thereby maintaining the continuity of food webs through time.

A prey organism that is harvested by a predator often is not entirely ingested. Many carnivores do not eat fur, feathers, or skeletons, even though these materials contain energy and nutrients. Even if carnivores ingest these parts, most cannot digest them; nor can herbivores digest all of the fibrous portion of plant tissue. If consumed, indigestible materials are excreted and subsequently digested by detritivores or broken down by decomposers. The portion of ingested food energy that is absorbed by an animal's digestive system is called **assimilated food energy.**

Further energy losses occur when assimilated food enters the body's cells and is processed to liberate energy that can be used for maintenance, growth, and reproduction. This process of energy liberation, called **cellular respiration,**

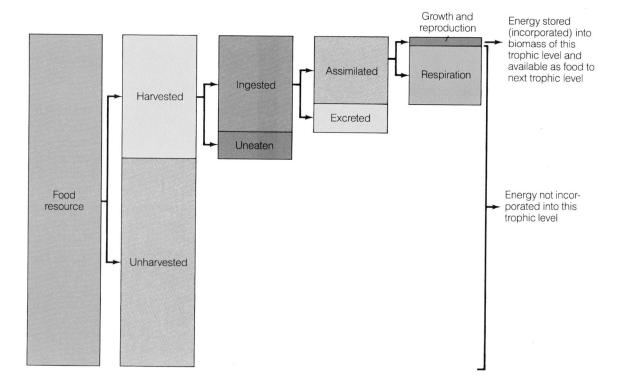

Figure 3.8   Generalized flow scheme of energy transfer and losses between and within trophic levels. The relative size of the boxes will depend on the trophic level, the nature of the ecosystem, and the type of animal involved.

Figure 3.10   A gray tree frog blends in with its surroundings thus enabling it to avoid detection by enemies.

© Don Blegen

Figure 3.9   Thorns on a fishhook cactus effectively ward off many potential predators.

© D. Wilder/Tom Stack & Associates

occurs in all living cells. However, as dictated by the second law of thermodynamics, cellular respiration is inefficient. Less than half the energy in sugars, which are the direct source of energy for cellular respiration, is converted to work. The rest is transformed to heat that cannot be used to perform work and is eventually lost to the organism's environment. Carbon dioxide and water, the other products of cellular respiration, are also released to the environment. The process of cellular respiration is summarized as follows:

Sugar + oxygen →
    carbon dioxide + water + energy for work + heat

$$C_6H_{12}O_6 + 6O_2 \rightarrow 6CO_2 + 6H_2O + \text{energy for work} + \text{heat}$$

Note that one of the raw materials essential for cellular respiration is oxygen. Without a continual supply of oxygen, cellular respiration ceases, and cells die. People die of asphyxiation, or oxygen starvation, when they are deprived of oxygen for longer than about five minutes. All plants and animals and most microorganisms require oxygen and thus are called **aerobic organisms.** Only a few species of microorganisms can survive in the absence of oxygen. These species are known as **anaerobic** (without oxygen) **microorganisms** and include bacteria that cause tetanus, gas gangrene, and some forms of food poisoning. We will consider the role of anaerobic bacteria in the treatment of organic waste in chapters 14 and 20.

Our discussion of energy transfer within food webs has demonstrated several factors that contribute to low efficiency. Agricultural food webs appear to be more efficient than natural food webs because people have developed strategies that partially compensate for inefficiencies. We evaluate these strategies in the next section.

Coping with low efficiency

In an effort to increase the supply of food, human ingenuity has developed several strategies to improve food web efficiency. These strategies include (1) replacing forests with cropland and grazing land, (2) developing high-yield crops and livestock, (3) harvesting at lower trophic levels, and (4) reducing competition. The success of these strategies is confirmed by the growth of the human population, which has increased from less than 5 million to over 5.4 billion—a thousand-fold increase—since the advent of agriculture some ten thousand years ago. But each strategy also has limits.

If you are lost in a forest, you cannot expect to survive by eating bark and twigs. Almost all the food energy in a forest occurs in unpalatable or indigestible materials. Accordingly, for centuries, humans have cleared away forests and replaced them with crops, thereby channeling a greater percentage of photosynthetic energy into the growth of seeds and fruits— materials that we and our domestic livestock can largely digest. During the last two hundred years, timberland in the United States has been reduced by some 20%, an area about the size of Texas. Worldwide, forest land has been reduced by 20% in just the past thirty years. Many of these forests were cleared to provide land for agriculture.

Clearing forests to enhance food production has its natural limits, however. Much of the forest land that is suitable for agriculture has already been cleared, and we need to define management goals for the forests that remain. If the objective is to convert additional forests to agricultural production, we must realize that much of the remaining forests grow in soil that is generally low in fertility and vulnerable to severe erosion. If the objective is to harvest forests for commercial products, we must realize that lumbering will reduce the recreational value of the land as well as its ability to support certain types of plants and animals. Given these conflicting demands on forest lands, how do we fulfill our needs for food and wood products and still maintain the quality of the environment for ourselves and the natural diversity of plants and animals?

Agricultural food webs appear to be many times more efficient than natural food webs, not because of any inherent differences between domesticated and wild species, but because of modern agricultural practices. For example, in the case of animal husbandry, commercially grown chickens in the United States no longer roam about the farmyard scratching the ground for seeds and insects. They are raised in climate-controlled buildings and fed a controlled diet of ground grain enriched with vitamins and minerals. Today's chicken feed contains little indigestible material; thus, chickens assimilate a greater percentage of the food they ingest. Because their food is brought to them, they expend little energy on movement, and climate control allows them to spend less energy on maintaining their body temperature. Chickens raised under such conditions spend less energy on maintenance, so they channel more energy into growth and reproduction.

If one looks only at the fraction of ingested food that goes into increased biomass via growth and reproduction, then modern animal husbandry does enhance energy efficiency. However, producers have only achieved these gains at the cost of a major energy subsidy. Modern animal husbandry consumes enormous quantities of energy in the form of petroleum, natural gas, and coal used in growing, harvesting, processing, and transporting animal feed; handling manure; and building and maintaining climate-controlled surroundings. These fuels do much of the work that chickens once did for themselves. Modern agriculture is highly dependent on the availability of these fuels, which are **nonrenewable resources**; that is, once they are used, they cannot be regenerated within a reasonable time. We consider the prospects for the continued availability of inexpensive supplies of coal, petroleum, and natural gas in chapter 18.

Similar energy subsidies are required for modern crop production. Although crop production per hectare has increased greatly in recent decades, we have not achieved this increase by developing plant varieties with a more efficient photosynthetic process. That is, production of sugars per unit of leaf area has not improved. Rather, we have increased the amount of leaf area per unit of ground area. In other words, the individual factory (leaf) is not more efficient, but because there are more factories (leaves) per unit of land, yields have

Figure 3.11 A new dwarf strain of high-yielding wheat (the crop with the shorter stalks) compared with a native strain. Development of dwarf strains has contributed significantly to increased wheat production in many nations.
FAO photo

Figure 3.12 In an effort to save their crops, African farmers attempt to beat back an invasion of locusts.
© A. Devaney, Inc.

risen. Just as building a new factory requires the appropriate resources, so the greater leaf area has required heavier applications per hectare of fertilizers and pesticides and also, in some areas, irrigation water. The manufacture, transport, and application of fertilizers and pesticides as well as the construction and operation of irrigation systems translate into a major energy subsidy from supplies of petroleum, natural gas, and coal.

Furthermore, plant breeders have developed crop varieties that channel more of the sugars produced by photosynthesis into the production of edible materials such as seeds and fruits. Consequently, these varieties use less of the sugars to produce leaves, stems, and roots, which are usually unpalatable and indigestible for humans. For example, new dwarf strains of wheat produce several-fold more grain per stalk than traditional tall strains (figure 3.11). The additional nutrients that are removed from the field by harvesting the high-yield grain must be replaced by the addition of fertilizers to maintain soil fertility.

Our understanding of food web dynamics tells us that only a small fraction of the energy at a particular trophic level is transferred to the next higher trophic level. Thus, if we ate less meat and more fruits, vegetables, and cereals, the amount of food energy available to us would increase automatically. A gram of grain yields about the same amount of energy as a gram of meat. So, according to the 10% rule, approximately ten times more food energy would be available to us if we ate the grain we now feed to livestock instead of eating the meat. A strict vegetarian diet is unappealing to many Americans, who are accustomed to a diet that contains approximately 40% animal products, but in many overcrowded areas of the world, people have been forced to rely on shorter food chains. For example, the diet of Southeast Asians consists almost entirely of rice or wheat and some green vegetables; an occa-

sional meal of fish for protein supplement accounts for only about 10% of their diet. Most of these people must take full advantage of existing food supplies and simply cannot afford the approximately 90% decline in energy availability that occurs between the trophic levels of the herbivore and the carnivore.

Because all organisms that occupy a particular trophic level compete for the energy resources of the next lower trophic level, humans can increase the available food supply by reducing the numbers of competing species. For example, farmers apply insecticides to kill insects that feed on crops, thus increasing our share of available food. If our competitors get a bigger share (which they do, for example, during locust plagues), then fewer people can be fed (see fig. 3.12). In these situations, famine and death may result unless food is imported from regions with a surplus. However, pesticides have some negative effects. Some insecticides kill valuable organisms such as honey bees, which play a major role in food production by pollinating nearly one hundred types of crop plants. Some pesticides also kill the natural enemies of the pests that farmers are attempting to control. Consequently, the pest population continues to grow despite the use of pesticides. In chapter 9, we discuss further the consequences of pesticide use as well as other strategies to reduce pest damage.

In spite of human ingenuity in coping with inefficiencies in food webs, nature continues to restrict our ability to exploit available food resources. In addition, our multiple demands on limited land resources raise other questions about our ability to feed a growing human population and still retain a rich diversity of plants and animals. We explore the issues related to feeding a rapidly growing human population in chapter 8 and those related to managing land resources, including maintaining biodiversity, in chapters 10, 11, and 21.

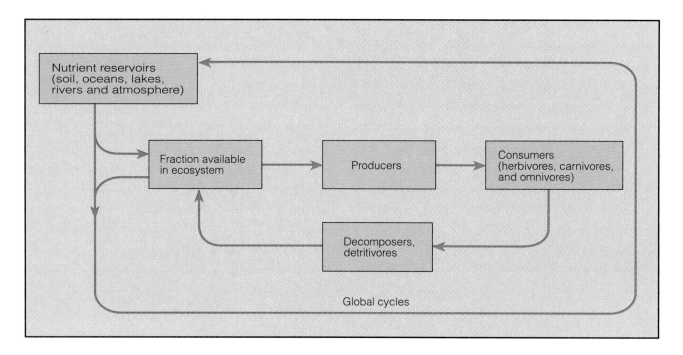

Figure 3.13   Generalized model of nutrient cycling.

The Cycling of Materials

Your cereal box provides information on the array of chemicals that are necessary for good nutrition. All organisms require certain chemicals for survival. Calcium, for example, is an essential component of bones, teeth, and shells in animals and of cell walls in plants. Phosphorus is essential for photosynthesis and cellular respiration. Nitrogen is a component of all the proteins that sustain living tissues. Protein deficiency may be the greatest single cause of human disease in the world. Both phosphorus and nitrogen are components of **DNA** (deoxyribonucleic acid), the cell component that contains the genetic code. In addition to these three nutrients, plants and animals require about twelve other elements, including iron, magnesium, and zinc. As shown in table 3.1, organic compounds (proteins, fats, and carbohydrates) make up over 35% of an organism's biomass. The basic elements of these organic compounds are carbon and hydrogen. In addition, oxygen is essential for cellular respiration, and carbon dioxide and water are necessary for photosynthesis. Since the availability of these chemicals has a direct impact on food and fiber production, we need to understand the sources of these chemicals and how they move within and among ecosystems.

Although the earth receives a continuous supply of energy from the sun, it has no comparable source of materials. For all practical purposes, the quantity of materials on earth is fixed and finite. Although all of the energy that flows through a food web is eventually lost as heat via cellular metabolism, nutrients flow in cycles (fig. 3.13). For hundreds of millions of years, life-sustaining materials have been continually recycled within and among the earth's ecosystems. Typical cycles involve the continual transfer of solids, liquids, or gases from one reservoir to another. Among the reservoirs involved are oceans, the atmosphere, soil, rocks, and organisms, including humans.

All cycles obey the **law of conservation of matter,** which states that matter can neither be created nor destroyed, although it can change in chemical or physical form. For example, when logs burn in a fireplace, a portion of them is converted to ash, and the rest goes up the chimney as carbon dioxide, water vapor, and creosote. We can look at this law in terms of accountability: All losses from one reservoir (e.g., the burning logs) must be accounted for as gains in other reservoirs (e.g., the ashes on the bottom of the fireplace and the gases that went up the chimney into the atmosphere.)

In the following sections, we describe four of the cycles that are essential for life: the **carbon, oxygen, nitrogen,** and **phosphorus cycles.** We consider another essential cycle, the **water cycle,** in chapter 13 to prepare for our examination of water pollution and the management of aquatic resources.

The Carbon Cycle

Materials such as carbon continually cycle around the globe. The winds, for example, may transport carbon dioxide exhaled by a person in California to a cornfield in Kenya, where it is taken up by a corn plant and incorporated into sugar to build an ear of corn. Insects eventually consume the ear of corn, and their cellular respiration releases carbon in the form of carbon dioxide back into the atmosphere. The carbon dioxide may then be carried by winds to the Pacific Ocean where it dissolves in the water and is transported great distances via ocean currents. In the following sections, we describe the many pathways that materials such as carbon can travel as they cycle within the environment.

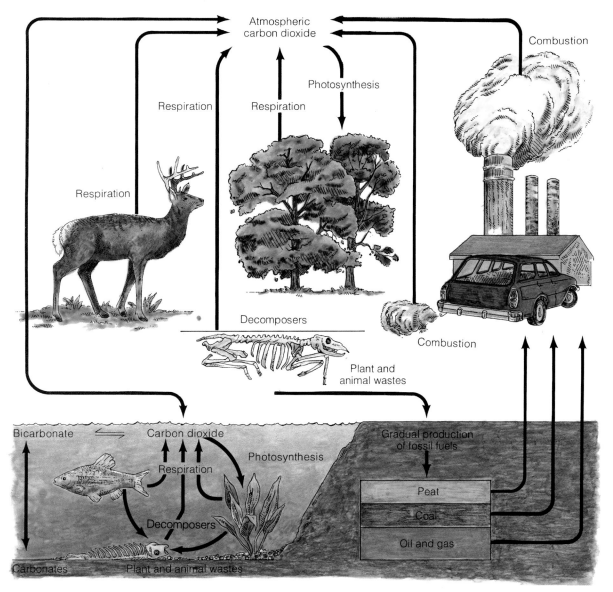

Figure 3.14   The carbon cycle.

Global cycles consist of interlocking subcycles. A major subcycle of the carbon cycle involves the interactions of organisms and the atmosphere (fig. 3.14). Photosynthesis cycles carbon dioxide (CO_2) from the atmosphere into green plants, where it is incorporated into sugar ($C_6H_{12}O_6$). Plants use sugar to manufacture a myriad of other organic compounds, including fats, proteins, and other carbohydrates. Both producers and consumers transform a portion of the carbon in these organic compounds back into carbon dioxide as a by-product of cellular respiration. In this way, plants and animals release carbon dioxide back to the atmosphere. Carbon in dead plants and animals is also returned eventually to the atmosphere through the cellular respiration of decomposers and other members of detritus food webs. If we compare the flow of energy through food webs with the cycling of carbon through ecosystems, we conclude that these two essential processes are intimately interwoven.

Some dead plant and animal materials become buried in sediments before they can be broken down completely by decomposers or consumed by detritivores. This process has been going on to a greater or lesser extent for hundreds of millions of years, but for us, it was particularly important during the Carboniferous Period, 280 to 345 million years ago, when trillions of metric tons of organic matter were buried. Through subsequent eons, much of these plant and animal remains were transformed into coal, oil, and natural gas as a result of heat and compression. Thus, luxuriant swamp forests that inhabited the earth hundreds of millions of years ago were transformed into the coal reserves of today (fig. 3.15). In certain marine environments, plant and animal remains were converted into oil and natural gas. (See chapter 17, which describes the geologic conditions under which fossil fuels form.) When we burn coal, oil, and natural gas, collectively known as **fossil fuels**, we are using energy that

(a)

(b)

Figure 3.15 (*a*) A diorama showing how a carboniferous swamp forest might have looked when it existed 280 to 345 million years ago. (*b*) Over the vast expanse of time since the growth of swamp forests, its organic remains were gradually converted to coal. This coal seam exposed in northern Alaska is about 4 meters (12 feet) thick.

(*a*) Field Museum of Natural History (*b*) U.S. Geological Survey/Photo by R. G. Ray

was locked in vegetation through photosynthesis hundreds of millions of years ago. In the combustion process, stored carbon combines with oxygen in the air to form carbon dioxide, which enters the atmosphere.

Humans have been consuming fossil fuels on a large scale for less than a century, yet most scientists contend that we will exhaust our available supplies of economically recoverable petroleum and natural gas within the next century. Although the precise time of depletion is uncertain, there is no doubt that we are consuming fossil fuels much more rapidly than natural processes can generate them. Hence, if we are to have sufficient supplies of energy to maintain our current standard of living, we must begin to develop alternative forms of energy. The status of our current energy supplies and the prospects for alternative energy sources are explored in chapters 18 and 19.

The exchange of carbon dioxide between the atmosphere and the oceans is the primary regulator of carbon dioxide levels in the atmosphere. At the interface between air and water, carbon dioxide continually moves between the two reservoirs. Some atmospheric carbon dioxide dissolves in the water, and some dissolved carbon dioxide escapes from the ocean surface to the atmosphere.* If atmospheric carbon dioxide levels increase, more carbon dioxide dissolves in ocean water. Conversely, if atmospheric carbon dioxide levels decline, more carbon dioxide escapes from the oceans to the atmosphere.

The carbon dioxide that dissolves in water is available for photosynthesis by aquatic plants. Some of the carbon dioxide that is incorporated into sugars via photosynthesis subsequently cycles through aquatic food webs.

Another subcycle of the carbon cycle involves the formation of limestone (calcium carbonate, $CaCO_3$), dolomite [calcium magnesium carbonate, $CaMg (CO_3)_2$] and other carbon-containing rocks. These rocks form in shallow marine ecosystems. As marine organisms die, their shells and skeletons settle to the bottom. Over long periods of time, these remains accumulate, are compressed by their own weight, and are gradually transformed to solid, carbonate rock. Subsequently, geological processes expose these rocks to the atmosphere, where chemical and physical processes decompose and fragment them. One decomposition product is carbon dioxide, which escapes to the atmosphere. The rate of release of carbon dioxide by these processes is so slow that it has little

*It may be surprising to learn that gases dissolve in water. A simple experiment illustrates this fact. When we heat a pan of water, tiny bubbles begin to appear. These bubbles are composed of atmospheric gases, including carbon dioxide, that were dissolved in the water.

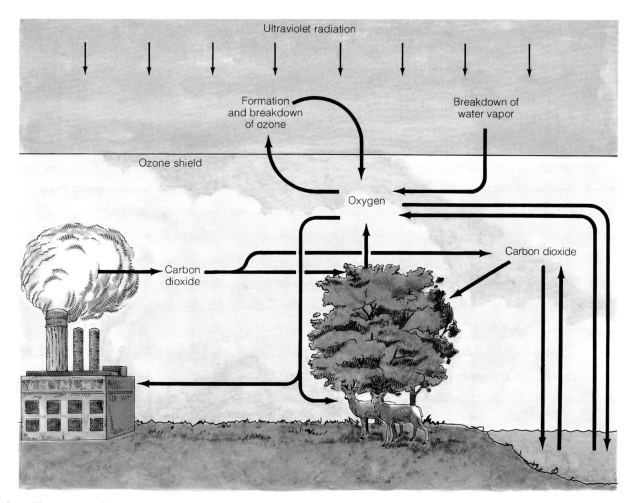

Figure 3.16   The oxygen cycle.

immediate significance for us. However, over the millions of years that constitute geologic time, the weathering of carbon-containing rocks has had an important influence on atmospheric carbon dioxide levels.

The Oxygen Cycle

Oxygen is found almost everywhere on earth. As a gas, oxygen (O_2) makes up nearly 21% of the air we breathe; it dissolves in water and occurs in pore spaces of soils and sediments. Oxygen combines with a multitude of other elements to form essential life-sustaining substances, including water (H_2O), carbon dioxide (CO_2), plant nutrients such as phosphate (PO_4^{3-}) and nitrate (NO_3^-), and organic compounds such as sugars ($C_6H_{12}O_6$), starch, and cellulose. Oxygen is also a component of many commercially important rocks and minerals, including limestone ($CaCO_3$) and iron ore (Fe_2O_3). Clearly the oxygen cycle interacts with many other cycles. We outline here only a few of its more important aspects. Figure 3.16 is a schematic model of the oxygen cycle.

The oxygen and carbon cycles have much in common. For example, an important subcycle of the oxygen cycle involves photosynthesis and cellular respiration, two processes that are also important in the carbon cycle. Furthermore, organic materials such as wood, coal, gasoline, and manure

(all of which contain considerable carbon) will burn only in the presence of oxygen, and they release carbon dioxide as a by-product.

The cycling of oxygen between the atmosphere and bodies of water such as streams, lakes, and oceans is another important subcycle. Atmospheric oxygen dissolves in surface waters, where aquatic organisms use it for cellular respiration. In chapter 14, we describe the consequences when contamination of waterways reduces dissolved oxygen levels.

A vital subcycle of the oxygen cycle is responsible for the production of ozone (O_3) in the upper atmosphere, where intense ultraviolet radiation (UV) from the sun converts oxygen (O_2) to ozone (O_3). While this is going on, other interactions decompose ozone to oxygen. These natural processes filter out most of the potentially harmful ultraviolet radiation. If not for this so-called **ozone shield,** life as we know it would not have evolved. The potential damage from exposure to ultraviolet light is described in the special topic entitled "The Hazards of Sunbathing."

The Nitrogen Cycle

Although 79% of the air we breathe is composed of nitrogen gas (N_2), plants and animals cannot use it directly to construct proteins and other nitrogen-containing compounds such

The Hazards of Sunbathing

A rise in the popularity of sunbathing has put people at greater risk for contracting skin cancer.
© Bob Coyle

Many people believe that a dark tan is attractive and a sign of good health, but mounting evidence indicates that too much sun can lead to several health problems. One of the most noticeable effects is premature aging of the skin, that is, the skin develops a leathery texture, and wrinkles and dark spots appear. The sun also contributes to certain types of cataracts—opaque regions in the lens of the eye that interfere with vision and, if untreated, may cause blindness. Most worrisome is the role of the sun in skin cancer, the most common form of cancer in the United States. More than 400,000 new cases of skin cancer are diagnosed each year; about one in seven Americans, then, will develop this disease during their lifetime. The culprit is overexposure to the ultraviolet portion of solar radiation—specifically, the portion of ultraviolet known as *UVB*. How can you protect yourself from this dangerous radiation?

The body has some natural defenses against UVB radiation. At the base of the *epidermis,* the outer layer of skin, lie *basal cells,* which regularly divide and produce new cells, as shown in the drawing on page 48. Some of these new cells migrate toward the skin surface, while others remain behind and continue to divide. The migrating cells, called *squamous cells,* die on their way to the

surface. Eventually these dead squamous cells reach the surface, where they become part of the *stratum corneum,* the outermost sublayer of the epidermis. This layer of tightly packed dead cells, usually about twenty cells thick, shields the underlying layers of skin by absorbing potentially damaging UVB radiation.

The stratum corneum gradually erodes as dead cells are shed from the skin. You can observe this by vigorously rubbing your arm in a bright light; the tiny particles of flaking skin resemble fine dust particles. To maintain the stratum corneum, squamous cells migrating from the base of the epidermis replace lost cells from below. The sequence from cell formation to shedding typically takes three to four weeks. UVB radiation speeds up the cell-formation process, so that a thicker layer of dead skin forms that better protects the underlying layers.

The body's second line of defense consists of specialized cells called *melanocytes,* which lie among the basal cells (see the drawing on page 47). Melanocytes respond to UVB by producing *melanin,* a pigment that absorbs UVB and is responsible for darkening the skin. Melanin spreads to surrounding squamous cells, which then migrate upward, die, and become part of the stratum corneum. The migration and death of squamous cells explain why a tan gradually fades, since melanin is lost during shedding.

Melanin does not provide total protection against UVB, however. Typically, several days elapse between the skin's initial exposure to UVB and the development of a protective tan. In the meantime, considerable skin damage can occur, especially if the initial exposure causes sunburn. Also, melanin does not absorb all the incident UVB, so that no one is immune to the hazards of overexposure. Although the risk of contracting skin cancer is greater for fair-skinned people, whose melanocytes produce less melanin, people with dark complexions, whose melanocytes produce more melanin, can also develop skin cancer.

Overexposure to UVB is linked to three forms of skin cancer: basal cell carcinoma, squamous cell carcinoma, and malignant melanoma. In each case, UVB damages the DNA of the respective cells, resulting in uncontrolled growth and the subsequent formation of tumors. A *basal cell carcinoma* grows slowly and usually does not spread to other parts of the body. This cancer accounts for about 75% of all skin cancers. Fortunately, it is easily treatable if caught early. A *squamous cell carcinoma* grows more quickly and is more likely to invade underlying skin structures. About 20% of skin cancers are squamous carcinomas. By far the most dangerous form of skin cancer is *malignant melanoma,* which begins in the melanocytes. These tumors grow quickly and can spread rapidly throughout the body. Malignant melanomas often appear as dark brown or black spots on the skin because of the increased production of melanin. If the melanoma is detected before it spreads, the patient is often cured. Once the tumors have begun to migrate, however, the cure rate plummets. Thus, malignant melanomas account for 75% of the deaths from skin cancer, even though they account for only 5% of total occurrences of skin cancer.

Like other forms of cancer, skin cancer usually takes 20 or more years to develop. Thus, the average age for the first discovery of skin cancer is 50. Unfortunately, that age is declining as skin cancer becomes increasingly more common among younger people. If our natural defenses are insufficient to protect us from developing skin cancer, what more can we do?

We must first recognize that we are exposed to much more UVB than we realize. Ultraviolet radiation penetrates clouds more readily than does visible solar radiation; thus, even when the sky is completely overcast, some protection from UVB is

The epidermis, the outer layer of skin, has several natural defenses against UVB radiation.

necessary. At the beach, a person may believe that the best way to avoid too much sun is to sit under a beach umbrella, jump in the water, or wear a T-shirt. Such strategies, however, provide inadequate protection. Sand reflects up to 50% of the incident UVB, exposing a person to dangerous radiation even in the shade of a beach umbrella. Water transmits UVB to a depth of a meter or so, and a wet T-shirt allows 20%-30% of incident UVB to reach the skin. The beach is not the only place where a person is likely to be exposed to high levels of UVB. Skiing at high mountain elevations, where UVB is more intense than at sea level, results in significant exposure. In addition, snow is even more reflective of UVB than is sand.

Common sense can help us to avoid overexposure to UVB. Initial exposure to the sun should be brief until a protective tan develops. Whenever possible, we should avoid the sun at its highest intensity, which is between 10 A.M. and 3 P.M. A good rule of thumb is that if your shadow is longer than you are tall, there is no problem; if your shadow is shorter than you are tall, then you should apply a sunscreen. Because most skin cancers appear on commonly exposed surfaces such as the forearms, face, and neck, wearing protective clothing (e.g., a wide-brimmed hat) is a good strategy. Applying a sunscreen to all exposed skin is always wise. Sunscreens have ingredients that selectively block UVB wavelengths.

When purchasing a sunscreen, be aware of the degree of protection offered: the sunscreen's sun protection factor (SPF). The SPF is a measure of the time that the skin can safely be exposed to the sun. The higher the SPF value, the longer the protection lasts. Experts recommend a minmum SPF of 15. Note that suntan lotions are not sunscreens. Suntan lotions merely help to keep the skin moist: they do not provide any protection from UVB.

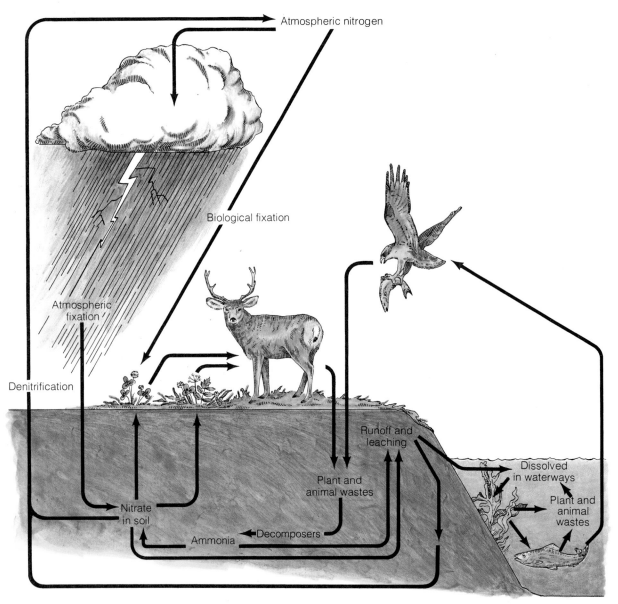

Figure 3.17 The nitrogen cycle.

as DNA. Thus, despite the abundance of nitrogen in the atmosphere, insufficient soil nitrogen often limits plant growth. Two natural processes, biological fixation and atmospheric fixation, convert nitrogen gas into forms that plants can utilize in their metabolism. Both processes form essential parts of the nitrogen cycle (fig. 3.17).

Biological fixation involves the activities of specialized microorganisms that convert nitrogen gas to ammonia (NH_3) by a series of chemical changes.* Some of these nitrogen-fixing microorganisms are specialized bacteria that live in the root nodules of leguminous plants, such as peas, beans, alfalfa, and clover (fig. 3.18). A buildup of ammonia is toxic to cells; thus, the ammonia produced by these bacteria is transported into nearby legume root cells where it is converted to

*Ammonia (NH_3) and ammonium (NH_4^+) are forms of nitrogen that are readily interconverted. Which one predominates depends upon the acidity of the soil or solution.

Figure 3.18 Pea plants have root nodules that contain nitrogen-fixing bacteria.

© Hugh Spencer/Photo Researchers, Inc.

Figure 3.19   Lightning is an important factor in the planet's nitrogen cycle.
Photo by: Dave Baumhefner, NCAR

other compounds of nitrogen that are transported throughout the plant. Biological fixation is also carried out by other species of bacteria that live free in the soil, rivers, lakes, and oceans. These microorganisms also convert nitrogen gas into ammonia, which they utilize in metabolism. Biological fixation contributes roughly 90% of the total nitrogen that is fixed each year.

Atmospheric fixation occurs during thunderstorms. Extremely high temperatures associated with lightning (fig. 3.19) cause nitrogen and oxygen to combine in the atmosphere. Eventually, extremely dilute nitric acid (HNO_3) forms and is washed to the ground by rain and snow. In the process, nitric acid is converted to nitrate (NO_3^-), which is taken up by plant roots.

Plants convert nitrate to ammonia, which is then readily incorporated into a variety of organic compounds, including amino acids, proteins, and DNA. As plants are eaten by herbivores that subsequently fall prey to carnivores, nitrogen in the form of these organic compounds cycles through food webs. The diet of many animals includes far more nitrogen-containing compounds than their bodies can utilize. Many mammals, including humans, convert the excess nitrogen to urea (CH_4N_2O) and excrete it in urine.

When plants, animals, and microorganisms die, a variety of decomposers convert the nitrogen in organic compounds (proteins and DNA as well as urea) back to ammonia, which returns to the soil. Some ammonia may be changed to nitrate by microbial action, a process known as **nitrification.** Either form of nitrogen can be taken up again by plants and recycled through food webs.

Not all of the fixed nitrogen remains within a terrestrial ecosystem. Nitrogen can be lost from a terrestrial ecosystem via two major pathways: denitrification and transport by water. **Denitrification,** carried out by a specialized group of anaerobic bacteria, converts nitrate back to nitrogen gas, which eventually escapes from the soil to the atmosphere. Note that nitrification and denitrification are opposing processes. While nitrification makes usable nitrogen available to food webs, denitrification removes usable nitrogen.

Although both nitrate and ammonia are highly soluble in water, ammonia tends to adhere to soil particles. Hence, it can be lost from a terrestrial ecosystem via soil erosion. In contrast, nitrate interacts little with soil particles and is readily carried away by surface runoff. Nitrate also seeps through soil into groundwater, where (as described in chapter 14) it may pose a significant health threat. Nitrate and ammonia normally remain dissolved in lakes, rivers, and oceans and can be taken up by aquatic plants and subsequently cycled through aquatic food webs, including detritus pathways. Some nitrogen is returned to terrestrial ecosystems by land animals that feed upon aquatic organisms. For example, eagles swoop down, catch fish in their talons, and carry them to their nests to feed their young (fig. 3.20).

The Phosphorus Cycle

Soil is the major source of many nutrients, including phosphorus, calcium, magnesium, potassium, and sulfur. The phosphorus cycle illustrates how these nutrients cycle within and between ecosystems (fig. 3.21). Plant roots take up phosphorus from the soil, mainly in the form of phosphate (PO_4^{3-}). Plants subsequently transport phosphate to their growing parts, and incorporate it into a variety of organic compounds, including DNA and certain types of fats. Phosphorus then cycles through food webs in these forms as herbivores consume plants and are in turn consumed by predators. When plants, animals, and microorganisms die, decomposers invade the remains, and their actions eventually return phosphorus to the soil, where it can be taken up again by plant roots.

Retention of phosphorus within a natural terrestrial ecosystem is critically important for organisms, because relatively little is available in the soil. Although phosphorus is added to the soil through the weathering (decomposition and fragmentation) of rocks that contain it, this process occurs very slowly. Thus, the most effective means of providing an adequate supply of phosphorus in an undisturbed terrestrial ecosystem is through retention and recycling by organisms.

Ecosystems have a variety of ways to conserve nutrients. For example, plants serve as a protective covering for the soil. This covering plus extensive root systems anchor the soil and thereby retard erosion by rain and wind. Another natural

Figure 3.20   A bald eagle catching a salmon illustrates a linkage between aquatic and terrestrial ecosystems.
© Tom Mangelsen/Peter Arnold, Inc.

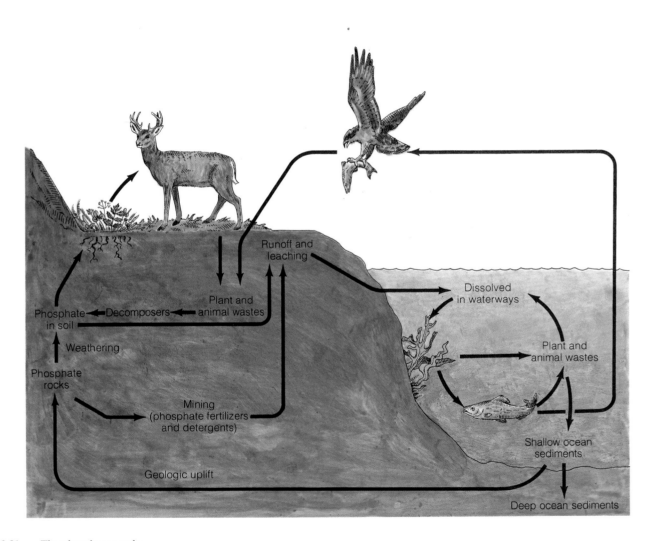

Figure 3.21   The phosphorus cycle.

Figure 3.22 Bracket fungi contribute to the slow decay of a tree stump.
© Craig Newbauer/Peter Arnold, Inc.

Figure 3.23 Estuaries, such as this salt marsh, are highly productive ecosystems.
© Andrew J. Martinez/Photo Researchers, Inc.

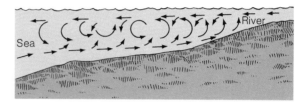

Figure 3.24 Circulation in an estuary, mixing lighter fresh water with denser sea water to form a nutrient trap.
Source: After E. Odum, *Fundamentals of Ecology*, 3d ed. Copyright © 1971, W.B. Saunders Company.

nutrient-conserving mechanism is the channeling of some nutrients into long-term storage reservoirs. For example, billions of metric tons of nutrients are stored in wood, where they are not subject to loss. When a tree dies, decomposers slowly release its store of nutrients to the soil; many decades may elapse before a large log decomposes fully (fig. 3.22). On the other hand, leaves constitute a shorter-term storage reservoir; leaves on deciduous trees live for only one growing season and subsequently decompose within a year or two. Additional nutrient conservation occurs just before trees lose their leaves, when some nutrients are transported back from the dying leaves to the branches and roots, where they are stored until growth begins anew in the spring. As much as 30% of the phosphorus in leaves is reabsorbed before the leaves fall.

Still, even an undisturbed terrestrial ecosystem loses some nutrients. Just as we saw in the nitrogen cycle, small amounts of phosphate, which binds tightly to soil particles, are carried off via erosion to lakes and other waterways. Once in an aquatic ecosystem, some of the phosphate is taken up by algae and rooted aquatic plants and again cycles through food webs. Phosphate-laden soil particles also sink to the bottom and mix with other sediments. If the body of water is shallow, wave action stirs up bottom sediments, making the phosphate available once again for uptake by algae. Just as in the nitrogen cycle, some phosphorus is returned to terrestrial ecosystems by land animals that feed on aquatic organisms.

Rivers transport significant amounts of phosphorus and other nutrients (including detritus) to coastal areas such as estuaries. **Estuaries** are exceptionally productive ecosystems that are transitional between land and sea (fig. 3.23). The high productivity of estuaries results, at least in part, from a special circulation pattern that retains and recirculates nutrients (fig. 3.24). Estuaries support many of the world's commercially important marine fisheries. We discuss the importance of estuaries and other wetlands in chapters 8, 10, and 14.

Some phosphorus is deposited in off-shore sediments and thus removed from circulation until geologic processes

reexpose them. The reappearance of these deposits may require hundreds of thousands or even millions of years. Uplift and weathering of exposed rock that contains phosphorus slowly add phosphorus to the soil of terrestrial ecosystems. The best estimates indicate that under natural conditions, the phosphorus added to a terrestrial ecosystem from weathering is about equal to what is lost by erosion, but this estimate remains to be experimentally verified.

We have now described the vital ecological processes of energy flow and nutrient cycling. Although the basic mechanisms of these processes are similar throughout the world, rates of energy flow (productivity) and patterns of nutrient cycling vary greatly from one region to another. The special topic entitled "Biomes of the World" compares and contrasts these two ecological processes in the world's major communities of plants and animals.

Humans, the Cycling of Materials, and Pollution

As the human population has grown and our technology has advanced, the impact of our activities has spread in rapidly widening circles. Today, such phenomena as potential global warming and the thinning of the ozone shield demonstrate that our impact is no longer local: We are now able to disrupt our environment on a global scale. In the following sections, we consider how human activities disrupt the natural cycling of materials.

Biomes of the World

Very large communities of plants and animals, called *biomes,* cover all of the continents except Antarctica. Biomes are named for the types of plants that dominate each landscape. For example, deciduous forests cover most of the eastern United States, while vast grasslands carpet the midsection of North America. The accompanying table lists the world's major biomes and some of their distinguishing characteristics.

The geographic distribution of the world's major biomes is illustrated in the accompanying map on pages 54–55. Because climate is often the principal control of the type of vegetation that can grow in a region, locales with similar climates usually support the same type of biome. Thus, boreal forests and tundra form nearly concentric bands around the North Pole, where the climate is very cold and dry. Deciduous forests grow in eastern North America, western Europe, and northeastern China. Extensive grasslands are present in the Great Plains of North America, the pampas of South America, the steppes of Eurasia, and the African veld. The world's major deserts include the Sonoran in North America, the Sahara in North Africa, the Gobi in Mongolia, the Atacama along the west coast of South America, and the great desert that stretches across the interior of Australia.

Although a biome is dominated by particular vegetation, we should not jump to the conclusion that plant and animal species are the same throughout a biome. In reality, interactions of such factors as soil, altitude, slope exposure (for example, a south-facing slope is warmer and drier than a north-facing slope), drainage patterns, and local climate mean that a myriad of combinations of plant and animal species exist within a biome. Disturbances by humans and other species as well as such physical forces as fires, hurricanes, and volcanism add to the incredible variety of communities that occur within the boundaries of a specific biome. In the following paragraphs, we compare and contrast two of the most important characteristics of biomes: their productivity and their patterns of nutrient cycling.

Because temperature and precipitation strongly influence productivity, plant production varies greatly among biomes, as shown in table 3.2 in the text. Abundant moisture, favorable temperature, and a year-round growing season combine to make tropical rain forests the most productive terrestrial biome on an annual basis. Although temperatures are favorable all year round for plant growth, many tropical rain forests experience a short dry period, during which many trees become dormant and lose their leaves. When the rains return, plants begin to grow again.

As we move poleward from the tropics, the length of the growing season and temperatures generally decline. Such changes contribute to significant reductions in annual productivity. The productivity of temperate deciduous forests is some 40% lower than that of the tropical rain forests. Moving farther poleward, we find that the annual productivity of the boreal forest is 30% lower than that of the temperate deciduous forests.

In temperate grasslands, the growing season is shortened by freezes in spring and autumn as well as by drought in summer. Summer dry periods are often so severe that prairie plants cease to grow and become dormant for a significant portion of the growing season.

Desert and tundra biomes are at the low end of the productivity spectrum, as the table shows. The impact of extreme environmental conditions on productivity is evident in these regions. Precipitation obviously limits productivity in deserts. Rainfall is not only sparse—less than 25 centimeters (10 inches) annually—but also sporadic, often occurring as brief, intense downpours in thunderstorms. Plants and animals that live in desert biomes burst into activity when infrequent deluges occur. When sufficient precipitation falls, for example, the seeds of many desert plants germinate rapidly, producing seedlings that flower and set seed within a span of only a few weeks. The appearance of that new plant growth, in turn, triggers reproductive behavior in such desert herbivores as grasshoppers and rabbits. However, once the water has evaporated or has been absorbed by the plants, activity slows down, and production comes to a virtual halt again.

Productivity in the tundra illustrates the impact of light precipitation, low growing-season temperatures, and a short growing season (sixty days or less). Vegetation grows slowly and consists of low plants, such as grasses, sedges, and such dwarf woody plants as willows and birches. Few animals (such as caribou, musk oxen, Arctic fox, Arctic hare, lemmings, and ptarmigans) make the tundra their year-round home. In the summer, however, large numbers of migratory birds, mostly geese and ducks, arrive, nest, and raise their young before migrating again toward the equator.

We now turn to a consideration of patterns of nutrient cycling. A fundamental distinction exists in the nutrient cycling patterns of terrestrial biomes between oligotrophic and eutrophic ecosystems. The basic distinguishing factor is soil fertility. The soils in *oligotrophic ecosystems* are low in fertility, whereas in *eutrophic ecosystems,* soils are high in fertility. Oligotrophic ecosystems include tropical rain forests, which grow on old, highly-leached soils; pine forests, which grow on sandy, acidic, well-leached soils; and chaparral (a shrubland that grows in a moderately dry, temperate, maritime region, with little or no summer rain), which grows on shallow, coarse-textured soils. Eutrophic ecosystems include grasslands, which grow on soils with a great deal of organic matter and nutrients; deciduous forests, which grow on soils with intermediate to high availability of nutrients; and those rain forests that grow on relatively young soils derived from nutrient-rich volcanic rocks.

Because soils in oligotrophic ecosystems contain few plant nutrients and often have a low nutrient-storage capacity, plant uptake and storage is the major mechanism for the conservation of essential nutrients. Perhaps the most important adaptation for retaining nutrients is the mat of humus and fine rootlets that are located at or near the soil surface in pine forests and tropical

Earth's major biomes and their characteristics

Biome	Dominant Growth Form	Representative Plants (Mostly Northern Hemisphere)	Climate
Forests			
tropical rain forests	trees, broad-leaved evergreen	many species of evergreen, broad-leaved trees, vines, epiphytes	125 to 1,250 cm annual rain; temperatures from 18° to 35° C
tropical deciduous forests	trees, both evergreen and deciduous	mahogany, rubber tree, papaya, coconut palm	Marked dry season, generally lower precipitation
temperate deciduous forests	trees, broad-leaved deciduous	maple, beech, oak, hickory, basswood, chestnut, elm, sycamore, ash	60 to 225 cm precipitation; droughts rare; some snow; temperatures from −30° to 38° C
northern coniferous (Boreal) forests	trees, needle-leaved	evergreen conifers (spruce, fir, pine), blueberry, oxalis	35 to 600 cm precipitation; evenly distributed; much snow; temperatures from −54° to 21° C; short growing season
Reduced forests			
scrubland chaparral	shrubs, sclerophyll evergreen	live oak, deerbrush, manzanita, buckbush, chamise	25 to 90 cm precipitation; nearly all during cool season; temperatures from 2° to 40° C
Grasslands			
tropical savannah	grass (and trees)	tall grasses, thorny trees, sedges	25 to 90 cm precipitation during warm season; thunderstorms; dry during cool season; temperatures from 13° to 40° C
temperate grasslands	grass	bluestem, Indian grass, gramma grass, buffalo grass, bluebunch wheatgrass	30 to 200 cm precipitation; evenly distributed or high in summer; snow; temperatures from −46° to 60° C
Tundra	diverse small plants	lichens, mosses, dwarf shrubs, grass, sedges, forbs	10 to 50 cm precipitation; snow drifts and areas blown free of snow; temperatures from −57° to 16° C
Deserts			
warm	shrubs, succulents	creosote bush, ocotillo, cacti, Joshua tree, century plant, bur sage (in U.S.A.)	0 to 25 cm precipitation; very irregular; long dry seasons; temperatures from 2° to 57° C with high diurnal fluctuations
cold	shrubs	sagebrush, saltbush, shadescale, winterfat, greasewood (in U.S.A)	5 to 20 cm precipitation; most in winter; some snow; long dry season; temperatures from −40° to 42° C with diurnal fluctuation

Sources: Data from W. A. Jensen and F. B. Salisbury, *Botany: An Ecological Approach*, copyright © 1972 Wadsworth Publishing Company; and climatic data from W. D. Billings, *Plants, Man, and the Ecosystem*, 2d ed. copyright © 1970 Wadsworth Publishing Company.

rain forests. As the leaves fall to the mat, they are quickly enveloped by a mass of fine rootlets. Those rootlets, in association with specialized fungi, reduce nutrient losses by efficiently taking up nutrients from decomposing leaves and twigs. Very few nutrients are stored in the soil in oligotrophic ecosystems.

An important difference between oligotrophic and eutrophic ecosystems is their response to a disturbance. Because most of the nutrients in an oligotrophic ecosystem are incorporated in leaves, stems, and the root mat, most of the ecosystem's nutrients are removed when the vegetation is

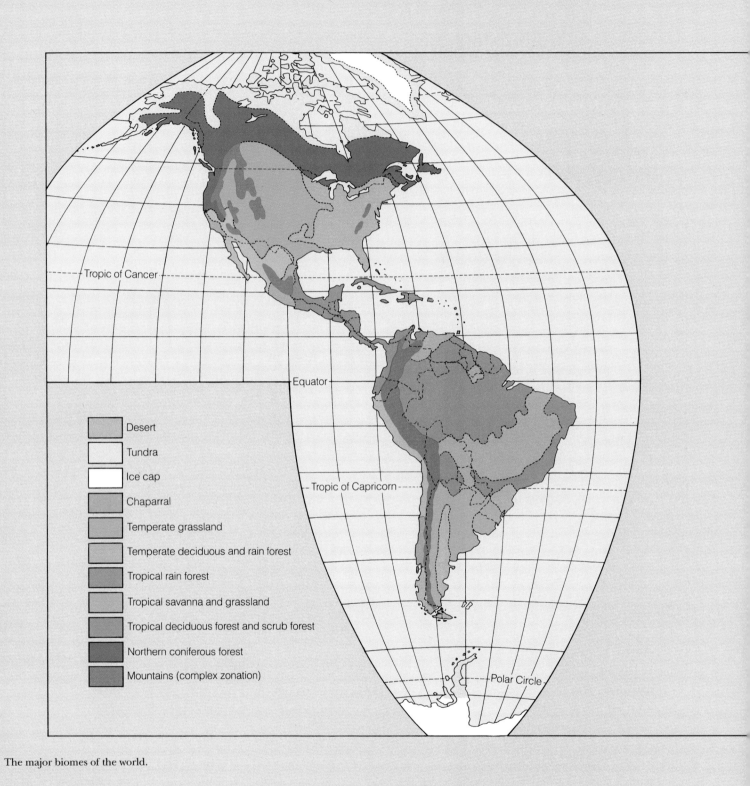

Desert

Tundra

Ice cap

Chaparral

Temperate grassland

Temperate deciduous and rain forest

Tropical rain forest

Tropical savanna and grassland

Tropical deciduous forest and scrub forest

Northern coniferous forest

Mountains (complex zonation)

Tropic of Cancer

Equator

Tropic of Capricorn

Polar Circle

The major biomes of the world.

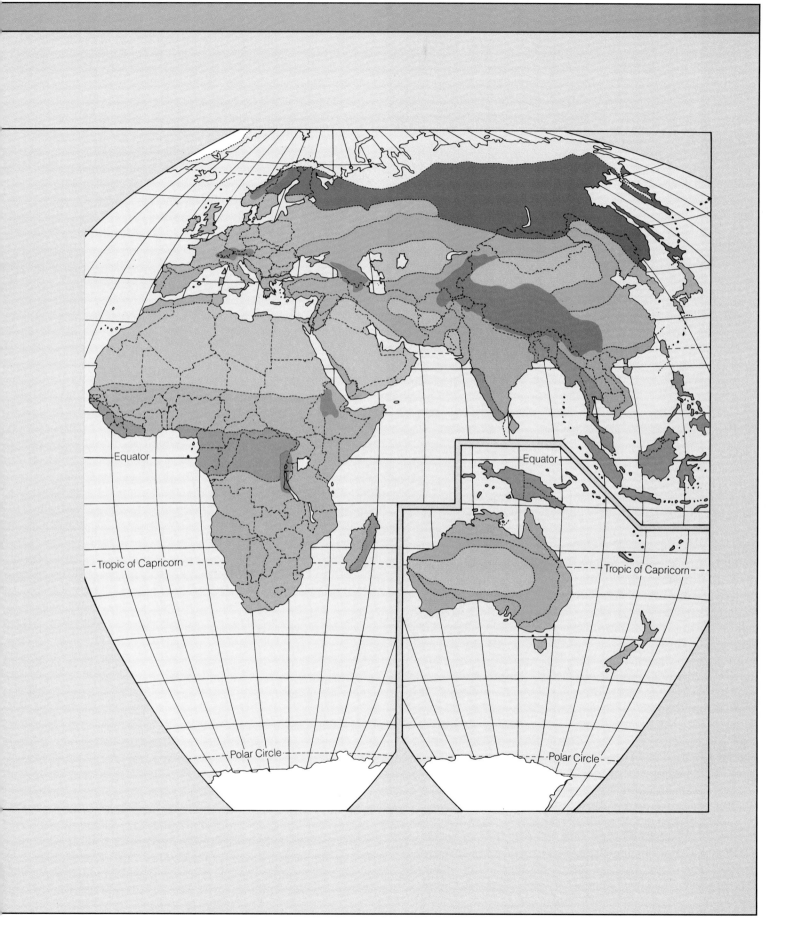

Equator

Tropic of Capricorn

Polar Circle

Equator

Tropic of Capricorn

Polar Circle

cleared and the root mat is destroyed by cultivation. Furthermore, because those soils often have little nutrient-holding capacity, application of fertilizers does little to improve crop yields. As a result, productivity usually falls off quite rapidly; within a few years after the original forest vegetation is cleared, agricultural yields fall to the point at which further cultivation becomes futile.

In contrast, most of the nutrients in a eutrophic ecosystem are in the soil, and the nutrient-conserving mechanisms of plants are relatively less important. Thus, the loss of vegetative cover does not greatly diminish the nutrient content of the ecosystem. Furthermore, because these soils have substantial nutrient-storage capacity, they respond well to the application of fertilizers. Thus, following a disturbance, a eutrophic ecosystem retains much of its productive capacity, and agricultural activities are usually successful (barring severe erosion).

If the disturbance is short-lived, such as a wildfire, both types of ecosystems usually undergo rapid revegetation. Often, colonization is particularly rapid in oligotrophic ecosystems after a fire, because many species of shrubs and trees that grow in such ecosystems resprout from underground stems and roots that are not harmed by fire. In eutrophic ecosystems, however, recolonization takes longer because the new vegetation is initiated, in most cases, by seeds, which take longer to become established. Revegetation that occurs from resprouting plants is quicker because sprouts already have a well-developed root system. Nutrients contained in the ashes of a fire also spur rapid regrowth. In oligotrophic ecosystems, however, that burst in regrowth soon slows as the ash nutrients are depleted, and the plants become dependent again on nutrient-poor soil.

Cycling Rates and Human Activities

Over time—at least until recently—the cycling rates of materials among various environmental reservoirs have been relatively stable. A **cycling rate** is defined as the amount of material that moves from one reservoir to another within a specified period of time. Cycling rates often reach an equilibrium, or balance; that is, the rate at which a material enters a reservoir equals the rate at which it leaves that reservoir. But an increase or decrease in a cycling rate may disrupt an equilibrium, so that the cycling rate into a reservoir no longer equals the cycling rate out of it. Figure 3.25 shows a model of the three basic patterns of flow into and out of a reservoir. In recent years, human activities have increasingly altered cycling rates and consequently disrupted the equilibrium of many cycles. In this section, we consider some of the human impacts on the phosphorus, nitrogen, and carbon cycles.

In the phosphorus cycle, the rate of phosphorus loss from an undisturbed ecosystem is usually low. If the protective vegetative cover is removed, however, as happens when a grassland is plowed to grow crops, the loss of phosphorus by water and wind erosion may increase tremendously. The phosphorus cycle is also disturbed when we mine phosphate rock and process it into fertilizers and detergents. The runoff of phosphate from agricultural lands and its inefficient removal by municipal sewage treatment plants enhance the transfer of phosphate to lakes and rivers.

Human activities have also altered the nitrogen cycle. Approximately sixty years ago, the development of the **Haber process** revolutionized agriculture because it provided an abun-dant and inexpensive source of nitrogen fertilizer. This industrial process synthesizes more than 60 million metric tons of ammonia annually from natural gas and atmospheric nitrogen—an amount roughly equal to a third of the ammonia produced annually by biological fixation in terrestrial ecosystems. Ammonia is injected into the soil, where nitrifying bacteria transform some of it to nitrate. Because agricultural lands are subject to erosion, some ammonia and nitrate runs off into lakes and rivers.

Today, nitrogen and phosphorus are entering waterways faster than they are being cycled out. Elevated concentrations of phosphorus and nitrogen act as fertilizers, accelerating the growth of algae and rooted aquatic plants. The enhanced plant growth, in turn, contributes to declining water quality in many waterways. In chapter 14, we explore the factors that control the cycling of nutrients and their impact on water quality.

An examination of the carbon cycle (fig. 3.26) also illustrates the impact of our activities on global cycles. Without human intervention, the flow of carbon dioxide between the atmosphere and terrestrial ecosystems would be essentially at equilibrium. That is, the amount of carbon dioxide incorporated into plants through photosynthesis each year would balance the carbon dioxide emitted to the atmosphere by the cellular respiration of plants, animals, and microorganisms. In recent years, however, human activities have begun to disturb the global carbon cycle. For example, atmospheric carbon dioxide levels increased by roughly 25% from 1850 to 1991 (fig. 3.27) and are expected to continue rising well into the next century. The principal causes of this upward trend

Input equals output:
no change in water level
(equilibrium)

Output greater than input:
water level drops

Input greater than output:
water overflows

are the burning of fossil fuels (about 80%) and deforestation (about 20%).

Since the dawn of the Industrial Revolution, the burning of coal, oil, and natural gas has emitted tremendous quantities of carbon dioxide to the atmosphere. As figure 3.26 illustrates, at least 5 billion metric tons of carbon dioxide are currently released each year as a result of the combustion of fossil fuels. Another source of atmospheric carbon dioxide is human clearing of forests, which may be releasing over a billion metric tons each year. Primeval forests, which contain massive trees, are significant reservoirs of carbon (fig. 3.28). In the nineteenth and twentieth centuries, huge amounts of carbon dioxide were released when forests in temperate regions were cleared and burned to create pastures and croplands. This process continues today with the burning of forests and savanna grasslands in tropical and subtropical regions. The clearing of forests, either for lumber or agriculture, contributes to elevated atmospheric carbon dioxide in another way. Because the trees are no longer there, they can't remove carbon dioxide from the air for photosynthesis. Thus, the amount of carbon dioxide that remains in the atmosphere increases as more and more forests are cut.

Figure 3.25   Three possible patterns of transfer rates into and out of a reservoir.

Figure 3.26   Annual rates of cycling of carbon dioxide among various reservoirs, expressed in units of billions of metric tons per year. Via the combustion of fossil fuels and deforestation, 6 to 7 additional billions of metric tons of carbon dioxide are added to the atmosphere each year.
Based on work by Bert Bolin of the University of Stockholm

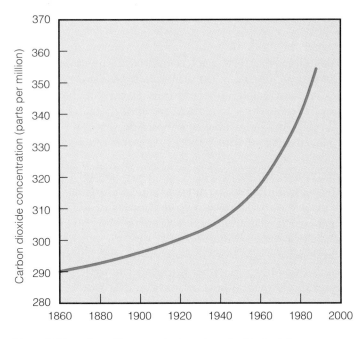

Figure 3.27   Trend in atmospheric carbon dioxide level due to increased combustion of fossil fuels and deforestation.
Source: Data from S.H. Schneider, "The Changing Climate," *Scientific American* 261:74, September 1989.

Figure 3.28   Massive tree trunks, such as these Douglas fir trees on Vancouver Island, British Columbia, store significant quantities of carbon.
© Thomas Kitchin/Tom Stack & Associates

While the cutting and burning of tropical rain forests continue to add carbon dioxide to the atmosphere, some ecologists suggest that the establishment of young, fast-growing, intensively managed forests should reduce the carbon dioxide content of the atmosphere. For example, some investigators suggest that reestablished timberlands in temperate regions such as the southeast United States are now slowly beginning to store carbon. Others counter that the continued degradation of temperate forests by acid rain and other air pollutants is reducing photosynthetic rates, thereby impairing the forests' capacity to remove carbon dioxide from the atmosphere. Hence, the future role of temperate-zone reforestation in the carbon cycle remains questionable. Nevertheless, the best current estimates suggest that in nontropical regions, the rate of carbon dioxide uptake essentially balances the rate of carbon dioxide release.

Natural processes have removed only about half of the carbon dioxide released into the atmosphere by human activities. Even though some compensating removal of carbon dioxide by the oceans is occurring, the rate of human-induced release of carbon dioxide into the atmosphere is still greater than the rate of absorption into the oceans. If world consumption of fossil fuels and the burning of tropical forests continue to accelerate, the carbon dioxide content of the atmosphere could double within sixty years.

The principal reason that humans have been able to have a significant impact on the carbon cycle is the relatively small amount of carbon dioxide in the atmosphere. Human-induced transfer of carbon dioxide into the atmosphere (6 to 7 billion metric tons per year) is a significant fraction (1% per year) of the total atmospheric carbon dioxide. Thus, it is well within our power to double the carbon dioxide concentration in the atmosphere within our lifetime. This chapter's issue describes some of the potential consequences of our disturbance of the global carbon cycle.

The Nature of Pollution

Pollution may be defined as an environmental disturbance that adversely affects the well-being of an organism directly or the natural processes upon which it depends. The type of material or the form of energy that is involved in the disturbance is called a **pollutant.**

Inevitably, all organisms (including humans) disturb their environment by utilizing resources and producing waste products. As we have seen in this chapter, humans have been particularly pervasive in disturbing their environment. In addition, physical forces such as hurricanes, floods, volcanic eruptions, and earthquakes are major perturbers of Earth's environment.

Anyone who has survived a hurricane or a strong earthquake would agree that we can do nothing to prevent such catastrophes. We can, however, reduce the toll of lost lives and damaged property by planning for such disasters, as described in chapter 21. On the other hand, we can do some-

The Potential Environmental Impacts of Elevated Atmospheric Carbon Dioxide

*B*ecause carbon dioxide is a resource for photosynthesis, we would expect that elevated levels of atmospheric carbon dioxide would trigger increases in plant production. Research on this question, however, has been less than conclusive. Although elevated levels of carbon dioxide enhance the yield of some crops, such as wheat and barley, and some trees, such as sugar maple and beech, other crops, such as corn, and other trees, such as birch and white pine, show little response to carbon dioxide enrichment. In some instances, the enhancement of plant growth rates has actually fallen off with continued exposure to high carbon dioxide levels. Perhaps most important, plant yields only improve significantly if other growing conditions, such as soil moisture and air temperatures, remain favorable. Thus, any significant improvements in crop production as a result of elevated carbon dioxide levels depend, to a large extent, upon future climatic conditions in the world's major agricultural regions.

It is important to realize that atmospheric carbon dioxide also influences the planet's climate. Because carbon dioxide helps to warm the lower atmosphere, scientists are asking whether or not the rapid increase in atmospheric carbon dioxide will lead to global warming and other climatic changes. Researchers have tried to answer this question by using numerical models of the atmosphere and electronic computers. These models predict that Earth's average surface temperature could rise some 2 to 6 degrees Celsius (4 to 11 degrees Fahrenheit) with a doubling of atmospheric carbon dioxide, which could occur by the middle of the next century. Warming of this magnitude could match the total post-Ice Age temperature rise—but at a rate that would be 10 to 100 times faster. Furthermore, the magnitude of warming is likely to vary from region to region. In fact, models predict that while tropical regions will experience little change in temperature, polar warming will be up to three times the average global change. In addition, models generally agree that annual rain and snowfall will increase poleward from about 30° latitude and within about 5° of the equator, and decrease elsewhere. Significant summer drying is predicted for the continental interiors of the middle latitudes, where much of the world's food is grown.

If the carbon dioxide level actually doubles by the middle of the next century, the consequent global climatic change could be greater than any experienced since the last major retreat of the glaciers some ten thousand to fifteen thousand years ago. Agriculture would be seriously disrupted. Although

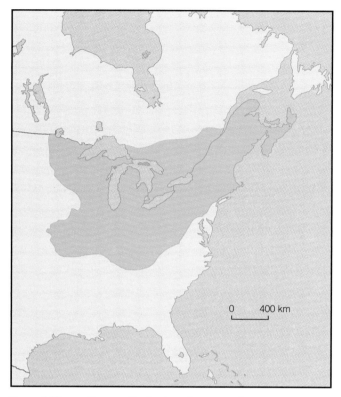

Figure 3.29*a*   Current distribution of sugar maple.

Source: Data from Margaret B. Davis and Catherin Zabinski, University of Minnesota.

elevated levels of carbon dioxide enhance the yield of some crops under favorable conditions, heat and moisture stress associated with global warming would actually cut crop yields. In many regions, traditional farming practices would have to change. For the North American Midwest, for example, a greater frequency of drought would force farmers to switch from crops that require lots of water, such as corn, to crops that require less water, such as wheat. Where sufficient, inexpensive water reserves are available, more land would have to be irrigated.

Forestry would suffer many of the same effects as agriculture. Higher temperatures and drier soils would reduce growth rates of many species of trees. These same conditions would

thing about disturbances for which humans are responsible. For example, we can reduce the contamination of surface waters by untreated sewage, cut emissions from automobiles, and clean up hazardous waste.

If we are responsible for disturbances that reduce the quality of our environment, then it makes sense to take corrective action. However, we do not all agree on whether ac-

tion is needed, nor on what specific steps should be taken if it is needed. Some people argue that any disturbance that harms plants, animals or humans in any way is unacceptable. But how seriously must organisms be impaired before the disturbance becomes unacceptable? Are we concerned about all species, or just a few, or just us? If we are not concerned about all species, what criteria do we use to select which species we

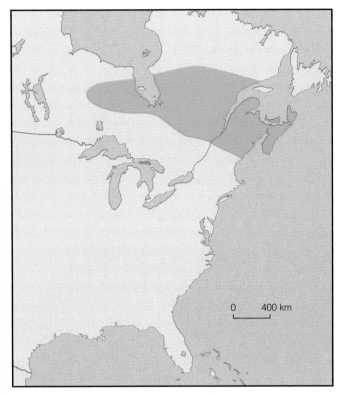

Figure 3.29*b* One projected distribution following a doubling of atmospheric carbon dioxide.
Source: Data from Margaret B. Davis and Catherin Zabinski, University of Minnesota.

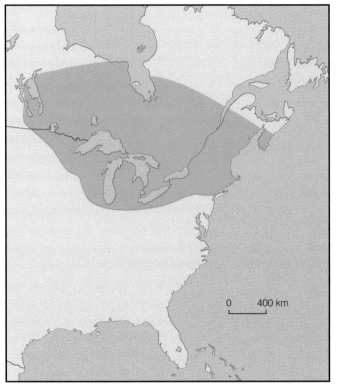

Figure 3.29*c* Another projected distribution following a doubling of atmospheric carbon dioxide, based on a different climate model.
Source: Data from Margaret B. Davis and Catherin Zabinski, University of Minnesota.

probably contribute to an increased frequency of forest and grassland fires.

Changes in the geographic distribution of plant and animal species are also likely. Figure 3.29 illustrates two scenarios for the future distribution of sugar maple, one of the dominant tree species of the eastern deciduous forest. Although the projected

distributions vary with the basic assumptions used in the two climate models, sugar maple distribution could shift several hundred kilometers to the north. Many plant species in the eastern United States could follow a similar pattern of change in their geographic range. There is no guarantee, however, that sugar maple or any other plants would advance to their

save? Finally, what are the economic and political ramifications of our choices?

Rarely will there be just one "right" answer to any of these questions. For example, suppose that a government regulatory agency has determined that the air pollution from an industrial plant presents a health hazard to people living downwind of the plant and requires the company to install expensive pollution-control equipment. After a cost-effectiveness analysis, however, the company feels that the equipment is too costly and decides to close its gates. Whether this disturbance (the air pollution) is acceptable or unacceptable depends upon your perspective—whether you are employed by

the company and face losing your job, or you live downwind and face losing your health.

Conclusions

In this chapter, we confront two related concepts that reoccur continually in environmental science: interrelationship and balance. We find that we cannot study an ecosystem without describing the interactions among the organisms, chemical substances, and physical conditions that characterize it. We cannot describe the pathways taken by the flow of energy and the cycling of materials without showing how each pathway

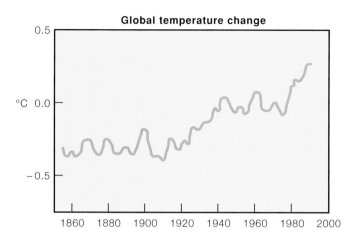

Global temperature change

Figure 3.30 Global mean temperatures show a warming trend over the past century, relative to the average for 1951–1980.

potential northern limits. Species vary greatly in their ability to disperse. Ecologists are also concerned about the ability of migrating species to adjust to changes in the length of the day and in soil type. In addition, obstacles such as large urban areas, agricultural fields, and highways pose significant barriers to the successful migration of many species.

The predicted amplification of the global warming trend in polar latitudes has prompted much speculation on the fate of the Greenland and Antarctic ice sheets. Some studies contend that the warming would melt enough ice to cause the sea level to rise and inundate densely populated coastal areas such as the East and Gulf Coasts of the United States. Warming also causes ocean water to expand, and this would contribute to a higher sea level. The net effect of glacial melting and expansion of ocean water would raise sea level an estimated 0.3-1.5 m (1-5 ft) within sixty years.

Some researchers hasten to point out that some benefits may come from warming caused by increased carbon dioxide. For example, crop yields may increase in far northern regions where cool temperatures and a short growing season now limit plant growth. In addition, warmer winters would reduce the demand for heating fuel in the middle and high latitudes. In their 1988 summary of potential socioeconomic impacts in the Great Lakes region due to a doubling of atmospheric carbon dioxide, S. J. Cohen and T. R. Allsopp of Atmospheric Environment Service of Canada reported that in Ontario, milder winters would reduce the energy demand for space heating by 45%, more than offsetting an estimated increase in summer air-conditioning demand of 7%. Warmer winters would also lengthen the navigation season on lakes, rivers, and harbors where ice cover is a problem.

Has the warming begun? The decade of the 1980s witnessed the highest global mean temperatures in over 120 years (fig. 3.30). Yet a note of caution is warranted. Even though some recent climatic trends are consistent with the changes that are predicted to accompany carbon-dioxide-induced warming, there is still no evidence of a direct cause-effect relationship. Climate controls other than carbon dioxide may well be responsible for these observed trends. Furthermore, the atmosphere is a highly interactive system, so that other processes may compensate for any carbon-dioxide-induced warming. For instance, the ocean has great thermal stability, which for a time may slow the warming. Warmer conditions in high latitudes may also mean more snowfall and eventually more, not less, glacial ice.

In spite of current scientific uncertainty, a growing number of experts argue that because so much is at stake, we should plan now for the worst future climate scenario. They argue for a sharp reduction in fossil fuel consumption and a switch to alternative fuels in conjunction with reduced deforestation and a massive effort at reforestation. As we shall see in succeeding chapters, such actions have other benefits as well.

affects and interacts with the others. And we really cannot describe environmental balance without referring to the dependence of each component or process in an ecosystem on the smooth functioning of all its interrelated parts.

These concepts are particularly important when we try to measure the effect that we have on the environment. When a raindrop strikes a pond, the effects of that disturbance are felt in ever-widening circles. Unlike these natural ripples, which are transient, human-made ripples may continue to disturb the environment for decades and perhaps centuries. One reason for studying environmental science is to make ourselves more sensitive to the effects of human-made ripples on

the functioning of ecosystems and thus on our own future well-being.

What are the consequences of disturbances for organisms, including ourselves? What are the limits of an organism's ability to adjust to environmental change? What happens when these limits are exceeded? We explore these and related questions in the next chapter.

Key Terms

El Niño
ecosystem

biotic
abiotic

producers
consumers
herbivore
carnivore
omnivore
detritus
decomposers
detritivore
inorganic chemicals
organic chemicals
energy
food chain
food web
trophic level
photosynthesis
cells
grazing food web
detritus food web
zooplankton
phytoplankton
law of energy conservation
first law of thermodynamics
energy efficiency
second law of thermodynamics
net primary production

biomass
assimilated food energy
cellular respiration
aerobic organisms
anaerobic microorganisms
nonrenewable resources
DNA
law of conservation of matter
carbon cycle
oxygen cycle
nitrogen cycle
phosphorus cycle
water cycle
fossil fuels
ozone shield
biological fixation
atmospheric fixation
nitrification
denitrification
estuary
cycling rate
Haber process
pollution
pollutant

Summary Statements

Ecosystems consist of living and nonliving components. Organisms comprise the living component. They are distinguished by their major functions: producers (green plants) manufacture food, whereas consumers (herbivores, carnivores, omnivores, decomposers, and detritivores) consume other organisms to obtain food. Chemical substances and physical characteristics comprise the nonliving portion of an ecosystem.

Within ecosystems, food energy moves only in one direction—from producers to consumers—in networks of pathways known as food webs. Ecologists recognize two types of food webs: grazing and detritus. The relative importance of each depends on the type of ecosystem in question.

Energy flow through food webs is governed by the first and second laws of thermodynamics. For several reasons, only a small percentage of food energy is transferred from one trophic level to the next higher trophic level. Because energy efficiencies are low and each organism requires a certain amount of energy to survive, only a limited number of organisms can survive at a particular trophic level.

With an understanding of food-web dynamics, we have increased the quantity of food available to humans by (1) reducing the number of organisms that compete for the same food, (2) converting forests and rangeland into cropland, (3) using fossil fuel energy subsidies to increase the efficiency of energy use by livestock and to increase crop yield per hectare, and (4) eating less meat and more fruits, vegetables, and cereals. In spite of our ingenuity in coping with food-web inefficiencies, nature continues to restrict our ability to exploit food resources.

Although energy is supplied continually by the sun, the quantity of materials on Earth is fixed and finite. The carbon, oxygen, nitrogen, and phosphorus cycles illustrate the many pathways along which essential materials are recycled.

All cycles obey the law of conservation of matter. All losses from one environmental reservoir must be accounted for as gains in other environmental reservoirs.

Over past millennia, the cycling rates of materials among various environmental reservoirs have become relatively constant; in many instances, they reach an equilibrium. Recently, however, human activities have increasingly altered cycling rates and thereby disrupted the equilibrium of many cycles. Some of the consequences include a decline in air and water quality and possible changes in climate.

Pollution is an environmental disturbance that adversely affects the well-being of organisms directly or the natural processes on which they depend. All organisms, including humans, disturb their environment by utilizing resources and producing waste products. If we are responsible for disturbances that reduce the quality of our environment, then it makes sense to take corrective action.

Questions for Review

1. List the biotic and abiotic components of an ecosystem.
2. Compare and contrast grazing and detritus food webs. How does their relative importance vary among different ecosystem?
3. Explain the importance of the second law of thermodynamics for understanding energy inefficiency in food webs.
4. List four reasons why energy efficiency in natural food webs can be as low as 1%.
5. The percentage of digestible food in the diet helps to determine the amount of energy that is transferred to the next higher trophic level. For each kilogram of food intake, which herbivore would obtain the most usable energy—one that eats seeds, wood, young foliage, or mature foliage? Explain your answer.
6. Compare and contrast the roles of photosynthesis and cellular respiration in the flow of energy through food webs.
7. Describe the role of energy subsidies in raising the energy efficiency of livestock to 20%.
8. Describe several advantages that accrue to an animal that can occupy more than one trophic level.
9. Define the law of conservation of matter. Describe its importance for understanding global cycles of nutrients.
10. Trace the route that carbon might follow as it cycles through a terrestrial ecosystem. Include at least four different organisms in the cycle.
11. Trace the route that phosphorus might follow as it cycles through a terrestrial ecosystem. Include at least four different organisms in the cycle.
12. How are the cycles that you constructed in questions 10 and 11 similar? How do they differ?
13. We burn fossil fuels for energy. What is the ultimate source of the energy found in fossil fuels?
14. Describe the roles of cellular respiration and photosynthesis in both the carbon cycle and the oxygen cycle.
15. How does biological fixation of nitrogen differ from atmospheric fixation of nitrogen? Which process provides the greatest amount of fixed nitrogen?
16. Describe the role of decomposers in the phosphorus and nitrogen cycles.

17. Microorganisms play many essential roles in the nitrogen cycle.
   Identify at least five points in the nitrogen cycle at which microorganisms are essential.
18. Describe how the disruption of the nitrogen cycle and the phosphorus cycle have affected water quality.
19. Describe several potential environmental impacts of elevated atmospheric carbon dioxide.
20. Define *pollution*. List several criteria that people might use to determine if an environmental disturbance is acceptable or not.

Projects

1. Visit several types of ecosystems in your region. Identify the major producers, herbivores, carnivores, decomposers, and detritivores, and describe evidence for their interactions. Also determine whether grazing or detritus food webs dominate. Describe the types of evidence that you used to make your determination.
2. Take a tour of your region and note examples of pollution. Does your community judge these incidences of pollution to be acceptable, or is something being done to control the pollu tion? What factors play a role in a community's decision either to ignore or to deal with a pollution problem?

Selected Readings

Aber, J. D., K. J. Nadelhoffer, P. Steudler, and J. M. Melillo. "Nitrogen Saturation in Northern Forest Ecosystems." *Bioscience* 39 (1989):378-86. An analysis of a proposal that excess nitrogen from fossil fuel combustion may be disrupting the activities of northern forests.

Bell, R. H. V. "A Grazing Ecosystem in the Serengeti." *Scientific American* 226 (July 1971):86-93. Describes how migrations of zebra, wildebeest, and Thomson's gazelle are synchronized with the availability of specific food sources.

Bergon, M.E., J. L. Harper, and C.R. Townsend. *Ecology: Individuals, Populations, and Communities.* 2d ed. Cambridge, Mass.: Blackwell Scientific Publications, 1990. An up-to-date and thorough treatment of ecological principles with abundant examples.

Berner, R. A., and A. C. Lasaga. "Modeling the Geochemical Carbon Cycle." *Scientific American* 260 (March 1989):74-81. A look at the long-term processes that transfer carbon among land, sea, and atmosphere and how people are disrupting the long-term cycle.

Cohn, J. P. "Gauging the Biological Impacts of the Greenhouse Effect." *Bioscience* 39 (1989):142-46. Projections concerning the impact of global warming on ecosystems and populations.

Houghton, R. A., and G. M. Woodwell. "Global Climatic Change." *Scientific American* 260 (April 1989):36-44. A consideration of the role of carbon dioxide in global warming and the steps that must be taken to halt further change.

Mooney, H. A., B. G. Drake, R. J. Luxmore, W. C. Oechel, and L. F. Pitelka. "Predicting Ecosystem Responses to Elevated CO_2 Concentrations. *Bioscience* 41 (1991):96-104. An examination of what is known about and what needs to be done to predict the responses of terrestrial ecosystems to rising carbon dioxide levels.

Mooney, H. A., P. M. Vitousek, and P. A. Matson. "Exchange of Materials between Terrestrial Ecosystems and the Atmosphere." *Science* 238 (1987):926-32. An examination of recycling processes.

Schneider, S. H. "The Changing Climate." *Scientific American.* 261 (Sept. 1989):70-79. An examination of the impact of people's disruption of the carbon cycle and strategies to slow global warming.

Smith, R. L. *Ecology and Field Biology.* 4th ed. New York: Harper & Row, 1989. A well-written, informative examination of ecological principles.

Spences, C. N., B. R. McClelland, and J. A. Stanford. "Shrimp Stocking, Salmon Collapse, and Eagle Displacement." *Bioscience* 41 (1991):14-21. A look at the cascading impact in a major food web following introduction of a seemingly harmless species of shrimp.

Chapter 4

Ecological Responses to Environmental Change

A unique combination of environmental conditions makes possible the growth of giant sequoia trees in the Sierra Nevada Mountains of California.

© Tom Stack/Tom Stack & Associates

*T*he environment is dynamic. Weather varies from one day to the next, the seasons come and go, rivers and lakes rise and fall, and forests and grasslands go up in flames and are replaced by new plant growth. The very face of the planet changes as wind and water sculpt the land, volcanoes erupt, mountain chains rise, plains subside, glaciers flow, the seas advance and retreat, and even the continents themselves drift over the face of the globe. Organisms die and are succeeded by their offspring; their populations wax, wane, and eventually disappear, to be replaced by those of other species; their remains build the soil and shape the seabed. People, too, are agents of change. We dam rivers, drain or fill in wetlands, clear forests, blanket the land with farms and cities, and foul the air, water, and land with our wastes.

We often fail to appreciate how much the natural variability of our environment affects our personal planning and that of commerce, industry, and government. For example, figure 4.1 illustrates the year-to-year variation in total snowfall for December in a typical community in the midwestern United States. The red line represents the average (normal) total snowfall for the month. We might conclude that the considerable variation in snowfall from year to year makes planning difficult. If you own a ski shop, for example, your sales depend to a great extent upon the depth of the snow cover during December. People are more likely to purchase ski equipment when a deep mantle of snow covers the ground than when the ground lies bare. Hence, when only 3.5 centimeters (1.4 inches) of snow fell in December of 1979, your sales were quite poor, and you were stuck with a large inventory. As a result, when you ordered your stock for the next winter, you decided to maintain a smaller inventory than you did in 1979, so as not to tie up as much of your capital. However, over 38 centimeters (15 inches) of snow fell that December. Such snowy conditions generated a strong demand for ski equipment, but because you had reduced your inventory, you sold out of skis early and had to turn away prospective buyers.

Figure 4.2 shows that considerable year-to-year variation also occurs in the average December temperature. Variations in temperature and precipitation influence such plans as municipal budgets for snow removal, electric utility plans for stockpiling fossil fuels, and personal plans regarding when to take a vacation and where to go.

Humans and other organisms can usually adjust to environmental change. For example, changes in the season initiate various physical and behavioral responses among animals. Foxes and wolves adjust to the coming of winter by growing a thicker coat of fur; deer congregate in small, sheltered areas and conserve their energy by greatly reducing their movements and foraging activities; chipmunks and groundhogs hibernate; and many birds and some insects migrate to warmer climates. We adjust to colder weather by wearing insulated clothing, staying indoors more often, and heating our homes, (and some people follow the birds south). The ability to adjust to change is essential for survival in an environment where change is constant.

However, some changes may be so great or occur so abruptly that some organisms cannot adjust to them. For ex-

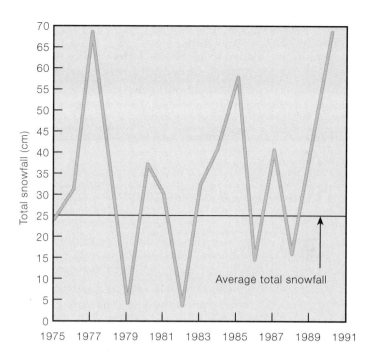

Figure 4.1 Variation in December total snowfall in Green Bay, Wisconsin.

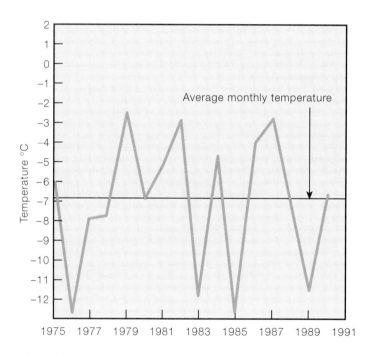

Figure 4.2 Variation in December mean temperature in Green Bay, Wisconsin.

ample, searing heat and prolonged drought during the summer of 1988 afflicted crops to the point that U.S. soybean production declined by one-fifth, corn by one-third, and spring wheat by one-half. Environmental change can cause not only the death of individual organisms, but also, with time, the death of populations and eventually even the disappearance of entire species. The fact that 99% of all species that have existed on this planet are now extinct attests to the limitations of adjustment to environmental change.

In this chapter, we examine the factors that govern an organism's ability to respond to and survive changes in its environment. We also consider the responses of ecosystems to change. An understanding of these processes helps us to appreciate the varying impacts of disturbances and the resulting challenges of resource management, which we discuss in subsequent chapters.

Organisms and Environmental Change

Among the components of the environment that change with time are temperature, precipitation, light levels, wind speed, the availability of food, water, and mineral nutrients and the presence or absence of various species. As these characteristics change, they influence the well-being and survival of organisms. To understand the response of organisms (including ourselves) to such changes, we need to consider two laws: the law of tolerance and the law of the minimum.

The Law of Tolerance

An organism's ability to respond to a change in its environment is described by the **law of tolerance.** This principle states that for physical conditions and chemical substances in an environment, minimum and maximum limits exist, called **tolerance limits,** beyond which no members of a particular species can survive.*

To illustrate the law of tolerance, let us return to the problem of phosphorus enrichment and excessive growth of algae and aquatic plants in lakes that we described in chapter 3. We can perform a simple experiment to demonstrate a cause-and-effect relationship between algal growth and phosphorus concentration. If we place algae in a water-filled aquarium that contains all essential nutrients except phosphorus, the algae will die; like all organisms, they require a minimum amount of phosphorus to survive. If we add phosphorus little by little to the aquarium, eventually it will reach a concentration that favors the survival of a few algae. That concentration represents the **lower tolerance limit** of algae for phosphorus (fig. 4.3). As we continue to add phosphorus, the algae multiply until they reach a maximum population. The phosphorus concentration at that point is called the **optimum concentration.** Above the optimum concentration, any additional phosphorus is toxic to some algae, and the population declines. At a still higher level, the phosphorus concentration is lethal even to the most phosphorus-tolerant algae. The concentration above which no algae can survive is called the **upper tolerance limit.**

In some respects, the law of tolerance supports the adage, "moderation in all things." Algae require phosphorus to function properly, yet excessive amounts can be lethal. Likewise, we require many minerals (including phosphorus) and

*The law of tolerance does not apply to toxins that are poisonous at extremely low concentrations.

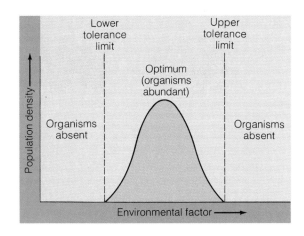

Figure 4.3 The law of tolerance illustrates the ability of organisms to respond to environmental change.

vitamins to maintain good health, but too much or too little of these nutrients can weaken, sicken, or even kill us. For example, an insufficient supply of vitamin D leads to a condition known as *rickets,* in which the bones and teeth fail to develop normally. An excess of vitamin D, however, leads to nausea, diarrhea, and weight loss; in extreme cases, it causes irreversible damage to the kidneys. It is interesting to note that although natural foods are generally poor in vitamin D, our skin produces it when exposed to ultraviolet light from the sun. Hence, rickets was once rather common in human populations that lived in northern regions where exposure to solar ultraviolet light was limited during long, dark winters. Recently, however, we have become aware that overexposure to ultraviolet light can lead to skin cancer. Either too little or too much of any required factor, such as food, vitamins, minerals, water, oxygen, sunlight, or heat can threaten the survival of individual organisms and even an entire species.

The genetic makeup of an organism (the information contained within its DNA) ultimately determines its tolerance limits for each environmental component. It is important to note that each individual in most populations of plants and animals is genetically unique. This is obvious as we look around the classroom. Each person looks and behaves differently from all the others. (Even identical twins, who have the same genetic makeup, will differ slightly in appearance and behavior.) Because each person has a different genetic makeup, each person also has slightly different tolerance limits. You may have a friend who catches a cold or the flu every winter and another friend who appears never to have been sick a day in her life. Remember that a tolerance curve represents the collective response of a population composed of many individuals, each with different tolerance limits. Hence, as a population is exposed to any vital chemical substance or physical condition that is above or below the optimum for that population, the least tolerant individuals may die, while the more tolerant individuals may survive.

As figure 4.4 illustrates, tolerance limits also vary among species. Note that some animals, such as humans and rats, are

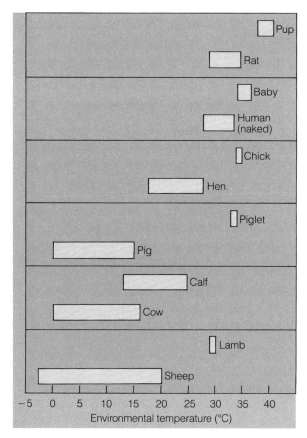

Figure 4.4 Tolerance to temperature varies with species. Each bar represents the temperature range in which the animal can maintain a constant temperature without resorting to special temperature-regulating mechanisms. Note that the young have considerably narrower limits than the adults.

Sources: Data from G. W. Cox and M. D. Atkins, *Agricultural Ecology,* © 1979 W. H. Freeman & Company; and H. Precht, J. Christopherson, H. Hensel, and W. Larcher, *Temperature and Life,* © 1973 Springer-Verlag.

less stressed by high temperatures than are other animals such as sheep and cows. Hence, we would expect changes in temperature to have differing impacts on different species.

The Law of the Minimum

Suppose that one day when we visit our favorite lake, we find that a population explosion of algae has turned the lake pea green. We might conclude that the addition of phosphorus-rich sewage and agricultural runoff to the lake has raised its phosphorus concentration to a level that approaches the optimum level for algal growth. But many chemical substances and physical characteristics simultaneously influence algal growth: water temperature, light penetration, and concentrations of other nutrients, such as nitrogen. How can we determine which factor actually accelerated algal growth in the lake?

To answer this question, we first consider another ecological principle: the **law of the minimum.** According to this principle, the growth and well-being of an organism is limited by the essential resource that is in lowest supply relative to what is required. This most deficient resource is called the

limiting factor. Suppose that we visit a clear mountain lake and find only a few algae growing. Analysis may show that the lake is poor in phosphorus, and we might assume at first that the lake contains too little phosphorus to support algal growth. Actually, some other factor, perhaps excessive acidity, might be more limiting. Thus, more observations and perhaps some experiments would be needed to determine which factor is primarily responsible for limiting the growth of algae in the lake.

Farmers expend a great deal of energy to reduce the impact of limiting factors on crop yields. For example, fertilizers replenish soil nutrients and thereby spur plant growth; irrigation reduces the stress of deficient soil moisture on crops; and pesticides diminish the yield-limiting impact of weeds, insects, and pathogens. Of course, adding fertilizer does little good if deficient soil moisture is the factor that is actually limiting crop growth. In chapters 8 and 9, we evaluate current efforts to reduce the impact of limiting factors on agricultural production and thereby increase the food supply for a hungry world.

Determining Tolerance Limits

Determining the tolerance limits of a given organism for a particular environmental factor is often challenging because so many things influence tolerance limits. For example, age strongly affects an individual's tolerance limits. In all species, the very young usually have the least tolerance for stress. A child whose diet is deficient in essential amino acids (the building blocks of protein) runs a high risk of suffering permanent brain damage, whereas an adult, whose brain is already fully developed, probably would not suffer this effect from a similar diet. The young also often have considerably narrower tolerance limits than adults for many environmental factors, including air temperature (see fig. 4.4).

Gender also influences tolerance limits. Women are nearly 50% more likely to die during hospitalization for a heart attack than men. One reason for this disparity may be that women have a higher incidence of heart failure, which is a measure of cumulative damage. Growing evidence suggests that men and women may also differ in their response to AIDS (Acquired Immune Deficiency Syndrome) and the onset of certain forms of cancer.

Another complication in determining tolerance limits is the genetic capacity of an organism to adjust to changes in its surroundings. For example, visitors to mountainous areas may initially experience drowsiness, headaches, nausea, and shortness of breath. The air at high altitudes contains less oxygen per unit volume than it does at lower altitudes, and these symptoms indicate that people are approaching the lower tolerance limit for oxygen. Over time, however, symptoms gradually disappear. The process of adjustment to low oxygen levels at high altitude and other stresses is known as **acclimatization.** The human body adjusts to low atmospheric oxygen concentrations by several mechanisms. For example, on initial exposure to low oxygen levels, the body increases the inflow and outflow of air between the atmosphere and the lungs. If a person remains at a high altitude, his production of red blood cells (carriers of oxygen in the bloodstream) begins to increase within a week or two. Some people are better able to adjust than others because of genetic differences. Still, the ability of the human body to acclimatize to low oxygen levels is limited. The highest permanent human settlement is at 5,300 meters (17,500 feet) in the Peruvian Andes.

Despite such complications as acclimatization and the varying effects of age and sex, we can determine tolerance limits relatively easily if we can isolate one species and change only one environmental factor at a time. However, we can only do this in a laboratory; in a natural, uncontrolled environment, many factors operate and interact simultaneously. Sometimes these interactions are relatively straightforward. For example, we are aware that diet and general health are related. If we eat the proper foods, we are usually less likely to catch a cold or the flu.

Sometimes interactions have unforeseen consequences. One such interaction is **synergism.** The interaction of two or more factors is said to be *synergistic* if the total effect is greater than the sum of the two or more effects acting independently; that is, the total effect is greater than the sum of its parts. For example, sulfur dioxide fumes attack and impair the lungs. Particulates (tiny particles such as soot and ash) also damage the lungs. However, when laboratory animals are exposed to air that contains both sulfur dioxide and particulates, their lungs sustain much greater damage than we would expect if they were exposed to each pollutant separately and in sequence. Thus, the interaction of sulfur dioxide with particulates is a synergistic interaction.

Two or more factors may also interact to produce a total effect that is less than the sum of the two effects acting independently; that is, the total effect is less than the sum of its parts. Such an interaction is known as **antagonism.** For example, nitrogen oxide is an air pollutant with effects similar to those of sulfur dioxide, but when nitrogen oxide and particulates interact, they somehow counteract each other, and their harmful effect on the lungs is much less severe than we would expect.

In the larger context of an ecosystem, the determination of tolerance limits is very difficult. An ecosystem may be composed of hundreds or thousands of species, each of which is influenced by other species and the physical and chemical characteristics of its environment. Moreover, the number and kinds of species that are present and such physical conditions as light levels, air temperature, and soil moisture fluctuate daily and seasonally.

The many and ever-changing interactions that occur in nature lead us to conclude that few simple cause-and-effect relationships exist in our environment. Most predictions about the impact of a particular environmental change, therefore, must be qualified. In most situations of interest, so many interactions occur that scientists find it difficult to identify all of them, let alone predict their outcome. This complexity often leads to disagreements about the causes of (not to mention the solutions to) environmental problems. A case in point is the dramatic decline and dieback of forests described in this chapter's issue.

Ecological Succession

A habitat may be so disturbed that conditions exceed the tolerance limits of many organisms, so that relatively little plant and animal life remains on the site. The causes of such severe disturbance include fires, hurricanes, volcanic eruptions, agricultural cultivation, the cutting of forests, construction (dams, highways, towns, etc.), and industrial pollution.

When the disturbance subsides or ends, plants and animals usually begin to recolonize the seemingly barren site within a few weeks or months. You may have witnessed such an event in a vacant lot or an abandoned garden that is quickly taken over by weeds. This first stage in recolonization is called the **pioneer stage,** and the successful species are called **pioneer species.** Over a period of months and years, these pioneer species are gradually replaced by other types of colonizers. The process of replacement, or succession, continues until an assemblage of species that is able to maintain itself on the site becomes established (barring another major environmental disturbance). This relatively stable stage of recolonization is called the **climax stage,** and the species that comprise this stage are called **climax species.** The continual sequence of colonization and replacement of species on a site is known as **ecological succession.** Hence, we can view ecological succession as a series of recovery stages following the disturbance of an ecosystem.

Given the dynamics of ecological succession, we can conclude that early successional plants and late successional plants differ considerably in their survival strategies. Early successional species must germinate and grow rapidly in the rather hostile conditions of an exposed landscape. In addition, as succession occurs, these species need to escape the site and colonize suitable areas in early stages of succession elsewhere. In other words, they must be well suited for dispersal and seedling establishment. Late successional species, on the other hand, must possess characteristics that allow them to compete well with other species and thereby persist longer on a site. Further discussion of these two sets of strategies appears later in this chapter.

The climax vegetation that eventually emerges on a site depends on the interaction of many factors, including climate, soil type, and drainage conditions. For example, the natural climax ecosystem of the Great Plains is grassland (which we have converted to fields of grain), primarily because grasses are more tolerant than trees of the relatively dry climate and because frequent thunderstorms ignite fires that kill woody vegetation but promote the growth of grasses.

Ecologists recognize two basic types of succession: primary and secondary. Which one takes place depends on the soil conditions after the disturbance has passed.

Why Are the Trees Dying?

*I*n once-thriving forests across large areas of central and eastern Europe and portions of eastern North America, tree growth is slowing, and in many locations, trees are either dying or are already dead (fig. 4.5). In Europe, eleven species are affected, including the most important conifers (Norway spruce, silver fir, Scotch pine, and European larch) and several deciduous species, including oak, maple, and birch. In North America, only conifers, including several species of pine, two species of fir, and red spruce, have been affected thus far.

Recent research has revealed that forests on both continents have been under continuous stress for at least the past two decades.* Compared to the growth rates of previous decades, the growth rates of trees have slowed significantly during the last twenty years. Unfortunately, this slowed growth occurs long before symptoms of deteriorating health become visible. These symptoms include yellowing and dying of leaves, and crooked and out-of-place shoots. Uprooted trees display stunted and deformed root systems.

Why are so many trees dying? No single cause can explain all the symptoms of damage that have been observed. Air pollution is considered to be the primary causal agent, but several air pollutants have been implicated, including acid rain, ozone, sulfur dioxide, and nitrogen oxides. In addition, insects, disease, and such climatic stresses as drought probably play a

*See the paper by Hinrichsen in the readings listed at the end of this chapter.

secondary role. Emerging evidence suggests that these factors are operating synergistically and antagonistically as well as by simple additive interactions. Furthermore, the importance of each factor appears to vary from one region to another. Although no single hypothesis can explain all the observed symptoms, for illustrative purposes we briefly consider the proposed role of three air pollutants that contribute to forest decline.

Ozone (O_3) is known to cause major damage to crops and forests. One proposed mechanism of ozone action is that it injures leaves and causes their premature loss. These effects reduce photosynthetic capacity, which in turn leads to reduced growth. Such weakened trees are more vulnerable to insect infestations, root rot, and ultimately death.

Another suspected culprit is airborne nitrogen fertilizer. Studies have found that 10% of the nitrogen that is applied to soils in fertilizer escapes to the atmosphere. (Refer to the discussion of the nitrogen cycle in chapter 3.) Some of this nitrogen fertilizer is eventually deposited onto forests. Although some additional nitrogen probably stimulates forest growth, some scientists believe that excess nitrogen poses a problem for trees. By stimulating growth, excess nitrogen prolongs the growing period. Consequently, when winter arrives, the trees have not had sufficient time to prepare for the rigors of winter weather. Growing tissues are much more vulnerable to winterkill than dormant, winterized tissues.

Acid rain may affect vegetation in several ways. Some research suggests that acid rain accelerates normal soil

Primary Succession

Primary succession occurs on solidified lava flows, newly formed sand dunes, rocks exposed by a retreating glacier, or waste heaps left by mining operations. Such sites have one thing in common: No soil is present. Because most plants cannot grow without soil, primary succession proceeds slowly until a soil forms. For example, soil development on bare rock often begins with the establishment of lichens and mosses (fig. 4.6) in the shelter provided by small cracks and depressions. These species can withstand the temperature and moisture extremes that characterize such habitats. Within such microsites, they are also able to trap tiny bits of detritus and soil particles. Consequently, they slowly grow outward. Concurrently, the acids produced by these organisms coupled with physical weathering (e.g., the freezing and thawing of water in rock crevices) slowly fragment and break down the

rock. Gradually, soil forms and thickens, which leads to greater moisture-holding capacity and the inclusion of more organic matter, which eventually makes possible colonization by grasses and herbs. With further soil development, shrubs may successfully invade the pockets of soil. Hence, the initial stages of primary succession are extremely slow. Hundreds of years may be needed for a cover of soil to develop that is deep enough to support shrubs and trees. As a general rule, primary succession occurs most rapidly in humid, tropical locations and most slowly in dry, polar locations.

Once soil has formed, however, the subsequent stages of primary succession are usually similar to those of secondary succession in comparable environments. **Secondary succession** occurs when soil is present but major disturbances, such as agriculture or fire, have removed most, if not all, of the vegetation. In other instances, even the toppling of a large acidification processes. As soil becomes more acidic, essential nutrients such as calcium and magnesium are leached out of the root zone. Deficiencies in these nutrients limit tree growth, which renders trees more susceptible to climatic stresses. In addition, soil acidification increases aluminum concentrations in the lower portion of the root zone. Higher concentrations of aluminum damage a tree's finer (feeder) roots, which absorb water and nutrients from the soil. Although damaged trees respond by producing masses of fine roots in the upper soil, these trees are now more vulnerable to other stresses. For example, superficial rooting increases a tree's vulnerability to wind throw and drought.

Trees are able to form fine roots in the upper soil layer because it is usually rich in organic matter, which binds with aluminum and reduces its toxicity. In some parts of Europe, high-elevation spruce/fir forests grow on soils with low levels of organic matter. In contrast, high-elevation forests in the eastern United States grow mostly on soils that are rich in organic matter. Thus, while soil acidification and aluminum toxicity appear to be major contributing factors to forest decline in some parts of Europe, acid rain appears to be less important in the United States.

Given the numerous interacting factors, much research and time are required to analyze the complex cause-and-effect relationships that are contributing to the decline and dieback of European and North American forests. In the meantime, what can we do to prevent further deterioration of these valuable resources? Chapter 16 presents more information about the sources of air pollutants and how they can be reduced.

Figure 4.5 Air pollution has been implicated in the die back of Fraser fir stands high in the southern Appalachian Mountains.
© John Shaw/Tom Stack & Associates

Figure 4.6 Lichens contribute to the very slow process of soil formation on bare rock.

tree in a dense forest sets the stage for secondary succession below the gap in the canopy. Because the soil usually remains after such disturbances have ended, revegetation normally begins within a few weeks.

Secondary Succession

Succession that occurs in abandoned agricultural fields, known as **old-field succession,** illustrates secondary succession. In the Piedmont region of North Carolina, European settlers cleared the native oak/hickory forests and farmed the land. In subsequent years, improper farming practices accelerated soil erosion and the loss of soil nutrients. Eventually, soils became so infertile that farmers had to abandon their fields and migrate westward to more fertile lands. We can reconstruct the stages of old-field succession by examining fields that have been abandoned for different lengths of time.

(a)

(b)

(c)

Figure 4.7 Various stages of old-field succession. (*a*) A marginal farm such as this one is often abandoned after several years of cultivation. (*b*) This field is in the broom-sedge stage of succession. (*c*) The climax stage of old-field succession in this area is an oak-hickory forest.
(*a*) and (*b*) © Breck P. Kent/Earth Scenes (*c*) Tennessee Valley Authority

Abandoned fields in the Piedmont region did not revert immediately to the original oak/hickory forests, even though oak and hickory trees in nearby woods were a source of seeds. Rather, a series of plant communities successively occupied the fields before an oak/hickory forest was eventually reestablished. Figure 4.7 shows three stages in old-field succession. In the year following abandonment, a field is colonized by a mixture of pioneer species that includes crabgrass, pigweed, and horseweed. The next year, the field is dominated by asters, which are succeeded in the following year by a nearly pure stand of a grass commonly called broom sedge. The pace of succession then slows; the broom sedge may persist for as long as twenty years, and during that time the field is gradually invaded by pine seedlings. Eventually, the pines mature and shade out the broom sedge. Later, oak and hickory seedlings slowly establish themselves beneath the pine trees, and as the pines mature and die, they are replaced by these hardwoods. Eventually (some 150-200 years after abandonment) the oak/hickory forest is reestablished. Because oak and hickory seedlings can grow in the shade of mature oak and hickory trees, the oak/hickory forest persists until it experiences a major disturbance.

As the vegetation changes, so too do the types of shelter and food available for animals. Consequently, the numbers and types of animals also change as succession proceeds. Figure 4.8 shows the changes in wildlife that occur during old-field succession in upstate New York.

At any time during primary or secondary succession, another disturbance such as fire or drought may again kill the vegetation and reexpose the landscape. In fact, ecologists have found that natural disturbances occur rather frequently. For example, major fires or hurricanes occur every thirty years or so in the New England states. If succession requires 150-200 years to reach climax, the probability is high that many sites will be disturbed again before they reach the climax stage. Thus, ecologists increasingly believe that climax communities are relatively rare in nature.

Although natural disturbances occur rather frequently, they occur in different regions at different times. Furthermore, even within a single region, the severity of a disturbance can vary dramatically from one place to another. For example, forest fires often leave an irregular patchwork of burned and unburned areas in their wake (fig. 4.9). Thus, a landscape is often occupied by a mosaic of communities at differing stages of ecological succession, which provides for a rich diversity of plant and animal life. In addition, this pattern of disturbance more or less continually provides favorable sites on which early successional species can become reestablished.

Old-field succession and other successional patterns demonstrate that the composition of plant and animal species continues to change in an area long after a disturbance has ceased. The search for the mechanisms by which one species replaces another is challenging. As ecologists learn more about the process of succession, they are finding that

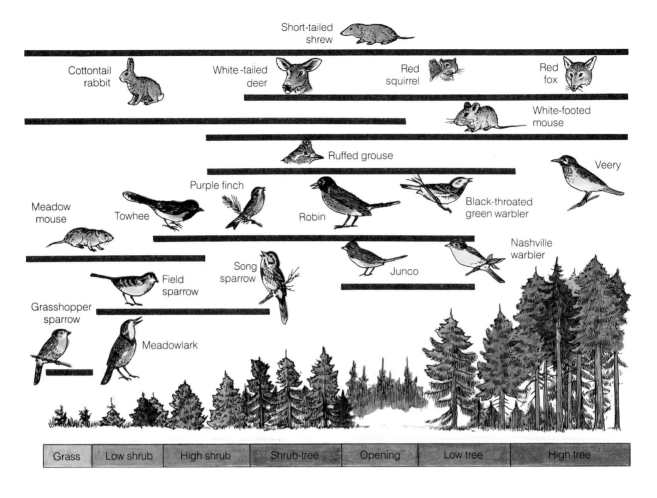

Figure 4.8 Wildlife succession in central New York. Note that some species appear and others disappear as the vegetation changes; a few species are common to many habitats. Horizontal bars indicate the range of habitats for indicated species.

Figure 4.9 Like most major fires, the Yellowstone fires of 1988 produced a mosaic of burned and unburned areas.
© Wendy Shattil and Bob Rozinski/Tom Stack & Associates

Effects of Human Activities on Succession

Although natural disturbances such as fire, flood, drought, hurricanes, insect infestations, and plant diseases occasionally disrupt ecosystems, humans constantly inflict major disruptions on their environment. In temperate regions around the globe, we have replaced species-rich forests and grasslands with highly simplified farms composed of only a few species such as corn, wheat, soybeans, and alfalfa. On a smaller, but growing scale, multispecies forests are being cleared and replaced by single-species tree plantations. For example, in the southeastern United States, row after row of pine trees now stand where oak/hickory forests once grew. In tropical areas, species-rich rain forests are being chopped down, burned, and replaced by much simpler ecosystems, such as banana, rubber, and cocoa plantations. Other major disturbances of human origin include air pollution, water pollution, and strip mining.

Not only are humans major perturbers of their environment, they frequently continue to disturb the same sites year after year. For example, the cultivation of farmland prevents old-field succession by continually disrupting the soil. In the southeastern United States, controlled burning prevents oaks and hickories from invading pine plantations (fig. 4.10). When

the importance of a particular mechanism of species replacement may vary from one stage to the next and from one site to another. Special Topic 4.1 presents a discussion of several mechanisms that have been proposed for species replacement.

How Do Species Replace One Another?

Ecologists have proposed three different models that describe how species replace one another during succession. They are the facilitation model, the tolerance model, and the inhibition model.

The *facilitation model* was first proposed in the early 1900s, when the concept of ecological succession was initially formalized. Essentially, the model states that organisms in each stage of succession modify their physical environment in such a way that the site becomes less suitable for themselves and more suitable for another group of species. Thus, pioneer species facilitate colonization of an area by less hardy species. This model is based on the common observation that plants and animals appear to modify their physical environment. As succession proceeds, humus and nutrients enrich the soil. As a weedy field gradually reverts to a forest, the climate beneath the developing tree canopy changes: Fluctuations in air temperature diminish, and more moisture is retained in the soil and air beneath the canopy. Recently, however, a careful reanalysis has shown that except for the early stages of primary succession, when pioneer species first begin to build up the soil, pioneer species generally do not make a site more favorable for later species.

The *inhibition model* and the *tolerance model* both assume that any species that can survive on the site as adults can establish themselves early in succession. Because of their ability to reproduce rapidly, produce many seeds, disperse widely, and germinate and grow readily, most of the early colonizers are likely to be pioneer species. If we look carefully, however, we can usually also find a few seedlings of the species that dominate later successional stages.

The tolerance model further proposes that although species that dominate in later successional stages will grow more slowly, they can tolerate the difficulties encountered in an environment dominated by pioneer species. As the slower growing, more tolerant species that characterize later stages of succession continue to grow and mature, the faster-growing, less-tolerant pioneer species will die out, and dominance will pass on to the more tolerant species of later stages. This sequence continues until no species can replace the already established ones. Note that in this model, the dominant pioneer species neither inhibit nor facilitate the establishment of later species.

In the inhibition model, the early occupants modify the site so that it becomes less favorable for colonization by both early and late successional species. Thus, unless they are damaged or killed, the established residents will exclude all newcomers. When a plant dies and a spot opens for colonization, the replacement plant may be from the same stage or a later successional stage, depending on the conditions of the site and the characteristics of the species available for replacement at that time. If the new colonizer is a member of the same successional stage, the existing successional stage will continue, but if the newcomer is from a later stage and can tolerate the environmental conditions, succession will progress to that later stage.

At present, no consensus exists among ecologists as to which model (inhibition or tolerance) better explains how secondary succession actually takes place. Perhaps both processes occur at the same time. As usual in ecological studies, several factors complicate the picture. For example, even sites that appear quite uniform, such as a recently abandoned cornfield, usually exhibit considerable spatial heterogeneity with respect to slope direction and angle, soil fertility, soil moisture, and distance and direction from potential seed sources. Another complication is that species vary in their ability to germinate and establish themselves. Then, too, there is the element of chance, such as the direction and strength of the wind when seeds disperse. Thus, the initial mix of seeds that germinate on differing sites is likely to be quite variable. Finally, the mechanisms that explain species replacement in early successional stages may not apply to later stages. Thus, we still have much to learn about the process of species replacement.

---

these climax species are allowed to invade a pine plantation, pine-tree production is reduced, subsequent pine regeneration becomes difficult, and the presence of oaks and hickories increases the fuel that is available for a destructive wildfire. To overcome these difficulties, forest managers periodically set controlled ground fires to prevent the fire-sensitive climax species from invading the area. (Pines are fire-resistant and usually remain undamaged.)

Each stage of succession is characterized by different patterns of nutrient cycling and energy flow; thus, as table 4.1 indicates, later stages of succession may be fundamentally different from early stages. Several consequences therefore accrue when humans replace later successional stages with earlier successional stages. These consequences include (1) an increase in productivity (yield), (2) an increase in the rates of nutrient cycling, (3) a reduction in species diversity, and (4) a change in species composition.

Increased productivity

As we saw in chapter 3, increased production of food is a major goal of our replacement of late successional communities with pioneer communities (agricultural fields). Crop plants grow faster than trees. Furthermore, as described in chapter 3, crop plants channel more of their energy into the production of seeds and fruits that are edible either directly by us or by our livestock. The situation is much the same in forestry. Many early successional trees are more productive than late successional species, and some are more desirable for wood

Figure 4.10 Controlled burning of a pine stand prevents the invasion of fire-sensitive climax species and the buildup of plant litter that would likely increase the chances of a destructive wildfire.
© Kirtley & Perkins/Photo Researchers, Inc.

products, such as lumber for construction. Commercially desirable successional species include loblolly pine in the Southeast and Douglas fir in the Pacific Northwest. Intermediate-stage habitats are also the preferred homes of many popular game species, such as white-tailed deer, elk, moose, bobwhite quail, and pheasants.

Increased nutrient cycling

As table 4.1 indicates, disturbance of later successional stages also alters the patterns and rates of nutrient cycling. Besides interrupting the uptake and retention of nutrients by plants, the loss of vegetative cover also raises the temperature and increases aeration of the soil surface. These changes speed the breakdown of **humus,** organic materials that are generally slow to decay. As a consequence, more nutrients are released into the soil, so that a temporary surge in available nutrients often occurs following a disturbance.

At the same time, the loss of humus and plant cover impairs the site's ability to retain water. Humus acts as a sponge that soaks up water and releases it slowly. The removal of plant cover ends **transpiration,** which is the loss of water vapor from leaf pores to the atmosphere. (Perhaps as much as 98% of the water taken up by plant roots is released to the atmosphere via transpiration.) Transpiration reduces soil moisture and thereby increases storage space in the soil for rainfall, so that runoff is greatly reduced until the soil storage space is refilled by precipitation. Hence, a loss of vegetative cover and its associated transpiration increases runoff, because the soil moisture remains higher and the soil cannot store as much of the rainfall. The combination of greater runoff and free soil nutrients often results in a significant loss of nutrients from a disturbed ecosystem.

Changes in nutrient cycling following a disturbance have been well documented by investigations at the Hubbard Brook Experimental Forest in New Hampshire.* For one study, researchers cut all the trees in a small watershed that were larger than 2 centimeters (1 inch) in diameter, and then inhibited subsequent regrowth for two years by periodically applying herbicides. As figure 4.11 indicates, nutrient losses were substantial. In the second year following the clearing of trees, stream-water concentrations of nitrate were 56 times higher than prior to deforestation. These concentrations exceed the recommended levels for drinking water. Clearly, significant amounts of this important plant nutrient were being lost from the watershed via runoff. Substantial increases in the concentrations of other plant nutrients were also measured in the stream water during the same period of time (for example, 16 times as much potassium, 4 times as much calcium and magnesium, and 2 times as much sodium).

*See the work by Borman and Likens in the list of readings at the end of the chapter.

Table 4.1 Contrasting characteristics of early and late successional stages

Characteristic	Early Successional Stages	Late Successional Stages
net biomass production (yield)	high	low
mineral cycling	open	closed
nutrient exchange rate	rapid	slow
number of species	low	higher
food chains	linear	weblike

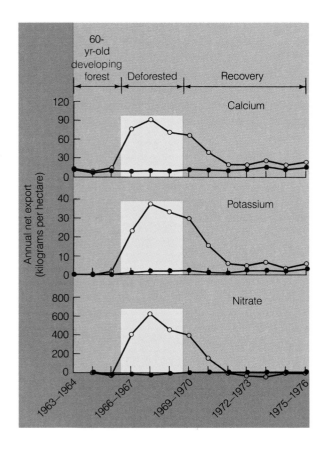

Figure 4.11 Deforestation can significantly increase the concentration of nutrients found in stream water, indicated by open circles. Subsequent revegetation reduces their loss and thereby lowers their stream-water concentration. Solid circles represent stream flow from a similar forest that was not cut.

The Hubbard Brook study illustrates the potential magnitude of nutrient losses from a disturbed ecosystem, especially if it remains disturbed for several years. Usually, however, new plant growth appears within only a few weeks following a disturbance. Recolonization quickly reduces nutrient losses by reversing the processes that accelerate their loss in the first place.

We have already learned that early successional (pioneer) species tend to grow rapidly. They also rapidly take up and retain soil nutrients; the tissues of many pioneer species contain even greater concentrations of nutrients than those of late successional species. Thus, early successional plants conserve substantial pools of nutrients. Figure 4.12 illustrates the relationship between increased storage of nutrients in vegetation and their reduced loss to stream water. Recolonizing vegetation also shades the ground, thereby lowering temperature and airflow at the surface, which in turn slows the release of nutrients from the decomposition of humus. Moreover, transpiration by the reestablished vegetation also begins to reduce runoff. At Hubbard Brook, transpiration reduced annual stream-water discharge by as much as 25%. If a site is not disturbed on a continual basis, then, recolonization can restore cycling processes quickly to rates that approximate predisturbance conditions.

Reduced number of species

Early and late successional stages also differ in the number of species present; there are usually fewer species in early successional stages than in late stages. (A true climax ecosystem is often somewhat less diverse than the immediately preceding stages.) Also, simple linear food chains tend to characterize early successional stages, whereas more complex food webs are present in late successional stages. These differences are particularly evident in agricultural regions where forests and prairies have been replaced by comparatively few species of crop plants, livestock, and associated weeds and pests (rodents and insects). What are the consequences of this reduction in species diversity?

The traditional view is that the greater the number of species and the number of linkages between them, the greater the stability of the ecosystem. **Ecosystem stability** is a measure of the ability of a community of plants and animals either to resist or to recover quickly from a disturbance. Intuitively, the traditional view makes sense. Imagine a very simple ecosystem, in which mice are the only important food source for a population of foxes. If a disturbance (e.g., disease) drastically reduces the population of mice, we might predict that some of the foxes would either have to migrate to another region in search of food or else die of starvation. Thus, intuition tells us that the energy flow in a very simple food web may fluctuate tremendously, depending on the well-being of only a few species.

Now imagine a more complex ecosystem, where several species, such as mice, chipmunks, rabbits, and various ground-dwelling birds, are available as prey for the foxes. In that situation, if an environmental change causes one prey species to become scarce, the foxes can turn to a more abundant prey population. Consequently, the energy flow from one trophic level to the next should remain roughly unchanged. Thus, the more complex ecosystem remains stable despite changes (assuming that not all of the species of prey become scarce at the same time). We can also find parallels in commerce. For example, a nation that has a variety of natural resources, manufacturing processes, and export markets tends to be more stable economically than a nation whose economy is based on only one or a few products.

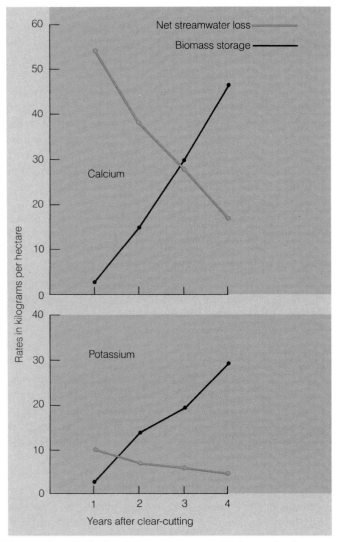

Figure 4.12 Following deforestation, regrowth of vegetation increases the storage of nutrients. This causes a decline in nutrients lost from the ecosystem.

But is our intuition actually confirmed when ecologists carefully examine the relationships between diversity and stability in natural ecosystems? Initial observations that associated increased stability with increased species diversity have not held up under close scrutiny and experimentation. In fact, most studies indicate an overall tendency for stability to decrease as complexity increases.* These studies also suggest, however, that the precise relationship between stability and complexity is elusive and that no single relationship would be correct for all ecosystems at all times.

The relationship appears to vary with the specific trophic level that is perturbed; that is, increasing complexity generally leads to increasing stability at the producer level, but it generally leads to decreasing stability at the top predator level. The relationship also appears to vary with the nature of the environment; that is, complex ecosystems of relatively stable environments (e.g., the tropics) are less stable and, therefore,

*See the work by Bergon, Harper, and Townsend listed in the readings at the end of the chapter.

Table 4.2 Contrasting characteristics of early and late successional plant species.

Characteristic	Early Successional Plants	Late Successional Plants
photosynthetic mechanisms adapted to high light levels	yes	no
growth rates	high	low
resource acquisition rates	fast	slow
recovery from resource limitation	fast	slow
seed dispersal	well dispersed	poorly dispersed

Reproduced, with permission, from the *Annual Review of Ecology and Systematics,* Vol. 10, © 1979 by Annual Reviews, Inc.

more susceptible to outside disturbances than less complex, but more stable, ecosystems of more variable environments, such as the temperate regions of the midlatitudes.

Changes in species composition

Major changes in species composition usually accompany disturbances of late successional stages. As we briefly pointed out earlier in this chapter, the characteristics of early successional species differ greatly from those of late successional species (see table 4.2). The significance of these differences becomes evident when we consider that agricultural areas are continually prone to invasion by unwanted pioneer species, which we call *weeds* if they are plants, and *pests* if they are insects or other animals. Weedy plants are particularly hardy species. Their ability to grow rapidly, to take up soil nutrients rapidly, and to recover rapidly from stress makes them formidable competitors for crop plants. They can produce large numbers of hardy seeds that are widely dispersed by wind or animals (fig. 4.13). Thus, weedy plants readily reestablish themselves during the following growing season in both the same fields and adjacent ones. Likewise, pioneer animals (mostly rodents and insects) have an innate genetic capacity for proliferation and wide dispersal; as a result, they can inflict heavy crop damage. Thus, in preparing the way for a profitable crop, a farmer is also providing opportunities for invasions by large numbers of pests and weeds that are well adapted to the rigors of life in cultivated fields. We derive little or nothing in the way of food, clothing, or shelter from these plants and animals, but we spend a great deal of time, money, and energy trying to eradicate them.

In contrast, while early successional organisms are proliferating in so many areas of the world, the species that require late successional conditions for survival are disappearing. In dozens of cases where plant and animal species have become extinct, or nearly so, habitat destruction is the single most important contributing factor. The fate of Attwater's prairie chicken (fig. 4.14) illustrates the problem. That species once thrived in the lush, tall-grass prairie along the Gulf Coast in

(a)

(b)

Figure 4.13 Two means whereby seeds of weedy plants are dispersed are illustrated by (*a*) the plumes of milkweed seeds that are carried off by the wind and (*b*) the hooks of burdock that cling to animals that brush against them.
(*a*) © Ed Reschke/Peter Arnold, Inc. (*b*) © Herbert Schwind/Photo Researchers, Inc.

Figure 4.14 Attwater's prairie chicken, a species endangered by destruction of its habitat.
© C. Allen Morgan/Peter Arnold, Inc.

southwestern Louisiana and southeastern Texas. But the plowshare turned the prairie into cropland, reducing the prairie chicken's range by 90% and its population by 99%. Today, Attwater's prairie chicken is found in only a few isolated pockets in southeastern Texas. Habitat loss threatens the extinction of numerous other animal species, including the San Joaquin kit fox, the Florida panther, and the African elephant. Loss of habitat also threatens many plant species, including the Tennessee purple coneflower, persistent trillium, dwarf

lake iris, and several species of orchids (fig. 4.15). In chapter 11, we discuss further the problems encountered in trying to preserve the diversity of plants and animals on the planet.

Limits to Succession

Nature's healing process takes a long time. Even under favorable circumstances, it takes 150-200 years for a climax oak/hickory forest to become established on abandoned farmland in the southeastern United States. Restoration requires even more time when a disturbance (e.g., a glacier) removes the soil or covers it (e.g., a lava flow). In some instances, sustained, improper land use produces a landscape where succession essentially does not occur at all. Consider the following examples.

Throughout most semiarid regions of the world, improper irrigation practices lead to **soil salinization,** an accumulation of salts at the soil surface. (When irrigation water evaporates, dissolved salts are left behind in the soil.) If salinization is not corrected, increasing soil salinity leads to declining crop yields. In severe cases, agriculture becomes impossible. Today, millions of hectares of formerly irrigated farmland have been abandoned throughout the world—from the Middle East to South America to the western United States. Since few plants can survive highly saline soils, salinization often leads to the formation of deserts. In fact, some human-made deserts in the Middle East have persisted for centuries. As described in chapter 8, soil salinization continues today as a major problem in semiarid regions throughout the world.

Figure 4.15 Dwarf lake iris, a species threatened by loss of its habitat.
© John Gerlach/Earth Scenes

In other instances, improper land use has contributed to excessive erosion and the complete loss of soil cover. Along the east coast of the Mediterranean Sea, for example, the mountainous landscape was once covered with the famous cedars of Lebanon. After these forests were cut down, some 5,300 years ago, farmers moved in and attempted to cultivate the steep mountain slopes that were vulnerable to severe soil erosion. Today, few cedars remain and the soil is gone. The bare limestone slopes bear mute testimony to the failure of succession to restore a cedar forest (fig. 4.16). Today, as a consequence of attempts to feed the world's rapidly growing human population, the soil in many regions is being washed and blown away faster than it is being replaced by natural soil-forming processes.

Time may be the most significant limiting factor on our efforts to restore disturbed landscapes. Natural succession is not a viable means of restoring much of the earth's disturbed land to its original state within our lifetime. To speed up restoration, we must employ such management practices as adding fertilizer to the land and planting hybrid plants that do well under the poor growing conditions that characterize many disturbed sites. Even these strategies meet with natural limitations, however. Soil building remains a slow process, and most trees take many decades to reach maturity.

Slow as it is, ecological succession is the natural recovery response of ecosystems to disturbance. In some cases, the best procedure for restoring a disturbed habitat may be to lower our expectations and allow succession to occur at its own rate.

Our evaluation of the consequences of replacing late successional stages with early successional stages provides mixed results. In agriculture, forestry, and wildlife management, humankind has derived considerable benefits. Yet continued poor land-use practices have accelerated soil erosion and soil salinization, contributed to the pollution of waterways, and led to species extinction. Without greater worldwide implementation of land-use practices that conserve land resources, the land will become increasingly less productive. We discuss these issues in more depth in chapters 8, 10, and 11.

Adaptation to Environmental Change

Organisms do not necessarily die when their environment changes. Most organisms have the capacity to adjust, at least in part, to environmental changes. For example, our bodies adjust to uncomfortably low temperatures by shivering. In response to falling skin temperature, the brain initiates an increase in muscle activity that requires the release of energy by cellular respiration. As we know, a by-product of cellular respiration is heat, which helps maintain normal body temperature. But our ability to adjust to changes in air temperature is limited. Under some conditions, we are unable to maintain normal body temperature, and death may result. For more on human responses to changing air temperatures, see Special Topic 4.2, "Temperature and Human Comfort" on page 80.

Many other species are able to adjust to low temperatures. Chipmunks, ground squirrels, and woodchucks adjust to the rigors of winter by hibernating. **Hibernation** involves a series of body changes, including the slowing of heart and breathing rates. As a result, the animal's body temperature declines to just a few degrees above freezing. These sharp decreases in body activity diminish an organism's energy needs, and it can rely on stored body fat to provide the energy to prevent freezing of the body. Thus, a hibernating animal can live through the winter without having to leave its den to search for food in hostile weather.

The ability to adjust to environmental change is an example of adaptation. An **adaptation** is a structural, functional, or behavioral characteristic of an organism that helps it to survive and reproduce in its environment. Individual organisms are born with a set of adaptations that is the result of the evolution of past generations. An overview of the process of evolution will help us to understand (1) how adaptations to environmental change come about, (2) why some organisms adapt better than others, (3) what limits the ability of organisms to adjust to environmental changes, and (4) what types of adaptation can occur.

Evolution: The Driving Force of Adaptation

Evolution is defined as a change in the genetic makeup of a population with time. The process of evolution involves two steps: (1) the production of genetic variability, and (2) the ordering of that variability by natural selection. (Some evolutionary biologists argue that other agents besides natural

Figure 4.16 The barren slopes of Lebanon illustrate well the limits of succession in restoring the landscape where human intervention causes a significant loss of soil. This grove of some 400 trees is all that remains of the fabled cedars of Lebanon.
© Anne E. Hubbard/Photo Researchers, Inc.

selection work on genetic variability, but most agree that natural selection is by far the most important agent.) We now consider how these two steps work together in evolution.

We have already noted that in most populations of plants and animals, each organism is genetically unique. The human population now numbers over 5.4 billion. Discounting the relatively low number of identical twins, each of us is special. No one else looks exactly like you or responds to environmental changes exactly as you do. That is a lot of genetic variability! But even more remarkable is that we can find billions of insects of a single species in just a few square kilometers. What is the source of all this variability?

Mutations, inheritable changes in the DNA sequences that make up the chromosomes, are the ultimate source of genetic variability. Mutations are random events; where and when they occur in chromosomes is a matter of chance. As genetic variability arises through mutations, an almost endless variety of genetic combinations becomes possible through sexual reproduction. In the formation of **gametes** (eggs and sperm), the genetic deck is, in effect, reshuffled. Every cell of the body, except the gametes, contains two sets of chromosomes—one set is inherited from each parent. When gametes are formed, the number of chromosomes in each cell is reduced by half. Each gamete receives one complete set of the chromosomes that originally existed in pairs. (One member of each pair was inherited from the mother and the other

member was inherited from the father.) As figure 4.17 shows, however, individual chromosomes are randomly distributed when gametes are formed. Taking the simplest example, an organism that has two pairs of chromosomes can produce four possible chromosome combinations in its gametes. The two chromosomes in the gametes may both be of paternal origin, or both may be of maternal origin, or one may be inherited from each parent. The latter case allows two possibilities. In our illustration, a gamete can contain the dot maternal chromosome and the oblong paternal chromosome, or it can contain the opposite combination. The number of possible chromosome combinations rises rapidly as the number of chromosome pairs increases. Since we have twenty-three sets of chromosomes, the number of possible chromosome combinations in the gametes of just one person is 8,388,608.

Additional genetic variation is introduced during the formation of gametes when pieces of chromosomes break off and become attached to other chromosomes (fig. 4.18). This process of genetic interchange, called **crossing-over,** is also a random event.

The greatest source of genetic variation occurs with the union of an egg and a sperm. For example, because a human male and female can each potentially produce 8,388,608 different chromosomal combinations in their respective gametes, the potential number of different chromosomal

Temperature and Human Comfort

To appreciate how air temperature influences human comfort, we must first understand that humans are *homeothermic*. This means that we regulate our internal, or core, temperature within 2 Celsius degrees (3.6 Fahrenheit degrees) of 37° C (98.6° F), despite much greater variations in the temperature of the *ambient air* (surrounding air). The term *core* refers to those regions of the body that contain vital organs such as the brain, heart, lungs, and digestive tract. If vital organs are not maintained at a nearly constant temperature, they do not function properly. Other parts of the body, such as the legs and arms, however, may undergo much greater temperature changes without ill effects.

At ambient air temperatures of 20° to 25° C (68° to 77° F), someone who is fully clothed and indoors will feel comfortable at rest. In this temperature range, the body readily maintains a core temperature of 37° C without having to resort to special temperature-regulating mechanisms.

When we are exposed to ambient air temperatures above or below the 20°–25° C range, the body must initiate processes that maintain the core temperature at 37° C. For example, if you stand in the sun on a day when air temperatures hit 30° C (86° F), your core temperature will begin to rise. In response, you perspire. The heat required to evaporate the perspiration is provided by your skin; therefore, your skin cools. You experience the same cooling effect of evaporation as you step out of a shower or climb out of a swimming pool. *Evaporative cooling* reduces skin temperature, which in turn normally leads to a drop in the body's core temperature.

If you are exposed to air temperatures below 20° C (68° F), and your core temperature begins to fall, you will start to shiver. This increased muscular activity produces additional heat, which helps to raise your core temperature

back to 37° C. Perspiring and shivering are examples of *thermoregulation,* natural physiological mechanisms that assist in maintaining a nearly constant core temperature regardless of the ambient air temperature.

Another thermoregulation process involves blood flow. Heat transfer from the body core to the skin occurs via the circulatory system. When you are exposed to air temperatures below 20° C, your body can limit heat loss by restricting blood flow to the skin. Under the direction of the nervous system, many of the tiny blood vessels in the skin constrict. As ambient air temperatures drop, additional blood vessels constrict, further reducing blood flow to the body surface. You can observe this phenomenon by immersing your hand in a container of ice water; the skin becomes paler. The reduced blood flow produces a thicker insulating layer between the heat-producing core and the skin surface.

In contrast, ambient air temperatures above 25° C (77° F) trigger increased blood flow to the body surface. Blood vessels near the skin surface dilate, giving the skin a flushed, or reddish, appearance. Greater flow of blood to the body surface raises skin temperature. As a consequence, the body-to-air temperature gradient increases, and cooling by radiation, conduction, and convection is enhanced. Greater blood flow to the skin also increases evaporative cooling by supplying more water for perspiring.

In addition to physiological changes, behavioral responses assist in thermoregulation. For example, if we feel hot, we shed clothing, seek shelter from the sun, or turn on a fan or air conditioner.

Under some conditions, thermoregulation is insufficient to maintain a core temperature of 37° C. For example, if a person hiking in the woods is drenched by a cold rain and then overexerts herself and becomes exhausted,

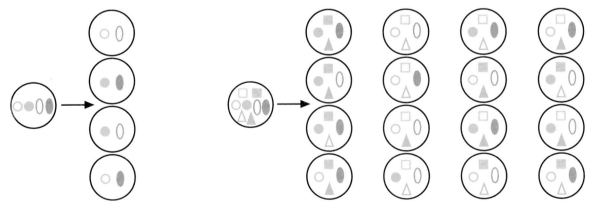

Figure 4.17   Schematic representation of the possible distribution of chromosomes during the formation of gametes, using 2 and 4 pairs of chromosomes. The dark chromosomes were originally of paternal origin and the light chromosomes of maternal origin. Two pairs of chromosomes permit only 4 possible different distributions, but 4 pairs permit 16 possible different distributions, and the number of possible combinations doubles with the addition of every chromosome pair. Thus the chromosome number of our own species (23) makes possible over 8 million chromosome combinations in each of our gametes.

thermoregulation may not be able to compensate for heat loss. Consequently, the core temperature drops, and hypothermia may ensue.

The term *hypothermia* refers to those responses that occur when the human core temperature drops below 35° C (95° F). Initially, shivering becomes violent and uncontrollable. The victim begins to have difficulty speaking and becomes apathetic and lethargic. If the core temperature falls below 32° C (90° F), shivering is replaced by muscular rigidity, and coordination deteriorates. Mental abilities are impaired, and the victim is generally unable to help himself. At a core temperature of 30° C (86° F), the person may drift into unconsciousness. Death may occur at core temperatures below about 24 Celsius degrees (75 Fahrenheit degrees) because the heart rhythm becomes uncontrollably irregular (ventricular fibrillation) or uncontrollably halted (cardiac arrest).

Once hypothermia begins, the victim is in serious trouble. With only a 3 Celsius degrees (5.7 Fahrenheit degrees) drop in core temperature, the body's ability to regulate its core temperature is already greatly impaired. If the core temperature drops to 29° C (85° F), thermoregulation is essentially ineffective. The first signs of hypothermia should never be ignored; one should take action immediately. Treatment takes two forms: prevention of further heat loss and the addition of heat. Further heat loss can be prevented by replacing wet clothing with dry clothing, finding shelter, and insulating the person from the ground, so that body heat is not conducted to the colder ground surface. The body can be heated by an external source, such as a space heater or other human bodies. Administering hot nonalcoholic drinks, if the victim is conscious, helps warm the core from the inside. In any event, medical attention should be sought as soon as possible.

In some situations, thermoregulation may be unable to prevent a rise in core temperature. For instance, a person exposed to hot desert conditions with an inadequate supply of water will eventually experience an increase in core temperature. If the core temperature continues to rise, hyperthermia may ensue. *Hyperthermia* refers to those responses that take place when the core temperature rises above 39° C (102° F). As the core temperature climbs to 41° C (105° F), thermoregulation breaks down, and a person may suddenly and quite unexpectedly collapse. The victim also experiences muscle cramps or spasms and slips into unconsciousness. Sweating ceases, although it is not known if this is a cause or a result of hyperthermia. With serious heat stress, the individual may die within a few hours unless the core temperature can be lowered artificially. These responses are collectively identified by various names, including *heatstroke, sunstroke,* and *heat apoplexy.*

The victim of hyperthermia must be treated promptly because, once thermoregulation fails, the core temperature rises rapidly. To save the victim, the core temperature must be lowered from outside the body. The victim should be moved to a cooler environment, and if possible, the body should be placed in cold water. Alternatively, sponging the body with alcohol enhances cooling as the alcohol evaporates. Again, one should seek medical attention as soon as possible.

The human body has a remarkable capacity to adjust to changing air temperatures and remain comfortable. This capacity is limited, however. If our core temperature begins to deviate from normal, we or our companions must take corrective action promptly. Failure to do so may be fatal.

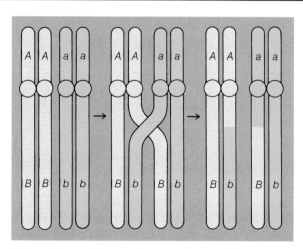

Figure 4.18 The crossing-over of segments between like chromosomes greatly increases genetic variability within a population.

combinations formed by the union of a sperm and an egg is 70,368,177,664, and this number probably increases by millions when crossing-over is taken into account. Furthermore, these numbers represent the possible combinations of only one man and one woman, considering their unique genetic makeup. When the entire human race is considered (each person having a unique genetic constitution) the number of possible genetic combinations is truly mind-boggling.

It is important to remember that the variation produced by mutations and sexual reproduction is random. The variation is not caused by and is not related to the present needs of the organisms or the nature of their environment.

We have now arrived at the second step of evolution: natural selection. In a population of genetically unique organisms, some individuals will have sets of **genes** (portions of a chromosome that code for particular characteristics) that produce traits that are better suited to current environmental conditions than others. Such individuals have a greater probability of surviving and of having offspring than members of the population who do not have those genes. Furthermore, their offspring are likely to inherit the same genes, so that they will also have a greater probability of surviving and having offspring. Thus, if the environment remains generally unchanged, most members of the population will eventually possess the beneficial set of genes. That sequence of events, in which a particular gene or set of genes is favored by the environment, is called **natural selection.** The result of natural selection is a population that contains a greater number of organisms that are better suited (adapted) to a given environment.

Natural selection is illustrated by the classic example of the peppered moth, which lives in the forests of England. Before 1850, most peppered moths were light in color; only occasionally was a dark variant observed. The first dark specimen was caught in the industrial town of Manchester in 1849. By 1895, however, the dark variant made up about 98% of the peppered-moth population in the Manchester area.

It was not clear at the time what caused this dramatic shift in the predominant color of the peppered moth. The following hypothesis was proposed in the 1930s, based on observations that the peppered moth tends to rest during the day on tree trunks and rocks, and their major predators were birds who located the moths by sight. Prior to the Industrial Revolution, the surfaces of tree trunks and rocks were mostly light in color, and they were often covered with light-colored lichens. Against the light-colored surfaces, the darker moths were much more visible to their predators than were the lighter moths (fig. 4.19). As a result, most of the dark variants were eaten by birds (that is, they were selected against). However, with the growth of industry, soot from factory chimneys blackened the rocks and tree trunks and killed the lichens. At that point, the light moths were at a disadvantage. Because they showed up more clearly against the soot-darkened surfaces on which they rested, they were selected against (eaten by birds) in far greater numbers than were the dark variants. As a result, over time, the dark variants increased in the population, while the number of light-colored moths declined. Eventually, the dark variants dominated.

This hypothesis seemed plausible to some biologists, but it was disputed by others. Finally, in the 1950s, H. B. D. Kettlewell of Oxford University put the matter to an experimental test. He released roughly equal numbers of the light and the dark forms of peppered moths in a rural area of England, where the tree trunks had not been blackened by industrial pollution and where the light moths naturally outnumbered the dark ones by almost twenty to one. Kettlewell and his colleagues found that birds did indeed prey on the moths by sight: of the 190 moths that were observed to be eaten by birds, 164 were dark, and only 26 were light.

Next, Kettlewell repeated the experiment under the opposite environmental conditions. In another part of England, where soot had discolored the tree trunks and the dark variants naturally dominated, he released approximately equal numbers of light and dark moths. Once again, birds ate the moths, but this time they ate roughly three times as many light moths as dark ones.

In the years since Kettlewell's experiments, air-pollution controls have been implemented in England. Today, there is less airborne soot, and tree trunks are becoming lighter. Thus, light coloration in peppered moths should once again be increasingly selected for. In fact, as expected, the light variants are beginning to dominate again.

Several important points in this example need emphasis. First, environmental change did not produce the dark form of the moth; the dark trait was already present in the population, probably as the result of a mutation. Natural selection can only act on the genetic variations that are already present in a population. It favors only those members that already possess a trait that makes them better suited to the changing environment.

Second, the period of time that is required for an almost complete shift in a dominating factor in a population varies according to its reproductive capabilities. The change in the coloration of peppered moth populations occurred relatively quickly because moths, like most insects, reproduce in large numbers several times a year. Table 4.3 presents some reproductive capabilities for comparison. The dark forms multiplied rapidly when bird predation was no longer an important limiting factor on their numbers. Also, the time that is necessary for a generation to mature and reproduce is a major determinant of the length of time required for a population to adjust to an environmental change. For example, populations of some bacteria can double in only twenty minutes. Hence, if some members of a bacterial population are resistant to a particular antibiotic, that population can adapt to the drug within hours or a few days. In marked contrast, long-lived organisms that reproduce slowly, such as ourselves, may require hundreds or thousands of years to undergo an evolution of comparable magnitude.

Types of Adaptation

The products of evolution are adaptations that are either structural, functional, or behavioral. For example, adaptations to cold temperatures include the growth of fur (structural), reduction of the flow of warm blood to the skin (physiological), and the ability to fly south for the winter (behavioral). Adaptations include not only relatively static characteristics, such as our opposable thumbs, but also the capacity to adjust to an environmental change, such as shivering when our body temperature is lowered.

Essentially, every organism is a complex package that contains an immense number of interacting adaptations. For example, organisms adapt in many ways to obtain food and oxygen from their environment, to retain nutrients and water, to avoid becoming a meal for other organisms, to ward off disease-causing organisms, to compete with other organisms for limited resources, and to adjust to both physical and biological changes in their environment. To illustrate the fascinating array of adaptations that organisms possess, consider some defensive adaptations.

(a)

(b)

Figure 4.19 Light and dark forms of peppered moths at rest on tree trunks. (*a*) The light variant is much less visible than the dark variant on this lichen-covered tree. (*b*) It is much easier to spot the light variant on this soot-covered tree.
© Michael Tweedie/Photo Researchers, Inc.

Table 4.3	Some measures of reproductive capabilities of different species	
Species	**Mean Length of Generation***	**Time Required to Double Population**
rice weevil	6.2 weeks	0.9 weeks
brown rat	31.1 weeks	6.8 weeks
human beings (1992 world rate)	15–20 years	40 years

Source: After E. P. Odum, *Fundamentals of Ecology.* Copyright © 1971 W. B. Saunders Company.

*Mean length of generation is the average time between the birth of individual members of a species and the birth of their first offspring.

As we briefly mentioned in chapter 3, many adaptations reduce an organism's vulnerability to attack. One such set of adaptations is the ability to produce defensive chemicals, such as the pungent chemical that is released by skunks when they are agitated. A less familiar example is the noxious chemical that is sprayed by the bombardier beetle when it is attacked by ants. That spray, which is emitted from the tip of the abdomen, can be aimed so precisely that the beetle can repel an ant from attacking any of its six legs.

Many chemical defenses are more subtle. Monarch butterflies, for example, contain a chemical that induces vomiting when it is ingested (fig. 4.20). Such a chemical is adaptive, because a predator that suffers from eating a monarch butterfly is less likely to attack and eat another one. Also, as mentioned in chapter 3, many plants produce chemicals that reduce attacks by herbivores and microorganisms.

Another array of defenses involves deceptive appearance. Many animals blend so well with their surroundings that they are nearly undetectable. We have already seen how differences in coloration make peppered moths more or less vulnerable to predation by birds. Natural selection favors

protective coloration in many populations in which a highly visible individual is in danger of being eaten. Many desert animals, particularly lizards, are light-colored and blend in with the light-colored background of desert soils. Other animals change colors with the seasons. For example, the Arctic fox is white during the winter snow season, but in the spring it sheds its white fur and becomes brown, thus blending in better with the tundra vegetation (fig. 4.21).

Rather than matching the color of the general background, some organisms mimic, or closely resemble, inanimate objects that are commonly found in the habitat. Walking sticks, for instance, look so much like sticks, and leafhoppers look so much like leaves that predators are often unable to tell the difference and thus pass up many nutritious meals. A particularly interesting type of **mimicry** is found in species that appear to be very similar to another species. The monarch butterfly and the viceroy butterfly, for example, are almost identical (fig. 4.22). The monarch butterfly can induce vomiting in its predator, but the viceroy cannot. Nevertheless, laboratory experiments have shown that birds that have had bad experiences with the monarch butterfly tend to avoid the viceroy butterfly. The viceroy butterfly therefore suffers little predation, because predators cannot distinguish it from its unpleasant look-alike.

Although many species are deceptive in their appearance, a few boldly advertise their presence (fig. 4.23). The bright colors and bold patterns displayed by these species contrast sharply with their surroundings. At first thought, we might question the survival value of such bold coloration, but further observation shows that most of these animals either taste bad, smell bad, sting, or produce toxic chemicals. After one or two unpleasant encounters with these animals, a predator will usually avoid them. Their conspicuous appearance allows the potential predator to recognize and avoid them easily. Thus, although a few individuals may be eaten, the population as a whole benefits because predators will no

(a)

(b)

Figure 4.20 A freshly eaten monarch butterfly (*a*) induces vomiting in a blue jay (*b*). After this experience, the blue jay is less likely to eat another monarch butterfly.
© Lincoln Brower

(a)

(b)

Figure 4.21 An Arctic fox. (*a*) Its white winter coat allows it to blend into the snow-covered landscape. (*b*) Its darker summer coat provides similar camouflage after melting snow exposes tundra vegetation.
(*a*) © James Simon/Photo Researchers, Inc. (*b*) © Steve Krasemann/Photo Researchers, Inc.

longer feed on members of a species that has caused them problems.

Limits to Adaptation

It is tempting to believe that adaptations will solve all of our environmental problems. Perhaps our lungs will become more resistant to air pollutants, or perhaps our skin will become more resistant to the damaging impact of ultraviolet radia-

tion. But nature imposes strict limits on adaptations—limits to which all organisms are subject. We have already mentioned some major natural limits on adaptation. First, each organism's capacity for adjustment to a particular change is limited by its genetic makeup; an organism cannot adjust to all environmental changes. Second, a changing environment selects only for characteristics that are already present in the population. And third, even if a heritable adaptive trait is present, the rate at which a population as a whole inherits the

Figure 4.22 The monarch butterfly (bottom) is mimicked by the viceroy butterfly (top).
© Breck P. Kent/Animals Animals

Figure 4.23 The colorful passion-vine butterfly (*Heliconius erato*) brilliantly displays its presence.
© Michael Fogden/Animals Animals

trait is limited by its reproductive capabilities (both the number of offspring per generation and the length of time required for each generation to mature and reproduce).

As a species, we have developed many technologies that appear to protect us from a hostile environment. Our climate-controlled dwellings protect us from inclement weather. Our agricultural technology helps to ensure us of a reliable food supply. Water treatment facilities reduce our chances of contracting waterborne diseases such as dysentery and hepatitis.

Our technological adaptations, however, are also limited. The development and large-scale implementation of most technological advances are expensive and time-consuming, and often, as we point out many times, our advances in technology create as many problems as they solve.

Conclusions

In this chapter, we have examined some of the ways in which individuals, populations, and ecosystems respond to changes in the environment. We have seen that environmental change initiates a continual process of selection, which determines the prevailing characteristics of a population or ecosystem by favoring the continued existence of some individuals over others. The net effect of environmental change, then, is a change in the type of organisms that live in an area.

Human activities are causing significant environmental changes. Some observers foresee a period of greater environmental change than we have yet experienced, as world population soars and people continue to strive for a higher standard of living. So far, our technological adaptations have perhaps made us the most successful species in the history of the Earth. Yet these same adaptations have also fouled the planet and given us the ability to destroy ourselves and most life on Earth. We also have the ability to contemplate the future and predict the consequences of our actions. It remains to be seen whether we will use this ability responsibly and successfully in the years to come.

Key Terms

law of tolerance
tolerance limits
lower tolerance limit
optimum concentration
upper tolerance limit
law of the minimum
limiting factor
acclimatization
synergism
antagonism
pioneer stage
pioneer species
climax stage
climax species
ecological succession
primary succession

secondary succession
old-field succession
humus
transpiration
ecosystem stability
soil salinization
hibernation
adaptation
evolution
mutations
gametes
crossing-over
genes
natural selection
protective coloration
mimicry

Summary Statements

The environment is changing continually. Some organisms are able to adjust to changes, others succumb to them. The law of tolerance and the law of the minimum help us understand the influence of environmental change on organisms, including ourselves.

The variables of age, sex, ability to adjust to change, and factor interaction all make it difficult to determine a specific organism's tolerance limits to a specific environmental change. Because so many interactions are possible among species and their environment, few simple cause-and-effect relationships exist in nature.

If an essential factor in the environment lies outside the tolerance limits of an organism, then that organism must be able to avoid the stressful conditions, or it will die.

When many plants and animals in an ecosystem are killed by a disturbance, the functioning of that ecosystem is severely disrupted. If the stress is removed, the damaged ecosystem may or may not eventually return to its original structure and species composition through a sequence of changes known as ecological succession. Because disturbances occur relatively frequently, climax stages are relatively rare.

Early and late successional stages differ in several ways. The early stages are more productive, but they are also more subject to nutrient loss via erosion. Although the early stages exhibit less species diversity than some later stages, ecologists do not yet understand clearly the complex relationships between species diversity and ecosystem stability. Because many late successional ecosystems have been replaced by ecosystems that are frequently disturbed, weedy and pest species and many species of game animals are more common today. But many other species of plants and animals that require late successional habitats are threatened with extinction.

In most situations, succession from pioneer to climax communities requires a century or longer. Sometimes a disturbance is so severe that the climax ecosystem may never be reestablished.

Most organisms can adjust, at least in part, to environmental change. That ability to adjust arises as a consequence of evolution, which involves two steps: (1) the production of genetic variability, and (2) the ordering of that variability by natural selection. The products of evolution are adaptations.

Each organism's ability to adapt to an environmental change is determined, therefore, by its genetic makeup. A changing environment selects for (favors) only those characteristics that already exist in a population. If a trait is already present, the rate at which a population can adapt to any given change is limited by the population's reproductive capabilities.

Questions for Review

1. Describe several ways in which natural variability in the environment affects how you plan your daily activities.
2. Describe the law of tolerance. Why is knowledge of this principle needed to understand the response of organisms to environmental change?
3. Consider how your ability to tolerate certain environmental stresses compares with the ability of your friends. Such stresses may include the flu, hot and muggy weather, final exam week, waiting in line, and a congested highway or store.
4. Describe the law of the minimum. What is a limiting factor?
5. How does the process of acclimatization make it more difficult to predict the impact of an environmental change?
6. What are synergistic and antagonistic interactions? How do they influence assessments that deal with the threat of environmental contaminants to human health?
7. Why is it so difficult for scientists to determine the causes of forest decline in Europe and North America?
8. List the ways in which early successional species and late successional species differ.
9. Describe the major differences between primary and secondary succession.
10. Why are climax ecosystems considered to be rare in nature?
11. Describe the mechanisms whereby early successional species quickly slow the high rates of nutrient loss that commonly follow the removal of vegetation.
12. Why are early successional species proliferating today, while many late successional species are in danger of extinction?
13. Describe, with examples, the limits on ecological succession.
14. Describe how sexual reproduction increases the genetic variability of a population.
15. By our activities, we have selected for bacteria that are resistant to antibiotics, insects that are resistant to insecticides, and larger cows that produce more milk. Compare and contrast this human-caused selection with natural selection.
16. After Kettlewell had run his peppered-moth experiment in a pollution-free area of England and found what he was looking for, why did he consider it necessary to rerun the experiment in a polluted area?
17. List the types of adaptations that reduce an organism's vulnerability to attack. Which type(s), if any, is (are) better adaptations than the others? Defend your answer.
18. What are the natural limitations on a population's ability to adapt to environmental change? How does the rate of environmental change influence a population's ability to adapt?

Projects

1. Contact the local office of your state's Department of Natural Resources or Conservation to find out where ecological succession is being used for vegetation or wildlife management. If possible, visit several of those sites and assess the success of the management practices.
2. List some of the environmental changes that occur in your area daily, seasonally, and annually. How are animals, plants and people adapted to survive these changes? Take a field trip to observe some of these adaptations.

Selected Readings

Barrett, S. C. H. "Mimicry in Plants." *Scientific American* 257 (Sept. 1987):76-83. A look at many ways in which mimicry helps plants to survive, from deterring predators to attracting pollinators.

Bergon, M. E., J. L. Harper, and C. R. Townsend. *Ecology: Individuals, Populations, and Communities.* 2d ed. Cambridge, Mass: Blackwell Scientific Publications, 1990. An up-to-date and thorough treatment of ecological principles, with abundant examples.

Bishop, J. A., and L. M. Cook. "Moths, Melanism, and Clean Air." *Scientific American* 232 (Jan. 1975):90-92. An examination of the processes that shaped the selection for wing color of the peppered moth in England.

Borman, F. H., and G. E. Likens. *Pattern and Process in a Forested Ecosystem.* New York: Springer-Verlag, 1979. An analysis and summary of the Hubbard Brook studies.

Colinvaux, P. A. "The Past and Future Amazon." *Scientific American* 260 (May 1989):102-8. An examination of the ability of the Amazon rain forest to tolerate human exploitation.

"Evolution." *Scientific American* 239 (Sept. 1978). A special issue devoted to many aspects of evolution and adaptation.

"The Dynamic Earth." *Scientific American* 249 (Sept. 1983). A special issue that discusses various aspects of change in and on the planet Earth over time.

"Fire Impact on Yellowstone." *Bioscience* 39, no. 3 (1989). An issue containing six articles that interpret the Yellowstone fires of 1988 and examine their impact on vegetation, animals, and streams.

French, A. R. "The Patterns of Mammalian Hibernation." *American Scientist* 76 (1988):569-75. An interesting look at how changing body temperatures promote winter survival and springtime reproductive success.

Grant, P. R. "Natural Selection and Darwin's Finches." *Scientific American* 265 (Oct. 1991):60-65. A fascinating account of how a single drought can significantly alter a population.

Hinrichsen, D. "The Forest Decline Enigma." *Bioscience* 37 (1987): 542-46. An examination of the many factors that underlie the extensive dieback of forests in North America and Europe.

Horn, H. S. "Forest Succession." *Scientific American* 233 (Nov. 1975):90-98. An examination of the many interaction that occur during ecological succession in forests.

Repetto, R. "Deforestation in the Tropics." *Scientific American* 262 (April 1990):36-42. An examination of government policies that encourage the exploitation and destruction of tropical forests.

Romme, W. H., and D. G. Despain. "The Yellowstone Fires." *Scientific American* 261 (Nov. 1989):36-46. An examination of role of fire as a necessary mediator of ecological change in Yellowstone National Park.

Schukla, J., C. Nobre, and P. Sellers. "Amazon Deforestation and Climatic Change." *Science* 247 (1990):1322-25. An analysis suggesting that reestablishment of the rain forest may be particularly difficult.

Smith, R. L. *Ecology and Field Biology.* 4th ed. New York: Harper & Row, 1989. A well-written, informative examination of ecological principles.

Chapter 5

Population Growth and Regulation

A mixed herd of zebra and wildebeest on the African plains. The population size of these grazing animals is determined by many factors, which vary in importance with time and location.

© Philip Kahl, Jr./Photo Researchers, Inc.

Rabbits were brought into Australia from England in 1859. In the absence of natural enemies (except for human hunters), the rabbits flourished and spread rapidly across the continent. By 1928, they had invaded nearly two-thirds of the Australian continent and ruined much of the grassland that supported sheep, Australia's main industry. Poisons, predators, and fences were ineffective against them, but in 1950 a solution was found—a virus that causes myxomatosis, a disease related to smallpox. This virus was discovered in South American rabbits, which were resistant to it. European rabbits, however, were not resistant.

The virus was introduced into Australia in 1951, and hopes soared as myxomatosis spread quickly and killed hundreds of thousands of the animal pests. Mosquitos accomplished the rapid dispersal of the virus; they bit infected rabbits and then spread the virus among healthy rabbits by biting them also. Over time, however, fewer and fewer of the remaining rabbits succumbed to the virus, and today rabbits are a major problem again (fig. 5.1).

This event illustrates several interactions among populations of humans, rabbits, mosquitos, and a virus. A **population** is a group of individuals of the same species occupying the same geographic area. The interaction of human populations with those of other species is a two-way street; they affect our well-being, and we affect theirs.

Plants, animals, and microorganisms provide services that are essential to our well-being. They may be sources of energy and essential nutrients, help purify the air we breathe and the water we drink, or play a major role in the recycling of essential nutrients. On the other hand, populations can become so large that they interfere with our well-being. For example, an overgrowth of algae can foul our source of drinking water, a swarm of locusts can devastate crops, or a large rat population in an urban area can endanger human health by harboring fleas that transmit typhus and bubonic plague. We define any species whose population is large enough to interfere significantly with human well-being as **pests.**

Because our interrelations with other species affect us so profoundly, we often attempt to regulate their populations. We regulate game animals (e.g., deer and rabbits) and waterfowl (e.g., geese and ducks) in the name of wildlife management. One goal of such efforts is to keep the size of a game population within the capacity of its habitat to support them. For example, a deer population that grows too large for a particular habitat will overbrowse and damage the vegetation, reducing the habitat's capacity to support the deer herd. Some people, however, are more interested in short-term goals than in managing for the long term. Many hunters, for example, have little regard for the ecological consequences of allowing wildlife populations to exceed the habitat's capacity. To improve their chances for a kill, they pressure wildlife managers to allow deer herds to proliferate. Many people in the tourist industry, too, favor the growth of deer herds because deer in the wild attract visitors who spend their money in nearby resorts. Wildlife managers must therefore deal not only

Figure 5.1 Despite numerous attempts to control their numbers, rabbit populations in Australia remain a major problem.
Australian Information Service

with difficult wildlife population issues, but also with economic and political pressures.

Although management of wildlife and pests poses many problems, the most worrisome population problem is the explosive worldwide growth of our own species. This growth, together with the soaring per capita demands of the world's affluent and developing societies, is stressing all of the earth's ecosystems. In response to our unrelenting demands for food and timber, we have produced vast regions of greatly simplified ecosystems, which have created new conditions that favor population explosions of weeds, insects, rats, and other pests. Our continued efforts to create these artificial environments disrupt the natural interactions among many species by polluting and destroying their habitats. Consequently, thousands of species are being pushed to the brink of extinction.

In this chapter, we examine how populations grow, the factors that promote and inhibit their growth, and the implications of these factors for the management of biological resources. We then apply the principles drawn from this examination in chapter 9 ("Pesticides and Pest Management"), chapter 10 ("Management of Forests, Rangelands, Parks, and Wetlands"), and chapter 11 ("Managing for Biodiversity"). The same ecological principles apply to the growth and regulation of all plant and animal populations, including our own. Human population growth, however, is also greatly affected by social and economic factors, so we defer our consideration of the growth and regulation of our own species to chapters 6 and 7.

Population Growth

Populations are dynamic. One may be growing while another is declining, but all populations fluctuate. Rarely is a population static, that is, not changing in size. The growth rate of a population depends on the net difference between additions to it and subtractions from it. Additions are made by **natality** (the production of new individuals by birth or germination) and **immigration** (migration into the population from elsewhere). Subtractions occur as a result of **mortality** (deaths) and **emigration** (migration out of the population). If additions exceed subtractions, the population grows; if subtractions exceed additions, the population shrinks.

Natality varies tremendously from one species to another. For example, a single female codfish in one season produces as many as nine million eggs. In marked contrast, female salmon lay fewer than one thousand eggs in the same period. Birds lay far fewer eggs than fish—from one (for the albatross) to fifteen (for the quail) eggs per clutch. Although most bird species lay one clutch each year, some (e.g., robins and bluebirds) may produce two or three clutches. Small mammals, such as mice, may have litters of four to six young as often as four times a year, but larger mammals characteristically have only one or two offspring per year. In general, the number of offspring is small in those species whose young require the most care.

Mortality rates, or chances of survival, also vary considerably from one species to another. In most species of fish, for example, fewer than 5% of the eggs that are produced actually develop into young and survive for a year. In most bird species, fewer than 25% of the eggs that are laid produce young that survive the first season. Those species whose young receive parental care have a greater chance of surviving, but the young of all species (as pointed out in chapter 4) are quite vulnerable to hostile weather, inadequate nutrition, predators, and disease. For many species, if a member survives its risk-laden first year, the chances that it will survive for its normal life span remain more or less constant.

Exponential Growth

The dynamics of population growth become clear if we follow the growth pattern of a population as it enters a new region that contains abundant resources. Population growth is slow at first, but eventually, it is much more rapid (fig. 5.2a). Such a pattern illustrates the exponential nature of population growth.

Exponential growth (also called **geometric growth**) occurs when a population (or some other factor) increases by a constant percentage of the whole over a specific period of time. For example, assume that your savings account is growing (accruing interest) at the rate of 6% per year. Next year, you will receive interest on both the principal and this year's interest. Thus, the total amount that you receive in interest next year will be more than a 6% increase on the amount originally deposited. Most of us are more familiar with **linear growth** (also called **arithmetic growth**), which is an increase

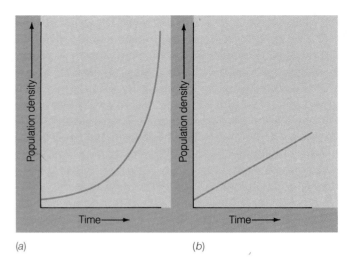

Figure 5.2 (*a*) An exponential, or geometric, growth curve. In a population that is growing exponentially, the rate of increase remains constant, but the growth in the number of individuals accelerates even more rapidly as the size of the population increases. (*b*) A linear, or arithmetic, growth curve. In a population that is growing arithmetically, the size of the population increases by a constant amount in a given period of time. Such growth, when plotted on graph paper, is a straight line, as shown.

by a constant amount during a specific period of time (fig. 5.2*b*). For example, a youth who grows at a constant rate of 5 centimeters (2 inches) per year is growing linearly. The contrast in the slopes of the curves in figure 5.2*a* and 5.2*b* implies that factors such as savings accounts and populations that grow exponentially increase much more rapidly than factors that grow linearly.

One difference between exponential and linear growth is illustrated by the following tale. Many years ago, a clever citizen presented an exquisite chess set to his king. The king was so pleased that he asked how he could reward his subject. In reply, the man asked the king to give him one grain of wheat for the first square on the chess board, two grains of wheat for the second square, four grains for the third, and so on; that is, with each square, the number of grains would double (increase by 100%). Thinking that this was a small price to pay for the chess set, the king readily agreed. Table 5.1 shows what happened. At first, the number of grains needed for each succeeding square appeared to increase slowly; but as the exponential growth curve in figure 5.2*a* illustrates, the number of grains required for each succeeding square soon began to grow rapidly. The tenth square required 512 grains, the twentieth 524,288 grains, and the fortieth, more than 549 billion. The cost for all sixty-four squares would have equaled the world's present annual production of wheat for the next two thousand years! Needless to say, the king ran out of wheat long before the sixty-fourth square. Subsequently, he was dethroned by his hungry subjects.

Obviously, the king was thinking about the more familiar linear growth when he agreed to the price. At the linear growth rate, only one extra grain would be added for each square, that is, 1 + 2 + 3 + 4 ... + 64. The total cost would have been much less than a bushel of grain. With exponential

Table 5.1	An illustration of exponential growth; Wheat grains on a chess board	
Selected Squares on Chessboard	**Number of Grains of Wheat Required**
1	1
2	2
3	4
4	8
5	16
10	512
20	524,288
40	549,755,809,568
64	9,223,372,036,854,775,808

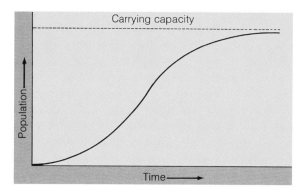

Figure 5.3 During the early phase of a sigmoid growth curve, the population shows an accelerating rate of growth. At a population density called the *inflection point*, the environment becomes limiting, and the growth rate begins to slow. This slowing continues as the population approaches the carrying capacity, at which no further growth will occur.

growth, however, factors increase by deceptively huge proportions in a relatively short period of time.

Because populations also grow exponentially, any population has the potential for explosive growth. A single pair of houseflies, for example, could have more than 6 trillion descendants in only one summer if all their eggs hatched and if all the young survived to reproduce. Obviously, this predicament never actually occurs; we are not buried under a deluge of flies or, for that matter, any other species. No population continues to grow exponentially for an indefinite period; at some point, something in the environment limits its growth.

The **sigmoid** (S-shaped) **growth curve** (fig. 5.3) illustrates one potential pattern when an environmental factor limits population growth. In this model, population growth is initially exponential, but at a particular population size, a limiting factor in the environment causes the growth rate to slow. Eventually the population levels off at or below the **carrying capacity,** that is, the population size that the total resources of the habitat can support on a sustained basis.

Although laboratory populations may follow a simple sigmoid growth pattern, most populations in nature frequently overshoot the carrying capacity. Then, as some environmental factor becomes limiting, many members of the population either die or migrate in search of more favorable conditions. Given the stresses associated with overshooting the carrying capacity, why don't most populations simply stop growing before they reach the carrying capacity?

The basic reason for the failure of most populations to level off at or just below the carrying capacity is the built-in momentum of exponential growth. Once a population begins to grow, it tends to continue growing. Consider for illustrative purposes the following modern-day version of an old French riddle. Assume that you are the proprietor of a plush resort, situated on the shore of a beautiful, pristine lake. One day your groundskeeper informs you that a small mat of algae has appeared on the far side of the lake. Your employee has determined that the algal mat is doubling in size daily, and that it will take about thirty days for the bloom to cover the entire lake. He asks you what to do about it. You tell him that you are too busy to bother with it now, and besides, the mat is

too small to worry about anyway. You will decide what to do when the pond is half covered, believing that you will have plenty of time to stop the algal growth. At stake, of course, is the attractiveness of the lake for your guests; an extensive algal mat will surely drive them away.

The riddle: What day must you decide to take action to stop the algal growth? The answer: the twenty-ninth day. Because the bloom is growing exponentially, by the time it covers half the pond, you will have just one day to control it. In a single day, you will have to decide upon and implement the proper control procedures and hope that they will be effective that day. If you are not successful, your guests will leave. Hence exponential growth has another deceptive characteristic—it is explosive and can reach the carrying capacity within a surprisingly short period of time.

This feature of exponential growth can cause a population to overshoot its carrying capacity because, in many cases, it initially grows faster than its control agents. For example, when predators are important in controlling the numbers of a rapidly increasing prey population, time is required for the predator population to reproduce or for other predators to immigrate into the area. Thus the prey population is likely to overshoot the carrying capacity before the number of predators is sufficient to control them.

In addition, a growing population typically contains a large percentage of young members who are not yet making full demands on their habitat for food, water, space, and other resources. When those young members mature and begin to demand an adult share of essential resources, the habitat cannot supply them adequately. Meanwhile, the population continues its exponential growth and rises above the carrying capacity. Consequently, increasing numbers of the population succumb to predators and parasites, perish because of inadequate resources, or migrate in search of more favorable conditions. Thus, the population falls below the carrying capacity. This pattern of population growth and decline is often repeated, and the long-term effect is an oscillation about the carrying capacity.

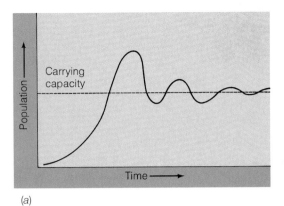

(a)

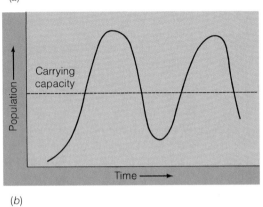

(b)

Figure 5.4 (*a*) Model of population growth characteristic of *K*-strategists. (*b*) Model of population growth characteristic of *r*-strategists.

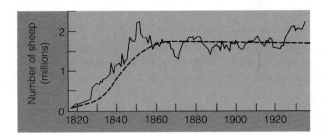

Figure 5.5 Growth of the sheep population during the more than one hundred years following their introduction on the island of Tasmania. Because sheep are large, long-lived, and reproduce slowly, the sheep population is relatively insensitive to environmental change. The sheep population usually showed less than 25% variation in year-to-year numbers after the population leveled off near the carrying capacity. The dotted line is the hypothetical sigmoid curve, about which the population fluctuates.

Figure 5.4 illustrates two models of population growth that commonly occur in nature. Note that the degree of oscillation about the carrying capacity varies greatly. Ecologists have found that the extent to which a population fluctuates in size depends upon how it allocates its resources for growth, maintenance, and reproduction. Each population employs a strategy to balance the use of resources required for reproduction now against their use to increase the probability of survival for reproduction in the future. We examine these strategies in the next section.

Reproductive Strategies

We can conclude from figure 5.4 that two quite different basic types of **reproductive strategies** occur in nature. We classify species into two basic types: *K*-strategists and *r*-strategists.

K-strategists (which include humans) maximize their probability of surviving to reproduce in the future at the expense of current reproduction. Accordingly, *K*-strategists allocate considerable energy to producing relatively large and long-lived adults. To help ensure a long adult life, *K*-strategists have mechanisms that allow them to adjust to environmental changes. For example, many animals that are *K*-strategists are able to maintain a nearly constant body temperature even though environmental temperatures fluctuate. This process, known as *thermoregulation*, enables them to remain active

throughout much of the year (see Special Topic 4.2 on page 80). Such animals also have an immune system that allows them to fight off infections. Acquired Immune Deficiency Syndrome (AIDS) illustrates what happens when the immune system is unable to ward off life-threatening diseases.

Maintaining a constant internal temperature and an immune system requires energy. Because only so much energy is available and because considerable energy is needed to build and maintain large adults, *K*-strategists have less energy available for reproduction. Thus, animals that are *K*-strategists produce few offspring over a long time period, but their offspring are usually large and require considerable parental care.

The reproductive strategies of trees are similar to those of large animals. They do not usually produce seeds until they are at least 15-20 years old, when they are 5-10 meters tall (15-30 feet) and have a well-developed root system. Once sexual reproduction begins, however, trees produce seeds for many decades or even, for some species, centuries. Because seeds from trees are often contained in fruits that are larger and more obvious than those produced by annual plants, trees may appear to produce more seeds than annuals, but the jungle of weeds that quickly take over an abandoned lot attests to the prodigious seed production of annual plants. In fact, a dense growth of small annual plants produces thousands more seeds than a stand of larger, but fewer, trees.

As figure 5.5 indicates, populations of *K*-strategists are usually relatively stable. For large organisms, day-to-day variations in the environment have a relatively minor impact on the size of a population. Even if climate or essential resources (e.g., food, water, or shelter) become limiting and diminish natality during a given year, many adults usually survive to reproduce the following year. For example, sheep reproduce during about 90% of their average ten-year lifespan, and they norm-ally produce one or two offspring each year. Such reproduc- tive characteristics tend to reduce the degree of variation in the size of a population.

Their lower reproductive capacities and their more limited ability to disperse make *K*-strategists poor colonizers.

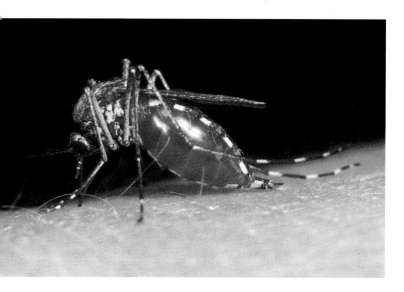

Figure 5.6 This mosquito, filled with blood, is a familiar *r*-strategist.
© Hans Pfletschinger/Peter Arnold, Inc.

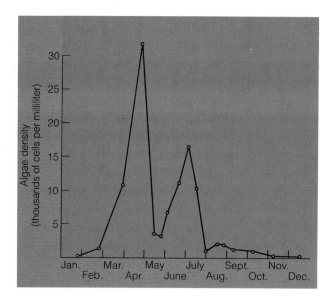

Figure 5.7 Variations in number of algae in the bay of Green Bay, Wisconsin. Because they are small and short-lived, and because they reproduce rapidly, algae are quite sensitive to environmental changes. Hence, in contrast to the sheep in figure 5.5, the number of algae vary by more than thirtyfold within a year.

Hence, they are more common in the late successional stages than in the early successional stages of terrestrial ecosystems (chapter 4).

At the other end of the spectrum, the ***r*-strategists** maximize their current rate of reproduction at the expense of surviving long enough to reproduce in the future. Unlike *K*-strategists, these organisms produce many offspring in a short period of time and thus may produce several generations in a single year. Because they allocate most of their resources to reproduction, they have little energy available for growth and maintenance. Thus, adults are usually small and short-lived (fig. 5.6). Many *r*-strategists are unable to maintain a constant body temperature and have poorly developed immune systems, so they are vulnerable to environmental change.

Species that are *r*-strategists are opportunistic; they reproduce rapidly when conditions are favorable or when a new habitat becomes available. Unfavorable habitat conditions, however, can cause the population to decline rapidly, or crash. Hence populations of *r*-strategists typically follow boom-and-bust cycles, as illustrated in figure 5.7.

Algae are *r*-strategists. A clear lake may develop the appearance of pea-green soup in only a few days when an initially small population of algae reproduces at an explosive rate. Often, however, the algal bloom disappears quickly as the population rapidly exhausts the available nutrient supply. Also, persistent cloudy weather can cause an algal population to crash because of a significant reduction in solar energy for photosynthesis.

Other *r*-strategists include insects, rodents, and annual plants. Because annual plants live only a single year, they allocate most of their energy to seed production rather than to growth. Although they rarely grow more than a meter in height, each annual plant commonly produces hundreds of seeds. Because *r*-strategists usually disperse widely and are good colonizers, they are common in pioneer and other early successional stages of terrestrial ecosystems.

In agriculture, we raise both *r*-strategists (crops) and *K*-strategists (livestock). In crop cultivation, we obviously select for *r*-strategists. Many crop plants, such as corn, wheat, and soybeans, are annual plants that put most of their energy into the production of seeds and fruits (much of our food). Ironically, however, by cultivating these plants, we also create conditions that are favorable for pest species, which are also *r*-strategists. In raising livestock, animal husbandry practices represent a trade-off between both ends of the spectrum of reproductive strategy. Although cattle, swine, and sheep are *K*-strategists and thus less productive than *r*-strategists, their larger size makes them easier to manage, and they are considered to be far more desirable as food sources than insects or rodents.

Many endangered species are extreme *K*-strategists. California condors produce only one chick every two years, and a female gorilla gives birth to only one infant every five or six years (fig. 5.8). Although *K*-strategists are more efficient than *r*-strategists in raising young, they do have limits to their survival rates. Illness, predation, and accidents take their toll on all populations, but their impact is particularly severe on populations with low reproductive capabilities. Today the gorilla, like many other endangered species that are *K*-strategists, survives in perilously small numbers and only in the most favorable habitats. Some ecologists suggest that such species as the gorilla and condor have pushed the *K*-strategy too far and that they would approach extinction regardless of human activities. There is no doubt, however, that humans are accelerating the demise of such species by poaching and destroying their habitats. We discuss further the characteristics and management of endangered species in chapter 11. Table 5.2 summarizes the contrasting characteristics of *r*-strategists and *K*-strategists.

Figure 5.8 Although extreme *K*-strategists, such as this lowland gorilla, take good care of their young, their relatively low rate of reproduction threatens the future of these species.
© Ron Austing/Photo Researchers, Inc.

Table 5.2 Contrasting characteristics of *r*-strategists and *K*-strategists	
***r*-Strategist**	***K*-Strategist**
adults small	adults larger
many small young	fewer, larger young
little or no care of young	more care of young
early maturity	later maturity
short life	longer life

Population Regulation: Biological Factors

We have seen that a population cannot continue to grow indefinitely. Sooner or later, some limiting factor will cause the population to stop growing or even decline. Ecologists

Table 5.3 Types of two-species interactions	
Type of Interaction	**General Nature of Interaction**
predation	the predator benefits at the expense of the prey
parasitism	the parasite benefits at the expense of the host
competition	each species is inhibited by the presence of the other
mutualism	each species is favored by the presence of the other

have found that both biological and physical factors play a role in limiting population size. Biological factors consist of interactions within and between populations and include predation, parasitism, and competition. Physical factors include fire and weather stresses, such as hurricanes, floods, droughts, and temperature extremes.

Factors that control the size of a population are either density-dependent or density-independent. The influence of a **density-dependent factor** strengthens as the density of a population (the number of individuals per unit area) increases. Biological factors are usually density-dependent. In competition, for example, the more the demand (the population size) exceeds the supply (the available resources), the more significant competition becomes in regulating a population.

A physical force is usually a **density-independent factor.** Unlike the impact of a biological factor, the impact of a physical factor varies little with population size. For example, a severe freeze in late spring might kill all the annual weeds in a field, whether there were 10 or 1,000 plants.

Often both types of factors interact to influence the size of a population. A late spring freeze (a density-independent factor), for instance, kills many oak flowers, which in turn reduces the acorn crop. The following winter, therefore, the squirrel population faces severe competition (a density-dependent factor) for the meager food resources that are available, and many will starve.

In the following sections, we consider more fully the influences of density-dependent and density-independent factors on population growth. We first consider how various biological factors may be involved in controlling the size of any given population.

Ecologists recognize several types of interactions among species. As table 5.3 illustrates, some of these interactions help regulate population size by producing a negative impact on one population (such as reducing the size of the population by predation and parasitism) or on both populations (as a result of competition). Some interactions, however, actually benefit both populations. In many cases, the survival of each interacting species depends on the presence of the other. That kind of relationship is called **mutualism** (see Special Topic 5.1). Our focus here, however, is on the role of negative interactions: predation, parasitism, and competition.

Mutualism: Species Helping Each Other

Mutualism—an interaction between species that is beneficial to each—occurs widely and is important to many populations. Such an interaction can benefit a species in several ways, including (1) nutrition, either through the digestion of food for the partner or the synthesis of vitamins or proteins for the partner; (2) protection, either from enemies or from environmental change, or (3) transport, such as the dispersal of pollen or the movement of materials from unsuitable to suitable environments.

Interactions between ruminants (e.g., cows and sheep) and the microorganisms that inhabit their stomachs are good examples of nutritional mutualism. A ruminant's diet consists mainly of cellulose, but like most animals, ruminants do not produce the necessary enzymes to break down cellulose. Unlike most animals, however, they have microorganisms in their stomachs that can digest cellulose. Furthermore, these microorganisms can also synthesize proteins from ammonia and urea that are present in the stomach. Some of these microorganisms are subsequently digested, thus providing the ruminant with protein. They can also synthesize all of the B-complex vitamins and vitamin K. Hence microbial activity enables ruminants to gain additional energy and nutrients. The microorganisms, in turn, benefit from free food and good housing—the warm, wet interior of the stomach provides a favorable environment for their growth and reproduction.

In their mutualistic relationship, these red ants obtain nourishment from aphids and protect the aphids from predators.
© Hans Pfletschinger/Peter Arnold, Inc.

We, too, have a complex assemblage of mutualistic microorganisms residing in our digestive systems. Unlike those of ruminants, our microbes cannot digest cellulose, but they do synthesize the B-complex vitamins and vitamin K.

Many organisms provide protection from enemies and in turn may be provided with food, protection, housing, or all of these. In East Africa, for instance, certain ants live in the swollen thorns of acacia trees. The thorns provide an ideal shelter and a balanced and almost complete diet for all stages of ant development. The ants, in turn, serve as protectors. As soon as a branch is touched by a browsing animal or some other organism, the ants feel the vibrations and respond by pouring out of their holes in the thorns and racing toward the ends of the branches. Along the way, they release a repulsive odor that usually drives the enemy away.

Ants are also involved in protective mutualisms with sap-feeding insects, such as aphids. Plant sap often contains too much sugar for the aphids to utilize. Consequently, aphids extrude a sugary solution, called *honeydew*, which serves the ants as a major source of food. In return, ants protect the aphids by attacking and driving away predators (left photo).

Flowering plants and their pollinators exemplify transport mutualism. A wide variety of animals (e.g., butterflies, beetles, hummingbirds, and bats) visit flowers to obtain food (right photo). Flowers produce pollen, which is rich in protein, and nectar, a nutritious sugary fluid. As the visitors feed, they brush against the male structures (which produce pollen), and their bodies are dusted with pollen. When they visit another flower,

The head of this bronzy hermit hummingbird is about to be dusted by pollen as it feeds upon nectar from a passion flower.
© Michael Fogden/Animals Animals

the pollen from the first flower is transferred inadvertently to the female structures of the second flower. This pollination process may then lead to fertilization and the production of seeds and fruits. Several advantages accrue to flowering plants from pollination by animals. Because the animals move directly from one flower to another, less pollen is wasted than in pollination that depends on the random movements of the wind.

Furthermore, when plants are relatively few and widely scattered, the food-seeking activities of pollinators make pollination more likely than it is with wind pollination.

How do mutualistic interactions form? Are the mutualistic microbes that live in us the descendants of microbes that were once intestinal parasites in our ancestors? Are mutualistic interactions stable, or do they frequently break down? For example, instead of protecting their partners, do aphid-tending ants sometimes turn on the aphids and devour them? Ecologists still have much to learn about these fascinating population interactions.

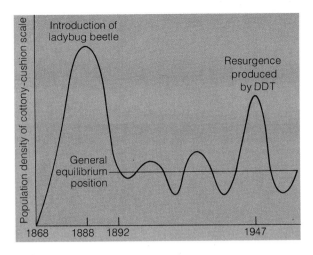

Figure 5.9 The effect of ladybug beetles on population size of cottony-cushion scale.

Predation

Predation is an interaction in which one organism (the predator) derives its sustenance by killing and eating another (its prey). The role of predation in keeping the predator species alive is clear; determining its effect on the prey population is more difficult. Most studies of the effect of predation have focused on wild game and pests. These studies indicate that the significance of predation in regulating a population depends on the types of predators, the types of prey, and the conditions of the habitat.

Predation is a major agent in limiting the population size of small animals, such as insects and rabbits. In fact, **biological control,** the regulation of pests by natural enemies, including predators and parasites, is an important alternative to pesticides.

A classic example of pest control by a predator involves ladybug beetles and the cottony-cushion scale, an insect that infests citrus plants. In the early 1870s, cottony-cushion scale insects were accidentally introduced into California from Australia. They spread rapidly, and within a few years, the infestation of that insect threatened the citrus industry. A resourceful government scientist journeyed to Australia to seek out the natural predators of the scale. He turned up a species of ladybug beetle, which, when introduced into California citrus orchards, quickly brought the scale under control. Figure 5.9 illustrates the ladybug's impact on the scale population. Overall, the ladybug has kept the scale population down, so it causes little damage to citrus crops. Ironically, during the late 1940s, the scale population exploded when the pesticide DDT was applied to control another species of scale. While DDT almost eliminated the ladybugs, it had little adverse impact on the cottony-cushion scale. Once the use of DDT was halted, the ladybug population recovered and soon had the cottony-cushion scale population under control again.

In the United States, biological control regulates other insect pests, including bark beetles and sawflies on ponderosa pine, spotted aphids on alfalfa, and codling moths on apple trees. Today, scientists can cite nearly a hundred examples from around the world in which insect pests have been at least partly controlled by the introduction of natural predators. We have more to say about biological control in chapter 9.

Although predation is often important in controlling insect populations, the role of predators in regulating populations of large animals is much less clear. The complex interactions between large predators and prey are well illustrated by the wolf and moose populations that reside on isolated Isle Royale in Lake Superior. When wolves first crossed the ice from Canada to the island in 1949 (one of the few years when an ice-bridge formed), it was generally believed that the island's thriving moose population was doomed. Because of widespread acceptance of the "big, bad wolf" myth, any mention of wolves creates the image of a marauding pack that slaughters every living thing in sight. In fact, wolves on Isle Royale bring down only one of every two dozen or so moose that they attack, and their victims are usually calves or moose that have aged past their prime (fig. 5.10). As moose age, arthritis and a lungworm make them more vulnerable to attack. Thus, in many instances, wolves function as scavengers by killing prey that would have died soon anyway.

Such interactions between wolves and moose persisted until the early 1970s, when several consecutive winters of heavy snow made it difficult for the moose to find sufficient browse. Scarcity of food created a whole generation of weak moose. In response to a greater supply of harvestable prey, the wolves prospered, and their numbers grew to a record fifty by 1980—double their 1960s average. Meanwhile, the wolves had reduced the moose population from a mid-1970s peak of 1,000 down to 600. Not only had the wolves significantly diminished their prey population, but more important, most of the remaining moose were too strong for the wolves to kill. With little harvestable food and no place to go, the wolf population plummeted from fifty to fourteen in only two years. Subsequently, with fewer wolves competing for a growing moose population, the wolf population recovered to twenty-three, close to the 1960s average.

In recent years, the wolf population has mysteriously declined to only about ten members. Because no pups appeared outside dens during the late 1980s, a suspected culprit is canine parvo virus, which infects domestic dogs and to which wolf pups are particularly susceptible. Another probable contributing factor is **inbreeding**—mating between close relatives. Presumably no new wolves have reached Isle Royale since the founder population (perhaps only a single pair) arrived in 1949. As a result, most, if not all, of the wolves on the island today are close relatives, which favors the buildup of deleterious characteristics such as increased susceptibility to disease. A recent study found that about 50% of the genetic variability has been lost in this population since their arrival 40 years ago. To ensure the survival of wolves on Isle Royale, wildlife managers are considering a reintroduction of wolves from another population.

Figure 5.10 A pack of wolves stalking a moose on Isle Royale in Lake Superior.
© Rolf Peterson

A particularly important finding of the Isle Royale study is that interactions between predators and prey are highly variable. At times, predators are able to harvest a large number of prey, so that the prey population declines while the predator population grows. At other times, the number of harvestable prey becomes a limiting factor, so that the predator population declines. In addition, the importance of predation as a regulator of prey population size is often influenced by habitat conditions, which may change dramatically. Furthermore, interactions may be influenced by parasites (in this case, lungworms and canine parvo virus).

In the past, predators were often killed in the belief that game populations would then prosper. But mounting evidence of the complex relationships between predator and prey populations makes it difficult for game managers to decide whether controlling predators is really wise. In addition, wildlife managers must contend with pressure from special-interest groups, such as hunters, who think that killing predators will save more wildlife for them, and from others who argue that many predators are endangered species and that perhaps all predator species should be protected.

The role of large predators in controlling the numbers of prey is not yet fully understood, but it is clear that predators contribute to the stability of prey populations in several ways. For example, as we have seen, large predators tend to keep prey populations healthy by killing the weak, diseased, and old members. If diseased animals lived longer, the disease would be more likely to spread to other members of the population.

Many prey species have adaptations that help them to escape predation, but not all species are so equipped. Many insects do not run or hide from predators or produce defensive chemicals. Aphids, for example, merely sit on leaves, sucking up plant juices; hence, they are easy prey (fig. 5.11). Such insects compensate for their inherent vulnerability by producing offspring by the hundreds of thousands. Thus, although predators consume them in great numbers, their high natality favors the population's survival. This balancing effect demonstrates the fundamental concept that population size results from the interaction of many factors. Some factors are genetically inherent to the population itself, while other factors stem from the nature of the ecosystem and the other populations that reside within it.

Figure 5.11 Aphids are "easy pickings" for the predatory lacewing fly. © Stephen Dalton/Photo Researchers, Inc.

Parasitism

Parasitism is an interaction in which one organism (the parasite) obtains its nourishment by living within or on another living organism (the host). Parasites usually do not kill their hosts immediately for nourishment, as predators do, but they may eventually kill their hosts in the process of obtaining their meals. When a host dies, the parasite frequently also dies, because the parasite loses its own habitat in the process. Thus, infection of a low-resistance host population by a highly virulent parasite population could lead to the extinction of both populations. As a result of natural selection, however, the interactions of a host population and a parasite population sometimes result in less virulent parasites and more disease-resistant hosts. Such changes exemplify the process known as **coevolution.**

A classic example of the coevolution of a parasite and a host is the short-lived reduction of the rabbit population in Australia. Recall from our earlier discussion that although the introduced virus initially killed hundreds of thousands of rabbits, eventually fewer and fewer of the remaining rabbits died of myxomatosis. Why did the virus no longer effectively control the rabbits?

In their attempt to determine what had happened, investigators found that, through natural selection, the virus had lost its "punch," and that the rabbit population had become less susceptible to myxomatosis. The rabbits that were infected by the most virulent strains died almost immediately, which had two effects: (1) early death of the hosts inhibited the spread of the most virulent forms of the virus, and (2) it drastically reduced the reproduction rate in the most susceptible rabbits. The most virulent forms were not spread widely, because mosquitos (the carriers of the virus) were not attracted to dead animals. Hence, if the virus killed its host quickly, there would be less time for mosquitos to bite a host that was infected with the virulent strain of virus and to transmit it to another rabbit.

Meanwhile, the rabbits that were infected with the less virulent strains lived longer, so that the mosquitos transmitted the less virulent strains to more rabbits. Thus, the relatively weak strains become more numerous in the viral population. At the same time, natural selection continued to favor rabbits that had some resistance to the virus. Eventually, the virus was no longer a significant regulating factor, and because the rabbits are r-strategists, the rabbit population began to increase rapidly again, virtually unchecked.

Thus, we see that some parasite-host relationships evolve toward a stable interaction in which the host tolerates a widespread, but not life-threatening infection. Consequently, the parasite is no longer an important regulator of the size of the host population. But conditions can change. If the habitat for the host population becomes less favorable, the consequent stress may render some individuals less tolerant of the parasite, thereby increasing host mortality; or a mutation in a few parasites may lead to a more virulent parasite population, which will also increase host mortality.

Coevolution between a parasite and a host is by no means inevitable, but it is more likely to occur with r-strategist hosts than with K-strategists. (By their very nature, parasites are usually r-strategists.) Species that are r-strategists usually have larger populations with considerable genetic variability. Thus, the probability is greater that some members of the population have some resistance to the parasite. In addition, the ability of r-strategists to reproduce rapidly in large numbers improves the chances that the resistance characteristic will occur with increasing frequency in succeeding generations.

In contrast, coevolution is less probable if the host is a K-strategist. Populations of K-strategists are usually smaller and thus usually have less genetic variability. Some members may be resistant to the parasite, but because they produce fewer young over a longer time period, they may die from other causes before they can pass on the resistant trait to significant numbers of offspring.

(a)

(b)

Figure 5.12 The effects of Dutch elm disease, an example of the destruction that can follow the introduction of a new parasite. (*a*) Elm-lined Gillett Avenue in Waukegan, Illinois, as it appeared in the summer of 1962. (*b*) Gillett Avenue in 1969, after the elms were destroyed by Dutch elm disease.

Elm Research Institute, Harrisville, New Hampshire

For example, consider what happened when chestnut blight was introduced into the United States. Before the turn of the century, the American chestnut tree (a *K*-strategist) was a dominant member of the Appalachian forests. In 1904, Asiatic chestnut trees were brought to New York, carrying a parasitic fungus to which they were resistant. American chestnuts, however, had no resistance to the new parasite. The fungus quickly spread, and by the early 1950s, the American chestnut had been virtually eliminated from the Appalachian forest ecosystem. In a few places, the roots of destroyed trees are still alive and continue to send up new shoots, but these, too, eventually are killed by the fungus.

Dutch elm disease (fig. 5.12) also demonstrates how a new parasite can essentially eradicate a nonresistant, long-lived host. Partly because it usually takes twenty years or longer for a tree to produce its first seed, the probability of coevolution between a very susceptible tree species and a new, virulent parasite is much reduced. Extinction of the host is the more likely result. In addition, if the parasite can survive only on that particular host, it will also become extinct.

Parasitism illustrates well the nature of density-dependent regulation of population size. The greater the density of the host population, the more likely it is that an uninfected member will come into contact with an infected member. Thus, crowding increases the probability that a parasite will spread throughout a host population and kill a greater portion of it. We are much more likely to catch a cold in a crowded school or subway than if we stay at home. In addition, if a growing host population exceeds the carrying capacity of its habitat, competition for limited resources may force some members into marginal areas. There, weakened by a scarcity of resources or an inhospitable climate (density-independent factors), they may become particularly susceptible to predators or to infection by parasites and, consequently, succumb.

Human populations, too, have always been afflicted by parasites. In fact, human history has been shaped partially by the impact of parasitism. In chapter 6, we examine the historical role of parasitism in limiting human populations.

Competition

The resources that are essential for life, such as food, water, and space, are finite, and sometimes the demand for them exceeds the supply. The interactions among organisms to secure a resource that is in short supply constitute **competition.** Recall that competition often involves the interaction of density-dependent and density-independent factors. The degree of competition depends on both the availability of resources, which is often determined by such factors as the weather (a density-independent factor), and the size of the populations (density-dependent factors) that depend on these resources.

Ecologists recognize two types of competition: intraspecific and interspecific. **Intraspecific competition** occurs among members of a single species, whereas **interspecific competition** takes place among populations of different species. Both types of competition are important in regulating population size.

Competition among animals

Intraspecific competition among animals is often a consequence of certain types of social behavior. Many animals establish territories, which they defend against intruders of their own species. For example, during the breeding season, each male robin establishes and defends the territory within which he and his mate build their nest. The male robin's song informs other robins of his territorial limits. If another male robin enters his territory, he will attempt to drive the intruder out (fig. 5.13). This sort of aggressive defense, which is an example of **territorial behavior,** separates breeding animals and tends to ensure an adequate supply of resources for members within each territory. Thus, territorial behavior favors population growth by partitioning resources so that members occupying and defending a territory are likely to have a sufficient supply of essential resources.

However, the same behavior pattern also tends to limit populations, because the number of breeding pairs cannot

Figure 5.13 Two male robins fighting over territory.
© Phillip Strobridge, National Audubon Society/Photo Researchers, Inc.

exceed the number of available territories. Although the size of each breeding territory may shrink as a population increases, a minimum size exists for every species. The territory of a red-winged blackbird is about 0.3 hectares (0.7 acres). In contrast, the bald eagle has a territory of 250 hectares (617 acres). Once the minimum is reached, animals without territories of their own are forced to migrate to marginal areas. Since the areas beyond the prime territories have limited resources, many outcasts fail to breed. Some succumb because of inadequate food, water, or shelter, and others, in their weakened condition, fall victim to predators and disease. Occasionally, an outcast will replace a member of the breeding population who has died, but as a rule, outcasts cannot force their way into a fully occupied habitat. Territorial behavior, therefore, appears to regulate population size by limiting the number of animals that can breed in a favorable habitat.

Social hierarchies, or pecking orders, may also play a role in intraspecific competition. In some species (e.g., wolves and chickens), encounters between animals result in dominant-submissive relationships, which involve the entire population. One animal may dominate all the other animals in the population, whereas another, though submissive to the first one, dominates the remaining animals, and so on.

Once a pecking order is established, greater stability and order prevail within the population. Because each animal knows its place, less fighting occurs. Therefore, animals do not injure one another, and the population as a whole is healthier than it would be if its members were continually struggling for dominance. Thus, social hierarchies foster the growth of a population by maintaining its overall health, but they also limit it as well, since low-status animals often fail to breed. During times of stress, submissive animals are the first to be deprived of adequate food, water, and shelter

(fig. 5.14). Even in the best of times, low-status animals may be so harassed by dominant animals that they are driven into marginal habitats, where they may die.

Interspecific competition occurs among animals of different species that vie for the same resources. The results of this type of competition vary, depending on many factors. Laboratory experiments suggest that if two species need similar resources, one will cause the extinction of the other. Such **competitive exclusion** occurs, for example, if we introduce two species of *Paramecium* (unicellular animals) into a tube that contains a fixed amount of bacterial food. Although both species thrive when grown separately, *Paramecium caudatum* cannot survive when grown in the same culture with *Paramecium aurelia* (fig. 5.15).

Convincing evidence of competitive exclusion in the field, where habitat conditions are considerably more complex than they are in the laboratory, is rare. In fact, many cases of localized extinction once thought to be a result of competition were actually caused by habitat changes. Consider, for example, a recent reexamination of the impact of the European cabbage butterfly on its American cousin. Following the arrival of the European species in New England in the 1860s, local butterfly collectors began to notice that the native butterfly species was disappearing from towns and the surrounding countryside. They blamed the extinction of the local species on competition from the European invader. Today, however, in Vermont, adults of both species still occur in large numbers in the same hay fields, mate in the same areas, and during the summer, lay their eggs on many of the same species of food plants.

Despite this large overlap in resource utilization, which creates a great potential for competition, recent studies show no evidence for competition between the two species. However, if competition does occur, each species has a refuge from the other. While both species are found in hay fields and roadsides, only the native species flies in the wooded areas. In addition, although the larvae of both can feed on some of the same plants, only the larvae of the European species can feed on a weedy mustard species, which was also introduced from Europe and has successfully invaded hay fields and roadsides. Significantly, this mustard plant is lethal to the larvae of the native butterfly.

If these two species of butterflies can coexist today, why did the native species disappear from urban areas in the late nineteenth century? No one knows for sure, but our improved understanding of the ecological relationships of the native butterfly suggests an explanation other than competitive exclusion. In May and early June, the only abundant plant on which the eggs and the larvae of the native species can develop is a species that is restricted to woodlands. Furthermore, the major summer larval food plant persists only where the land is cultivated each year. In nineteenth-century New England, woodlands were cut to supply railroads and industries. Cleared areas were often used as pastures, but they were abandoned later when cities expanded. Consequently, suitable sites for the food plants on which the larvae of the native butterfly could feed declined dramatically. Thus, localized

Figure 5.14 The dominant members of this pride of lions feed upon a wildebeest while the submissive animals wait their turn to feed on the remains.
© F. S. Mitchell/Tom Stack & Associates

extinction of the native butterfly in the urban areas may well have resulted from a loss of food sources rather than from competition with the introduced species.

Other studies have shown how closely related species are able to coexist in the same region. In a classic study, the noted ecologist R. H. MacArthur studied the feeding behavior of five species of Maine warblers. He observed that although all five species feed on basically the same type of insects, which are found in spruce trees, competition is greatly diminished by differences in feeding habits. As figure 5.16 shows, the hunting activities of each species are confined to a specific part of a tree. Although feeding zones overlap to some extent, other behavior patterns further reduce competition. For example, one species of warbler captures insects at the tips of spruce needles, while another hunts insects hidden underneath the needles. Also, because the five species all have different nesting times, their times of greatest food need differ as well. Hence, these five species are able to coexist with a minimum of competition.

Another factor that reduces competition among species is differences in the hunting periods of predators. For example, hawks and owls both feed on similar types of animals, but because hawks capture their prey during the day and owls are nocturnal, they usually do not compete with one another for the same prey.

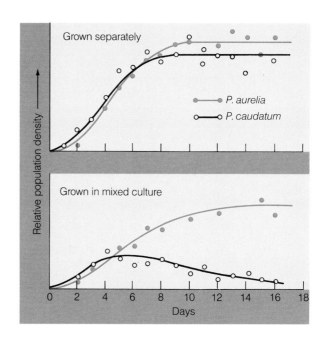

Figure 5.15 *Paramecium caudatum* and *Paramecium aurelia* thrive when grown separately in a controlled environment with a fixed food supply. When grown together, however, the *Paramecium caudatum* population becomes extinct—an example of competitive exclusion.

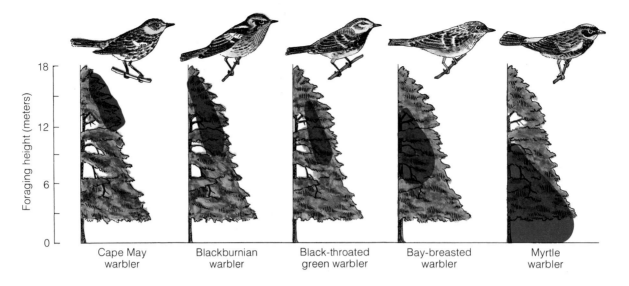

Figure 5.16 The foraging areas used by five species of warblers in spruce forests in Maine.

Structural adaptations also reduce competition between species. The downy woodpecker and the hairy woodpecker, for example, occupy similar habitats in North America. As figure 5.17 shows, these species are similar in appearance, but the hairy woodpecker is larger and has a larger bill. These structural differences allow the hairy woodpecker to eat larger insects and to reach insects that are hidden deeper within tree trunks.

With respect to nesting sites, we find again that different species of birds use different resources. Some birds nest on the ground; others nest on a tree branch or in a cavity in a tree trunk. A few species use a burrow in the ground; others employ a crevice in the side of a cliff.

Thus, we see that numerous mechanisms allow animals to avoid competition with their neighbors. Although the result of competition, which is popularly referred to as the "survival of the fittest," conjures up images of animals fighting tooth and nail for survival, such violent encounters are rare. After all, as we noted in chapter 4, the most fit are those whose young make up a greater percentage of the next generation. When competition is lessened because of adaptations that reduce confrontation, animals can direct more energy to reproduction and rearing their young. The term *fittest*, therefore, often applies best to those organisms that avoid competition.

Competition among plants

Intraspecific competition also plays a significant role in controlling the size of plant populations. Generally, most seeds fall close to the parent plant. When these seeds germinate, they produce a dense growth of seedlings. Unlike animals, plants are unable to make compensating adjustments in their spacing to reduce competition, and so the whole mass of seedlings competes for sunlight, water, and soil nutrients in the same space.

As figure 5.18 indicates, the impact of competition depends on the density of seedlings (the number of individual plants per square meter, for example). At extremely low densities, plants may grow vigorously, because there is little competition. As plant density increases, the biomass yield for each unit of area also increases. At moderate densities, plants respond to increased competition with a reduction in growth rate. Thus, the plants are smaller and individually produce fewer or smaller seeds or both. Despite the reduced size of individual plants, the total biomass produced on each unit area continues to increase. Eventually, however, biomass production approaches a constant value, which is equivalent to the carrying capacity. At higher plant densities, competition becomes so severe that some plants die. The mortality associated with severe competition is termed **self-thinning.** Despite this mortality, yield per unit area remains little changed from yields produced at moderate plant densities.

These relationships are important in the functioning of natural plant communities, and they are crucial in agriculture and forestry. Experienced farmers and foresters know that by increasing the planting density, they can only increase their yield for each unit of area up to the carrying capacity of the site. A further increase in planting density will only initiate self-thinning; it will not improve the total yield.

Interspecific competition in the plant kingdom (in contrast to that in the animal kingdom) commonly leads to competitive exclusion. Take, for example, the competitive interactions of bluebunch wheatgrass and cheatgrass on the rangelands of the Columbia River Basin in Washington. Before the advent of domestic grazing, that rangeland was dominated by bluebunch wheatgrass, along with a rich diversity of other plants, but heavy overgrazing destroyed the native vegetation, creating a habitat that was open to invasion. Cheatgrass, a native of Europe, soon dominated several million hectares

(a)

(b)

Figure 5.17 Differences in the structural adaptations of (*a*) the hairy woodpecker and (*b*) the downy woodpecker enable them to exploit somewhat different food sources. This reduces competition between the two species.
(*a*) © John Gerlach/Animals Animals. (*b*) © Rod Planck/Photo Researchers, Inc.

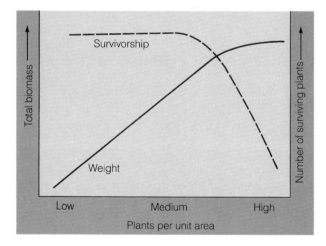

Figure 5.18 As plant density increases, total biomass production increases until a maximum production level is reached. Increasing plant density also eventually results in lowered survivorship.

that were previously occupied by bluebunch wheatgrass. Although ranchers prefer to graze cattle on wheatgrass, it has been almost impossible to reestablish wheatgrass, even where grazing has been eliminated.

Field and laboratory experiments indicate that cheatgrass can exclude wheatgrass for several reasons. The seeds of both species germinate at approximately the same time, during the fall, when the soil is moist. Then during the winter, wheatgrass becomes dormant, while cheatgrass continues to grow. As a result, the extensive root system of cheatgrass grows deeper and gains control of the soil before the wheatgrass seedlings can become better established. Also, cheatgrass matures four to six weeks earlier than wheatgrass, so it withdraws most of the water from the soil before the wheatgrass has a chance to tap it. During the typically hot and dry summers, wheatgrass seedlings usually succumb to severe soil moisture deficits. Hence in places where summers are relatively dry, cheatgrass excludes wheatgrass by out-competing it for soil moisture. In areas where plentiful rain falls in the summer,

Figure 5.19 Model of stratification of vegetation in a deciduous forest.

may find the forest floor carpeted with a spectacular array of spring wildflowers. These spring ephemerals renew their growth three to five weeks before the trees leaf out. In that period, ephemerals flower, set seed, and die back. Because they bloom early in the spring and their growth cycle is relatively brief, these plants have access to sunlight before the tree canopy closes and shades them.

The fact that plants grow to differing heights reduces competition for space. In forests, a vertical stratification of vegetation occurs, as shown in figure 5.19. In mature forests of the southern Appalachians, tall trees (e.g., beech, sugar maple, basswood, and tulip trees) constitute the upper canopy. Beneath that layer are smaller trees (e.g., dogwood, magnolia, and ironwood) that seldom or never reach the upper canopy. Shrubs (e.g., spicebush, witchhazel, and pawpaw) grow beneath the smaller trees; below the shrubs, wildflowers and a ground layer of mosses and lichens flourish. Vertical stratification is most apparent in tropical rain forests, which have three to five overlapping strata of trees. Little light penetrates the multiple tree strata of tropical forests; thus, ground vegetation (e.g., shrubs, herbs, and ferns) is relatively sparse. Consequently, most herbivores in a tropical rain forest live in the tree canopies, where the food is.

Population Regulation: Physical Factors

Recall that physical factors are density-independent agents in influencing population size. The most dramatic factors are catastrophic events, such as hurricanes, tornadoes, floods, and wildfires. Less dramatic are year-to-year variations in seasonal patterns of temperature and precipitation, as well as the length of the freeze-free period.

Weather can directly affect population size. An extended period of hot, dry weather can produce habitat conditions that exceed the tolerance limits of plants in a hay field as well as many associated species of herbivorous insects. Extreme weather may also lower population size indirectly by reducing food resources. For example, pregnant female kangaroo rats in the Mojave desert need an adequate supply of herbaceous (nonwoody) plants in January and February to provide them with water, vitamins, and energy. If winter rainfall is too scant to stimulate sufficient growth of those plants, kangaroo rat natality suffers.

The impact of a dynamic environment on any given species usually varies with the sensitivity of that species. As we noted earlier, species that are small and short-lived, and that reproduce rapidly (*r*-strategists), usually are sensitive to weather extremes. Hence, insect and rodent populations often fluctuate dramatically in response to major changes in the weather. Large, long-lived organisms that reproduce slowly (*K*-strategists), such as large game animals and trees, are less sensitive to weather extremes and hence usually exhibit smaller fluctuations in population size.

however, cheatgrass does not compete as well and grows only on recently disturbed sites.

Another type of competition in plants occurs when one plant species produces chemicals that inhibit the germination or growth of another species. That type of competition occurs on abandoned farmland during secondary succession. Recall from chapter 4 that broom sedge invades farmland within three years following abandonment, quickly replaces the pioneer weeds, and then remains for fifteen to twenty years, often in almost pure stands. Investigators have found that the roots and shoots of broom sedge contain chemicals that inhibit the growth of other plant seedlings. When broom sedge enters a field, it releases those chemicals into the soil, retarding the growth of pioneer weeds. Since the chemicals retard invasion by other plants, the rate of secondary succession slows considerably.

Although competitive exclusion sometimes eliminates a plant species from a particular habitat (where it might otherwise survive), abundant observational data suggest that plants, like animals, have evolved distinct strategies that reduce competition. If we visit a deciduous forest in the early spring, we

Because weather is so variable, most weather-related stresses are only short-term events. Favorable conditions usually return within a few weeks, months, or years. Thus, extreme weather often only temporarily reduces the size of a population; it usually recovers. For example, when deep snow prevents bobwhite quail from scratching for food, their population declines drastically. During the following spring and summer, however, the quail's reproductive success is often unusually high—probably as a result of reduced intraspecific competition. Thus, by fall, their population is restored to near-normal numbers.

Plant populations, too, can be severely reduced in the short run by such stresses as drought, freezes, or fire. But through a variety of mechanisms, most plant populations avoid extinction. The seeds of many species remain viable in the soil for many years, despite such stresses as drought or fire, which might kill the parent plants. Other plant species die back yearly into underground structures (e.g., tulip bulbs and the underground stems of many grasses), thus avoiding exposure during seasons when stressful conditions occur. By such mechanisms, some plant species are able to withstand stressful conditions for many years.

How Are Populations Actually Regulated?

The natural world is a more uncertain place than most ecologists once believed it to be. With their current understanding of the bewildering complexity of ecosystems, ecologists now realize that the size of a population at any one time is likely to be governed by interactions among several factors. Furthermore, the relative importance of a particular factor depends on the nature of the species and the ecosystem. The relative importance of a factor also often changes with time.

Interspecific competition is likely to be more important as a population regulator in particular circumstances—for example, at higher trophic levels in food webs rather than at lower levels, for species-rich ecosystems rather than for species-poor ecosystems, and in ecosystems that are subject to only modest physical disturbances. Physical factors are likely to have a greater influence in other circumstances—for example, for r-strategists rather than K-strategists, and in temperate environments, where weather is more variable, rather than in a tropical rain forest, where the weather is more stable.

Changes that occur from one year to the next can shift the relative importance of population control factors. Recall that wolves killed more moose on Isle Royale when the moose were weakened by several consecutive winters of heavy snow. When more favorable winters returned and the moose had access to adequate browse, the wolves could only kill calves and infirmed adults.

Future research will further clarify the relative influence of competition, predation, parasitism, catastrophic events, and weather extremes on population size and how the importance of these factors varies with species, types of habitat, and time.

Conclusions

The well-being of any population, including our own, depends on its physical environment and its interactions with other populations. In the long run, many density-independent and density-dependent factors influence the size of a population and tend to keep it more or less in balance with its habitat. Although fluctuations in size may be quite large, population growth rarely gets out of control for long, nor do many populations become extinct.

In recent decades, human intrusion has dramatically altered population-regulating mechanisms. We have greatly simplified many ecosystems, and throughout the world we have introduced new predators, parasites, and competitors. One consequence of these activities has been the extinction or near extinction of native species. Many scientists now view this global loss as a major threat to the well-being of our own species. We explore further the issues associated with loss of biodiversity in chapter 11.

As with the European rabbit in Australia, another consequence of our introduction of new species is that alien species may flourish to the point that they become pests. In other cases, we have removed natural control agents (e.g., predators or competitors) of native species and thereby triggered population explosions. Thus, populations that were once controlled by their natural enemies have become pests. We consider further the nature of pests and strategies for controlling their populations in chapter 9.

No other single species has had a greater impact on the planet than our own. As our numbers and our demands on resources continue to increase, our activities further threaten the survival of other species as well as the natural support systems upon which we depend. Many experts believe that control of our own population is the most important environmental problem facing us today. In the next two chapters, we apply the principles of population growth and regulation to our own species.

Key Terms

population	density-dependent factor
pest	density-independent factor
natality	mutualism
immigration	predation
mortality	biological control
emigration	inbreeding
exponential growth	parasitism
geometric growth	coevolution
linear growth	competition
arithmetic growth	intraspecific competition
sigmoid growth curve	interspecific competition
carrying capacity	territorial behavior
reproductive strategies	social hierarchies
K-strategists	competitive exclusion
r-strategists	self-thinning

Summary Statements

The size of a population at any given time is determined by the difference between additions to it (natality and immigration) and subtractions from it (mortality and emigration). Both natality and mortality vary with environmental conditions and the type of organism.

Populations exhibit exponential growth. In contrast to linear growth, exponential growth generates huge numbers in a very short period of time. Because of exponential growth, a population can reach a habitat's carrying capacity in a surprisingly short period of time.

The size of a population usually fluctuates about the habitat's carrying capacity for that species. The range of fluctuation depends on the habitat conditions and the nature of the species. Species that are *r*-strategists generally experience larger fluctuations than those that are *K*-strategists.

Both biological and physical factors play a role in regulating population size. Biological factors include predation, parasitism, and competition. Physical factors include fire and weather-related stresses, such as floods, droughts, hurricanes, and temperature extremes.

Predation, parasitism, and competition are density-dependent factors; their regulating influence increases with the density of the population. In contrast, fire and weather-related stresses are density-independent factors; their impact on population size does not vary greatly with the size of the population. Both density-dependent and density-independent factors usually help to determine the size of a population.

Predation is often important in regulating the population of small animals, such as insects. The impact of predators on populations of large animals appears to vary considerably with time and often depends on habitat conditions.

Establishment of a new parasite-host relationship often has a devastating impact on the host population, particularly if the host is a long-lived species. If the host reproduces relatively rapidly, natural selection may lead to a stable interaction through development of a less virulent parasite and a more immune host. Such an interaction is known as *coevolution*.

Territorial behavior and social hierarchies (pecking orders) are examples of intraspecific competition (competition among members of the same species) that may regulate animal population size. If interspecific competition (competition between species) is reduced, many similar species of animals can coexist in the same habitat. Factors that reduce interspecific competition include differences in timing of hunting, location of feeding activities, and types of food.

Self-thinning illustrates the impact of severe intraspecific competition among plants. Although competitive exclusion is a more common consequence of interspecific competition between plant species than it is of such competition between animal species, vertical stratification and differences in the timing of growth and reproduction reduce competition among plant species.

Short-term, weather-related stresses may temporarily reduce population size, but populations usually recover quickly as the weather again becomes favorable. *K*-strategists are usually less sensitive to variable environmental stresses than are *r*-strategists.

The relative importance of biological and physical factors in regulating a population's size depends, in part, on the nature of the habitat and the nature of that population as well as on the characteristics of other populations that reside in or move through an ecosystem.

Questions for Review

1. How do natality and mortality influence a population's growth rate?
2. Describe the differences between exponential growth and linear growth. Why must a wildlife manager or pest manager be concerned with these differences?
3. Why do populations commonly overshoot the carrying capacity of their habitat? What might be their impact on the habitat during the time of excess population?
4. Contrast the reproductive strategies of an *r*-strategist and a *K*-strategist.
5. Why are pest species more likely to be extreme *r*-strategists than extreme *K*-strategists?

6. Are humans *K*-strategists or *r*-strategists? Support your answer.
7. How do density-dependent factors and density-independent factors differ in the ways that they limit population size?
8. What is biological control? Describe its importance in pest management.
9. Explain how the importance of predation in regulating prey populations varies with the type of prey and habitat conditions.
10. Although predators may not always be the primary agents in controlling the size of prey populations, they do contribute to the stability of these populations. Explain.
11. What circumstances will normally favor coevolution of a parasite and its host? Under what conditions is coevolution less likely to occur?
12. Using examples, describe how parasitism is a density-dependent regulating mechanism.
13. Describe how territorial behavior and pecking orders function in population regulation. Identify animals in your area that demonstrate these types of social interaction.
14. What is competitive exclusion? Why is competitive exclusion more common between plant species than between animal species?
15. Compare and contrast the means whereby plants and animals reduce interspecific competition.
16. Why are short-term weather stresses relatively insignificant in the long-term regulation of population size?
17. Describe how physical and biological factors may interact to control the population size of a species. Cite examples from your area.
18. Many ecologists argue that we can no longer say that one particular population regulator is the dominant means by which population numbers are controlled. What is the basis for that assertion?

Projects

1. Visit a park or natural area in your region, and observe several species of plants and animals. Identify and describe some of the adaptations that reduce competition for resources among these species.

2. Check with the local office of your state Department of Natural Resources, Department of Conservation, or Department of Agriculture to learn whether it is involved in projects to control pests or excess populations of wildlife. If it is, determine which population control measures are being used. How successful are those measures in reducing overpopulation?

Selected Readings

Bergerud, A. T. "Prey Switching in a Simple Ecosystem." *Scientific American* 249 (Dec. 1983):130-41. A study of a prey-species population crash and the subsequent impact on its major predator and other prey populations.

Bergon, M. E., J. L. Harper, and C. R. Townsend. *Ecology: Individuals, Populations, and Communities.* 2d ed. Cambridge, Mass.: Blackwell Scientific Publications, 1990. An up-to-date and thorough treatment of ecological principles with abundant examples.

Cochran, M. F. "Chestnuts—Making a Comeback?" *National Geographic* 177 (Feb. 1990):128-40. A chronicle of the efforts to save the survivors of the blight and to breed resistant strains.

Gilbert, L. E. "The Coevolution of a Butterfly and a Vine." *Scientific American* 247 (Aug. 1982):110-21. An interesting account of the coevolution of a parasite and its host.

Moehlman, P. D. "Social Organization in Jackals." *American Scientist* 75 (1987):366-75. An examination of how a complex social system permits the successful rearing of dependent young.

Morse, D. H. "Milkweeds and Their Visitors." *Scientific American* 253 (July 1985):112-19. A look at the numerous interactions of the nectar feeders, herbivores, predators, and parasites that gather on milkweed plants.

O'Brien, S. J., M. E. Roelke, L. Marker, A. Newman, C. A. Winkler, D. Meltzer, L. Colly, J. F. Evermann, M. Bush, and D. E. Wildt. "Genetic Basis for Species Vulnerability in the Cheetah." *Science* 227 (1985):1428-34. An investigation into the cost of genetic uniformity in a small population.

Rennie, J. "Living Together." *Scientific American* 266 (Jan. 1992):104-13. A fascinating look at the latest research into parasite-host relationships.

Smith, R. L. *Ecology and Field Biology.* 4th ed. New York: Harper & Row, 1989. A well-written, informative examination of ecological principles.

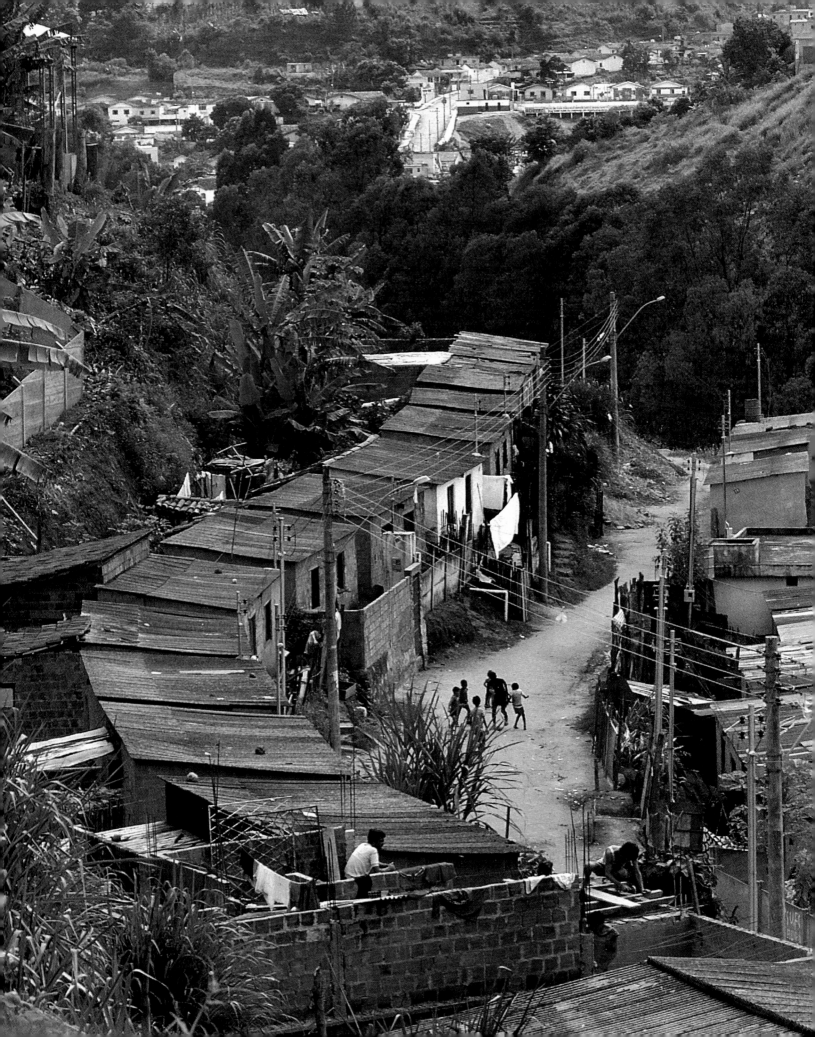

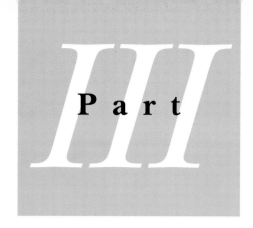

Part III

Human Population

*I*n this section, we consider a fundamental problem facing humanity: our rapidly growing numbers. Probably no other species has a greater impact on the planet than our own. As our numbers and demands on resources rise, our activities increasingly threaten the survival of other species as well as the natural support systems upon which all life depends. Consequently, many experts argue that controlling human population growth must be humanity's top environmental priority. In chapter 6, we examine some of the reasons for the recent population explosion and explore the marked contrasts in population growth among nations. In chapter 7, we consider the motivational factors that influence family size and then evaluate fertility-control methods. We conclude by projecting future human population growth.

In most less-developed nations, a large fraction of the population is mired in poverty and must live in shantytowns such as this one on the outskirts of Juiz de Fora, Brazil.
© Steve Maines/Stock Boston

Chapter 6

Human Population Dynamics

In the United States, the system of highways cannot keep up with increasing affluence and the demands of a growing number of people. Rush hour traffic crawls on this Los Angeles freeway.

© Tom McHugh/Photo Researchers, Inc.

*I*ncreasingly people are beginning to understand that the resources of the planet are finite and limited. Environmental economist Herman Daly (of the World Bank) has observed that throughout most of human history, natural resources were abundant, while human-made capital (e.g., sawmills and fish nets) was the limiting factor. Today the balance has shifted dramatically. Fishing boats and timber-harvesting equipment often lie idle because of insufficient natural resources to sustain their use. What then is the capacity of Earth to sustain the human population?

This question is most difficult to answer. Unlike other species, human populations vary greatly in their per capita demand for resources. By world standards, the average American has extravagant demands. We have (1) a diet that regularly includes fresh vegetables and fruits and high-quality meats; (2) more clothing than needed to keep warm and dry; (3) spacious, climate-controlled housing; (4) excellent medical care; (5) transportation that often includes more than one auto; and (6) a variety of home entertainment options. In stark contrast, an average Ethiopian is extremely impoverished and has no access to any of these resources. An average American obviously requires many times more resources to support his or her life-style than an average Ethiopian (fig. 6.1). In fact, although Americans make up less than 5% of the world's population, we consume more than 30% of the world's resources.

If everyone adopted an Ethiopian standard of living—that is, if all humans consumed resources at the bare subsistence level—then the planet could support several times today's 5.5 billion people. On the other hand, most experts believe that the planet could not support even the present world population at the level of consumption of the average American. It is important to realize, then, that the human carrying capacity of the planet hinges on the quality of life selected.

(a)

(b)

Figure 6.1 Considerably more resources are required to support our standard of living than are required by most citizens of less-developed nations. The contrast between (*a*) a suburban home in the United States and (*b*) a typical stone dwelling in Ethiopia is striking.

(*a*) © J. L. Shaffer. (*b*) © Kazuyoshi Nomachi/Pacific Press/Photo Researchers, Inc.

Human Population Dynamics   111

(a)

(b)

Figure 6.2 As the population of the desert Southwest soars, competition for scarce fresh water also increases. Is it wise to use such an essential resource to irrigate (*a*) cropland or (*b*) golf greens?
(*a*) © William E. Fergusen. (*b*) © M. Long/Visuals Unlimited

Even in the United States, the pressures of population growth are forcing us to make difficult choices. Should land near a large urban area be used for a park, a housing tract, or a landfill? Should a limited water supply be used for commercial and industrial development, or for irrigation of cropland (fig. 6.2)? Should forest lands be managed for timber production or to protect endangered species (fig. 6.3)? Television and newspapers frequently describe conflicts over limited resources.

We are also influenced by population growth in other nations. For example, the United States annually receives twice as many immigrants as all other nations combined. During the 1980s, about 700,000 people legally immigrated to the United States in the hope of achieving a higher standard of living. Furthermore, at least another 200,000 to 400,000 enter the country illegally each year in a desperate search for jobs. Experts predict that these numbers will rise in the 1990s because of continuing worldwide population and economic pressures. Population pressures sometimes prompt nations to invade neighboring nations to gain access to essential resources. Such unrest often draws in other nations that want to protect their economic and political interests in the region. Such conflicts always lead to death, destruction, social disruption, and severe economic hardships.

More people means more environmental disruption, because they require more resources and produce more waste products. Environmental threats such as acid rain, thinning of the ozone shield, and potential global warming suggest that the human population is now disrupting the environment on a global scale. Contamination and disruption of the environment by any nation eventually influences the quality of the environment of other nations.

Population growth, however, is not the only factor that contributes to our environmental problems. Rising affluence, inappropriate technologies, and mismanagement of resources also contribute directly or indirectly to our difficulties. But unless world population growth is slowed significantly, few nations will be able to meet the needs of future generations, much less provide adequate food, water, shelter, health care, and education for those who are already here.

In this chapter, we examine some of the causes of the recent human population explosion and explore more fully the contrasts in population growth among nations. In the next chapter, we evaluate strategies for slowing the growth of the human population.

Figure 6.3 As the human population grows, habitat for rare species such as the northern spotted owl are increasingly lost to human demands for raw materials such as timber.

© Wm. Grenfell/Visuals Unlimited

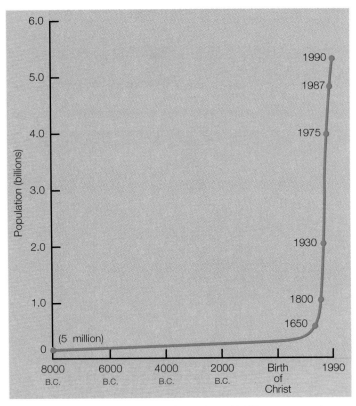

Figure 6.4 Growth of human population from 10,000 years ago to the present.

Historical Overview

As figure 6.4 indicates, we are living in a unique period. Through much of human history, the global human population remained quite small and, like that of other *K*-strategists, typically fluctuated only slightly. By the seventeenth century, however, the human population entered a period of explosive growth which continues today. Given our understanding of the nature of population growth (chapter 5), we would expect that some time in the future, the human population will reach the planet's carrying capacity. In this section, we examine the historical events that have thus far shaped this pattern of population growth.

Early on, sustained growth was impossible because our ancestors were particularly vulnerable to a hostile environment. Food was often scarce, so famine was common; protection against harsh climates was limited because clothing and shelter were primitive; and frequent outbreaks of disease such as the bubonic plague and smallpox claimed many lives. Population growth rates were very low because high birth rates were matched by high death rates.

Then, about ten thousand years ago, humankind began a gradual transition from hunting and gathering to subsistence agriculture. As people learned to cultivate crops and domesticate animals, they gradually abandoned their traditional customs of gathering seeds, fruits, and roots and hunting indigenous species for food. Consequently, the food supply became more reliable, and human populations began to grow at a somewhat faster rate. Nonetheless, crops remained vulnerable to the vagaries of weather and to attacks by pests such as insects and rodents. Thus, famines continued to claim many lives. In addition, disease continued to take a heavy toll, as sanitation and other living conditions remained primitive.

Historically, epidemics of disease have severely limited the growth of human populations. During the fourteenth and fifteenth centuries, the bubonic plague (a disease caused by a bacterium carried by fleas that feed on rats) killed hundreds of thousands of people in Europe and Asia, reducing the population of some countries by half. People were terror-stricken because no one knew the causes of the epidemics, and thus little could be done to prevent calamity.

By the early 1800s, however, medical science began to control many of these diseases by means that we now take for granted. One of the first great advances occurred in 1796, when Edward Jenner demonstrated that smallpox could be prevented by inoculating human beings with pus that was extracted from cowpox lesions. Today, smallpox, which once killed hundreds of thousands of people each year, has been virtually eradicated.

Medical advances gradually dispelled ancient beliefs that demons and noxious vapors caused illness. Once people accepted the fact that germs cause many diseases, measures to protect public health (such as chlorination of drinking water and protection of food from contamination) became more common. Also, antiseptic techniques were developed and eventually became commonplace in doctors' offices and hospitals. (Surprising as it may seem today, as recently as 150 years ago, enlightened doctors who advised their colleagues to wash their hands between examinations of patients were often ridiculed and sometimes even banished from their profession.) Personal hygiene was also promoted; people bathed more often, and washable clothing became popular.

Today, medical advances continue to reduce the number of premature deaths and to lengthen human life spans. In the United States, for example, heart disease and strokes are major causes of death. Yet, in the past decade a combination of better medical treatment and improved life-style (including regular exercise, no smoking, and a better diet) has reduced the number of deaths due to heart disease by nearly 30%. Medical advances have also increased dramatically the survival rate following the treatment of certain types of cancer, particularly if detected early. Thus, we can be confident that future medical advances and improvements in life-style will further reduce the number of premature deaths and lengthen life spans.

Another factor that contributes to lowered death rates is the continued increase in the food supply. Major advances in agricultural technology in recent decades have more than doubled the per hectare crop yield in many parts of the globe. Better storage and transportation have reduced spoilage and ensured that more food is moved from farm to market to home. Furthermore, advances in worldwide communication have softened the blow of famines by providing access to emergency food supplies from other nations.

In summary, advances in medicine, sanitation, food production, transportation, and communication have led to major reductions in human death rates worldwide. However, birth rates remain quite high in many nations; thus, the human population is growing at the rate of some ninety million per year.

The explosiveness of the recent era of human population growth is illustrated by considering some major mileposts. It took several hundred thousand years (approximately 99% of human history) for the human population to reach one billion, which occurred sometime around 1800 (see fig. 6.4). By 1800, the population was growing so fast that it doubled to two billion people in only 130 years, and then in only 45 years, it doubled again. Hence, we are now living in a period of unprecedented human population growth.

Major Factors Affecting Human Population Growth

In chapter 5, we described some principles of population growth and regulation. These same principles apply to human populations because humans are biologically similar to other species. Hence, we know that the population growth rate of a country depends on the net difference between additions (inputs) and subtractions (outputs). Recall that inputs are births and immigration, while outputs are deaths and emigration. Although migration among nations may reduce population pressures temporarily, the planet's isolation forces us to consider the numerous human populations as one global human population.

We now examine how birth rates, death rates, and several other critical factors influence human population growth.

Birth Rates, Death Rates, and Exponential Growth

Crude birth rates and crude death rates are major determinants of changes in human population size. The **crude birth rate** is the annual number of live births per 1,000 people in a population. The **crude death rate** is the annual number of deaths per 1,000 people in a population. These rates are used to determine the **rate of natural population change,** which is the rate at which a population is increasing or decreasing in a given year due to a surplus (or deficit) of births over deaths.

The rate of natural population change is given by the following formula:

Rate of natural population change =
(crude birth rate – crude death rate) × 100%

In the United States in 1991, the crude birth rate was 17 per 1,000 (0.017), and the crude death rate was 9 per 1,000 persons (0.009). Thus, in the United States,

Rate of natural population change =
(0.017 – 0.009) × 100% = 0.8%

The significance of the rate of natural population change is easier to visualize when we consider its corresponding **doubling time,** that is, the length of time required for a

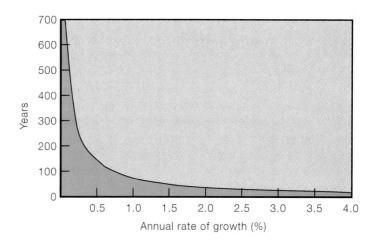

Figure 6.5 Growth rates and their doubling times in years.

Table 6.1	Doubling times associated with rates of natural population change in selected nations	
Nations	**Rate of Natural Population Change**	**Number of Years for Population to Double**
Romania, Poland	0.5%	140
Sri Lanka	1.0%	70
Singapore, Thailand	1.5%	47
Fiji, Columbia	2.0%	35
Chad, Morocco, Malaysia	2.5%	28
Mali, Congo, Pakistan, Guatemala	3.0%	24
Uganda, Yemen, Solomon Islands	3.5%	20
No examples	4.0%	18

Source: Data from the *1991 World Population Data Sheet* (Washington, D.C.: Population Reference Bureau, Inc., 1990).

population to double at a fixed rate of growth (fig. 6.5). Doubling time is given by the following formula:

Doubling time in years = 70/ X,
where X = rate of natural population change

For example, if the population of the United States continues to grow at the current rate of 0.8%, its population will double in approximately ninety years. In contrast, if the population of Nicaragua continues to grow at its current rate of 3.4%, it will double in just twenty-one years. That is, within the next twenty-one years, Nicaragua will have to double its food supply, housing units, schools, hospitals, and job opportunities just to maintain its current standard of living. Table 6.1 lists the doubling times associated with the rates of natural population change for a number of nations.

A rate of natural population change greater than zero indicates exponential growth. Recall from chapter 5 that exponential growth occurs when a factor increases by a constant percentage of the whole during a fixed period of time. Also recall that exponential growth has two very important characteristics. First, it can generate enormous numbers in a short period. For example, the world's population is nearly 5.5 billion and is doubling (experiencing a 100% increase) about every forty years. Assuming that this doubling time persists, the world's population in 2030 (about the time when today's college students will be retiring) will be nearly 11 billion people. World population would double again to 22 billion by 2070. That is, shortly before the third centennial of the United States, the world would need to support over four times as many people as it does today. Maintaining the same growth rate for another forty years would mean that by 2110, the world's population would stand at 44 billion people, eight times greater than it is today.

We also need to realize that seemingly small differences in exponential growth rates can produce surprisingly great differences in population size within just a few generations. South Korea, for example, has nearly a 1% rate of natural population change, and Syria's rate is over 4%. Today, South Korea has a population of nearly 43 million people, which is essentially 3.5 times that of Syria. But if each nation were to continue to grow at its current rate, their populations would be equal in about thirty-six years. Furthermore, within one hundred years, Syria's population would be nearly five times larger than that of South Korea. Thus, even small declines in exponential growth rates contribute significantly to efforts to keep the human population from exceeding the planet's carrying capacity.

Exponential growth also means that populations can reach a fixed upper limit (the carrying capacity) within a surprisingly short period of time. Recall from chapter 5 that exponential growth often causes animal and plant populations to overshoot the carrying capacity of their habitats. Human populations also overshoot their nation's carrying capacity for certain resources. For example, the recent history of production and consumption of petroleum in the United States shows how growth in consumption of a resource can outstrip growth in its production.

During the 1950s and 1960s, increased production of petroleum in the United States generally kept pace with increased consumption associated with a growing population and economy. Consequently, imports of crude oil increased by only about 25%, from 372 million barrels in 1960 to 483 million barrels in 1970. But in the early 1970s, the increase in domestic petroleum production slowed, while consumption continued to grow at a relatively high rate. To make up the deficit, more petroleum was imported from other nations. Between 1970 and 1977, the amount of imported crude oil soared by 500%! The most troublesome impact was economic. Demand caused the price of petroleum imports to soar from $2.7 billion in 1970 to $44.5 billion in 1977, a sixteenfold increase. Although conservation measures and an economic recession temporarily reduced our petroleum imports in the

1980s, our reliance on imports has increased once again, while domestic production remains essentially at late 1960s levels. To remain above the carrying capacity for petroleum, the United States must continue to pay an economic price, reflected in a strongly negative balance of trade, and a political price, indicated by a military presence in the Middle East. In chapters 17 and 18 we revisit this problem in the context of today's energy resources and future energy alternatives.

Total Fertility Rate

A useful measure for estimating future population growth is the **total fertility rate (TFR),** that is, the average number of children that a woman bears in her lifetime. Currently the TFR in the United States is 2.1, equal to the replacement level fertility. **Replacement level fertility** is the number of children that a couple must have to replace themselves. Because some children die before they reach reproductive age and because some individuals are physically unable or choose not to have children, the average family size in the United States must be 2.1 children, rather than 2.0, to ensure replacement. In nations that experience higher infant mortality and shorter life expectancies, the replacement level fertility may be as high as 2.7 children per family.

As figure 6.6 illustrates, the TFR can fluctuate considerably with time. Several factors influence the total fertility rate, including average age of marriage, infant mortality rates, educational and employment opportunities for women, and the availability of reliable methods of contraception. The role of these factors in determining population size is described further in chapter 7.

Total fertility rates vary widely from one region to another, as shown in table 6.2. Europe and North America are currently below replacement level fertility; parents in these regions currently are not having enough children to replace themselves. Other regions are above replacement level

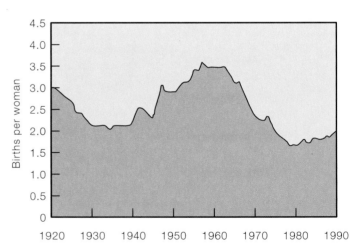

Figure 6.6 Variation in total fertility rates of American women over the past 70 years.
Source: Data from the U.S. Census Bureau and the Population Reference Bureau.

fertility. Extreme values occur throughout most of Africa, where couples are having two to three times as many children as needed to replace themselves. If such high total fertility rates persist, these nations face continued high rates of population growth.

Age Structure

To predict future population growth, we also need to know the **age structure** of a population, that is, the percentage of people at each age level in a population. Understanding a

Table 6.2	Total fertility rates for various geographical regions, 1991
Region	**TFR**
Africa	6.1
Latin America	3.5
Asia	3.3
Oceania	2.6
North America	2.0
Europe	1.7

Source: Data from the *1991 World Population Data Sheet*, Population Reference Bureau, Inc., Washington, D.C.

population's age structure is important because not all individuals are equally likely to bear children or to die. Only those women who have reached reproductive age can give birth (the reproductive age of most women is between 15 and 44 years). Also, people in the reproductive age group are less likely to die than those who are very young or very old. Hence, a country's female population can be subdivided into three major age groups: prereproductive (0-14 years old), reproductive (15-44 years old), and postreproductive (45 years old and older). By classifying the citizens of a country into these three categories, we generate an age-structure diagram for that nation. Then we refine the diagram by subdividing it into age groups, or *cohorts*, of 5-year intervals (0-4 years, 5-9 years, and so on). Furthermore, we differentiate between males and females by placing them on opposite sides of the diagram. Age-structure diagrams for Kenya, the United States, and Denmark appear in figure 6.7.

Age-structure diagrams help us to assess the current growth status of a population and predict what might happen in the future. For example, an age-structure diagram that has the shape of a broad-based pyramid profiles a population with a large percentage of people who are quite young and indicates a recent history of high birth rates. The population of Kenya forms such a broad-based pyramid (see fig. 6.7). A broad base also indicates that a large number of people will

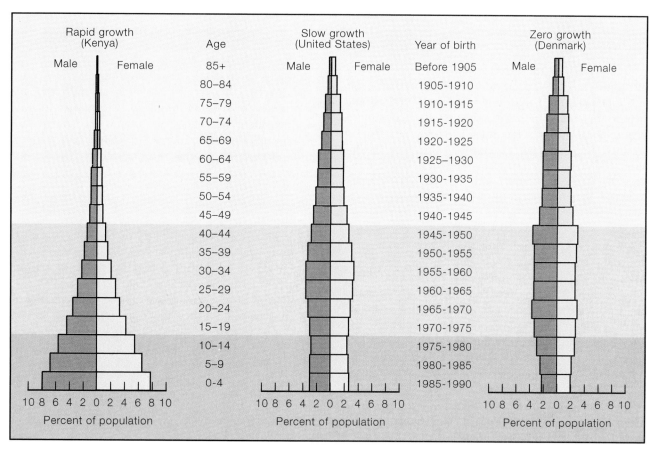

Figure 6.7 Population age-structure diagrams for nations with fast (Kenya), slow (United States), and essentially zero (Denmark) population growth rates.

Source: Data from the U.S. Census Bureau, the United Nations, and Population Reference Bureau.

soon enter the reproductive age group. Consequently, a significant increase in the total number of births appears to be inevitable in the future. In contrast, the upper portion of the pyramid indicates that a small percentage of the population is at or approaching old age. Hence, we expect the total number of deaths over the short term to remain relatively small. When the total number of births is increasing and the total number of deaths remains low, a population will grow rapidly.

The relatively large percentage of young people provides a built-in momentum for explosive population growth. Recall from chapter 5 that a growing population tends to continue to grow; and that momentum tends to make it climb above the carrying capacity.

A contrasting age-structure diagram is illustrated for Denmark (see fig. 6.7). Because birth rates in Denmark have been low and relatively stable for a long time, its age-structure diagram looks more like a rectangle than a pyramid. The prereproductive age group is approximately the same size as the reproductive age group. Thus, we can expect little change in the total number of births as the prereproductive group reaches reproductive age (assuming that the preferred family size remains the same for this generation). Furthermore, the reproductive group is approximately the same size as the postreproductive group. Thus, we can expect the total number of deaths to change little in the future. The age-structure diagram of Denmark therefore describes a population that is likely to grow little, if at all, in the years to come and may eventually decline.

The age-structure diagram for the United States illustrates a situation that is intermediate between those of Denmark and Kenya (see fig. 6.7). The base of the age-structure diagram is much narrower than that of Kenya, but we should not conclude that our population will soon stop growing. A closer examination of the diagram shows a definite bulge in the reproductive age group. This cohort is the "baby boom" generation that was born during the two decades following World War II. In contrast, the number of persons who are older than sixty is smaller than the number of people in the reproductive age group.

Even if the total fertility rate remains at 2.1, the population will continue to grow for another fifty years because so many new families are forming. The total number of births will continue to be greater than the total number of deaths until about 2040. At that time, all age groups should be about the same size, and the U.S. age-structure diagram will resemble that of Denmark. If migration is disregarded, the size of the population will be unchanging or stationary. The U.S. Census Bureau projects that by 2040, the population in the United States will be 302 million people—nearly 50 million more than today.

Zero population growth (ZPG) means that a population ceases to grow; that is, if we disregard migration, the birth rate equals the death rate. The projection that the United States will reach zero population growth by the year 2040 is based on the assumption that during the coming decades, the total fertility rate will remain essentially unchanged. However, we are aware that the TFR for women in the United States has varied in the past (see fig. 6.6). Thus, if the TFR increases again at any time during the next 50 years, the United States will not achieve ZPG until well beyond 2040.

The age profile of the United States implies that once a population starts to grow, it will continue to grow for some time. This behavior of populations is termed **population momentum.** Once a human population begins to grow, it usually includes a relatively large percentage of people in the prereproductive age group. As these people mature and have families, they force the population to grow for some time, even if each couple (on average) only replaces itself. Thus, the U.S. baby boom (which peaked in the late 1950s) is providing the momentum that will mean continued population growth until at least 2040.

Population momentum is a crucial factor for many nations in South and Central America, Africa, and Asia that are in the midst of unprecedented rates of natural population increase. In these countries, nearly 40% of the people are younger than 15 years old and will soon be raising families. Hence, even if we assume that replacement level fertility can be achieved immediately and then maintained (both highly unlikely events), it will take more than a century for these nations to achieve zero population growth, unless the death rate climbs.

Variations Among Nations

We are now aware that the pattern of population growth shown in figure 6.4 is an oversimplification. In reality, population growth rates vary considerably among nations, as figure 6.8 shows. With some important exceptions, most countries are classified into one of two groups according to their rates of natural population increase. The first group comprises the **more-developed nations,** which include the nations of Europe and North America, the former Soviet Union, Japan, Australia, and New Zealand. The second group comprises the **less-developed nations,*** which include most of the nations of Central and South America, Africa, and Asia.

The more-developed countries usually have a low rate of natural population increase—less than 1% per year. The populations of several European nations, including Denmark, Austria, Germany, Greece, and Italy, are essentially not growing at all, while the population of Hungary is actually declining. In contrast, the less-developed countries have an average annual rate of natural population increase that is higher than 2%, which means that their populations double in thirty-five years or less (see table 6.1). Many African nations (e.g., Kenya, Togo, Zambia, and Tanzania) and Middle East nations (e.g., Syria and Jordan) have growth rates that exceed 3.5%; thus, their populations double in about twenty years.

---

*Refer to page 24 for the distinction between more-developed and less-developed nations.

Annual population growth rate and approximate doubling time	
4.0%–18 years	Syria (3.8%), Kenya (3.8%) Tanzania (3.7%)
3.5%–20 years	Yemen (3.5%), Saudi Arabia (3.4%) Iran (3.3%), Ghana (3.2%), Libya (3.1%)
3.0%–23 years	Africa (average 3.0%), Kuwait (3.0%), Egypt (2.9%), El Salvador (2.8%), Bolivia (2.6%)
2.5%–28 years	Ecuador (2.4%), Mexico(2.3%) Lebanon (2.1%), Less-developed nations (average 2.1%) Latin America (average 2.1%)
2.0%–35 years	India (2.0%), Brazil (1.9%) Chile (1.8%)
1.7%–40 years	World average (1.7%)
1.5%–47 years	Sri Lanka) (1.5%), China (1.4%) Singapore (1.3%), Taiwan (1.1%)
1.0%–70 years	South Korea (0.9%), Canada (0.7%), Ireland (0.6%)
0.5%–140 years	More-developed nations (average) (0.5%), France (0.4%) Sweden (0.3%), Europe (average 0.2%), Italy (0.1%)
0.1%–170 years 0% −0.5%	Denmark (0.0%), Hungary (−0.2%)

Figure 6.8 Examples of growth rates among nations.

Source: Data from *World Population Data Sheet 1991*, Population Reference Bureau.

What accounts for these dramatic differences in the rates of population growth among nations? As we saw earlier, through much of human history, death rates and birth rates were high and nearly equal. Hence, human populations grew slowly. However, death rates began to fall in western Europe in the seventeenth and eighteenth centuries as health care, sanitation, and food production improved. Since high birth rates continued, rates of natural population increase climbed. Within a few decades, however, birth rates also began to decline. Several factors, including industrialization, urbanization, improved per capita income, and the improved socioeconomic status of women contributed to a lowering of birth rates.

Today, both birth rates and death rates are low and nearly equal again in Europe. The transition from historically high birth and death rates to current low birth and death rates is described as a **demographic transition** (fig. 6.9). A similar transition also occurred in the other more-developed countries. Consequently, the more-developed nations are now growing at an average rate of 0.5% per year—a doubling time of approximately 128 years. Although this growth rate may appear to be quite low, it is actually high in historical terms. For most of human history, the doubling time of the global population has been on the order of 30,000 years.

While a demographic transition was occurring in Europe and North America, most of the rest of the world was still experiencing both high birth and death rates. Improvements in health care, sanitation, and food production did not occur in many of the less-developed countries until after World War II. Then their death rates began a steep decline, but birth rates remained high—as they still are—in most less-developed nations, so the population in these nations continues to swell. The less-developed nations are now growing at an average rate of 2.1% per year—a doubling time of only thirty-three years.

In addition to major differences in population growth rates between more-developed and less-developed nations,

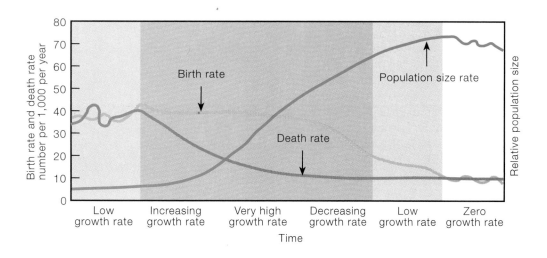

Figure 6.9 Generalized model of the demographic transition.

Table 6.3 Comparisons of more-developed and less-developed nations in 1991

Features	More-Developed Nations	Less-Developed Nations
crude birth rate	14/1,000 population	30/1,000 population
crude death rate	9/1,000 population	9/1,000 population
doubling time	137 years	33 years
total fertility rate	1.9	3.9
infant mortality	14/1,000 births	75/1,000 births
life expectancy at birth	74 years	62 years
wealth (per capita GNP in 1988 U.S. dollars)	$16,990	$750

Source: From *1991 World Population Data Sheet,* (Washington, D.C.: Population Reference Bureau, Inc., 1991).

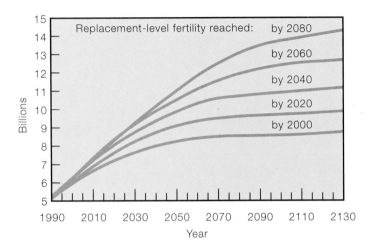

Figure 6.10 Ultimate world population size depends upon when replacement-level fertility is reached.

several other striking contrasts are shown in table 6.3. Experts disagree over whether some or all of these contrasts result from differences in population growth or are merely coincidental. These disputes may never be resolved, but everyone agrees that the quality of life differs markedly between the more-developed and less-developed nations.

During recent decades, several less-developed countries (e.g., Singapore, Taiwan, South Korea, Cuba, Costa Rica, Thailand, and Indonesia) have impressively reduced their fertility. During the last decade, the rate of population growth has declined faster in the People's Republic of China (which accounts for 25% of the world population) than in any other large less-developed nation. But in many nations, particularly in Africa and western Asia, fertility rates have declined little from their historically high levels. In chapter 7, we explore the reasons why fertility rates are declining rapidly in some nations while they remain near historical highs in other nations.

What, then, are the projections for human population growth? Disregarding the question of carrying capacity for the moment, we are aware that the lowering of the total fertility rate to replacement-level fertility will ultimately lead to zero population growth. Thus, the time required to arrive at replacement level fertility determines to a large extent whether the world's ultimate population reaches nine billion, fifteen billion, or some other number (fig. 6.10).

Conclusions

We are living in an unprecedented time in human history. The earth has never witnessed a period of such explosive growth in human population. While the more-developed nations are experiencing fertility rates that are near their replacement level fertility, many less-developed nations are experiencing fertility rates that are two to three times greater than their replacement level fertility. Given the limitations on the earth's ability to sustain the human population, many population experts are concerned that fertility rates in many less-developed nations will not decline far enough soon enough to prevent grave economic and ecological consequences. In the next chapter, we consider strategies to facilitate reductions in fertility.

Key Terms

crude birth rate	age structure
crude death rate	zero population growth (ZPG)
rate of natural population change	population momentum
doubling time	more-developed nations
total fertility rate (TFR)	less-developed nations
replacement level fertility	demographic transition

Summary Statements

An average American requires many times more resources to support his or her lifestyle than the average citizen of a less-developed nation. The carrying capacity for the world human population depends upon the quality of life that is selected. Population pressures occur both in the United States and abroad. The more affluent the population, the greater is its impact on the global environment.

Because our ancestors were vulnerable to a hostile environment, the human population for thousands of years remained quite small and grew very slowly. Development of agriculture, advances in medicine, and improved sanitation have contributed to the recent period of unprecedented population growth.

A population's natural rate of change is a function of its crude birth rate, crude death rate, total fertility rate, and age structure. A population will probably continue to grow if (1) its crude birth rate remains higher than its crude death rate, (2) its total fertility rate remains higher than its replacement level fertility, and (3) a large percentage of its population is in the prereproductive age group.

Because of population momentum, even a population that has reached its replacement level fertility will continue to grow if its reproductive and prereproductive age groups are large. A nation with zero population growth will have nearly equal numbers in the prereproductive, reproductive, and postreproductive age groups.

The more-developed nations have gone through a demographic transition involving a decline in death rates followed by a decline in birth rates. Because death rates and birth rates are again nearly equal, their populations are growing slowly. Most less-developed nations have experienced a recent decline in death rates, but birth rates remain high. Hence these nations are growing rapidly.

Questions for Review

1. How does the quality of life differ between the more-developed and less-developed nations? How is the quality of life of a nation's citizens related to that nation's carrying capacity?
2. How do high population growth rates in other nations influence the well-being of American citizens? How does the high level of U.S. affluence affect citizens of less-developed nations?
3. List five factors that have contributed to the growth of the human population during the last ten thousand years.
4. Define crude birth rate and crude death rate. How are these two rates used to calculate the rate of natural population change?
5. Compare doubling times among several more-developed and less-developed nations. Describe the importance of doubling time for determining the future resource needs of a country.
6. What is the difference between the total fertility rate and replacement level fertility?
7. Describe how the total fertility rate for the United States has varied over the past seven decades. What factors may have contributed to these rate shifts?
8. Compare the shape of the age-structure diagram for the United States with that of Kenya. What do the differences in shape tell us about past differences in population growth in these two nations? What do the differences tell us about future population growth in each country?
9. What is population momentum? Why is it necessary to understand the effects of population momentum before one can realistically predict when a country will achieve zero population growth?
10. Is there a relationship between population momentum and the fact that populations sometimes overshoot the carrying capacity?
11. Describe the process known as the *demographic transition*.
12. Why didn't the demographic transition begin at the same time in the less-developed nations and the more-developed nations?

Projects

1. Prepare an age-structure diagram for your community. Interpret the diagram relative to past and future population growth.
2. Talk with family members to determine the life spans of the past five or six generations of your family. Has the age at death changed appreciably in recent generations? What would account for these changes? How have the causes of death changed?

Selected Readings

Ehrlich, P. R., and A. H. Ehrlich. "Population, Plenty and Poverty." *National Geographic* 174 (Dec. 1988):914-45. A well-illustrated and informative consideration of how growing populations and increasingly affluent life-styles are straining the planet's resources.

Goliber, T. J. "Africa's Expanding Population: Old Problems, New Policies." *Population Bulletin* 44 (Nov. 1989). Washington D. C.: Population Reference Bureau. An examination of the history of population growth in Sub-Saharan Africa, with a consideration of policies to slow population growth.

Keyfitz, N. "The Growing Human Population." *Scientific American* 261 (Sept. 1989):118-26. An overview of the many dimensions of world population growth.

McEvedy, C. "The Bubonic Plague." *Scientific American* 258 (Feb. 1988):118-23. An account of the plague: how it kills, how it spreads, its toll on human populations, and its control today.

McFalls, J. A. "Population: A Lively Introduction." *Population Bulletin* 46 (Oct. 1991). Washington, D.C.: Population Reference Bureau. Reviews the dynamic forces that cause human populations to grow or decline and prospects for future growth.

Merrick, T. W. "World Population in Transition." *Population Bulletin* 41 (April 1986). Washington, D.C.: Population Reference Bureau. A status report on world population growth, its history, its regional trends, and the debate over its effects.

Population Reference Bureau. *Annual Population Data Sheet.* Washington, D. C.: Population Reference Bureau. A concise summary of population data, compiled for more than 125 nations, that is updated yearly.

Van de Kaa, D. "Europe's Second Demographic Transition." *Population Bulletin* 42 (March 1987). Washington D. C.: Population Reference Bureau. An examination of the factors behind the continuing population decline in Europe.

Van de Walle, E., and J. Knodel. "Europe's Fertility Transition: New Evidence and Lessons for Today's Developing World." *Population Bulletin* 34 (Feb. 1980). Washington, D.C.: Population Reference Bureau. A reexamination of the causes of the demographic transition in Europe and the implications for slowing population growth in less-developed nations.

Chapter 7

Controlling Human Population Growth

A congested New York City street underscores
some of the implications of a soaring human
population.
© Steve Elmore/Tom Stack & Associates

With an estimated 1.1 billion citizens, the People's Republic of China accounts for over 20% of the world's population. In recent decades, the government of China has come to view population growth as a major obstacle to improving its citizens' standard of living. Given China's population growth rate, relative scarcity of arable land, crowded urban conditions, and high unemployment, the Chinese government decided that population control was an imperative and embarked on a vigorous campaign to limit family size. During the 1970s, the Third Birth Planning Campaign encouraged fertility reduction with the slogan "later, longer, and fewer," advocating later marriage, a longer interval between births, and fewer children.

The year 1979 saw the emergence of a much more stringent plan termed the *one-child campaign* (fig. 7.1). Couples pledging to have only one child could receive free hospital delivery of their child, free medical care and education for the child, preferential treatment in housing, grain allotments, as well as monetary bonuses. Government campaigns during the early 1980s supported the one-child policy by encouraging abortion, sterilization, and the use of IUDs (interuterine devices). Birth-control devices and surgical sterilizations were provided free to married couples. As a consequence of these initiatives, the total fertility rate declined dramatically from about 6 children per woman in the late 1960s to 2.1 children per woman by 1984.

As one might expect, however, controversy surrounded the one-child policy from the beginning. In some provinces, infanticide was practiced when the firstborn was a girl. In many other instances, a birth was simply not reported to the authorities if it was a girl. Among China's twenty-nine provinces, acceptance of the one-child policy ranged from only 10% in impoverished, remote, rural provinces to 75% in large urban areas and better-off coastal provinces.

In the mid-1980s, some government officials realized that while most couples did not want five or six children, neither did they want to have just one child, particularly if it was a female. Many Chinese couples are poor and live under harsh conditions. Labor needs (sons to cultivate the land) and the need for old-age security (sons to take care of them) are high priorities. For these people, the lack of sons is perceived as a greater personal threat than the government-imposed fines for having a second child without official approval. Many are so poor that they could never pay the fine anyway.

In response to widespread dissension and resistance, the government gradually eased its policies. A new policy of "opening small holes," created more than a dozen categories of exemptions to allow a couple to have a second child. Even if a couple did not fall into an exemption category, nearly 50% of rural couples were officially allowed to have another child

Figure 7.1   A billboard on a main street in Beijing, People's Republic of China, encourages a one-child family.
© Owen Franken/Sygma

after an appropriate interval should the first child be a girl. In addition, the government became increasingly reluctant to pursue confrontational policies.

These new policies, however, changed again in 1989 as a result of the Tiananmen Square uprising. Some experts suggest that continuing discord over population policies contributed to that confrontation between China's citizens (mainly students) and the central government. Afterwards, political hard-liners regained control of the central government and began to invoke again the one-child policy and its stringent provisions. Because most of the hard-liners are in their 80s, many experts believe that China's population policies will soon be eased again as more moderate leaders take positions of power.

China's experience with its strict one-child policy teaches several important lessons. Even in a highly structured society with a strong totalitarian government, people will resist population control if they believe that doing so is in their best interests. Even strong economic incentives such as free medical care and education for a child may not be sufficient to sway people, particularly impoverished agrarian citizens. China's experience also illustrates the long lead time required to meet population goals. Although the goal of 1.2 billion people by the year 2000 was set thirty years in advance, and the goal has been vigorously pursued for over two decades, China will probably overshoot its target population by at least 10%.

Increasingly, more nations, like China, are realizing that the human population must come into balance with available resources. Overcrowding, unemployment, pollution, and conflicts over dwindling resources suggest that the human population is approaching the planet's carrying capacity. Already, hundreds of millions of people live in extreme poverty without adequate food, shelter, clean water, health care, or education. How can we restrain human population growth?

A population will stop growing when either (1) the death rate rises to equal the birth rate, or (2) the birth rate declines to equal the death rate. Recall from chapter 5 that the first scenario is nature's solution to overpopulation. As a growing population approaches and overshoots the carrying capacity, the death rate increases (coupled with outward migration, if possible) until the population drops below the carrying capacity to a level that the resource base can sustain. To let nature take its course, however, is to permit the human misery and suffering that would accompany a rise in famine, pestilence, and civil strife. Few people would consider such a decision humane or ethical, especially considering that most deaths would occur among children. Thus our only choice is to lower the birth rate.

In this chapter, we examine strategies for reducing human fertility. We first consider the motivational factors that influence family size and then evaluate fertility-control procedures. Next, we examine the special challenges that nations such as the United States face as they approach zero population growth. We conclude with a consideration of the future of human population growth.

Motivational Factors

A growing body of evidence (as well as common sense) suggests that fertility rates decline when parents find it desirable to have smaller families. Thus, a country that wants to curb its population growth must motivate its citizens to have fewer children. Recent surveys of desired family size, however, reveal that in many less-developed countries, most couples still prefer large families. For example, the ideal family size among sub-Saharan Africans ranges from 5.3 children in Ghana to 8.8 children in Mauritania. Most people who live in more-developed countries are members of much smaller families, and they may wonder why these people want large families, especially in view of the bleak living conditions that often afflict such areas of the world. What motivates couples to have large families?

We need to remember that until recently, a society could only survive in a hostile environment by coping with historically high death rates. To compensate for high mortality, many societies encouraged early marriages and an early start to child-bearing, on the assumption that at least a few out of a large number of children would survive to adulthood and maintain the society. Large numbers of children not only benefited society, but also provided many rewards to parents. These benefits included continuing a family lineage, providing a means to hand down property, guaranteeing family labor, and securing economic well-being in old age. In addition, large families satisfied the need for social approval. If a society believes that large families are important, most of its citizens will have large families in order to gain the approval of others, particularly members of their extended family. On the other hand, the failure of a couple to meet the expectations of family and society often brings rejection or ostracism.

Although death rates have declined substantially everywhere in the world, the social pressures that encouraged high fertility rates still remain in many societies. Why? Almost everywhere in the world, high fertility rates are associated with poverty, illiteracy, and a rural life-style. These three factors largely explain the persistence of many traditional cultural beliefs and practices that favor high birth rates (fig. 7.2).

Many population experts suggest that three conditions must be met before a population will experience a substantial decline in fertility: couples must (1) accept the belief that they alone have control over their bodies and the number of children that they have, (2) they must want to have smaller families, and (3) they must understand and master effective techniques of fertility control. Once people believe that they can make their own decisions about family size, they must want to have smaller families. What motivates couples to have fewer children?

As we saw in chapter 6, the demographic transition is essentially complete in most of the more-developed countries. In addition, desired family size has declined dramatically in recent years in rapidly developing nations such as Taiwan, South Korea, and Singapore. We conclude, therefore, that as a general rule, desired family size falls as economic development progresses. Thus, many population experts argue that industrialization, urbanization, and relatively high per capita incomes are necessary prerequisites to reducing family size.

Recent events in such nations as Indonesia, Thailand, and India suggest, however, that motivation to reduce fertility can occur in regions where the economy is not yet highly industrialized, the population is still largely rural, and per

Figure 7.2   Women and children are the principal agricultural labor pool in sub-Saharan Africa. The need for people to work the land has led to a social structure that resists fertility control.
© Jane Thomas/Visuals Unlimited

capita income is low. Declines in family size in these nations appear to be associated with an improvement in the status of their female citizens. Women tend to gain status by obtaining an education, working in nonagricultural and nondomestic jobs, and participating in social and political organizations (fig. 7.3). Researchers suggest that these activities tend to compete with child-bearing and child-rearing and thus strengthen the motivation to limit family size. Researchers also suggest that women who have a higher social status and better education also tend to marry later, which also reduces the fertility rate. In contrast, early marriage, high fertility, and large family size persist in societies where the status of women remains low.

For a while active debate took place between those who espoused economic development and those who favored improving the status of women as to which strategy was more effective in reducing family size. It soon became evident, however, that the two strategies are intertwined. Although improving the status of women does not necessarily require

economic development, a stronger economy certainly provides more resources to educate women and more job opportunities for them outside the home. In turn, a woman's job status and level of income are determined to some extent by her level of education. Furthermore, better economic conditions permit more free time for involvement in social and political issues.

Each nation has its own unique blend of economic, social, and cultural characteristics that influence the choice of family size. To motivate families to reduce their fertility level, policymakers are best advised to tailor a combination of policies to the characteristics of each nation.

Fertility-Control Methods

Only when people believe that they can control their fertility and are motivated to do so will they seek a means of birth control. Of course, the means must then be available to them. Table 7.1 lists and briefly describes various methods of birth

Figure 7.3 Employment opportunities for women outside the home, such as this garment factory in Uzbekistan, are associated with declining family size.

© Dieter Blum/Peter Arnold, Inc.

control. Special Topic 7.1 on page 128 describes some of the basic processes of the male and female reproductive systems and how contraceptives such as the pill function to control fertility.

Fertility-control procedures vary in effectiveness, safety, convenience, and compatibility with local cultural, religious, and sexual attitudes. Thus, the popularity of each procedure also varies. Table 7.2 compares and contrasts the estimated use of different forms of contraception in the United States and the world. While the frequency of sterilization is about the same in the United States as it is worldwide, there are some significant differences between the United States and the world in contraceptive use. For example, health concerns and legal action have almost removed the IUD from the U.S. marketplace, but it remains the second most popular form of contraception elsewhere in the world.

The choice of an appropriate fertility-control procedure depends upon many factors. Each person is genetically unique and has a unique health history. Hence, the choice should be made in consultation with a physician to avoid unforeseen complications. Furthermore, the preferred form of birth control will probably change with the differing phases of a couple's life span. For example, because sterilization is considered to be irreversible, a couple should not consider it until they are sure that they do not want any more children. Cultural and social conditions also influence the choice. Because there are so many variables, no single fertility-control procedure has been universally accepted. An ideal method, however, would have the following characteristics: It should be effective, safe, inexpensive, convenient, free of side effects, and compatible with local cultural, religious, and sexual attitudes.

Effectiveness

The effectiveness of fertility-control devices is clearly of primary importance. Failure rates vary considerably, however. Table 7.3 lists the percentage of women who experienced contraceptive failure during the first twelve months of use. Sterilization is almost 100% effective. **Sterilization** is the surgi-

Table 7.1 Methods of birth control

Operation of Method	Description of Method
Methods that prevent entry of sperm	abstinence: choosing to refrain from sexual intercourse
	coitus interruptus: the withdrawal of the penis from the vagina prior to ejaculation
	condom: rubber or latex sheath that fits over the penis and prevents release of sperm into the vagina
	spermicides: chemicals that kill sperm and also serve as a barrier to the passage of sperm into the cervix
	diaphragm: thin rubber dome that covers the cervix and prevents entry of sperm into the uterus
	sterilization: surgical alteration of either the female (tubal ligation) or male (vasectomy) reproductive system to prevent pregnancy
Methods that avoid or suppress the release of the egg	natural family planning: timing of intercourse to avoid the transient period of a woman's greatest fertility
	oral contraceptive (pills): combination of female hormones that suppresses the release of eggs from the ovary
Methods that prevent implantation	intrauterine device (IUD): small metal or plastic object that is placed inside the uterus to prevent the implantation of a fertilized and developing egg
Methods that prevent birth in case of pregnancy	abortion: voluntary termination of an established pregnancy by medical procedures before the fetus has attained the ability to survive outside of the uterus

Table 7.2 Use of different contraceptive methods in the United States and worldwide

Method	Estimated Use (percent)	
	United States	World
sterilization	33	36
oral contraceptives	32	15
condom	17	10
IUD	3	19
diaphragm	4–6	—

Sources: Data from *Developing New Contraceptives: Obstacles and Opportunities,* National Academy Press, 1990; and "Levels and Trends of Contraceptive Use as Assessed in 1988," *Population Studies,* No. 110, United Nations, 1989.

Table 7.3 Percentage of women experiencing contraceptive failure during the first twelve months of use

Method	Percent Failing
vasectomy	0.2
female sterilization	0.4
IUD	6.0
oral contraceptive	6.2
condom	14.2
diaphragm	15.6
natural family planning methods	16.2
other*	22.2
spermicides	26.3

Source: J. A. Ross, "Contraception: Short-term vs. Long-term Failure Rates," *Family Planning Perspectives* 21: 275–277, 1989.
*Mainly withdrawal and douche

cal alteration of either the female (tubal ligation) or male (vasectomy) reproductive system to prevent pregnancy. Users of the pill or the IUD also experience relatively low failure rates. The **contraceptive pill** is a combination of female hormones that suppresses the release of eggs from the ovary (ovulation). To use it successfully, a woman must remember to take it every day, because each pill that is forgotten increases the chances of pregnancy. An **IUD** is a small metallic or plastic object that is placed inside the uterus to prevent the implantation of a fertilized and developing egg. The undetected expulsion of an IUD is probably the leading cause of the failure of this procedure. Couples who use other methods experience two to four times greater failure rates. Historically, users of natural family planning methods experienced the highest failure rates. **Natural family planning methods** comprise a variety of procedures that involve abstinence during the transient period of a woman's greatest fertility. Recent studies, however, show that natural family planning methods

are no longer the least successful procedure.* A shift toward more sophisticated forms of the method that are also more effective has improved their success rate. In addition, today's users of natural family planning methods generally are more highly motivated and more capable of practicing the procedures successfully.

The probability of contraceptive failure declines with the age of a woman. As figure 7.4 on page 132 illustrates, the specific relationship between age and failure rate varies somewhat among methods, but women twenty-four and younger experience greater failure rates for all methods than women over twenty-four. Although many factors probably contribute

*J. A. Ross, "Contraception: Short-Term vs. Long-Term Failure Rates," *Family Planning Perspectives* 21 (1989):275-77.

Human Reproductive Systems and Fertility Control

*T*o answer the questions we may have about fertility control, we need to understand the structure and function of the human reproductive systems. First, we examine the male reproductive system and determine why a vasectomy is an effective form of fertility control. Next, we describe the female reproductive system, particularly the menstrual cycle, and then examine how the contraceptive pill works in conjunction with the menstrual cycle. We also describe several other forms of contraception.

The structure of the male reproductive system is illustrated in the drawing below. This system produces sperm (the male gamete), which are deposited during intercourse within the reproductive tract of the female. Sperm originate within the testes, which are located in the scrotum. Sperm production begins at the onset of puberty and continues into old age. A young male adult produces several hundred million sperm per day.

From the testis, immature sperm are carried by fluids to the epididymis, a mass of coiled tube overlying the testis. As the sperm move through the epididymis, they continue to mature and gain the ability to swim and to fuse with an egg. Most sperm are stored temporarily in the epididymis.

A tube known as the *vas deferens* runs from the epididymis into the abdominal cavity, where it eventually merges with the urethra, which extends the length of the penis. The urethra, which emerges from the urinary bladder, thus serves as the passageway for the excretion of urine as well as for the expulsion of sperm. Ducts from three accessory glands—the seminal vesicles, the prostate, and the bulbourethral gland—empty into the vas deferens/urethra tract. These glands produce seminal fluids that provide a medium for the transport of sperm.

With continued stimulation during erection, muscles in the walls of the epididymis and vas deferens contract, thereby forcing sperm into the urethra. Concurrently, the three accessory glands release fluids into the vas deferens/urethra tract. These fluids and the sperm constitute semen. Ultimately, contractions of muscles in the walls of the urethra propel semen from the penis, a process known as *ejaculation*.

We can now understand what a vasectomy is and how it prevents conception. As the name implies and the drawing at the top of page 129 illustrates, a vasectomy involves surgically cutting the vas deferens. Small sections of both vas deferens are removed, and the ends are surgically tied so that sperm can not be propelled into the urethra. Conception is impossible because

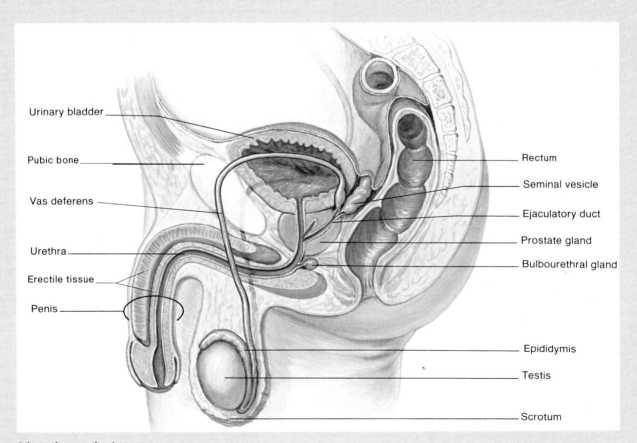

Organs of the male reproductive system.

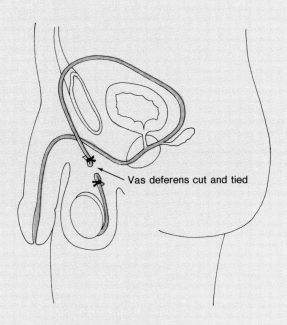

Vas deferens cut and tied

A vasectomy is a surgical procedure that prevents normal movement of sperm.

the ejaculated semen contains no sperm. Although sperm accumulate in the vas deferens behind the vasectomy, they are removed either by white blood cells within the epididymis or by leakage through the epididymal wall. Because blood vessels and nerves between the testes and the rest of the body are left intact, a vasectomy does not physically influence sexual performance.

Condoms are another common form of male contraception. Fitting over an erect penis, the rubber or latex sheath traps the semen and prevents it from entering the vagina. Improved strength, lubrication, and design have considerably enhanced their durability and reliability. Latex condoms are the only form of birth control that provides some protection against sexually-transmitted diseases, particularly AIDS and gonorrhea.

The female reproductive system is more complex than the male's because it performs more functions. These functions include production of eggs (the female gametes), reception of semen, and if fertilization occurs, implantation and development of the fetus. Egg production occurs in the ovaries (see the drawing below). Unlike males, who do not produce gametes until puberty, females begin to develop gametes during the third month of fetal development. At birth, each ovary contains about one million undeveloped eggs. Females also differ from males in that no additional gametes form after they are born; at

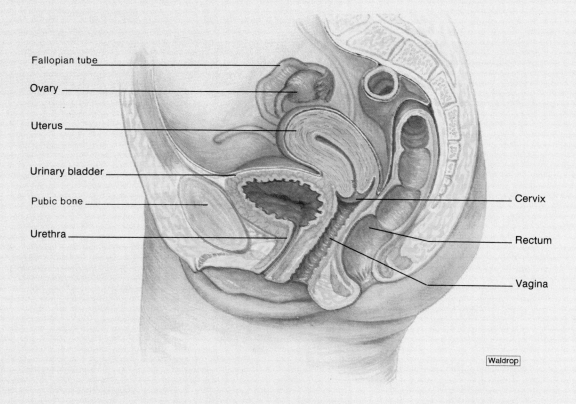

Fallopian tube

Ovary

Uterus

Urinary bladder

Pubic bone

Urethra

Cervix

Rectum

Vagina

Waldrop

Organs of the female reproductive system.

birth, females have all the undeveloped eggs that they will ever have.

Egg development does not proceed until puberty. At that time, eggs begin to complete their development, usually one at a time, about every twenty-eight days until a woman reaches menopause, which typically occurs at approximately age fifty. When an egg has matured, it bursts through the ovarian wall, a process called *ovulation*. The egg is then swept into the fallopian tube—a long tube that leads to the uterus. The journey from the ovary to the uterus normally takes about three days. An egg, however, is capable of being fertilized for only about twenty-four hours. (In contrast, sperm can remain viable within the female tract for up to three days.) Fertilization, if it is to occur, takes place in the fallopian tube. If the egg is fertilized, the embryo moves into the uterus, where it may become implanted in the endometrium (the lining of the uterus). The drawing below illustrates the pathway of fertilization and implantation. If the embryo does not implant successfully, or if the egg is not fertilized, the endometrium is shed.

At its base, the uterus narrows to form a muscular ring called the *cervix*, which controls the size of the opening between the uterus and the vagina. This opening through the cervix allows the passage of sperm to the uterus and ultimately the fallopian tubes. Dilation of the cervix at the time of birth permits the expulsion of the fetus from the uterus into the birth canal (vagina). The vagina also serves as the receptive organ for the penis and the initial repository for semen following ejaculation.

We can now understand how tubal ligations, IUDs, and diaphragms work in fertility control. Tubal ligations are surgical changes in the fallopian tubes that result in sterilization (see drawing below). Several different procedures are used to interrupt a fallopian tube. One method is to cut each fallopian tube in two and surgically tie each end. Alternatively, an instrument may be inserted into each tube that cauterizes and thereby seals it. The consequence of each of these methods is that an egg cannot continue its journey toward the uterus, nor can sperm that enter a fallopian tube from the uterus reach an egg. Although tubal ligation is a somewhat more difficult procedure and involves more risk than a vasectomy, the end result of each is the prevention of further migration of a gamete through the reproductive tract. Both procedures are generally considered to be irreversible.

An intrauterine device (IUD) is a copper or plastic coil that a physician inserts into the uterus. They are designed to remain in the uterus for an extended period. IUDs produce a low-grade, chronic inflammation of the uterine lining, which apparently inhibits the implantation of a fertilized egg.

The diaphragm, which is a soft rubber or plastic cap that fits over the cervix, acts as a barrier that prevents the entry of sperm into the uterus. To be effective, a diaphragm should be inserted into the vagina no more than six hours before intercourse. Furthermore, a diaphragm should be used in conjunction with a spermicidal cream or jelly because it is impossible to determine if there are any gaps between the diaphragm and cervix that would allow sperm to enter the uterus. In addition, a diaphragm should be left in place for at least six hours after sexual intercourse.

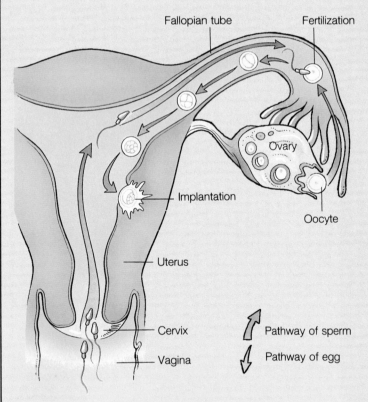

Pathway of fertilization of an egg by sperm and subsequent implantation.

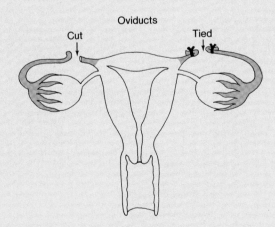

A tubal ligation is a surgical procedure that prevents normal movement of an egg.

To understand how contraceptive pills control fertility, we must first examine the menstrual cycle, which is divided into two phases (see drawing below). Phase 1 involves the maturation of the egg and the initial preparation of the endometrium for possible implantation of the embryo. The endometrium thickens and develops a richer blood supply. Phase 1 ends with ovulation. Phase 2 involves the continued development of the endometrium. If fertilization and implantation do not occur, the outer layers of the endometrium are shed during a bleeding process called *menstruation.*

These two phases are controlled and coordinated by varying levels of hormones (see drawing). During phase 1, the ovary in which an egg is maturing synthesizes a variety of hormones, particularly estrogen. This hormone helps to direct the further development of the egg as well as induce the development of the endometrium. Estrogen secretion peaks just before ovulation. After ovulation, estrogen secretion declines, while the secretion of progesterone (another hormone) greatly increases. Progesterone converts the growing endometrium into a tissue that can accept and nourish a developing embryo.

If fertilization and implantation occur, the developing placenta (a combination of fetal and endometrial tissues) soon secretes a hormone that signals the ovary to continue to release progesterone until the placenta is fully functional. If fertilization does not occur, or if a fertilized egg does not successfully implant, no message is sent to the ovary. Hence, progesterone levels eventually decline to a point that the endometrium is not

maintained, and all but the permanent inner layer is shed. Menstruation follows, and the cycle begins anew.

Users of oral contraceptives take a pill that contains synthetic preparations of both estrogen and progesterone. The pill is taken for twenty-one days followed by a seven-day break. The high levels of both hormones maintained throughout the twenty-one-day period act to suppress ovulation. Meanwhile, the endometrium responds positively to high levels of these hormones by enlarging and developing a rich blood supply. Thus, when a woman stops taking the pill after twenty-one days, reduced hormonal levels result in menstruation.

In recent years, the controversial drug RU 486 has been used in France to interrupt pregnancies. This chemical blocks the action of progesterone, which (as we have seen), is critical to the development and maintenance of the endometrium following fertilization and implantation. RU 486 can be taken as a tablet in conjunction with another hormone (prostaglandin) that increases the frequency and strength of uterine contractions. In France, this drug combination is approved for terminating pregnancies of up to forty-nine days' duration (counting from the first day of the last menstruation).

New fertility-control measures will continue to be developed. A basic understanding of the functioning of human reproductive systems allows us to evaluate for ourselves the effectiveness, safety, and ethical considerations of these procedures.

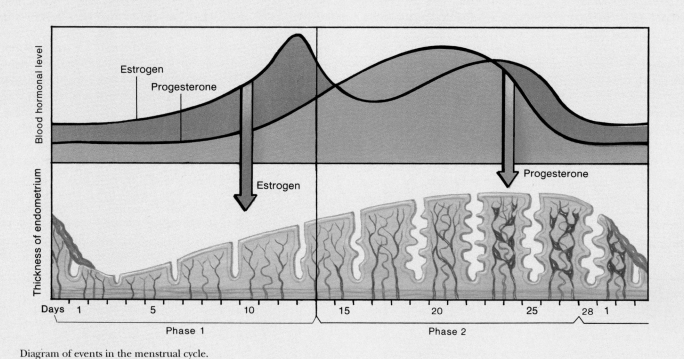

Diagram of events in the menstrual cycle.

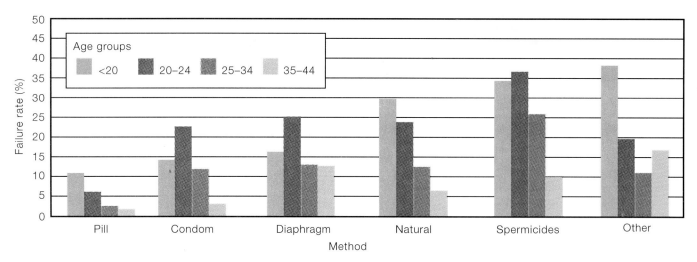

Figure 7.4 Percentage of women experiencing contraceptive failure during first twelve months of use.

to this difference, younger couples tend to be less diligent in following correct procedures. Other contributing factors include variations in attitudes about the method or about pregnancy, motivation to avoid pregnancy, knowledge of the proper use of the method, and experience with the method.

Safety

The safety of fertility-control methods is also very important. In assessing their safety, however, we must keep in mind that pregnancy itself poses risks to the mother. Generally, mortality rates among pregnant women are considerably greater than mortality rates among women who are using a fertility-control method. Complications during pregnancy and childbirth are responsible for about 23 deaths per 100,000 expectant mothers each year in the United States and Great Britain. In less-developed nations, the death rate is much higher: about 60 deaths per 100,000 expectant mothers each year.

The contraceptive pill continues to raise more health-related questions than any other method. Because the pill (1) changes the body's normal hormonal balance, (2) continues to be a popular form of contraception, and (3) has been on the market only since 1960, the potential long-term health effects of the pill remain the focus of ongoing research. Specifically, research is aimed at discovering whether women who take oral contraceptives face an increased risk of heart attack, stroke, and cancer.

Studies have shown that former users of the pill do not have an increased risk of heart attack or stroke. Current users, however, are about 2.5 times more likely to experience a cardiovascular disease as nonusers, but most of this risk is confined to older women who are also heavy smokers.

Some findings are quite encouraging regarding the relationship between oral contraceptives and the incidence of common forms of cancer. Studies clearly indicate that oral contraceptives protect women from uterine and ovarian cancer. Women who take the pill for four years have a 60% lower risk of uterine cancer and a 50% lower risk of ovarian cancer than women who have never taken it.

The most questionable and thus most troublesome potential relationship is that between the pill and breast cancer.* One difficulty in assessing this relationship is that breast cancer has a long latency period and frequently does not appear until women reach the age of sixty or older. Most women who have used the pill have not yet reached age sixty. Consequently, most studies on a possible link between oral contraceptives and breast cancer have been conducted on younger women whose risk is small to begin with and who make up a small fraction of the population of women who contract breast cancer.

Some studies have shown an increased incidence of breast cancer in younger women who have taken oral contraceptives; other studies have not. A major reason for the varying conclusions is the abundance of confounding factors, which include duration of use, use before or after a first pregnancy, age when use began, length of time since use was initiated, whether or not the woman was ever pregnant, and the formulation used. (Older formulations of the pill contained higher dosages of hormones than the pills currently marketed.) Thus, researchers are finding it difficult to assess which particular set or sets of factors actually increase the risk of contracting breast cancer.

Given the conflicting results and numerous scientific shortcomings of these studies, the U.S. Food and Drug Administration (FDA) continues to argue that the findings thus far are too equivocal to make any definitive new rulings. At the time of this writing, the patient information insert that accompanies oral contraceptives contains a warning that "some studies have shown an increased risk of breast cancer in some women."

Inserting an IUD frequently causes some bleeding and discomfort for a short time. If these symptoms persist or are excessive, the IUD should be removed. Thousands of women who used an early form of IUD called the Dalkon Shield contracted pelvic inflammatory disease (PID), a disease of the fallopian tubes and surrounding tissues caused by the same bacterium responsible for gonorrhea. Because

*See the article by J. H. Johnson listed at the end of this chapter.

severe cases can lead to sterility, the Dalkon shield was soon taken off the market.

Although the chances of contracting PID while using today's IUDs are much smaller, the actual level of risk remains controversial. So many risk factors (e.g., a history of gonorrheal infection, frequency of intercourse, and number of recent sexual partners) have been implicated in the contraction of PID that the actual risk associated with the use of an IUD is difficult to determine. Some experts, however, contend that women who have only one sexual partner and no history of sexually-transmitted diseases run little risk of PID as a result of IUD use.

The **diaphragm,** a thin rubber dome that covers the cervix and prevents entry of sperm into the uterus, is nearly free from side effects and complications. In fact, the death rates associated with the use of diaphragms (and condoms) actually refer to deaths of women in childbirth as a result of pregnancies involving contraceptive failure. A **condom** is a rubber or latex sheath that fits over the penis and prevents the release of sperm into the vagina.

When abortions are performed in medical settings, serious postoperative complications are rare. **Abortion** is the voluntary termination of an established pregnancy before the fetus has attained the ability to survive outside the uterus. When performed during the first trimester of pregnancy, less than 1% of legal abortions in the United States result in such major complications as hemorrhaging or a pelvic infection. The chances that an induced abortion might reduce the ability to have a child later is quite small.

Moral and Ethical Considerations

Cultural, religious, and sexual attitudes greatly influence a person's choice of birth-control measures. Undoubtedly, the most controversial method of preventing birth is abortion. The primary goal of those who support legalized abortion is not to reduce fertility rates; their aim is to ensure women's right to control their own fertility. Furthermore, proponents of abortion argue that abolition of legal abortions would result in more unwanted births and recourse to potentially dangerous illegal abortions. Experts estimate that over 100,000 maternal deaths occur each year in nations where illegal abortions are commonly performed by unqualified practitioners.

Opponents of legalized abortion champion the rights of the unborn child. They believe that from the moment of conception, every fetus is a human being with a right to life. However, conception and fetal development consist of a sequence of stages; each stage blends into the next without any clear demarcations between them. Although scientists can describe these events, personhood is not a state that can be evaluated by scientific experiments or observed under a microscope. There is nothing that a scientist can measure to determine whether the entity that is developing in the uterus has become a person who has the right to life. Therefore, the question of precisely when personhood begins during fetal development is not a scientific question. Rather, the controversy about abortion is a complex philosophical and ethical

question that is beyond the scope of this textbook. (For a presentation of the various philosophical viewpoints on abortion, the reader is referred to the collection of papers edited by Joel Feinberg, which is listed among the readings at the end of this chapter.)

About 40% of the world's population now lives in nations where induced abortion is permitted on request. While most of the more-developed nations allow abortions, most of the less-developed nations severely restrict it. In most western European and English-speaking nations, about half of all abortions are obtained by young (less than twenty-five years of age), unmarried women who wish to delay the first birth. In marked contrast, in eastern Europe and less-developed nations, most abortions are obtained by women who are older, married, and already have two or more children. These women often use abortion as a means of ending an unintended pregnancy because effective methods of contraception are not available.

Current Status and Future of Fertility-Control Procedures

Our evaluation of fertility-control procedures leads to one very important conclusion: Every procedure has advantages and disadvantages. For example, although sterilization is highly effective, poses little threat to health, and is a one-time procedure, it must be considered irreversible. Furthermore, many people balk at the idea of having surgery unless they are faced with a life-threatening condition. Although oral contraceptives are highly effective, a woman must remember to take a pill every day, and they may pose some long-term health hazards. Although natural family planning methods pose no direct health threat and satisfy the expectations of some religions for a "natural method," they have a relatively high failure rate. Hence, today's procedures are all inadequate in one way or another.

What are the consequences of these inadequacies? A 1990 report by the National Academy of Sciences estimates that contraceptive failure is responsible for about two million unintended pregnancies a year in the United States. The Academy also suggests that one measure of the inadequacy of today's fertility-control methods is the high incidence of abortion. They estimate that contraceptive failure is responsible for one-half of the 1.6 million abortions performed each year in the United States. The association between high abortion rates and inadequate fertility-control procedures is evident as well in the former Soviet Union, which has the highest per capita abortion rate in the world and provides poor contraceptive assistance to its citizens (e.g., oral contraceptives are essentially unavailable there).

Given the inadequacies of current methods and the need for more and better choices, what should be the priorities for research and development of new contraceptive methods? Some small improvements can be made in current birth-control methods. For example, the FDA recently approved the use of small silicone-rubber rods that slowly release hormones that act as contraceptives. These rods are implanted below the skin and are designed to prevent conception over a five-year period. Perhaps more important, this procedure delivers contraceptives at a more constant and lower rate than oral contraceptives.

Some experts, however, believe that bolder contraceptive strategies are required. The following list presents one such set of priorities.*

1. A new spermicide with antiviral properties
2. A "once-a-month" pill, effective as a menses inducer
3. A reliable ovulation inducer
4. Easily reversible and reliable male sterilization
5. Antifertility vaccine

The top priority on this list (a new spermicide with antiviral properties) is obviously a response to the AIDS epidemic as well as the need for better contraceptives. Other priorities address many of the shortcomings of current procedures that we have described. Although most fertility-control procedures are directed toward females, two on the list address males. A vasectomy (male sterilization) is simpler and safer than a tubal ligation (female sterilization). Hence, an easily reversible vasectomy would be a major breakthrough. A male contraceptive pill would also be a major advance, but there are significant problems to overcome. Assuring long-term safety is even more difficult for males than it is for female users of oral contraceptives because males are fertile over a longer portion of their life span.

Unfortunately, the outlook for development of new contraceptives is dim. Several factors have prompted the pharmaceutical industry to drastically reduce its commitment to research and development in this area. Among these factors are the time and money needed to develop and market a procedure. After the catastrophic effects of the sedative thalidomide on fetal development became clear in the 1960s (chapter 12), birth-control procedures have been more strictly regulated and monitored. Thus, regulations regarding contraceptive drugs are now more stringent than for most other classes of medication. Also, new federal legislation includes more safety regulations for nondrug methods of birth control as well, such as IUDs. More tests for the effectiveness and safety of all new birth-control procedures are also required, including tests for side effects, reversibility, and potential carcinogenic effects. Consequently, approval of a new contraceptive procedure may require twenty years or more of research, development, and testing. Pharmaceutical companies must therefore make a large, long-term financial commitment at great risk. Even after testing is complete, there is no guarantee that a new contraceptive will be approved by the FDA.

Even if a procedure wins FDA approval, the major problem of product liability remains. Many Americans today expect a "risk-free" environment (chapter 12). Furthermore, the legal system seems to favor substantial settlements for

*See the article by C. Djerassi listed at the end of this chapter.

anyone injured by a product or procedure, particularly where drugs and medical practice are involved. Awards or out-of-court settlements in excess of one million dollars to a single person are not uncommon.

Indisputably, contraceptive procedures have harmed some women and their unborn, and it is just and reasonable for them to be compensated, but this issue has other ramifications. These forces also limit the number of available contraceptives. For example, the pharmaceutical industry has withdrawn all forms of IUDs from the U. S. market except two. The sole remaining manufacturer requires users to sign a twenty-page product liability waiver form. Furthermore, without significant changes in FDA approval processes and product liability laws, the public must continue to rely on a three-decade-old contraceptive technology well into the future.

Public Policy

To speed up the demographic transition, many policymakers believe that governments must make a stronger commitment to fertility reduction. We have noted the success that several less-developed countries have had in reducing their fertility rates. These declines took only a few decades, which is considerably more rapid than the reduction that occurred in more-developed countries. Although the approaches to fertility reduction were different for each of these less-developed nations, all had the following features: (1) some combination of social and economic incentives to encourage small families, (2) educational campaigns concerning the negative impacts of continued population growth, and (3) widespread availability of family-planning services. We now examine these approaches to reducing fertility rates.

Incentives

Governments have used a variety of incentives in their attempts to reduce family size. For example, recall that to encourage acceptance of its one-child policy, the People's Republic of China provided such incentives as free hospital delivery, free medical care for the child, preferential treatment for housing and food allotments, and a monetary bonus. Other nations, such as India, make a one-time payment to people who become sterilized or use contraceptives.

In a few cases, incentives that improve the welfare of the community have been offered. In a northern province of Thailand, piglets were given to women who agreed not to become pregnant while the animals were being fattened for market. The local family-planning group then marketed the pigs and shared the profits with the women who raised them. During one three-year period, no woman who contracted to raise a pig became pregnant.

Recall that early marriages and early child-bearing tend to promote large family size. Hence fertility-reduction policies can focus on incentives to encourage couples, particularly women, to delay marriage and subsequent childbirth. Such incentives include programs that provide education for young women and create career opportunities for women.

Recall also that high infant mortality rates tend to encourage high birth rates. Hence, programs targeted at improving the nutrition and health care of infants are likely to trigger a decline in fertility rates. One way of achieving this goal is to inform people that breast feeding improves infant survival. In addition to being highly nutritious for an infant, breast milk contains immune factors that help a child ward off disease.

Educational Programs

Successful educational programs link fertility reduction to future benefits for individual families and for society. By emphasizing future relationships between population and resources, governments can shift the focus of child-bearing from the well-being of parents (such as having children for old-age security) to the well-being of children. Policy planners in the People's Republic of China noted that each Chinese person subsisted on barely one-tenth of a hectare of land, and they projected that if each family had an average of two children, the population would swell by another 300 to 400 million people. Thus, each member of the next generation would have to exist on 40% less land. Furthermore, the planners showed that if the People's Republic of China grew beyond 1.2 billion persons (its current population is about 1.1 billion persons), it would probably have a very difficult time feeding itself and maintaining its current standard of living. Thus, government officials used population and resource projections to build support for their one-child family program.

Family-Planning Services

Family planning is a voluntary program of fertility control that a couple uses to achieve the family size of their choice. Family-planning services provide information on birth control (fig. 7.5) and often distribute contraceptive devices. To be effective, family-planning agencies must make special efforts to reach the poor and people who live in rural areas. In addition, successful programs provide a wide array of contraceptive procedures, in order to serve people with differing needs and different sociocultural backgrounds.

Transition to a Stationary Population in the United States: Problems and Prospects

During the early 1970s, the population of the United States reached a major milestone—the total fertility rate of American women dropped below the replacement level fertility of 2.1 children per woman. The total fertility rate has hovered between 1.8 and 2.1 since 1974. As a consequence, the rate of natural population increase continues to decline. In fact, projections by the U. S. Census Bureau show that by 2040, the

Figure 7.5   Providing information on birth control methods is an essential part of services provided by a family planning clinic.
© E. F. Anderson/Visuals Unlimited

nation will arrive at zero population growth. Figure 7.6 shows the projected age profile in the United States for 2040. This transition toward a stationary population already has begun to affect our social and economic systems. Let us examine these effects.

To better appreciate the consequences of the shift toward zero population growth, we must first realize that current low fertility in the United States was preceded by the extraordinary baby boom of the 1950s and 1960s. During that time, men and women married earlier and had children sooner than couples do today. In addition, the total fertility rate increased to a peak of 3.7 births per woman in 1957. Between 1955 and 1964, nearly 42 million babies were born in the United States—more than during any other ten-year period, before or after, in the nation's history.

Subsequently, the fertility rate declined rapidly, and the baby-boom generation gave way to the birth-dearth

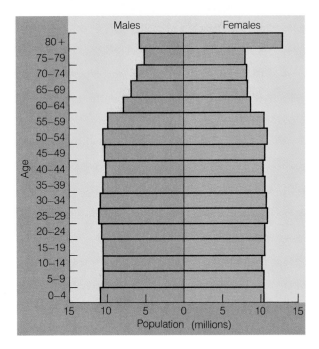

Figure 7.6 The approximate age structure of the United States if it achieves zero population growth by the year 2040.

Source: Data from L. F. Bouvier, "America's Baby Boom Generation: The Fateful Bulge." *Population Bulletin* 35 [April], 1980.

Figure 7.7 Many entry-level jobs went unclaimed when the birth-dearth generation reached working age.

© George E. Jones III/Photo Researchers, Inc.

generation. As total births declined from over 4 million annually during the baby-boom years to 3.1 million by 1975, the painful task of retrenchment followed. Schools closed, and teachers were laid off. In addition, as the birth-dearth generation entered the job market in the late 1980s and early 1990s, many part-time and entry-level jobs went unfilled (fig. 7.7).

Meanwhile, the total number of births per year began to rise again and reached four million by 1989, despite an essentially unchanging fertility rate. There were two reasons for this increase: (1) the women of the baby-boom generation had reached their child-bearing years, and (2) most of the many women who had delayed child-bearing to start their careers were now having children. Thus, we can expect the total number of births to remain relatively high during the early 1990s and then begin another decline.

One consequence of this recent increase in births is that elementary school enrollment is growing again and will continue to do so until about 1996. The enrollment pattern in secondary schools is similar, but it lags behind the elementary school pattern (fig. 7.8). Hence, communities that laid off teachers and closed schools 10 to 15 years ago are now hiring more teachers and building more schools. Furthermore, these baby-boomlet children will force adjustments in other facets of U.S. social, economic, and political systems as well.

Despite the current baby boomlet, the long-term trend of declining fertility and longer life span in the United States translates into an aging population. As figure 7.9 shows, the number of people at least sixty-five years old will nearly double between now and 2030. The number of elderly will also double over the next forty years in Japan and most European nations. The populations of South Korea, Singapore, and Malaysia

will age even faster because of the rapid demographic transition in these nations. Thus many countries are becoming increasingly concerned about labor shortages, the solvency of pension plans, and adequate health care for the elderly. The issue on page 137 explores strategies to cope with graying populations in the United States and other nations that are completing their demographic transition.

World Population: What of the Future?

Some of our insights concerning the future give us hope; others cause deep concern. Population experts can point to many positive changes that have taken place in recent decades. The world's annual rate of natural population increase, which peaked at 2% in the early 1960s, has since slowly declined to about 1.7%. Most of the more-developed nations are approaching zero population growth, and rates of natural population increase have declined significantly in some less-developed nations, particularly in Asia. Furthermore, there

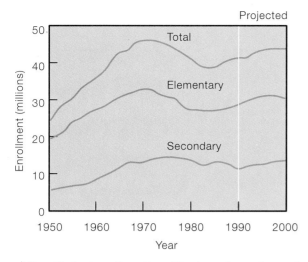

Figure 7.8 Student enrollment in public elementary and secondary schools from 1950 to 2000.

From Jeanne E. Griffith, Mary J. Frase, and John H. Ralph, "American Education: The Challenge of Change," *Population Bulletin*, vol. 44, no. 4 (Washington, D.C.: Population Reference Bureau, Inc., 1989), p. 7. Used by permission.

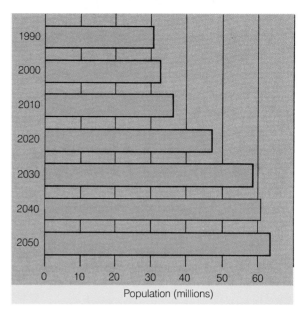

Figure 7.9 A doubling of the number of Americans who are 65 years old or older will occur within the next four decades. But their increased numbers will be offset by a reduction in the size of the population that is younger than 18 years old (assuming that birth rates remain low). This shift in population structure will necessitate a major reallocation of resources from the young to the elderly.

Source: Data from L. F. Bouvier, "America's Baby Boom Generation: The Fateful Bulge." *Population Bulletin* 35 [April], 1980.

Issue

How to Cope with Aging Populations

As populations age, many questions arise. How can a nation maintain its economic vitality if there are not enough workers? How can a nation provide sufficient health care for its rapidly aging population? There are no easy answers to these questions; the answers proposed usually generate considerable controversy as nations attempt to adjust to the new realities of completing their demographic transition. Policies to cope with aging can be divided into two categories: (1) policies to reverse population aging and (2) policies to accommodate population aging.

One set of policies to reverse population aging encourages an increase in births. Singapore, for example, has initiated policies to promote childbearing, principally among its dominant Chinese population (fig. 7.10). Incentives for having a third child include a tax rebate of $10,000, and for working mothers, an additional rebate of 15% of earned annual income. Sweden provides generous maternity leave with pay for working mothers as well as readily accessible day-care facilities. Although the fertility rate of both nations has risen to replacement level, there is controversy about the long-term success of such incentives. Evidence from some eastern European countries suggests that although incentives may encourage couples to have children sooner, their ultimate family size remains at or below the replacement level.

To reverse population aging and to alleviate labor shortages, many nations encourage immigration of young workers. In the United States, for example, immigration accounts for almost one-third of the annual population growth. Some European nations have recruited hundreds of thousands of workers (many non-European) since World War II. The war decimated the European working-age population and the postwar baby boom was too small to meet the growing need for workers.

The presence of foreign workers, however, raises many sensitive issues. Should foreign workers receive the same rights and privileges as citizens? Should they be allowed to bring their families and establish permanent residence? Political and social unrest often grow as a nation's ethnic composition changes. Fear and misunderstanding frequently lead to discrimination in housing, wages, and hiring practices (fig. 7.11), and requirements for achieving citizenship become very restrictive. Thus, although a more-developed nation may need foreign labor to maintain its economic vitality, its citizens may resist government policies to encourage their immigration.

Policies to accommodate population aging focus on maintaining the viability of public pensions and the availability of long-term health care. In the United States today, there are approximately 3.2 workers for each beneficiary in the social

There are precious moments only families can share.

Picture yourself taking home a new baby. The awe on your children's faces as they meet their new brother or sister. The unique precious moments that follow as baby grows and learns from the family. The surge of pride and joy you experience as you both watch your children achieve more. Warms the heart, doesn't it?

You can benefit from the measures introduced which make it easier to have children now. Medisave can be used for the birth of the first three children. You can make use of the subsidised child-care services. Your children get priority in school registration when you have three. Tax relief and rebates for children are generous, especially for the 3rd and 4th child. If you are a working mother, you can claim a special tax rebate for the birth of your third child. You will also get priority for upgrading your HDB flat.

Children. Life would be empty without them.

Figure 7.10   Given an aging population, Singapore's government has initiated policies to encourage childbearing.

security system. In 2030, however, there will probably be only 2 workers for each beneficiary. To ensure the future fiscal viability of the social security system, the U.S. government has reduced benefits by raising the age of eligibility for receiving benefits in the next century. Likewise, Japan has changed its funding formula so that future retirees will receive lower benefits. In addition, the United States recently raised the contribution rates (social security tax rates). Although changing the rules in the middle of the game is always risky, most people are usually more concerned about making ends meet today. Thus, politicians frequently can reduce future benefits, particularly twenty to thirty years down the road, without incurring significant backlash from their constituents.

Given the rapidly rising costs of health care in more-developed nations, a major concern for the future is providing adequate long-term health care for an aging population. In West Germany, for example, health-related expenses quadrupled between 1970 and the late 1980s, largely because of the

Figure 7.11   Foreign workers and their families are not readily assimilated into European societies.
© C. Parry/The Image Works

graying of its population. Japan has been forced to redesign its national health policy for the elderly in order to contain costs. Efforts to enact a national health insurance plan in the United States always trigger considerable debate. With rising costs and an aging population, the passage of such a program is even more problematic.

Traditionally, when elderly people can no longer care for themselves, the children (usually a daughter) provide the care. Many couples who will enter advanced age over the coming decades, however, have few or no children. Furthermore, given the realities of two-income families and single-parent families, fewer adult children have the time or financial resources to care for both their children and their parents. In the United States, this problem has been recognized in proposed family-leave policies that would allow time off for both child and parent care.

In the future, the costs of elderly care will probably fall more and more on the government. Japan and Singapore already provide tax rebates to children who support their elderly parents. For the elderly who do not have children to care for them, governments may need to subsidize home care programs and day nursing homes.

Slowing population growth and population aging are new phenomena in world history. We face many choices as we try to balance the needs of the elderly with the needs of the rest of the population. Setting priorities will not be easy, but we understand the issues, and we have time to make orderly adjustments. In addition, we can continue to learn from the successes and failures of other countries that are undergoing similar transitions.

are signs that the demographic transition will occur more rapidly in the less-developed countries, once it begins, than it did in the more-developed nations. Contraceptive use has grown from 9% (during 1960-1965) to nearly 55% today (fig. 7.12). Four out of five people in the less-developed world today live in countries that at least have an official policy for slowing population growth. Furthermore, the total fertility rate in less-developed nations has fallen from 6.1 in the 1960s to 3.9 in the early 1990s. These developments lead population experts to predict a continued decline in the global rate of natural population increase.

Optimism about the future must be tempered, however, by several factors, including the realities of exponential growth and population momentum. Although the natural rate of population increase may be declining, multiplying a larger population base by a smaller rate of natural increase still produces large increases in the total number of people added to the population each year. Thus, as figure 7.13 shows, world population will continue to grow by well over eighty million people annually for at least thirty more years.

Another disquieting reality is the marked contrast between the more-developed nations and many of the less-developed nations. We can conclude from the two age-structure diagrams in figure 7.14 that the less-developed nations have a tremendous potential for continued growth, while the more-developed nations have essentially arrived at ZPG. In addition, many married women of reproductive age in less-developed nations do not use contraceptives, particularly in Africa (fig. 7.12), and many women in less-developed nations still prefer a family size that is well above replacement level fertility. Thus, as figure 7.13 indicates, almost all global population growth will take place in the less-developed nations.

What, then, are the projections for world population growth? Assuming that policymakers continue to create an environment that motivates people to limit the size of their families, that parents decide to limit the size of their families to the replacement level, and that all people have easy access to good family-planning programs, the United Nations projects that the world's population will exceed eleven billion before it stabilizes in approximately one hundred years. Two all-important questions remain. (1) Can these assumptions be realized? (2) Can food, water, and energy supply keep pace with the demands of a growing population?

We have evidence that perhaps the race is already being lost in the forty-one nations that the United Nations defines as least developed. Table 7.4 contrasts some characteristics of these nations, which include Afghanistan, Bangladesh, Haiti, and most of the sub-Saharan countries, with those of the more-developed nations and the other less-developed nations. Poverty, hunger, disease, and premature death haunt the citizens of these nations, which lack the resources to meet the needs of their current population. Furthermore, any increase

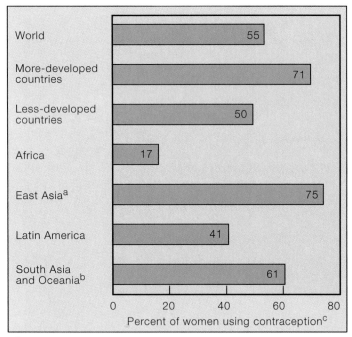

aExcludes Japan
bExcludes Australia and New Zealand
cMarried women of childbearing age

Figure 7.12   Contraceptive use in world regions (1990 estimates).

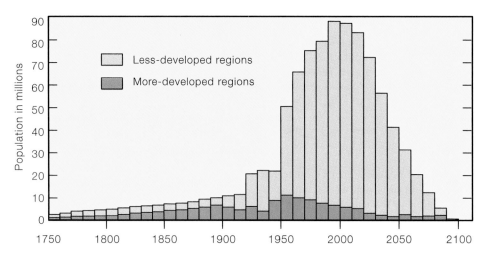

Figure 7.13   Average annual increase of world population per decade.

in resources such as food, water, energy, schools, and health care that might improve the plight of today's citizens will be consumed by the additional millions that will be born in the coming decades. Their future appears even more grim when we consider that 45% of the population in these nations is under the age of fifteen and has yet to start having children.

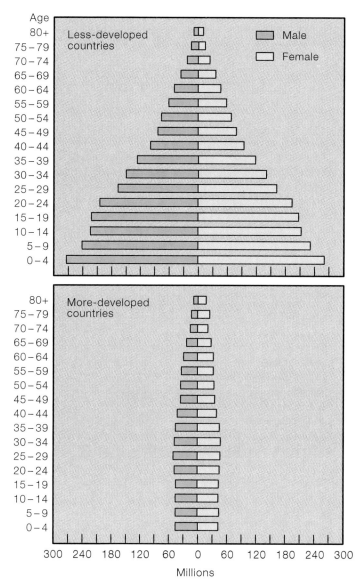

Figure 7.14   Age-structure diagrams for the less-developed and more-developed countries, 1989.

But even these events are not the whole story. Stressed by overpopulation, some regions are evolving into ecological disaster areas. Deforestation, overgrazing, and the resulting severe soil erosion are reducing the land's productivity. Thus, many subsistence farmers and nomads are losing their livelihood. The more-developed countries also face similar problems. For example, accelerated soil erosion in the United States is reducing the fertility of the world's richest agricultural land. Tragically, the planet's carrying capacity for humans appears to be diminishing, while the human population continues to grow.

Conclusions

Considerable differences in rates of natural population increase exist today among nations on our planet. Associated with these differences are other major inequalities in affluence, diet, shelter, health care, education, and other factors that influence the quality of life. Furthermore, future global population growth will occur mostly in those nations that are least able to provide resources and services. Although we who live in more-developed nations may believe that overpopulation is a problem only in less-developed nations, we need to remember two essential points. First, each of us places much greater demands on the environment than does the average citizen of a less-developed nation because of our high per capita resource consumption and consequent disposal of wastes. Second, in a sense the more-developed nations are also overpopulated because they must import considerable quantities of resources in order to maintain a high standard of living.

In later chapters, we examine and evaluate efforts to increase the carrying capacity of the planet for humans by improving the supply of such resources as food, water, energy and minerals. We also consider the impact of human numbers and technology on environmental quality. Our overriding conclusion is that, from a global perspective, the human population is approaching the earth's limited capacity to sustain it. Thus, at some point the human population must cease to grow. Nature dictates that over the long-term, our numbers (like those of all other species) must more or less balance the supply of needed resources. However, the question remains whether we will achieve that balance by rational and orderly efforts to lower the birth rate or reach it through chaos and a rise in the death rate.

Table 7.4   Population and other characteristics of least-developed, less-developed, and more-developed nations

Status of Nation	Annual Population Change (percent)	Life Expectancy at Birth (yrs)	Infant Mortality (per 1,000)	Children under 15 (percent)	1987 GNP per Capita (U.S. $)
least-developed	2.4	47.1	133	45	$200
less-developed	2.1	57.6	89	37	$700
more-developed	0.7	72.3	16	22	$12,000

Source: Data from *Population Newsletter*, No. 46, December 1988; and *World Bank Atlas*, 1989.

Key Terms

sterilization

contraceptive pill

IUD

natural family planning
methods

diaphragm

condom

abortion

family planning

Summary Statements

A population will cease to grow either when the death rate rises to the level of the birth rate, or the birth rate declines to the level of the death rate. A reduction in the birth rate is the more rational and humane way to bring the human population into balance with the planet's limited resources.

The greatest incentives for fertility reduction result from socioeconomic development (such as education, employment opportunities, and health care) that is aimed at improving the status of women.

A variety of procedures to control fertility are available. Although no method is perfect, key characteristics include effectiveness, safety, low cost, convenience, and compatibility with cultural, religious, and sexual attitudes.

Sterilization is the most effective fertility-control procedure. The contraceptive pill and the IUD are also highly effective. The probability of contraceptive failure declines with a person's age.

The form of fertility control that continues to raise the most health-related questions is the contraceptive pill. Although use of the pill reduces the risk of contracting uterine and ovarian cancer, its role in the onset of breast cancer remains unanswered. Today's IUDs and other forms of fertility control pose little health risk.

In terms of cultural and religious attitudes, abortion is the most controversial method of controlling the birth rate. The issues are philosophical and ethical, and cannot be addressed by scientific research.

No current fertility-control method is totally adequate in meeting a couple's needs for birth control. Although small improvements can be made in current procedures, the outlook for major breakthroughs in contraceptive methods is dim. Regulatory procedures and litigation have prompted the pharmaceutical industry to abandon contraceptive research and development.

To speed up the demographic transition in less-developed countries, governments need to make stronger commitments to the allocation of resources to fertility reduction. Successful public policies include a combination of social and economic incentives, educational programs that link population growth with future resource availability, and widespread availability of family-planning services.

As many nations approach ZPG, their populations become older. This shift in the age structure has many social, political, and economic ramifications, but because the shift will be gradual, there is time to plan for a smoother transition.

Some of our insights into the future give us hope; others cause deep concern. The world's population growth rate has begun to decline, the demographic transition is occurring relatively rapidly in some less-developed nations, and more couples want smaller families. Hence, natural rates of population increase are expected to continue to decline.

On the other hand, continued high rates of natural population increase in many nations (particularly in sub-Saharan Africa), limited use of contraceptive procedures, and the fact that 33% of the current world population is less than fifteen years old will create serious population pressures in the future.

Already, citizens of the least-developed nations must contend daily with poverty, hunger, disease, and early death. Evidence suggests that the planet's carrying capacity is diminishing, while the human population continues to grow. Just how the human population will reach a balance with global resources remains unclear.

Questions for Review

1. What factors motivate people to have large families?
2. Name the three conditions that must be met before a population will experience a significant decline in fertility.
3. What factors motivate people to have smaller families? How do these factors vary in importance from one culture to another?
4. List the various forms of contraceptive procedures in order of decreasing effectiveness. Why are some procedures more effective than others?
5. Why does the incidence of contraceptive failure decline as people age?
6. Summarize the long-term health implications of using contraceptive pills. On balance, is taking the pill an acceptable risk?
7. How does the incidence of abortion vary among more-developed and less-developed nations?
8. List the characteristics of an ideal birth-control method.
9. Describe the priorities proposed for the development of new contraceptive methods. What is the outlook for the development of these methods?
10. Describe several economic and social incentives for encouraging a reduction in fertility rates.
11. List the elements of a successful family planning program.
12. The transition in the United States to zero population growth will require many adjustments. Describe some of the expected social, environmental, and economic changes. How might your life be influenced by these adjustments?
13. Describe several strategies that other nations use to cope with aging populations.
14. List the recent population developments that suggest that the global natural rate of population change will continue to decline.
15. What significance do exponential growth and population momentum have for the future of global human population growth?
16. List several differences in characteristics between the least-developed nations and the more-developed nations. Is the magnitude of these differences expected to increase or decrease in the coming decades?

Projects

1. Visit a Planned Parenthood office in your community. What are the major problems that the agency is attempting to solve? What are the prevailing attitudes of the agency's clients about their desired family size? Also visit a Right-to-Life office. How do the attitudes of the Right-to-Life employees differ from the

Planned Parenthood employees? How are they similar? Could these two groups work together toward a common goal?

2. Design a survey to assess attitudes concerning family size. Administer the survey to several diverse groups, such as your class, a religious organization, and your neighbors. Compare the responses and identify the reasons for similarities and significant differences among the groups.

Selected Readings

Caldwell, J. C., and P. Caldwell. "High Fertility in Sub-Saharan Africa." *Scientific American* 262 (May 1990): 118-25. A look at the socioreligious system that accounts for the continuing high fertility rates in sub-Saharan Africa.

Djerassi, C. "The Bitter Pill." *Science* 245 (1989): 356-61. An overview of the recent history of contraceptive development and use and a prognosis for future research and development of contraceptive procedures.

Feinberg, J. *The Problem of Abortion.* 2d ed. Belmont, CA: Wadsworth, 1984. Presents a variety of perspectives on abortion.

Hellig, G., T. Buttner, and W. Lutz. "Germany's Population: Turbulent Past, Uncertain Future." *Population Bulletin* (December 1990). Washington, D.C.: Population Reference Bureau. A history of German population growth, with a major section on policies for coping with a declining population.

Henshaw, S. K. "Induced Abortion: A World Review, 1990." *Family Planning Perspectives* 22 (1990):76-89. A worldwide review of trends in abortion patterns.

Johnson, J. H. "Weighing the Evidence on the Pill and Breast Cancer." *Family Planning Perspectives* 21 (1989):89-92. An evaluation of recent studies linking oral contraceptives and breast cancer.

Jones, E. F., and J. D. Forrest. "Contraceptive Failure in the United States: Revised Estimates from the 1982 National Survey of Family Growth." *Family Planning Perspectives* 21 (1989):103-9 An in-depth evaluation of the failure rate of various contraceptive procedures.

Martin, L. G. "Population Aging Policies in East Asia and the United States." *Science* 251 (1991):527-31. A comparison of policies for coping with aging populations.

Ulmann, A., G. Teutsch, and D. Philibert. "RU 486." *Scientific American* 262 (June 1990):42-48. An examination of how this controversial drug is used to terminate unwanted pregnancies.

Volume II

Environmental Science

Managing Biological Resources

Part IV

Managing Biological Resources and Environmental Quality

*I*t is evident that certain natural laws govern humans as a species and our environment. Our surroundings, for example, have a limited capacity to absorb wastes. When this limit is approached or exceeded, natural functions are impaired. Likewise, when pollutants enter our bodies, our health often suffers. In this section, we examine the management of biological resources. Mounting concerns regarding the supply of critical resources such as food and timber underscore the need for management practices that protect the environment and conserve the planet's limited resources. More and more people are calling for

sustainable development of biological resources. In chapter 8, we consider the challenge of producing enough food for a burgeoning global population. Chapter 9 evaluates procedures to control pests that threaten our well-being by causing disease or competing with us for food and fiber. Chapter 10 examines conflicts over the limited availability of forests, rangelands, parklands, and wetlands. In chapter 11, we describe a significant problem that is increasingly attracting public attention: the loss of the planet's plant and animal species. Finally, chapter 12 describes the effects of toxins (poisons) on the human body.

Johnson Creek in central Idaho, a remote area that so far has been little modified by human activity.
© David R. Frazier Photolibrary

147

Chapter 8

Food Resources

Although the adoption of resource-intensive agriculture has contributed to a significant increase in food production, its future is clouded by the realities of environmental degradation and limited resources.
© Victoria Hurst/Tom Stack & Associates

148

Over the centuries, famines have caused much human misery and taken the lives of millions of people. One of the worse famines in recent history struck the African continent in 1983 and 1984 as a severe drought affected nearly thirty African nations. One particularly hard-hit area was the Sahel, an east-west band along the southern fringe of the Sahara Desert from the Atlantic Ocean to Ethiopia (fig. 8.1), where much livestock died, and many nomadic people lost their livelihood. As the drought intensified, large numbers of people with no food and no means of support were forced to walk many kilometers to relief stations in search of food. Some people died on the way, and others arrived at these camps in such an emaciated condition that they soon died.

Although United Nations officials warned repeatedly that a widespread famine was imminent, the world failed to notice the desperate plight of these people until November of 1984. Television vividly depicted the horror of starving children and spurred an outpouring of aid from several more-developed nations. Food was rushed to a few African nations that appeared to be suffering the most. In the meantime, famine had killed as many as 500,000 people, and the lives of hundreds of thousands of others were in jeopardy.

Exploring the reasons underlying the Sahel tragedy will give us an understanding of the complexities involved when the human population exceeds the capacity of the land to feed it. The Sahel is located in the transition zone, between the hot, steamy equatorial region to the south and the arid Sahara Desert to the north. Normally, the Sahel experiences a rainy summer and a dry winter. Beginning in the late 1960s, however, and generally continuing through much of the 1980s, the monsoon winds that bring rain to the region arrived later than usual. When the monsoons did appear, they lasted for only a short period and produced below-average rainfall. The average annual precipitation for the Sahel in the late 1960s declined by 45%, and 1983 and 1984 were the driest years of the century (fig. 8.2)—a disastrous turn of events for a region that has only marginal rainfall at best. As

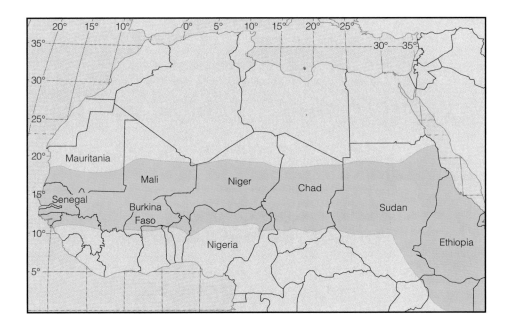

Figure 8.1   The Sahel lies between the Sahara Desert to the north and the tropical rain forest to the south. The region experiences dry winters and wet summers.

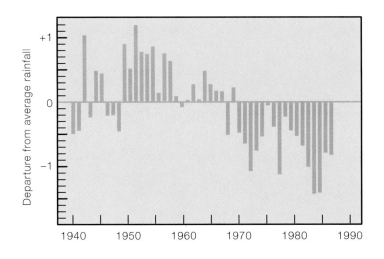

Figure 8.2   An index of the yearly average April-to-October rainfall for twenty sub-Saharan locations. The unusually wet weather in the 1950s gave way to an extended period of extreme drought. Rainfall has improved in recent years.

Figure 8.3 Desolation in the African Sahel caused by an extended period of extreme drought.
© Carl Purcell/Photo Researchers, Inc.

the drought intensified, livestock overgrazed the land, stripping it bare of vegetation. As a result, the soil dried up and blew away, livestock perished, and people starved (fig. 8.3).

More than just a change in the weather contributed to this tragedy, however. Rainfall was relatively abundant in the 1950s (fig. 8.2), and during that time, the Sahel received aid from more-developed nations, which made possible marked improvements in living conditions: better medical care lowered death rates, and imported technology spurred food production. Consequently, both human and livestock populations experienced a burst of growth. Therefore, even though prolonged droughts were not uncommon, when drought returned to the Sahel in the late 1960s, its impact was intensified by greater populations of humans and livestock.

Ironically, while the Sahel suffered through a devastating famine, much of the world experienced vastly improved harvests. In fact, the enormous growth in world grain production that occurred between 1950 and the mid-1980s has no precedent. Since 1950, West European wheat yields and U.S. corn yields have nearly tripled. Furthermore, rice yields in China and wheat yields in India have more than doubled since the early 1950s. Crop yields improved significantly in many nations around the globe, and world grain production increased 2.6 times during this period (fig. 8.4). More importantly,

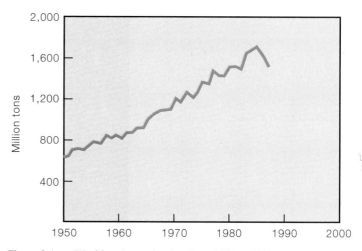

Figure 8.4 World grain production from 1950 to 1988.
Sources: Data from U.S. Department of Agriculture and Worldwatch Institute.

growth in food production outpaced population growth as world per capita production of grains actually rose by nearly 15% from 1950 to 1984. The term **Green Revolution** was coined to describe these unprecedented events.

Some people believed that the Green Revolution would solve many of the world's food problems, but as figure 8.4

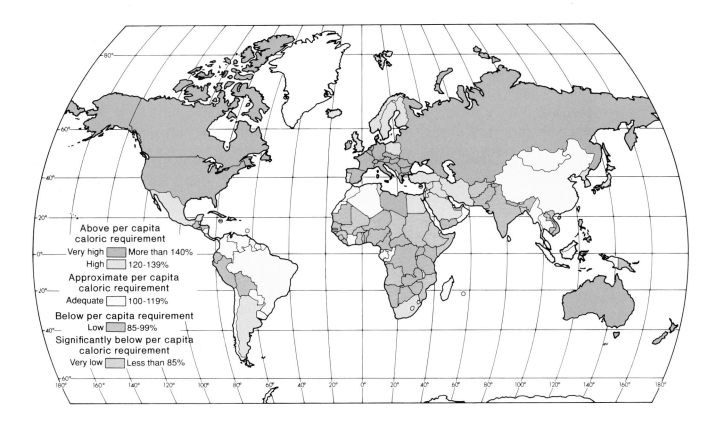

Figure 8.5   The geography of hunger, 1978 through 1980. Many less-developed nations barely receive an adequate supply of calories.
Sources: Data from the World Bank and *1983 World Population Data Sheet.*

also indicates, grain production has leveled off since 1984. Although drought-reduced harvests in India in 1987 and in the United States, Canada, and China in 1988 contributed significantly to the downturn in global grain production during those years, grain harvests have essentially plateaued since 1984. Meanwhile the number of additional mouths to be fed continues to grow by more than ninety million people each year.

Pushing agriculture to produce more food is causing other pressures. The enormous growth in world food production has been accomplished largely by greater reliance on fertilizers, pesticides, fossil fuels, and irrigation. In turn, these inputs are increasingly stressing our environment. Groundwater contamination by pesticides and nitrates from fertilizers is becoming more common, and consumers are more wary of eating pesticide-contaminated food. In addition, mismanagement of cropland has accelerated soil erosion and increased soil salinity in many areas of the globe. Ironically, many attempts to increase production are actually counterproductive, and in some localities are reducing crop yields. The future of agriculture is being shaped increasingly by the realities of environmental degradation and limited resources.

In this chapter, we consider some of the many dimensions of global food resources. We look first at how an inadequate diet affects human health. We then explore and evaluate various strategies for improving global food production. Finally, we draw some conclusions about prospects for feeding a growing world population.

Forms of Hunger

Inadequate nutrition can take many forms. One basic nutritional requirement is energy; without a sufficient intake of energy, we cannot adequately perform our daily activities. The human body converts all three principal types of food molecules—carbohydrates, fats, and proteins—into glucose. Recall from chapter 3 that glucose is a required raw material for cellular respiration, the process whereby energy is liberated from food.

The energy content of food is commonly measured in **calories.** Although the required minimum caloric intake varies with age, body size, and activity level, the United Nations recommends an average dietary intake of 2,350 calories per person per day. The diets of hundreds of millions of people in less-developed nations routinely fall below the U.N. recommended minimum caloric intake (fig. 8.5). These people suffer **undernutrition.** In addition to calories, the body also requires various materials that it cannot synthesize from food. These chemicals include amino acids, vitamins, and minerals that must be present in food. People whose diets are inadequate in one or more of these nutrients suffer **malnutrition.**

The human body utilizes twenty different amino acids as building blocks to synthesize proteins. Unlike plants and bacteria, however, we are unable to synthesize all twenty amino acids. Eight amino acids, known as **essential amino acids,** must be present in our diet if we are to synthesize all our required proteins.

Table 8.1   Sources and deficiency symptoms of some important vitamins

Vitamin	Major Sources	Deficiency Symptoms
A	egg yolk, green or yellow vegetables, fruits, liver, butter	night blindness; dry, flaky skin
B_1 (thiamine)	pork, whole grains	beriberi: fatigue, heart failure, indigestion, nerve damage
B_2 (riboflavin)	milk, eggs, liver, whole grains	inflammation and cracked skin, including lips; swollen tongue
B_6 (niacin)	whole grains, liver, and other meats	pellagra: cracked skin, digestive disorders, mental changes
Folic acid	liver, leafy vegetables	anemia, digestive disorders
C	citrus fruits, tomatoes, green leafy vegetables, potatoes	scurvy: bleeding of gums and skin, slow healing of wounds
D	fish oils, liver, fortified milk and other dairy products, sunlight on skin	rickets: defective bone formation
K	synthesis by intestinal bacteria, leafy vegetables	bleeding, internal hemorrhaging

Figure 8.6   Children in Nigeria suffering from a form of protein-calorie malnutrition called *kwashiorkor.*
© Dourdin/Photo Researchers, Inc.

Protein malnutrition is perhaps the greatest single cause of human disease in the world. **Kwashiorkor** (a form of protein malnutrition) occurs in many less-developed nations where young children receive insufficient protein to support their rapidly growing bodies. Symptoms of this devastating illness include lethargy, depressed mental abilities, failure to grow, hair loss, and swelling caused by accumulation of fluids beneath the skin (fig. 8.6). Kwashiorkor commonly develops when a child is switched from highly nutritious mother's milk to a diet of mostly cereals or roots, such as sweet potato and cassava—foods rich in carbohydrates, but poor in essential amino acids. Fortunately, permanent retardation can be avoided if a child is put on a balanced diet soon enough.

To maintain good health, humans also require about twenty vitamins, most of which the body's cells cannot manufacture. Deficiencies of each vitamin are associated with specific metabolic effects. Table 8.1 lists some of the essential vitamins, their major food sources, and associated deficiency symptoms. Vitamin A deficiency, for example, results in dry, flaky skin that is susceptible to invasion by microorganisms. Vitamin A deficiency also reduces the ability of the eye to respond to light. Mild deficiencies usually lead to an inability to see at night (night blindness). In extreme cases, blindness may occur. Vitamin A is also known as an anti-infection vitamin. People with insufficient Vitamin A are more prone to infections in the eyes, kidneys, or respiratory tract. Severe deficiencies in children dramatically increase deaths due to diarrhea, measles, and other maladies.

Recall from chapter 3 that we also require a variety of minerals for proper nutrition. Among these are iron, a component of hemoglobin, which carries oxygen in the blood; calcium, a component of bones; and potassium, required for transmission of nerve impulses and for muscle contraction. Table 8.2 lists these and other minerals and their sources. Iodine is an example of an essential mineral that is lacking in the diet of some regions. Iodine is a component of the hormone thyroxin, which is produced by the thyroid gland. Without sufficient iodine, thyroxin production declines, and the thyroid gland, located in the neck, swells to form what is known as a *goiter.* Because thyroxin stimulates metabolism, other symptoms of low thyroxin production include mental and muscular sluggishness and extreme lethargy. A victim often sleeps 14–16 hours a day. In extreme cases, people develop deafness and arteriosclerosis (hardening of the arteries). Goiters and associated symptoms are common in mountainous regions of Asia, Africa, and Latin America, where

Table 8.2 Functions and sources of some important minerals

Mineral	Functions	Sources
calcium (Ca)	structure of bones and teeth; essential for nerve impulse conduction, muscle fiber contraction, and blood coagulation	milk, milk products, leafy vegetables
iron (Fe)	component of hemoglobin; essential for cellular respiration	liver, meats, egg yolk, whole grains, legumes, nuts, dark green vegetables
magnesium (Mg)	constituent of bones and teeth; essential role in carbohydrate and protein metabolism	milk, dairy products, legumes, nuts, leafy green vegetables
potassium (K)	helps maintain water balance and regulate pH; promotes metabolism; needed for conduction of nerve impulse and contraction of muscle fibers	avocados, dried apricots, meats, nuts, potatoes, bananas
phosphorus (P)	structure of bones and teeth; component in nearly all metabolic reactions; constituent of chromosomes and many proteins	meats, poultry, fish, cheese, nuts, whole grain, cereals, milk, legumes
sodium (Na)	helps regulate water balance; needed for conduction of nerve impulses and contraction of muscle fibers; aids in regulation of pH and in transport across cell membranes	table salt, cured ham, sauerkraut, cheese, graham crackers
sulfur (S)	essential part of various amino acids and vitamins	meats, milk, eggs, legumes

insufficient levels of iodine in the soil result in crops that are deficient in iodine. Adding potassium iodide to table salt is a simple, inexpensive means of preventing iodine deficiency.

Most people in more-developed nations have access to a balanced diet of meat, dairy products, cereals, vegetables, and fruit. As a consequence, relatively few of them suffer any of the forms of malnutrition that we have described. But hundreds of millions of impoverished people in less-developed nations depend almost exclusively on a diet that typically consists of carbohydrate-rich cereals or roots or both. Relatively few can afford adequate amounts of meat, dairy products, fruit, and green vegetables, even if they were available in the marketplace. Thus, it is no surprise that many of these people suffer from deficiency diseases.

In the remainder of this chapter, we explore and evaluate various strategies to overcome hunger in the face of an ever-growing human population. Because nearly one-half of the world's annual food production is lost to our competitors, such as insects, birds, rodents, bacteria, and fungi, a critical strategy to increase the food supply is to cut crop losses to these pests. (Because such species also affect our well-being in other ways, we consider the more general topic of pest control in chapter 9.) Another strategy to increase food production is to cultivate more land. A third strategy is to improve yield on land that is already under cultivation. Finally, we evaluate the prospects of the world's fisheries as sources for much-needed protein.

Cultivating New Land

From the time that agriculture began, perhaps ten thousand years ago, until approximately 1950, the major strategy for increasing world food production was to cultivate new land. Today, however, no readily available reserves of arable land are left in the world except in humid, tropical regions. In this section, we examine the potential of expanding agriculture into tropical rain forests.

The Amazon Basin of South America and the rain forests of equatorial Africa are the only remaining large tracts of arable land where rainfall is sufficient to support intensive agriculture. The lush vegetation of tropical rain forests suggests high soil fertility and an excellent potential for agriculture, but in this case, appearances are deceiving. Soil in most tropical rain forests is actually quite low in nutrients and humus; almost all the mineral nutrients are contained in the vegetation, not in the soil. Special Topic 8.1, "Soil Properties and Plant Production," explains the partitioning of nutrients between soil and vegetation in tropical rain forests.

The native people of the tropical rain forests have adapted to this pattern of nutrient distribution by practicing **slash-and-burn agriculture.** Farmers cut down and burn vegetation on small plots, usually less than 1 hectare (2.5 acres) in area (fig. 8.7). The ashes of the burned vegetation, which contain some of the nutrients that were in the plants, serve as fertilizer. Within three to five years, the harvesting of crops and their incorporated nutrients, coupled with the considerable runoff and leaching associated with heavy rainfall, depletes the soil of most nutrients. As a result, the soil becomes too infertile to produce adequate crops, and farmers must move on to slash and burn new plots. Within 20-25 years, the original plots are overgrown again with vegetation, and the farmers return and start the cycle again.

Although crop yields from slash-and-burn agriculture are low in comparison to yields from resource-intensive agriculture, this farming technique was a successful and effective adaptation to tropical soils as long as populations remained small. Today, however, rapidly growing human populations are endangering the future of tropical rain forests as agriculture and logging operations expand ever deeper into them. To improve crop yields and to save more of the remaining forests, research in recent years has focused on developing resource-intensive agriculture.

Research has demonstrated that some soils in the Amazon Basin can sustain high crop yields if adequate fertilizers

Soil Properties and Plant Production

Soil is a dynamic habitat whose processes are vital for plant production. A fertile soil is a hospitable place for plant growth. Air circulates freely through some of its pore spaces, while other pore spaces retain water long after a rain. A tablespoon of soil contains millions of particles to which plant nutrients adhere. The same tablespoon also contains billions of microorganisms, including bacteria and fungi. Most of them are decomposers that break down organic matter, forming humus and releasing plant nutrients. Other microbes improve soil fertility by participating in the nitrogen cycle (chapter 3). Still others produce a sticky substance that binds soil particles together and helps to reduce erosion.

A vertical profile of most soils reveals a sequence of distinct layers, called *soil horizons,* produced by the gradual upward transition from undisturbed bedrock or sediment to surface vegetation. Although no single set of soil horizons can illustrate the wide variety of soil types, the accompanying drawing shows a generalized soil profile. Horizons are classified on the basis of physical and chemical characteristics, and these, in turn, are determined—for each soil type—by the degree of leaching to which each layer is subjected. *Leaching* is the process by which downward-seeping water dissolves, transports, and redeposits soluble soil components.

The sequence of major horizons that may be found in a soil profile are the O-horizon, A-horizon, E-horizon, B-horizon, C-horizon and R-horizon. The *O-horizon* is the surface layer; it is composed of fresh or partially decomposed organic matter (*litter*). The topsoil, called the *A-horizon,* is the major location of mineral nutrients for plants and usually contains most of the plant roots. The A-horizon also accumulates humus. Leaching transports some of the soluble soil constituents of the A-horizon, such as calcium carbonate ($CaCO_3$), to the subsoil, which is called the *B-horizon.* In some soils, leaching is severe enough to form a leached horizon, the *E-horizon,* which separates the A- and B-horizons. Because the B-horizon contains less organic matter, it is less fertile than the A-horizon. The B-horizon may also be a zone of accumulation of clay, aluminum, or iron. Below the B-horizon lies the *C-horizon,* a mineral layer that is made up of the partially decomposed, underlying bedrock or sediment. Because its fertility is low, the C-horizon is penetrated by few plant roots. The *R-horizon* is the underlying layer of consolidated bedrock such as granite, limestone, or sandstone.

Some soils form from the breakdown of the underlying bedrock. Others develop from sediments that have been deposited on the bedrock. Such sediments include windborne

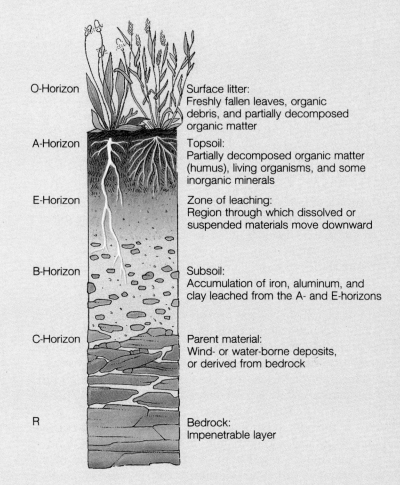

O-Horizon — Surface litter:
Freshly fallen leaves, organic debris, and partially decomposed organic matter

A-Horizon — Topsoil:
Partially decomposed organic matter (humus), living organisms, and some inorganic minerals

E-Horizon — Zone of leaching:
Region through which dissolved or suspended materials move downward

B-Horizon — Subsoil:
Accumulation of iron, aluminum, and clay leached from the A- and E-horizons

C-Horizon — Parent material:
Wind- or water-borne deposits, or derived from bedrock

R — Bedrock:
Impenetrable layer

silts and sand, waterborne deposits, volcanic ash, and glacial drift (deposited by a glacier).

Five factors play a role in the development of a soil profile and its properties, such as moisture-holding capacity, degree of aeration, and nutrient availability: (1) composition of parent bedrock or sediment, (2) climate (influences moisture and temperature), (3) topography (which controls water drainage), (4) types of plants and animals, and (5) time.

Consider how these factors interact to influence soil development. Prairie soils, such as those in Iowa and Illinois, are exceptionally fertile. Most of them are derived from *loess,* a windborne, silty parent material that has a high moisture-holding capacity and is rich in plant nutrients. In some areas, loess reaches a depth of several meters. These deep, loess-derived soils, along with moderate precipitation and high summer temperatures, promote the growth of grasses and other nonwoody plants, such as prairie clover, sunflowers, and goldenrods. Many of these plants send roots several meters into the soil—a growth pattern made possible by the depth of loess. When the roots die and begin to decay, they help to increase the thickness of the soil humus layer. Because these soils are relatively young, leaching has removed relatively small amounts of nutrients from the topsoil. The highly productive prairie soils of Iowa and Illinois are thus the products of a particular combination of factors— parent materials, climate, organisms, and time.

In contrast to prairie soils, some tropical rain forest soils in the Amazon Basin are quite infertile. A major contributing factor to low nutrient levels in tropical rain forests is climate. High temperatures and frequent heavy rains accelerate the activity of decomposers, so the remains of dead plants and animals are broken down rapidly. However, essentially no nutrients are released directly into the soil. For the most part, decomposition is carried out by fungi that grow in living plant roots and extend into the soil. As these fungi decompose the remains of dead plants and animals, they take up mineral nutrients that are released and transport them directly into plant roots. The few nutrients that are released directly into the soil are quickly washed away by heavy rains. Hence, soils are infertile.

Another factor that contributes to low soil fertility in the Amazon Basin is the geologic stability of that region. Relatively little large-scale geologic activity, such as volcanic eruptions and glaciation that would provide new sources of nutrient-rich parent material, has occurred for hundreds of millions of years. Although rivers have deposited some sediment, most soils are derived from sandstones that are nutrient-poor. That geologic stability, coupled with a warm, wet climate, has produced deep, highly-leached, nutrient-poor soils.

A rich legacy of fertile soils is a major reason why the United States is one of the major breadbaskets of the world; 64% of the U.S. is covered with moderately to highly fertile soils. The world as a whole does not fare nearly as well; with a corresponding value of 45%. With only about 5% of the world's population and a much greater percentage of the world's fertile soil, the United States is capable of producing large quantities of surplus food.

Figure 8.7   Slash-and-burn plot in the tropical rain forest of Costa Rica. © Jack Swenson/Tom Stack & Associates

are applied. In fact, fertilization means that three crops can be grown each year on the same plot, yielding several times more food than can be produced by the slash-and-burn method alone. Furthermore, a rotation of rice, corn, and soybeans produces the highest yields. In contrast, continuous planting of the same crop results in much lower yields because of pest buildup. If people can apply this model of intensive agriculture more widely in the Amazon Basin, they will require less land to support more people, and many hectares of tropical forest can be spared from clearing.

Although agricultural technology that is better adapted to the dynamics of rain forest ecosystems is encouraging, it faces several limitations. For example, its success is based largely on the availability of fertilizers; without them, yields decline to zero after the third consecutive crop. Furthermore, continuous cropping leads to severe soil erosion in hilly regions unless hillsides are terraced. Although farmers can control weeds and pests by crop rotation, pesticide application, and selection of pest-resistant varieties, researchers expect that continuous cultivation will increase pest problems. Modifications such as the use of different crop varieties, crop rotation, and variable fertilizer application rates will probably be needed to adapt to local conditions. Hence, successful use of this new technology will require skilled personnel to train farmers in the new techniques, money to provide needed resources and to support skilled personnel, receptive farmers, and time.

The time necessary to adapt new agricultural technology to local regions and to transmit knowledge of it to local farmers may be running out. Because of growing population

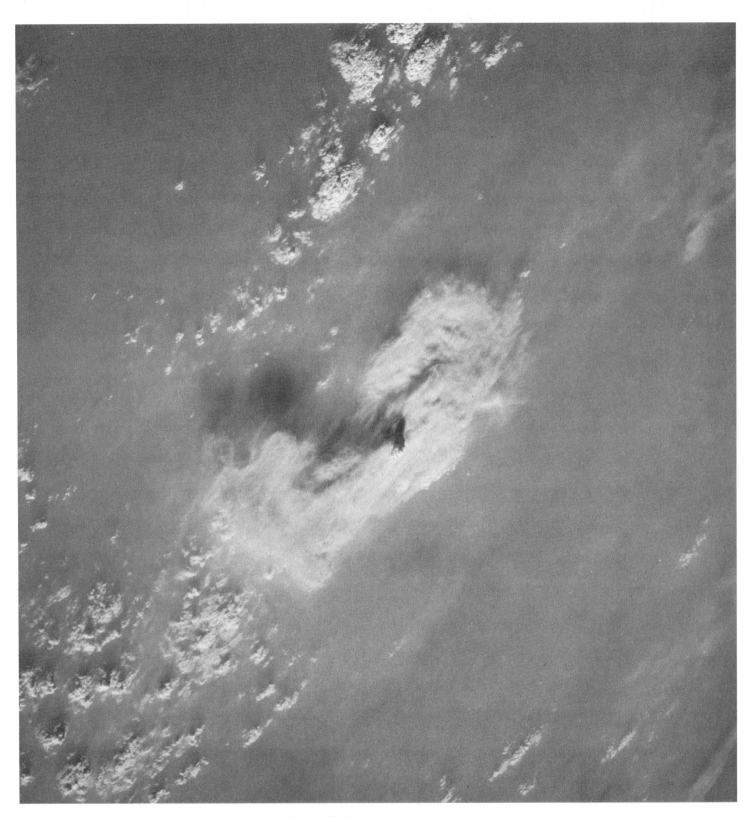

Figure 8.8 Slash-and-burn agriculture has become so widespread in the Amazon basin that a persistent pall of smoke obscures the ground. In this satellite photograph, smoke rises from a forest fire and adds to the smoke pall that already obscures the ground. The individual fire emitting this plume is estimated to be about the size of those that burned in Yellowstone National Park during the summer of 1988.
NASA

Figure 8.9   Gully erosion in this plowed field has made it unsuitable for future crop cultivation.
© Frank W. Hanna/Visuals Unlimited

pressures in these areas, the rate of deforestation was 90% greater in 1989 than it was in 1979. Slash-and-burn agriculture accounts for 60% of the region's deforestation (fig. 8.8). Clearings are larger and closer together, and the time allowed for a forest plot to regenerate is shorter. More than 2% of the Amazon Basin's forests are cleared every year for agriculture and forest products. Only about one out of every ten trees that are cut each year is replaced by reforestation projects. At this rate, the great Amazonian rain forest will vanish in about forty years.

Most deforestation is carried out by displaced, small-scale farmers who are victims of pervasive poverty. Some people view deforestation as necessary for providing food and forest products to impoverished people. Others are concerned that the loss of the vast genetic reservoir that lies within the disappearing forest will have serious long-term ramifications, as described in chapter 11.

An analysis of potential new arable lands shows that not much more can be done in semiarid regions. Large irrigation projects in recent decades have greatly expanded agriculture into semiarid lands. Today, essentially all of the economically attractive sites for large irrigation reservoirs have been devel-

oped. Some people are concerned about the environmental effects of additional dam construction. Even when new water sources are tapped, the economic advantages of using the water for residential, commercial, and industrial purposes far outweigh the return from agricultural uses. The challenges of enhancing the supply of fresh water for agriculture and other sectors of the economy are explored further in chapter 13. Given these difficulties, agricultural experts expect no significant expansion of irrigated land in the near future.

In summary, the worldwide potential for profitably expanding cropland is limited. Some tropical nations, such as Brazil, will be able to add new cropland, but the successful application of new and appropriate agricultural technologies in the tropics will be costly and slow.

In evaluating the future availability of cropland, we must also examine the other side of the coin; that is, how much land is being taken out of crop production? Each year, millions of hectares of cropland are lost. Some farmland is abandoned because it is so severely eroded that cultivation is no longer profitable (fig. 8.9). Urbanization—the construction of new homes, shopping malls, office buildings, factories, and highways—also takes over agricultural lands, often

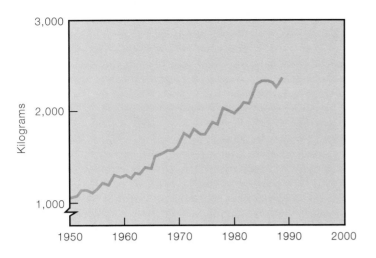

Figure 8.10 World grain yield per hectare from 1950 to 1989.
Source: U.S. Department of Agriculture.

consuming the most productive farmland. Such losses of crop-land have been most pronounced in the densely populated, rapidly industrializing countries of East Asia, including Taiwan, South Korea, Japan, and China.

The net growth in cultivated land has currently slowed to a crawl as gains and losses largely offset each other. In the near future, however, the total land under cultivation may actually begin to decline under the twin pressures of further land degradation and demands that cropland meet other needs of a growing human population. In the meantime, the world's population continues to grow by over ninety million people each year. Clearly, we cannot supply the amount of food required to feed the growing world population by expanding the cropland base. Thus, the prospects for increasing the food supply fall almost exclusively on our ability to improve per hectare productivity.

Increasing Production on Cultivated Land

Since 1950, roughly 80% of the growth in world food output has come from improved yields on land already under cultivation. Higher yields have been achieved not only in nations with advanced agricultural technology, but also in many less-developed countries, which have imported American and European technology. Such significant improvement in yields, however, has not occurred in all nations, particularly in Africa.

As figure 8.10 illustrates, the unprecedented surge in per hectare crop yields that occurred from the 1950s to the 1970s leveled off in the 1980s. At this time, the significance of this leveling-off is not clear. Perhaps we are seeing merely a brief interruption of a long-term trend, and per hectare yield will soon resume its climb. Indeed, a closer examination of figure 8.10 reveals that production per hectare has plateaued for periods of two to three years several times since 1950 and that production was especially erratic during the 1970s and 1980s. On the other hand, the current plateau may indicate that future increases in food production are less certain in

nations where the yield per hectare may be approaching its upper limit and in a world where environmental degradation may be reducing per hectare yields.

Many factors contribute to successful crop production. Plant productivity—whether in a natural ecosystem or an agricultural one—is influenced primarily by the weather (mainly moisture and temperature) and soil fertility. Other significant factors that affect plant yields are a plant's genetic makeup and populations of competitors and herbivores. Agriculture can modify these factors to some extent by irrigation, crop breeding (biotechnology), and the application of fertilizers and pesticides. Furthermore, economic considerations usually govern the availability of these resources to individual farmers. In chapter 9, we describe efforts to control pests that limit crop yields. We now explore several other factors that influence the production of food on land—climate variability, soil fertility, water supply, genetically-improved crops and livestock, and environmental degradation.

Adapting to Climatic Variability

Climate is by far the most important factor that controls food production. Climate regulates soil development, the length of the growing season, the heat supply during the growing season, and the amount of rain and sunshine that is available for plants. Climate also influences growing conditions for animals. Thus, climate governs the types of plants (including crops) and animals (including domesticated ones) that thrive in particular regions and determines both plant yields and animal production. Climate also influences the number of crops that are grown each year. In regions with a year-round growing season and adequate water, two or three crops can be grown in the same field each year.

Although it may come as a surprise in this age of advanced agricultural technology, our food supply remains largely at the mercy of the weather. This vulnerability to weather was evident in the United States three times during the 1980s as the nation experienced major drought-reduced harvests in 1980, 1983, and 1988. The most severe and widespread drought occurred in 1988, when searing heat and meager rainfall afflicted most states but was especially brutal in the midwestern farm belt. The combination of record-breaking heat and drought reduced corn production by 30%, soybean production by 20%, and spring wheat production by nearly 50% from 1987 levels (fig. 8.11). Perhaps for the first time in U.S. history, domestic grain production fell below consumption. Fortunately, carryover stocks were large, and 1989 brought considerably more favorable growing weather.

Drought was not confined to the United States in 1988. All or parts of most nations in the temperate latitudes were stricken by record drought and heat. Droughts devastated crops in China, Argentina, Canada, and many nations that border the Mediterranean Sea. In stark contrast, the summer and fall months were wet and stormy for most of the tropics from Latin America eastward to Africa and southern Asia. The Sahel experienced its wettest growing season in twenty years, while India and southern China recorded torrential

Figure 8.11   Record-breaking heat and drought in 1988 dramatically reduced crop yields in the United States.
© John S. Flannery/Visuals Unlimited

rains. Water submerged more than ten million hectares (twenty-five million acres) in southeastern China. Cool, rainy weather in Japan cut the rice harvest to its lowest level in five years.

Such spatial and temporal variations in the weather are to be expected (see chapter 15). Weather is inherently variable from one place to another and from one year to the next. Good years are followed by bad years, which are eventually succeeded by good years. Unfortunately, as soon as favorable weather returns, people tend to forget bad weather and their concern for coping with bad times. For example, for twenty years the Sahel was a focus of global attention as persistent drought and famine generated an international outpouring of assistance. But when the rains returned in the late 1980s, the tendency was to assume that drought would never again return to the region, and world attention focused on presumably more pressing problems. In fact, if policymakers were to treat drought as a recurrent event, they could deal with it in ways that would provide a better chance of stabilizing a region's food production during both the good and lean years. The same can be said for other weather hazards such as floods and late spring and early fall freezes that recur in certain regions of the world.

In many ways, the world's agricultural system is becoming more vulnerable to the vagaries of weather. Increased susceptibility stems from (1) expansion of agriculture into climatically marginal land where the growing season is short or rainfall is marginal, (2) increased reliance on new high-yielding hybrids that are more vulnerable to weather extremes than traditional native varieties, (3) greater use of agricultural technology (e.g. irrigation), and (4) shrinking fresh water re-

sources. Thus, world agriculture is developing a production base that will generate excellent harvests during favorable weather but sharply lower and potentially disastrous harvests during unfavorable weather.

The heat and aridity of 1988 raised renewed concerns about global warming. Atmospheric models predict that in the event of global warming, midcontinent regions (many of the world's bread baskets) will experience more frequent episodes of dry, hot summers. The same models also suggest that torrential rainfalls will occur more frequently in tropical regions. Although the 1988 drought and other weather events suggest a warming trend, most scientists are unsure that the planet has entered a prolonged period of global warming. They do believe, however, that the devastating events of 1988 would recur more frequently if global warming materializes. We have more to say on the potential causes and other impacts of global warming in chapter 15. Even if the earth's climate does not undergo significant future warming, the inherent variability of weather will certainly continue to play havoc with food production.

How can agriculture better cope with climatic variability? First, we need to recognize that humankind cannot control climate to ensure a more dependable food supply. Weather-modification projects such as rainmaking have met with limited success, partially because scientific understanding of precipitation processes is incomplete (see chapter 13). Dependable weather-modification technology is still at least several decades away (assuming that such technology can ever be developed at all). Hence, our only practical alternative is to learn to adapt to the inherent variability of climate.

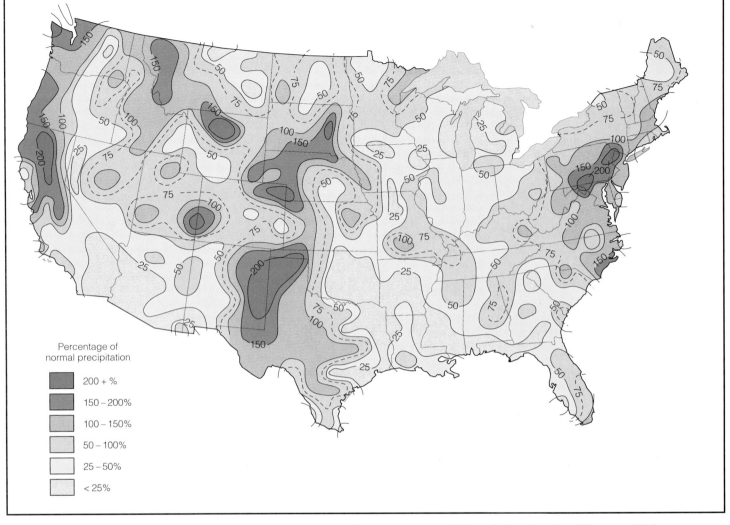

Percentage of normal precipitation

- 200 + %
- 150 – 200%
- 100 – 150%
- 50 – 100%
- 25 – 50%
- < 25%

Figure 8.12 Precipitation for May 1988, expressed as a percentage of the long-term average, varied greatly, from less than 25% to over 200%.
Sources: NOAA and USDA Agricultural Weather Facility.

One strategy for adapting to a variable climate is to take advantage of **agroclimatic compensation,** which occurs when good growing weather in one area or growing season offsets poor growing weather in another area or growing season. As 1988 illustrated, the magnitude and direction of climatic departures from long-term averages vary with geographic location. For example, while parts of the central Midwest received less than 25% of their average May rainfall, many regions on the High Plains received over 200% of their average May rainfall (fig. 8.12). In an area as large as the United States, we would not expect all regions to suffer poor growing-season weather simultaneously. In fact, there could even be an abundant harvest of a particular crop, depending on the region and its season of growth. By the time the heat wave and drought struck the Midwest in 1980, for example, much of the wheat crop had already matured. Thus, although yields of the later-maturing corn and soybean crops declined significantly, the 1980 wheat harvest set a record high for total production.

Regional agroclimatic compensation, however, also has its limitations. Favorable weather conditions in a region of

low soil fertility (which already limits crop yields) probably would not compensate for a decline in yields caused by unfavorable weather where soils are highly fertile. For example, much of Illinois is covered by soils derived from loess, a highly fertile, silty sediment. During waning phases of the Ice Age some 14,000-20,000 years ago, winds blowing across river floodplains lifted silt and deposited it downwind as loess. Loess was not deposited evenly over the landscape. Soils that developed in thicker loess are considerably more fertile than those that developed in thinner loess. Even when thinner loess soils receive a good deal of fertilizer, corn yields are only 50%-60% of those on the thicker loess soils. Therefore, good weather in an area of thin loess soils will not produce enough corn to compensate for lower corn yields due to poor growing weather in a region of thick loess soils.

Another limitation on regional agroclimatic compensation stems from the geography of crop cultivation. For example, over two-thirds of U.S. oranges are produced in Florida, which has recently experienced several severe freezes. For three days around Christmas of 1983, an arctic air mass surged into Florida, dropping temperatures well below freezing throughout much of the state. Damage to the citrus crop exceeded $1 billion. In mid-January of 1985, another severe cold snap caused further damage to surviving citrus trees. Losses from this double blow reduced citrus-producing acreage by almost 90% in Lake County, formerly Florida's second largest citrus-producing county. Subsequently, a number of growers replanted, and the region was recovering when another deep freeze over Christmas weekend 1989 again ruined much of Florida's citrus crop. Thus, while regional agroclimatic compensation may provide some flexibility, it also has serious limitations, especially for crops that are grown only in a few areas.

We can take advantage of year-to-year variability in crop production by storing grain in bountiful years to provide food for lean years. However, only a few nations, such as the United States and Canada, consistently harvest enough food to have carryover stocks every year. Even when a less-developed nation experiences a bumper harvest, it may lack the storage facilities to protect the excess grain from weather and pests such as rodents and insects.

Another strategy for coping with changeable weather is to modify agricultural practices to fit current weather conditions. During times of drought, crops that use less water (e.g., wheat, oats, and barley) can be substituted for crops that require more water (e.g., corn and rice). Irrigation systems that lose less water to evaporation and to percolation through the soil can replace less efficient systems. In some instances, farmers can turn to groundwater supplies by drilling new wells or reconditioning old ones when surface water resources are inadequate for irrigation.

Although such strategies are often successful in more-developed nations, they may not be applicable in many less-developed nations. Farmers in more-developed nations have access to credit (loans), modern machinery, energy, and information that enables them to adapt relatively quickly to change. In most less-developed nations, however, such resources are not available, and farmers must rely on their own meager resources to cope with adversity. Hence, hostile weather in these regions often leads to an extreme scarcity of food and starvation.

Maintaining Soil Fertility

After climate, soil fertility is probably the most significant factor that influences per hectare crop productivity. Thus, the most important driving force for the increased grain yields of recent decades has been growth in fertilizer use. From 1950 to 1988 world fertilizer application rose by almost 900%, while per capita use quintupled (fig. 8.13). In recent years, however, annual application rates have become somewhat erratic as the yield response to fertilizers has diminished, and weak farm economies have forced farmers to cut costs by reducing their use of fertilizer.

Much of the fertilizer applied today is **inorganic fertilizer,** which is composed of the three major plant nutrients: nitrogen, phosphorus, and potassium. Phosphorus, in the form of phosphates, and potassium, in the form of potash, are ex-

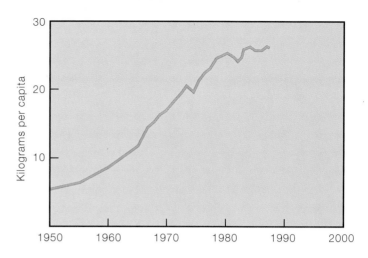

Figure 8.13   After a thirty-year trend of increasing per capita use of fertilizer, applications leveled off in the 1980s.
Source: U.S. Department of Agriculture.

tracted from rich deposits in the ground and granulated for commercial sale. Nitrogen, in the form of ammonia, is synthesized by the Haber process, which uses natural gas as an energy source to synthesize ammonia (NH_3) from atmospheric nitrogen (N_2) and hydrogen (H_2). Because inorganic fertilizers produce high yields, are easy to handle and apply, and are relatively inexpensive, they are widely used. They also allow farmers to obtain high crop yields without having to raise livestock to produce manure.

Although the United Nations projects that worldwide supplies of commercial inorganic fertilizers will be adequate for some time, a major limitation to fertilizer use is the unequal geographic distribution of potash and phosphate deposits, and natural gas reserves (needed for ammonia production). The world's largest reserves of phosphate are in Morocco, the People's Republic of China, the United States, the former Soviet Union, Jordan, and Mexico. Large deposits of potash are found in Germany, Canada, the United States, and the former Soviet Union. Mexico and the Middle East nations, which have large petroleum reserves, have the greatest potential for manufacturing ammonia. In marked contrast, these resources are scarce in most less-developed countries. Furthermore, these nations usually lack the financial resources to develop the deposits that they do have or to import sufficient quantities of inorganic fertilizers.

Such regional differences cause fertilizer application rates to vary markedly among nations. Fertilizers are used most heavily in Europe and least in Africa. On the whole, fertilizer use per hectare of arable land in less-developed nations is only about 15% of that in more-developed nations. Such a disparity is tragic. Additional fertilizer applied to land that is already heavily fertilized brings relatively small improvements in yields. However, when fertilizers are applied to nutrient-depleted soils, such as those in many less-developed nations, dramatic increases in yield usually follow (unless other factors, such as weather and pests, are even more important as limiting factors).

Figure 8.14 A pile of manure chips next to a cooking fire in a village in the sub-Saharan nation of Mali. Manure is an important fuel in the least-developed nations.
© Wolfgang Kaehler

Animal manure (an example of an organic fertilizer) is sometimes suggested as a real alternative to commercial inorganic fertilizers. However, over 90% of the manure that is produced annually in the United States is already worked into the land. In some less-developed countries, where fossil fuels are too expensive and firewood is scarce, manure is often the only fuel available for cooking and space heating, uses that are given higher priority than improving soil fertility (fig. 8.14).

Researchers are investigating means to improve soil fertility without the need for large annual investments in commercial fertilizers. These research efforts include improving the nitrogen-fixing capabilities of legumes (e.g., alfalfa and soybeans) and inoculating the soil with species of free-living, nitrogen-fixing, microorganisms. (See chapter 3 for an overview of the nitrogen cycle.)

As we assess the future availability of fertilizers, we must also realize that today's fertile soils are being lost at an alarming rate. Although soil erosion occurs naturally (chapter 3), improper agricultural practices have greatly accelerated the process. Soil erosion today is undermining the productivity of about one-third of the world's cropland. Each year, rains wash nearly 4 billion metric tons of topsoil into waterways in the United States, and three-quarters of that soil originates on farmlands. Furthermore, wind erosion accounts for another 1 billion metric tons of lost soil annually. On average, a midwestern farm field loses 2.5 centimeters (1 inch) of topsoil every 10-20 years, while it takes 200-300 years for nature to regenerate this lost soil. Hence, soil from American farmland is being eroded up to thirty times faster than it is being produced by natural soil-forming processes. Studies in the Midwest show that a loss of 2.5 centimeters of topsoil reduces corn yields by about 6%. Special Topic 8.1 describes some factors that influence soil formation.

Soil erosion in less-developed countries is considerably more severe than it is in the United States, and it worsens as the demand for food increases. As population pressures force people to move onto marginal land, hillsides erode, and deserts expand. Farmland is classified according to the degree of its erosional degradation. Land where the yield potential has

Table 8.3 Estimated land degradation from erosion

Continent	Percent of Land Surface Degraded			
	Slight	*Moderate*	*Severe*	*Total*
Africa	60	23	17	100
Asia	56	28	16	100
Australia	38	55	7	100
Europe	69	25	6	100
North America	70	23	7	100
South America	73	17	10	100

Sources: *World Agriculture Situation and Outlook Report*, Economic Research Service, U.S. Department of Agriculture, Washington, D.C., June 1989, based on data compiled by Harold E. Dregne.

been reduced by less than 10% is described as slightly degraded. Moderate degradation includes land where the yield potential has been reduced by between 10% and 50%, while severely degraded land has lost more than 50% of its yield potential. Table 8.3 shows that while all regions have some severely degraded cropland, less-developed regions have more land in this category than do more-developed regions.

Many methods of controlling soil erosion are available to farmers. For example, when farmers plow furrows up and down rolling hills, runoff flows rapidly downhill and accelerates soil erosion. However, farmers who plow parallel to land contours, called **contour farming** (fig. 8.15), and flatten slopes into terraces can reduce soil erosion significantly. Although such procedures require an increase (albeit small) in cultivation time and fuel use, the savings in valuable soil make up for the added cost.

Crops differ in their ability to anchor soil against erosion; thus, a planting technique known as **strip cropping** (fig. 8.15) is effective in reducing soil erosion. In strip cropping, a crop that is planted in widely spaced rows, such as corn or soybeans, is alternated with a crop such as alfalfa, which forms a more complete ground cover. Furthermore, leaving plant residues on the soil surface by delaying plowing until spring also protects the land from excessive erosion. Most farmers, however, prefer to plow in the fall so that they can plant earlier in the spring. Delays in spring planting lead to shorter growing seasons and reduced harvests.

Recent research has focused on **no-till cultivation.** In this strategy, loosening of the surface soil, planting, and weed control are combined into one operation through the use of specialized machinery that plants seeds and applies herbicides and fertilizers to unplowed soil. Combining these activities minimizes disturbance of the soil surface and greatly reduces the potential for soil erosion. But there is a trade-off: Reducing tillage increases weed and insect populations, which necessitates greater use of herbicides and insecticides.

Finally, the planting of trees to form **shelterbelts** reduces wind erosion (fig. 8.16). Following the Dust Bowl days of the 1930s, many shelterbelts were established in the upper Great Plains. Shelterbelts also trap snow, thereby increasing soil moisture in the spring. In addition, they provide habitats for animals that help control pest populations in adjacent fields.

Figure 8.15   Strip cropping and contour cultivation help reduce soil erosion on the rolling terrain of Pennsylvania.
© Breck P. Kent/Earth Scenes

Figure 8.16   Acting as windbreaks, these rows of trees and bushes significantly reduce soil erosion.
© John Cunningham/Visuals Unlimited

Figure 8.17 In California's Imperial Valley, irrigated fields to the left of the canal contrast sharply with the nonirrigated dry lands to the right. © Tom McHugh/Photo Researchers, Inc.

Soil erosion is a multifaceted problem. Even though many methods of controlling it are available, relatively few farmers in the United States employ them. The land is a farmer's principal economic asset, and we would expect most farmers to protect it, but recent economic studies have shown that the short-term costs of implementing soil-conservation measures often exceed the short-term benefits. Thus, although the long-term benefits of conserving valuable topsoil are obvious, few farmers believe that they can afford to consider these benefits when their short-term economic survival is at stake.

Another major difficulty in dealing successfully with soil erosion is that most farmland is privately owned. A long-standing American tradition is that landowners are free to manage (or mismanage) their land as they please. Thus, voluntary government programs, such as those administered by the U.S. Soil Conservation Service (SCS), have had only limited success. This situation is slowly changing, however, because participation in some federal agricultural aid programs requires compliance with certain SCS standards.

Better Management of Irrigation Systems

Inadequate soil moisture is a common limiting factor for crop production. In areas where rainfall is too unreliable to grow crops profitably, water must be provided by irrigation. Without supplemental water supplies, California's Central Valley (fig. 8.17) and the Aral Sea Basin—major food producing regions of the United States and the former Soviet Union respectively—could hardly be cultivated. Without irrigation, harvests from the major grain-growing areas of the Great Plains, northern China, and northwestern India would decline by 30%–50%. Today, one-third of global food production comes from the 17% of the world's cropland that is irrigated; thus, irrigation is clearly critical for world food production.

Irrigation is facing many challenges. For example, the efficiency of irrigation systems averages less than 40%; that is, less than 40% of the water drawn for irrigation is actually taken up by plants. Figure 8.18 shows the losses of water in a typical irrigation system. Much of the water seeps through the bottom and sides of unlined canals before reaching crops. Additional amounts run off the land or infiltrate the soil below the root zone because farmers apply water excessively, unevenly, or at the wrong times. Furthermore, the high temperatures and abundant sunshine common in irrigated regions promote high rates of evaporation.

Low efficiency not only means that much of the irrigation water is not available to improve yields, but major losses of water also often lead to the degradation of irrigated land. For example, without adequate drainage, seepage from unlined canals and overwatering of fields can be so substantial that the groundwater table rises. Eventually the soil's root zone becomes saturated with water, and plant roots can not obtain adequate oxygen. Such waterlogged soils are no longer suitable for cultivation. Although methods exist to prevent waterlogging and to rehabilitate waterlogged soils, they are often expensive.

When irrigation water evaporates, salts dissolved in it are left behind (chapter 13). Thus, high rates of evaporation frequently lead to a buildup of salts at the soil surface. This process, known as **soil salinization,** is illustrated in figure 8.19. As salts accumulate, they become increasingly toxic to plants. If they are not flushed away, crop yields decline as the salinity of the soil increases. Flushing away the excess salts, however, requires large quantities of additional fresh water that would otherwise be available to enhance crop production. Furthermore, flushed-out salts degrade the quality of water in streams and rivers that receive the flushing water. Although no one knows for sure the extent of the damage caused by soil salinization, estimates suggest that, worldwide, harvests on about 25% of the irrigated land are reduced by salt buildup. A similar amount of damage is estimated for irrigated land in the United States.

Thus, rehabilitation of irrigation systems should be a high priority. Siltation of canals, waterlogging, and excess salinity can often be controlled with proper management. However, management is costly: Irrigation systems must be continually maintained, and drainage, groundwater, and soil salinity must be monitored closely. Nevertheless, if faulty systems are allowed to continue functioning, they will produce millions of hectares of salt flats or water-logged soils and once-productive cropland will have to be abandoned.

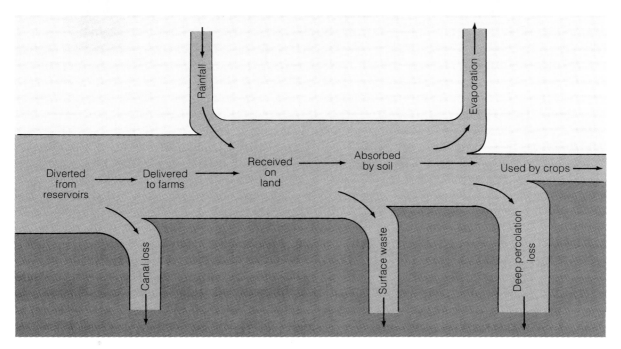

Figure 8.18 Water is gained and lost at several points in irrigation operations. Gains and losses at various points are roughly proportional to the widths of the shaded areas.

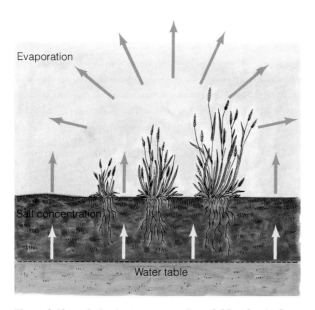

Figure 8.19 Irrigation water contains soluble salts. As the water evaporates, it leaves the salts behind at the soil surface. As the soil's salinity increases, crop yields decrease.

Another major problem facing irrigation in many regions is unsustainable fresh water resources. In the United States, roughly one-fifth of the country's irrigated land is watered by pumping rates that exceed the natural groundwater recharge rates. Perhaps the most significant challenge facing irrigation is competition among different users for a limited resource. Today, irrigated lands are often in regions that are also experiencing rapid population growth and urbanization. Because agriculture may account for as much as 80% of water usage in these regions, competition between farmers and thirsty, growing cities over a scarce supply of water is becoming increasingly intense. We discuss this important topic further in chapter 13.

Excessive withdrawal of water for irrigation has lead to one of today's greatest environmental disasters—the dying of the Aral Sea. The issue on page 166 explores the causes of the potential demise of the Aral Sea and the chances for its recovery.

The Role of Biotechnology

A major reason for the success of the Green Revolution was the development of high-yielding hybrids of major crops such as corn, wheat, and rice. These hybrids came about by traditional methods of plant breeding, the painstaking task of transferring pollen from one selected variety to another and assessing the potential of the newly created variety. In the 1980s, a new green revolution may have begun with the development of techniques to transfer beneficial genes from the cells of one organism into the cells of another. Much has been claimed for the potential contributions of **transgenic plants** and **transgenic animals** to improved agricultural productivity. Certainly the ability to introduce genes into crop plants that would confer greater drought tolerance, salt tolerance, resistance to pests, and ability to fix nitrogen could raise per hectare yields as well as permit expansion into marginal agricultural lands. Similarly, transgenic livestock with improved growth rates and feeding efficiencies would enhance the supply of badly needed animal protein. However, as we are well aware, each technology has its trade-offs. In this section, we evaluate the potential as well as the limitations of biotechnology for increasing food production.

The greatest successes in biotechnology so far have been in livestock production. For example, a hormone that is naturally produced in dairy cows (bovine growth hormone, BGH) can now be mass-produced by genetically-altered bacteria.

The Vanishing Aral Sea: Can It Be Saved?

Once larger than any of the Great Lakes except Lake Superior, the Aral Sea (located in the desert of the former Soviet Union's central Asian republics of Uzbekistan and Kazakhstan) is now shrinking and appears doomed. During the past thirty years, the water level of the Aral Sea has dropped more than 10 meters (33 feet), which translates into a loss of about 40% of its surface area. Fishing villages that once sat at the lake's edge are now more than 30 kilometers (20 miles) from the shoreline (fig. 8.20). As the volume of water in the lake has declined, its salinity has risen. Thirty years ago, twenty-four native species of fish thrived in the lake. Today there are no fish, because of the lake's high salinity, and the once-prosperous commercial fishing industry is dead.

Vast salt flats are left behind as the lake shrinks. Although some salt flats have been exposed for twenty years or more, the salt-encrusted sediments are so toxic to plants that ecological succession has not occurred, and the exposed lakebed remains a barren wasteland. Each year the wind sweeps away as much as 39 million metric tons (43 million tons) of salty grit. Long-term exposure to airborne salt particles has brought marked increases in respiratory and eye diseases to the region's residents, and infant mortality rates are the highest in the former Soviet Union. Furthermore, as the lake recedes, its moderating influence on the region's climate diminishes. Summers are becoming hotter and winters colder, increasing the stresses on the region's inhabitants. What has caused this environmental disaster?

Two large rivers (the Amu Darya and the Syr Darya) once flowed into the Aral Sea. Over time, input from the rivers more or less matched the output due to evaporation (the Aral Sea has no surface outlet). This equilibrium was disrupted in the 1960s as the consequence of massive irrigation projects that diverted huge quantities of water from the two rivers. The amount of water used to irrigate cotton, rice, and melons doubled in the last thirty years. Consequently, the flow of water into the lake slowed to a trickle. In fact, during droughts the rivers completely dry up before reaching the Aral Sea. Without the inflow of river water to compensate for evaporative losses, the Aral Sea is vanishing into the atmosphere.

Although there are grand schemes to save the Aral Sea by diverting water from Siberian rivers, 2,400 kilometers (1,500 miles) away, environmentalists argue that such diversions will only create greater problems. A more realistic solution is to

Figure 8.20 These fishing boats sit stranded in a sea of salt flats some 30 kilometers (20 miles) from the present shoreline of the Aral Sea.
© David Turnley/Black Star

increase river flow to the lake, but that would require a corresponding reduction in irrigated land. Even to maintain the lake at its present level would require nearly a 50% reduction in the 7 million hectares (18 million acres) now under irrigation in the region. But then what would happen to the people who depend upon irrigation for work and food?

No doubt irrigation will receive higher priority than restoring the Aral Sea. Without sufficient water to maintain even the current level of the Aral Sea, restoration efforts are likely to focus on the deltas where the two rivers merge with the Aral Sea. Increasing the flow of water to the small lakes and marshes that dot the two deltas could lead to a commercial fishery that would at least partially replace that lost in the Aral Sea. Currently, however, the deltas are also degraded. For example, large populations of boar, deer, and other wildlife that once inhabited the wetlands are now mostly gone, the victims of overhunting and habitat destruction.

There appears to be little hope for the Aral Sea and the people who once depended upon it for food and livelihood. The sea will continue to shrink. Deeper portions of it will probably remain as small, isolated lakes, but they will be four to five times more saline than ocean water. Even these small remnants of the Aral Sea will be dead, victims of human competition for a limited supply of fresh water.

When this hormone is injected into dairy cows, milk production increases by 10%–25% without any apparent ill effects on the cows or consumers who drink the milk they produce. In some situations, however, livestock may experience both beneficial and adverse effects. For example, when the gene that induces the synthesis of BGH was introduced into pigs, two successive generations of pigs expressing the gene showed significant increases in both feed utilization efficiency and daily weight gain. But they also experienced a high incidence of arthritis, stomach ulcers, and kidney disease. Hence, more research is needed to develop methodologies to overcome these adverse effects.

The practical benefits from plant biotechnology have so far been quite limited. The most important accomplishments have been the development of resistance to herbicides and insects (chapter 9). A major reason for the slow progress is that the genetic makeup of plants is complex, and the insertion of just one or two genes will probably not produce a superior variety. For example, chromosomes in each corn cell contain about twenty thousand genes and at least thirty genes (located at currently unknown positions on the chromosomes) influence yield. Cereals pose another problem because they are generally not compatible with the bacterium most commonly used to introduce alien genes. Furthermore, once genes are transferred, conventional plant-breeding techniques must be used to evaluate the altered plants and to make them available to farmers. Normally 10–15 years are required to create a new hybrid and to produce a marketable quantity of its seed.

There are also many potential political, social, and economic obstacles to the development and widespread application of biotechnology to food production. For example, in the United States transgenic plants and animals come under the jurisdiction of several federal agencies. The regulatory actions of these agencies greatly influence whether transgenic plants and animals and their products ever proceed beyond the experimental stage to the market place. In addition, the level of public acceptance of foodstuffs from genetically-altered organisms remains unknown. Regulation and public acceptance are issues that arise in nations with relatively abundant food supplies; they are of no concern to hungry people. However, the future of biotechnology in less-developed nations faces other major problems. Currently, most less-developed nations do not have the financial resources to develop and maintain an infrastructure (that is, laboratories, equipment, trained personnel, etc.) to support widespread implementation of biotechnology.

Some scientists raise a fundamental concern with all efforts to develop more productive crops and livestock, whether by traditional methods or by biotechnology. Today's high-yielding varieties of crops such as rice, for example, are the products of breeding experiments that utilized thousands of varieties, both native and domestic. These new varieties are initially resistant to insects and disease, but sooner or later insects and disease-causing organisms evolve and successfully attack the new crop varieties. (See chapter 5 for a review of evolutionary processes.) Plant breeders then scurry in search of other varieties with the genes that confer natural resistance to the newly evolved pests. Unfortunately more and more crop varieties are disappearing, replaced by improved crops that are genetically uniform. For example, India once grew over thirty thousand varieties of rice. Today, more than 75% of rice production in India comes from fewer than ten varieties.

To save the genes of the remaining varieties, scientists are establishing **seed banks.** For example, the U. S. Department of Agriculture's National Seed Storage Laboratory at Fort Collins, Colorado has a cache of over 225,000 samples of seeds. Yet native and traditional varieties of crops are disappearing much faster than they can be located and collected. Thus, as agriculture struggles to feed a hungry world, essential genetic resources to support that struggle are disappearing. We have more to say about the value of genetic resources in chapter 11.

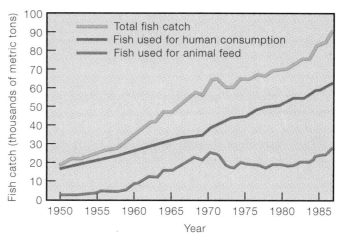

Figure 8.21   Global fish catch and use between 1950 and 1986.

There is little doubt that biotechnology will ultimately result in more productive livestock and crop plants. Progress will continue to be slow, however, and many technological, political, social, and economic problems must be solved if biotechnology is to produce another green revolution comparable in impact to that of the 1960s and 1970s. In the meantime, humankind should mount a massive effort to save the genetic diversity that is an important key to solving today's and tomorrow's food production problems.

We have seen that modern agriculture relies heavily on the massive consumption of resources such as fertilizers, pesticides, irrigation water, and fossil fuels. Furthermore, modern agriculture places increasing stress on the environment through wind and water erosion of soil, salinization and waterlogging of soil, and contamination of surface water and groundwater by fertilizers and pesticides. Given this huge demand on resources and the resulting environmental degradation, more people are concluding that modern agriculture is unsustainable over the long term and are calling for a switch to sustainable agriculture. For more on this topic, see the issue on page 168.

Harvesting the Oceans

We now turn our attention from terrestrial food production to fish and shellfish production in oceans and lakes. Fish and shellfish are more important as sources of protein than as sources of calories. The United Nations lists more than thirty countries in which fish protein represents 40% or more of the total supply of animal protein. In many of these nations, particularly in Asia and western Africa, people rely so heavily on diets of grain or starchy roots that without fish protein, many people would suffer from severe protein deficiency.

Today, the future yield of the world's fisheries is uncertain. Although the global fish catch has shown steady growth since 1950 (fig. 8.21), the United Nations Food and Agricultural Organization (FAO) believes that it is approaching the maximum sustainable harvest from the oceans. Hence, in the near future, the global fish catch is expected to plateau, and it may even begin to decline.

Sustainable Agriculture: Can It Succeed?

What is meant by sustainable agriculture? Definitions of **sustainable agriculture** depend upon the people involved. A small-scale organic farmer who uses only natural products on the land does not define sustainable agriculture in the same way as a large-scale farmer who wants to cut costs by reducing the use of inorganic fertilizers and commercial pesticides. We will use the definition of sustainable agriculture proposed by Gordon K. Douglass of Pomona College.* Professor Douglass views sustainable agriculture in terms of long-term food sufficiency, wise stewardship, and community.

Sustainability in terms of long-term food sufficiency requires that agricultural systems be more ecologically based, so that they do not destroy their natural resource base. Instead of depending upon commercial products such as inorganic fertilizers and synthetic pesticides, a sustainable farm relies upon resources found on or near the farm. To improve soil fertility, for example, a key component is crop rotation, that is, growing a planned sequence of various crops in a field. A crop rotation that includes leguminous plants such as soybeans and alfalfa, as well as grains, such as corn and oats is beneficial in several ways. Nitrogen-fixing bacteria in the nodules of leguminous plants improve soil fertility for future grain crops. In addition, soybeans and alfalfa require less nitrogen to grow than do grain crops such as corn. Another key component is the addition of organic materials such as manure to the soil, or the plowing-under of alfalfa (green manure). Such organic materials improve the structure and fertility of soil as well as its water-storage capacity. Because water readily infiltrates soils that are relatively rich in organic matter, surface runoff and soil erosion are greatly reduced. Likewise, alternative methods of pest control are available that both significantly reduce environmental degradation and decrease reliance on commercial pesticides.

Today only a small percentage of farmers practice sustainable agriculture as defined by Douglass. Proponents cite political and economic policies as major obstacles to widespread acceptance of these alternative farming practices. For example, current government policies favor large-scale corporate farms, discourage crop rotation, and encourage overproduction. Furthermore, government subsidies of irrigation systems result in unrealistically low prices for fresh water that encourage waste of this very limited resource.

These concerns lead us to Douglass' second perspective on sustainable agriculture: that of wise stewardship. This viewpoint requires that agricultural systems be based upon a conscious ethic of responsibility to future generations and to other species. This ethic suggests that humankind is a part of and dependent upon nature and we should not view nature as something to dominate and exploit.

The third perspective on agricultural sustainability is that of community. This viewpoint suggests that we cannot sustain agricultural systems if there are major inequities in the distribution of power, land, and wealth among the world's people. In many nations, government policies and market structures favor large landowners at the expense of peasant farmers with regard to the benefits of modern agricultural techniques. Government policies also tend to favor the growing of exportable cash crops (which benefit the large landowners) rather than the production of foodstuffs that will improve the nutrition of impoverished small-scale farmers. In the United States, communities of small family farms once supported thriving small towns, but as these farms are increasingly taken over by large corporate farms, many rural communities are withering and dying. Today, a prosperous agricultural economy certainly does not imply a prosperous farming community. Proponents of sustainable agriculture argue that economic and political policies must be aimed at revitalizing rural communities.

*G. K. Douglass, Agricultural Sustainability in a Changing World Order (Boulder, Colo.: Westview Press, 1984).

The major source of pessimism is **overfishing,** the harvesting of more fish than are produced in a given period. Almost all important stocks of cod, haddock, flounder, and hake are either fully exploited or overfished. In addition, many stocks of highly-valued species such as salmon and tuna are depleted.

One reason for such overexploitation is technological advancement. Modern fishing vessels are equipped with sonar and helicopters that increase the chances of locating large schools of fish. Probably the most important reason, however, is a lack of compliance with fishing quotas. From the 1950s to the mid-1970s, the principal fishing grounds in international waters were managed by a series of agreements among coastal and fishing nations. When quotas were established, it was up to each nation to enforce limits on its own fishermen. Such a system was obviously vulnerable to abuses, and overexploitation of fisheries resulted.

Political conditions changed somewhat in the mid-1970s, when the Law of the Sea Treaty allowed coastal nations to take control of fisheries by extending their jurisdiction to 200 kilometers (124 miles) offshore. This law, however, has done little to change the picture. Off the New England coast, for example, a reequipped and larger U.S. fishing fleet easily took up the slack in pressure on fish stocks as foreign fishing ships were denied the right to fish within the United States' 200-km zone. Concerned about their own livelihoods, U. S. fishermen continue to resist suggestions that quotas be established on fish harvests and number of boats. Meanwhile the

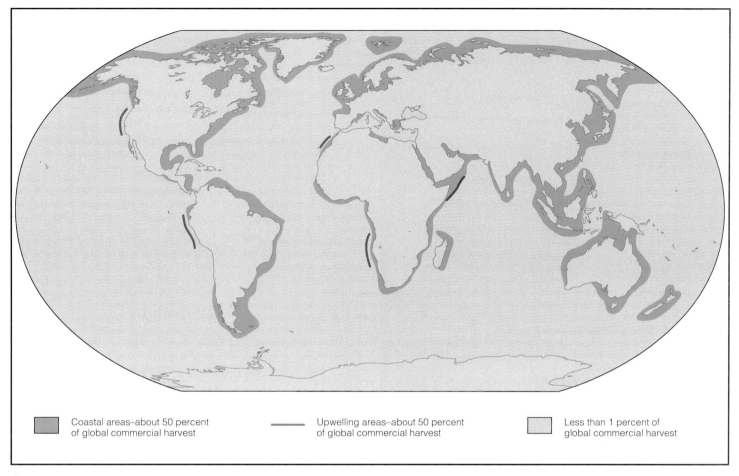

■ Coastal areas–about 50 percent of global commercial harvest	— Upwelling areas–about 50 percent of global commercial harvest	■ Less than 1 percent of global commercial harvest

Figure 8.22   Distribution of the world's fisheries.

Source: Data from *Patterns and Perspectives in Environmental Science*, Washington, D.C.   Natural Science Foundation, 1972.

fisheries are not recovering, and the situation is much the same elsewhere.

Another threat to sustained yields from the ocean is pollution and destruction of habitats that are vital for fish production. The most productive areas of the ocean (estuaries, reefs, and regions of upwelling) are adjacent to continents (fig. 8.22). Unfortunately, these zones are becoming seriously polluted by eroded sediments, dredge spoils, pesticides, and domestic and industrial wastes. For example, in the United States some 30% of the shellfish beds are now closed because of possible contamination by disease-causing microorganisms originating in untreated domestic wastewater (chapter 14). More oil spills and tanker accidents in recent years pose additional threats to coastal marine life. Finally, commercial, industrial, and recreational development threaten to eliminate fish and shellfish habitats altogether (see chapters 10 and 21). If coastal zones are not managed properly, fish production throughout the world may actually decline.

One way to meet the growing demand for fish protein is to exploit new regions and species. Possibly the only large oceanic region that remains underutilized is the southwest Atlantic, particularly off the coast of Argentina. Three species regarded as underutilized in several oceanic regions are octopus, squid, and krill (zooplankton). Although large amounts of krill exist off the coast of Antarctica, no one knows what fraction can be harvested without upsetting the food webs of marine animals such as whales (including the endangered blue whale), seals, and penguins.

Several other options exist for expanding the supply of fish for human consumption. One strategy is to develop markets for the less desirable fish that are usually discarded at sea when caught along with more desirable species. Most of these fish die anyway after being released. Another option is to improve refrigeration and processing facilities to reduce spoilage, which accounts for a 40% loss in the catch by some less-developed nations. A third strategy is to use some of the fish currently sold as animal food for human consumption. As much as 20,000 metric tons (40% of the total fish catch) of such fish as sardines and anchovies are ground up each year and fed to livestock and domestic pets.

All these strategies to increase world fish harvests are forms of hunting and gathering. Thus, another approach to improving the fish harvest is **aquaculture:** the growing and harvesting of fish for human consumption in confined areas such as ponds, lakes, irrigation ditches, and fenced-in portions of estuaries. Today, aquaculture accounts for 15% of all fish consumed worldwide. Aquaculture is particularly important in Asia, where it is a major source of protein for the poor.

Figure 8.23 Man harvesting tilapia fish in a pond in northern Thailand.
© Wolfgang Kaehler

The most successful form of aquaculture has been fresh water fish farming. To achieve high yields, managers have developed several practices that are based on ecological principles described in chapters 3-5. For example, to improve the yield of fish without increasing competition, ponds are stocked simultaneously with several species that require different types of food. Fish that are commonly used in such polycultures include Chinese, Indian, and European carp, various tilapia species, and channel catfish. To enhance production of organisms that are consumed by fish, nutrients are added in the form of inorganic fertilizers, agricultural wastes, or human sewage. Such methods have produced relatively high harvests in such diverse locations as India, Israel, Thailand (fig. 8.23), and the United States.

Marine fish farming (*mariculture*) has focused on the raising of shellfish and salmonid fishes (salmon and trout). Yields of shellfish (such as oysters and mussels) are enhanced by using racks, rafts, or trays. Shellfish that grow on these submerged structures rather than on the bottom have access to more food and are protected from bottom-dwelling carnivores and burial by sediments. In addition, many shellfish and fish can be mass-produced rather easily in hatcheries and then released to enhance old fisheries or establish new ones.

As we saw for agriculture, some scientists forecast an increasingly important role for biotechnology in the improvement of fish production by aquaculture. Already transgenic carp and rainbow trout have been developed that grow 20%-40% faster than ordinary specimens. Other efforts are underway to enhance viral immunity of carp and to increase resistance in Atlantic salmon to freezing of body tissues. Applying biotechnology to aquaculture, however, faces many of the same challenges that we identified for agriculture. High costs and regulatory obstacles are likely to slow the use of biotechnology in fish and shellfish production.

If the FAO is correct, the world's fish harvest is rapidly approaching its maximum sustainable level. Some experts suggest that even the current harvest level cannot be sustained unless fisheries and coastal zone management improve around the globe. Furthermore, the potential contribution of aquaculture is limited; it will probably never produce anywhere near the quantity of fish that is produced naturally in the oceans.

Conclusions

We are left with a very difficult question: Can enough food be produced to feed an ever-growing human population? We can be reasonably certain that the area of cultivated land is not likely to change appreciably in the near future. Thus, assessing prospects for food production is primarily a matter of predicting how quickly and how much per hectare yields will rise.

Clearly many things can be done to improve crop yields. Many current strategies still focus on increasing the availability of such resources as fertilizers, irrigation systems, pesticides, and genetically-improved crops and livestock. Although these strategies have improved food production in the short-term, people are questioning the long-term sustainability of modern agriculture. Increasingly, people are calling for methods of food production that are more environmentally friendly. For example, massive international efforts are needed to protect soil from further degradation by water and wind erosion, waterlogging, and salinization. Major efforts are also required to conserve fresh water and to restore land that is already degraded. Without these efforts, the amount of cultivated land as well as yields per hectare could actually decline in the future with devastating implications for the hungry people of the world.

Although the greatest potential gains are possible in less-developed countries, food production in most of these nations is limited by inadequate resources, including an infrastructure that could provide economic and technical assistance to farmers. Achieving significant gains in food production in any nation, more-or-less-developed, depends to a great extent on government policies that encourage agricultural research and ensure the availability of needed resources at prices that farmers can afford. Also, governments must ensure that farmers receive a fair price for their crops and livestock. Today, however, many governments focus heavily on industrial and commercial development. Furthermore, pricing policies strongly favor urban consumers at the expense of farmers. Even farmers in more-developed nations increasingly face similar problems as these nations become more urbanized. Public policies that reduce a nation's agricultural capabilities must change if the world is to have any chance of feeding its growing population.

Undoubtedly, food production will continue to increase for some time. But history warns that we should expect years of poor harvests and declining production. In the long run, whether or not the global population will level off before we exceed the earth's sustainable capacity to feed us is still an open question.

Key Terms

Green Revolution	contour farming
sustainable agriculture	strip cropping
calorie	no-till cultivation
undernutrition	shelterbelts
malnutrition	soil salinization
essential amino acids	transgenic plants and animals
kwashiorkor	seed banks
slash-and-burn agriculture	sustainable agriculture
agroclimatic compensation	overfishing
inorganic fertilizer	aquaculture

Summary Statements

Hunger can take the form of undernutrition, malnutrition, or both. Insufficient protein in the diet is the most common form of malnutrition and can lead to serious illness and sometimes permanent disability or death.

Although people in humid tropical regions are able to cultivate new lands, many tropical soils presently limit agricultural expansion because of low fertility. Meanwhile, land is being lost for crop production; either it is too degraded, or it is being appropriated for urbanization and industrialization. Hence, the amount of land under cultivation soon may begin to decline.

The Green Revolution produced unprecedented growth in per hectare yields during the 1960s and 1970s. During the 1980s, however, grain production throughout the world plateaued.

Climate is the most important factor that controls food production. Crop failures during recent years in both more-developed and less-developed nations illustrate how vulnerable the world's food supply is to the weather. Furthermore, the world agricultural system is becoming even more vulnerable to the vagaries of weather. Agroclimatic compensation and other strategies need to be developed and implemented to cope with the impact of climatic variability on the food supply.

Next to climate, soil fertility is the most important environmental factor that influences per hectare crop production. Although supplies of inorganic fertilizers seem to be adequate for the near future, unequal distribution of fertilizer resources around the world severely limits the growth of global food production. Furthermore, fertile soil is being lost by erosion at an alarming rate. Many procedures are available to reduce soil erosion, but because of economic hardship, few farmers are employing these techniques.

Irrigation is necessary in regions where rainfall is insufficient for growing crops. However, competition between urban and rural areas for a limited supply of water is becoming more severe. Problems caused by irrigation, such as soil salinization and waterlogging, are removing irrigated lands from cultivation.

Proponents claim that biotechnology will generate more productive livestock and crop plants. Although few successes have been achieved thus far, most experts believe that biotechnology will eventually contribute to increased food production. Progress will be slow, however, and many problems will have to be solved if biotechnology is to usher in a new green revolution.

Fish are an important source of protein for people in many countries, but the FAO believes that the world catch size is approaching the maximum sustainable harvest. The major limitations on increasing the fish harvest are overfishing and the degradation of habitats that are vital for fish production.

Prospects for feeding the world's growing population are uncertain. Strategies such as increasing the availability of fertilizers, pesticides, irrigation water, and genetically-improved crops and livestock will improve food production in the short term, but we must also implement strategies to ensure the long-term sustainability of agriculture. These strategies include massive international efforts to conserve water and to protect land from further degradation by wind and water erosion, salinization, and waterlogging. Providing farmers with sufficient resources will require solving many difficult political and economic problems in all nations.

Questions for Review

1. What are the differences between undernutrition and malnutrition?
2. Although the vegetation in tropical rain forests is usually very lush, the soils are often quite infertile. Explain this apparent paradox.
3. Describe the process of slash-and-burn agriculture. What are its advantages and disadvantages?
4. Evaluate the potential for expanding agriculture into tropical rain forests.
5. Even if agriculture continues to expand into new lands, the total land in cultivation may actually decline over the coming decades. Explain.
6. List five reasons why climate is the most important controlling factor in crop production.
7. How is modern agriculture becoming increasingly vulnerable to climatic variability?
8. What is agroclimatic compensation? How can it help us cope with weather variability? What are its shortcomings?
9. How do inorganic fertilizers differ from organic fertilizers such as manure?
10. How do leguminous plants improve soil fertility?
11. Why is manure not returned to the land as a fertilizer in some less-developed nations?
12. What are strip-cropping and no-till cultivation? How do they help to reduce soil erosion?
13. List several reasons why less than 40% of the water is actually taken up by plants in many irrigation systems.

14. What is soil salinization? Why is it often a problem on irrigated lands?

15. Describe the potential of biotechnology to improve world food production. Do you have any reservations about eating food that comes from transgenic plants and animals?

16. Describe why many of the world's fisheries are overharvested.

17. Describe why pollution and destruction of habitats in coastal zones may lead to a decline in the world fish catch.

18. What is aquaculture? Evaluate its potential for increasing the world supply of protein.

Projects

1. Survey the nutritional status of people in your community. Do any of them suffer from undernutrition or malnutrition? If so, what are the reasons? What is being done to alleviate their hunger?

2. Visit your local planning office and determine how much cropland in your region has been lost to urbanization in recent years. Does your community or state have policies that promote preservation of cropland?

3. Visit a farm that practices sustainable agriculture. How do cultivation practices on this farm differ from those of surrounding farms? Does the farmer believe that sustainable farming provides a sufficient profit?

Selected Readings

Brown, L. R., ed. *State of the World.* New York: Norton. An annual update developed by the Worldwatch Institute that examines a variety of environmental issues, including world food production.

Crossen, P. "Climate Change and Mid-Latitude Agriculture: Perspectives on Consequences and Policy Responses." *Climate Change* 15 (1989):51-73. A consideration of the potential impact of global warming and how nations can respond.

Crosson, P., and N. J. Rosenberg. "Strategies for Agriculture." *Scientific American* 261 (Sept. 1989):128-35. An examination of some social and economic changes that will be necessary to improve food production without further degrading the environment.

Dahlberg, K. A. "Sustainable Agriculture—Fad or Harbinger?" *Bioscience* 41 (1991):337-40. An evaluation of the forces that support sustainable agriculture and those that oppose it.

Ellis, W. E. "The Aral: A Soviet Sea Lies Dying." *National Geographic* 177 (Feb. 1990):73-93. A sobering look at the impact of irrigation on the Aral Sea and the people who depended upon it for food and livelihood.

Gasser, C. S., and R. T. Fraley. "Genetically Engineering Plants for Crop Improvement." *Science* 244 (1989):1293-99. A look at some achievements in developing transgenic agricultural plants and an evaluation of key issues that affect the commercial introduction of these plants.

Glantz, M. H. "Drought in Africa." *Scientific American* 256 (June 1987):34-40. An examination of strategies for dealing with recurrent droughts in sub-Saharan Africa.

Juskevitch, J. C., and C. G. Buyer. "Bovine Growth Hormone: Human Food Safety Evaluation." *Science* 249 (1990):875-84. An evaluation of the risk posed to consumers who drink milk produced by cows injected with bovine growth hormone.

Miller, R. W., and R. L. Donahue. *Soils: An Introduction to Soils and Plant Growth.* Englewood Cliffs, N.J.: Prentice Hall, 1990. A basic examination of soil properties and management.

Pursel, V. G., C. A. Pinkert, K. F. Miller, D. J. Bolt, R. G. Campbell, R. D. Palmiter, R. L. Brinster, and R. E. Hammer. "Genetic Engineering of Livestock." *Science* 244 (1989):1281-92. An evaluation of the beneficial and adverse consequences of developing transgenic livestock.

Reganold, J. P., R. I. Papendick, and J. F. Parr. "Sustainable Agriculture." *Scientific American* 262 (June 1990): 112-20. A description of farming methods used in sustainable agriculture and an evaluation of environmental and financial outcomes.

Repetto, R. "Deforestation in the Tropics." *Scientific American* 262 (April 1990):36-42. An examination of government policies that encourage exploitation of the tropical rain forests.

Rhodes, R. E. "The World's Food Supply at Risk." *National Geographic* 179 (April 1991):74-105. A fascinating look at the race to rescue wild and domesticated relatives of today's major crop varieties.

Scrimshaw, N. S. "Iron Deficiency." *Scientific American* 265 (Oct. 1991):24-30. A look at the impact of one of the world's greatest nutritional problems and how it can be treated.

World Resources. New York: Basic Books. An annual international report on a wide range of environmental issues, including world food production.

Pesticides and Pest Management

A helicopter crop dusting a sugar beet field in California. Pesticide application is a popular strategy to eliminate competitors for our food supply.

© Inga Spence/Tom Stack & Associates

courge to humans and livestock alike, the tiny tsetse
fly has been a major force in the history of Africa
(fig. 9.1). Tsetse flies harbor microscopic parasites
called trypanosomes that cause sleeping sickness in humans
and a similar disease known as nagana in livestock. A thousand years ago, the tsetse fly and the trypanosome blunted the
advance of Islam into central Africa. Horses and camels of
invading Muslims inexplicably died upon reaching the tsetse
belt south of the Sahara Desert (fig. 9.2). The fly and its
companion also slowed European exploration and colonization of Africa in the 1800s. Today, they still stifle economic
progress. Each year, over twenty thousand people fall victim
to the parasite, but many more cases probably go unreported.
At least fifty million Africans reside within the tsetse belt and daily
face the threat of being bitten by a trypanosome-infected tsetse.

Tsetse flies pick up trypanosomes with the blood that
they ingest from infected hosts. (Like mosquitoes, tsetse flies
feed on the blood of humans and many other animals.) They
transmit the parasites as they feed on other victims. Trypanosomes multiply rapidly in the human bloodstream. Early symptoms include severe headaches and high fever. If treated early,
the disease is curable, but if it is left untreated, the parasite
enters the spinal cord and the brain, where treatment can be
as toxic to the patient as it is to the parasite. As the infection
spreads, the victim's speech becomes slurred and he or she
becomes more and more apathetic. Eventually the person
lapses into a coma and often dies.

Trypanosomes also use wild game animals as hosts, but
coevolution has lead to a parasite-host interaction that leaves
native animals unaffected. (See chapter 5 on coevolution.)
Thus, these animals are an ever-present source of trypanosomes that can be transmitted to humans and livestock by
tsetse flies. Early attempts to control the tsetse often focused
on the eradication of game animals. Fortunately, more enlightened means are used today. Still, despite seventy years of
research and control efforts, the tsetse and the trypanosome
reign over a large region of Africa.

In addition to the tsetse fly and the trypanosome, an
amazing variety of species interferes with our well-being. Like
the trypanosome, **pathogens** (disease-producing organisms)
such as viruses, bacteria, protozoa, and parasitic worms bring
illness and death to hundreds of millions of people worldwide
each year. Human diseases caused by viruses include the common cold, AIDS (Acquired Immune Deficiency Syndrome),
chicken pox, measles, and infectious hepatitis. Bacteria cause
considerable human suffering and death through such diseases as whooping cough, syphilis, gonorrhea, cholera, dysentery, tetanus, and diphtheria. Debilitating diseases caused by
protozoans include malaria, amoebic dysentery, and African
sleeping sickness. Parasitic worms that reside in such vital
organs as the stomach, intestines, liver, heart, and lungs infect
hundreds of millions of people in less-developed nations.
Tapeworms, flukes, hookworms, and other parasitic worms
cause diseases such as schistosomiasis, onchocerciasis (river
blindness), trichinosis, and elephantiasis (fig. 9.3).

Some species like the tsetse fly play a role in human
diseases by transmitting pathogens from one person to an-

Figure 9.1   The tsetse fly, a carrier of parasitic trypanosomes, spreads
sleeping sickness and, for that reason, has played a major role in the
history of Africa.
© J. A. L. Cooke/Animals Animals

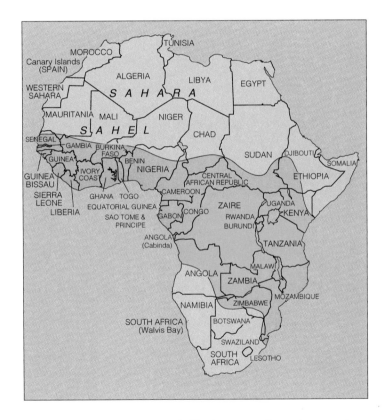

Figure 9.2   Their ranges bounded by deserts, nearly two dozen species
of tsetse flies occupy a wide belt that spans the continent of Africa.

other or from an animal to a person. Such disease-carrying
species are called **vectors.** Mosquitoes function as vectors for
such diseases as malaria, encephalitis, and yellow fever. Fleas
carry the bacterium for bubonic plague from rats to humans,
and deer ticks carry the bacterium for Lyme disease from
deer and field mice to humans.

Other species dramatically influence our well-being by
playing havoc with the production of food and fiber. In fact,
the activities of these species result in a loss of about half of

Figure 9.3 This woman from Mozambique is suffering from elephantiasis. This disease develops when a parasitic worm (filaria) blocks vessels in the body's lymph system. As a consequence, fluid accumulates and causes extreme swelling in parts of the body and hardening of the surrounding skin.

© Thomas S. England/Science Source/Photo Researchers, Inc.

Figure 9.4 The tumorlike galls of corn smut (a fungus) will destroy this ear of corn.

© Michel Viard/Peter Arnold, Inc.

A major worldwide endeavor in recent decades has been to improve the human condition by controlling the populations of pest species. Such efforts have had mixed results. Although some pest populations in certain regions have been greatly reduced, many victories have come at the expense of environmental disturbances. Some pest control programs have inadvertently reduced populations of valuable species (honeybees, for example) and contaminated food and water. Furthermore, many pest populations remain more or less uncontrolled.

In this chapter, we examine and evaluate present measures for pest control and consider future directions for pest management. Our focus is on the control of agricultural pests and disease vectors. Control of human diseases is generally the province of medical science, which has made relatively few attempts to apply ecological approaches to the improvement of human health.

The Nature of Pest Species

We label a species a pest if its population grows large enough to influence our well-being adversely. (For some people, however, a single mouse or cockroach constitutes a significant pest problem.) In terms of our discussion in chapter 5, a typical pest is an *r*-strategist. (Recall that *r*-strategists produce many offspring in a short period of time, and adults are usually small and short-lived.) Because *r*-strategist populations can explode when environmental conditions are favorable, such species frequently inflict considerable damage. If we were to list our "Top Ten Pests," mice, cockroaches, termites, and ants (all *r*-strategists) would probably rank high.

the world's food production each year. Some species of insects and rodents consume crops; other insects, bacteria, and fungi weaken or kill crop plants or domestic animals. Still others induce rotting of crops, either in fields before harvesting or in storage facilities (fig. 9.4). Furthermore, weeds reduce food production by competing with crop plants for light, soil nutrients, and water.

Species that cause disease, transmit pathogens, compete with humans for food or fiber, or otherwise threaten human health or well-being are called **pests.** (Pathogens are sometimes not included with the other pest species.) Weeds are considered to be pests when they threaten human welfare by competing with plants that humans value for food, timber, or other uses.

Figure 9.5   Although relatively few in number, an elephant herd's voracious appetite can consume a village's crops in a matter of hours.
© Leonard Rue Enterprises

Our discussion in chapter 5 would probably not lead us to believe that *K*-strategists are pests. (Recall that *K*-strategists produce few young over a long period of time, and the adults are large and long-lived.) However, *K*-strategists can also be pests; they can inflict damage that is disproportionately great given their population size. For example, most of us think of African elephants as an endangered species, not as a pest species. Farmers in eastern Africa, however, consider elephants to be pests, because even a small herd can easily consume a village's crops overnight (fig. 9.5).

Whether or not we consider a plant a weed depends upon its habitat. Recall from our discussion of secondary succession in chapter 4 that when we cultivate the soil (whether for a small flower or vegetable garden or for a wheat field covering hundreds of hectares), we create bare soil, a site that is prime for the successful invasion of *r*-strategists. In contrast, when we grow a lush lawn or an immaculate putting green, we create a habitat that is perhaps better exploited by *K*-strategists that can compete in the dense cover of grasses. A prime example of such a *K*-strategist is the nearly ubiquitous dandelion, whose deep taproot and ground-level rosette of large leaves makes it an extremely effective competitor in mowed landscapes.

The Goal of Pest Control

Policymakers, the media, or both may claim that the goal of a particular pest-control program is total eradication of the pest. This may be laudable when speaking of such dreaded diseases as AIDS or malaria, but such a goal is very difficult to achieve in nature. Pest populations are usually large and widely dispersed over a region composed of many different habitat types. Thus, some members of a pest population usually escape the control agent. Even if the entire population in a region is exterminated, before long, new pests usually migrate into the region from an outside population. A more realistic goal of pest control, therefore, should be an economic one, that is, to reduce the pest population to a level at which no further reductions are profitable.

Table 9.1   Use of synthetic pesticides in the United States

Land-Use Category	All Pesticides		Herbicides		Insecticides		Fungicides	
	Treated hectares (× 10⁶)	Quantity (× 10⁶ kg)	Treated hectares (× 10⁶)	Quantity (× 10⁶ kg)	Treated hectares (× 10⁶)	Quantity (× 10⁶ kg)	Treated hectares (× 10⁶)	Quantity (× 10⁶ kg)
agricultural lands	114	341	109	199	34	74	5	68
government and industrial lands	28	55	30	44	—	11	—	—
forest lands	2	4	2	3	1	1	—	—
household lands	4	55	3	26	3	25	1	4
Total	148	455	144*	272	38	111	6	72

Source: Adapted from D. Pimentel and L. Levitan, "Pesticides: Amounts Applied and Amounts Reaching Pests," *Bioscience* 36: 86–91. Copyright © 1986 American Institute of Biological Sciences, Washington, D.C. Used by permission.

*Same land may be treated with several classes of chemicals.

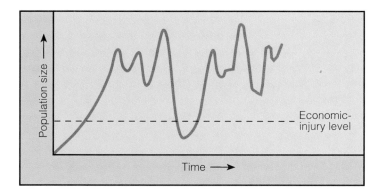

Figure 9.6   The population fluctuations of a hypothetical pest. When the pest population exceeds the economic-injury level, the goal of pest control should be to bring the population down again to the economic-injury level.

The concept of an **economic injury level** for a hypothetical pest is illustrated in figure 9.6. If the population of a pest is below the economic injury level, then the costs of control outweigh the benefits of further population declines. Note, however, that a pest population is often above the economic injury level, so that economic benefits would accrue from a pest-control program that reduced the population to the economic injury level. Let us emphasize again, however, that total eradication of a pest species is not a justifiable goal on purely economic grounds; that is, the cost of eradication (assuming that it is possible) outweighs the economic benefits of higher crop yields.

To reduce a pest population to the economic injury level, an ideal pest-control program should consist of procedures that are effective, environmentally safe, and profitable. Let us now examine how various pest-control procedures measure up to these criteria.

Synthetic Pesticides

Synthetic pesticides are chemicals formulated to kill certain types of pest species. Major groups of pesticides include insecticides, fungicides, and herbicides. **Insecticides** are chemicals directed at killing insects. Synthetic insecticides include **chlorinated hydrocarbons** (such as DDT and dieldrin), **organophosphates** (such as malathion and parathion) and **carbamates** (such as carbaryl). **Fungicides** are directed at fungi such as molds, blights, and rusts. A common synthetic fungicide is captan. **Herbicides** kill certain weed species. Common synthetic herbicides include 2,4-D and atrazine.

Although chemical control of pests began some 2,500 years ago, pest-control technology developed slowly until World War II, when it underwent a major revolution. The pressing need to control the insect vectors for many tropical diseases that were afflicting military personnel spurred the screening of hundreds of synthetic chemicals for insecticidal properties. The three major groups of insecticides used today were developed during this time. Following World War II, these new insecticides were adopted worldwide by farmers, ranchers, and health teams charged with insect control. The discovery of other chemicals for the control of rodents, fungi, and weeds soon followed. Of the approximately 500 million kilograms (1.1 billion pounds) used in the United States today, about 60% are herbicides, 24% insecticides, and 16% fungicides (table 9.1).

Agriculture uses the greatest share of pesticides (75%) (table 9.1). Note, however, that households account for 12% of all pesticide usage. Furthermore, the National Academy of Sciences found that homeowners often apply four to eight times more synthetic pesticides per hectare than do farmers. In 1990, homeowners spent $1.2 billion on pesticides. Pesticides are clearly big business: The retail value of pesticides sold in the United States annually tops $6 billion.

Positive Aspects of Synthetic Pesticides

Synthetic pesticides are heavily used worldwide because they are effective, relatively inexpensive, and easy to apply. One measure of their effectiveness is the reduction in food losses to pests during the raising and storage of crops. Total worldwide food losses to pests amount to nearly 45% of total food production. (Preharvest losses account for about 30%, while postharvest losses amount to about 15%.) In the United States, nearly one-third less food is lost to pests than is lost on average worldwide; our widespread use of pesticides is responsible for a 35% increase in yield. Viewed from an economic perspective, for the $3 billion invested each year in pesticides to control agricultural pests, about $12 billion are returned.

Other benefits have also accrued from the use of pesticides. Because pesticides improve per hectare crop yields, U.S. farmers do not have to cultivate as much land. Hence, more land is available for forests, recreation, and wildlife preserves. Furthermore, agricultural expansion would necessitate the tillage of land that is highly erodible. Thus, the use of pesticides indirectly reduces soil erosion and contributes to water quality.

Pesticides have also contributed to the control of human diseases spread by vectors. Chlorinated hydrocarbons, especially DDT, have controlled outbreaks of malaria, plague, yellow fever, and louse-borne typhus, four of the most important epidemic diseases in human history. Perhaps the single most important benefit of pesticides has been protection from malaria. Today over one billion people (approximately one-fifth of the world's population) no longer live with the threat of malaria.

Negative Aspects of Synthetic Pesticides

The benefits of using synthetic pesticides in public health and agriculture have been many. Unfortunately, however, many of them also have significant environmental costs and risks. Synthetic pesticides can kill nontarget species, trigger pesticide resistance in pest populations, and contaminate surface and ground waters.

Killing nontarget species

Some pesticides, such as DDT, kill a wide variety of species and thus are called **broad-spectrum pesticides.** In addition to killing target pests, most synthetic pesticides are toxic to other species, including humans. As many as 67,000 cases of human pesticide poisoning occur worldwide each year, some of which are fatal. Improper use of pesticides by poorly trained workers frequently leads to poisoning during application. In other instances, workers who enter fields too soon after pesticide application to weed or harvest crops pick up pesticides by brushing against contaminated foliage. We have more to say about the effects of pesticides and other toxins on humans in chapter 12.

Honeybees are beneficial insects that are particularly vulnerable to insecticides. When honeybees forage for pollen and nectar, they frequently come in contact with insecticides, which may kill them. Insecticide poisoning has contributed to significant declines in wild and domestic honeybee populations during recent decades in the United States. This decline has serious ramifications, because honeybees pollinate about $20 billion worth of crops each year. Without insect pollination, there would be no apples, cherries, almonds, cucumbers, melons, or strawberries and no seeds to grow such important forage crops as alfalfa and clover.

Killing nontarget species may also disrupt natural pest-control mechanisms, such as parasitism and predation (discussed in chapter 5). Such a disruption occurred in efforts to control insect pests on cotton growing in the Rio Grande Valley. Initially, the boll weevil, pink bollworm, and cotton fleahopper were the major pests that had to be controlled each year to achieve a profitable yield of cotton. Two potential pests of cotton—the tobacco budworm and the bollworm—were usually kept in check by their natural enemies. However, when the new chlorinated hydrocarbon pesticides, such as DDT, became commercially available shortly after World War II, farmers commonly applied these insecticides to their cotton crop 10-20 times during each growing season. As a result, cotton yields soared as the major insect pests were decimated. Within a few years, however, populations of the three former major pest species plus previously-checked populations of tobacco budworm and bollworm exploded, and cotton yields plummeted. The insecticides applied to control the three major pest species had also devastated the natural enemies of the tobacco budworm and bollworm. Without their natural population-control agents, these two minor pests became major pests. In an attempt to control three pests, the cotton growers had produced two more.

Development of pesticide resistance

A closer look at insecticide usage in the Rio Grande Valley cotton fields reveals another frequently encountered problem. With continual application of pesticides, some pest populations evolve that are no longer harmed by the pesticide. This phenomenon is known as **pesticide resistance.** By 1960, the bollworm and tobacco budworm were becoming more and more resistant to DDT. Although growers increased the dosage and frequency of DDT application, by 1965 these pests were no longer controlled by DDT or any other chlorinated hydrocarbon pesticide. Next, the growers tried the organophosphate pesticide, methyl parathion, which worked temporarily. By 1968, however, the tobacco budworm had also developed resistance to that pesticide. In subsequent years, cotton yields declined dramatically even though fields were treated 15-20 times each growing season with every conceivable combination of pesticides. Although pesticides contributed to bumper cotton crops in the early 1950s, within 20 years, pesticide-resistant populations and destruction of the pests' natural enemies threatened the livelihood of cotton growers. Unable to meet expenses, farmers cut the acreage planted to cotton by over 60%.

Although development of pesticide resistance by insects in the Rio Grande Valley is an extreme case, the phenomenon is common. A 1990 report of the American Chemical Society disclosed that genetic strains of 504 species of insects and mites, 273 species of weeds, and 150 species of plant pathogens (mainly fungi) are resistant to one or more pesticides.* Pesticide resistance is not restricted to agricultural pests; widespread and growing resistance to insecticides is appearing among populations of the *Anopheles* mosquito (the vector for malaria).

We can apply the principles of evolution and population dynamics described in chapters 4 and 5 to understand how

---

*See the article by Green et al., listed at the end of the chapter.

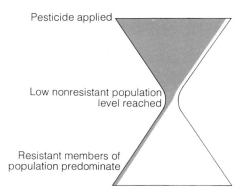

Pesticide applied

Low nonresistant population
level reached

Resistant members of
population predominate

Figure 9.7 A model of the selection for a pesticide-resistant population. The unshaded area represents pesticide-resistant members of the population; the shaded area represents pesticide-sensitive members. The passage of time goes from the top to the bottom of the figure.

pesticide resistance develops. We first must realize that (1) the population of a specific pest species inhabiting a hectare of land commonly numbers in the thousands to millions, and (2) such a huge population typically shows considerable genetic variability. Therefore, some members *may* have genes that provide natural resistance to a particular pesticide; they do not die when a pesticide is applied. After several applications of the pesticide, most nonresistant members of the population have been selected against (that is, killed), and a small resistant population survives. In the absence of intraspecific competition (between resistant and nonresistant members), subsequent generations (composed almost entirely of pesticide-resistant members) often thrive, and the population soars. A schematic representation of selection for a pesticide-resistant population is shown in figure 9.7.

Growers who apply pesticides to their crops are, in effect, taking natural selection into their own hands, and they unintentionally select for members of the pest population that are most resistant. Instead of eliminating a pest, they are responsible for the delayed explosion of a pest population that is composed primarily of resistant members. When growers react to the situation with increased dosages, more frequent applications, and the use of other pesticides, they usually continue the selection process and make it increasingly difficult to control the pest.

Water contamination

Another problem with some pesticides, particularly chlorinated hydrocarbons, is that they degrade slowly and thus persist in the environment for many years. Some pesticides readily infiltrate soils, particularly sandy soils, and eventually accumulate in the groundwater. In 1989, the EPA detected pesticides in the groundwater of twenty-six states. Fortunately most concentrations were below the health advisory level established by the EPA. However, as long as persistent pesticides are used, the potential remains for them to contaminate groundwater, perhaps at concentrations that would adversely affect human health. Meanwhile, the United States spends $1.3 billion annually to monitor groundwater for pesticides.

Some persistent pesticides also adhere to soil particles that are washed by runoff to lakes, rivers, and estuaries. Some

of them, such as DDT, are taken up by aquatic food webs and may subsequently accumulate in fish and shellfish to levels that make them unfit for human consumption (chapters 12 and 14). In addition, the accumulation of pesticides has contributed to population declines among several endangered species, such as the bald eagle and brown pelican, which occupy the higher trophic levels in aquatic food webs (see chapter 11).

Thus, we see that using synthetic pesticides involves both benefits and costs in terms of pest control and environmental quality. Assessing the impact of synthetic pesticides is complicated by the enormous variety on the market. More than one thousand different chemicals are used as pesticides in more than thirty thousand different formulations. Furthermore, each formulation has different properties and behaves somewhat differently in the environment.

Response to problems

During the 1970s, as the public became more aware of the problems associated with pesticide usage, these chemicals came under much closer scrutiny. Chlorinated hydrocarbons offered the greatest threat to the environment, so the EPA banned the widespread use of several of those insecticides, including DDT, aldrin, dieldrin, and heptachlor. The agency may permit their use in an emergency, however, such as during an epidemic of a disease that is transmitted by an insect vector. Furthermore, these insecticides are still being sold in many other nations. Although most European nations have also banned one or more pesticides that contain chlorinated hydrocarbons, virtually all nations in Central and South America, Africa, and Asia still apply them in large quantities.

When some of the chlorinated hydrocarbons were removed from general use, other insecticides such as organophosphates became more popular. Organophosphates are more compatible with the environment than chlorinated hydrocarbons, because they are generally less persistent and less likely to accumulate in food webs. However, some organophosphates, such as ethyl parathion, are more toxic to mammals than are chlorinated hydrocarbons (table 9.2); thus, they present a greater direct health hazard to humans.

In recent years, pesticide research has taken new directions. Chemical companies are now focusing research efforts on the development of pesticides that are less persistent and more selective. In addition, more precise methods of application are being studied. For example, 20%-25% of pesticide applications are by aircraft (crop dusters). Even under ideal weather conditions for aerial application, only about 50% of the pesticide lands on the target fields (fig. 9.8).

Although industry and government are undertaking programs to minimize risks to human health and the environment, a growing number of people (including farmers) are becoming more dissatisfied with the heavy (often total) reliance on synthetic pesticides in many pest-control programs. They cite increasing incidences of (1) surface and groundwater contamination, (2) population disruptions of nontarget species, (3) contamination of food webs, and (4) development of pesticide-resistant pests. They call for alternative

Table 9.2	Toxicity to nontarget species and the persistence of selected synthetic insecticides. (Ratings range from a minimum of 1, which may include zero toxicity, to a maximum of 5.)				
			Toxicity		
	Rat	Fish	Bird	Honeybee	Persistence
DDT (organochlorine)	3	4	2	2	5
lindane (organochlorine)	3	3	2	4	4
ethyl parathion (organophosphate)	5	2	5	5	2
malathion (organophosphate)	2	2	1	4	1
carbaryl (carbamate)	2	1	1	4	1

Source: Adapted from M. Begon, J. L. Harper, and C. R. Townsend, *Ecology: Individuals, Populations, and Communities*, 2d ed., Blackwell Scientific Publications.

Figure 9.8 Although aerial spraying is a popular form of application, much of the pesticide does not hit the target areas.
© Inga Spence/Tom Stack & Associates

pest-control measures that are more environmentally compatible and that better assure human safety. Many people have come to realize that synthetic pesticides, like many other products of modern technology such as chlorofluorocarbons (CFCs) and the internal combustion engine, have both positive and negative aspects. The issue on page 181 describes the continuing controversy over aerial spraying to control the Mediterranean fruit fly in Los Angeles County, which illustrates many concerns raised by synthetic pesticides today.

Alternative Pest-Control Methods

Ideally, alternative methods of pest control should (1) affect only target pests, (2) control pests permanently by reducing their populations to the economic injury level, and (3) not select for resistance. Because much of the annual production of plant biomass survives the onslaught of enemies such as pathogens and herbivorous insects, we can conclude that there are many natural means of pest control. We now consider how potential pests are controlled in nature and how we can use this knowledge to control agricultural pests and vectors that transmit diseases to humans.

Biological Control

In chapter 5, we described the process of **biological control,** the use of natural predators and parasites to control pest populations. Recall that in the example given, an insect (a ladybug beetle) was used to control populations of another insect (cottony-cushion scale) that threatened to wipe out the California citrus industry. In other situations, weeds have been controlled by the introduction of herbivorous insects. For example, the prickly pear cactus was introduced into Australia in the 1830s as an ornamental plant. The cactus escaped

from cultivation and rapidly produced dense stands that covered over 24 million hectares (60 million acres) of grazing land (fig. 9.9). To combat the cactus and reclaim the land, the cactoblastis moth was introduced in 1935 from South America, where it controlled the native populations of the prickly pear cactus. The female moth lays its eggs on the cactus. When the caterpillars emerge from the eggs, they bore into the pads (stem) of the cactus and devour tissue, thereby damaging the plant. Within a few years, the moth devastated the Australian prickly pear cactus population (fig. 9.9) and has continued to control it. Today the cactus is found only in sparse, widely distributed populations.

Biological control has several advantages. It is nontoxic and usually targets a specific pest. Furthermore, biological control is often self-perpetuating; that is, once a population of predators or parasites is established, it does not have to be reintroduced. Also, the possibility that the pest might develop a resistance to the control agent is minimal as compared with synthetic pesticides, largely because both the pest and its biological control agents often continue to coevolve and thereby maintain a stable interaction. Worldwide, approximately seventy pests have been successfully controlled by the introduction of their natural enemies, including the spruce sawfly in Canada, the coffee mealy bug in Kenya, and the spotted alfalfa aphid in the United States.

Biological control does have limitations. One of them is time: following its introduction, a control population requires time to reduce the pest population. But to maximize production and profits, farmers usually want immediate results. In addition, considerable research is required for a clear understanding of a potential control agent—how it interacts with a pest, its natural enemies, and its physical environment. Finding the best control agent usually requires fifteen to twenty years of research. Even after a control agent has been

The California Medfly Controversy

The California Department of Food and Agriculture (CDFA) argues that it must be stopped before it reaches the state's Central Valley, the vast agricultural region that is one of the nation's premier breadbaskets. Residents of Los Angeles County respond by vigorously protesting the aerial spraying of pesticides over their homes to kill it. (In 1990, over 2,600 square kilometers [1,000 square miles] were sprayed from helicopters.) The "it" at the center of this continuing controversy is the medfly (Mediterranean fruit fly) which appeared in Los Angeles County in eight of ten years during the 1980s. This fruit fly originated in Africa and is now widespread throughout tropical regions. Its larvae feed on and damage a great variety of vegetables and fruits.

Agriculturalists strongly support the spraying because they fear severe economic losses if the medfly breaks out of Los Angeles County and infests the state's rich agricultural regions. California raises over 250 types of fruits and vegetables that have a combined annual worth of more than $2 billion. Because these crops are susceptible to medfly larvae, an outbreak of this pest would inflict significant economic damage. Residents of Los Angeles County, however, are angered by the repeated dousings from aerial spraying because they fear that they are being poisoned by the pesticide malathion. State authorities, however, contend that the malathion treatments are safe in the small doses used. Some independent scientists agree. They suggest that citizens of the Los Angeles basin daily face much greater health and safety risks by breathing the region's polluted air and driving on the freeways.

A more recent and probably more pertinent controversy involves some scientists and CDFA authorities concerning the reasons why medflies continually reappear. State authorities contend that the spraying program eradicates the medfly population but that subsequently the medfly is reintroduced from places like Central America and Hawaii. Pointing to interceptions by USDA authorities of illegal shipments of fruit from Hawaii to California, CDFA personnel argue that medflies are repeatedly introduced into the Los Angeles region in smuggled fruit.

Other scientists, such as James Carey of the University of California at Davis, have a different opinion. Citing maps that show that the medflies continue to appear year after year in essentially the same areas, Carey contends that the chances of repeated introduction to the same places are slim. Rather, he thinks that medfly populations have resided in the Los Angeles basin at low levels and thus undetected since the 1970s. The pesticide sprayings have knocked the medfly populations down, but not out.

Determining the reason(s) for the repeated appearances of the medfly is critical for effective management of this pest. If, in fact, the medfly is continually being reintroduced, then the current spraying program is effectively eradicating the pest population each year. (It may be wise, however, to reevaluate the current means of eradication.) If, on the other hand, the medfly is now established in California, the chances of ever eradicating it will have to be critically evaluated. If scientists and public policymakers determine that medfly populations are likely to persist, then perhaps the current spraying program should be abandoned and efforts directed toward the development of integrated pest management programs (described later in this chapter), especially in agricultural regions. At the time of this writing, no consensus has been reached on the reason for the continued reappearance of the medfly.*

*For more on opposing viewpoints, see M. Barinaga, "Entomologists in the Medfly Maelstrom," *Science* 247(1990):1168-69.

Figure 9.9 Although it once produced dense stands that covered over 24 million hectares (60 million acres) in Australia, the prickly pear cactus was reduced to sparse, widely-distributed populations following introduction of the cactoblastis moth.
© Australian Information Service

Figure 9.10 Since its introduction to the United States about a century ago, over one hundred natural enemies of the gypsy moth have also been introduced. Nevertheless, gypsy moth caterpillars continue to defoliate large numbers of trees each year. These are oak trees on Maryland's eastern shore.

© James L. Amos/Photo Researchers, Inc. ; © John M. Burnley/Photo Researchers, Inc. (inset)

identified, there is no guarantee that it will control the pest. Only about 40% of all attempts to control insect pests by introducing their natural enemies have produced significant reductions in the pest populations. Nearly one hundred natural enemies have been introduced to control the gypsy moth in the United States, so far without any major success. Thus, this major pest continues to defoliate a great variety of trees, including birches, oaks, larch, and aspen (fig. 9.10).

Sometimes the control agent becomes a pest itself. An example of such a turnabout occurred in Jamaica, where the Indian mongoose was introduced to kill rats in sugarcane fields. At first, the mongoose did reduce the rat populations, but the rats subsequently took to the trees where the mongoose could not reach them. The mongoose's diet then expanded to include poultry and ground-nesting birds. Furthermore, the mongoose carries rabies. Ironically, the mongoose, first introduced as a control agent, is now a pest in Jamaica.

Natural Pesticides

In the coevolution that occurs between plants and their pests, plants have developed a vast arsenal of **natural pesticides,**

chemicals that are injurious to their enemies. Some researchers suggest that chemicals produced by plants may be the most important natural control of insects. For example, citrus plants produce **antifeedants,** chemicals that prevent certain insects from feeding by destroying their sense of taste. Other plants produce chemicals that interfere with an insect's ability to produce chitin, the hard, protective material that is the primary component of an insect's outer covering (exoskeleton).

Some plants produce nerve poisons, such as pyrethrum. Pyrethrum is a powerful contact insecticide that induces rapid paralysis or knockdown. It is produced by the flowers of a daisylike plant (fig. 9.11). Several pesticide companies extract pyrethrum from these flowers and use it in formulations sold in the United States and elsewhere to control flies and other household and garden insects. Another powerful natural poison is rotenone, a chemical produced in the roots of certain species of the legume (bean) family. It is a powerful inhibitor of cellular respiration. Rotenone is effective against such pests as aphids, red spider mites, and some beetles (Colorado potato beetle, for example). Scientists have just begun the task of screening thousands of plant

Figure 9.11 Harvesting pyrethrum flowers in Kenya. The flower heads are ground into a powder and used as a commercial insecticide in the United States.
© Yoram Lehmann/Peter Arnold, Inc.

species for chemicals that are harmful to their natural enemies. More successes are expected.

Recently, considerable attention has focused on the use of natural toxins produced by pathogens to control insect pests. A bacterium, *Bacillus thuringiensis* (*B.t.*) is commercially available throughout the world. This bacterium occurs naturally in the soil and produces crystals composed of powerful toxins. If insect larvae ingest these crystals, their digestive juices release the toxins. Death ensues within three days. *B.t.* crystals have proven effective against mosquitoes and many species of beetles that are agricultural pests. The chemicals are quite toxic to target species, but they appear to have no effect on nontarget species, including the pest's natural enemies and humans. Resistance to particular *B.t.* toxins has already evolved in a few insect populations as a consequence of unusually heavy reliance upon them to control these insects. Different strains of the bacterium produce distinct toxins, and in many strains, each bacterium produces a number of distinct toxins. Such a large arsenal facilitates the substitution of an effective toxin for one that is no longer effective.

Disease-Resistant Crops

Crop geneticists have imitated the natural resistance of native plants to some of their enemies by developing pest-resistant crops. For example, at one time, the Hessian fly annually caused several hundred million dollars in damage to the wheat crop. By breeding strains of wheat that are resistant to this pest, crop geneticists have virtually eliminated the Hessian fly as a significant pest. Other pests, such as the corn borer, aphid, and loopworm, have also been controlled through the development of resistant crop varieties.

The new technologies of gene transfer (genetic engineering) hold great promise for the development of pest-resistant crops. Recall from chapter 8 that **gene transfer** is the process whereby a beneficial gene from one species is introduced into the cells of another species. The gene that codes for the production of *B.t.* toxins has already been directly introduced into the genome of such crop plants as cotton, potato, and tomato. Furthermore, the structure of the transferred gene has been modified so that plant cells produce larger amounts of the *B.t.* toxin, thereby enhancing the plant's resistance to its insect enemies. In one field trial, tomato plants treated with such modified *B.t.* genes suffered little damage, while plants without the gene experienced total defoliation. Many other advances in biotechnology are on the horizon.

Pest populations will almost certainly develop resistance eventually to the toxins produced by transgenic plants, as they have to other synthetic and natural toxins. To slow down the evolutionary responses of pests to our assaults, evolutionary

Figure 9.12　An agricultural landscape consisting of a variety of crops with intervening fencerows favors long-term biological control of pests.
Bureau of Reclamation, U.S. Department of the Interior

biologists and agricultural scientists are teaming up to develop new pest-control strategies. One such strategy is the *multiple-toxin* approach. Simply put, the more barriers placed in the path of a pest, the more difficult it is for the pest to adapt. Thus, efforts are underway to develop transgenic plants that can produce more than one *B.t.* toxin. A second strategy stems from observations that plants do not naturally produce equal amounts of a toxin all season long or in all of their parts. Hence, it may be possible to produce genetically engineered plants that restrict pests to feeding on their less-important, nontoxic parts. Such a strategy would preserve food production while reducing selection pressures for resistance in the pest population.

Although these strategies to slow the evolution of pest populations hold great promise, much research and development remain to be done, both in the laboratory and the field. No doubt the success of these and other strategies will vary greatly among pest species and crop plants. Furthermore, as we noted in chapter 8, many questions remain about regulating large-scale use of genetically engineered plants and animals. Another unknown is how the public will respond when food products derived from genetically engineered plants reach supermarket shelves.

Control by Cultivation

Field studies strongly suggest that natural enemies are more effective controllers of pest populations when the environment is heterogeneous. **Environmental heterogeneity,** sometimes called **environmental patchiness,** is a measure of the degree of diversity of habitats within a region. A landscape consisting of patches of varying vegetation provides refuges for natural enemies as well as for pests. The availability of refuges usually leads to a long-term, stable interaction between the pest and its enemies. Consequently, environmental heterogeneity favors long-term biological control below the economic injury level.

Pest control in agricultural areas, therefore, can be better achieved, at least in part, by maintaining fencerows and by planting a variety of crops (fig. 9.12). In reality, most agricultural areas are **monocultures;** that is, only one crop is planted over a vast area. In addition, few fencerows remain today, because they stand in the way of the large machinery that farmers use to provide a better economy of scale. Large machinery also encourages the planting of large fields to a single crop, rather than a variety of crops. Thus, the modern farm landscape exhibits little or no heterogeneity (fig. 9.13).

Heterogeneity in time, which can be achieved by crop rotation, can also limit outbreaks of pests. **Crop rotation** involves planting each field to a different crop every one to three years. For example, a field may be sown to alfalfa for a few years and then planted in successive years to wheat, soybeans, corn for a couple of years, and then oats. The next growing season, the cycle begins anew with alfalfa.

Some pests, such as corn rootworm, feed only on a specific crop. Thus, crop rotation helps to prevent pest populations from building up over the years in a particular field by eliminating their food source at that site. For example, by simply alternating corn with alfalfa, wheat, or soybeans from one year to the next, farmers can substantially limit the size of

Figure 9.13 A simplified agricultural landscape provides few or no refuges for pests and their enemies. Thus, biological control is usually ineffective in such areas.
© Bob Coyle

the corn rootworm population in the soil. However, most farmers today plant the particular crop each year that they believe will produce the greatest monetary return for their efforts. As a result, the benefits of crop rotation are typically given little weight in decision-making.

Control by Insect Sex Attractants and Growth Regulators

In some cases, an insect's own chemical make-up can be turned against it. For example, females of many insect species release chemicals known as **sex attractants** that make it easier for males to locate them. Traps baited with a sex attractant are sometimes set to catch males and prevent their mating with females. Populations of gypsy moth and the Japanese beetle (a major pest on fruit trees and crops) have been partially controlled by this strategy. In other cases, a sex attractant is sprayed into the air, causing males to become so disoriented that they fail to locate the females. Populations of the grape berry moth, the most serious pest of grapes grown east of the Rocky Mountains, have been significantly reduced by this procedure. A major barrier to the wider application of sex attractants is their relatively high cost.

Like humans, insects have a variety of chemicals that regulate their normal growth and development. **Juvenile hormones** constitute one such group of chemicals. Changes in concentration of a juvenile hormone initiate natural changes in an insect's life cycle (fig. 9.14). If we apply a juvenile hormone in the proper concentration at the right time, we can disrupt an insect pest's life cycle. If larvae cannot develop into their adult stage, they cannot reproduce. Consequently, the population size of the next generation is reduced, and the pest causes considerably less economic damage. Juvenile hormones are usually harmless to humans, other mammals, and birds, but they may affect the pest's enemies as much as the pest itself.

Control by Sterilization

The concept of controlling a pest species by **male sterilization** is relatively simple. Large numbers of males are raised in a laboratory and subsequently sterilized, usually by exposure to radiation. When these males are released, they mate with females, which lay infertile eggs. The lowered birth rate dramatically reduces the pest population.

Although simple in concept, this technique works only for particular insect species. Its success depends first upon sterile males mating with most or all of the fertile females. To accomplish this, scientists must swamp the target area with sterile males, which must often outnumber fertile males by at least a factor of 100. Hence, the pest must lend itself to mass rearing in a laboratory. Second, the female must mate infrequently and preferably only once. Third, the target area must be well isolated, so that the control program is not undermined by immigration of fertile males.

The most successful case of pest control by sterilization was the virtual eradication of the screwworm fly from the southern United States and northern Mexico in the 1970s.

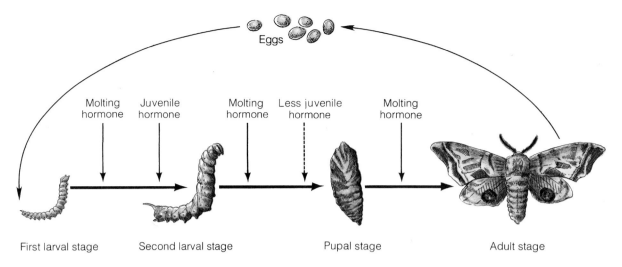

Figure 9.14 The role of juvenile hormone in the development of a moth. When the molting hormone plus large amounts of juvenile hormone are present, the organism will remain in the larval state. When the juvenile hormone is present in very small concentrations or is absent, the larval form will develop into the pupal stage. When no juvenile hormone is present, the adult stage develops.

The female screwworm fly lays its eggs directly on fresh wounds in livestock and wild animals. The eggs hatch into larvae that eat the flesh of the host animal. Untreated, the victim dies. The sterile-male method worked well in this instance because female screwworm flies mate only once. In addition, population densities were low; thus, released sterile males easily outnumbered wild fertile males. A similar program conducted by the United Nations in the early 1990s has largely eliminated a major screwworm infestation in Libya. Male sterilization has also been used with limited success against other pests, such as the oriental fruit fly and the gypsy moth.

A major disadvantage of male sterilization is expense. A large and properly equipped facility to raise and sterilize the males, special systems to release the sterile insects at the proper time, and a well-trained and coordinated staff are all necessary. In the case of the screwworm, cattle were judged to be valuable enough to justify the cost.

Integrated Pest Management

As scientists learn more about population regulation in nature, it is becoming obvious that various control agents work at different times in different places (chapter 5). This recognition has led to the development of integrated pest-management programs. **Integrated pest management (IPM)** is the coordination of all suitable procedures that can be used in as environmentally compatible a manner as possible to maintain a pest population below the economic injury level. Such programs do not necessarily exclude synthetic pesticides, but their use is minimized.

As an illustration of IPM, we return to the plight of the cotton growers in the Rio Grande Valley. In the early 1970s, when these growers were no longer able to control the insect pests with synthetic pesticides, a new system of pest management was developed. That system included three basic components:

1. Cotton stalks were shredded and plowed under by mid-September, which reduced the number of weevils by reducing their overwintering habitat.
2. A rapid-fruiting, short-season cotton variety was cultivated, which could be harvested before the weevils would normally attack the cotton bolls.
3. A limited application of insecticides was carefully timed in the spring to kill overwintering adult weevils and to minimize the impact on the insect enemies of the bollworm and tobacco budworm.

When these procedures were followed carefully, there was usually no need for additional insecticide applications. As a result of this IPM program, production costs have declined in the Rio Grande Valley, and cotton yields have more than doubled since the late 1960s.

Integrated pest-management programs have also been successful with several other crops, including sorghum, peanuts, alfalfa, apples, and peaches. Approximately 12 million hectares (30 million acres), or about 8% of U.S. farmland, are now being managed with IPM programs. Many researchers believe that IPM holds great promise for economic and environmentally sound pest control. However, many difficulties must still be overcome before this practice is widely accepted. Integrated pest management requires a great deal of research and testing, and the resulting procedures usually need to be developed to fit each crop and geographic region. For example, the program that controls cotton pests in the Rio Grande Valley is not effective in the San Joaquin Valley of California, because the two regions differ in many ways. The insect pests and their behavior in terms of how and when they attack cotton plants are different, and the climate and soil conditions are not the same. Thus, IPM programs are time-consuming and expensive to develop and evaluate.

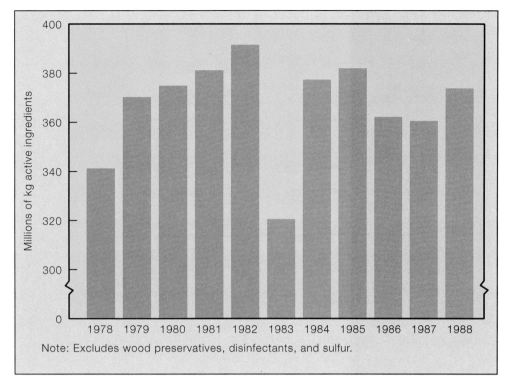

Figure 9.15 Agricultural use of pesticides since late 1970s shows relatively little change in the amounts applied.
Source: Environmental Protection Agency

Conclusions

It is clear that the need for pest control will become even greater in the future. Because the human population continues to grow rapidly, more effort must be given to the continuing battle to reduce food losses to agricultural pests. Furthermore, because most of the human population growth will occur in less-developed nations where hundreds of millions of people already suffer from diseases that are spread by vectors, more effort must also be directed at controlling these pests. How will these battles be fought?

As figure 9.15 indicates, agriculturalists still depend heavily on synthetic pesticides—but times are slowly changing. More and more farmers are switching to alternative forms of pest control. They are becoming convinced that synthetic pesticides involve real health and environmental risks. They are increasingly eager to protect their lands and their groundwater, as well as the well-being of their workers and people who buy their products. But even more powerful forces are at work. With continued research and development, some procedures such as the application of *B.t.* crystals work as well as synthetic pesticides at equal or lower cost. Furthermore, farmers know that consumers are demanding food that is free of unnecessary chemicals. Thus, some fruit and vegetable growers have switched to alternative control methods because they sell their produce to processors who insist on minimal chemical residues. In some states, such as California, strict environ-

mental legislation requires massive amounts of paperwork by growers who use synthetic pesticides. Such administrative costs are persuading some farmers to switch to alternative methods of control. Thus, the more-developed nations are slowly shifting to pest-control techniques that are more environmentally sound and more compatible with natural pest control. Table 9.3 describes several environmentally safe procedures that we can follow to control pests in our own homes and gardens.

Table 9.3 Things we can do to control pests

1. To prevent entry of such insects as ants and roaches, caulk windowsills, door thresholds, and baseboards.
2. To avoid attracting insects, don't leave out food and dirty dishes. Clean up spilled food.
3. Control ants and roaches by sprinkling borax around baseboards, appliances, and ducts.
4. Get rid of aphids by spraying plants with a mixture of soap dissolved in water.
5. Create heterogeneity in your lawn and garden by planting a variety of plants.
6. If there are just a few insects or weeds, remove them by hand.
7. If possible, purchase only alternative means of pest control.
8. If you must use synthethic pesticides, use them in the smallest amounts possible.

Pest populations pose particularly severe problems in less-developed countries, where few people can afford to lose food and hundreds of millions of people are afflicted with vector-borne diseases. Many of these people work at the subsistence level and cannot afford pesticides even when they are available. These impoverished, technologically unsophisticated countries are in great need of inexpensive, simple-to-use alternatives.

Control programs for agricultural pests in less-developed nations should probably focus on cultivation of pest-resistant crop varieties and improvement of local farming practices. Also, crops must be protected from pests after harvesting, since approximately one-third of the world's food loss to pests occurs after the crops are harvested. In particular, rat- and bird-proof structures are needed that can be fumigated with nonpersistent pesticides to eliminate insects and fungi. In many areas today, even bumper crops do little to alleviate hunger because storage and transportation facilities are inadequate. Surplus crops must be stored outside, where they either rot or are devoured by rats and birds.

Control of vector-borne diseases may be shifting from the control of the vector to simpler methods of controlling the pathogen. Vaccines are under development that battle the malaria-causing parasite after it is injected by mosquitoes into the human blood stream. In addition a drug called Mectizan is available that destroys the larvae of the parasitic worm that causes onchocerciasis (river blindness). The drug is taken orally once a year.

As we have seen, many of our pest-control successes have lasted only a few years, as pest populations have not only survived our assaults, but often have resurged to previous population levels. Although we are beginning to develop control strategies that are based upon a growing understanding of evolutionary and ecological processes, we should not underestimate the evolutionary potential of pest populations. The biological evolution of pests will probably continue to keep pace with the evolution of our strategies to control them. Thus, although pest-control strategies are slowly becoming more environmentally friendly, our battles with pests will continue to be hard-fought.

Key Terms

pathogen	biological control
vector	natural pesticides
pest	antifeedant
economic injury level	gene transfer
synthetic pesticide	environmental heterogeneity
insecticide	environmental patchiness
chlorinated hydrocarbons	monoculture
organophosphates	crop rotation
carbamates	sex attractant
fungicide	juvenile hormone
herbicide	male sterilization
broad-spectrum pesticide	integrated pest management (IPM)
pesticide resistance	

Summary Statements

An amazing variety of pests threatens our health or interferes with our production of food and fiber. Most pest species are *r*-strategists.

The goal of most pest-control programs should be to reduce the pest population to the economic injury level.

Synthetic pesticides are formulated to kill specific types of pests. In more-developed countries, the use of pesticides has reduced crop losses significantly, making more land available for recreation, forests, and wildlife preserves. Synthetic pesticides also reduce the threat of vector-borne diseases for more than a billion people.

On the negative side, some synthetic pesticides kill beneficial species and disrupt natural pest-control mechanisms. Pesticide resistance has developed in populations of over 900 species of insects, weeds, and plant pathogens. Because some synthetic pesticides are persistent, they may accumulate in food webs and groundwater and reach toxic concentrations.

Each synthetic pesticide must be evaluated on its own merits. Insecticides that are most harmful to the environment are the persistent chlorinated hydrocarbons.

Alternatives to synthetic pesticides include biological control, natural pesticides, development of pest-resistant crop varieties, crop cultivation practices, sex attractants, juvenile hormones, and male sterilization. Each alternative has its advantages and disadvantages.

Integrated pest management is an attempt to simulate nature by combining several pest-control procedures that produce effective control while minimizing environmental damage. Although very effective programs have been developed for some crops in some regions, IPM programs remain quite time-consuming and expensive to develop and evaluate.

Given the continued growth of human populations, particularly in less-developed nations where food is often scarce and vecto-borne diseases are prevalent, the need for effective pest-control programs will become even greater in the future. Although synthetic pesticides continue to be the mainstay of most pest-management programs, economic, political, and social forces are beginning to stimulate a switch to alternative strategies of pest control.

Questions for Review

1. Describe the characteristics of a pest species. Why are most pest species *r*-strategists?
2. What is the economic threshold level? Describe its importance for pest-management programs.
3. Who is the biggest user of synthetic pesticides? How much pesticide do homeowners apply to their lawns?
4. Describe the benefits of synthetic pesticide usage.
5. List the types of environmental problems that often result from the use of broad-spectrum pesticides.

6. Describe how an insect population develops resistance to an insecticide. Relate this process to the observation that some antibiotics no longer control the growth of certain types of bacteria (such as the bacterium that causes gonorrhea) in people.
7. List the characteristics of an environmentally sound pesticide.
8. What is biological control? List its advantages and disadvantages.
9. Present several examples of natural pesticides and how they control pests.
10. Define environmental heterogeneity. What is its importance for effective biological control?
11. What is an insect sex attractant? How is it used to control a pest population?
12. Describe the conditions required for successful control of a pest by male sterilization.
13. Present an example of an integrated pest-management program. Why are IPM programs difficult to develop?
14. What are some of the political, economic, and social forces that appear to be stimulating a switch from synthetic pesticides to alternative means of pest control?

Projects

1. Interview people in the agriculture, forestry, and health sectors to find out the five pest species that most threaten the well-being of people in your region. What common characteristics do these species share? What is being done to control them? Are these pest-control programs effective? Why or why not? What, if any, are the environmental consequences of these control procedures?
2. Visit a nursery or a garden and lawn supply business. Determine what fraction of their products consists of synthetic pesticides versus natural chemicals. Find out if the owner has detected an increased interest in alternative means of pest control.
3. Visit the buildings and grounds office of your campus. Find out what insect and plant species your campus has judged to be pests. What methods are currently used to control these pests? Are they effective? Discuss with your class whether these species are truly pests. Also evaluate the pest-control procedures your campus uses. Are there better alternatives?

Selected Readings

Adkisson, P. L., G. A. Niles, J. K. Walker, L. S. Bird, and H. B. Scott. "Controlling Cotton's Pests: A New System." *Science* 216(1982):19-22. A description of the integrated pest-management program that controls pests of cotton in the Rio Grande Valley.

Barrett, S. C. H. "Waterweed Invasions." *Scientific American* 261(Oct. 1989):90-97. An account of how studies of the water hyacinth, one of the world's most widespread aquatic weeds, are leading to new programs for weed control.

Dahlsten, D. L., and R. Garcia, eds. *Eradication of Exotic Pests.* New London, Conn.: Yale Univ. Press, 1989. An analysis of the concept of eradication and an interesting examination of the attempts to eradicate eleven of the world's major pests.

Dover, M. J., and B. A. Croft. "Pesticide Resistance and Public Policy." *Bioscience* 36(1986):78-85. Examines the history of pesticide resistance and describes strategies for coping with this growing problem.

Gerster, G. "Tsetse: Fly of the Deadly Sleep." *National Geographic* 170(Dec. 1986):814-33. A lucid description of the continuing war to control the insect that is a scourge to humans and cattle alike in Africa.

Gould, F. "The Evolutionary Potential of Crop Pests." *American Scientist* 79(1991):496-507. An account of the fascinating ability of crop pests to adapt to pest-control procedures and an examination of newly evolving pest-control strategies.

Green, M. B., H. M. LeBaron, and W. K. Moberg. *Managing Resistance to Agrochemicals: From Fundamental Research to Practical Strategies.* Washington, D. C.: American Chemical Society, 1990. An examination of pesticide resistance as well as prescriptions for implementing management responses.

Habicht, G. S., G. Beck, and J. L. Benach. "Lyme Disease." *Scientific American* 257(Dec. 1987):78-83. A fascinating account of the discovery of Lyme disease and the vector that transmits it to humans.

Horn, D. J. *Ecological Approach to Pest Management.* New York: The Guilford Press, 1988. An examination of the principles and practices that support effective and balanced pest-control programs.

Lambert, B., and M. Perferoen. "Insecticidal Properties of *Bacillus thuringiensis.*" *Bioscience* 42(1992):112-22. An examination of the characteristics of this bacterium and its future for controlling insect pest populations.

Pimentel, D., and L. Levitan. "Pesticides: Amounts Applied and Amounts Reaching Pests." *Bioscience* 36(1986):86-91. Examines the routes taken by excess pesticides as they move throughout the environment.

Reganold, J. P., R. I. Papendick, and J. F. Parr. "Sustainable Agriculture." *Scientific American* 262(June 1990):112-20. Includes a review of alternative pest-control practices.

Strobel, G. A. "Biological Control of Weeds." *Scientific American* 265(July 1991):72-78. Reviews a wide variety of environmentally sound approaches to weed control.

Chapter 10

Management of Forests, Rangelands, Parks, and Wetlands

Harvesting redwood trees on the north coast of California.

© Inga Spence/Tom Stack & Associates

*I*n the early 1990s, a major battle was brewing between wilderness and energy. At the heart of the confrontation is the Arctic National Wildlife Refuge (ANWR), a primordial landscape populated primarily by musk oxen, caribou, and bear. With only about two hundred permanent human residents, this 7.5 million-hectare (19 million-acre) refuge is about as unaltered by human activities as any wilderness can be today. ANWR's 160-kilometers (100-mile) coastline is the only remaining segment of Alaska's 1,900-kilometer (1,200-mile) Arctic Ocean coastline that is still off-limits to economic development (fig. 10.1). Beneath this coastal plain lies a pool of oil that may be the largest remaining oil reserve in the United States. Hence, the conflict.

The United States consumes enormous quantities of oil (chapter 18). Because domestic supplies are unable to meet demand, it must import from other nations about half of the petroleum it consumes each year. Furthermore, America's largest oil field, Prudhoe Bay (also in Alaska), is past peak production. Within ten years, production will decline by one-half of its current output. Federal officials argue that without further exploration for domestic oil, the national economy will be further weakened as billions of dollars leave the country to pay for imported oil. Hence, the petroleum industry, an oil-thirsty nation, and the state of Alaska (which receives over 80% of its government revenues from taxes, fees, and royalties associated with oil) in the early 1990s turned to ANWR. The coastal plain sits atop the same geological formations

that underlie Prudhoe Bay. The oil reserves under ANWR may be as large as those at Prudhoe Bay; on the other hand, the odds are 4 to 1 against finding any exploitable oil at present prices. Only drilling will tell how much oil this formation holds.

Government and industry officials estimate that only 5,200 hectares (13,000 acres) of the coastal plain are required for exploration activities, but conservationists argue that more land than this would actually be disrupted. Oil production would require a pipeline connecting the coastal plain and the trans-Alaska pipeline (fig. 10.1). Furthermore, additional land would be needed for drilling pads and waste pits, production facilities, water treatment plants, housing for workers, and other structures.

Conservationists also argue that the nation must protect its few remaining wilderness areas. ANWR's coastal plain, for example, is the seasonal home of hundreds of thousands of migratory birds. Each summer, 30 species of shore-birds, tundra swans, and as many as 325,000 lesser snow geese arrive to nest and feed upon the abundant tundra vegetation and insect life. In addition, the Porcupine herd of caribou, which numbers about 180,000, migrates each summer across the Porcupine River to ANWR to bear their young and feed on tundra plants (fig. 10.2). Concerns have been voiced over the impact of oil exploration on the migratory birds, caribou, and other wildlife such as grizzly and polar bears, moose, musk oxen, wolves, and wolverines that inhabit the coastal plain

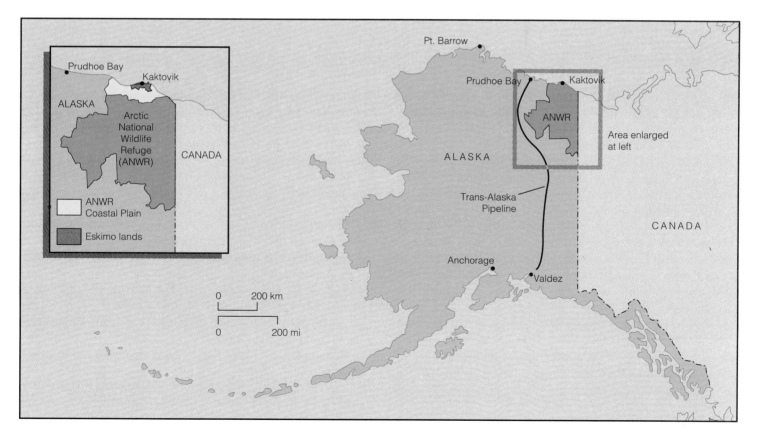

Figure 10.1   The ANWR coastal plain harbors both oil and pristine wilderness.

Figure 10.2   Each year, some 180,000 head of the Porcupine caribou herd migrate about 560 kilometers (350 miles) between their winter range along the Porcupine River and their summer calving grounds on the ANWR coastal plain.

© Johnny Johnson/Animals Animals

during critical stages in their life cycles. Furthermore, conservationists point out that arctic tundra ecosystems are especially vulnerable to disturbance. Something as simple as tractor tracks can scar the landscape for decades.

Some government and petroleum officials downplay the potential impact of oil production. They contend that the oil industry has learned much from its past mistakes and that with new technology, oil development has become a much safer and cleaner endeavor. They also point out that the central arctic caribou herd, which gives birth to most of its calves near the Prudhoe Bay oil wells, has tripled in number in recent years. Although they do not suggest that petroleum production promotes the growth of wildlife populations, they argue that oil production and wildlife can coexist.

Conservationists counter that caribou herds worldwide have grown over the past decade. They suggest that the growth of the central arctic herd may be due to natural population cycles or changing weather patterns. Other possible explanations include restrictions on caribou hunting and sharp declines in natural predators (particularly wolves) since oil development began.

Conservationists also urge renewed emphasis on energy conservation and alternative energy sources instead of exploitation of the oil reserves that lie beneath the coastal plain (chapter 19). Energy officials counter that existing alternatives are inadequate to meet U.S. energy needs over the next twenty years. However, if or when exploration begins, another ten to fifteen years will elapse before the first oil from ANWR reaches the trans-Alaska pipeline. Thus ANWR's oil, even if reserves are large, will not solve either today's or tomorrow's energy problems. Why then disturb one of the nation's few remaining wilderness areas? After all, other proposals to enhance U.S. energy security (such as damming the Colorado River in the Grand Canyon National Park and tapping Yellowstone National Park's geothermal potential) have been dismissed as unwarranted. As of this writing, the question to drill or not to drill in ANWR is unresolved.

The ANWR controversy is just one of many land-use conflicts that we face today. We are constantly modifying and reshaping the landscape for housing sites and industrial and commercial purposes. We terrace hillsides, drain and fill wetlands, cover the ground with concrete and asphalt, dam rivers and streams, and carve out campsites and parks. We grow our food and fiber on land, and we build our cities on it. Furthermore, we tear up the land, extract its rock, mineral, and fuel resources, and use it as a dumping ground for our wastes.

Figure 10.3 In the past, timber was harvested with little regard for the scars left behind. The remaining slash often fed wildfires, and the land left unprotected by any cover was subject to severe erosion.
© B. H. Brattstrom/Visuals Unlimited

At one time there was so much unoccupied land in the United States that there seemed to be more than enough space and resources to meet human demands. But times are changing. More people are realizing that the land and its resources are finite. As our population and its demands continue to grow, so do land-use conflicts. Urban sprawl is threatening nearby open spaces, farmers are draining wetlands, miners compete to exploit minerals and fossil fuels in natural areas, loggers are harvesting forest land that harbors endangered species, and sheep and cattle are competing with indigenous wildlife on rangelands.

In this chapter, we survey U.S. land-use patterns and examine efforts to manage land-use conflicts. We focus primarily on federal **public lands,** that is, forests, rangelands, parks, and wilderness areas owned and managed by the federal government. Our reasons for concentrating on these landscapes are that (1) the federal government manages over one-third of the total land area of the United States; (2) federal lands are increasingly the sites of major land-use conflicts; and (3) an individual citizen has a much greater chance to influence land-use patterns on federal land than on private land. Although we focus on federal public lands, it is important to note that similar land-use conflicts occur on public lands owned and managed by local and state governments.

Historical Perspective

Several federal agencies, including the U.S. Forest Service, National Park Service, Fish and Wildlife Service, and Bureau of Land Management regulate activities on public lands. However, these lands were not always regulated. During the first one hundred years of U.S. history, developers exploited natural resources with virtually no restrictions. During that period, the federal government handed over enormous parcels of land to anyone who was willing to exploit its minerals, timber, or water. Consequently, the nation got the resources that it needed, and the economy strengthened, but the environment frequently suffered considerable damage (fig. 10.3). In the late nineteenth century, however, some individuals began to argue that the consequences of such give-away policies were becoming intolerable. They called for measures to conserve natural resources and preserve national treasures. More and more people urged the federal government to exercise more prudent stewardship over public lands for the common good.

The first major step in the direction of resource conservation on public lands was the establishment of Yellowstone National Park in 1872. Later that same year, California took steps to protect Yosemite Valley. Twenty years later, the first federal forest reserves (later called *national parks*) were set aside, and by 1900, six national parks were established.

In the early part of this century, President Theodore Roosevelt championed the conservation movement; he was concerned about the mounting cost of widespread mismanagement of the nation's natural resources. In 1908, Roosevelt called a White House Conference on National Resources, and one outcome was the appointment of a fifty-member National Conservation Commission. Gifford Pinchot (who had been appointed Chief Forester of the Department of Agriculture in 1898) and Roosevelt persuaded Congress to enlarge the nation's forest reserves nearly fivefold. That major goal was accomplished over the vigorous objections of ranchers, miners, and others who were accustomed to having free rein over public lands.

Pinchot's conservation philosophy was the guiding principle for federal management of public lands throughout much of this century. While he favored resource development over preservation, he advocated development of the nation's resources for the public good with a minimal impact on the environment whenever and wherever possible. Other individuals, such as John Muir, differed sharply with Pinchot. Muir (one of the founders of the Sierra Club) argued that some areas should be kept totally free of development or resource exploitation. To this day, differences of opinion among individuals and groups supporting these and other philosophies on the use of the nation's resources create the basis for innumerable land-use conflicts.

Today, public lands under federal management total almost 325 million hectares (800 million acres)—approximately 1.2 hectares (3 acres) per citizen. More than half of Idaho and Utah and more than 80% of Alaska and Nevada is federal land (fig. 10.4). The law permits a wide array of activities on these lands (fig. 10.5). Based on federal law, federal public land falls into two general categories: (1) regions set aside primarily for preservation, and (2) areas intended for regulated multiple use. Wilderness areas, wild and scenic rivers, national parks, and wildlife refuges are designed primarily to protect and preserve invaluable natural resources. In national forests and on lands regulated by the Bureau of Land Management, a variety of activities are permitted. These areas are usually subject to regulations intended to prevent multiple uses from interfering with one another and to ensure that uses are compatible with environmental quality. Although these management goals are simply stated, federal land agencies must in fact cope with numerous conflicts. In the following sections, we examine the nature of these conflicts and attempts to resolve them.

National Forests

The U.S. Forest Service manages the National Forest System (NFS) in trust for the American public. This vast system of forests and grasslands spreads over forty-two states and Puerto Rico. It encompasses 77 million hectares (191 million acres) including 204,000 kilometers (128,000 miles) of streams and

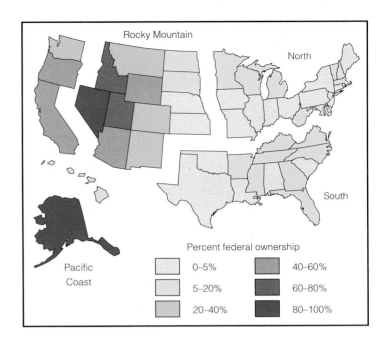

Figure 10.4   Large portions of the western states are owned by the federal government.

Source: Based on data from *1984 State of the Environment and Assessment at Mid-Decade*, The Conservation Foundation.

890 thousand hectares (2.2 million acres) of lakes. All told, the NFS comprises 8.5% of the nation's total land area and nearly 15% of the total forestland. The Multiple Use–Sustained Yield Act of 1960 mandated that the Forest Service manage these public lands to provide a sustained production of numerous valuable resources, including timber, livestock forage, minerals, water, fish and wildlife habitats, and recreation.

Historically, however, the Forest Service has had considerable difficulty in balancing multiple uses of the public land that it oversees. The agency has focused primarily on watershed protection, timber production, livestock grazing, and mineral extraction, and neglected other resources. In addition, some critics accuse the Forest Service of maximizing grazing and timber and mineral production while ignoring the environmental consequences of such practices. In response, Congress enacted the Forest and Rangeland Renewable Resources Planning Act of 1974 (RPA) to assure long-term, sustainable management of renewable resources and to increase public involvement in resource policy and budget debates. The RPA charged the Forest Service to give greater emphasis to ecologically sensitive management of NFS lands and to focus more attention on fish and wildlife resources, wilderness areas, recreation areas, and endangered species.

Despite the intent of the RPA, most management plans for individual national forests continued the traditional emphasis on timber production until the late 1980s. At that time, a resurgence of public concern for the environment coupled with growing demand for leisure-time resources began to force the Forest Service to alter its perspective. These changes became evident in 1990 when the Forest Service released

(a)

(b)

(c)

its five-year strategic plan for national forest and rangeland management programs as mandated by the RPA. Four new high-priority themes reflect a greater concern by the Forest Service for other resource management values. These four themes are (1) recreation, wildlife, and fisheries resource enhancement, (2) environmentally acceptable commodity production, (3) improved scientific understanding of natural resources, and (4) responses to global resource issues.

Many of the specific objectives associated with the four priority themes will improve the long-term, sustainable management of NFS lands. One objective is to reduce the level of timber harvests to provide habitats for threatened and endangered species as well as to preserve scenic and recreational sites. Reductions in the timber harvest will also reduce the number of below-cost timber sales, which are particularly irritating to conservationists. To enhance timber sales, the Forest Service builds logging roads to reach the trees, surveys the tracts that contain the trees, and draws up legal documents for the sale. In many sales, large, mature trees sell for as little as $1 per tree. When costs of preparation are compared with receipts from sales, one finds that the Forest Service frequently loses money. In fact, over the past ten years, the Forest Service has averaged an annual total loss of $343 million on timber sales. Taxpayers, of course, must absorb these losses.

Another objective of the 1990 strategic plan is a reduction in clear-cutting and an increase in partial cutting. **Clear-cutting** involves the removal of all the trees in a plot in a single cutting. Foresters have relied heavily on clear-cutting, especially since the 1960s. Generally done in plots of 8-16 hectares (20-40 acres) scattered throughout a forested region, clear-cutting can lead to several forms of environmental degradation. A close clustering of clear-cut plots can cause considerable forest fragmentation (fig. 10.6). Such habitat fragmentation often has a severe impact on species that require older-growth forests (see chapter 11). In addition,

Figure 10.5 A variety of activities are permitted on federal (public) lands including (*a*) recreation, (*b*) grazing, and (*c*) mining. Public lands must be managed so that these activities do not conflict with one another.

(*a*) © Michael Fredericks/Earth Scenes (*b*) © Tom Stack/Tom Stack & Associates (*c*) BLM

Management of Forests, Rangelands, Parks, and Wetlands   195

Figure 10.6 A close clustering of large, clearcut plots can lead to a loss of species diversity and accelerated soil erosion.
© Jack Swenson/Tom Stack & Associates

Figure 10.7 Partial cutting of a forest stand leaves behind trees that serve as seed sources.
© Milton H. Tierney, Jr./Visuals Unlimited

closely clustered clear-cut plots on steep slopes are subject to severe soil erosion and stream sedimentation.

After clearing, a clear-cut plot may reseed naturally. Increasingly, however, the plot is planted with seedlings selected for high growth rates. Clear-cutting thus eventually leads to establishment of an even-aged stand, often composed of only one or two species of trees. Such plots are more vulnerable to pest outbreaks, fires, and perhaps even to climate change.

Partial cutting covers a continuum of activities that range from selectively cutting only a few trees (by desired size and species, for example) to harvesting most but not all (typically 85% to 90%) of the trees on a site (fig. 10.7). In the latter situation, the remaining trees serve as seed sources and as refugia from which species can invade the surrounding developing forest as it becomes established and matures. Aggregating partially cut sites reduces habitat fragmentation and thus helps maintain species diversity. Furthermore, soil erosion is usually less of a problem than on clear-cut sites.

The Forest Service is now officially planning to make the transition from a timber/grazing/watershed/mining orientation to one with more emphasis on fisheries, wildlife, recreation, and environmentally friendly resource use. However, transferring such strategic planning to actual forest management will be most challenging. These new initiatives require significant additional funding, which will be difficult to obtain from Congress. The forest-products industry remains a powerful economic and political force, and many people within the Forest Service remain strongly committed to timber production. Furthermore, if timber harvests are to be reduced, retraining should be made available to those in the timber industry who will lose their jobs. There is a consensus that forestry practices must be more ecologically based and that the Forest Service must achieve a better balance in its management of the diverse resources on NFS lands. However, progress will be slow as research and management evolve to enhance the compatibility of multiple resource uses and to ensure a long-term, sustained supply of resources from National Forest Service lands.

National Rangelands

Rangelands are essentially unsuited for rain-fed crop cultivation or forestry. Grasses or shrubs such as sagebrush (fig. 10.8) dominate these arid and semiarid tracts. Essentially all of the rangeland in the United States is west of the Mississippi River. About one-third of it is publicly owned, most of that land is managed by the Forest Service or the Bureau of Land Management (BLM).

Federal rangelands are primarily managed for livestock production. In 1989, ranchers leased nearly 108 million hectares (267 million acres) of federal land for grazing their livestock. Although grazing has long been the traditional use of federal rangelands, management of public rangelands primarily for this single purpose is coming under increasing fire from conservation groups.

Figure 10.8 Sparse precipitation limits plant production on this sagebrush-dominated rangeland in Arizona.
© William E. Ferguson

Figure 10.9 Overgrazing often leads to severely degraded rangeland such as shown to the right of the fence in this photograph.
© Dennis Paulson/Visuals Unlimited

From an economic perspective, conservationists view grazing leases as large subsidies to a few privileged ranchers. Only about 2% of private ranchers graze their cattle on public lands, and they pay leasing fees that are only about one-sixth to one-eighth of the fees for leasing equivalent tracts of private grazing lands. From another perspective, grazing fees usually cover less than one-half of the federal costs of range management; the balance is borne by taxpayers.

Environmental costs are even greater. In 1987, a BLM survey of its grazing lands reported that nearly 60% were in unsatisfactory condition. The major source of degradation is **overgrazing:** exceeding the land's carrying capacity by raising more cattle or sheep than the available forage can sustain. Degradation of rangeland begins with a decline in the palatable perennial grasses. This loss in plant cover allows less-palatable, weedy annual grasses to invade a rangeland. If overgrazing and associated trampling of vegetation and soil continue, vegetative cover of all species begins to diminish, making the land more vulnerable to erosion by water and wind (fig. 10.9).

Overgrazing has a particularly dramatic impact on **riparian areas,** that is, lands adjacent to streams and rivers, where vegetation is strongly influenced by the presence of water. In 1988, the BLM reported that 90% of the streamside habitat that it manages in Colorado was in unsatisfactory condition, and 80% of BLM's riparian lands in Idaho were degraded. Overgrazing of riparian habitats leads to erosion and stream sedimentation (fig. 10.10). As livestock trample denuded stream banks, channels widen, and water temperatures rise, threatening native species of fish. The loss of cottonwood trees along these streams results in population declines of such birds as bald eagles and hawks, which require large trees for roosting and nesting sites. Tree-cavity nesters such as bluebirds and woodpeckers also decline as riparian vegetation disappears. Overgrazing has also severely affected other riparian and upland species, including gamebirds and songbirds.

Although rangelands that are currently managed by federal agencies were once home to large numbers of grazing herbivores, populations of many species such as bison, mule deer, and pronghorns have declined by as much as 95% since the advent of livestock grazing. The future for these species will not improve as long as federal agencies allocate essentially all of the annual plant production on rangelands to livestock grazing.

Resolving the conflicts associated with sustained management of rangelands appears to be relatively simple: Reduce livestock herds to a level that the rangelands can sustain without degradation. Conservation groups also propose that a better balance can be achieved between wildlife and livestock populations. Even though the carrying capacity of a particular site varies with weather, season, slope exposure, and soil and vegetation type, the scientific and management expertise is available to make reasonably sound estimates of how many livestock an area can sustain. Furthermore, restoration of rangelands in general and riparian habitats in particular is often successful (fig. 10.11).

Impediments to resolving these conflicts, however, are numerous. Ranchers see any reduction in grazing allotments as a threat to their livelihood. Few admit that rangeland degradation is a problem, or if they do, they claim that grazing is not the cause. Cattlemen's associations exert considerable political pressure in the West and traditionally have had their way in the management of federal rangelands. They view conservationists as outsiders who are threatening a way of life that has prevailed for generations. Meanwhile, livestock grazing has always been a central component of the management policies of federal land agencies, particularly the BLM. Obviously, these resource managers are reluctant to compromise the basis of their professional careers. Thus, federal agencies have been slow to initiate any major reforms. Meanwhile conservation groups are asking why public rangelands are not being managed primarily to preserve species diversity, watersheds, recreation, and scenic qualities. They also ask why federal agencies continue to subsidize grazing while permitting further deterioration of these public lands.

Figure 10.10 Excessive grazing by livestock can seriously degrade riparian habitats. Riparian zones are narrow strips of land that border rivers, streams, or other bodies of water.
Wayne Elmore and Robert L. Beschta/BLM

Figure 10.11 Reduction of grazing and reestablishment of vegetation can restore degraded riparian habitats. This is the same creek as shown in figure 10.10 after 8 years without grazing. The steep cutbank in the background is no longer being eroded by the stream.
Wayne Elmore and Robert L. Beschta/BLM

(a)

Figure 10.12 Some of the many treasures protected in the National Park System: (*a*) Yosemite Falls in Yosemite National Park, California.
(*a*) © Charles Mayer/Photo Researchers, Inc.

National Parks

The world's first national park was founded in 1872 when the U.S. Congress set aside over 800 thousand hectares (2 million acres) to establish Yellowstone National Park. Today the National Park Service (NPS) manages more than 350 units, including national parks, monuments, lakeshores, seashores, historic sites, and battlefields. The NPS manages these units to preserve a vast array of natural and historic features as well as to meet public recreational needs (fig. 10.12). Through nature walks, campfire talks, visitor center displays, numerous pamphlets, and other educational services, National Park Service employees provide nearly 270 million annual visitors with a better understanding of how nature works.

As National Park Service land has increased and the U.S. population and its resource demands have grown, land-use conflicts have also multiplied. Examples of conflicts are as varied as the national treasures that the Park Service is mandated to preserve. In this section, we present only a sample of the complex management issues that the NPS faces. We first focus on internal stresses and then consider external stresses on the park system.

Internal Stresses

There is little debate that the greatest stress on many national parks is overcrowding (fig. 10.13). Americans are loving their national parks to death. The most popular national parks in 1991 included Great Smokey Mountains (8.7 million annual visitors), Grand Canyon (3.9 million), Yosemite (3.4 million), and Yellowstone (2.9 million). During the height of the tourist season, certain regions within these and many other parks are overwhelmed with congested traffic, vandalism of natural and cultural features, noise, air pollution, and litter. As more and more people flock to popular scenic attractions, they cause ever-widening circles of destroyed vegetation and compacted soil. Millions of feet are wearing down the hiking trails

(b)

(c)

Figure 10.12 continued   (b) A grove of sequoias in Sequoia National Park, California; (c) hoodoos in Bryce Canyon National Park, Utah.
(b) © Richard Weiss/Peter Arnold, Inc. (c) © Ned Haines/Photo Researchers, Inc.

and making them more subject to erosion (fig. 10.14). The greatest threats to human safety are not the steep cliffs and wild animals, but traffic accidents. Park rangers spend considerably more time managing people than natural resources.

Recent events in Yosemite illustrate the opportunities and the difficulties of curbing human impacts on national parks. In 1980, the NPS released a management plan for Yosemite that called for greatly reducing park traffic and overnight accommodations in the park, and relocating some nonessential NPS and concession functions outside the park. Much of the park's spectacular Yosemite Valley has become a virtual town, complete with tent cabins, motels, restaurants, liquor stores, and even a video outlet. In the past ten years, shuttle bus service has expanded, reducing traffic in some areas, and a golf course has been removed from the valley floor. Otherwise, progress has been scant; NPS and concession buildings and functions remain in place. In fact, pressures on the park have increased dramatically, as the annual visitors have risen by over 35% since the management plan was initiated in 1980.

Figure 10.13   Visitors crowd a footpath in Glacier National Park, Montana.
© John Gerlach/Visuals Unlimited

Management of Forests, Rangelands, Parks, and Wetlands   199

Figure 10.14 Heavy use of high country trails in national parks leads to serious soil erosion.
© Hank Andrews/Visuals Unlimited

Reasons for limited progress are numerous. Some conservation groups suggest that the NPS has not been vigorous enough in pursuing solutions; in particular, they argue that the NPS needs to develop more innovative alternatives to the complex problems of people management. In their defense, NPS personnel argue that Congress has not appropriated sufficient funds to make the needed changes. In fact, the National Parks and Conservation Association suggests that the Park Service has inadequate funds to maintain its buildings, roads, and trails (a maintenance backlog of more than $2 billion) and to fill hundreds of needed ranger positions. (The ratio of rangers to visitors is currently at an all-time low.)

Additional funds may become available in response to a report issued in the spring of 1990. In that report, the Department of the Interior (of which the NPS is an agency) noted that in 1988, concessionaires in national parks sold $490 million in goods and services, but paid the federal government only $12.5 million in licensing fees, an average rate of return of 2.5%. Obviously, business is booming, and the Department of the Interior, believing that the federal government (the taxpayers) is not receiving adequate compensation in licensing services, is asking Congress to raise the licensing fees. In addition, conservation groups are requesting that these proceeds go to the National Park Service. Currently, they go to the U.S. general treasury. In its early years, the NPS negotiated favorable contracts to attract concessionaires in the belief that they would, in turn, attract more visitors to the largely underutilized parks. But times have changed; there are now fifteen times more visitors to U.S. national parks than in 1950. The NPS no longer needs to offer incentives to concessionaires to attract visitors.

The greatest challenge to the NPS may be on the horizon. As visitor numbers continue to soar, park managers are finding it increasingly difficult to preserve resources while providing for visitor needs. The NPS may soon have to tackle the concept of carrying capacity and make the controversial decision to limit admission to the most crowded parks.

Another internal stress arises from management of wildlife populations within the parks. Conditions in Yellowstone illustrate some of the conflicts. As in many parks, the boundaries of Yellowstone were established to preserve natural features such as the geysers. Wild animals, however, do not recognize political boundaries. Thus, the historic winter migration patterns of elk and bison in Yellowstone frequently take them outside park boundaries, where they are unprotected (fig. 10.15). Furthermore, most parks are surrounded by private or public lands that are not managed for wildlife preservation. Thus, bison in Yellowstone run afoul of local ranchers who graze their cattle on National Forest lands adjacent to the park.

Ranchers fear the bison will transmit brucellosis to livestock, a disease that causes cattle to abort. Although no such cases have been documented, Montana law allows ranchers to kill any bison that crosses the northern park boundary. In most years, relatively few bison are shot. In 1989, however, large numbers of bison left the park. A series of mild winters and the lack of natural predators allowed the herd to grow dramatically. Then a severe winter with heavy snows drove as many as 800 of the estimated 1,000 bison that make up the northern herd onto grazing lands outside the park. National attention focused on the herd as hunters killed more than 550 bison. Although the herd is in no danger of disappearing, many conservationists contend that better management policies must be developed. At this time, the Park Service, Forest Service, and the State of Montana are developing a regional plan to manage the bison on an ecological basis rather than a political one.

In some parks, the NPS must contend with environmental damage caused by nonnative species. For example, when wild boars escaped from a North Carolina game preserve in the 1920s, they spread into the Great Smokey Mountains National Park and thrived. By the 1940s, they were a menace. Wild boars compete with native species such as black bear for food. In addition, they root up large areas, destroying the beautiful array of native wildflowers that flourish on the forest floor. In thirteen years, through a combination of trap-and-transfer and shooting, over six thousand boars have been removed from the park, so that now the boar population numbers only a few hundred. Nonnative species are an

Figure 10.15 A bison grazing near geysers in Yellowstone National Park.
© Farrell Grehan/Photo Researchers, Inc.

especially severe problem in Hawaiian national parks, where feral goats, pigs, and cats devastate the islands' unique vegetation and wildlife.

External Stresses

Because many NPS lands are islands in a sea of developed lands, they are frequently subject to severe external pressures. An example is the Everglades National Park, which is located at the southern end of the Florida Everglades, a 160-kilome-

ter-long (100-mile-long) region of sawgrass prairie, palm hammocks, and mangrove swamps. An 80-kilometer-wide (50-mile-wide) sheet of water once flowed unhindered from Lake Okeechobee south through the Everglades to Florida Bay, sustaining a unique mix of subtropical and temperate vegetation and wildlife. Over 700 plant species and 300 animal species reside in the park, including the endangered manatee and Florida panther. As the population of south Florida exploded in recent decades, however, over 2,250 kilometers (1,400 miles) of canals and levees have diverted large quantities of water to thirsty urban areas and expanding farmland (fig. 10.16). Vast irrigated farms have spread to the park's boundaries. During times of drought, water is diverted to cities and farm fields, while the park withers. In contrast, water is dumped into the park during times of heavy rainfall to avoid flooding of farmland and residences. This yo-yo effect imperils the future of the park's wildlife and vegetation. Populations of nesting wading birds have declined dramatically from 300,000 sixty years ago to fewer than 15,000 today.

To combat these problems, managers have developed a model that simulates seasonal rainfall patterns and monitors the natural pattern of water delivery to the northern boundary of the park. But, an improved water delivery system has brought other problems. Much of the land upstream from the park is now either intensively farmed for vegetables or supports large herds of cattle. As a result, water flowing into the park now carries a burden of nitrogen and phosphorus compounds that is 10–20 times greater than a few decades ago. These nutrients spur the growth of cattails, which replace the saw grass and shade out the algae that form one of the bases of the Everglades food web. To cope with these problems, farmland on the northern boundary of the park is being converted to filtration marshes, and efforts are underway to reduce contamination by manure and commercial fertilizers.

Time will tell if these management strategies will enable the park to recover from decades of misguided management of south Florida's water resources. One other troublesome problem remains. South Florida continues to attract over 360,000 new residents each year. Eight of the ten fastest growing cities in the United States are in Florida. Will there be

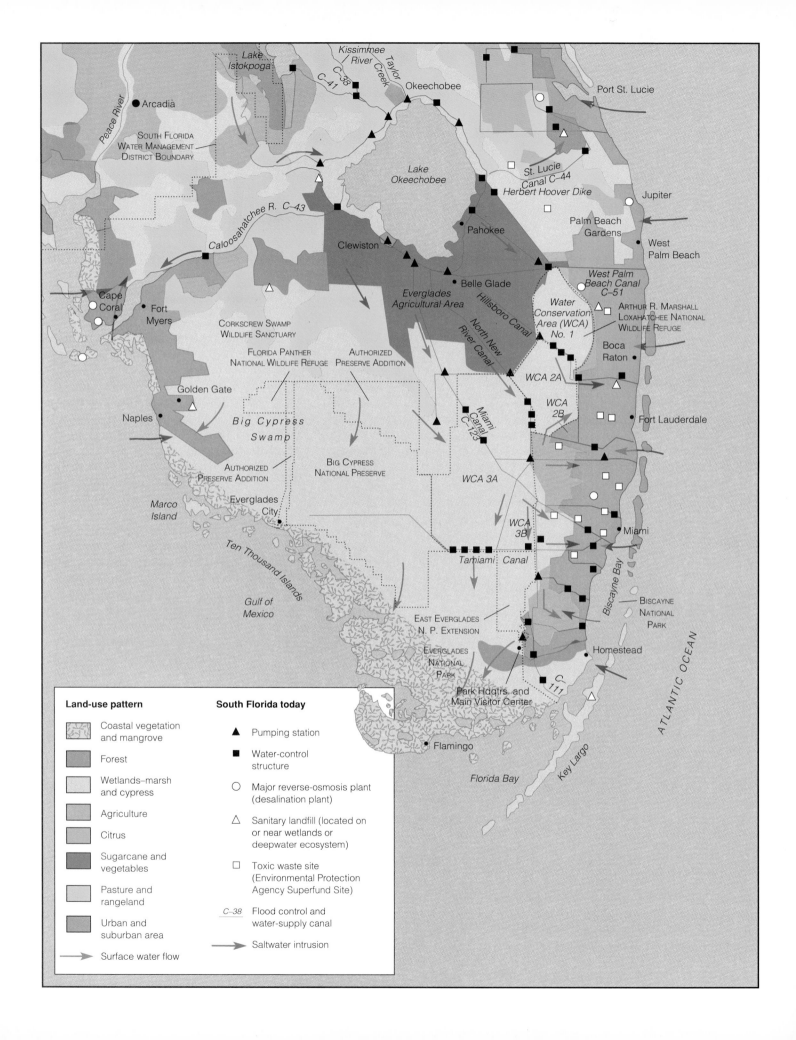

Land-use pattern

- Coastal vegetation and mangrove
- Forest
- Wetlands–marsh and cypress
- Agriculture
- Citrus
- Sugarcane and vegetables
- Pasture and rangeland
- Urban and suburban area
- → Surface water flow

South Florida today

- ▲ Pumping station
- ■ Water-control structure
- ○ Major reverse-osmosis plant (desalination plant)
- △ Sanitary landfill (located on or near wetlands or deepwater ecosystem)
- □ Toxic waste site (Environmental Protection Agency Superfund Site)
- C–38 Flood control and water-supply canal
- → Saltwater intrusion

Lake Istokpoga

Kissimmee River

Taylor Creek

C–38

C–41

Arcadia

Peace River

SOUTH FLORIDA WATER MANAGEMENT DISTRICT BOUNDARY

Okeechobee

Lake Okeechobee

Port St. Lucie

St. Lucie Canal C–44

Herbert Hoover Dike

Jupiter

Palm Beach Gardens

West Palm Beach

Caloosahatchee R. C–43

Clewiston

Pahokee

Belle Glade

Everglades Agricultural Area

Hillsboro Canal

West Palm Beach Canal C–51

Water Conservation Area (WCA) No. 1

ARTHUR R. MARSHALL LOXAHATCHEE NATIONAL WILDLIFE REFUGE

Boca Raton

North New River Canal

WCA 2A

WCA 2B

Fort Lauderdale

Cape Coral

Fort Myers

CORKSCREW SWAMP WILDLIFE SANCTUARY

FLORIDA PANTHER NATIONAL WILDLIFE REFUGE

AUTHORIZED PRESERVE ADDITION

Golden Gate

Naples

Big Cypress Swamp

AUTHORIZED PRESERVE ADDITION

BIG CYPRESS NATIONAL PRESERVE

Miami Canal C–123

WCA 3A

WCA 3B

Miami

Everglades City

Marco Island

Ten Thousand Islands

Gulf of Mexico

Tamiami Canal

Biscayne Bay

BISCAYNE NATIONAL PARK

Homestead

EAST EVERGLADES N. P. EXTENSION

EVERGLADES NATIONAL PARK

Park Hdqtrs. and Main Visitor Center

C–111

Flamingo

Florida Bay

Key Largo

ATLANTIC OCEAN

Figure 10.16 A massive system of canals and levees designed to control flooding and divert water to urban and agricultural regions threatens the future of the Everglades National Park.

Source: Data from N. Duplaix, "South Florida Water: Paying the Price," *National Geographic* 178(1): 88–113.

enough water for people, crops, cattle, and the wildlife and vegetation of Everglades National Park? If not, which sector will experience the most harm?

Even when national parks are surrounded by other federal lands, differences in management priorities can lead to conflicts. National forests often border national parks. While the NPS promotes preservation of natural resources, the NFS promotes the use of resources such as wildlife, timber, coal, and metals. Thus, conflicts are inevitable. For example, NPS officials became alarmed when a coal mining company announced its intent to strip-mine an area in the Dixie National Forest adjacent to Bryce Canyon National Park. NPS personnel and conservation groups argued that 24-hour flood lights, dust, and blasting from strip-mining would obscure scenic views from the park, pollute the park's pristine air, and shatter its silence. (Acoustic studies have shown that the natural silence in this park rivals that of a sound recording studio.) Mining would also scar thousands of hectares that are visible from park overlooks. After many years of negotiations, this battle may be coming to an end. At this time, both conservation groups and coal interests are supporting a congressional bill that would give the coal company financial credits toward the purchase of new leases in central Utah in exchange for withdrawal of leases from the Dixie National Forest.

Other resource development threats continue to appear. For example, in 1989 exploratory wells were drilled in the Winema National Forest within 2 kilometers (1.2 miles) of the edge of Crater Lake National Park, in Oregon, to assess the region's potential for geothermal energy development. At this time, no one knows what the ramifications of extracting geothermal energy might be for the lake.

Pollution also threatens many parks. Raw sewage has been found in the underground rivers of Mammoth Cave National Park, where endemic species of blind fish and shrimp live. Tests show that sewage contamination in the caves' groundwater exceeds Kentucky's water-quality standards 20-25% of the time. To solve the problem, NPS personnel and local authorities have developed a regional water treatment system, although its construction is currently on hold because of insufficient funds.

A more pervasive pollution threat to national parks is declining air quality. Although many parks were once noted for their pristine vistas, NPS air-quality monitoring indicates that at some locations (such as Grand Canyon National Park), visibility is reduced as much as 90% of the time by human-caused haze (fig. 10.17). In addition, vegetation in many parks is showing signs of decline, caused in part by acid rain, ozone, and sulfur dioxide. The amendments to the Clean Air Act of 1990, described in chapter 16, attempt to address many of these problems.

Perhaps the most visible and controversial conflict that has embroiled the National Park Service in recent years is its fire management policies. This issue is discussed on page 205.

National Wilderness Areas

In 1964, Congress established the National Wilderness Preservation System, in which congressionally designated land was to be left "unimpaired for future use and enjoyment as wilderness." Furthermore, Congress stipulated that these lands shall be "devoted to the public purposes of recreation, scenic, educational, conservation, and historical use." By 1990, almost 37 million hectares (92 million acres) of U.S. public land were designated as wilderness. Approximately 60% of this land is in Alaska. The 474 units that comprise the National Wilderness Preservation System are managed by the National Park Service, National Forest Service, Fish and Wildlife Service, or the Bureau of Land Management.

In the 1960s, voices were also raised to protect the remaining few wild rivers from development. Most of the rivers in the United States had already been dammed to provide flood control, water for irrigation, and/or to generate electricity. In response, Congress authorized in 1968 the National Wild and Scenic Rivers System. The objective of the system is to prevent development on or along selected segments of rivers that have particular aesthetic or recreational value or that are still in a natural, free-flowing state. Today the System protects over 75 rivers that account for about 0.2% of the nation's 5.6 million kilometers (3.5 million miles) of rivers.

Although each person may have a somewhat different idea of what constitutes wilderness, the Wilderness Act calls for designated lands to be "affected primarily by the forces of nature, with the imprint of man's work substantially unnoticeable." Achieving this lofty goal presents many difficult challenges. Furthermore, federal agencies cannot protect irreplaceable wilderness resources by simply leaving them alone. Active management programs are absolutely necessary, because these special public lands face many of the same pressures that we have already described for other lands managed by federal agencies.

As we would expect, the greatest conflicts arise from human pressures. The Indian Peaks Wilderness in Colorado, for example, is one of the most frequently visited in the Rocky Mountains. On a typical summer day, several hundred visitors begin their hikes at the main trailhead. About three million people live within a ninety-minute drive. A rising number of visitors threatens to diminish further the solitude and natural surroundings. In the future, officials may have to disperse visitors more widely and limit the number of campsites. Limits on permits have already been imposed in a few wilderness areas.

Several wilderness areas are increasingly affected by surrounding urban and recreational development. Some homeowners need only jump their back fence to enter the Twin Peaks Wilderness near Salt Lake City. Although prohibited in wilderness areas, bicycles, snowmobiles, and all-terrain vehicles often cross the boundaries. Because it is too costly to fence and patrol wilderness boundaries, stopping

Figure 10.17 Vistas in the Grand Canyon National Park are often obscured by haze and smog.
© Bob McKeever/Tom Stack & Associates

Fire Management Policies in National Parks

The year 1988 brought national attention to the National Park Service in general and to Yellowstone National Park in particular. The spectacular Yellowstone fires that summer produced such headlines as "A Legacy in Ashes," "Yellowstone Destroyed," "Valuable Timber Lost," and "Survival of Yellowstone in Jeopardy." These headlines along with sensational video footage of wildfires roaring through Yellowstone gave the public the erroneous impression that the entire park had been reduced to cinders. Such media exposure also reinforced the popular notion that all fires are destructive. Because Yellowstone National Park had a fire management policy that under certain conditions allowed lightning-started fires to burn out on their own, the Park Service was blamed for the conflagration. In fact, many of the major fires were human-caused, and some originated on adjacent U.S. Forest Service lands (fig. 10.18). Nevertheless, federal, state, and local politicians, business people of the gateway towns on the park's perimeter, and some members of the media severely criticized the Park Service for its fire management policy. Let us evaluate the ecological basis for this policy.

The legislative mandate of the National Park Service is to maintain, as near as possible, a primitive ecological situation. Ecologists have known for some time that naturally caused fires are among the factors that have shaped the evolution of vegetation and wildlife for millennia. The aftermath of the Yellowstone fires of 1988 supports this perspective.

Contrary to popular opinion, less than 1% of the park experienced the high-intensity fires that left little more than ashes. In fact, as the fires progressed, they produced in their wake a mosaic of burned, partially burned, and nonburned areas. As a result, new habitats were created that will follow various pathways of ecological succession. The fires increased the heterogeneity (patchiness) of the landscape, which sustains greater species diversity.

Plant growth began almost immediately after the fires. Less than 1% of the soil was subjected to heat extreme enough to penetrate more than 2.5 centimeters (1 inch). Thus, the fires left most seeds unharmed, and many germinated when soil moisture conditions became favorable. Furthermore, the fires released billions of seeds of the lodgepole pine, a species that provides nearly 80% of the forest cover in Yellowstone. Lodgepole pines produce two types of cones. Some shed seeds when they reach maturity. Others sealed with a resin coating and will not open until they are exposed to high temperatures from a fire. Hence, they may remain unopened for years, but they release millions of seeds within a few days after a fire. Following the 1988 fires, densities of lodgepole pine seeds in some areas of the park ranged from 50,000 to 100,000 seeds per acre. While many of these seeds nourished birds, mice,

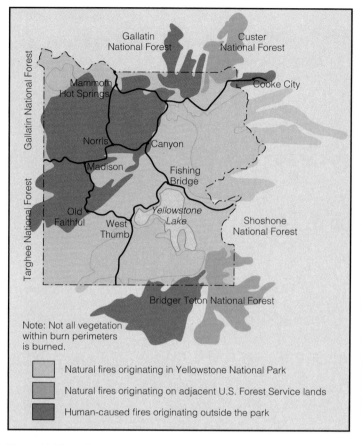

Figure 10.18 Extent of natural and human-caused fires in the Greater Yellowstone Area in 1988. Not all vegetation within burn parameters was burned.
Source: *Yellowstone Fire*, 1988, National Park Service.

and squirrels, some escaped predation, germinated, and will eventually produce forests that are much like the ones that burned. Other plants, such as willows and grasses, quickly resprouted from underground structures that were not harmed by the fires. Thus, within only one or two growing seasons, the charred forest floor was replaced with herbaceous plants, shrubs, and the seedlings of tomorrow's forests (fig. 10.19).

Rather than destroying the vegetation of the park forever, the fires actually spurred a renewal of life in Yellowstone. Prior to the fires, plant growth had slowed considerably on some sites where the soils are inherently infertile. The relatively dry and cool summer weather in Yellowstone, which slows decomposer activity, also contributes to low soil fertility. Without fire, most of the nutrients are tied up in living and dead vegetation. The fires of 1988 triggered a resurgence of plant growth because they released plant nutrients in the form of ash and because they exposed the previously shaded forest floor to the sun. Herbivores flourished after the

Figure 10.19 Following the Yellowstone fires of 1988, vegetation such as these pink fireweeds rapidly recolonized many burned areas.
© Jeff Foott/Tom Stack & Associates

fires as they benefited from a diet of highly nutritious vegetation. In turn, predators experienced a bonanza because the fires caused small herbivores to concentrate on islands of unburned vegetation, where they were particularly vulnerable.

What happened to the famed wildlife of Yellowstone during the fire? For those of us whose memory of the terrified Bambi escaping a forest fire is still vivid, the results may be surprising. Wildlife skillfully avoided the fires, except where fires moved rapidly along a broad front. A November 1988 survey of carcasses in Yellowstone Park revealed that less than 1% of the park's approximately 30,000 elk (summer population count) were killed by fire or smoke. Likewise, only 5 bison (park population, 2,700) and 2 moose were killed. Even these

resources were not wasted; they provided abundant food for bears, coyotes, ravens, and other scavengers throughout the following winter. Even such tiny critters as chipmunks and ground squirrels escaped the fires as they sought shelter in their well-insulated, underground burrows.

We can conclude that, contrary to popular opinion, fire is a regenerative force in Yellowstone National Park, as in many other ecosystems. Figure 10.20 illustrates this cycle of regeneration. The Park Service's fire management policy of allowing certain lightning-started fires to burn out on their own has a solid ecological foundation.

Since this policy was implemented in 1972, scientists have learned that almost all lightning-induced fires simply fizzle out before burning even a single hectare of land. Why, then, did such massive fires roar through Yellowstone in 1988? Fire specialists cite abundant fuel, favorable weather conditions, and several sources of ignition as key contributing factors to a large forest fire. None of these factors alone is usually sufficient to trigger a major fire. In Yellowstone in the summer of 1988, all three were present.

Decades of fire-suppression prior to the early 1970s contributed to a buildup of fuel as dead wood and litter accumulated on the forest floor, but such an accumulation often only supports a **ground fire,** which merely creeps along the ground, burning only the litter, shrubs, and small trees. Recall from chapter 4 that forest managers often follow a policy of **prescribed burns,** setting ground fires to reduce fuel buildup or to prevent the invasion of climax species or both.

Major fires occur when the forest crown catches fire. For a **crown fire** to develop, a fire must have a way to move from the ground to the crown. In Yellowstone, such a means develops when shade-tolerant species such as spruce and fir invade a lodgepole pine stand. Such invasions usually occur if there are no fires on a lodgepole pine site for one or two hundred years (fig. 10.20). Because spruce and fir do not self-prune (the lower branches do not fall off as the tree grows), the highly flammable lower branches serve as a perfect ladder for the ground fire to reach the crown of lodgepole pines (fig. 10.21). Once a fire reaches the crown, it may quickly become uncontrollable. In the summer of 1988, many lodgepole pine stands not only contained considerable fuel, but also supported an understory of spruce and fir "ladder" trees. The stage was set for a major blaze when the right weather conditions came along.

In 1988, practically no rain fell in June, July, and August, an event unprecedented in the park's 112-year weather record. The moisture content of the vegetation thus became extremely low. All that was needed was a spark. That summer Yellowstone recorded twice the average number of lightning strikes. By early fall, about 50 lightning-caused fires had occurred in the park. (Recall also that many fires were caused by accidents or careless campers.) Furthermore, a series of dry weather fronts triggered unusually high winds, which fanned the flames into fire storms. At times the fire storms had all the fury of hurricanes. It is no

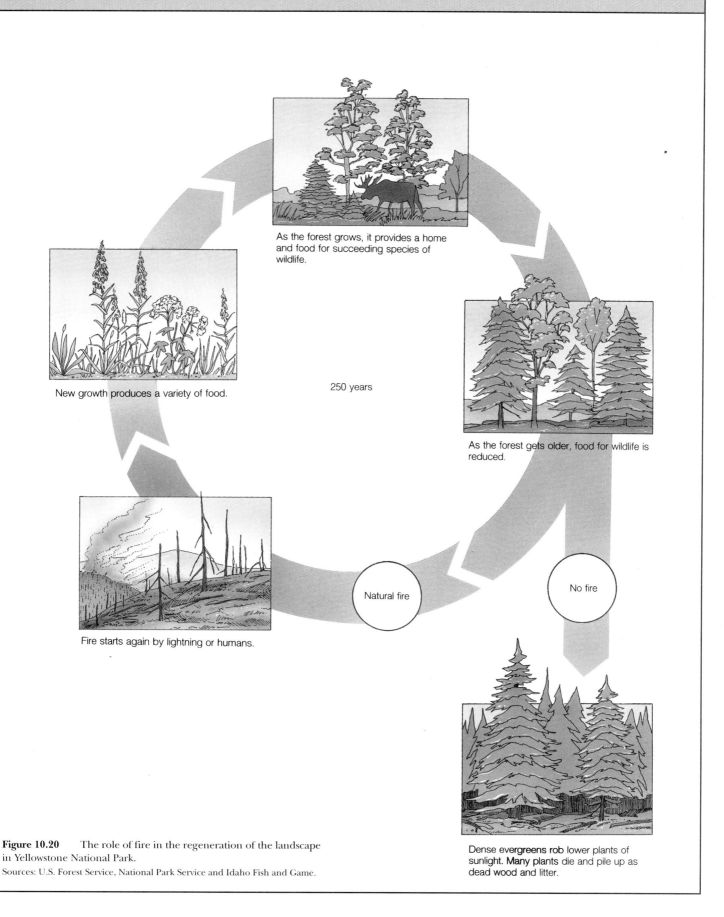

As the forest grows, it provides a home and food for succeeding species of wildlife.

250 years

New growth produces a variety of food.

As the forest gets older, food for wildlife is reduced.

Natural fire

No fire

Fire starts again by lightning or humans.

Dense evergreens rob lower plants of sunlight. Many plants die and pile up as dead wood and litter.

Figure 10.20   The role of fire in the regeneration of the landscape in Yellowstone National Park.
Sources: U.S. Forest Service, National Park Service and Idaho Fish and Game.

Figure 10.21   The unpruned, lower branches of spruce and fir trees provide a ladder for a ground fire to reach the canopy of lodgepole pines.
© Chase Swift/Tom Stack & Associates

wonder that 25,000 well-equipped firefighters had no chance of controlling them. Thus, the major blazes resulted from the coincidence of several extreme events. Although such fires are unlikely to happen again in Yellowstone during our lifetime, there is abundant evidence that such conflagrations have recurred in Yellowstone every 200-300 years for millennia. A September snowfall eventually snuffed out the fires—one natural process gave way to another.

The fire management plan of Yellowstone Park was not irresponsible. Not all lightning-caused fires were allowed to burn. The plan included a provision to suppress wildfires if they threatened human life or property, historic and cultural sites, or endangered species. Furthermore, the policy called for suppression of all human-caused fires by cost-effective, safe, and environmentally sensitive means.

Today, Yellowstone Park is buzzing with life. Not only are vegetation and wildlife flourishing, but record numbers of human visitors swarm over the park to see its geysers and thermal pools as well as the aftermath of the fires. The gateway towns are booming with business. But while the wilderness setting of the park survives, its fire management plan does not. Despite its firm ecological foundation, the policy was rescinded in December of 1988 by an interagency Fire Management Policy Review Team. The team, composed of representatives of five federal agencies, stated that "the agencies reject as impractical and unprofessional that fires be allowed to burn free of prescriptions or appropriate suppressive action." At this time, Yellowstone Park personnel are rewriting the park's fire management policies. Although they hope to have most of the old policies reinstated, allowing nature to take its course under certain circumstances remains an issue.

Meanwhile, in 1990, wildfires in Yosemite National Park forced officials to close that park for the first time in history. Public attention was once again drawn to the role of fire in natural landscapes such as national parks. Although some people may demand that federal agencies adopt management policies to stop all fires, everyone should realize that sooner or later, ecosystems such as the forests of Yellowstone and Yosemite will burn. Nature frequently cannot be controlled, despite our best-laid management plans. Furthermore, fires are one of nature's ways to regenerate and renew the landscape. To prevent all fires is ultimately to invite the decline of the same landscapes we want to preserve.

these incompatible uses requires education and cooperation. In addition, alternative sites for such recreational vehicles should be provided.

The Wilderness Act also makes provisions for certain nonconforming uses such as access to state or private land within a wilderness area, grazing, and control of fire, insects, and disease. Achieving the proper balance (which differs greatly depending upon a person's concept of wilderness) between protection of wilderness and these permitted uses can be quite challenging and perplexing. For example, people who own inholdings sometimes request that a road be built across the wilderness area so that they can drive to their property. But no motorized vehicles (including the landing of airplanes and helicopters) are permitted in wilderness areas. (An exception was made in the designation of the Frank Church River of No Return Wilderness.) To reduce such conflicts, efforts are underway to purchase inholdings.

Managing fire, insects, and disease poses special problems. Although the ecological roles of these natural phenomena are slowing being recognized, controversy rapidly escalates when these forces threaten resources and properties beyond the wilderness boundary. Currently lightning-caused fires may be permitted to burn under certain circumstances if they are part of an approved management plan. When fires must be controlled, fire crews are instructed to use motorized equipment and disturb the soil only when such actions are absolutely necessary to protect private property and safety.

As a special place to experience solitude, natural surroundings, and unconfined recreation, wilderness is an invaluable resource. Major educational efforts are required, however, to let the public know why such areas must be protected from the stresses of a growing human population and its activities. Even wilderness visitors need to be reminded that they have a special responsibility not to degrade the resource and not to affect adversely the experience of other visitors. Certainly wilderness-area managers face formidable challenges in this rapidly developing world.

In the preceding sections, we have examined some of the land-use conflicts that arise on federal public lands as management agencies attempt to balance the various demands of U.S. citizens. Although we have focused on federal lands, similar conflicts occur on lands owned and managed by state and local governments. Furthermore, the questions that we have raised about sustainable and environmentally sound management practices apply to private lands as well as to public lands.

In the final section, we examine some of the conflicts that arise over the management of wetlands. As with other resource lands, wetlands fall under both public management and private ownership.

Wetlands

Wetlands are transitional between terrestrial and aquatic ecosystems; they include marshes (dominated by grasses and sedges), swamps (dominated by shrubs or trees), bogs, mudflats, and small ponds. Groundwater is at or near the surface, so that much of the soil and sediment is saturated with water.

A distinction is usually made between coastal and inland wetlands. The water in most coastal wetlands is salty or brackish (a mixture of salt and fresh water) and subject to daily tidal oscillations. Inland wetlands have fresh water, are nontidal, and usually occupy formerly glaciated lowlands. Wisconsin, for example, has about 2 million hectares (5 million acres) of wetlands (down from the original 4 million hectares) that date from the last recession of glaciers about ten thousand years ago. Some freshwater wetlands occur near the coast, protected from the sea by sand dunes and fed by precipitation or groundwater.

Historically, wetlands were viewed as worthless, and were often filled in or drained to make way for what were considered better uses. At the time of settlement by Europeans, what are now the lower forty-eight states had an estimated 87 million hectares (215 million acres) of wetlands. A 1991 report by the Fish and Wildlife Service estimates that 53% of these wetlands have been lost. Twenty-two states have lost at least half of their wetlands while ten states have lost more than 70% (fig. 10.22). Most wetlands occur along the coasts of the Atlantic Ocean and Gulf of Mexico, the Mississippi River, and the glaciated regions of the upper Midwest.

In the early days, wetland loss was mostly along the eastern seaboard, where marshes and swamps were drained and filled to accommodate new towns and cities. Many coastal cities, for example, are primarily sited on former wetlands. Wetlands continue to be transformed for suburban shopping malls and industrial parks, but today agriculture is responsible for perhaps 85% of total wetland loss. Wetlands are drained to grow crops.

In recent decades, people have become more aware of the value of wetlands to society and the need to stem their loss. Wetlands perform several important ecological functions:

1. Because nutrients and detritus accumulate in wetlands, they are very productive ecosystems and provide feeding and nesting habitats for a variety of wildlife, as well as spawning grounds for fish. Roughly half of the commercial fish catch in the North Atlantic depends on coastal wetlands for some part of its life cycle. Furthermore, the National Wildlife Federation estimates that about 25% of endangered plants and 45% of endangered animals depend directly or indirectly on wetlands.
2. Wetlands act as huge sponges by holding large amounts of water; thus, they regulate streams that flow through them, holding back water at times of flooding and releasing water during periods of low stream flow.
3. Wetlands filter out wastes, especially nutrients, carried by streams or tidal waters that flow through them.
4. Wetlands frequently recharge groundwater reservoirs (chapter 13).

Channelization of rivers and streams is particularly destructive to wetlands. **Stream channelization** is a controversial flood-control strategy that involves the straightening and ditching of a meandering stream channel. Changes in the

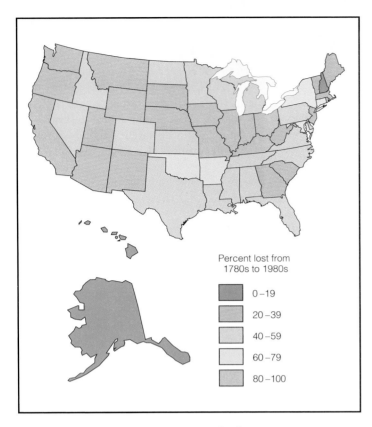

Figure 10.22 America's disappearing wetlands.
Source: U.S. Department of the Interior, 1990.

Percent lost from
1780s to 1980s

- 0–19
- 20–39
- 40–59
- 60–79
- 80–100

Figure 10.23 An aerial view of Florida's Kissimmee River prior to channelization.
South Florida Water Management District

Kissimmee River drainage basin in south central Florida illustrate well the effects of channelization. Back in 1962, the Kissimmee River was a gently flowing stream that meandered from Lake Kissimmee southward through thousands of hectares of wetlands to Lake Okeechobee (fig. 10.23), a river course of about 160 kilometers (100 miles). By 1971, stream channelization had drastically transformed the Kissimmee into a straight, swiftly flowing canal, 100 meters (330 feet) wide, 10 meters (33 feet) deep, and only 93 kilometers (58 miles) long (fig. 10.24). The purpose of channelization was to control seasonal flooding that damaged agriculture and hindered development south of Orlando. Ironically, subsequent studies showed that channelization of the river was not effective in controlling flooding. Also, more than 75% of the river's original wetlands, a total of 16,000 hectares (40,000 acres), were drained as a result of channelization, causing sharp declines in wildlife populations, including bald eagles (down 74% since channelization) and wintering and resident waterfowl (down more than 90%). Much of the drained wetland was converted to cattle ranches; thus, the river now carries a significantly greater load of such plant nutrients as nitrogen and phosphorus. Furthermore, an increased sediment load has greatly reduced aquatic life and severely affected recreational fishing.

Although wetlands first received federal protection with enactment of the Clean Water Act of 1972 (described in chapter 14), progress in wetland preservation has been slow. A major obstacle has been the inability of the four concerned federal agencies—the EPA, Soil Conservation Service, Fish and Wildlife Service, and the Army Corps of Engineers—to agree to a single set of criteria for determining whether a specific area constitutes a wetland. The future of wetlands became brighter in 1989 when the four agencies authored a joint manual containing a common definition of the term *wetland*. But efforts to preserve wetlands soon received a severe jolt.

The Bush administration in late 1991 proposed a series of significant changes in wetland preservation regulations. The most controversial change was a redefinition of *wetland*. Scientists, wetland managers, and conservation groups discovered that over 50% of today's wetlands would no longer qualify as wetlands under the new definition; consequently, they would be removed from federal protection and opened to commercial, industrial, and agricultural development. Furthermore, scientists from a wide range of disciplines argued that the administration's wetland criteria were sorely deficient; some components had no scientific basis, while other components would be nearly impossible to document. The latter deficiency would probably lead to additional wetland losses. Under proposed guidelines, if the determination was inconclusive after tests had been completed, the land would automatically cease to be designated as a wetland. Although public uproar prompted the Bush administration to back off from the issue, the question of what constitutes a wetland remains unresolved as of this writing.

While efforts to preserve wetlands were facing political barriers, the National Research Council published a report in late 1991 asking the federal government not only to save all wetlands, but also to initiate a new program to restore aquatic communities that had already been damaged. A few small wetland restoration projects around the nation have demonstrated that if the water source can be restored and the soil remains more or less undamaged, some wetlands can be returned to a condition that approximates their original composition and function.

Figure 10.24   Channelization of the Kissimmee River in Florida led to massive alteration of adjacent wetlands. Meandering segments of the river's former course prior to channelization are visible on either side of the channel.
© James Balog/Black Star

Perhaps the most ambitious restoration project proposed to date involves returning the Kissimmee River and its environs to pre-1962 conditions. As a key component of that effort, engineers reestablished the Kissimmee's original course along a 19- kilometer (12 -mile) segment during a 1984 testing phase. The success of that project indicates that large-scale restoration of the entire river is feasible. The restoration plan calls for returning almost 100 kilometers (62 miles) of the original channel as well as 14,000 hectares (35,000 acres) of the original floodplain wetlands to natural conditions. The estimated cost of the entire project is more than $500 million. (By comparison, the original channelization of the Kissimmee cost $30 million.) As of this writing, the focus is on whether the federal government and the state of Florida will authorize sufficient funds to reverse a major ecological catastrophe.

To this point we have focused on freshwater ecosystems. However, coastal wetlands are also in peril. The issue on page 212 describes the alarming loss of the largest coastal wetland in the United States, the Louisiana Bayou.

Loss of the Louisiana Bayou

While most wetland loss in the contiguous United States is inland and due to agriculture, the wetlands of the Louisiana bayou southwest of New Orleans are a glaring exception (fig. 10.25). Loss of these wetlands is a potentially serious problem for many reasons. The Louisiana wetlands provide (1) winter habitat for perhaps two-thirds of the Mississippi Flyway's migratory waterfowl (more than 5 million ducks and geese); (2) nurseries for most of the commercial fish and shellfish (oysters, shrimp, and crabs) and about half of the recreational fish taken from the Gulf of Mexico; (3) homes to the nation's second largest populations of wading birds, southern bald eagles, and shorebirds; and (4) a buffer against powerful tropical storms that track out of the Gulf. Since the turn of the century, more than 400,000 hectares (1 million acres) of Louisiana wetlands have been lost. By the late 1980s, the annual rate of loss was about 15,000 hectares (38,000 acres), up from an estimated 4,100 hectares (10,000 acres) in 1970. Louisiana now accounts for about 80% of all coastal wetland losses in the continental United States.

The main reason for the loss of Louisiana wetlands is the long-term effort to control the course of the Mississippi River. The bayou is part of the huge Mississippi delta (actually, a series of overlapping deltas), which has been built up over millions of years by sediment deposited by the river. Periodically, buildup of sediment chokes the main channel and forces the river to find a new outlet to the Gulf. The river has therefore repeatedly shifted course, swinging back and forth like a pendulum, depositing sediment, and building a delta that now stretches from near Cairo, Illinois, south to the Gulf of Mexico, a distance of about 1,600 kilometers (1,000 miles).

To make the river below New Orleans more navigable for commercial ships, engineers built levees that essentially confined the river to a main channel. Consequently, the river now flows rapidly into the Gulf, and the scouring action of its current reduces the need for costly dredging. On the other hand, the relatively strong current also delivers much of the river's sediment load directly into the Gulf, shutting off the sediment supply to the delta, the bayou, and beaches along the coast. Less sand for protective beaches has allowed storm waves to cut tidal inlets through them. In particular, the storm surge created by a hurricane (e.g., Andrew in 1992) causes considerable coastal erosion. Meanwhile, natural compaction and settling of delta sediments (locally enhanced by withdrawal of groundwater and petroleum) is causing a relative rise in sea level.

Further compounding the problem in the bayou are about 16,000 kilometers (10,000 miles) of pipeline ditches and navigation canals, many of which are used to reach oil and gas rigs (fig. 10.26). Canals channel salt water into the wetlands, where it kills the vegetation with its extensive root systems that bind and anchor delta sediments. The result of all this is accelerated erosion by sea waves and tides and transformation of the bayou to open water.

Costly remedial action is planned to slow if not halt the rate of wetland loss in Louisiana. This is likely to involve

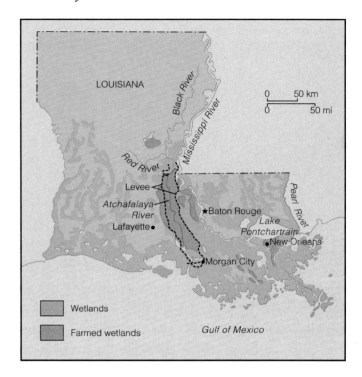

Figure 10.25   Louisiana's wetlands are falling victim to efforts to control the course of the Mississippi River.

Figure 10.26   This canal through a Louisiana bayou serves to transport oil crews to and from oil rigs.
© C.C. Lockwood/Earth Scenes

(1) diversion of Mississippi River water into the wetlands to increase sedimentation, (2) rebuilding of badly eroded offshore barrier islands to help hold back Gulf waters, and (3) development of alternatives to navigation canals (e.g., helicopters and hovercraft).

Conclusions

Conflicts over land use, particularly regarding public lands, are bound to become more frequent and more severe as the human population continues to grow and as we step up our efforts to locate new sources of fossil fuels and other resources. As population/resource pressures continue to rise, public lands are coming under increasing scrutiny by commercial resource developers and a variety of citizen groups. Resource developers, pointing to national needs for fuel, timber, minerals, and water resources, are pressing for less rigid land-use restrictions. Public lands hold perhaps half of the nation's undiscovered oil and gas; about 10% of domestic coal production is from public lands; and approximately 54% of the nation's softwood timber inventory is on public lands, mostly in the Pacific Northwest.

In contrast, many citizen groups are pressing for tighter land-use restrictions for large-scale development and the purchase of more private land to be placed under public protection. But perspectives on how to manage public lands vary greatly among these organizations. There are major conflicts even among users of the recreation resources as some press for more and better facilities such as roads, parking lots, campgrounds, and trails, while others seek to limit further loss of solitude and scenic beauty by limiting visitor access. Since each user group addresses real needs and concerns, public land policies will, no doubt, continue to be reviewed and modified. Many of these conflicts, like other public issues, will ultimately be resolved in legislatures and courts.

Key Terms

public lands	ground fire
clear-cutting	prescribed burns
partial cutting	crown fire
rangelands	wetlands
overgrazing	stream channelization
riparian areas	

Summary Statements

The land-use philosophies of Gifford Pinchot and John Muir continue to be major forces that shape resource management policies for federal lands.

Traditionally the Forest Service has focused primarily on watershed protection, timber production, livestock grazing, and mineral extraction. In response to the Forest and Rangeland Renewable Resources Act, this agency now places a higher priority on managing recreation, wildlife, and fisheries resources as well as environmentally acceptable timber production.

Overgrazing on rangelands is leading to significant degradation of grazing lands, especially riparian habitats. Conservationists question the low fees charged to ranchers for grazing rights on federal rangelands. Resolving the conflicts associated with sustained management of rangelands has proven to be particularly difficult.

The National Park System faces many internal and external stresses. Internal stresses include the impact of a growing number of visitors and the management of wildlife within park boundaries. External stresses include air and water pollution and conflicts over management policies with agencies that manage adjacent lands. Although fire plays a vital ecological role in many ecosystems, some of the Park Service's fire management policies have come under heavy criticism since the Yellowstone fires of 1988.

Managers of National Wilderness Areas face many of the same conflicts that trouble the Park Service and other resource agencies. Again the greatest conflicts arise from human pressures and the attempt to maintain wilderness.

Because they were once viewed as worthless, many wetlands have been lost through agricultural and urban development. Although wetlands perform many important ecological functions, they continue to be threatened by stream channelization and other activities that either drain or fill these valuable habitats.

Questions for Review

1. How did Gifford Pinchot and John Muir differ in their perspectives on land use and resource conservation?
2. In what ways have the management policies of the Forest Service changed in response to the Forest and Rangeland Renewable Resources Planning Act?
3. How do clear-cutting and partial cutting differ in their environmental impact?
4. What is a rangeland? Where in the United States are the rangelands located?
5. What are riparian habitats? How have they been affected by overgrazing?
6. Although overgrazing of rangelands has been an issue for decades, little progress has been made in resolving conflicts. Describe the nature of the conflicts from the perspective of a rancher, a BLM manager, and a conservationist. What would you do to resolve the conflicts?
7. Present several reasons why managing wildlife populations in national parks is particularly challenging.
8. Describe how outside pressures are threatening the Everglades National Park.
9. Construct a diagram that illustrates the role of periodic fires in regenerating vegetation in Yellowstone National Park.
10. How did the Yellowstone fires benefit resident herbivores, carnivores, and scavengers?
11. How does a ground fire differ from a crown fire?
12. Describe the coincidence of events that resulted in the Yellowstone fires of 1988. What are the chances that such fires will recur within the next hundred years?
13. List five types of internal stresses and five types of external stresses that contribute to management conflicts in national parks.
14. According to Congress, what are the characteristics of a wilderness area?
15. Describe some of the similarities and differences among the resource management policies of the Park Service, Forest Service, and Bureau of Land Management.
16. What are the differences between coastal wetlands and inland wetlands?

17. List four benefits of wetlands.
18. Describe several reasons why stream channelization is often so destructive of wetlands.

Projects

1. Obtain a topographic map of your region and identify local, state, and federal lands that are managed for their natural resources. Prepare an inventory of these lands by level of government and management priorities. Do you believe that adequate public land has been set aside in your region to meet future resource needs?
2. Visit nearby public lands that are managed for recreation, timber, wildlife, mining, grazing, and so on. What conflicts do resource managers face? What can you do personally to help resolve these conflicts?
3. Discuss the following question in your classroom: Can a wilderness area be used by human visitors and still truly be a wilderness?

Selected Readings

Duplaix, N. "South Florida Water: Paying the Price." *National Geographic* 178 (July 1990):89-113. A description of the numerous water management problems that plague the Everglades National Park.

Fege, A. S. "Wilderness Tomorrow." *American Forests* 95 (July/Aug. 1989):37-41. An examination of the many challenges that face wilderness managers and the actions needed to protect these irreplaceable resources.

Kenney, J. "Control of the Wild." *National Parks* 65 (Sept./Oct. 1990):21-25. A look at the difficult task of managing wildlife populations in national parks.

Leopold, A. *A Sand County Almanac.* New York: Oxford Univ. Press, 1949. A noted naturalist eloquently describes his land-use ethic.

Loftin, K. "Restoring the Kissimmee River." *Geotimes* 36 (Dec. 1991):14-17. An examination of efforts to restore the Kissimmee River and the cost.

Pritchard, P. C. "The Best Idea America Ever Had." *National Geographic* 180 (Aug. 1991):36-59. A fascinating look at the varied activities of Park Service personnel as they try to preserve national park resources in the face of an unending wave of visitors.

Romme, W. H., and D. G. Despain. "The Yellowstone Fires." *Scientific American* 261 (Nov. 1989):36-46. An examination of the perspective that major fires are not only inevitable, they are also necessary agents of ecological change.

Schueler, D. "That Sinking Feeling." *Sierra* 75, no. 2 (1990):43-50. Describes the loss of the Louisiana bayou.

"Wilderness America: A Vision for the Future of the Nation's Wildlands." *Wilderness* 52, no. 184 (Spring 1989). An entire issue commemorating the twenty-fifth anniversary of the establishment of the National Wilderness Preservation System.

Wuerthner, G. *Yellowstone and the Fires of Change.* Salt Lake City: Haggis House, 1988. A well-illustrated and authoritative account of the Yellowstone fires of 1988.

Wuerthner, G. "The Price Is Wrong." *Sierra* 75 (Sept./Oct. 1990):38-43. An examination of the conflicts over the grazing of livestock on public lands.

Managing for Biodiversity

Chapter 11

A lush tropical rain forest in Tikal, Guatemala.
Such ecosystems house the greatest diversity of
plant and animal life on the planet.
© Philip Sze/Visuals Unlimited

Enormous flocks of passenger pigeons, sometimes more than two billion strong, once darkened the skies. Today they are gone forever. The last passenger pigeon, an aging female named Martha, died in the Cincinnati zoo in 1914. The birds' savory flesh and gregarious nature led to their demise; they were good to eat and easy to kill. With the advent of railroads, which could transport harvested birds to major markets before they spoiled, thoughtless commercial harvests of the pigeons began. In the 1840s, several thousand people made their living by hunting passenger pigeons and other birds. Pigeons were shot (fig. 11.1), clubbed to death while in their nests, or trapped. In Michigan alone, more than one billion passenger pigeons were killed in a single year. Within only a few decades, the highly prized species had become extinct.

Few people mourn the passing of the passenger pigeon, and certainly the planet's forests, prairies, and seas appear to be teeming with life. But today hundreds of thousands, perhaps millions, of species are facing the same fate that befell the passenger pigeon. The current rate of extinction of birds and mammals may be as much as 100 to 1,000 times greater than during recent centuries.

Species that are considered to be in immediate danger of extinction are called **endangered species.** In 1992, the U.S. Fish and Wildlife Service listed as endangered 277 U.S. species (table 11.1) and 489 foreign species of animals (fig. 11.2). Others are classified as **threatened species** because they are still relatively abundant in some parts of their ranges, but their numbers have declined significantly in other areas. For example, the grizzly bear remains relatively secure in Alaska and western Canada, but its numbers have fallen dramatically in the Rocky Mountain states. The U.S. Fish and Wildlife Service lists 97 U.S. species and 36 foreign species of animals as threatened (table 11.2).

The potential demise of such animals as the bald eagle, peregrine falcon, gray wolf, and grizzly bear prompted preservation efforts by federal and state agencies in the late 1960s.

The plight of rare plant species, on the other hand, was not officially recognized until the late 1970s. Given the later start, the U.S. Fish and Wildlife Service lists relatively few plant species as endangered or threatened. Although the Center for Plant Conservation (located at the Missouri Botanical Garden in St. Louis) reported in 1990 that nearly 700 species of native plants may become extinct within the next decade, the Service in 1992 only listed 243 species as endangered and 64 species as threatened (fig. 11.3). The cactus family is

Table 11.1	Some endangered animals of the United States	
Mammals		**Birds**
Ozark big-eared bat		Masked bobwhite (quail)
Eastern cougar		Hawaiian coot
Key deer		Mississippi sandhill crane
San Joaquin kit fox		Hawaiian creeper
Florida manatee		Hawaiian duck
Utah prairie dog		Snail kite
Sonoran pronghorn		California clapper rail
Gray wolf		Red-cockaded woodpecker
Reptiles		**Amphibians**
American alligator		Desert slender salamander
American crocodile		Texas blind salamander
Blunt-nosed leopard lizard		Houston toad
San Francisco garter snake		Santa Cruz long-toed salamander
Green sea turtle		
		Insects
Fish		El Segundo blue butterfly
Humpback chub		Mission blue butterfly
Mohave chub		Oregon silverspot butterfly
Maryland darter		San Bruno elfin butterfly
Pecos gambusia		Smith's blue butterfly
Comanche Springs pupfish		
Colorado River squawfish		
Gila trout		

Table 11.2	Some threatened animals of the United States and other nations
Mammals	
baboon	
black howler monkey	
grizzly bear	
red kangaroo	
Southern sea otter	
Birds	
bald eagle	
arctic peregrine falcon	
Reptiles	
Mona ground iguana	
Eastern indigo snake	
loggerhead sea turtle	
desert tortoise	

Figure 11.1 Passenger pigeons being hunted for market.
American Museum of Natural History

particularly imperiled, with over twenty species on the list. Another highly threatened family is the orchids. No worldwide listing of endangered/threatened species of plants is currently available. Nevertheless, conservationists suggest that at least 10% of the world's estimated 250,000 species of flowering plants are endangered.

In recent years, the large-scale destruction of tropical rain forests (which contain at least half of the planet's species) has expanded concern from the potential extinction of a few hundred species to the potential loss of millions of species. Thus, attention is now focused on preserving **biodiversity,** the total number of species and the genetic diversity within each species. No one knows exactly how many species still remain on this planet. Although 1.4 million species have been given scientific names, estimates of the number of unidentified species range up to 100 million. Many biologists agree, however, that there are at least 10 million species. Many also contend that unless the rampant alteration of

(a)   (b)   (c)

Figure 11.2   Examples of foreign animals on the endangered species list: (*a*) giant panda; (*b*) Japanese crane; (*c*) the Galápagos tortoise.
(*a*) © Stan Wayman/Photo Researchers, Inc. (*b*) © George Holton/Photo Researchers, Inc.(*c*) © Miguel Castrpo/Photo Researchers, Inc.

(a)   (b)   (c)

Figure 11.3   Examples of plants on the endangered species list:
(*a*) persistent trillium; (*b*) Tennessee purple cornflower; (*c*) San Clemente larkspur.
(*a*) © Jeff Lepore/Photo Researchers, Inc. (*b*) © Paul Somers (*c*) © T. Oberbauer

ecosystems such as wetlands and tropical rain forests is halted, perhaps 20% of all species now living will disappear within the next thirty years.

A rapidly growing human population, with its need for more food, water, energy, space, and other resources, threatens the survival of our fellow species. In this chapter, we explore reasons to preserve biodiversity, the forces that threaten biodiversity, and strategies for preserving it.

The Value of Biodiversity

What good are all these species? After all, we seem to be getting along fine without the Carolina parakeet, the California grizzly bear, and the eastern elk, all of which are now extinct. Although most people are at best lukewarm about saving endangered species, others argue that the loss of biodiversity is one of the most serious problems that humankind faces in the coming decades. Why are opinions so divided?

Biodiversity and the Genetic Reservoir

Recall from chapter 4 that every organism is a unique combination of genetic characteristics that enable it to adjust to a variety of environmental conditions. All the genes of all the individuals in a population collectively form a **gene pool.** Scientists are confident that gene pools include undiscovered traits that are potentially important for the future of our own species. Consider some examples of the value of gene pools.

Even in today's high-tech society, about 45% of the medical prescriptions written in the United States contain at least one product of natural origin. Numerous lifesaving drugs have been isolated from flowering plants. Quinine (from the cinchona tree) is used to treat malaria; digitalis (from the foxglove plant) is used to treat chronic heart trouble; and morphine and cocaine (from the poppy plant and the coca shrub respectively) are used to reduce pain. Other drugs that have been extracted from plants are used to treat leukemia, several forms of tumorous cancer, various heart ailments, and hypertension (high blood pressure). A recent exciting find is taxol, a chemical isolated from the bark of the Pacific yew tree. Taxol's unique properties give it the potential to damage cancer cells without harming normal cells. Long considered worthless by loggers, the Pacific yew grows only in the rapidly disappearing ancient forests of the Pacific Northwest. In spite of considerable research activity, only about 5,000 of the earth's 250,000 species of flowering plants have been screened thoroughly for possible medical applications.

Many useful drugs have been derived from unlikely sources. The fungus *Penicillium notatum* produces the lifesaving antibiotic penicillin (chapter 1), and certain species of bacteria produce the important antibiotics tetracycline and streptomycin. Thanks to these antibiotics and more than a 1,000 others, diseases such as typhoid fever, scarlet fever, bubonic plague, diphtheria, syphilis, and gonorrhea can be treated more effectively.

Maintenance of large gene pools is also valuable for agriculture. All crops and livestock are domesticated native plants and animals. Agriculture still needs wild populations of native species to provide new genetic characteristics in order to solve present and future food-production problems. Recall from chapter 8 that the new varieties of wheat and rice, which have significantly increased food production in many nations, are products of breeding experiments that utilized thousands of native and domesticated plant varieties. Because pests, pathogens, and technology evolve over time, many challenges that we cannot predict are certain to present themselves—challenges that are likely to require the genetic resources of other varieties and breeds.

Despite their great value, agricultural gene pools are being swept away in the wake of human activities. As recently as fifty years ago, 80% of the wheat grown in Greece consisted of native strains that were well adapted to the local environment and resistant to local pests. Today, only 5% of those wheat varieties still exist.

Industries also use a wide variety of plant and animal products and thus benefit from a rich diversity of species. As many as 2,000 plant species throughout the world have economic importance. American building, furniture, and papermaking industries use more than one hundred different species of trees. Cotton, flax, hemp, jute, and agave provide fiber for the manufacture of textiles, rope, and twine. Natural substances from a multitude of plant species are used to manufacture rubber, dyes, tanning agents, perfumes, lotions (e.g., aloe), resins, gums, insecticides (e.g., pyrethrum and rotenone), oils (e.g., turpentine in paints and varnishes), and other products.

What if the wild ancestors of rice, wheat, and corn had been eradicated before they were domesticated? These three grains account for more than one-half of the world's food production. What if the natural sources of medicines and industrial products that we now use had not been discovered before the source species became extinct? These examples make it clear that natural gene pools form an essential foundation of human civilization.

The Value of Biodiversity in Maintaining Ecosystem Functions

All species participate in the numerous ecological processes that occur within and between ecosystems. Each one contributes to one or more functions, such as production, decomposition, water and nutrient cycling, soil generation, erosion control, pest control, and climate regulation. These functions are not only vital to the survival of an ecosystem; collectively, they are essential to human survival.

Some species, of course, appear to be more important than others in the functioning of ecosystems. The American alligator, for example, is a dominant member of some subtropical wetland ecosystems, such as the Florida Everglades (chapter 10). Depressions excavated by alligators (gator holes) hold fresh water during the relatively dry winters. These reser-

Figure 11.4   Alligators maintain depressions, known as "gator holes," which retain water during dry periods and serve as reservoirs for aquatic organisms.
© Larry Lipsky/Tom Stack & Associates

voirs preserve aquatic life and provide fresh water and food for birds and mammals (fig. 11.4). If the alligators were eliminated, gator holes would fill quickly with sediment and aquatic plants, and wildlife could not depend on them for survival during the dry periods. The alligators' feeding habits also influence the Everglades' ecosystem. By eating large numbers of gar, a major predator of such game fish as bass and bream, alligators indirectly provide more game fish for other Everglades predators.

The American alligator is an example of a **keystone species,** that is, a species that plays a central role in controlling the relative abundance of other species in an ecosystem. Another keystone species is the sea otter, whose presence or absence greatly affects other sea animals along the Pacific coast of North America.

Although relatively few species function as keystone species, many function as critical-link species. **Critical-link species** play an essential role in ecosystem functions, regardless of their trophic level, magnitude of biomass, or possible role as keystone species. Many decomposer microorganisms are critical-link species. Although relatively small in size and total biomass, they function as essential links in recycling nutrients within ecosystems (chapter 3). Some critical-link species may provide food for a network of species (such as a plant species that supplies food for several herbivorous insects); others may play essential roles as pollinators of flowers or dispersal agents

for seeds or fruits. Although relatively little is known about coevolution among species (chapter 5) in tropical rain forests, the high degree of animal-dependent pollination and dispersal suggests that tropical ecosystems may be particularly rich in critical-link species relationships.

Economic Value

Economic value is often cited in arguments favoring species preservation. The vast multitude of natural products used in industry and agriculture add trillions of dollars to the world's economy. In addition, tourism and recreation generate billions of dollars, and the most popular tourist attractions are generally not human-made marvels, but national and state parks and forests. Sport hunters and fishermen spend hundreds of millions of dollars each year; thus, they and their commercial suppliers have a vested interest in the preservation of their target species. For reasons that are basically economic, many local and national conservation groups have worked toward habitat and species preservation; although their efforts are often restricted to only a few species of game mammals, upland game birds, water fowl, and fish. A few nongame animals, such as eagles, hawks, cranes, and wolves, have received considerable attention (mainly by state and federal agencies and citizen groups), but most nongame species are virtually ignored.

Figure 11.5 The Texas blind salamander is an endangered species, but its appearance fails to raise much human concern over its fate.
© Joseph T. Collins/Photo Researchers, Inc.

Aesthetic Value

Some proponents of species preservation are guided principally by aesthetic values. The thrill of watching an eagle soar silently overhead, an afternoon spent listening to the capricious chatter of chipmunks, the taste of wild blackberries, the refreshing fragrance of spring wildflowers, and the softness of a bed of moss conjure up pleasant feelings in most of us. Still, those who base their arguments on aesthetics alone must recognize that most plants and animals hold little aesthetic value for most people. Many people may be thrilled by the image of a wolfpack roaming the wild, but very few become excited at the thought of the Texas blind salamander resting in a cave (fig. 11.5). Human nature tends to embrace the warm and fuzzy (birds and mammals) and reject the cold and sometimes slimy (reptiles, amphibians, fish, insects, clams, and plants). Furthermore, when people are hungry, poor, and unemployed, aesthetic concerns are not generally among their priorities.

Intrinsic Value

The arguments we have presented to justify the preservation of biodiversity have a common thread: They are all based on its value to people. Using such a foundation, however, presents many practical problems. One such problem is that many species have little or no direct value in the conventional sense. More than one hundred species have become extinct in the United States alone since the 1600s without causing any well-documented impact on the economy or general health of our society. Thus it is difficult to convince people to be concerned about an obscure species that may become extinct if, for example, a dam is built or a forest is cut to provide jobs and lumber. Even when a threatened species has significant economic value, many people will reason that a dam or a shopping mall would provide even greater economic return.

Problems in assigning value to biodiversity also arise when we consider the ecological roles of species. Although ecologists know little about the specific functions of most of the globe's 1.4 million identified species, relatively few species appear to have value as keystone or critical-link species. Fur-

thermore, there is some evidence that the ecological roles of many species overlap. For example, although the chestnut blight killed essentially all of the American chestnut trees that once dominated the forests in Appalachia (chapter 5), various species of oaks have filled in the gaps left by the fallen chestnut trees. Thus, Appalachian forests continue their services in production, decomposition, nutrient and water cycling, soil generation, erosion control, pest control, and climate regulation. (We hasten to add that such ecological services are also of great economic value.)

Thus, assigning conventional values to biodiversity has some serious shortcomings. Furthermore, as David Ehrenfeld of Rutgers University argues, assigning such values to biodiversity merely legitimizes the very processes that are wiping it out.* That is, the short-term economic gain that accrues from cutting down tropical rain forests, killing wildlife for commercial products, draining wetlands for agriculture, and so on is the driving force that is propelling thousands of species to the brink of extinction. Rather than basing arguments for the preservation of biodiversity on economic or ecological values, Ehrenfeld reasons that the value of a particular species does not depend on the properties of that species; rather, value is an intrinsic part of biodiversity. Ehrenfeld argues that "For biological diversity, value *is*. Nothing more or nothing less." In agreement were noted naturalists John Muir and Aldo Leopold who pointed out that the very existence of biodiversity constitutes its right to survive. Furthermore, many theologians argue that we have a responsibility to be faithful stewards of the earth and its resources, including other organisms with whom we share the planet. They suggest that during our brief journey on this planet, we have no right to destroy its inhabitants.

A Historical Perspective on Species Extinction

Extinction is not a new phenomenon. Species have been vanishing from the face of the earth since life began about 3.5 billion years ago. Furthermore, extinction is the rule, rather than the exception. Over 99% of the species that ever lived on this planet are now extinct. Mounting evidence suggests that the history of life on earth consists of relatively long episodes of biological stability (perhaps hundreds of thousands of years) when relatively few species became extinct. On occasion, these stable periods have been punctuated by short bursts of massive species extinctions. During such mass extinctions, biodiversity declined precipitously; in some cases 50% or more of all living species disappeared. Five or six such mass extinctions have occurred over the past 600 million years.

These brief periods of mass extinction appear to be associated with times of massive geological upheaval. Such major global events as continental drift (Special Topic 17.2), volcanic eruptions, and bombardment by massive asteroids may

*D. Ehrenfeld, "Why Put a Value on Biodiversity?" in *Biodiversity*, ed. E. O. Wilson (Washington D.C.: National Academy Press, 1988).

Figure 11.6 A reconstruction of a scene from the Pleistocene era at Rancho La Brea Tar Pit, near present-day Los Angeles. Illustrated are vultures, saber-toothed tigers, mammoths, giant ground sloths, and dire wolves. These species became extinct about 11,000 years ago.
Charles R. Knight, American Museum of Natural History

have contributed to episodic losses of thousands of species. A mass extinction that has captured the imagination of many people occurred 65 million years ago when the dinosaurs vanished.*

A more recent, but much smaller mass extinction took place in North America at the close of the last Ice Age (Pleistocene Epoch) some eleven thousand years ago. This extinction apparently only involved many large prey and their associated predators—horses, camels, mammoths, mastodons, ground sloths, giant beavers, jaguars, and saber-toothed tigers (fig. 11.6). Only about 30% of the large game species that lived at that time, including moose and caribou, survived. Hence this extinction is often called the Pleistocene *megafaunal extinction.*

Nobody knows for certain what caused this mass extinction. We do know that the climate changed dramatically during this period. As the North American ice sheet shrank, a corridor opened along the MacKenzie River between the Canadian Rockies and the retreating continental ice front. This allowed extremely cold air to flow out of the Arctic Basin and onto the Great Plains (fig. 11.7). R. Dale Guthrie of the University of Alaska argues that a shift to colder conditions on the Great Plains set in motion a series of changes that ultimately led to the megafaunal extinction. He reasons that the growing season shortened and seasonal contrasts in temperature and moisture increased. In response, the mosaic of grasses, shrubs, and trees that supported a rich diversity of large herbivores and their predators was replaced by simpler vegetational patterns; the woody plants could no longer tolerate the more extreme climate. Thus, as both the quantity and quality of their essential resources declined, the large mammals began to disappear locally and eventually became extinct over their entire range.

*See the article by Alvarez and Asaro listed at the end of this chapter.

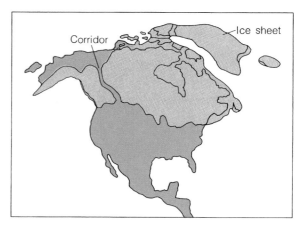

Figure 11.7 The opening of a corridor between the Rocky Mountains and continental glaciers some eleven thousand years ago allowed cold air to flow out of the Arctic onto the Great Plains. It also allowed people from Siberia to invade the southern portion of North America. The blue area is glaciated; the brown area is unglaciated.

Paul S. Martin of the University of Arizona has a different explanation. He points out that the Pleistocene megafaunal extinction closely coincides with the arrival of humans in North America. To explain how the appearance of relatively few human hunters could bring about the extinction of so many species of large animals, Martin proposes the following scenario. These early immigrants crossed the Bering land bridge from Siberia into Alaska, traveled down the MacKenzie corridor, and eventually spread south and east across North America. As these hunters advanced across the continent, they killed animals in large numbers (e.g., by setting fires to drive large herds over cliffs or into rivers where they drowned). When the hunters moved on, carnivores eliminated the few large prey animals that were left behind. Eventually, as the hunters advanced across the continent and large herbivores became extinct, the large, specialized carnivores, such as the saber-toothed tiger, became extinct as well. Other prey animals were apparently too wary and fleet (e.g., deer) or had group defenses (e.g., musk oxen, which form a protective circle) that prevented the specialized carnivores from preying on them effectively.

Quite likely, the combination of changes in the environment, a more extreme climate, and the arrival of humans stressed the megafaunal populations of North America. Perhaps without the additional impact of human hunting, however, these species could have survived the change in climate. This megafaunal extinction may have been a harbinger of the tremendous impact of modern humans on global biodiversity.

Modern Threats to Biodiversity

The rate of extinction is accelerating rapidly today, and the sharp upturn can be directly linked to the unprecedented growth of the global human population and the accompanying exploitation of the earth's resources. Humankind is contributing to the loss of biodiversity in several ways: (1) by destroying natural habitats, (2) by hunting for food and commercial products (e.g., furs, hides, and tusks), and (3) by introducing alien species. Although the relative significance of each type of stress varies among species, alteration of natural habitats has had by far the greatest adverse impact on biodiversity.

Destruction of Natural Habitats

Until a few centuries ago, people lived mostly in small, isolated communities surrounded by a sea of wilderness. This pattern of settlement has changed drastically in the last two hundred years. Throughout the world, people have transformed complex forest and prairie ecosystems into farms and ranches where a few crop and livestock species are the dominant life forms. Today, natural prairies are all but gone, and the remaining ancient forests are quickly disappearing. Tropical rain forests have already been reduced to perhaps 50% of their original area. At the current rate of destruction, little of the rain forest will remain in thirty years. Many Americans decry the loss of tropical rain forests, but the destruction of ancient forests is also occurring within the United States (fig. 11.8). Over 24,000 hectares (60,000 acres) of ancient (old-growth) trees are cut each year in the Pacific Northwest, mostly for lumber that is exported to Japan and other Pacific rim nations. In Alaska's Tongass National Forest, about half of the most productive forest land has been logged since 1950.

Expansion of human settlements has also severely affected wetlands (chapter 10). Marshes and swamps have been filled in for housing developments, shopping centers, and industrial parks. Other wetlands have been drained to provide fertile soils for crop production or livestock ranching. For example, glacial prairie potholes in the upper Great Plains have long been nesting sites for migratory waterfowl. Over recent

Figure 11.8 Ancient forests in the Pacific Northwest, like tropical rain forests, are being rapidly lost to timber cutting.
© Tom Kitchin/Tom Stack & Associates

decades as many as 50% of the prairie potholes have been drained to increase acreage for crops or commercial development. Loss of nesting habitat plus recurrent drought account for a 30% decline in duck populations over the past two decades. Of all wetlands that were present when European settlement began in what are now the lower forty-eight states, less than half remain.

Over the past two to three centuries, humanity has largely conquered and tamed the wilderness that once surrounded its communities. As people have dominated the landscape, suitable habitat for wild species has not only declined in total area, but the remaining habitat fragments are more isolated. Today, many wild species are found mainly in isolated habitat fragments (e.g., parks and reserves) in a sea of humanity. Figure 11.9 illustrates the reduction and fragmentation of woodland in Wisconsin between 1831 and 1950. Because these habitat fragments are surrounded by dissimilar habitat, they are often referred to as **habitat islands.**

We can gain valuable insight into the effects of the formation of habitat islands on species diversity by first considering the history of human impact on actual islands surrounded by

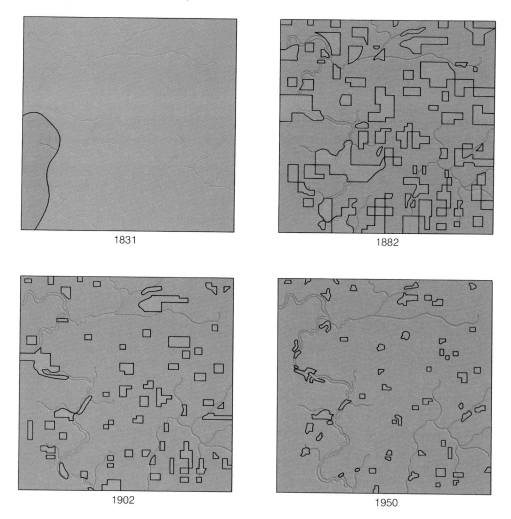

1831

1882

1902

1950

Figure 11.9 The reduction and fragmentation of woodland (shaded areas) in Cadiz Township, Wisconsin, (6 miles by 6 miles) between 1831 and 1950. The fragmentation created small, isolated habitat islands.

water. Because island-dwelling species generally have small populations, restricted genetic diversity (gene pools), and limited available habitat, they are usually quite vulnerable to changes in their environment. For example, no large herbivorous mammals lived on the Hawaiian Islands until the 1700s, when Europeans introduced domestic livestock. Without grazing pressure from native herbivores, there had been little natural selection for thorns or toxic chemicals in plants to ward off herbivores. Native plants were therefore vulnerable to the impact of introduced livestock. Furthermore, the limited size of habitats favored overgrazing. Humans also affected the native flora through habitat destruction (deforestation and cultivation) and the introduction of weedy plants that competed successfully against native species. Thus, species that had not been threatened previously were devastated by humans and their domesticated animals. Perhaps 1,800 native plant species once grew on the Hawaiian Islands, but today more than 200 endemic species are thought to be extinct and another 800 are considered to be endangered.

Human impact has not been restricted to plants alone; nor has extinction been perpetrated only by people of European ancestry. About two thousand years ago, just prior to Polynesian colonization, at least 86 species of land birds lived in the Hawaiian Islands. Forty-five species disappeared after Polynesian colonization, and another 11 have disappeared since Europeans arrived. Several more species of birds are on the verge of extinction. The U.S. Fish and Wildlife Service lists more endemic species of Hawaiian birds as endangered or threatened than it lists for the entire continental United States.

The numerous extinctions on the Hawaiian Islands are not an isolated phenomenon. Similar large-scale reductions in native species have occurred on isolated islands in all oceans, including the famous Galápagos Archipelago in the Pacific Ocean and the two large islands of New Zealand. Will a similar fate befall species on habitat islands? Partially in response to this question, ecologists have been investigating the dynamics of biodiversity on islands. Using mathematical models and field studies, ecologists have developed a theory of island biogeography that they can use to predict the impact of habitat fragmentation on biodiversity.

The **theory of island biogeography** holds that biodiversity on an island remains more or less constant (essentially at equilibrium) because the extinction rate for species on the

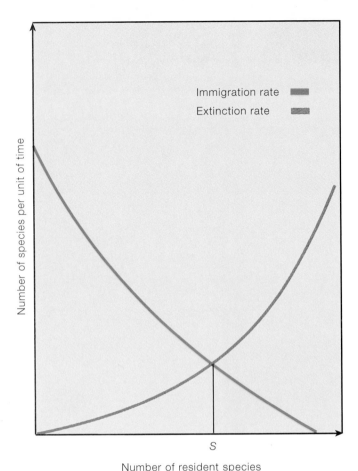

Figure 11.10   The balance between immigration and extinction on an island results in a particular equilibrium value (S) for species richness.

island is balanced by new species migrating to the island (fig. 11.10). At first glance, these findings suggest that the current level of biodiversity on habitat islands is rather secure. But the theory is based upon islands that are in equilibrium; that is, they are not changing. Many habitat islands around the globe, however, continue to shrink and become more isolated. Herein lies the major problem for the future of biodiversity on habitat islands.

If the area of a particular habitat, such as a tropical rain forest, is reduced by a particular amount, the number of species living in that habitat will decline to a new, lower equilibrium. The theory of island biogeography suggests a general rule of thumb: If the area of an island diminishes tenfold, the number of species will be cut in half. Today the rich Brazilian rain forest along the Atlantic coast has been cleared to the extent that less than 1% of its original area remains. Even if the total remnant could be protected forever, the theory of island biogeography predicts that biodiversity will decline by as much as 75%.

Even relatively large habitat islands may not be able to support all of their original biota. Consider, for example, the future of the grizzly bear in Yellowstone. By worldwide standards, the Greater Yellowstone ecosystem (encompassing 53,500 square kilometers or 20,650 square miles and comprising a national park and six national forests) is a large habitat island. But, the average grizzly bear has a home range of 250 to 1,000 square kilometers (100–400 square miles). Ecologists estimate that 50-90 animals are required if there is to be a 95% probability that the population will survive for one hundred years. Hence, survival of the grizzly population may require a reserve of up to 90,000 square kilometers (35,000 square miles). A network of roads throughout much of the Greater Yellowstone ecosystem poses additional problems for the grizzlies. Roads permit easy human access, which enhances opportunities for poaching. Since 1975, more than fifty grizzly bears have been killed illegally within the Greater Yellowstone ecosystem. Thus, continued survival of grizzlies on this habitat island, even though it is relatively large, remains in doubt.

Reduction in the size of habitat islands not only makes survival less likely for resident organisms, but the accompanying isolation also reduces the chances for successful migration to such islands. Highways, farms, and cities serve as effective barriers for plant and animal migration. Any observant highway traveler who notices the road kill of small animals, is well aware of the impact of highways on animal movement. Even some species of birds that could easily fly between habitat islands rarely do so because they instinctively avoid flying over certain types of habitat. Other migrating animals may not be able to survive the conditions of the intervening habitats. Studies of chipmunks, for example, suggest that they can survive only a few hours in the open sun. Hence, successful migration from one forest fragment to another would be impossible if the intervening distance required more than a few hours of travel time. Finally, the smaller and more isolated the island, the greater the probability that a migrating organism will miss the island; the smaller the target, the less likely migrants are to hit it.

Figure 11.11 summarizes the significance of island biogeography theory for biodiversity. We can expect that large islands close to other islands will support the greatest number of species. Small, isolated islands will support the fewest species. These generalizations, however, are subject to qualification. For example, a collection of small habitat islands that offers considerable habitat heterogeneity may support greater biodiversity than a single large island that is more homogeneous. In addition, a single large island is vulnerable to the spread of disease that could eliminate a species, whereas several isolated islands would be much less susceptible to an epidemic.

In previous chapters, we have described the value of environmental heterogeneity (patchiness or fragmentation) in stabilizing population interactions (chapter 9) and enhancing species diversity (chapter 10). These concepts appear to contradict our present discussion, but the key difference between nature's activities and our activities is the subsequent degree of isolation. For example, we saw in chapter 10 that fires create a mosaic of burned, partially-burned, and totally burned areas. These habitat fragments, however, are usually only a short distance from similar fragments, so that the chances for successful migration among them are relatively high. Furthermore, the intervening habitat may not pose a significant barrier to migration because it pro-

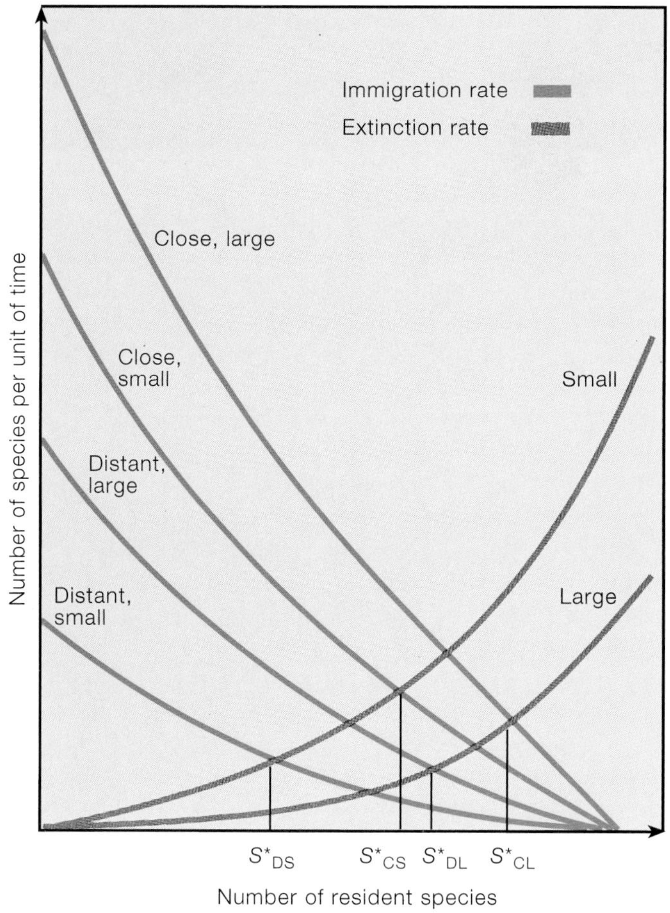
Immigration rate
Extinction rate
Close, large
Close, small
Distant, large
Distant, small
Small
Large
Number of species per unit of time
S*DS
S*CS
S*DL
S*CL
Number of resident species

Figure 11.11 The effects of island size and degree of isolation on equilibrium species richness. S^* is the equilibrium species richness (D = distant, C = close, L = large, S = small).

vides some shelter (e.g., standing or fallen charred timber) from the weather and predators. Even if such shelter is not immediately available after a fire, ecological succession soon provides it.

In marked contrast, human destruction of habitats continually increases the distance between fragments while creating effective barriers to dispersal. Highways, agricultural fields, and urban areas are essentially permanent (ecological succession is not permitted in such habitats) and insurmountable barriers to the migration of many species of plants and animals.

In addition to the conspicuous destruction just described, human activities alter natural habitats in many less obvious ways. Acid rain increases the concentration of aluminum in water that drains into streams, where it may become toxic to frogs and salamanders (chapter 16). Pesticides such as DDT reduce the thickness of the egg shells of eagles and peregrine falcons, disrupting successful reproduction (chapter 12). Global warming might force the relocation of many species (chapter 3), but such relocations would be greatly hampered by human barriers to migration.

Habitat alteration is a by-product of many human activities. Thus, if we are to slow the accelerating trend toward a loss of biodiversity, we must give greater priority to species and ecosystem preservation when we make decisions regarding land use. We discuss strategies for habitat preservation later in this chapter.

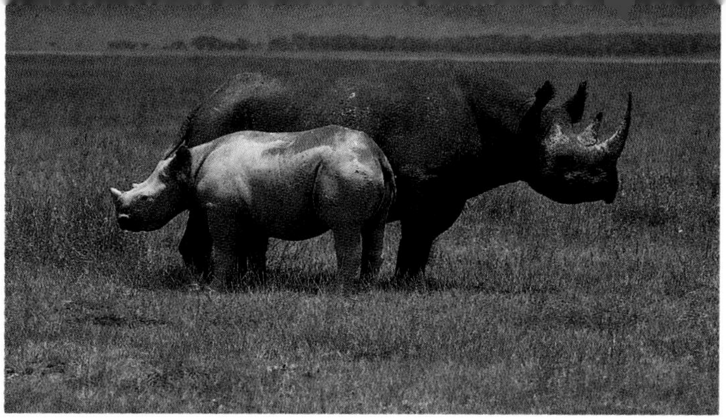

Figure 11.12 Poachers have brought the black rhinoceros to the brink of extinction in its native habitat.
© W. Wisniewski/Photo Researchers, Inc.

Commercial Hunting

Humans hunt animals for food, sport, or commercial products. In the past, a significant number of species either became extinct or endangered because of their food value. These species include the passenger pigeon (extinct), great auk (extinct), heath hen (extinct), wild turkey, sandhill crane, and trumpeter swan. Today the few species that are endangered because of their food value include the giant tortoise (for its flesh and liver) and the green sea turtle (for its flesh and eggs). Sport hunting is usually well regulated, and game species are only in danger where governments ineffectively thwart poachers; however, commercial hunting threatens worldwide biodiversity. International trade in plants and wildlife is big business, annually generating more than $5 billion, of which up to 30% may be illegal. Such trade includes live specimens, skins, furs, and reworked pieces such as jewelry.

Public attention has recently focused on the slaughter of large animal species such as the elephant and the rhinoceros. Several species of rhinoceros face extinction because of human demand for their horns. Poachers have eliminated 95% of the black rhinos since 1970 (fig. 11.12). The world's second largest land mammal faces a hostile world where a single rhino horn can sell for as much as $6,000. In North Yemen, rhino horns are made into handles for fancy daggers. In Asia, rhino horns are ground into a powder that is sold to relieve headaches, heart problems, and liver ailments. The powder is also used as a skin ointment. Modern medicine, however, has not been able to confirm the effectiveness of powdered rhino horn for any of these uses.

During the 1980s poachers killed well over half of Africa's elephants for their ivory tusks, which were carved into statues, necklaces, rings, and various other ornamental products sold primarily in the affluent nations: Japan, the United States, and the European community. In 1989, a worldwide ban on international trade of African elephant products was approved. Consequently, the price of ivory dropped from $30 a kilogram (2.2 pounds) in 1989 to $3 in 1991. The decline in profits has become a major disincentive for poachers; the loss of elephants to poachers has declined as much as 90% in some nations such as Kenya, Uganda, and Tanzania. However, the future of wild elephants remains in doubt. Some African nations are pressing for an end to the ban so that they can regain badly needed financial resources from the ivory trade. In addition, a rapidly growing human population is appropriating more and more of the African elephant's habitat.

Illegal hunting threatens many species in the United States. Middle East oil sheiks will pay as much as $10,000 for a living peregrine falcon. In a sixteen-month period, two North American poachers netted $750,000 from smuggling live peregrine falcons and goshawks to Europe and the Middle East. Black bears are killed by the hundreds because their gall bladders sell for $3,000 per pound in foreign markets, where they are used in herbal medicines. In less than ten years, the average age of legally killed bears in the Smokey Mountains has dropped by half. Trophy hunting is accelerating because collectors are anxious to get a prize specimen (e.g., the head of a big-horn sheep or a polar bear skin) before those animals are all gone. The U.S. Fish and Wildlife Service estimates that illegal profits from U.S. wildlife total $200 million annually, and such tax-free income continues to grow. Nationwide poaching of wildlife equals or exceeds the legal kill.

The pet trade also consumes a huge number of wild animals. Each year, millions of live mammals, birds, reptiles, amphibians, and fish are imported into the United States primarily for sale as pets (fig. 11.13). Such sales are often a frivolous waste, because most of these animals make very poor pets. When they lose their novelty, many are flushed away by their owners or abandoned to fend for themselves in a foreign habitat. The total waste of animals in the pet trade is even more extreme, because perhaps five to ten times the number of animals that make it alive to the United States die during capture or shipment.

Commercial trade also endangers many species of plants, particularly cacti and orchids. The saguaro cactus symbolizes the deserts of southern Arizona (fig. 11.14), but it is quickly disappearing from its native habitat, a victim of cactus rustlers. This king of the southwest cacti is highly prized by homeowners and landscapers, but because it grows so slowly, nurseries rarely attempt to cultivate them. Thus, poachers find a lucrative black market where a prime stolen plant can bring up to $1,000.

Commercially hunted species are frequently caught in a vicious circle. The more they are hunted, the rarer they become, and the rarer they become, the more intensely they are hunted. People will pay more for a rare species, and as the financial stakes soar, poachers are more willing to take the risk of bagging an endangered organism. Potentially huge profits spur the formation of highly sophisticated international poaching and smuggling rings that deal in endangered species or commercial products made from them.

Introduction of Alien Species

As long as people have traveled, they have introduced (accidentally or intentionally) plant and animal species into new

Figure 11.13 Trading of these parrots in a marketplace in Brazil threatens their future existence in the wild.
© Martin Wendler/Peter Arnold, Inc.

Figure 11.14 The saguaro cactus is rapidly disappearing from its native habitat in the desert Southwest.
© Dale E. Boyer/Photo Researchers, Inc.

geographical areas. Sometimes the new environment had an opening, and the alien became established without seriously interfering with the well-being of native species. Often, however, the alien is a superior predator, parasite, or competitor, and causes the extinction or near extinction of native species.

Unquestionably, introduced domestic livestock have dramatically changed local biotas. Goats, sheep, and cattle often overgraze native ecosystems, reducing the native plant cover and creating conditions that favor the invasion of highly competitive and herbivore-tolerant weedy species. Recall from chapter 5 that before the advent of domestic grazing in the Columbia River Basin of Washington, these rangelands were dominated by native bluebunch grass and a rich diversity of other native plants. Heavy overgrazing destroyed the native vegetation, creating a habitat open to invasion, and cheatgrass, a native of Europe, soon dominated several million hectares. This pattern of change in plant species composition has been repeated often around the world wherever people have introduced domestic livestock.

Domestic livestock often successfully compete with native herbivores. For example, in 1957 fishermen released goats on one of the Galápagos Islands where tortoises also lived. When a research team from the Charles Darwin Station arrived on the island in 1962 to study the tortoise population, they found only empty tortoise shells, many of them wedged among the rocks that covered the upland slopes. Investigation of the lowlands indicated that the flourishing goat herd had consumed all the vegetation that had previously been within reach of the tortoises. The hungry tortoises apparently were forced to seek food on the rocky slopes, where they either became trapped between rocks or fell over precipices and starved to death.

Introduced predators can wreck havoc on native prey. Rats and feral cats, for example, have eliminated ground-nesting birds in many areas, including Jamaica, Australia, New Zealand, and the Galápagos Islands. In one infamous case, pigs played a major role in the extinction of a ground-nesting bird, the dodo, which lived only on the island of Mauritius in the Indian Ocean. The dodo was flightless and laid its eggs on the ground. When sailors introduced pigs to the island to provide a source of fresh meat as they sailed between the East Indies and Europe, the pigs devoured the dodo's eggs. Such predation coupled with hunting by humans soon led to the demise of the dodo.

Even aquatic species have felt the impact of alien invaders. During the 1940s and 1950s, the sea lamprey devastated populations of lake trout and whitefish in the western Great Lakes. After attaching to a fish, the lamprey opens a hole in the side of its victim (fig. 11.15) and secretes an anticoagulant that keeps the wound open. Thus, the parasite can feed up to several weeks until it is satiated or its host dies. Although the lamprey had long been in Lake Ontario, its access to the other Great Lakes was blocked by Niagara Falls. The lamprey was able to circumvent this barrier when the Welland Canal was constructed to allow seagoing ships to sail between the western Great Lakes and the Atlantic Ocean. With no natural enemies and abundant hosts, the lamprey flourished and nearly wiped out the major fish species. It continues to be a concern to fish managers on the Great Lakes.

Characteristics of Endangered Species

Some species are more vulnerable to human activities than others. Table 11.3 lists characteristics shared by many endangered species. For example, many species are endangered because they are specialists; that is, they have narrow habitat

Figure 11.15   Two lamprey feeding on a brown trout.
© Gary Milburn/Tom Stack & Associates

Table 11.3	Some characteristics that increase vulnerability to extinction
Characteristic	**Examples**
specialized feeding habits	giant panda (bamboo) snail kite (apple snail)
limited or specialized distribution	green sea turtle (lays eggs on only a few beaches) Kirkland's warbler (nests only in six- to fifteen-year-old jack pine stands)
occupies top trophic levels	bald eagle, gray wolf
low reproductive potential (extreme *K*-strategist)	California condor, gorilla
hunted for commercial products	rhinoceros, African elephant

tolerances and restricted food sources. In addition, as described in chapter 5, many endangered species are extreme *K*-strategists; that is, their reproduction begins at a late age, they produce few young, and long intervals occur between births of offspring. If the number of young that reach reproductive maturity continues to be insufficient to offset adult mortality, a species is doomed to extinction. Today, hundreds of thousands of species are vulnerable to extinction simply because they live in ecosystems like wetlands and tropical rain forests that are bearing the brunt of habitat destruction.

Attempts to save an endangered species utilize the concept of a **minimum viable population (MVP).** What is the smallest number of individual members required to ensure the survival of an isolated population? Although this question is important, it has two unfortunate implications. The first is that there is a number, the MVP, that serves as an accurate lower limit below which a population is vulnerable to extinction and above which it is safe. In reality, of course, the risk of extinction varies continuously with population size. The second implication is that the MVP value applies to all populations of a species; however, we know that the viability of a local population depends upon the particular conditions in

Genetic Problems of Small Populations

As populations decline in size, they often experience a loss of genetic variability that can further increase their vulnerability to extinction. As the number of individuals declines, the probability that mates will be closely related increases. The mating of close relatives, called *inbreeding*, increases the likelihood that deleterious characteristics will appear in the population. In addition, an inbred population often experiences a decline in overall vigor.

The hazards of inbreeding are well illustrated by the cheetah. Although it ranged over Africa, Asia, Europe, and North America twenty thousand years ago, the cheetah today is restricted to isolated populations in South Africa and East Africa, with overall numbers as low as 1,500 to 5,000. Investigators recently found that cheetahs have dramatically lower levels of genetic variation than other cats and mammals. Concurrently, they found high mortality among young cheetahs and high susceptibility to viral infections at all ages. Researchers also discovered that the sperm count (sperm per unit volume of ejaculate) was ten times lower than it is in related cat species. In addition, 70% of the sperm were malformed. Investigators believe that lack of genetic variability, vulnerability to viral diseases, and sperm abnormalities are results of inbreeding.

Genetic drift also becomes more important as population size declines. Genetic drift occurs when chance events such as accidental death lead to a loss of genetic variability within a population. Consider, for example, a small population in which only two individuals have a particular trait. Although this trait may be especially beneficial, it would be lost if both individuals were killed by lightning or swept away in a flood or shot to death by hunters. Thus, genetic drift leads to a net loss in genetic variability within a population.

Reduced to small, isolated populations, cheetahs may be experiencing the negative effects of inbreeding.
© Arthus-Bertrand/Peter Arnold, Inc.

Lack of genetic variability limits a population's ability to respond to inevitable environmental change. Consider again the cheetah. Today it is a highly successful hunter, but as the fastest animal on four legs, the cheetah is also highly specialized. Being so specialized and having so little genetic variability, the cheetah probably has a poor chance of adapting to a significant change in its environment.

What about the thousands of other endangered species? How much genetic variability remains in their populations? What are their chances for adapting to environmental changes?

---

the local habitat. Nevertheless, although a specific MVP value for an entire species does not really exist, MVP analysis remains one of the few tools that resource managers can use to approximate the risk of extinction.

What forces influence a small population's risk of extinction? Obviously environmental uncertainty plays a major role in the long-term survival of any small population. Temporal changes in the availability of essential resources and in the activities of predators, prey, parasites, hosts, competitors, and mutualists all can threaten a population with extinction (chapters 3, 4, and 5). Natural disturbances such as fire, drought, heat waves, hurricanes, and floods can mean extinction to small, isolated populations. Human-caused disturbances (e.g., clearing a forest, polluting a lake) can also take a toll on local populations. Field biologists and resource managers are also becoming increasingly aware of the genetic problems that small populations face as their numbers dwindle. Special Topic 11.1 considers the implications of **inbreeding** (mating between close relatives) and **genetic drift** (a change in gene

frequency determined by chance rather than by evolutionary advantage) for small populations.

The extinction of the heath hen illustrates the stresses experienced by a small population. Once common from New England to Virginia, the number of heath hens began to decline as European settlement progressed. By 1900, there were fewer than 100 survivors—all on Martha's Vineyard, an island off Cape Cod, Massachusetts. To save the remaining birds, a predator-control program was instituted, and in 1907 a portion of the island was set aside as a refuge. In response, the heath hen population rose to 800 birds by 1916. That summer, however, a fire destroyed most of their nests, and that setback was followed by unusually heavy predation by goshawks. The following spring, the birds numbered between 100 and 150. By 1920, the population had increased again to about 200, but disease took its toll, and the population fell below 100 again. Although the species endured a while longer, the last survivor had died by 1932. In the last stages of decline, the proportion of males to females increased,

Figure 11.16 These five species received top priority in 1989 in the allocation of financial resources for recovery of endangered species: (*a*) bald eagle; (*b*) grizzly bear; (*c*) red-cockaded woodpecker; (*d*) American peregrine falcon; (*e*) gray wolf.

(*a*) © Johnny Johnson/Animals Animals (*b*) © Brian Milne/Animals Animals (*c*) © David and Hayes Norris/Photo Researchers, Inc. (*d*) © Tom McHugh/Photo Researchers, Inc. (*e*) © Tom and Pat Leeson/Photo Researchers, Inc.

and more and more birds appeared to be sterile, perhaps as a result of inbreeding.

Sometimes, small populations are able to survive environmental stresses because immigration from other colonies compensates for losses. Populations that produce excess individuals are termed **source populations,** whereas populations that would go extinct without immigration are termed **sink populations.** Long-term survival of the sink population(s) depends upon the continued well-being of the source population(s) and the availability of corridors that permit dispersal between sources and sinks.

Efforts to Preserve Biodiversity

Efforts to preserve biodiversity have focused largely on individual species. Since enactment of the Endangered Species Act of 1973, the U.S. Fish and Wildlife Service has listed over 680 native species as endangered or threatened. Furthermore, the Service has approved over 310 **recovery plans,** specific plans of action for the recovery of species listed as endan-

gered or threatened. These plans identify the causes of a species' decline and specify actions needed to reverse the downward trend. The great majority of these plans focus on a single species.

Dealing with loss of biodiversity on a species-by-species basis, however, has some significant shortcomings. The listing process is cumbersome, labor-intensive, and costly and has not kept up with the forces that threaten biodiversity. Currently more than 3,600 U.S. species await consideration by the Service for endangered species status, but only about 50 are added to the list each year. In addition, financial resources for recovery continue to concentrate on only a few species. During 1989, federal and state agencies spent nearly $44 million for the conservation of 347 endangered and threatened species (approximately two-thirds of all listed species), but twelve species accounted for over half of those expenditures. The top five species were the bald eagle, grizzly bear, red-cockaded woodpecker, American peregrine falcon, and gray wolf (fig. 11.16). With the listing of additional species and prospects of limited tax dollars, few recovery plans are likely to receive enough financial support to have a chance of saving the endangered species.

Many conservationists also express concern that traditional species-level approaches do not adequately consider vital ecological relationships such as source and sink dynamics, habitat sizes, and migration corridors, environmental gradients of moisture, temperature and day length, and large-scale stresses such as global warming. Finally, the species-level approach inevitably sets up choices between *desirable* species that are likely to be protected and *insignificant* species that are likely to be ignored.

The landscape approach to preserving biodiversity consists of integrating a system of reserves. Such strategies provide many benefits. Preserving the landscape not only provides protection for a whole community of organisms, but also helps preserve the vast array of ecological relationships that are essential for long-term survival of the community. Furthermore, a naturally functioning landscape may require less management than does species-level conservation. Thus, the landscape approach may ultimately be more cost-effective.

No doubt a few highly *desirable* species such as the bald eagle and the grizzly bear will continue to receive special attention, but there is a broad consensus today that the future of most of the planet's biodiversity rests mainly on the preservation and/or restoration of large tracts of natural and seminatural habitat.

Landscape Approach: Habitat Preservation and Restoration

As noted earlier, habitat destruction is currently the chief threat to biodiversity; hence, habitat preservation and restoration are the most important strategies for preserving biodiversity. In tropical and subtropical nations, where rich biodiversity is particularly threatened by human activities, some large natural preserves have been established, but there is great need to preserve much more of the remaining forests, wetlands, and grasslands. Unfortunately, many tropical nations are burdened by enormous financial debts, which drive many of these countries to short-term and unsustainable exploitation of their natural resources. These activities include logging, farming, and mining in tropical forests as well as the draining of wetlands and the conversion of grasslands to agricultural land. Thus, these nations are caught in a vicious circle. Economic crises are spawning environmental crises which, in turn, exacerbate economic crises.

In recent years, a consortium of U.S. and international conservation organizations has developed an innovative solution to the escalating problem of resource degradation in less-developed nations, the **debt-for-nature swap.** Simply stated, this program converts unpaid notes of indebted nations into funds for conservation activities in the same countries. Costa Rica, the most aggressive debt swapper, has converted over $80 million of debt into funds for the protection of hundreds of millions of hectares of tropical forests. Swap monies have

Figure 11.17 Debt-for-nature swap funds are used to hire and train park workers such as this Costa Rican park guard who has discovered an illegally burned forest tract in La Amistad Biosphere Reserve.
© Chip and Jill Isenhart/Tom Stack & Associates

hired, trained, and equipped dozens of conservation workers and park managers as well as supported research for conservation and sustained development of Costa Rica's natural resources (fig. 11.17).

Debt-for-nature swaps in other nations, including Ecuador, Bolivia, Zambia, and Madagascar, have provided badly needed funds for land acquisition, environmental education, and the development of environmentally friendly tourism. Because only 5% of the less-developed world's total commercial foreign debt has been swapped and because numerous critical natural areas still badly need protection, the potential of this strategy remains largely untapped.

In the United States, federal and state agencies as well as many private conservation groups have also set aside millions of hectares of land. Here too, additional lands are needed to ensure that the landscape approach will protect biodiversity. Dr. Edward Grumbine of the Sierra Institute argues that greater ecosystems (landscapes aimed at biodiversity conservation) must include (1) enough habitat for minimum viable populations of all native species in the region, (2) areas large enough to accommodate natural disturbances such as fire, (3) habitats within which species and landscape processes may evolve over centuries, and (4) human occupancy and use at levels that do not cause ecological degradation. Although the Great Barrier Reef Marine Park off the east coast of Australia (fig. 11.18) may meet all of these criteria, few other preserves do.

Figure 11.18 A view of a portion of Great Barrier Reef Marine Park off the east coast of Australia.
© C. Seghers/Photo Researchers, Inc.

Additional efforts are needed to acquire new preserves, to enlarge present reserves, to provide migration corridors that connect preserves, and to modify uses of surrounding land so that they are more compatible with preserving biodiversity.

A promising example of what could be done at the landscape level is the reserve concept known as the Rothsay Prairie Landscape in Minnesota. Within an area of over 3,200 hectares (8,000 acres), several discontinuous protected areas totaling 1,600 hectares (4,000 acres) have been established (fig. 11.19). By themselves, these habitat islands will not provide the conditions necessary to preserve the rare species and ecological processes of the northern tall grass prairie ecosystem that once occupied most of western Minnesota. Connecting these small, isolated fragments, however, is a large corridor of disturbed native prairie and wetlands and formerly cultivated land under private ownership.

Model conservation programs enacted by the Minnesota legislature provide a strategy to consolidate the currently protected areas into a larger managed landscape. The present prairie preserves could be connected by restoring the intervening areas of degraded prairie and wetlands. The concept also calls for surrounding the core of remnant and restored natural areas with a buffer of low intensity agriculture where haying and grazing are permitted. Tax exemption opportunities could provide economic incentives for local landowners to participate in the Rothsay Prairie Landscape program.

At the time of this writing, no concrete plan had been formulated. Certainly, formal planning will have to involve the affected landowners. If the Rothsay Prairie Landscape concept becomes reality, several isolated fragments of native prairie would be integrated into a single landscape that would not only be large enough to protect the current native species, but would also allow the establishment of other native prairie species, perhaps even bison. The restored landscape would probably also accommodate natural disturbances, including fires, while permitting an economic return to farmers who live on the edge of the prairie.

Species Approach: Controlling Commercial Hunting

Establishing preserves does not always ensure adequate protection of endangered species. Poaching of animals and plants on preserves (as well as off preserves) by hunters, collectors, and traders is a serious problem worldwide. In an effort to curb poaching, the Convention on International Trade in Endangered Species of Wild Fauna and Flora (CITES) was negotiated in 1973. As of January 1992, 112 nations were members of CITES. This international treaty regulates the export and import of wild specimens and derived products. By vote of member nations, species may be placed on one of three (lists) appendices. Appendix I lists species in danger of extinction that are, or may be, affected by commercial trade. CITES prohibits all commercial trade in these species. Appendix II includes species that may become endangered if their trade is not brought under control. Commercial trade in species listed on Appendix II is allowed only if export permits are obtained stating that trade will not harm the species. Appendix III lists species that individual treaty nations identify as subject to domestic regulations to restrict or prevent overexploitation.

Although Appendix I is not identical with the U.S. Fish and Wildlife Service's list of endangered species, there is considerable overlap. Thus, in accord with CITES, it is illegal to import living specimens of endangered species (such as the orangutan) for sale in pet shops in the United States. Likewise, it is illegal to import tiger-skin coats or jewelry carved from ivory, or to export products that are made from American crocodiles.

Undoubtedly, many populations would be extinct today were it not for CITES. Yet each solution has its shortcomings; for CITES, it is enforcement. Because international trade in plants and wildlife is both monitored and regulated by the individual member nations, the effectiveness of the treaty depends upon the law enforcement and regulatory infrastructure of each country. Many nations do not have the financial

Native prairie and wetland in fair to good condition

Badly degraded pasture, some restoration potential; some areas probably old fields

Cropland active or recently retired

Land owned by the Minnesota Department of Natural Resources (DNR) or The Nature Conservancy

Private land

0   1 mi

0   1 km

Figure 11.19 The mosaic of private and public, natural and degraded land that potentially comprises the Rothsay Prairie Landscape. Courtesy of Robert Dana/Minnesota Department of Natural Resources.

resources to employ and train a sufficient number of wardens to protect highly-prized species such as elephants and rhinoceros. Even in the United States, law enforcement officials are often outnumbered, out-financed, and inadequately equipped to deal with commercial poachers.

Penalties for trafficking in illegal wildlife and plants and their products have recently been stiffened; a recent modification of the Endangered Species Act permits fines up to $100,000. Sentencing, however, is left to the discretion of judges. Some of them view poaching as a serious offense; others send a message of tolerance, even to repeat offenders. Economic incentives for poaching often far outweigh the risk of paying fines or serving a jail sentence, and the rarer a species becomes, the greater are the economic incentives. Thus, some conservationists argue that all nations, including the United States, must allocate more resources to reduce illegal hunting. Otherwise the future of many species with commercial value is indeed bleak.

Species Approach: Captive Propagation

Some species have been reduced to such small numbers that habitat preservation or reduction of poaching are not enough to save them; they require more intensive management. One strategy is **captive propagation,** that is, the breeding of animals in zoos and plants in botanical gardens for subsequent release back into the wild. To accomplish this goal, the American Association of Zoological Parks and Aquariums has approved survival plans for nearly forty species since 1980. Each plan is developed by a committee that judges how many captive animals are needed to ensure each species' survival and genetic diversity, how many animals can be kept in the available space, and which animals should be mated with one another.

Survival plans have been developed for the gorilla, rhinoceros, tiger, lemurs, and lesser-known endangered species. Already, zoo-born animals are replenishing a few depleted wild populations; examples include the release of red wolves to their natural habitat in North Carolina and the restoration of the Arabian oryx (a small, almost pure-white species of antelope) into preserves in Jordan and Oman. An exemplary case of restoration is the golden lion tamarin, a squirrel-sized monkey from Brazil's coastal rain forests (fig. 11.20). About fifteen years ago, the National Zoological Park in Washington D.C. and other U. S. zoos began a captive breeding program with approximately 70 tamarins. Today the number of animals in captivity has grown to over 400. The National Zoological Park has already sent nearly 50 tamarins to Brazil to be released in a reserve that was set aside to protect these popular monkeys.

Similar efforts are underway to save threatened plants. Since 1985, the Center for Plant Conservation has been working through its network of nineteen leading botanical gardens to conserve plants for the future. Each botanical garden

Figure 11.20 The golden lion tamarin, a participant in a successful captive breeding program.
© Larry Cameron/Photo Researchers, Inc.

to conserve plants for the future. Each botanical garden is responsible for growing plants native to its own region. Conservation techniques include research on traditional propagation methods (seeds and cuttings) as well as more modern methods such as tissue cultures. Another vital conservation effort is seed storage at regional gardens or at the U. S. Department of Agriculture (chapter 8).

Critics argue that captive propagation can save relatively few of the hundreds of thousands of species that are facing extinction. Furthermore, captive propagation curtails the natural evolution of species. For example, seeds of a threatened plant species may have to be stored for decades before circumstances in its former range become favorable for reintroduction. During this time, ecological conditions in its native habitat continue to change. Hence, with time, plants grown from stored seeds become less fit for survival in their native habitat.

Most proponents of captive propagation readily agree that landscape preservation should be the first priority. They also quickly point out, however, that some species are suffering such a precipitous decline that captive propagation might be the only way to save them. In addition, the conditions that caused some species to decline in the wild may persist for decades or longer. For example, there are few, if any, places in Africa where black rhinos could be released and kept safe from poachers. In other cases, there may be no habitat left to establish free-living populations. Several species of lemurs face extinction because their native forests in Madagascar are falling under the pressures of a rapidly-growing human population. In the future, forest remnants may be too small and isolated to sustain viable populations of lemurs on this island nation.

Conclusions

There is some good news! Several species have taken a step back from the brink of extinction. For example, since the American alligator was given protection from hunting, it has rebounded so well that it can again be legally hunted in many parts of its range. Curtailment of the use of DDT has contrib-

uted to a soaring eagle population. The number of nesting pairs in the contiguous United States has risen from an estimated 400 in the early 1960s to over 2,600 in 1989. The number of California gray whales has risen from 2,000 to more than 20,000. Some species (e.g., the red wolf, California condor, and the golden lion tamarin) have been reintroduced into their native habitats. On a much larger scale, debt-for-nature swaps are proving to be quite successful, and their potential remains largely untapped. The landscape approach to preserving biodiversity is gaining broader acceptance. Ecologists are learning more about populations, ecosystems, and landscapes. In addition, conservation biologists and resource managers are getting better at developing and implementing programs to preserve biodiversity.

Yet much work remains. Scientists and resource managers still know too little about the complex ecological interactions that sustain species, communities, and landscapes. Meanwhile, a rapidly growing human population and its resource demands continue to threaten the planet's rich diversity of life. Economic development will continue to stress native plants and animals.

As population/resource pressures intensify, conflicts over the management of endangered species also intensify. A prime example is the case of the northern spotted owl, listed as a threatened species in 1990. In late 1991, in accordance with the Endangered Species Act, the U.S. Fish and Wildlife Service designated 3.3 million hectares (8.2 million acres) of ancient (old growth) forests on federal lands in California, Oregon, and Washington as critical habitat for the owl. This action would ban logging on these lands in order to protect the continued existence of the owls. The timber industry strongly criticized this decision, arguing that over 32,000 jobs would be lost. In 1992, a cabinet-level committee revised the amount of land to be designated as critical habitat down to 2.2 million hectares (5.4 million acres). Some conservation groups subsequently argued that the protection of only 2.2 million hectares would prove to be insufficient to save the owls. As of this writing, the issue remains in the courts as conservation groups, government agencies, and logging interests continue to file suits and countersuits.

Although this continuing conflict is often described in the national media as a battle between loggers and the owl, the issues are much more complex. Perhaps the two most important facts are that (1) only 13% of the ancient forests remain in Oregon and Washington out of about 7.7 million hectares (19 million acres) that existed a century ago and (2) increased mechanization, exporting raw logs rather than milled lumber, and relocation of the timber industry from the Northwest to the Southeast have already contributed to a loss of almost 30,000 timber-related jobs. Thus, some observers suggest that the two real issues are how much of the ancient forest ecosystem should be maintained, and as the timber runs out, how should the economic transition from a timber-based economy to something else be managed?

On a worldwide basis, the forecast for biodiversity is bleak. Already short of food, land, water, and firewood, many less-developed nations face spectacular population growth in com-

What Will Be Required To Save the Salmon?

Some of nature's most magnificent creatures are about to become extinct. The salmon run in the Columbia River/Snake River Basin has dropped 85% in the past fifty years. Furthermore, only 3% of the remaining fish are wild stocks; the others are hatchery fish. The Idaho sockeye salmon has been proposed for listing under the Endangered Species Act. The runs of chinook salmon in the Snake River and coho salmon in the Columbia are also likely to be listed.

Although a variety of factors, including overfishing, deforestation, and pollution contribute to declines in salmon populations, hydroelectric dams account for about 75% of the losses. Salmon swim upstream to spawn. The young salmon (smolts) swim downstream after hatching to the Pacific Ocean, where they grow and mature and then return to the same streams to spawn. The fifty-six major dams on the Columbia River watershed are formidable obstacles to both upstream and downstream migration. Although dam operators have added fish ladders to help the salmon upstream and structures to guide them downstream around the turbines, many are killed in the turbines. (Imagine trying to swim through a giant food-processor.) Furthermore, smolts on their downstream passage are held up in reservoirs, where they are exposed to predators and pathogens and water that is too warm. Releasing hatchery fish was supposed to mediate the loss of native stocks, but more than half of them never make it to the first dam, raising questions about the hardiness of these stocks.

What can be done about the decline in salmon populations?

In one sense, the answers are relatively simple: (1) provide more water downstream to assist salmon migration, (2) build more protective devices around dams, (3) improve water quality in the Columbia River, and (4) place stricter limits on fish harvests. But implementation will be quite difficult. For example, more water for fish means less water for irrigation and the generation of electricity. Farmers will be especially hard hit; not only will they have less water for irrigation, they will also have to pay more for electricity to pump water from groundwater reservoirs. The Pacific Northwest has flourished in recent decades in part because abundant cheap hydroelectric power keeps utility bills the lowest in the nation, but economic and population growth has consumed all the surplus power. If that growth is to continue and if the salmon are to be saved by reducing hydrogeneration of electricity, electric rates will have to increase significantly. Everyone will be affected, from major industries to individual homeowners.

The problem of rising utility bills is exacerbated by pressure from the federal Office of Management and Power to force the Bonneville Power Administration (which markets Northwest hydropower) to pay off its debts. Bonneville owes billions of dollars because it defaulted on loans to build five nuclear power plants. (Two were abandoned before construction began, and two were mothballed; only one is operating.) How can Bonneville pay its debts without raising its rates?

As the human population approaches the earth's capacity to sustain its demands, there are fewer and fewer untapped resources. If other species such as salmon are to survive, more resources such as water must be allocated to support them. How willing are we to pay more and use less so that other species can survive?

ing years (chapter 6). In East Africa, the home of such famous wildlife areas as the Serengeti Plains and the Ngorongoro Crater, the human population is expected to double in less than twenty-five years. Hence, many conservationists urge immediate action to avoid the wholesale loss of species.

Resource managers are concluding that conservation and economic development must be closely coupled. Recall, for example, the incorporation of economic incentives for local farmers in the development of the Rothsay Prairie Landscape in Minnesota. This coupling of biodiversity preservation and economic development is essential in less-developed nations where high population densities and poverty threaten the last vestiges of wildlands. Few people are interested in saving wildlife and vegetation if they are hungry and do not have enough fuel to cook the little food they have.

To accomplish a union of conservation and economics in impoverished countries, nonendangered resources within nature preserves might have to be made more accessible to people as sources of food and fuel. In some instances, even limited harvesting of endangered species may be appropriate if populations attain a level that can sustain such activities. Some African nations currently allow controlled hunting of elephants despite their endangered status. The monetary return from even one elephant may be greater than the annual income of an entire village. These nations argue that villagers are more likely to take an active role in preserving biodiversity if the preservation of endangered species is linked to their own economic well-being.

Through its diverse and pervasive activities, our species is perhaps the single most important force in the process of natural selection today. Despite the value of biodiversity, we are threatening the survival of hundreds of thousands of species. It is true that few people care about the long-term survival of plants and animals until their own short-term needs for survival—adequate food, clothing, shelter, and health care—are met. Even when our basic needs are met and we are relatively prosperous, most of us are still more concerned about further improving our material well-being than we are about saving endangered species. How many of us would readily pay more for electricity, gasoline, or lumber so that endangered species could be saved? How many of us would idly stand by while we lost our jobs because a forest could not be cut or a road could not be built? Yet money could be allocated to retrain people who lose their jobs, and most people have enough discretionary income to pay more for essential goods and services. People in the more-developed nations can do much to preserve biodiversity if they only have the will to do so. The issue on page 234 entitled "What Will Be Required To Save the Salmon?" describes some of the sacrifices that would be needed.

Key Terms

endangered species
threatened species
biodiversity
gene pool
keystone species
critical-link species
habitat islands
theory of island biogeography
minimum viable population (MVP)

inbreeding
genetic drift
source population
sink population
recovery plan
debt-for-nature swap
captive propagation

Summary Statements

Biodiversity is threatened throughout the world as hundreds of thousands of species of animals and plants face extinction.

Each species has a unique set of adaptive characteristics that might be useful someday in medicine, agriculture, or industry. Each species also plays a role in maintaining the functioning of ecosystems.

Other arguments for species preservation include aesthetic and intrinsic values. In view of short-term economic considerations, however, including the need for jobs, housing, and food, preserving biodiversity is usually low among our priorities.

Over 99% of the species that have ever existed on this planet are now extinct. Many of them disappeared during short bursts of massive extinction associated with catastrophic geological events. One of the first major human impacts on biodiversity may have occurred about eleven thousand years ago when many species of large mammals became extinct in North America.

During the past two centuries, human activities have caused a sharp rise in the number of species that have become extinct. Humans are reducing biodiversity in several ways: the alteration of natural habitats, commercial hunting, and the introduction of alien species.

After two centuries of expanding human settlements, wild species are found only on habitat islands surrounded by a sea of humanity. As these islands become smaller and more isolated, resident species become more vulnerable to extinction.

Illegal commercial hunting threatens many species that are being overexploited for their value as clothing (furs and hides), food, ingredients in herbal medicines, pets, collectors' trophies, or other products.

Alien species such as domesticated animals have dramatically altered local biota. Overgrazing by livestock has allowed many landscapes that were once dominated by native vegetation to be replaced by alien weedy species. Introduced predators such as rats and feral cats often devastate native prey.

Endangered species often have one or more of the following characteristics: (1) they are specialists that require particular types of food and other habitat conditions; (2) they are extreme K-strategists; (3) they have a relatively large minimum viable population. Many species are vulnerable to extinction simply because they live in such ecosystems as wetlands and tropical rain forests that are bearing the brunt of human alterations.

The best way to preserve biodiversity is to preserve natural landscapes. Additional efforts are required to acquire new preserves, to enlarge present preserves, to provide migration corridors between preserves, and to modify uses of surrounding land.

Renewed efforts are needed to curb poaching of endangered species. Tougher enforcement of the U. S. Endangered Species Act and the Convention on International Trade in Endangered Species of Wild Fauna and Flora (CITES) would reduce illegal hunting. As human settlement pressures continue, however, zoos and botanical gardens may prove to be the only safe havens for some endangered species.

The outlook for biodiversity is mixed. Some species, such as the American alligator, bald eagle, and red wolf, appear to be making a comeback. The landscape approach to preserving biodiversity is gaining greater acceptance. Meanwhile, a rapidly growing human population and its resource demands continue to threaten the planet's rich diversity. To preserve biodiversity, we must make greater efforts to couple conservation and economic development in both more-developed and less-developed countries.

Questions for Review

1. What is an endangered species? What is a threatened species? How do they differ?

2. What is a gene pool? Why is it important to preserve as many gene pools as possible?

3. What is a keystone species? Describe how the alligator serves as a keystone species in some subtropical wetlands.

4. List five ways in which a plant or animal species may contribute to the well-being of humankind.

5. Describe how human beings may have contributed to the extinction of large mammals in North America eleven thousand years ago. How have human activities changed since that time? What is the significance of these changes for biodiversity?

6. Explain why habitat destruction is the most important threat to biodiversity.

7. Why are island populations particularly vulnerable to extinction?

8. Describe the major elements of the theory of island biogeography. What are the implications of this theory for the development of management strategies to preserve biodiversity?

9. Give three examples of species that are endangered by illegal hunting.

10. Describe how some commercially hunted species get caught in a vicious circle that could lead to their extinction.

11. How might overgrazing by livestock contribute to the successful invasion of alien weedy species?

12. List several characteristics that make some species particularly vulnerable to extinction.

13. Referring to the traits listed in table 11.3, speculate on the potential for survival of each of the following hypothetical animals: (a) a mouse-sized omnivore that gives birth to thirty young each year and inhabits weedy fields throughout the temperate portion of the northern hemisphere; (b) a predatory bird the size of an eagle that raises only two young every other year and is found only on a few remote islands in the South Pacific; (c) a weasel-sized predator that has valuable fur, raises five young every spring, and resides in logged forests of the Pacific Northwest.

14. Why is the minimum viable population size an important consideration in attempts to save endangered species? What are some of the shortcomings of relying on a particular minimum population value for a specific population?

15. If a choice has to be made, why is it more important to save the habitat of a source population than that of a sink population?

16. What is a recovery plan? What is its purpose?

17. Describe the importance of the debt-for-nature swap program for preserving biodiversity in less-developed nations.

18. List the advantages and disadvantages of captive-breeding as a strategy for preserving biodiversity.

Projects

1. Check with your state's Department of Conservation or Natural Resources to determine the plants and animals in your region that are on the endangered or threatened species list. What efforts are being made to preserve these species? How can you contribute to these efforts?

2. Visit your city or regional planning office and request a current land-use map as well as a map that is 30-50 years old. How much natural habitat has been lost to development? Can you detect any habitat islands? If so, are they being managed to preserve biodiversity? Are they large enough to sustain their native plant and animal populations? What are your community's land-use policies regarding habitat preservation?

3. Consider the following argument for a classroom discussion: "Conservationists are very naive; they are just not living in the real world. When it's a question of what's more important, me or the Houston toad, nobody is going to choose the toad." Do you agree or disagree? Defend your position.

Selected Readings

Alvarez, W., and F. Asaro. "What Caused the Mass Extinction? An Extraterrestrial Impact." *Scientific American* 263 (Oct. 1990):78-84. An account of the evidence that an asteroid or a comet triggered the extinction of the dinosaurs.

Chadwick, D. H. "Elephants: Out of Time, Out of Space." *National Geographic* 179 (May 1991):2-49. An examination of the impact of declining habitats on shrinking elephant herds.

Cohn, J. P. "Captive Breeding for Conservation." *Bioscience* 38 (1988):312-16. A review of success stories as well as controversies that surround this strategy for species preservation.

Devine, B. "The Salmon Dammed." *Audubon* 94 (1992):83-89. An evaluation of the impact of hydropower dams on the future of salmon in the Columbia River Basin.

Ehrlich, P. R., and E. O. Wilson. "Biodiversity Studies: Science and Policy." *Science* 253 (1991):758-62. An examination of key issues that link preservation of biodiversity, economics, and politics.

Endangered Species Update. University of Michigan, Ann Arbor, Mich. Also contains the Endangered Species Technical Bulletin published by the U.S. Fish and Wildlife Service. A monthly publication that describes the many international, national, and regional efforts to save endangered species; includes articles on the role of ecology and conservation biology in preserving biodiversity.

Findley, R. "Will We Save Our Endangered Forests." *National Geographic* 178 (Sept. 1990):106-36. An examination of issues surrounding the loss of the endangered virgin woodlands of the Pacific Northwest.

Gibbons, B. "Missouri's Garden of Consequence." *National Geographic* 178 (Aug. 1990):124-40. An account of the activities of the Missouri Botanical Gardens, one of the world's leading institutions in the battle to save tropical rain forests.

Grove, N. "Quietly Conserving Nature." *National Geographic* 174 (Dec. 1988):818-45. Insights into how the businesslike activities of the Nature Conservancy have set aside millions of hectares to save animals, plants, and their habitats from extinction.

Grumbine, E. "Protecting Biological Diversity Through the Greater Ecosystem Concept." *Natural Areas Journal* 10 (1990): 114-20. Explores the strategy of developing an integrated system of large nature reserves (landscapes) to protect biodiversity.

Leopold, A. *A Sand County Almanac and Sketches Here and There.* New York: Oxford Univ. Press, 1949. A noted naturalist eloquently describes the value of wild things.

Lord, J. M., and D. A. Norton. "Scale and Spatial Concept of Fragmentation." *Conservation Biology* 4 (1990):197-202. An examination of the types of habitat islands and the implications of habitat fragmentation for conservation.

May, R. E. "How Many Species Inhabit the Earth?" *Scientific American* 267 (Oct. 1992):42-48. A fascinating examination of a question that is crucial for efforts to conserve biodiversity.

Poten, C. J. "America's Illegal Wildlife Trade." *National Geographic* 180 (Sept. 1991):106-32. A troubling examination of how poachers are attacking wild animal populations to feed a growing global demand for American wildlife.

Raven, P. H. "The Politics of Preserving Biodiversity." *Bioscience* 40 (1990):769-74. A discussion of the political realities of preserving global biodiversity.

Soulé, M. E. *Conservation Biology: The Science of Scarcity and Diversity.* Sinauer Associates, 1986. A look at the emerging field of conservation biology and its role in preserving biodiversity.

Soulé, M. E. "Conservation: Tactics for a Constant Crisis." *Science* 253 (1991): 744-50. A thorough analysis of why biodiversity is endangered today, and potential solutions.

Terbough, J. "Why American Birds Are Vanishing." *Scientific American* 266 (May 1992):98-104. An examination of the causes of the population decline of many American songbirds.

Westman, W. E. "Managing for Biodiversity." *Bioscience* 40 (1990):26-33. An examination of ecological and public policy questions related to efforts to conserve biodiversity.

Wilson, E. O. "Threats to Biodiversity." *Scientific American* 261 (Sept. 1989):108-16. An informative overview of the role of habitat destruction, particularly in the tropics, in species extinction.

Wilson, E. O., ed. *Biodiversity.* Washington, D.C.: National Academy Press, 1988. An excellent collection of original articles that deal with the many facets of preserving biodiversity.

Environmental Toxicology

Chapter 12

Accidents such as this one involving a railroad tank car may release toxic fumes into the atmosphere or toxic chemicals into the soil and waterways.
AP/WideWorld Photos

F ish Unsafe to Eat! Drinking Water Contaminated! Industrial Accident Releases Toxic Cloud! Headlines such as these appear with some regularity. Reading further usually reveals conflicting opinions regarding whether or not public health is really threatened. Often the range of opinions of scientists and government officials is so broad that the public becomes totally confused. Some people wonder who to believe. If the so-called experts cannot agree on what is safe, is there really a problem after all?

Controversy is inevitable when someone in a responsible position warns the public that they are being exposed to harmful chemicals in their food, drinking water, or the air they breathe. For example, in the summer of 1989, scientists at the National Wildlife Federation warned women of child-bearing age not to eat any fish from Lake Michigan. Regulatory agencies from several of the Great Lake states and commercial fishermen were outraged at what they viewed as an exaggerated risk.

Some people believe that exposure to any and all chemicals is dangerous, but every substance that we encounter is made of chemicals; we ourselves are extremely complex organic chemical factories. Rather, the two basic questions are: To which chemicals should we limit our exposure, and to what extent should we limit our exposure? Neither question is easy to answer.

The need to provide answers to these questions was recognized as early as 1938 when the U.S. Food and Drug Administration (FDA) was created under the Food, Drug, and Cosmetic Act. This agency requires proof of the effectiveness and safety of drugs, food additives, and cosmetics. Today a variety of other federal laws seek to control chemical contamination of air, water, and land. For such matters, the U.S. Environmental Protection Agency (EPA) is the primary regulatory body. Both the EPA and the U.S. Department of Agriculture (USDA) are responsible for ensuring that the nation's "food basket" is safe.

Government officials base their decisions regarding our exposure to hazardous chemicals on the principles of **toxicology,** often called the *science of poisons,* or **toxins.** In this chapter, we first examine several aspects of toxicology, including routes of exposure, responses to exposure, and our natural defenses to toxins. Then we consider how to assess the relative risks to which we are exposed. Finally, we examine how scientists use principles of toxicology and risk assessment to estimate safe exposure levels and thereby protect the public.

Routes of Exposure to Toxins

Before a substance can injure an organism, it must contact or enter the organism (radioactive materials and explosives are exceptions). Foreign substances enter by four possible routes: inhalation, skin (or eye) absorption, ingestion, or injection (fig. 12.1). Often the mode of entry is significant for the degree of harm that a foreign agent causes. For example,

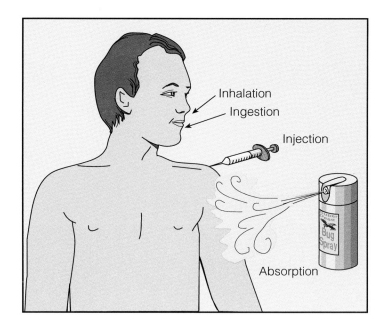

Figure 12.1 The routes of exposure to toxins: Inhalation, ingestion, injection, and absorption.

although rubbing alcohol (isopropyl alcohol) can be beneficial when applied to the skin, ingesting it causes serious illness. Thus, decisions regarding exposure must involve careful attention to the probable route of entry.

Inhalation

For humans, inhalation is the most frequent route of exposure to chemicals. A person inhales about 20,000 liters (700 cubic feet) of air per day. Thus, even moderate levels of air pollution can result in significant exposure. Furthermore, lung tissue is highly susceptible to the action of chemical agents such as acidic and caustic vapors, which can destroy cells. Even chemical agents that do not directly affect the lungs may pass through them into the bloodstream. For example, inhaled carbon monoxide (CO) does not damage lung tissue directly; rather, it enters the bloodstream, where it attaches to hemoglobin molecules and severely impairs the blood's oxygen-carrying capacity. The brain is particularly sensitive to reduced levels of blood oxygen. As a consequence, inhaling carbon monoxide leads to dizziness, and in sufficiently high concentrations can cause loss of consciousness and eventually death. Carbon monoxide is reported to be responsible for 44% of all deaths reported by coroner's offices across the United States. We have more to say on the impact of air pollution on the respiratory system in chapter 16.

We rely primarily on our noses to warn us of exposure to potentially hazardous gases, but our noses are not always trustworthy. Some toxic gases, such as carbon monoxide, defy detection because they are odorless, and their harmful effects are not immediate. Every day people are unknowingly exposed to and subsequently overcome by carbon monoxide through faulty furnaces and motor vehicle exhaust systems.

Another odorless gas is methane (CH_4), the primary component of natural gas. Although methane is not directly poisonous, it is explosive. To warn natural gas users of a possible leak, trace amounts of a quite odorous gas (ethyl mercaptan), are intentionally added to natural gas.

Absorption

A second route of exposure is direct contact with skin and eyes. Some chemicals such as strong acids (e.g., sulfuric acid [H_2SO_4] in automobile batteries) or alkaline materials (e.g., sodium hydroxide [NaOH] in drain cleaners) destroy (burn) tissue directly and are particularly painful. Other chemicals pass through the skin into the bloodstream and are transported to other organs, where damage may occur. Some pesticides (organophosphates) enter the body in this way and impair the ability of nerves to transmit messages. Death can result if enough of a chemical is absorbed. Because cosmetics, lotions, and deodorants have lengthy contact times with the skin, the FDA requires extensive toxicological studies to establish their safety. Skin abrasions and cuts allow a more direct entry for chemicals. The eye, unlike dry skin, is particularly vulnerable because many air pollutants readily dissolve in the fluid on the eye's surface.

Ingestion

A third route of exposure is ingestion through consumption of contaminated food or beverages or by accidental means. By entering food webs, environmental contaminants such as pesticides and heavy metals can eventually get into the food we eat. **Heavy metals** include the elements lead (Pb), mercury (Hg), chromium (Cr), nickel (Ni), cadmium (Cd), and arsenic (As). Humans and other species that occupy upper trophic levels can ingest significant quantities of toxins as a result of **bioaccumulation,** the process whereby certain chemicals become more highly concentrated as they are transferred to organisms at successively higher trophic levels. Special Topic 12.1 describes this important environmental phenomenon.

Most foods also contain low levels of natural toxins. Chocolate contains theobromine, a possible gene-altering substance; potatoes contain solanine and chaconine, which can reach lethal levels if the potatoes are damaged or exposed to light for extended periods; celery, parsley, parsnips, and figs contain psoralen derivatives, which become active cancer-causing agents when activated by light. Mushrooms, black pepper, mustard, horseradish, cottonseed oil, many herbal teas, and peanut butter all contain small amounts of at least one component with toxic properties. When we ingest these substances we knowingly or unknowingly encounter the small risk associated with exposure to them.

Most ingested toxins must be absorbed by the digestive tract before they initiate any toxic response, but a few chemicals burn or irritate the gastrointestinal tract directly. For example, accidental ingestion of strong alkalis or strong acids

Table 12.1	Common materials with hazardous properties
adhesives (airplane glue)	office-copier chemicals
ammonia solutions	paints
battery fluids	paint thinners
bleaches	pesticides
cleaning agents	photographic chemicals
degreasing solvents	printing inks
detergents	rubber chemicals
duplicating-machine fluids	shellacs
fuels (gasoline)	solvents (turpentine)
industrial oils	varnishes
lye	waste oil
lacquers	

severely burns the mouth and even the stomach. Swallowing toxic substances is the most common cause of poisoning among young children. On the other hand, many therapeutic drugs are administered orally because they are absorbed through the gastrointestinal tract and then distributed throughout the body.

Our homes and offices contain a variety of potentially hazardous chemicals (table 12.1). Of particular concern is preventing the accidental exposure of small children. One way to teach them to stay away from dangerous chemicals is by using Mr. Yuk stickers (fig. 12.2).

Injection

The fourth route of exposure to chemicals is injection through the skin by needle and syringe. Injection is often the preferred means of administering medications because it is most direct route to target organs. While accidental injection is an unlikely route of exposure for the general public, medical personnel are at considerably greater risk to such exposure. Accidental puncture wounds from used needles contaminated with the Acquired Immune Deficiency Syndrome (AIDS) virus or potent drugs (for example, drugs used to treat cancer) can expose doctors, nurses, paramedics, and sanitation staff to life-threatening agents.

Responses to Toxin Exposure

The body's response to a foreign chemical—whether inhaled, absorbed, ingested, or injected—depends on a host of factors. They include the frequency and duration of exposure, the dose, and the age, sex, and general health of an exposed individual.

Duration and Frequency

Chemical exposures are generally classified as either acute or chronic. **Acute exposure** is short-term exposure to a relatively

Bioaccumulation of Toxins

Recall from chapter 3 that plants are a source of energy and concentrated nutrients for all consumer organisms. In their normal functioning, however, plants can also take up pesticides such as DDT, industrial chemicals such as polychlorinated biphenyls (PCBs), or heavy metals such as mercury. These chemicals have one property in common: they persist in the environment. Persistent chemicals are not readily broken down by chemical or biological processes operating in soil, water, or organisms themselves. Because these substances are not degraded by the plant, they are retained and concentrated in its tissues. Some of them are passed from one trophic level to the next, building in concentration as one organism consumes another. This process of ever-increasing concentration is called *bioaccumulation* or *food-web accumulation*.

Bioaccumulation is a consequence of the transfer inefficiencies that exist between trophic levels—the biomass declines at each successively higher level (chapter 3). Furthermore, persistent toxic substances are not readily broken down or excreted by organisms. Thus, from one trophic level to the next higher trophic level, the amounts lost are small compared with the amounts that are transferred. Because most of the toxins that are ingested are retained and most of the biomass is lost, the toxins become more concentrated at successively higher trophic levels (see the accompanying drawing).

Bioaccumulation is especially pronounced in aquatic food webs, which usually involve four to six trophic levels rather than the two or three levels that are common in terrestrial ecosystems. In addition to becoming contaminated via bioaccumulation, some fish and shellfish can absorb certain toxins directly through their gills from the surrounding water. In these ways, a toxin may become from one thousand to over a million times more concentrated in upper-trophic-level

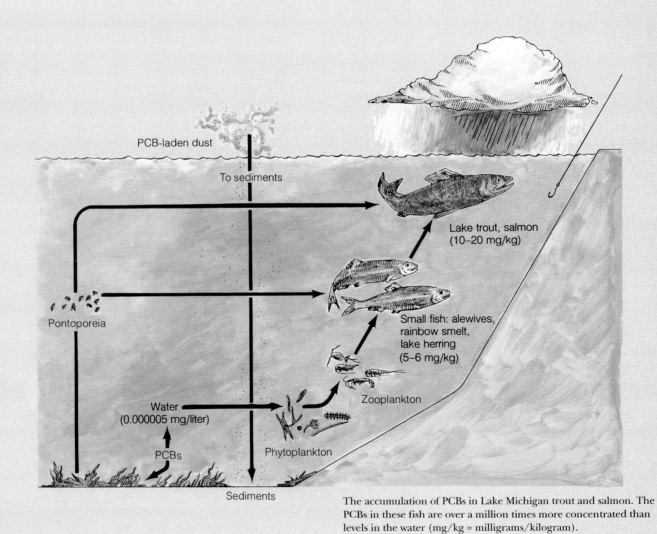

The accumulation of PCBs in Lake Michigan trout and salmon. The PCBs in these fish are over a million times more concentrated than levels in the water (mg/kg = milligrams/kilogram).

organisms compared to its concentration in the environment. Consequently, people and fish-eating birds such as eagles and osprey are at the greatest risk of consuming harmful amounts of toxic materials.

The significance of PCB bioaccumulation is illustrated by the level of contamination in Lake Michigan, one of the world's largest reservoirs of fresh water. The contamination process begins with seemingly insignificant amounts of PCBs in sediments and lake water, which are incorporated into bottom-dwelling organisms and algae. These slightly contaminated organisms are consumed regularly by small fish, such as alewives, smelt, and young salmon and trout, which are consumed by larger salmon and trout. PCB levels in the larger fish increase as the fish grow older, until the level of PCBs in their bodies makes them unsafe for human consumption. Most large predator fish in Lake Michigan exceed the level that the FDA considers safe for human consumption. To protect the public from the PCB hazard, the EPA has banned the commercial sale of Lake Michigan's highly prized salmon and trout, and sportsmen have been warned not to eat any large fish and not to eat more than one meal per week of medium-sized trout or salmon from the lake. Pregnant women and females in their prereproductive years are given more restrictive warnings concerning the amount of Lake Michigan fish they should eat. Many states now issue warnings concerning the levels of various toxins (e.g., PCBs and mercury) in fish from surface waters in their jurisdiction.

Bioaccumulation of PCBs in the environment also presents problems for fish-eating birds. Populations of many of these animals declined dramatically in the 1970s and 1980s. Investigators strongly suspect that PCBs are responsible for the poor survival rates of birds living in or near PCB-contaminated habitats. Some birds that do survive exhibit gross deformities, like those shown in the accompanying photograph. The propensity of such compounds as DDT, toxaphene, chlordane, dieldrin, and PCBs to bioaccumulate led to their being legally banned from most uses in the United States and many other developed countries. Consequently, the levels of many bioaccumulating substances in the environment are declining, and some fish-eating bird populations, such as eagles, are recovering. Other populations, such as the Caspian tern, however, have not yet responded to reduced contaminant levels in their food supply. The accompanying table lists some toxins that commonly bioaccumulate along with the U.S. Food and Drug Administration standards for these toxins in sport fish.

One of the physical deformities exhibited by fish-eating cormorants from the Great Lakes is a crossed bill.
U.S. Fish and Wildlife/Photo by K. L. Stromborg

United States Food and Drug Administration standards for contaminants commonly found in sports fish	
Contaminant	**Standard (max. level, parts per million)**
polychlorinated biphenyls (PCBs)	2
DDT	5
toxaphene	5
chlordane	0.3
dieldrin	0.3
mercury	0.5
dioxin	0.000010

Figure 12.2 Mr. Yuk warns children to stay away from dangerous substances in the home.
© Wm. C. Brown Communications, Inc./Doug Sherman

Figure 12.3 A transportation accident such as this may result in acute exposure of humans to toxic substances.
© Breck P. Kent/Earth Scenes

high concentration of a toxin. It usually occurs as a consequence of transportation or industrial accidents (fig. 12.3). Symptoms of acute exposure usually appear immediately and are often life-threatening. A particularly tragic incident occurred in Bophal, India, in December 1984 when one of the gaseous chemical ingredients used to manufacture a pesticide was released accidentally. A toxic cloud swept downwind from the chemical plant into a densely populated neighborhood and within a few hours killed more than two thousand people and injured over ten thousand others.

Chronic exposure involves repeated or continuous exposure to relatively low levels of toxins. For example, cigarette smokers and people who live or work with smokers daily inhale low amounts of tars and nicotine released by burning tobacco. Damage from chronic exposure often appears only after many years. Serious problems develop more frequently in certain industrial occupations. Long-term exposure to high levels of dust in occupations such as mining, metal grinding, and the manufacture of abrasives, for example, may eventually cause lung diseases such as black lung (from coal dust); byssinosis, or brown lung (from cotton dust); silicosis (from quartz dust generated during mining); and asbestosis (from asbestos fibers).

Dose-Response Relationships

We all know that a single aspirin tablet does not usually produce a harmful effect, but swallowing the entire contents of a small bottle of aspirin at once may be lethal. The size of the dose is an important determinant of the specific response. **Toxicity** is the type and intensity of response evoked by a chemical. To determine the response to chemicals, toxicologists administer controlled doses to laboratory test animals, usually rats or mice, and use the information they gain from these experiments to approximate the hazards for humans.

The toxic effects of chemicals are manifested in various ways. Some chemicals interfere with the ability of an organ (e.g., the liver or kidney) to function. Toxins may also interfere with blood-forming mechanisms, or the functioning of **enzymes** (proteins that accelerate specific chemical reactions), the central nervous system (including the brain), or the immune system. For example, dioxin, an extremely toxic compound, is thought to affect DNA and ultimately the immune system.

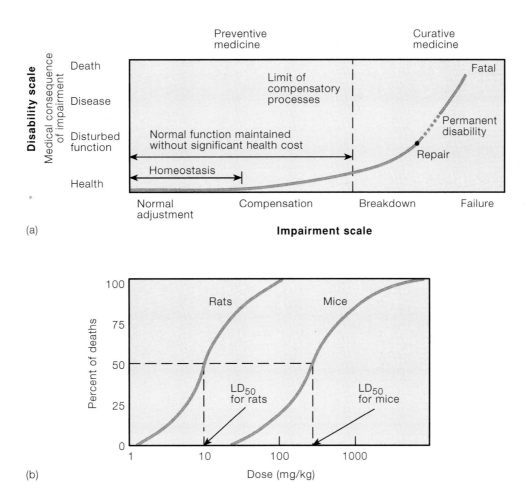

Figure 12.4   Two types of dose-response curves. (*a*) Response of an organism as the dose increases, and (*b*) points where death is the measured response in two different laboratory animals.

Toxicologists analyze both sublethal and lethal responses to determine toxicity. To measure sublethal responses, they monitor blood, urine, organ tissues, and the general health of animals. Sublethal responses may include changes in the number and type of blood cells, alterations in blood chemistry, changes in body weight, tumor formation, and retardation of growth.

Scientists usually find that low doses produce no response (fig. 12.4). The highest dose that produces no toxic response is called the **no-observed-effect level (NOEL).** Doses below this level are considered relatively safe for test animals. As the dose is increased above the NOEL, some test animals begin to show symptoms of poisoning. At some higher dose, a few animals die. At increasingly higher doses, more animals die, until eventually a dose is reached that kills all the animals in the test population.

From dose-response experiments, toxicologists determine a common measure of the relative toxicity of different toxins, the **LD$_{50}$** value, which is the dose that kills 50% of the individuals in a test population. Table 12.2 shows the wide range in toxicity among a variety of chemicals. Note that LD$_{50}$ values are expressed on a per-body-weight basis, that is the number of milligrams of toxin per kilogram (mg/kg) of body weight lethal to 50% of the test population. (The use of units such as mg/kg is one way of expressing concentration. For other ways, see Expressing Concentrations in Appendix B.)

When describing the danger associated with chemical exposure, toxicologists carefully select specific descriptors to communicate the relative hazard (table 12.3). *Super toxic* is used to describe botulinus toxin because a fraction of a milligram is lethal, whereas ethyl alcohol is described as *slightly toxic*. These descriptors are based on LD$_{50}$ values.

Physicians, pharmacists, toxicologists, and even wildlife biologists must base dose rates on body weight to avoid harming their subjects. Wildlife biologists frequently tranquilize animals so that they can examine or relocate them. The dose of the tranquilizing drug depends on two factors: the total weight of the animal and the recommended number of milligrams of tranquilizer per kilogram of animal. The recommended dose of phencyclidine hydrochloride, a common tranquilizer, is 1.5 mg/kg for both grizzly bears and white-tailed deer. Thus, a 500-kilogram grizzly bear requires

Table 12.2 Approximate acute LD_{50} of various chemical agents

Agent	LD_{50} (mg/kg)	Approximate Lethal Dose for Humans
ethanol	10,000	between pint and quart
sodium chloride (table salt)	4,000	between ounce and pint
morphine sulfate	1,500	
DDT	100	between teaspoonful and ounce
strychnine sulfate	2	few drops
nicotine	1	
d-tubocurarine (curare)	0.5	one drop
dioxin (TCDD)	0.001	1/500th of a drop
botulinus toxin	0.00001	1/50,000th of a drop

Source: John Doull, Curtis D. Klaasen, and Mary O. Amdur, editors, *Casarett and Doull's Toxicology: The Basic Science of Poisons*, 2d ed. Copyright © 1980 Macmillan Publishing Company, Inc.

Table 12.3 Toxicity ratings based on oral LD_{50} values for humans

Toxicity Descriptor	Approximate Lethal Dose (mg/kg)
practically nontoxic	15,000 or more
slightly toxic	5,000–15,000
moderately toxic	500–5,000
very toxic	50–500
extremely toxic	5–50
super toxic	less than 5

Table 12.4 Effects of alcohol on human behavior

Blood-Alcohol Level	Effects
0.02%–0.04%	judgment and reasoning somewhat affected; inhibitions lessened
0.05%–0.9%	coordination impaired; judgment and reasoning unreliable
0.10%–0.14%	slow reaction time, blurred vision, lack of coordination
0.16%–0.29%	staggering, slurred speech, visual impairment
0.30%–0.49%	marked lack of coordination, stupor, convulsions
0.50% and higher	coma and death

750 milligrams of the drug, whereas a 100-kilogram deer requires only 150 milligrams. A good estimate of an animal's weight is essential because an overdose may be lethal.

Lethal levels of air or water pollutants are used by scientists to predict harmful effects to people and the environment. These levels are stated in terms of LC_{50} values, that is, the concentration lethal to 50% of a test organism. In air, the LC_{50} value is expressed as the number of milligrams of a substance in a cubic meter (mg/m^3) of air that will kill 50% of a test population when inhaled over a specified period. In water, on the other hand, LC_{50} values refer to the number of milligrams per liter (mg/L), or parts per million (ppm), of a substance required to kill 50% of the population of an aquatic species—usually fish—in a specified time (often ninety-six hours, or four days). For example, the LC_{50} for chlorine dissolved in water is approximately 0.020 ppm for rainbow trout when exposed for ninety-six hours, whereas crayfish can tolerate at least 0.550 ppm. For humans, an exposure of a few minutes to air containing 1,000 ppm chlorine gas is fatal.

While LD_{50} values are important for all of us, the dose-response effects at lower levels of exposure are of even greater interest because organisms are more likely to encounter these concentrations. Although alcohol (ethanol, CH_3CH_2OH) is not usually considered an environmental contaminant, we have more medical information about this drug than almost any other chemical. We will use the human response to alcohol to illustrate the gradient of response to sublethal doses.

Alcohol level in humans is measured by analyzing the breath or blood (fig. 12.5). Blood-alcohol levels are expressed in milligrams of alcohol per 100 milliliters of blood. Drivers in most states are considered to be intoxicated if their blood alcohol equals or exceeds 100 milligrams per 100 milliliters. This level of intoxication is more commonly expressed as 0.10%. People with lower blood-alcohol levels exhibit some loss of coordination and impaired manual dexterity. An individual's response also depends on the amount of food in the digestive tract, other drugs in the bloodstream, and personal tolerance for alcohol.

Generally, as blood-alcohol levels rise, individuals lose visual acuity, experience a higher pain threshold (source of the expression, "feeling no pain"), and increasingly lose coordination (table 12.4). Relatively high blood-alcohol levels cause vomiting, and eventually an intoxicated person slips into unconsciousness. When blood-alcohol levels reach 350 to 400 mg/100 ml (0.35%–0.40%), alcohol functions as an anesthetic. Death ensues at levels only slightly higher than those required to anesthetize a person. Thus, people who become severely intoxicated run the risk of dying from alcohol poisoning.

As we can infer from our discussion of tolerance limits in chapter 4, the toxic response in humans and other animals varies with (1) the sex of the exposed individual, (2) the

Figure 12.5   A public safety officer administers a breath analysis for the presence of alcohol.
© Jim Shaffer

ability of the individual to develop a tolerance to toxins, and (3) factor interactions, including synergism and antagonism. Differences in the response of the two sexes to organophosphate pesticides have been shown experimentally. Some of these pesticides are more toxic to female rats and mice, whereas the opposite is true for other organophosphate pesticides. A first-time cigarette smoker typically experiences nausea, but tolerance develops after smoking only a few more cigarettes. Many drug prescriptions warn consumers not to drink alcohol when they use them because of potential adverse interactions.

Recall also from chapter 4 that the very young and very old are generally most sensitive to environmental stresses, including toxins. The young have not yet fully developed their capacity to detoxify toxins, and older individuals lose some of their body's ability to detoxify chemicals. Furthermore, prior to maturity, growing organs are particularly vulnerable to disruption by toxins. The human brain is not fully developed until six or seven years after birth. Thus, children under the age of seven are particularly susceptible to chemicals that adversely affect learning and behavior. We know that lead poisons children who eat lead-based paint chips from old houses. (Children with a disease called *pica*, crave nonfood items such as chalk, clay, or paint chips, and thus are particu-

larly susceptible.) Lead attacks the blood-forming mechanism, the gastrointestinal tract, and in severe cases, the central nervous system. Lead may also impair the functioning of the heart and the kidneys. In the early 1990s, scientists demonstrated that lead affects a child's ability to learn.* Long-term studies have shown that even children with blood-lead levels not high enough to cause clinical symptoms performed lower than average in school. In the United States, one child in six is estimated to have elevated blood-lead levels.

Another complication in assessing the effect of a toxin is that individual species vary in their response to a given toxin. For example, when dioxin, (an extremely toxic chemical) is administered to female rats and female guinea pigs, the guinea pigs die at 1/75 the dose required to kill the rats. When a chemical has widely ranging LD_{50} values for test animals and toxicologists have test results only from animal studies, they use a wider safety margin when setting allowable levels for human exposure to that chemical. Typically, allowable human exposure levels are set at one hundred times lower than the lowest experimental NOEL when only test data on animals are available.

*See the article by Needleman et al., listed at the end of this chapter.

Table 12.5 Some environmental agents that act as carcinogens, teratogens, and mutagens

Carcinogens	Teratogens	Mutagens
asbestos fibers	alcohol	benzo(a)pyrene
benzene	arsenic	lead compounds
benzo(a)pyrene	diethylstilbestrol (DES)	mercury compounds
certain chemotherapeutic agents	lead compounds	mustard gas
chromium compounds	lithium	ultraviolet light
diethylstilbestrol (DES)	methyl mercury	x rays
nickel	rubella (German measles)	
nitrosamines	thalidomide	
ultraviolet light	x rays	
vinyl chloride		
x rays		

Setting safe levels for human exposure to specific chemicals is clearly a challenging task. An interdisciplinary team from a number of medical, biological, and chemical disciplines usually makes such decisions. Medical evidence gathered from chronic occupational exposure, accidental exposure, and even suicides provide data for setting safe levels. For chemicals where no information on human exposure exists, the interdisciplinary team must base decisions on experiments with animals and use wide safety margins. A broad margin of safety is not always possible, however, as in some industries where it is neither economical nor technically feasible to attain these low levels of exposure. In other cases, levels of natural contaminants preclude the use of wide margins of safety. In the case of mercury (a heavy metal) in fish, natural levels are so high that much of this important food source would be judged unfit for human consumption if a hundred-fold safety factor were used to set the safe level. Thus, we should realize that if we regularly eat foods such as fish that have levels of mercury near the FDA limit, we are relying upon a relatively narrow safety margin.

Carcinogens, Mutagens, and Teratogens

It is particularly worrisome to many people that some forms of energy (X rays and ultraviolet light, for example) and some chemicals are carcinogens, mutagens, or teratogens, all of which can have catastrophic effects on health (table 12.5). In this section, we consider the health implications of these environmental factors.

Carcinogens are agents that cause cancer. They may be chemical (cigarette smoke), biological (cancer-causing viruses), or physical (ultraviolet light). Cancer is the uncontrolled growth of cells. Each one of us is made up of approximately 100 trillion cells, any one of which can be transformed to a malignant (cancerous) cell. Approximately one hundred different types of cancer have been identified. Figure 12.6 shows the incidence of cancer at various sites in the body.

In the United States today, cancer is second only to heart disease as a cause of death; more than 500,000 people die from it each year. Cancer eventually kills one out of about every four Americans. Although lung cancer has increased over 200% during the last thirty-five years, the incidence of all other forms of cancer has collectively declined by about 13% over the past thirty years.

Carcinogens trigger uncontrolled cell growth in many different ways; figure 12.7 shows one model of cancer formation. An early first step in the development of cancer occurs when an **initiator,** a type of carcinogen, structurally modifies a gene that normally controls cell growth. (A *gene* is a specific segment of a DNA molecule.) Such altered growth-regulating genes are called **oncogenes.** For a cell to begin to grow uncontrollably, at least two different growth-regulating genes must be altered.

Formaldehyde, a widely used chemical in industrial glues, is thought to be an initiator. Vapors of formaldehyde trigger the development of malignant tumors in the respiratory tracts of rats. To date, however, formaldehyde has not been shown to cause cancer in humans. In some cases, an initiator is a metabolic by-product. While benzo(a)pyrene, a natural product of the incomplete combustion of organic materials, including tobacco, is not an initiator, the body metabolizes it to a related chemical, benzopyrene-7,8-diol-9,10 epoxide, which is an initiator.

Oncogenes usually remain dormant until they are activated by another type of carcinogen, called **promoters** (fig. 12.7), which may act in several ways. For example, normal cells appear to prevent the activation of an oncogene in an adjacent initiated cell. A substance may act as a promoter by killing normal cells that surround an initiated cell. Lifelong exposure to promoters significantly increases the risk of cancer. People whose diets are high in salts or fats, for example, are more likely to develop stomach cancer and colon cancer respectively. Apparently, both dietary factors act as promoters by killing normal cells. Removing promoters from the area of an initiated cell that has not yet completed the promotion stage prevents the formation of a cancer cell. Thus, reducing the salt and fat content of one's diet greatly reduces the risk of contracting cancer.

Promoters may also activate oncogenes by inhibiting the action of **suppressor genes** (fig. 12.7), which prevent oncogenes from initiating uncontrolled cell growth. If suppressor genes are inactivated, oncogenes can then spur tu-

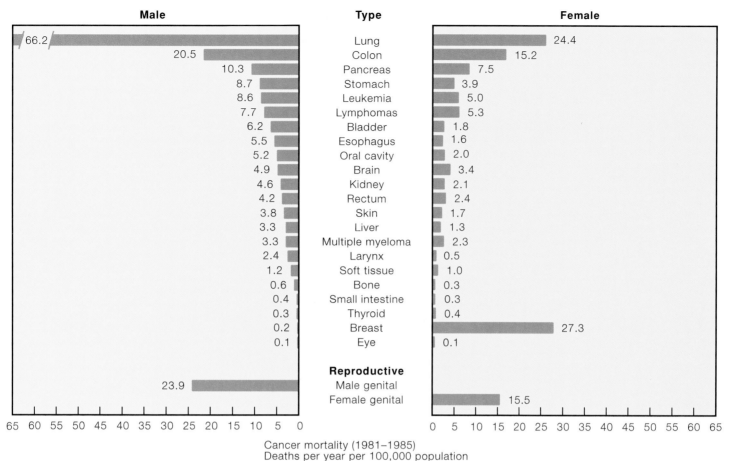

Figure 12.6 Cancer mortality in the United States by type and sex.

Source: *1987 Annual Cancer Statistics Review,* National Cancer Institute, U.S. Department of Health and Human Services.

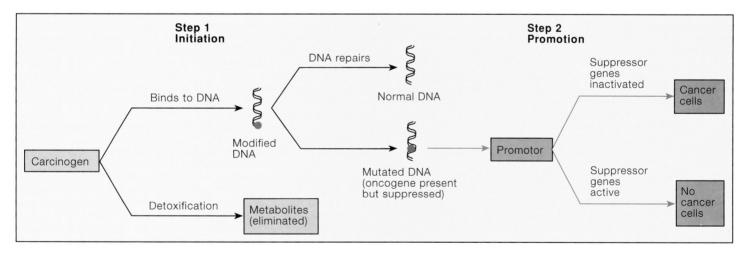

Figure 12.7 A possible model for cancer-cell formation. Cancer-cell formation often involves two agents: a carcinogen and a promotor.

mor formation. A better understanding of tumor suppressor genes may well provide the means for future anticancer therapy based on nature's own method of protecting against cancer.

Recall that for most toxins, a threshold dose is necessary before a response occurs. Many scientists believe, however, that exposure to any amount of a carcinogen poses some risk;

that is, there is no margin of safety. This property of carcinogens poses problems for scientists and public policymakers. For example, the chlorine added to public water supplies kills most pathogens (chapter 14), but its use also results in the formation of small amounts of carcinogens such as chloroform ($CHCl_3$). Because it is impossible to reduce chloroform

Human Reproduction and Toxins

*I*nadvertent exposure of women of reproductive age to chemicals in the workplace has become a major environmental issue. Until recently, management took specific action to prevent the exposure of women to teratogens. For example, battery manufacturers reassigned women from jobs that required handling hazardous lead-containing products to jobs where exposure to lead was minimal but where the pay was significantly lower. Some women, still of reproductive age, argued that they had the right to the higher paying jobs, because they had chosen not to bear any more children. Management, however, argued that these women might still become pregnant and then sue the company if their child developed a health problem. Battery manufacture is only one of many industries where women risk being exposed to teratogens.

In March of 1991, the U.S. Supreme Court ruled that employers could not discriminate against women in job placement on the basis of potential exposure to hazardous materials. Although women's rights groups praised the Supreme Court decision, industry still feared legal reprisals and threatened to move hazardous jobs to other nations. Although the decision reinforced the principle of equal employment opportunity for both men and women, scientists remain concerned about the impact of this decision on the health of future generations. This concern stems from two observations: (1) a developing fetus is inherently vulnerable to toxins, and (2) birth defects are relatively common. For example, 2-3% of the three million babies born annually in the United States, have major congenital defects that are manifested within the first year. The frequency of all malformations jumps to 16% of live births when all major and minor defects (e.g., hearing impairment) that become apparent later in life are included.

Scientists estimate that 23-25% of birth defects among live births can be traced to genetic factors associated with one or both of the parents, while 7-11% are explained by external environmental factors. No causal factor can be determined for 55-70% of birth defects. Because we cannot perform tests on humans, it is difficult to determine the relative contribution of genetic factors and teratogens to this last group of birth defects.

We now examine the development of the fetus from conception to birth to gain a better understanding of the impact of teratogens. During our discussion, we must keep in mind that medical specialists have a very limited knowledge of the causes of birth defects.

Fetal development involves a series of overlapping stages, some of which are more vulnerable than others to exposure to teratogens (fig. 12.8). Conception occurs when a sperm fertilizes an egg in a fallopian tube. During its journey down the fallopian tube to the uterus, the fertilized egg (initially a single cell) begins to divide. About three days after conception, the embryo arrives at the uterus, and it becomes implanted in the uterine wall within another three or four days. During the week between conception and implantation, the embryo divides repeatedly and develops into a multicellular mass (blastocyst). At this stage, the specific future role of each cell in the development of the fetus is not yet determined, and the cells are said to be *undifferentiated.* Exposure to a toxin at this stage may delay further cell division and development; a high concentration of a toxin kills the blastocyst.

By the end of the second week, placental tissue connects the mother and embryo. The placenta serves as the conduit through which exchanges of nutrients and waste occur between the blood of the mother and that of the embryo. By the third week, cells begin to differentiate; that is, they become specialized in both structure and function, and begin to develop into specific organs such as the heart, liver, and kidneys. During the first two months of cell differentiation, the newly forming organs are most vulnerable to disruption by teratogens. For example, if a pregnant women contracts rubella during this time, the embryo may suffer damage to the heart, lens of the eye, inner ear, or brain depending upon the specific time of exposure. Also, heavy alcohol consumption affects normal organ development. Alcoholic women are prone to miscarriages (spontaneous abortion) during the first months of pregnancy. By the end of the first trimester (approximately twelve weeks), differentiation is

levels in drinking water to zero, public policy regulators set the allowable concentration of chloroform and other potential carcinogens at a level that will cause one additional case of cancer in a population of one million people. Thus, they opt for a minimal risk for contracting cancer (with chlorination) rather than the greater risk of contracting a water-borne disease (without chlorination).

The incidence of cancer (number of cases per million population) is dose-dependent, while the severity of the response (cancer) is independent of dose. This means that as a population is exposed to higher levels of a carcinogen, we would expect more cases of cancer to develop. But once an individual has contracted cancer, the dose of the carcinogenic agent has little to do with the severity of the disease. This distinction between toxins and carcinogens is the reason that exposure regulations for substances classified as carcinogens are much more stringent than for toxins.

Mutagens cause **mutations,** which are inheritable changes in the DNA sequences of chromosomes. Recall from chapter 4 that all organisms have a package of chromosomes that are the result of evolution over numerous generations. Hence, most organisms are well adapted for survival in their particular habitat. Mutations are often harmful because they involve a random change in the natural functioning of chromosomes or their component genes; such changes (mutations) do not usually benefit the organism's offspring. If we use a screw-

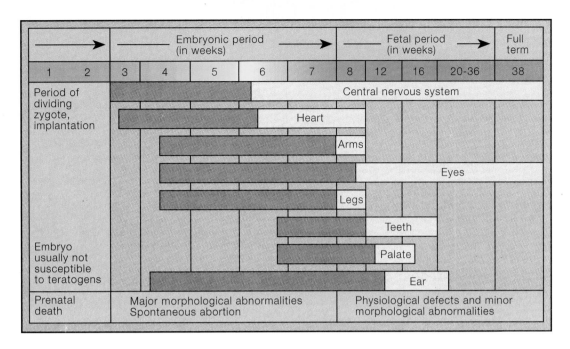

Figure 12.8 Human developmental stages and susceptibility of organ systems to reproductive toxins. Blue regions represent time periods when tissue is highly susceptible to environmental toxins.

essentially complete. Although the risks to the developing fetus are greatly reduced after this time, teratogens continue to pose a hazard as the organs develop through the remainder of the gestation period.

The placental linkage between a woman and the fetus allows many substances that enter a woman's bloodstream to reach the fetus. Each year an estimated 400,000 babies are born with symptoms of alcohol and/or cocaine addiction. These symptoms include low birth weight, poor muscle coordination, and brain damage. Many of these children are so severely handicapped that they may never be able to care for themselves.

Sometimes, the deleterious effect of a teratogen does not appear until many years after the mother has been exposed to it. For example, diethylstilbestrol (DES), was taken by pregnant women in the United States between 1945 and 1970 to prevent

miscarriages. In the late 1960s, however, medical specialists discovered abnormalities in the reproductive systems of daughters of women who took DES. Many have difficulty conceiving or carrying a fetus to full term. Overall, about 75% of all females exposed to DES developed benign tumors. Also many sons of those women have low sperm counts. In addition, the risk of vaginal or testicular cancer appears to be significantly greater in people whose mothers took DES.

One in six couples desiring to have children is unsuccessful, but the role played by teratogens in the infertility of offspring is uncertain. Infertility is caused both by genetic and external environmental factors such as disease, drugs, alcohol, radiation, and food contaminants. Of the 600 chemicals shown to cause reproductive abnormalities in laboratory animals, only 20 have been definitely linked to similar problems in humans.

driver to randomly alter the workings of a fine Swiss watch, how likely is it that the watch will run better after our tinkering? Every living organism is immeasurably more intricate than a watch. Some common diseases caused by mutations include cystic fibrosis, sickle-cell anemia, hemophilia, and Huntington's disease (a loss of neural muscular control in middle-aged adults). People who have a family history of these or other inheritable diseases are advised to seek genetic counseling before having children.

Teratogens cause abnormalities in a developing fetus. People frequently associate teratogens with severe deformities, such as the shortening or absence of arms or legs. Such was the impact of thalidomide, a sedative that was taken

by thousands of pregnant women in West Germany during the early 1960s. The effects of many teratogens, however, are usually more subtle. For example, rubella (German measles) contracted during the first trimester of pregnancy may produce cardiac defects and deafness in the offspring. Sometimes the deleterious effect of a teratogen does not appear until many years after the mother has been exposed. This was the case for DES (diethylstilbestrol), a drug taken by pregnant women in the United States for more than thirty years to prevent miscarriages. We discuss DES and other agents that influence human reproduction more broadly in the issue entitled "Human Reproduction and Toxins."

Many carcinogens, mutagens, and teratogens act in the same way by causing changes in DNA that eventually affect cell development. An examination of table 12.5 shows that many chemicals and forms of radiation appear in more than one category. Because of this relationship, initial screening for substances that may be carcinogenic, mutagenic, or teratogenic can be accomplished by testing their capacity to cause mutations in a particular strain of bacteria (*Salmonella typhimurium*). This unique strain of bacteria requires the essential amino acid histidine to grow. It can grow on histidine-free media only if it has first mutated. Thus, if the bacteria grow on histidine-free media after exposure to a test chemical, we can assume that the chemical caused the bacteria to mutate and that it is therefore likely to be a carcinogen. This test is called the Ames test after Dr. Bruce Ames of the University of California at Berkeley.

Natural Defense Mechanisms: Detoxification Routes

The body has a number of **detoxification mechanisms,** that is, means whereby the body metabolizes or excretes toxins. The primary focus in this section is on detoxification of ingested poisons.

Ingested toxins enter the body via the gastrointestinal tract, a long, highly specialized tube that extends throughout the length of the body's trunk (fig. 12.9). Ingested materials are processed in the mouth, stomach, small intestine, and large intestine. Glands in the lining of the stomach and small intestine together with the liver and pancreas, release fluids and enzymes essential for digestion. These fluids and enzymes convert the complex organic molecules of our food into simpler organic molecules that are absorbed (mainly by the lining of the small intestine) and taken up by the bloodstream.

The gastrointestinal tract has a variety of defense mechanisms against toxins. The same processes that digest food also metabolize ingested toxins into breakdown products called **metabolites,** most of which are not toxic. In addition, certain materials in the gut capture toxins and prevent them from moving into the bloodstream. For example, soluble fiber tends to capture chemicals such as cholesterol and carry them out of the body in feces. For this reason, a high-fiber diet (a diet rich in grains, fruits, and vegetables) lowers blood cholesterol and reduces the risk of colon cancer.

Dead cells that continually slough off from the lining of the gut capture toxins such as heavy metals. Because cells in the lining follow a three- to five-day sequence of dividing, dying, and sloughing off, the body continually renews this defense mechanism. But every defense has its limits. If the concentration of heavy metals is greater than the capacity of dead cells to remove them, some excess metals enter the bloodstream.

The body has still another line of defense. All of the blood from the stomach and small intestine flows directly to the liver (fig. 12.9), where a host of detoxifying enzymes me-

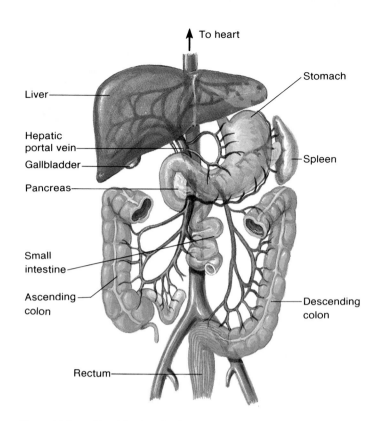

Figure 12.9 Blood leaving the digestive tract flows through the hepatic portal vein to the liver before being circulated to other parts of the body.

tabolize toxins. The liver is our most important internal defense against toxins. When alcohol in the bloodstream enters the liver, it immediately begins to detoxify (metabolize) alcohol by converting it to other chemicals, which are eventually eliminated from the body. The rate of detoxification, however, is relatively slow; blood-alcohol levels decrease at the rate of between 15 and 20 milligrams per 100 milliliters per hour. Thus, when alcohol is consumed faster than it is metabolized, blood-alcohol levels rise. Figure 12.10 indicates the approximate blood-alcohol levels in humans after varying numbers of alcoholic drinks. An average drink, which contains about 18 grams of alcohol, is assumed to be one 355-milliliter (12-ounce) can of beer, 120 milliliters (4 ounces) of wine, or one mixed drink.

Most metabolites formed in the liver are excreted in the bile and eventually released into the small intestine. If the small or large intestine does not absorb these metabolites, they are excreted in the feces. It is interesting to note that many therapeutic drugs, such as those used to treat cancer, are administered intravenously (through the veins), rather than orally. Thus, the drug is not metabolized by the liver before it reaches its target organ.

Enzymatic breakdown of toxins does not always produce a metabolite that is less toxic than the parent substance. The detoxification of two common alcohols, methanol (wood alcohol, CH_3OH) and ethanol (grain alcohol, CH_3CH_2OH) illustrate the differing toxicity of metabolites. Both methanol and ethanol are readily absorbed by the lining of the stomach and taken up by the bloodstream. In the liver, the same enzyme system breaks down the two alcohols, as illustrated in figure 12.11. The metabolites of ethanol, acetic acid and acetaldehyde, are relatively nontoxic, but the metabolite of methanol, formic acid, travels through the bloodstream and damages the light-sensitive region of the eye. Permanent blindness results from the ingestion of too much methanol. Wood alcohol is an example of a chemical that does not have to be directly toxic to cause harm; sometimes the enzymatic breakdown of a chemical to a more toxic metabolite is the actual cause of toxicity.

In the long run, even the liver has a limited ability to cope with continual exposure to toxins. A chronic abuser of alcohol, for example, eventually develops fibrous tissue in the liver, which constricts the flow of blood. Restriction of blood flow reduces the capacity of the liver to detoxify alcohol. Such degeneration of liver function, called *cirrhosis*, is eventually fatal. Although the liver is the principal defense against toxins, the kidneys also play a significant role. They filter the blood and subsequently excrete toxins from the body in urine.

We have noted that the body is able to detoxify some poisons, but the body's defenses are limited. Thus, our first line of defense against toxins is to minimize our level of exposure. How, then, do we assess our risk of exposure to toxins?

Analysis of Risk

To live is to take risks. But what is risk? How do we assess it? Can we avoid it? If not, can we reduce it? For an individual or for society, risk is often a factor in decision making. Do we choose to use a seat belt in a car, wear helmets when we cycle, drink before driving, live near a nuclear power plant, go to the beach for a tan, or regularly eat food high in fat and salt?

Assessing Risk

Risk is the chance that something harmful or undesirable will occur when we are exposed to a substance or engaged in some activity. The decision to take a risk may be our own, or it may be made for us. We may voluntarily take an **individual risk** by participating in skydiving, downhill skiing, automobile racing, jay walking, bungee jumping, or other risky activities. **Societal risks** are imposed on us, often without

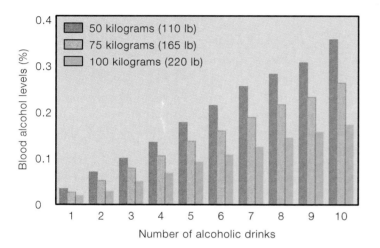

Figure 12.10 Blood-alcohol levels vary with the number of drinks consumed and the weight of the individual.

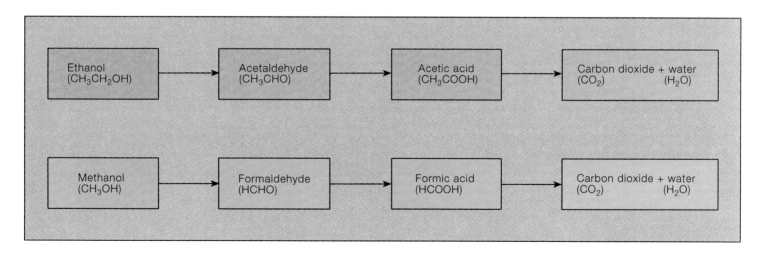

Figure 12.11 The metabolites formed when two common alcohols are ingested and metabolized.

Table 12.6 Comparative risks of death in a given year

Specific Risk	Annual Risk Probability	Annual Risk Odds (chances)
motor vehicle accident (total)	2.4×10^{-4}	2.4/10,000
pedestrian-motor vehicle accident	4.2×10^{-5}	4.2/100,000
home accidents	1.1×10^{-4}	1.1/10,000
electrocution	5.3×10^{-6}	5.3/1,000,000
air pollution (eastern U.S.)	2×10^{-4}	2/10,000
cigarette smoking (one pack per day)	3.6×10^{-3}	3.6/1,000
background radiation (sea level)	2×10^{-5}	2/100,000
all cancers	2.8×10^{-3}	2.8/1,000
four tablespoons peanut butter per day	8×10^{-6}	8/1,000,000
drinking water (EPA limit of chloroform)	6×10^{-7}	6/10,000,000
drinking water (EPA limit of trichloroethylene)	2×10^{-9}	2/1,000,000,000
alcohol, light drinker	2×10^{-5}	2/100,000
frequent flier	5×10^{-5}	5/100,000
mountaineering	6×10^{-4}	6/10,000

From R. Wilson and E. A. C. Crouch, "Risk Assessment and Comparisons: An Introduction," *Science* 236: 268. Copyright © 1987 AAAS.

our knowledge or direct consent. For example, the EPA allows low levels of certain pollutants (heavy metals and certain carcinogens) in public drinking water. Anyone drinking the water has, knowingly or unknowingly, accepted the drinking water standard used by the EPA as a reasonable measure of safety. But is the assumed level of risk low enough to protect the general public? Or are exposure levels so excessively restrictive that unnecessary costs (for control of contaminants) must be borne by consumers and taxpayers?

How then do we assess risk? **Risk assessment** is a process whereby we critically examine our activities with the objective of avoiding or reducing our exposure to dangerous substances or harmful activities. Risk assessment begins with the question, What can go wrong? Playing football could result in a pulled muscle, fractured leg, or spinal injury and paralysis. Unsafe sex may result in an unwanted pregnancy, syphilis, or AIDS. Drinking contaminated water could cause intestinal discomfort, hepatitis, or cancer. Next, we ask what the chances are that an undesirable outcome will occur. Is there a fifty-fifty chance or are the odds as low as one in a million? What are the consequences if we accept the risk and lose? Will the unsafe practice shorten our lives? Will the undesirable outcome permanently affect the quality of our lives (paralysis, for example)? Or will it just inhibit our mobility for a few days?

Most often we express risk in terms of the probability of death. For example, automobile accidents on average kill about 58,000 people in the United States per year. Since the nation's population is about 250 million, the probability of being killed in an automobile accident during a single year is about 2.4 chances in 10,000. Your chances of being killed by an automobile in your lifetime are about 2 in 100.

Stated in another way, auto accidents account for approximately two out of every one hundred deaths. Table 12.6 lists the probability (risk) of death during a single year from a variety of causes.

Another way of thinking about risk is to make comparisons. Professor Richard Wilson of Harvard University compares risks in terms of the "statistical person," a person thirty years old with a life expectancy of seventy-four years. He calculates that the life span of a statistical person is shortened by 9 seconds by drinking a single diet soda, 5 minutes for smoking a single cigarette, 1,800 days for being an unmarried male, and 2,700 days by being male rather than female. Risks can also be communicated by comparing activities that pose the same risk. The following all pose a one in a million chance of death: drinking a half liter of wine, paddling 6 minutes in a canoe, traveling 16 kilometers (10 miles) on a bicycle, 480 kilometers (300 miles) in a car, or 1,600 kilometers (1,000 miles) by commercial jet, spending 3 hours in a coal mine, and living 2 days in New York City.

In expressing risk quantitatively, scientists must be objective. Historical data provide the basis for most quantifications of risk, but these sources are sometimes insufficient or nonexistent. For example, industry introduces a few thousand new chemicals each year. The risk levels associated with these chemicals to people and the environment must be based on experiments with laboratory animals, as described earlier in this chapter. Estimating risk from animal studies is less reliable than direct evidence on risk for humans, but such studies are still important in making comparisons of relative levels of risk.

Factors Affecting Risk Level

One important criterion in decision making is to minimize risk. For example, take-offs and landings are the most hazardous phases of air travel. Thus, when we travel by air, most of us would choose a large commercial jet that gets us to our destination directly rather than a small commuter aircraft that makes many intervening stops. But the level of risk is not the sole criterion in decision making. Other factors include (1) convenience—it is easier to travel by car than by bus, even though a car is not as safe as a bus; (2) economics—a subcompact car is less expensive but riskier than a full-sized automobile; (3) psychological—although travel by car is less safe than by airline, we generally prefer to drive a car and have personal control than have an airline pilot in control; or (4) personal bias—some people prefer the personal exhilaration associated with a sports car versus a safer van or station wagon.

In general, when it comes to risk, people tend to be overly optimistic; that is, they underestimate real personal risks. In a recent study of a variety of hazards, people underestimated their own risk 78% of the time. People tend to be too optimistic when (1) dealing with hazards with which they have limited familiarity, (2) encountering risks where they feel they have control of the situation, and (3) participating in activities that have low risk. We often feel that an undesirable outcome always happens to the other person.

In other cases, people are too pessimistic about risk. This is generally true with regard to hazardous chemicals and the nuclear power industry. For example, the worst nuclear accident in history, the explosion of the nuclear power plant in Chernobyl, Russia, killed 31 people within 60 days primarily because of exposure to radiation. Within a 3-to 15-kilometer ring around the plant, 24,000 people were exposed to high levels of radiation, which eventually is likely to cause an additional 131 deaths due to cancer. This hardest hit region around the Chernobyl accident is expected to experience a 2.6% increase in the incidence of cancer. Considering the fact that 20% of the 24,000 people who lived near Chernobyl would normally be expected to develop cancer in their lifetime, the risks of living near a nuclear power plant are probably not as great as most people surmise, even in the case of a major accident that releases considerable radiation. While these statistics are not intended to downplay the potential hazard of a nuclear accident, they do put in perspective other serious risks such as cigarette smoking, which is implicated in perhaps 30% of all cancers. In general, people tend to be overly pessimistic about exposure to radiation and overly optimistic with regard to exposure to cigarette smoke.

Societal Risks

We have focused so far on how risk influences personal decision making. In many situations, it is more appropriate to assess risk for society as a whole. As a society, we are faced with many involuntary risks that government agencies attempt to assess and manage through regulation. Recall from our discussion in chapter 2 how public policymakers respond to concerns voiced by their constituencies. If a real or perceived risk to the public well-being is present, legislation is frequently passed that gives a government agency the legal authority to set performance or exposure standards designed to protect the public. In most cases, industry has a chance to argue for or against proposed standards before they become law. Typically, industry argues against strict regulatory standards on the basis of economic considerations (the cost of compliance exceeds the benefits), or it may argue that reasonable control technology is not available to meet the regulatory standard.

In cases of exposure to hazardous, but noncarcinogenic chemicals, regulatory standards are usually set at levels that are one hundred times lower than the no-observed-effect level (NOEL). For carcinogens, the acceptable level of exposure is the amount of material estimated to cause one additional case of cancer in one million people (a risk of 1.0×10^{-6}) when exposed at the maximum level allowed by a federal standard over a seventy-year life span. In actual practice, this goal is not being met. In a recent study of regulation of human exposure to carcinogens, 70% of actual regulations failed to meet this guideline. The least stringent regulations involving significant numbers of people have a lifetime risk of approximately 1.0×10^{-4} or one hundred times the federal standard. However, the risk of contracting cancer from such exposures is still substantially less than the risk from background sources (radiation from bedrock, ultraviolet radiation from the sun, and naturally occurring carcinogens in food), which is 2.5×10^{-1} (i.e., one in four people will contract some form of cancer during their lifetime).

Widespread public fear of cancer has spurred the federal government to develop strict regulatory policies concerning exposure to carcinogens through air (especially in the industrial setting), water, and food. Scientists are just now beginning to understand that we are also exposed to an overwhelming number of naturally occurring carcinogens.* Dr. Bruce Ames of the University of California at Berkeley estimates that our intake of natural carcinogens exceeds our exposure to synthetic carcinogens by a factor of ten thousand. Raw mushrooms, potatoes, peanut products, alcohol, natural root beer (sarsaparilla), bacon, some herbal teas, and brown mustard are commonly ingested substances that contain trace amounts of natural carcinogens.

*See the article by Ames et al., listed at the end of this chapter.

Molds that grow on seeds and fruits of food plants synthesize a variety of natural toxins, including carcinogens, that apparently help the molds to survive, and some of these carcinogens find their way into food. Aflatoxin, one of the most potent carcinogens in laboratory rats, is formed by molds that grow on corn, wheat, and nuts and thus enters the human food chain. For example, peanut butter marketed in the United States contains on average about 2.0 ppb aflatoxin. Because of their potency, mold toxins may be one of the most important sources of carcinogens in our food supply. The concern is even greater in less-developed nations, where poor storage facilities and warm, humid climates greatly enhance mold growth. In more-developed nations, modern storage facilities that use fungicides greatly reduce mold contamination.

From our discussion of risk assessment, it is clear that we still have much to learn about exposure to chemicals, especially carcinogens, mutagens, and teratogens. Research suggests that for society as a whole, pollution only contributes minimally to the cancer hazard compared to background levels. This does not mean that we should be complacent about pollution, but it does indicate that we need a more equitable balance between the fear of exposure and the actual risk. Currently, the possibility of exposure to even minute quantities of almost any chemical is stringently controlled by regulations that make compliance costly for industry. In many cases, these costs provide society with no tangible benefits. On the other hand, there still remain many occupations in the United States where lifetime exposure to the chemicals of the trade can result in permanent illness and a shortened life span.

Conclusions

We have examined human exposure to chemicals in an effort to understand the risks we face from continual exposure to synthetic and naturally occurring toxins. Some species are significantly more sensitive to toxins than we are, while other are more tolerant. For all organisms, we know that the severity of symptoms increases as the dose of noncarcinogenic toxins increases. The sensitivity of organisms varies because species differ in their ability to detoxify toxins.

When risks from activities such as driving a car, jaywalking, or sunbathing are compared to risks from chemical exposure, we see that today's society perceives risks from many toxins as being much higher than they actually are. The result has been strict regulation of activities that involve exposure to toxins. We will apply the fundamental concepts of toxicology again when we examine the impact of water and air pollutants (chapters 14 and 16) and the management of hazardous waste (chapter 20).

Key Terms

toxicology	oncogene
toxins	promoter
heavy metals	suppressor gene
bioaccumulation	mutagen
acute exposure	mutation
chronic exposure	teratogen
toxicity	detoxification mechanisms
enzymes	metabolites
no-observed-effect level (NOEL)	risk
LD_{50}	individual risk
LC_{50}	societal risk
carcinogen	risk assessment
initiator	

Summary Statements

Toxins enter organisms via four possible routes: inhalation, absorption, ingestion, and injection. Inhalation is the most frequent route of human exposure to toxins.

Normally, we ingest only minor amounts of toxins. The amounts we ingest from animals and plants contaminated with toxins through bioaccumulation are exceptions.

Short-term, or acute, exposure to toxins involves exposure to high levels of toxins with almost immediate symptoms. Long-term, or chronic, exposure to toxins involves exposure to low levels of toxins over lengthy periods, usually with long-delayed symptoms and illnesses.

The response to most toxins is dose-dependent; higher doses produce more severe symptoms.

The LD_{50} value, the dose required to kill 50% of a test population, is one way to express the toxicity of a substance.

A carcinogen is a substance or radiation that causes cancer. Carcinogens include initiators and promoters that act on DNA.

Mutagens are substances that cause heritable changes in genes.

Teratogens cause defects between conception and birth.

The body defends against toxins by excreting or metabolizing them. The liver is the primary organ for metabolizing toxins.

Risk is a measure of the likelihood of something going wrong. People vary in their response to risk; typically, they underestimate the risk associated with familiar activities and overestimate the risk associated with unfamiliar activities or highly publicized risks.

Toxicological data are used to determine the allowable levels (risk) of exposure of humans to toxins.

Questions for Review

1. List the four major routes of entry for toxins into the body. For each route, describe one of the body's defense mechanisms against a potential toxin.
2. Distinguish between acute and chronic exposure to toxic chemicals.
3. Do acute and chronic exposure to the same chemical result in similar symptoms? Explain, using an example.
4. The oral LD_{50} values for methanol, ethanol, and isopropanol, three common alcohols, in rats are 13,000 milligrams per kilogram, 21,000 milligrams per kilogram, and 5,840 milligrams per kilogram respectively. List these three alcohols in order of increasing toxicity. What would be an appropriate descriptive term for the toxicity of each alcohol?
5. From the graph in figure 12.10 determine the blood-alcohol level for a 75-kilogram person who has consumed five average alcoholic drinks. Would this person be considered legally intoxicated if caught driving?
6. A blood-alcohol level of 0.10% is considered the level of legal intoxication in most states. Express this concentration in parts per thousand (ppt), parts per million (ppm), and parts per billion (ppb). (Consult Appendix B.)
7. Paracelsus (1493-1541) is credited with the statement, "All substances are poisons; there is none which is not a poison. The right dose differentiates a poison and a remedy." Does this statement agree with modern toxicological principles? Explain.
8. Give four major reasons why the same dose of a chemical can produce different toxic responses in people.
9. Give an example of a metabolite that is more toxic than its parent substance.
10. Two major categories of carcinogens are initiators and promoters. How do they differ? Which type is considered to be the more potent carcinogen?
11. How do carcinogens, mutagens, and teratogens differ?
12. Why do many of the same substances appear on lists of carcinogens, mutagens, and teratogens?
13. During what weeks of the human gestation period are teratogenic effects most likely to be initiated? Why?
14. Speculate on why we can purchase some drugs over the counter, while others can be purchased only with a prescription.
15. What three federal agencies are primarily responsible for regulating levels of chemical contaminants in food? Give an example of a type of contaminant regulated by each agency.

Projects

1. Look around your home and prepare an inventory of hazardous and toxic materials. What properties of these products make them hazardous? Make suggestions for decreasing the risk of chemical-related accidents in the home.
2. Without consulting any references, rank the risk of dying for the following, from highest to lowest, on the basis of your own impressions: (a) driving a car, (b) smoking one pack of cigarettes a day, (c) traveling frequently by air, (d) drinking water with the EPA limit of chloroform, (e) exposure to background levels of radiation. Now consult table 12.6, and compare your perception of the relative risks with the actual risks. How do you account for the differences between your perception and reality?
3. Contact your local poison control center and determine which types of toxins are most commonly encountered in children. What treatments are usually required? Most poison control centers are located in hospitals.

Selected Readings

Abelson, P. H. "Incorporation of New Science into Risk Assessment." *Science* 250 (1990): 1497. Editorial that discusses the background for future legislative mandates.

Ames, B. N., R. Magaw, and L. W. Gold. "Ranking Possible Carcinogenic Hazards." *Science* 236 (1987): 271-80. A review of why animal cancer tests are of limited value in predicting human risks from carcinogens. The article also presents a ranking system for industrial and natural carcinogens that contaminate food and water.

Cohen, L. A. "Diet and Cancer." *Scientific American* 257 (Nov. 1987): 42-48. A review of some nutritional recommendations that may reduce the incidence of cancer.

Klaassen, C. D., M. O. Amdur, and J. Doull, eds. *Toxicology*. 3d ed. New York: Macmillan, 1986. An in-depth coverage of the science of poisons from the medical perspective.

LaFond, R. E., ed. *Cancer: The Outlaw Cell*. 2d ed. Washington, D.C.: American Chemical Society, 1988. A discussion of developments in cancer research.

Needleman, H. L., A. Schell, D. Bellinger, A. Leviton, and E. N. Allred. "The Long-Term Effects of Exposure to Low Doses of Lead in Childhood." *The New England Journal of Medicine* 322 (1990):83-88. An 11-year follow-up report on children exposed to various levels of lead, emphasizing effects on central nervous system function, including learning.

Ottoboni, M. A., *The Dose Makes the Poison*. 2d ed. New York: Van Nostrand Reinhold, 1991. A guide to the fundamental concepts of toxicology; includes many case studies.

Schmidt, K. F. "Dioxin's Other Face." *Science News* 141 (1992): 24-27. Overview of what is currently understood about the toxicity of dioxin.

Shane, B. S. "Human Reproductive Hazards." *Environmental Science and Technology* 23 (1989):1187-95. An examination of the effects of toxins on the developing fetus.

Slovic, P. "Perception of Risk." *Science* 236 (1987): 280-85. An examination of the judgments people make when asked to evaluate hazardous activities and technologies.

Steinmetz, G. "Fetal Alcohol Syndrome." *National Geographic* 181 (Feb. 1992): 36-39. A look at the irreversible effects of excessive alcohol consumption on fetal development.

Travis, C. C., and H. A. Hattermer-Frey. "Determining an Acceptable Level of Risk." *Environmental Science and Technology* 22 (1988): 873-76. An examination of the relative risk allowed by federal agencies for exposure to various carcinogens.

Weinstein, N. D. "Optimistic Biases about Personal Risks." *Science* 246 (1989): 1232-33. An examination of the biases concerning harmful activities that affect personal decision making.

Wilkinson, C. F. "Being More Realistic about Chemical Carcinogenesis." *Environmental Science and Technology* 21 (1987): 843-47. A discussion of cancer, carcinogens, and efforts to reduce the incidence of cancer.

Wilson, R., and E. A. Crouch. "Risk Assessment and Comparisons: An Introduction." *Science* 236 (1987): 267-70. Risk assessment is presented as a means of avoiding, reducing, and otherwise managing risks.

Volume III

Environmental Science

Managing Physical Resources

Part V

Managing Physical Resources and Environmental Quality

*T*his final section describes environmental issues that have major physical dimensions. Most of them stem from human disturbance of the planet's physical systems (water, land, and air) and our exploitation of physical resources such as minerals and fossil fuels. Such disturbance may adversely affect human health, or the health or habitats of other organisms. Although the emphasis here is on the physical, the ecological impact is wide ranging. Our typical approach is to describe the normal functioning of a physical system, how the system is disturbed, and what we are doing to manage the system to reduce our impact. We do this for water in chapters 13 and 14, air in chapters 15 and 16, and land in chapter 17. Chapters 18 and 19 focus on the consequences of our reliance on energy resources mined from the earth's crust (primarily fossil fuels) and prospects for alternative energy sources. Chapter 20 deals with current methods of waste disposal and strategies for waste management and resource recovery. In chapter 21, we consider how conflicts in land use affect the coastal zone, how land use is limited by natural hazards, and how community growth stresses the environment.

To meet a growing demand for electricity, the United States will probably increase its reliance on coal-fired power plants such as this one at La Grange, Texas.
© David R. Frazier Photolibrary

Chapter 13

The Water Cycle and Freshwater Resources

On earth, water occurs in all three phases: ice, liquid, and vapor. It continually cycles among oceanic, terrestrial, and atmospheric reservoirs. This global water cycle supplies us with a fixed quantity of fresh water.
© Spencer Swanger/Tom Stack & Associates

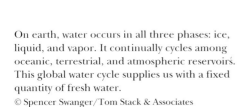

The year 1992 marked the sixth consecutive year of drought in California. Compounding the problem of recent water shortages are the state's rapid population growth (an average of 600,000 new residents per year during the 1980s) and huge agricultural industry (which uses 85% of the state's water supply to grow 25% of the nation's produce). Conflicting demands for dwindling water resources have pitted city residents against agricultural interests and north against south. Northern Californians have steadfastly resisted mounting pressures from thirsty southern Californians to divert more water to the south.

California receives the bulk of its precipitation during the winter. Winter rains and mountain snows have been well below long-term averages since the late 1980s, however, and reservoirs that they feed have shrunk (fig. 13.1). In March of 1991, the Hetch Hetchy reservoir that supplies San Francisco with its water was down to only 9% of capacity. At the same time, the snowpack of the Sierra Nevada Range was down 87%, and Lake Shasta, the northern California starting point for a series of canals that deliver irrigation water to the farms of the Central Valley, was only 35% full.

In response to water demands and mounting shortages, state and local officials set priorities for water use and imposed mandatory water conservation measures. The goal was to cut residential water use by 31% and agricultural use by 50%. Failure to adhere to water-conservation regulations means stiff fines. In San Francisco, filling swimming pools or hot tubs and washing cars at home were prohibited. In some areas, new home construction slowed because of regulations limiting the number of new hookups to the local water supply. As a consequence of the drought, most Californians have become keenly aware of their dependence on fresh water and the finite nature of that precious resource. As of this writing, the drought continues, and even more drastic measures have been proposed, including a complete cutoff of water to some seven thousand farms in the San Joaquin Valley.

Earth is often referred to as the watery planet for good reason; almost three-quarters of its surface is covered by water. Most of that water (97%), however, is too salty to drink or to use for irrigation or for most industrial purposes. Recall from chapter 3 that our bodies are about 60% water; without daily water intake, we would succumb to severe dehydration. Most areas of human settlement are near rivers, lakes, or other freshwater sources, demonstrating the dependency of civilization on an adequate supply of fresh water. We are familiar with the many important personal functions of water: replenishing our bodily fluids, bathing, cooking, washing, and flushing away wastes. Water is also essential for agriculture, food processing, transportation, power generation, and the manufacture of countless consumer goods.

Our concerns about water focus on both its quantity and quality. The global supply of fresh water is essentially fixed; thus, the per capita supply inevitably declines as the human population continues its rapid growth (chapter 6). Already, in some arid and semiarid regions of the world (e.g., the Sahel of Africa), water shortages are causing much hardship. Our

Figure 13.1 Six years of drought have lowered water reservoirs to the point that residents of southern California face mandatory water curtailments. This is what remains of Tamarack Lake in the Sierra Nevada Range. © Gerald & Buff Corsi/Tom Stack & Associates

reliance on a fixed supply of fresh water for food, power, and transportation is exacerbated by our reliance on water to dilute and flush away our wastes. Today, the waste load in some waterways exceeds the natural **assimilative capacity** of the system, that is, its ability to dilute and degrade waste without long-term disruption. While certain aquatic organisms decompose waste and thus cleanse waterways, this natural cleansing process can be overwhelmed by too much waste or destroyed by toxins.

In this first of two chapters on water in our environment, we examine the natural freshwater delivery system and its limitations and the management of our limited supply. In the next chapter, we focus on water pollution and how we can improve water quality.

The Water Cycle

We saw in chapter 3 that ecosystems depend on the continual cycling of materials and flow of energy. Water is an essential resource that cycles through various reservoirs within and among ecosystems. As water cycles, it plays a key role in the overall functioning of the environment.

Of all substances on earth, water is the only one that occurs naturally in all three phases: solid (ice and snow), liquid, and vapor. Water in its various phases is distributed among oceanic, atmospheric, and terrestrial reservoirs (table 13.1). The ocean, the largest of these reservoirs, contains 97.2% of the planet's water; most of the remainder is tied up as ice in Antarctic and Greenland glaciers. Relatively small amounts occur in rivers and lakes, and in the tiny open spaces within soil, sediment, and rock (soil moisture and groundwater).

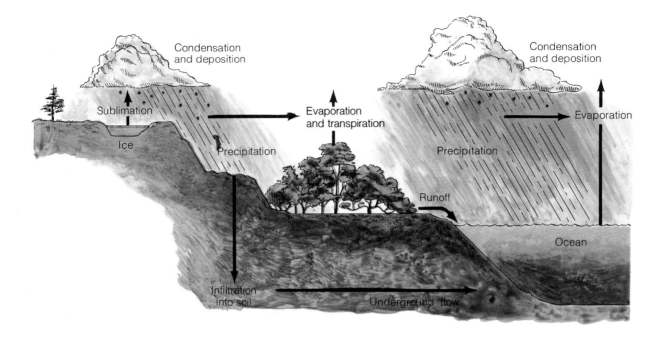

Figure 13.2   The water cycle. The world's water is distributed among its oceanic, terrestrial, and atmospheric reservoirs. Most of it occurs in the ocean basins.

Table 13.1	Water stored in global reservoirs of the water cycle
Reservoir	**Percent of Global Total**
world oceans	97.2
ice sheets and glaciers	2.15
groundwater	0.62
lakes (fresh water)	0.009
inland seas, saline lakes	0.008
soil moisture	0.005
atmosphere	0.001
rivers and streams	0.0001

The ceaseless flow of water among the various reservoirs, known as the **water cycle,** is illustrated in figure 13.2. Briefly stated, water vaporizes from sea and land to the atmosphere, where clouds form. From clouds, rain and snow fall to the earth's surface and resupply terrestrial reservoirs. From these reservoirs, water drips, seeps, and flows back to the sea. The endlessness of the global water cycle is expressed in a verse from Ecclesiastes: "Every river flows into the sea, but the sea is not yet full. The waters return to where the rivers began and starts all over again."

Virtually all the water we use must be fresh. Fresh water, by convention, contains less than 1,000 ppm dissolved salts. Sea water has an average salinity of 35,000 ppm,* so 97.2% of the water on the planet is not suitable. By U.S. standards, dissolved salts should not exceed 500 ppm for drinking water and 700 ppm for irrigation water. The cycling of water between the earth's surface and the atmosphere naturally puri-

*See Appendix B for an explanation of concentration units.

fies water that is too salty or polluted and ultimately governs the global supply of fresh water. In the next section, we examine this essential function of the global water cycle.

Air/Water Interaction

Water cycles from the earth's surface to the atmosphere by evaporation, transpiration, and sublimation. **Evaporation** is the process by which water changes from a liquid to a vapor at temperatures lower than its boiling point. Water evaporates from all open bodies of water as well as from soil and the damp surfaces of plant leaves and stems. Evaporation of ocean water is the principal source of atmospheric water vapor.

Transpiration is the process by which water absorbed by plant roots eventually escapes as vapor through tiny pores on the surface of green leaves. On land, transpiration is considerable and is often more important than direct evaporation from the surfaces of lakes, streams, and soil. For example, a single hectare (2.5 acres) of corn typically transpires 34,000 liters (8,800 gallons) of water per day. Measurements of evaporation and transpiration are usually combined as **evapotranspiration.**

Sublimation is the process by which water changes directly from a solid to a vapor without first becoming liquid. The gradual shrinkage of snowbanks, even though the air temperature remains well below the freezing point, results from both sublimation and settling. Solar radiation (chapter 15) provides the energy for evaporation, transpiration, and sublimation.

Water returns from the atmosphere to land and sea through condensation, deposition, and precipitation. **Condensation** is the process by which water changes from vapor to liquid (in the form of droplets). The formation of water

Figure 13.3 Clouds in their many forms are the visible products of condensation and deposition within the atmosphere. Clouds are composed of water droplets, ice crystals, or both.

droplets on the outside surface of a cold can of cola is an example of condensation. **Deposition** is the process by which water changes directly from vapor to solid (ice crystals), as when frost forms on an automobile windshield. When condensation and deposition take place in the atmosphere, water droplets and ice crystals are visible as clouds (fig. 13.3). **Precipitation** (e.g., rain, drizzle, snow, ice pellets, and hail) returns a major portion of atmospheric water from clouds to the earth's surface, where most of it vaporizes back into the atmosphere. If all water were removed from the atmosphere as rain and distributed evenly over the earth's surface, its depth would be only about 2.5 centimeters (1.0 inch).

Evaporation (or sublimation) followed by condensation (or deposition) purifies water. As water vaporizes, suspended and soluble substances like sea salts are left behind. Through these cleansing mechanisms, water from the salty seas eventually falls on land as freshwater precipitation that replenishes terrestrial reservoirs. This natural water purification process is known as **distillation.** Ultimately, this function of the water cycle supplies us with a fixed quantity of fresh water.

Humidity and Saturation

Precipitation is not possible without clouds, which do not form unless air is very nearly saturated with water vapor. Hence, we need to understand how air becomes saturated with water vapor.

Through evapotranspiration and sublimation, water molecules disperse rapidly and mix with the other gaseous molecules (mostly nitrogen and oxygen) that compose air. We can therefore specify the *actual* concentration of water vapor as so many grams of water vapor per kilogram of air. There is an upper limit to this concentration, however, and when air contains its *maximum* water vapor concentration, it is said to have achieved **saturation.**

As figure 13.4 shows, the saturation concentration increases as the air temperature rises, and vice versa. Thus, saturated, warm air is more humid (contains more water vapor) than an equivalent amount of saturated cold air. The temperature dependence of the saturation concentration is not surprising, because temperature regulates the activity of molecules. Water molecules are more active at higher temperatures and hence more readily escape from the surface of oceans, lakes, and rivers to the atmosphere as vapor.

Relative humidity is perhaps the most familiar way of describing the amount of water vapor in the air; it is a routine part of daily weather reports and forecasts. **Relative humidity** is defined as the ratio of the actual concentration of water vapor in the air to the concentration at saturation. Multiplying this fraction by 100% expresses relative humidity as a percentage. That is,

Relative humidity
= Actual concentration/Saturation concentration $\times$ 100%

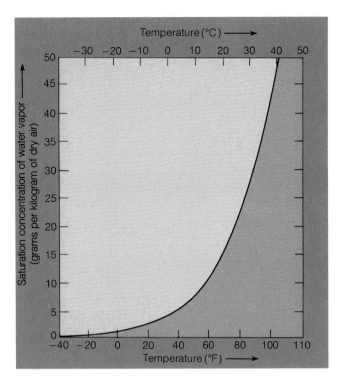

Figure 13.4 The concentration of water vapor at saturation increases with rising temperature. Hence, warm saturated air is more humid than cool saturated air.

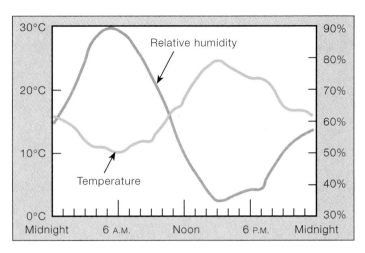

Figure 13.5 The variation of air temperature and relative humidity over the course of a 24-hour day. Note that the relative humidity is highest when the air temperature is lowest, and lowest when the air temperature is highest.

As an illustration, let us calculate the relative humidity for a muggy summer afternoon. Suppose that the air temperature is 32° C (90° F), and the water vapor concentration is 20 grams of water vapor per kilogram of dry air. We see from figure 13.4 that the saturation concentration of air at 32° C is about 30 grams per kilogram. Hence, from the formula, we compute the relative humidity to be 66.7%:

$$\text{Relative humidity} = \frac{20\ \text{gm/kg}}{30\ \text{gm/kg}} \times 100\% = 66.7\%$$

Suppose that we continue to monitor the relative humidity into the following night. By early morning, the air temperature drops to 25° C (77° F), while the water vapor concentration has remained essentially unchanged. Again consulting figure 13.4, we see that the saturation concentration of air at 25° C (77° F) is about 20 grams per kilogram. Hence, from the formula, we compute the new relative humidity to be 100%:

$$\text{Relative humidity} = \frac{20\ \text{gm/kg}}{20\ \text{gm/kg}} \times 100\% = 100\%$$

As the relative humidity approaches 100% (i.e., saturation), condensation or deposition of water vapor becomes more and more likely. Water vapor may condense on exposed surfaces as dew or, if the temperature is below freezing, as frost. Condensation or deposition within the atmosphere produces clouds. What, then, causes air to approach 100% relative humidity?

Achieving Saturation

The most common means whereby the relative humidity approaches saturation and clouds form is through cooling of the air. Note that (1) the saturation concentration of water vapor varies *directly* with temperature, and (2) the relative humidity varies *inversely* with the saturation concentration of water vapor. This means that as the temperature drops, the relative humidity increases, and vice versa. As figure 13.5 shows, the relative humidity is usually highest during the coolest time of day (around sunrise) and lowest during the warmest time of day (early afternoon). With sufficient cooling, the relative humidity increases to 100%.

One way that air cools is through expansion. In fact, expansional cooling is the most important means of cloud formation in the atmosphere. **Expansional cooling** occurs when the pressure on a parcel of air (e.g., a cubic centimeter) drops, as it does when air ascends within the atmosphere. We can think of air pressure as the weight of a column of air that stretches to the top of the atmosphere. Air pressure is greatest at sea level and drops rapidly with increasing altitude. Thus, as a parcel of air ascends, the pressure acting on it becomes less and less. Ascending air expands in the same way that a helium-filled balloon expands (and eventually bursts), as it drifts skyward.

If you have ever released air from a tire, you know that the escaping air feels relatively cool. Inside a tire, the air is under great pressure, and as it escapes through the valve and out into the atmosphere (where the pressure is less), it expands and cools. Conversely, when air is compressed, its temperature rises, and its relative humidity drops. Thus **compressional warming** occurs when air descends within the atmosphere.

In summary, rising air cools, increasing the relative humidity. With sufficient expansional cooling, the relative humidity approaches 100%, and condensation or deposition takes place—that is, clouds form. On the other hand, de-

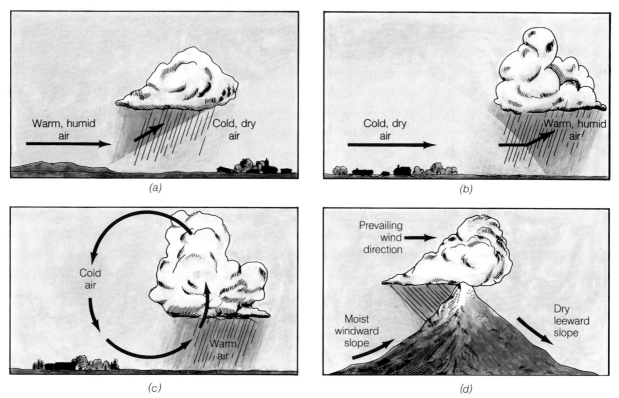

(a) *(b)*

(c) *(d)*

Figure 13.6   When air ascends within the atmosphere, it cools, and the relative humidity increases. With sufficient cooling, the relative humidity approaches saturation, and clouds form. Air ascends (*a*) along a warm front, (*b*) along a cold front, (*c*) in a convection current, and (*d*) on the windward slope of a mountain.

scending air warms, decreasing the relative humidity, so that clouds either vaporize or fail to form in the first place. What, then, causes air to rise? Air rises (1) along the surface of a warm or cold front, (2) as the ascending branch of a convection current, or (3) up the slopes of a hill or mountain.

Clouds and often precipitation are triggered by frontal uplift, which occurs when masses of warm and cold air meet. A front is a narrow zone of transition between two masses of air that differ in temperature or humidity. Warm air is less dense than cold air, thus, it displaces cold air by riding up and over it (fig. 13.6*a*). The warm air advances, and the cold air retreats. The leading edge of the advancing warm air is known as a **warm front.** On the other hand, cold air displaces warm air by sliding under it and forcing it upward (fig. 13.6*b*). The leading edge of the advancing cold air is known as a **cold front.** Along both warm and cold fronts, air rises, expands and cools, the relative humidity approaches 100%, and clouds form.

A **convection current** develops when the sun heats the earth's surface, which then heats the overlying air. The heated air rises, expands, and cools (fig. 13.6*c*). Eventually, the air becomes so cool and dense that it sinks back toward the surface, where it is heated again. Clouds may form where convection currents ascend, and the sky is cloud-free where convection currents descend (fig. 13.7). In general, the higher the altitude reached by convection currents, the greater the expansional cooling, and the more likely it is that clouds and precipitation will form. Convection currents that soar to great altitudes typically spawn thunderstorms (fig. 13.8).

Figure 13.7   The sky is partially covered by cumulus clouds. Where air ascends, clouds develop. Where the air descends, the sky is cloud-free.

Orographic lifting occurs when air is forced upward by topography, the physical relief of the land (fig. 13.6*d*). As winds sweep across the land, hills and valleys force the moving air to alternately ascend and descend. If relief is sufficiently great, the resulting expansional cooling and compressional warming of the air affects the distribution of clouds and precipitation. For example, a mountain range that intercepts a prevailing flow of humid air is cloudier and wetter on its windward slopes (facing the wind) than on its leeward slopes (downwind side). As air is forced to rise along the windward

Figure 13.8 Where convection currents surge to great altitudes, cumulus clouds billow upward and form thunderstorm (cumulonimbus) clouds.

(a)

(b)

Figure 13.9 These two scenes from the Pacific Northwest illustrate the contrast in vegetation between (a) the wet, windward slopes of a mountain range (temperate rain forest in the Olympic National Park, Washington) and (b) the dry, rainshadow region to the leeward side of the mountain range (grassland-scrub habitat along the Columbia River in Lincoln County, Washington).
(a) © Zig Leszczynski/Earth Scenes (b) © Jack Wilburn/Earth Scenes

slopes, it expands and cools, which increases its relative humidity. With sufficient cooling, clouds and precipitation develop. Meanwhile, on the mountain's leeward slopes, air descends and warms, which reduces the relative humidity, so that cloud formation and precipitation are less likely. In this way, a mountain range induces two contrasting climatic zones: a moist climate on the windward slopes and a dry climate (or rainshadow) on the leeward slopes.

The orographically induced contrast in precipitation is especially apparent from west to east across Washington and Oregon, where the north-south Cascade Range intercepts the prevailing flow of humid air from the Pacific Ocean. Exceptionally rainy conditions exist in the western portions of those states, and semiarid conditions in the eastern portions (see fig. 13.9). On Mount Waialeale, Hawaii, the contrast is spectacular. Annual rainfall varies from 11,700 millimeters (460 inches, the heaviest in the world) on the windward slopes of the mountain to only 460 millimeters (19 inches) on the leeward slopes.

In every case, markedly different plant and animal communities live on the windward versus leeward slopes of a mountain barrier. This climatic contrast also affects the domestic water supply, the crops that can be grown, and the type of human shelter required. For example, Denver and nearby communities on the leeward side of the Colorado Rocky Mountains must import water from the wetter, western side of the mountain range via a 21-kilometer (13-mile) tunnel cut beneath Rocky Mountain National Park.

Cloud and Precipitation Processes

Let us consider further the process of cloud formation. In their myriad forms, **clouds** are visible manifestations of condensation and deposition of water vapor in the atmosphere. They are composed of tiny water droplets, ice crystals, or both.

Condensation and deposition take place on tiny solid and liquid particles that are suspended in the atmosphere. These particles, called **condensation nuclei,** are the products of both natural and human activity. For example, forest fires, volcanic eruptions, wind erosion of soil, salt-water spray, and effluents from domestic and industrial chimneys are continual sources of nuclei.

Some condensation nuclei are **hygroscopic nuclei;** that is, they have a strong affinity (attraction) for water molecules. Condensation (or deposition) begins on those nuclei when the relative humidity is less than 100%. Many nuclei produced by industrial activity are hygroscopic and, as we will see in chapter 16, an abundance of hygroscopic nuclei (such as might occur downwind of an urban-industrial area) may locally enhance cloud and precipitation development.

Once clouds develop, however, there is no guarantee that it will rain or snow. In fact, most clouds—even those associated with powerful storm systems—do not yield any precipitation. A special combination of circumstances, not yet completely understood, is required for clouds to yield precipitation. Cloud particles (droplets and ice crystals) are so small that they remain suspended indefinitely within the atmosphere

Precipitation Processes and Rainmaking

Ultimately, virtually all of the fresh water on the planet originates as rain and snow that falls from clouds. However, the particles that make up clouds (water droplets and ice crystals) are too small to survive a journey to the earth's surface; they would vaporize in the unsaturated air below a cloud. In clouds that yield rain or snow, a growth process must take place by which cloud particles become large enough to reach the ground as raindrops or snowflakes. In understanding how precipitation forms, scientists have come up with techniques to stimulate its formation in clouds.

Most precipitation that falls in middle and high latitudes originates through the *Bergeron process,* named for the Scandinavian meteorologist Tor Bergeron, who first described it in 1933. The Bergeron process takes place in *cold clouds,* that is, clouds at temperatures below the freezing point of water. In such clouds, the Bergeron process requires the coexistence of water vapor, ice crystals, and supercooled liquid water droplets, in which the water remains liquid even though the temperature is subfreezing—possible within clouds even down to $-38°$ C ($-37°$ F).

Within cold clouds that are most likely to yield precipitation, supercooled water droplets initially far outnumber ice crystals. Very quickly, however, ice crystals grow at the expense of supercooled water droplets, primarily because there is a greater saturation concentration (of water vapor) surrounding water droplets than surrounding ice crystals.

At subfreezing temperatures, water molecules vaporize more readily from liquid water droplets than from ice crystals because water molecules are bonded more strongly in the solid phase than in the liquid phase. The saturation water vapor concentration is therefore greater over water than over ice. Hence, in a cloud composed of a mixture of ice crystals and supercooled water droplets, a relative humidity that is saturated (100%) for water droplets is actually supersaturated (higher than 100%) for ice crystals. Consequently, water molecules migrate (diffuse) from water droplets to ice crystals; that is, water droplets vaporize, and ice crystals grow.

As ice crystals grow larger and heavier, they fall faster and collide and coalesce with smaller ice crystals and supercooled water droplets in their path and grow still larger. Eventually, ice crystals become so large that they fall out of the cloud. If air temperatures are subfreezing at least most of the way to the ground, the crystals reach the earth's surface in the form of snowflakes. If the air below the cloud is above freezing, the snowflakes melt and fall as raindrops.

The objective of most cloud-seeding experiments is to stimulate the natural Bergeron process. Usually the cold clouds selected for seeding contain supercooled water droplets but are deficient in ice crystals. Hence, they are seeded with either silver iodide (AgI), a substance with crystal properties similar to ice, or dry ice pellets, frozen carbon dioxide (CO_2) at a temperature of about $-80°$ C ($-110°$ F). Silver iodide crystals act like ice crystals and grow at the expense of supercooled water droplets. On the other hand, dry ice pellets are cold enough to cause surrounding supercooled water droplets to freeze, and the frozen droplets then function as nuclei that grow into snowflakes. Cloud seeding is done from aircraft, either by dropping flares that disperse silver iodide crystals or by dispensing dry ice pellets from a hopper.

---

until they vaporize. They must somehow grow if they are to fall as raindrops or snowflakes. This is unlikely because about a million cloud droplets are required to form a single raindrop. We examine how cloud droplets and ice crystals grow into raindrops and snowflakes in Special Topic 13.1.

The Global Water Budget

The return of water from the atmosphere to the land and sea in the form of precipitation completes an essential subcycle in the global water cycle. This is a dynamic process; the average time that a water molecule remains in the atmosphere is only about ten days. To learn more about the water cycle, we compare the flow of water as it moves into and out of terrestrial reservoirs with the flow of water as it moves into and out of the sea. The balance sheet for the total inputs and outputs of water to and from the various global reservoirs is called the global water budget (table 13.2).

Table 13.2 Global water budget

Source	Volume in m^3 per Year
precipitation at sea	$+3.24 \times 10^{14}$
evaporation from sea	-3.60×10^{14}
net loss from sea	-0.36×10^{14}
precipitation on land	$+0.98 \times 10^{14}$
evapotranspiration from land	-0.62×10^{14}
net gain on land	$+0.36 \times 10^{14}$

Each year on land, the total amount of precipitation exceeds evapotranspiration by about one-third. At sea, however, annual evaporation exceeds precipitation. Hence, over the course of a year, the global water budget indicates a net gain of water on land and a net loss of water from the oceans, with the excess on land approximately equal to the deficit for the oceans. The land, however, is not getting soggier, nor are the world's oceans drying up, because the excess precipitation on land drips, seeps, and flows back to the sea.

The net flow of water from land to sea has important implications for the global distribution of water pollutants. The flow of water from land to sea means that the ultimate destination of contaminants dumped into waterways on land is the ocean. Today, considerable debate centers on the capacity of the ocean to assimilate waste—especially toxic and hazardous waste. We have more to say about ocean pollution in chapter 20.

Once precipitation reaches the land, it follows one or more routes. Some water vaporizes, and the remainder either seeps into the ground or runs off as rivers and streams to the sea. These latter pathways are respectively labeled the **infiltration** and **runoff** components of the water cycle. The ratio of the portion of water that infiltrates the ground to the portion that runs off depends on the intensity of rainfall and on the vegetation, topography, and physical properties of the land surface. Heavy rain that falls on frozen soil or city streets will mostly run off; on the other hand, light rain falling on a lawn or forest will mostly seep into the ground.

Water that infiltrates the ground replenishes soil moisture, some of which is taken up by plant roots and then mostly (98%) transpired to the atmosphere. Water that seeps to greater depths may fill the spaces between particles of soil or sediment or the fractures within rock, where it is known as **groundwater.** Groundwater flows very slowly in the subsurface toward points of discharge, including wells, springs, rivers, lakes, and the ocean.

On its journey from land to sea, runoff is stored for varying periods in one or more terrestrial reservoirs (rivers, streams, lakes, snowfields, or glacial ice). Those reservoirs along with groundwater are potential sources of fresh water for agricultural, industrial, and domestic uses. About 78% of U.S. freshwater supply comes from surface runoff, while about 22% is withdrawn from the ground. States that are most dependent on surface reservoirs for their water are Montana (98% of the total), South Carolina (97%), Alabama (96%), and West Virginia (96%). The states that are most dependent on groundwater are Kansas (85%), Mississippi (68%), Florida (64%), and Arkansas (64%). For the nation as a whole, the greatest demands for fresh water are for agriculture (42% of the total) and electric power generation (39%). Domestic and commercial purposes use 10% of the supply; industry and mining use about 9%.

So that we may better understand how contamination and overexploitation affect freshwater resources, we examine the physical characteristics of groundwater, rivers, and lakes in the following sections.

Groundwater

Although most U.S. residents draw their fresh water from surface reservoirs, globally, the volume of groundwater is about thirty-five times that of fresh water in lakes, rivers, and streams. Water that occurs within 750 meters (2,250 feet) of the earth's

Table 13.3 A consumer's guide to bottled water

Term Used for Water	Source
natural water	protected spring or well, as opposed to water from a municipal system or public source
spring water	underground formations from which water flows naturally to the earth's surface
well water	holes bored, drilled, or otherwise constructed in the ground, which tap aquifers
natural sparkling water	protected springs or wells that naturally contain carbon dioxide, making it bubbly
mineral water	government-approved and regulated underground source or natural spring; mineral content is not changed by the manufacturer
sparkling water	another name for carbonated water
seltzer	generally, tap water that has been filtered and carbonated
club soda	generally, tap water that has been filtered and carbonated with minerals and mineral salts added

Reprinted by permission from *The Christian Science Monitor,* © 1991 The Christian Science Publishing Society. All rights reserved.

surface is usually considered economically recoverable; in the United States its total volume is more than nine times that of the Great Lakes.

The general perception that water withdrawn from the ground is cleaner than surface waters has spurred consumer interest in bottled water, but not all bottled water is groundwater, as table 13.3 shows. Furthermore, in some regions human activities threaten both the quantity and the quality of subsurface water. Because the quality of much surface water has deteriorated, it is all the more important that we seek to maintain the quality of groundwater. In this section, we examine the physical properties of the groundwater reservoir so that we may better understand its vulnerability to pollution.

Water that infiltrates the ground occupies the tiny open spaces (pores) between soil particles or sediments or the fissures in bedrock. Downward-infiltrating water typically accumulates in two horizontal zones: the zones of aeration and saturation (fig. 13.10). In the upper one, the **zone of aeration,** pore spaces contain mostly air and some water. This zone includes in its upper reaches the belt of soil moisture that is tapped by plant roots. In the lower one, the **zone of saturation,** pore spaces and rock fractures are completely filled with water. This zone constitutes the groundwater reservoir. Wells must penetrate the zone of saturation.

The interface between the zones of aeration and saturation is the **water table.** As shown in figure 13.11, the water table roughly parallels the overlying topography; it rises under hills and sinks under valleys. In addition, the depth of the water table fluctuates in response to prevailing weather condi-

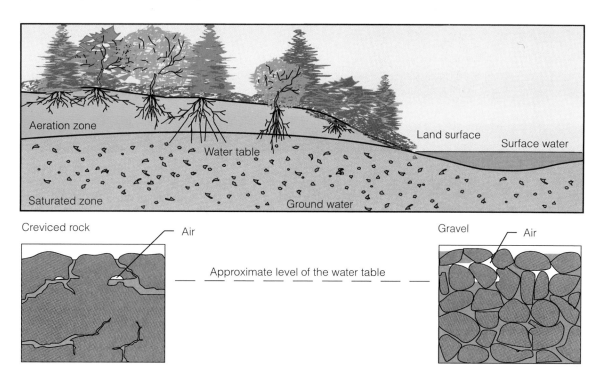

Figure 13.10 Downward-seeping water accumulates in the zones of aeration (where openings are mostly filled with air) and saturation (where openings are saturated with water).

Source: Adapted from *Ground Water and the Rural Homeowner*, U.S. Geological Survey, U.S. Department of the Interior.

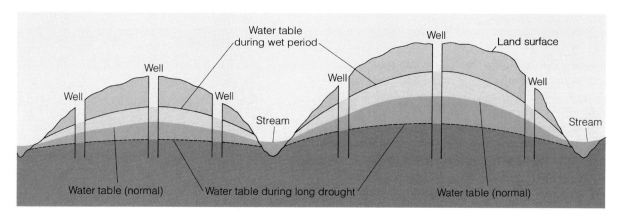

Figure 13.11 The elevation of the water table varies. Note that it tends to follow topography and that it rises or falls during extended wet or dry periods.

tions, dropping during dry periods and rising during moist ones. During droughts, shallow wells frequently go dry, and people must drill deeper wells. In most of the United States, the water table is less than 30 meters (100 feet) below the surface, but in some arid localities, wells must be several hundred meters deep before they intercept the water table.

The property of soil, sediment, and rock that enables these materials to transmit water within the zones of aeration and saturation is known as **permeability.** The degree of permeability depends upon the total volume of pore space, the **porosity,** and how well the individual pore spaces are interconnected. A layer of sediment or rock that is highly permeable and contains and transmits water is known

as an **aquifer** (from the Latin for water carrier). In general, layers of sand and gravel are good aquifers, whereas clay and most crystalline rocks such as granite (unless it is highly fractured), have low permeability and are not likely to yield much groundwater.

We have been describing **unconfined aquifers,** which are overlaid by permeable earth materials and recharged by water seeping down from above. Thus, local rainfall and snowmelt supply water to unconfined aquifers. **Confined aquifers,** on the other hand, are sandwiched between impermeable layers of rock or sediment and are recharged only where the aquifer intersects the land surface. In some cases, the recharge area is hundreds of kilometers away from the location of a well or

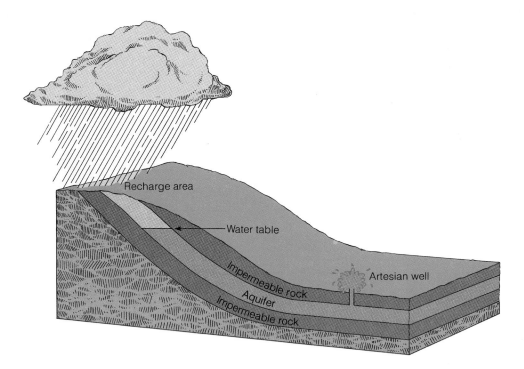

Figure 13.12 Artesian conditions develop when water in a confined aquifer is under pressure. When a well taps the aquifer, water flows to or even above the land surface.

other point of discharge. The water supply in a confined aquifer may therefore depend on weather conditions at a distant location that may be quite unlike local weather conditions.

Groundwater seeps downward under the influence of gravity. Hence, where a confined aquifer is inclined, the water is under pressure. If a well is drilled into such an aquifer, the water rises and may burst above the land surface (fig. 13.12). Such a free-flowing groundwater system is known as an **artesian well.** A major artesian aquifer dips southeastward under Florida from a broad recharge area in the northwestern quarter of the peninsula. Near Miami, wells must be drilled at least 250 meters (800 feet) to intercept the aquifer.

Flow Characteristics

Groundwater flows very slowly through aquifers from recharge areas to discharge areas such as wells, rivers, marshes, lakes, and seas. Flow is so extremely slow that the velocity is often measured in meters per year; in fact, a flow rate of 15 meters (50 feet) per year is typical. Hence, contrary to some popular opinion, groundwater is far from a surging underground river. Its slow pace is due to the frictional resistance offered by the soil particles and sediment that the water must seep between and around.

Slow movement has important implications for groundwater quality. Because it is slow, the flow is also nonturbulent; that is, the water describes smooth, gently curving streamlines toward points of discharge (fig. 13.13). Nonturbulent flow means that the mixing of contaminants with cleaner water

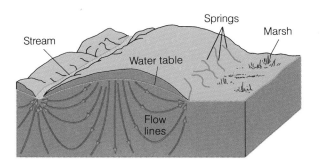

Figure 13.13 Groundwater flows along smooth streamlines through permeable soil, rock, or sediment from the recharge area to points of discharge, such as streams and springs.

(dilution) is very slow. In contrast, turbulent surface waters rapidly dilute pollutants. The slow rate of dilution in groundwater means that high pollutant concentrations can persist for many years.

A second and closely related result of the slow movement of groundwater is the relatively long residence time of contaminants. Once a contaminant enters an aquifer, it usually takes a long time for it to be flushed from the system. The U.S. Geological Survey reports that pollutants may remain in some aquifers for hundreds of years, fouling the water supply for generations to come. We consider other aspects of groundwater pollution again in chapter 14.

The typically slow and smooth flow of water in the subsurface also affects the pattern of withdrawal. When water is pumped out of a well, the vacated volume forms a **cone of depression** in the water table (fig. 13.14.) If wells are too

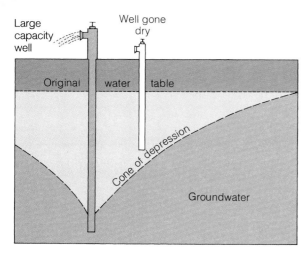

Figure 13.14 Pumping by a large, high-capacity well depresses the water table into a cone of depression so that nearby shallow wells run dry.

close together, heavy withdrawal from a deep well forms a cone of depression that may cause neighboring shallow wells to run dry.

Groundwater Mining

Aquifers are depleted if the rate of withdrawal exceeds the natural rate of recharge. This is known as **groundwater mining,** and it can cause coastal aquifers to turn salty and result in property damage due to ground subsidence.

In coastal areas where fresh groundwater is adjacent to salty groundwater, excessive withdrawal of fresh groundwater can allow salt water to migrate inland (fig. 13.15). As salt water replaces fresh water in the aquifer, wells begin to deliver salt water. This phenomenon, called **salt-water intrusion,** is most common in flat coastal plains like those of Florida and southeastern Georgia.

Where groundwater withdrawal exceeds recharge, the sediment that composes aquifers compacts and **ground subsidence** may occur as the overlying land surface adjusts to aquifer compaction by sinking. Millions of dollars in property damage can result from ground subsidence, which often causes structural damage to buildings, fractures utility pipes, reverses the direction of flow of sewers and canals, and increases tidal flooding along seacoasts. In late 1991, researchers at the University of Arizona reported that groundwater mining is causing the Tucson basin aquifer to compact and the ground surface to subside. The aquifer underlies the city of Tucson and its surroundings. Average subsidence rates of the ground surface currently vary between 1 centimeter and almost 5 centimeters per year, and in thirty years could exceed 1 meter in places.

In addition to causing salt-water intrusion and ground subsidence, groundwater mining affects the future of agriculture. In many arid and semiarid regions of the world, modern agriculture taps aquifers to irrigate millions of hectares of once powder-dry, but fertile soils. The benefit is lush,

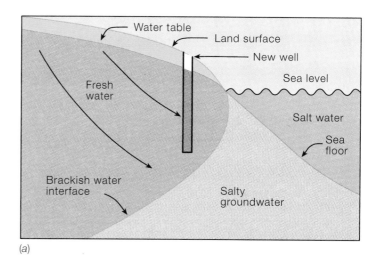

Figure 13.15 Cross-sectional diagrams of the position of the fresh-water/salt-water interface typical along flat coastal regions. (*a*) A new, unpumped system and (*b*) the same system after overwithdrawal.

highly productive cropland. The problem is that in such climates, irrigation usually removes water from the ground at a rate that far exceeds the natural rate of recharge. Groundwater mining is currently depleting the extensive High Plains aquifer and threatening the region's future economic viability. See the issue on pages 272–273.

Runoff: Rivers and Streams

Water that does not evaporate or infiltrate the ground constitutes the runoff component of the water cycle. Rainfall and snowmelt enter stream and river channels by overland flow. Precipitation that falls directly into the stream and groundwater seepage also contribute significantly to stream flow. Rivers and their tributaries drain water from a fixed geographical region called a **drainage basin** or **watershed.** Climate, vegetation, topography, and various activities, both natural and human, in the drainage basin affect the quantity and quality of runoff water (and groundwater).

Groundwater Mining and the High Plains Aquifer

The High Plains aquifer is the largest in the United States, supplying about 30% of the nation's irrigation water. It underlies portions of eight states from West Texas and eastern New Mexico north to South Dakota (fig. 13.16). The aquifer consists of sandy and gravelly rock layers averaging 65 meters (210 feet) thick at depths of less than 350 meters (1040 feet). The total water contained in the aquifer would flood the fifty states of the United States to an average depth of about 0.5 meters (1.5 feet).

Pumping of the High Plains aquifer for irrigation allows intensive agricultural cultivation that the region's usual rainfall could not sustain. The region is in the rainshadow of the Rocky Mountains, and receives an average annual precipitation of 40-70 centimeters (16-28 inches). Irrigation has increased crop yields by 600-800% over dry-land farming. In fact, the aquifer supports nearly half of the nation's cattle industry, one-quarter of the cotton crop, and much of its corn and wheat.

Exploitation of the High Plains aquifer began after the devastating drought of the 1930s, but really soared in the decades following World War II with installation of high capacity pumps. For example, the number of wells in Texas jumped from 8,356 in 1948 to 42,225 by 1957. In southwest Nebraska, the number of irrigated hectares grew nearly eightfold between 1953 and 1983. Today, an estimated 170,000 wells tap the High Plains aquifer, about one for every 260 hectares (640 acres).

Throughout the western plains, the rate of withdrawal for irrigation far exceeds the natural rate of recharge, and the aquifer is slowly being depleted. In portions of Texas, infiltration rates are estimated at only 0.6-1.5 millimeters per year. Water application rates in that area, however, range from 15 centimeters (6 inches) per year for wheat to 46 centimeters (18 inches) per year for corn. Between 1950 and 1980, the annual volume of water pumped from the High Plains aquifer tripled.

Since the 1930s, an estimated 11% of the aquifer has been pumped out, and this is expected to increase to 25% by 2020. In some localities, depletion rates have been much greater. By 1980, 38% of the water under Kansas had been withdrawn, and in Texas the water table had dropped by as much as 62 meters (200 feet).

Recent economic pressures have forced farmers who rely on the High Plains aquifer to place greater emphasis on water conservation. Rising energy prices beginning in the 1970s meant higher pumping costs. Where the water table is dropping, water must be lifted a greater distance, which increases the cost of irrigation. Some farmers have abandoned their former practice of spraying water into the air in favor of using nozzles that inject water directly into planted furrows thus reducing the loss of water by evaporation. In addition, some farmers have excavated runoff pits that capture and recycle irrigation water. Switching from water-thirsty crops (e.g., alfalfa and corn) to less profitable dry-land crops (e.g., cotton) also saves water.

Although conservation efforts have eased the problem somewhat in recent years, regions that continue to mine water will have to make some painful economic adjustments. A Texas Agricultural Experiment Station study of the Lubbock area predicts that by the year 2015, irrigated land area will have to be reduced by 95%. Between now and then, agricultural productivity is expected to decline by 70%, and other businesses in the area will therefore suffer. On the other hand, in Nebraska, where two-thirds of the total water in the aquifer are located, a crisis in supply will be delayed somewhat. In any event, as growing populations continue to make demands on groundwater reservoirs, problems of groundwater mining will force communities to seek other sources of water, to conserve water, or both.

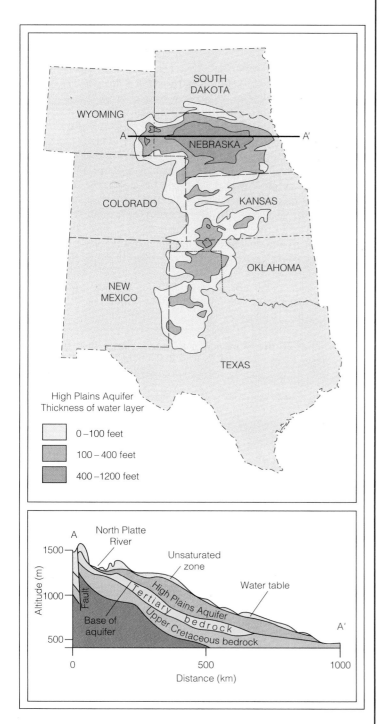

Figure 13.16 The High Plains aquifer is a source of water for irrigation in the area shown. Groundwater mining, however, is depleting this resource.

Gravity causes river water to flow downhill on its seaward journey. Where the gradient of the channel is steep, water flow is rapid and usually turbulent; where the channel slope levels off, the flow rate and turbulence slacken. Along a river's course, a portion of the available energy of motion (kinetic energy) is used in eroding and shaping the channel. Some energy is also used to transport sediment that is loosened from the channel along with other substances dumped (e.g., industrial waste) or washed (e.g., eroded soil) into the river.

Materials that are dumped or washed into a river from the land, together with the sediment that is eroded from the channel, constitute the river's *load*. A river transports its load in suspension, solution, and along the riverbed by the pushing and rolling action of the current. The types of substances that are dissolved in a river vary with climate, bedrock, and soil types, and with agricultural and industrial activities within the drainage basin. Soil erosion contributes suspended silt and clay particles and bits of detritus, along with dissolved substances, including plant nutrients and pesticides. Groundwater contributes a significant amount of the dissolved components of river-water; rain and snow supply a relatively small amount by washing pollutants from the air. Some rivers carry an added burden in the form of waste materials such as sewage and mine tailings (waste rock).

Because a river's ability to transport substances depends not only on the rate of water flow, but also on the quantity of water involved, stream discharge is the best measure of the transporting ability of a river. **Stream discharge** is defined as the volume of water that passes a fixed point along a river's course in a unit of time, usually expressed in cubic meters (or cubic feet) of water per second. Where and when discharge is great (e.g., in the upper Midwest during spring snowmelt), channel erosion accelerates, and the river can transport more and larger materials. When river discharge drops, however, as when water is diverted for irrigation, the river's ability to transport materials decreases, and suspended sediments settle out.

Weather in the drainage basin is the chief control of stream discharge. Discharge usually changes with the rainy and dry seasons. In some places it may even fluctuate significantly within a single day. Rapid discharge oscillations are typical of the arroyos of the American Southwest. Arroyos are dry streambeds (zero discharge) until an afternoon cloudburst over the watershed triggers a dramatic but short-lived gully-washing discharge. In contrast, rivers in some regions flow out of heavily watered mountains only to completely dry up in the desert plains beyond—never reaching the ocean directly.

Rivers are the lifeblood of many cities, and in rural areas they supply irrigation water essential for agriculture. Egypt, for example, is said to be a gift of the Nile River. Its 55 million citizens depend almost entirely on the Nile for their water supply. Just about all of Egypt's cropland must be irrigated. Likewise, the economy of the American Southwest is largely based on water diverted from the Colorado River. In areas

Water Laws

People have long recognized the benefits of having a reliable supply of fresh water, especially in arid or semiarid regions. Legal restrictions, however, regulate the redistribution of water resources in the United States. Two different types of water laws exist, which are administered by the individual states.

In nearly all states east of the Mississippi River, water use is governed by the *riparian* (bank or shoreline) *doctrine,* whereby the owner of land that adjoins a stream is entitled to receive the natural flow of that stream, undiminished in its quantity and quality. Riparian rights are transferred when the land is sold. During times of drought, all riparian owners have equal rights to reasonable use of water, although a priority use system may be imposed (domestic uses usually receive highest priority). The law in some states allows permit holders (riparian owners) to use water only under prescribed conditions. For example, if stream flow is diminished during dry periods, the permit holder may not be allowed to use any water. Also, the transfer of water from one watershed to another is prohibited.

In contrast, most western States are governed by *appropriation water laws.* Under these laws, only the owners of water rights may divert water from a watercourse, and then only for purposes that are deemed to be beneficial. Also, ownership of water rights is not tied to ownership of land that adjoins the watercourse. These laws came about after irrigation was introduced to arid and semiarid regions of the western United States. They were instituted to prevent anyone from buying land upstream and diverting the limited water supply away from dependent downstream users. Today, the law enables early water-rights holders to get their full share of water during a drought, while owners of more recently acquired water rights may be forced to do without. In addition, various treaties and compacts affect the quantity of water that various states can withdraw from the rivers of a particular region.

Laws that govern the use of groundwater distinguish between two types of underground water resources: subterranean streams (water in the zone of saturation) and percolating water. Subterranean streams are generally subject to the same water laws as surface water. Percolating water is owned by landowners, who are allowed to withdraw enough of it for their own use, as long as this does not infringe on the water rights of others. Furthermore, landowners may not pollute groundwater.

---

where water is limited, it is prudent to control community growth and avoid overexploitation of the natural discharge of rivers. Community growth control is not happening, however, in the Southwest. Cities are therefore forced to buy up farmland merely to obtain water rights. This increasingly popular practice of water-ranching is expected to bring an end to irrigated agriculture by 2020 in Pima County, Arizona, home of Tucson.

How do we arrive at a measure of the reliable river water supply? During extended dry periods when runoff into river channels is near zero, the river discharge originates from groundwater and water that is slowly released from wetlands. This so-called **base discharge** is considered to be the most dependable water supply. Many water management specialists suggest that the wisest water-use policy is to match our demands for river water with the base discharge. Still, the lower the average rainfall of a region, the greater the likelihood of recurrent droughts. In relatively dry areas such as the American Southwest, prolonged drought can cause river discharge to drop below base level, requiring strict conservation of the available water supply. For a description of laws regulating water use, refer to Special Topic 13.2.

Lakes

Lakes are temporary impoundments of water that occupy depressions in the landscape and receive water from runoff or groundwater or both. A variety of geological processes may be responsible for forming a lake basin: earthquakes (e.g., Lake Nyasa, East Africa), volcanic collapse (e.g., Crater Lake, Oregon), and glaciation (e.g., the Great Lakes). Water often also accumulates in abandoned rock quarries and other surface-mining operations. Furthermore, we create artificial lakes when we dam rivers and streams.

Although lakes exist in many different climates and contain waters that range widely in depth, temperature, salinity, and acidity, they are all relatively short-lived in terms of geologic time. Lakes gradually age as they fill with sediments that are washed from the watershed as well as undecayed plant and animal matter (detritus) produced within the lake. Portions of the shrinking lake are invaded by aquatic rooted plants such as cattails and bullrushes (fig. 13.17). Eventually the entire lake is covered by plants. This process of gradual change whereby a lake becomes a marsh or swamp is another example of ecological succession (chapter 4).

Figure 13.17 In time, a lake fills with inorganic sediments and partially decayed organic matter and is gradually transformed into a marsh or swamp. This is an example of succession.
© Dorothea Mooshake/Peter Arnold, Inc.

Lakes that are fed by springs or groundwater seepage may disappear if the regional water table drops. A climatic shift that produces greater evaporation, decreased rainfall, or both may reduce a lake's water volume or even cause it to dry up. About ten thousand years ago, a major climatic change from moist and cool to arid and warm conditions led to the disappearance of more than one hundred large lakes in the western United States, primarily California, Nevada, and Utah. Remnants of some of those lakes exist today; Great Salt Lake in Utah is what remains of the much more extensive prehistoric Lake Bonneville.

Enhancing Freshwater Supplies

So far we have examined some of the physical characteristics of our chief freshwater reservoirs: groundwater, rivers, and lakes. Confronted with increasing demands on a finite supply, human ingenuity has devised several strategies to increase the availability of fresh water, including desalination, cloud seeding, dams, watershed transfers, and conservation.

Desalination

Two widely separated events during the early 1990s, the Persian Gulf War and drought in California, heightened public awareness of desalination. **Desalination** is the production of potable water by treating saline water to remove or reduce its salts. During the Gulf War, large oil slicks threatened to disrupt desalination facilities that supply most of the fresh water for Saudi Arabia and other Gulf States. In California, years of persistent drought have forced some coastal communities to study the feasibility of desalination.

The state's first desalination plant opened in October 1990 on San Nicolas Island, off the coast of southern California. In March 1992, Santa Barbara opened a new $25 million desalination facility that is expected to meet one-third of the city's water needs.

Desalination involves either of two basic technologies: reverse osmosis or distillation. Most commercial desalination plants built or proposed for California will use **reverse osmosis,** whereby water under pressure is forced through a thin plastic membrane, leaving behind dissolved substances. Most Saudi Arabian desalination plants employ distillation, whereby high-temperature, high-pressure steam condenses into fresh water. The steam is recycled from electric power plants, where it drives turbines.

The principal drawback of desalination is the high cost of using technologies that are energy-intensive (distillation more so than reverse osmosis). In California, for example, desalinated water is expected to be at least twelve times more expensive than groundwater. It is not surprising, then, that desalination is most common in portions of the Middle East where energy is both abundant and inexpensive. Worldwide, about four thousand desalination plants produce about 13 billion liters (3.4 billion gallons) of fresh water daily. About 60% of this production takes place on the Saudi Arabian peninsula, but even there, costs are too high for agricultural applications of desalinated water.

Cloud Seeding

Since World War II, much research has gone into **cloud seeding,** an attempt to stimulate natural precipitation processes by injecting nucleating agents into clouds. Silver iodide crystals or dry ice pellets are the usual nucleating agents (see Special Topic 13.1). An example of a long-term cloud-seeding project has taken place on the windward western slopes of California's Sierra Nevada mountains. During winter, scientists seed clouds with the goal of enhancing snowfall and thereby thickening the mountain snowpack. The consequent increase in spring runoff is intended to help California meet its increasing agricultural and domestic water demands.

Although cloud seeding is probably successful in some instances, the actual volume of additional precipitation produced by cloud seeding and the advisability of large-scale seeding efforts are matters of considerable controversy. Some cloud seeders claim to increase precipitation by 15-20% or more, but the question always remains: Would the rain or snow that follows cloud seeding have fallen anyway? Even if apparently successful, cloud seeding on a large geographic scale may merely redistribute a fixed quantity of precipitation, so that an increase in precipitation in one area might mean a compensating reduction in another area. Thus, although rain-making might benefit agriculture on the High Plains of eastern Colorado, it might also deprive wheat farmers of rain in adjacent Kansas and Nebraska. Clearly, cloud seeding poses problems that may be just as complex as the ones that we are attempting to solve. Such conflicts have caused legal wrangles between adjoining counties and states.

Dams

Because precipitation varies with seasons and geography, dams have been built to conserve surface water and regulate its availability. Reservoirs that form behind dams collect water during wet periods and store it for use during dry periods. In many locations around the world, dams provide water essential for industrial, domestic, and agricultural use. Irrigation, for example, allows more land to be farmed, increases per hectare crop yields, and often enables farmers to grow more crops per year on the same land. Because dams reduce or eliminate variations in water supply, they stabilize the agricultural economy, especially in drought-prone regions.

Dams have other important functions as well. In mountainous areas, deep river valleys are dammed to create huge reservoirs for generating hydroelectric power (fig. 13.18). Many dams are built for flood-control; they make the fertile floodplains of major rivers habitable and reduce flood damage to crops, dwellings, roads, and businesses. Reservoirs also provide recreational opportunities, such as boating, swimming, and fishing.

The benefits of dams, however, must be weighed against the costs:

1. By interrupting the natural flow of a river, dams trap nutrients and sediments in their reservoirs. Nutrients accelerate the growth of algae and other aquatic plants and thereby reduce water quality. Sediments gradually fill the reservoir, so that it eventually loses its basic function (fig. 13.19).
2. Dams interfere with the migration and spawning activities of fish such as salmon (chapter 11).
3. Dams can disrupt estuarine ecosystems. Dams across rivers that flow into estuaries reduce the flow of fresh water that dilutes the sea water in an estuary. The increased salinity of estuarine waters can prevent the reproduction of some types of fish and shellfish. Dams also reduce the transport of nutrients into estuaries, limiting their productivity.
4. Flooding of land behind dams can destroy vast areas of valuable agricultural land, wildlife habitat, and scenic beauty.
5. Storage of water behind dams raises the water table, often waterlogging the soils of surrounding lands and reducing crop or commodity productivity.
6. Faulty construction, excessive rainfall, or earthquakes can cause dams to fail, and the floodwaters from a breached dam can take a terrible toll in lives and property (chapter 21).

Proposals for new dam projects are almost always controversial, some groups benefit much more from construction than others. Groups that support dam construction cite the economic benefits derived from increased water supply. Recreation developers also support dam construction. Opponents of such projects are typically naturalists, white-water boaters, and landowners whose homes and businesses would be inundated by such projects. Controversy is especially heated

Figure 13.18 At a hydroelectric facility, the kinetic energy of flowing water is tapped to drive turbines that generate electricity.
© Brian Parker/Tom Stack & Associates

Figure 13.19 A dam greatly slows the flow of a river, causing sediment to settle out in the reservoir behind the dam. The functional lifetime of a dam-reservoir system depends upon the volume of the reservoir plus the rate of erosion within the drainage basin.
© Francois Gohier/Photo Researchers, Inc.

if a scenic or wild river is proposed as a site for a dam. Such proposals always require an environmental impact statement (chapter 2).

Watershed Transfer

One of the most common techniques for making more fresh water available is watershed transfer, the diversion of water from one watershed to another. Many major U.S. cities, among them Los Angeles, Phoenix, Denver, and New York, rely on other watersheds in addition to their own for fresh water.

No doubt the nation's most exploited watershed is that of the Colorado River, the major source of fresh water in the arid Southwest.* The Colorado winds its way some 2,240 kilometers (1,400 miles) from its headwaters in the snow-capped Colorado Rockies to the Gulf of California in extreme northwest Mexico. By the time it reaches the sea, watershed transfers and evaporation have so depleted the river that its channel is dry.

Some thirty dams and huge reservoirs (including Lake Mead behind Hoover Dam, and Lake Powell behind Glen Canyon Dam) have greatly modified the flow of the river and its tributaries. Aqueducts and irrigation canals siphon

*See the article by Carrier listed at the end of the chapter.

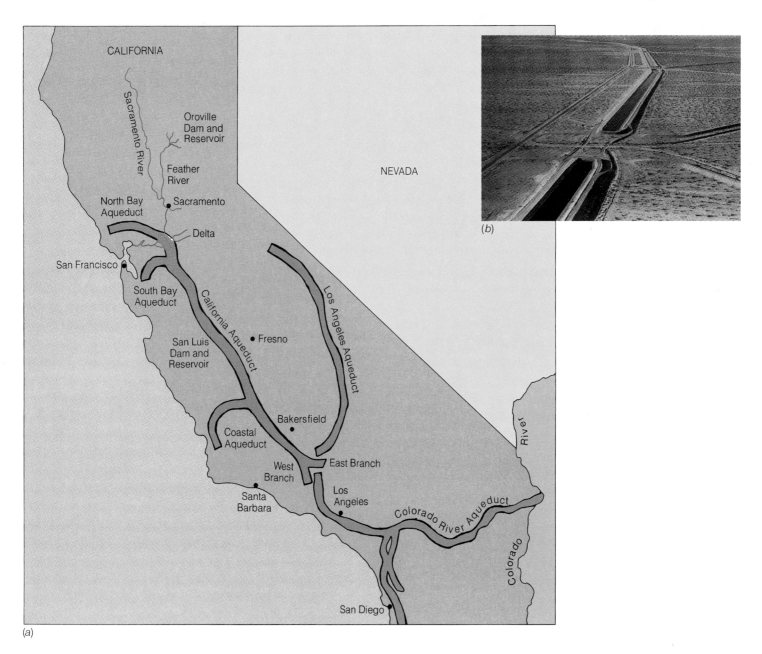

(a)

Figure 13.20 (a) Route of the Colorado River, Los Angeles, and California aqueduct systems. (b) A portion of the Colorado Aqueduct crossing the California desert.

Source: The Metropolitan Water District of Southern California.

off water for use in seven states and northern Mexico. For example, a 714-kilometer (444-mile) aqueduct system transports water from the Colorado River to Los Angeles and the irrigation systems of California's Central and Imperial Valleys (fig. 13.20). Also, the recently completed Central Arizona Project diverts Colorado River water from Lake Havasu (formed by Parker dam) on the Arizona/California border to the cities of Phoenix and Tucson.

As the water travels through the arid Southwest, perhaps 10% evaporates, which increases its salinity. In addition, irrigation water dissolves salts from the soil surface; thus, excess irrigation waters that return to the Colorado are more salty than they were to begin with. To meet drinking water standards, water from the Colorado that enters southern Califor-

nia must be diluted with less-saline water from northern California sources. By the time the river reaches Mexico, its flow is reduced to a mere trickle that is too saline for human consumption and most crops. Through an agreement with Mexico, however, the U.S. government guarantees that Colorado River water entering Mexico is of acceptable quality for both human consumption and irrigation. To fulfill that commitment, the U.S. was forced to construct and operate an expensive desalination facility. The desalinated water is about three times more costly than irrigation water north of the facility at Yuma, Arizona.

Allocation of Colorado River water to user states (and Mexico) is strictly regulated. If all the allocations were actually used, the water demand would exceed the usual river discharge by 18%. Because not all states are now using their full allotment, there are no shortages, but this is expected to change in the future as the population of the Southwest continues its rapid climb. Already, some conflicts over Colorado River water have had to be decided in the courts. In 1985, as the result of a U.S. Supreme Court decision (*California* v. *Arizona*, March 9, 1985), Arizona gained access to 2.8 million acre-feet of water that was previously allocated to California. (One acre-foot is the volume of water that would flood one acre of land to a depth of one foot—about the annual water demand of an urban family of four.) This allocation is part of the Central Arizona Project, mentioned earlier. The diverted water is intended to partially compensate for the region's rapidly depleting groundwater.

Water Conservation

Water conservation is the most practical means of extending available freshwater supply, and it begins at home, especially in the bathroom. Of the 380 liters (100 gallons) of water used each day by an average American, some 150 liters (40 gallons) are used to flush the toilet. A conventional toilet uses 19-26 liters (5-7 gallons) per flush, although new, low-flush toilets use only 4-6 liters (1-1.5 gallons) per flush. We can conserve significant quantities of water by inserting an object in a conventional toilet tank. A closed, weighted-down plastic container (such as a milk jug) or plastic-wrapped brick will do. With each flush, you save the volume of water displaced by the object. Table 13.4 lists other steps that you can take in and around the home.

In areas where voluntary water conservation is insufficient to reduce water consumption, communities have enacted water-use ordinances that include stiff fines. (Recall that this was one of the steps taken during the recent California drought.) In extreme cases, city water departments have installed a flow-restriction device in front of a household water meter. This device reduces the flow of water into the house to a mere trickle, forcing conservation.

Irrigation, which accounts for about 80% of water demand in the western United States, requires judicious conservation. Irrigation represents a *consumptive use* of water; that is, most irrigation water (more than 60%) is lost by evapotranspiration or infiltration and thus is not available for immediate reuse (chapter 8). (In contrast, municipal use of water is considered to be *nonconsumptive*; about 95% can be recovered for immediate reuse.) In our earlier discussion of the High Plains aquifer, we listed some practices that farmers use to conserve water. To this we add the need to monitor soil moisture, so that irrigation is used only when necessary.

Water conservation is also possible in industry through recycling and the introduction of processing that relies on less water. For example, many paper mills now employ a water treatment-use-treatment cycle that dramatically cuts water needs and also helps the facility meet water-quality effluent standards.

Treated wastewater and stormwater runoff are also important sources of fresh water that are usually discharged into

Table 13.4	Ten things you can do to conserve water

1. Turn off the tap while brushing teeth, shaving, or washing dishes.
2. Install low-flow faucet aerators and showerheads (reduces flow by 50% or more).
3. Keep your shower short or your bathwater low. (At a flow rate of 19 liters (5 gallons) per minute, a 5-minute shower saves about 95 liters (25 gallons) over a 10-minute shower.)
4. Run automatic dishwasher only when full and only on the short cycle.
5. Use a clothes washer only with a full load, or else turn down the water-level setting. The short or gentle cycle uses only about half the water of the full cycle.
6. Patronize car washes that recycle water.
7. Water the lawn only when necessary and only before 8 A.M. or after 6 P.M. to reduce evaporation.
8. Mulch outdoor plants to retain moisture.
9. Position sprinklers to avoid watering sidewalk and street.
10. Fix leaky faucets.

Source: Data from J. Getis, *You Can Make a Difference.* Copyright © 1991 Wm. C. Brown Communications, Inc., Dubuque, Iowa.

rivers or the ocean. This water can be conserved by routing it into recharge basins where it infiltrates and recharges the groundwater reservoir. In some low-lying coastal areas, recharge basins help prevent salt-water intrusion.

Conclusions

Although the ocean is by far the largest reservoir of water on earth, excessive salinity makes sea water unusable for domestic, agricultural, and most industrial purposes. The global water cycle, however, processes sea water naturally, and through distillation, makes some of it available as fresh water in lakes, rivers, glaciers, and the ground. All reservoirs of the global water cycle are interrelated, and the functioning of the system provides both fresh water and habitats for freshwater organisms. We must be sensitive to the way this cycling works, and recognize the need to protect the relatively small amount of fresh water it produces. Contaminants that enter the cycle at any point may eventually find their way into all parts of the system. We cannot expect natural cleansing processes to keep pace when we overload those reservoirs with contaminants.

Only desalination actually increases the global supply of fresh water. Cloud seeding and watershed transfer merely redistribute fresh water from one region to another, and dams create only temporary impoundments that do not enhance the total supply. Furthermore, as the world's human population continues to grow, the amount of fresh water available for each person inevitably declines. At present, on a global basis, we use only about 25% of the total freshwater supply, but in specific regions, the supply is very limited and is being stressed by drought and the demands of a rapidly growing population. These realities underscore the need for water conservation measures.

Even as the demand for fresh water continues to rise, we are degrading the quality of many freshwater sources, thereby creating even greater problems of supply and demand. In the next chapter, we explore the types and properties of pollutants that enter our water supply, their impact on water quality and on aquatic life, and methods of water pollution abatement.

Key Terms

assimilative capacity	runoff
water cycle	groundwater
evaporation	zone of aeration
transpiration	zone of saturation
evapotranspiration	water table
sublimation	permeability
condensation	porosity
deposition	aquifer
precipitation	unconfined aquifer
distillation	confined aquifer
saturation	artesian well
relative humidity	cone of depression
expansional cooling	groundwater mining
compressional warming	salt-water intrusion
warm front	ground subsidence
cold front	drainage basin
convection current	watershed
orographic lifting	stream discharge
clouds	base discharge
condensation nuclei	desalination
hygroscopic nuclei	reverse osmosis
infiltration	cloud seeding

Summary Statements

The global water cycle is the ceaseless circulation of water among the planet's oceanic, atmospheric, and terrestrial reservoirs. In the United States, the largest accessible freshwater reservoir is groundwater.

Water cycles from the earth's surface to the atmosphere via evapotranspiration and sublimation, and returns to the earth's surface via cloud formation (condensation, deposition) and precipitation.

Evaporation (or sublimation) followed by condensation (or deposition) purifies water. Distillation, a natural function of the water cycle, supplies the planet with fresh water.

When air contains its maximum water vapor concentration, it is saturated, and clouds develop. The saturation concentration varies directly with air temperature, so that warm saturated air is more humid than an equivalent volume of cold saturated air.

The relative humidity approaches saturation (100%) primarily through expansional cooling of air. Expansional cooling and consequent cloud development occur through lifting along fronts, in convection currents, and along the windward slopes of mountains.

Water vapor condenses (or deposits) on nuclei, tiny solid and liquid particles suspended in the atmosphere and produced by both natural and industrial activities. Hygroscopic nuclei promote cloud development at relative humidities under 100%.

The global water budget indicates a net flow of water from land to sea via infiltration and runoff.

The water table is a surface that roughly parallels the overlying topography and marks the upper boundary of the groundwater reservoir. An aquifer is a porous and permeable layer of sediment or rock that contains and transmits groundwater from recharge areas to points of discharge.

Groundwater is particularly vulnerable to contamination because its typically slow and nonturbulent flow (1) limits the rate of dilution, and (2) favors relatively long residence times for pollutants.

Groundwater mining occurs when the rate of withdrawal exceeds the natural recharge rate and can lead to salt-water intrusion, ground subsidence, and depletion of subsurface water.

Rivers drain a fixed geographical area known as a drainage basin or watershed. River discharge varies with climate and the physical and vegetational properties of the drainage basin.

Lakes are temporary impoundments of water that are formed by a variety of geological processes.

Attempts to enhance the local availability of fresh water include desalination, cloud seeding, dams, and watershed transfers. Each of these strategies has both benefits and drawbacks. Our most promising option is water conservation.

Questions for Review

1. Give a common illustration of each of the following phase changes: sublimation, deposition, evaporation, and condensation.
2. Little difference usually exists between the concentration of water vapor in the air over the southwestern United States and over northeastern areas. However, the northeastern area receives much more precipitation. What is the fundamental reason for this difference?
3. How does the saturation concentration of water vapor vary as air temperature changes? Why does the saturation concentration depend on temperature?
4. How is relative humidity computed?
5. Explain why the relative humidity is lower indoors during the winter than during the summer.
6. How is distillation a water purification process? What is the source of energy that drives natural distillation in the global water cycle?
7. Where in the United States would you expect the highest rates of evapotranspiration? Explain your response.
8. Describe four different processes that cause expansional cooling of air and eventual cloud development.
9. Explain why groundwater is particularly vulnerable to contamination.
10. Describe some of the negative consequences of groundwater mining.
11. Speculate on how the depth of the water table might vary seasonally in locations where the ground freezes during winter.

12. Distinguish between (a) confined versus unconfined aquifers, and (b) porosity and permeability. What types of earth materials are generally good aquifers?
13. What is the source of your tap water? Is water use in your home consumptive or nonconsumptive? Elaborate on your response.
14. What is a watershed? Describe the various factors that govern the discharge of a river.
15. In what sense is a lake a *temporary* impoundment of water?
16. Describe some of the ways whereby lakes come into existence.
17. Describe some ways to conserve irrigation water.
18. List and critically evaluate the various methods of freshwater enhancement.
19. List some strategies to conserve water around the home.
20. What happens to the dissolved salts in a lake that experiences high evaporation rates? What happens to dissolved salts in irrigation water?

Projects

1. How has water played a role in the location and historical development of your community?
2. Cite examples of specific projects in your area that modify the water cycle for the purpose of public water supply, electric power generation, or flood control.
3. Determine the source of the fresh water supply for your region of the country. Is there a problem with groundwater mining? How is the drinking water treated in your area?

Selected Readings

Anderson, T. L., ed. *Water Rights: Scarce Resource Allocation, Bureaucracy, and the Environment.* San Francisco: Pacific Institute for Public Policy, 1986. Focuses on the distribution of water in water-short regions.

Carrier, J. "The Colorado: A River Drained Dry." *National Geographic* 179 (June 1991):4-35. A trip down the Colorado River Valley reveals the environmental impacts of increasing demands for the river's water for irrigation, hydropower, municipal use, and recreation.

Cole, G. A. *A Textbook of Limnology.* Prospect Heights, Ill.: Waveland Press, 1988. Covers the physical, chemical, and biological characteristics of lake ecosystems.

Dolan, R., and H. G. Goodell. "Sinking Cities." *American Scientist* 74(1986):38-47. Examines the impacts of water, oil, and gas removal from beneath some of the major cities of the world.

Dunne, T., and L. B. Leopold. *Water in Environmental Planning.* New York: W. H. Freeman, 1980. Deals with water-related problems, particularly those involving runoff.

Golubev, G. N., and A. K. Biswas. *Large-Scale Water Transfers: Emerging Environmental and Social Experiences.* Oxford, England: Tycooly, 1985. Reviews regional impacts of large-scale dam projects.

Heath, R. *Basic Ground-Water Hydrology.* Denver: United States Geological Survey, 1983. Discusses fundamental concepts regarding movement and storage of groundwater.

Moran, J. M., and M. D. Morgan. *Meteorology, the Atmosphere and the Science of Weather.* 3d ed. New York: Macmillan, 1991. Covers principles of weather plus the basic properties of the atmosphere; includes three chapters on water in the atmosphere.

Plummer, C. C., and D. McGeary. *Physical Geology.* Dubuque, Iowa: Wm. C. Brown, 1991. A basic text that includes well-illustrated chapters on surface and groundwater.

Postel, S. "Saving Water for Agriculture." *State of the World, 1990.* New York: W. W. Norton, 1990. Reviews the status of the worldwide demand for irrigation water.

Sutherland, P. L., and J. A. Knapp. "The Impacts of Limited Water: A Colorado Case Study." *Journal of Soil and Water Conservation* 43 (1988):294-98. Analysis of the growing water demand in the Arkansas River Basin near Colorado Springs.

Water Quality

New fiberglass gasoline tanks are installed to
replace metal tanks that corroded, leaked, and
polluted groundwater.
© Craig Newbauer/Peter Arnold, Inc.

The five interconnected Great Lakes form a vast drainage basin that ultimately feeds the St. Lawrence River (fig. 14.1). The lakes account for about 20% of the world's fresh water and 95% of all surface fresh water in the United States. The drainage basin is home for one of three Canadians and one of eight Americans—a total of 60 million people. The lakes supply drinking water for 26 million people. Half of Canada's industry and 40% of the United States' industry are located in the massive watershed. Much of the economy of this region owes its existence to the Great Lakes. The concentration of human activity in the region, however, has created serious water pollution problems from urban and agricultural runoff and toxic chemicals.

International cooperation is obviously needed to stem water pollution problems in a drainage basin that involves more than one nation. For the United States and Canada such cooperation began in 1909 with the Boundary Water Treaty, which established the International Joint Commission (IJC) on the Great Lakes. The IJC's charge is to provide advice to Canada and the United States on water-quality problems, including monitoring, and to demonstrate progress in cleanup efforts. Working through this commission, the two countries signed a Water Quality Agreement in 1978 to eliminate point sources (e.g., drain pipes) of approximately 350 different chemicals. In 1987, this agreement was amended so that forty-two *pollution hot spots* (areas where toxic chemicals and polluted runoff are most severe) would be the primary focus of pollution abatement efforts.

This chapter examines the various types of water pollutants, their effects on human health and aquatic ecosystems, and strategies to reduce their negative impacts.

Drainage Basin Activities

We gain a useful perspective on water-quality problems by identifying the various activities—both natural and human—that occur in a drainage basin. Most water pollution problems stem from land-based activities such as farming and manufacturing (fig. 14.2) rather than water-based activities such as shipping, boating, and swimming. We can identify three general categories of land use within a drainage basin: (1) natural and seminatural areas that are relatively undisturbed, (2) rural-agricultural areas, and (3) urban-industrial areas. Each land use introduces a different mix of natural and human-produced contaminants into surface and groundwater.

Water-quality specialists find it useful to distinguish between two types of water pollution sources within a drainage basin—point and nonpoint. **Point sources** consist of industrial drain pipes or discharge pipes from sewage treatment plants. In contrast, **nonpoint sources** are broad areas, such as agricultural fields or lawns. Common examples of pollutants from nonpoint sources include sediments eroded from construction sites, and fertilizers, pesticides, and animal wastes washed from farm fields and city lawns. In contrast to water pollutants from point sources, nonpoint-source pollutants are much more difficult and expensive to control because their concentration is quite low and the volume of water is enormous.

Natural Landscape Interactions

As water flows over the land and seeps through soils, sediments, and rock crevices, small amounts of soluble materials dissolve in it (chapter 3). Substances that most frequently dissolve in surface and groundwater include calcium (Ca^{2+}), magnesium (Mg^{2+}), potassium (K^+), bicarbonate (HCO_3^-), and chloride (Cl^-). Concentrations of these substances in surface and groundwater vary considerably, depending to a great extent on the bedrock geology. Water that comes in contact with granite or sandstone has a relatively low dissolved-substance content, because the minerals composing such bedrock are highly insoluble. Water that flows through fractured limestone or dolomite, on the other hand, acquires a relatively high content of dissolved substances. Limestone ($CaCO_3$) and dolomite ($CaMg[CO_3]_2$) are examples of bedrock that readily dissolve in water. Such water is referred to as **hard water** because of its relatively high concentrations of calcium and magnesium. (Laundry detergents do not clean as well when these substances are present; hence, homeowners must install water softeners that remove them.)

Water with a low to moderate dissolved-substance content is considered to be of high quality. If the concentration becomes too high (above 500 milligrams/liter), public health officials consider the water to be unfit for human consumption.* When dissolved-substance levels top 700 milligrams per liter, the water is toxic to plants and cannot be used for irrigation. Because ocean water contains approximately 35,000 milligrams per liter of dissolved salts, it far exceeds standards for drinking or irrigation water.

Contrary to popular opinion, not all water that comes from natural areas is of high quality. For example, streams that originate in or pass through swamps and marshes are usually somewhat acidic and discolored, because they dissolve organic acids generated by the decomposition of detritus—leaves, bark, and twigs. Streams that drain melting glaciers are choked with fine sediment, an inhospitable environment for most forms of aquatic life. Groundwater in some regions is naturally contaminated with excessive amounts of fluoride, radium, and even arsenic.

Human Landscape Interactions

Agricultural areas degrade water quality in several ways. Cultivation often accelerates soil erosion; thus, fertilizers and pesticides become part of runoff and in some cases reach groundwater. Organic waste from barnyards and feedlots also washes into surface waters during rainstorms and spring snowmelt.

Rivers that flow through urban-industrial areas also carry a great number and variety of pollutants. Individual point sources can contribute disease-causing organisms (pathogens), plant nutrients, organic wastes, and toxic and hazardous wastes.

---

*Concentrations of dissolved substances in water are usually expressed in terms of mass (or weight) per volume. Concentrations expressed in milligrams per liter have the same magnitude as parts per million (ppm).

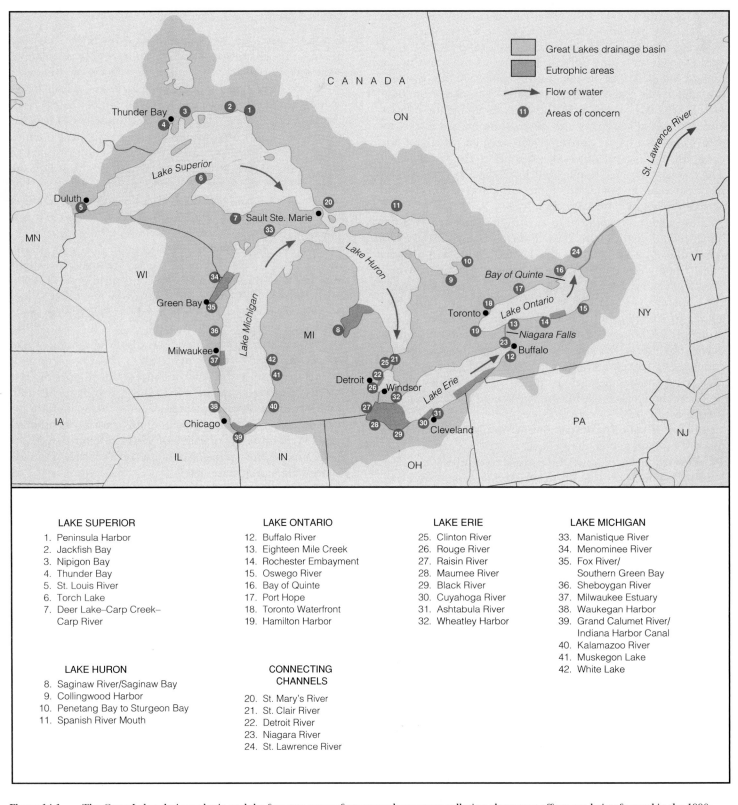

Figure 14.1 The Great Lakes drainage basin and the forty-two areas of concern where water pollution abatement efforts are being focused in the 1990s. Sources: Data from Great Lakes Water Quality Board and the Environmental Protection Agency.

LAKE SUPERIOR

1. Peninsula Harbor
2. Jackfish Bay
3. Nipigon Bay
4. Thunder Bay
5. St. Louis River
6. Torch Lake
7. Deer Lake–Carp Creek–
   Carp River

LAKE HURON

8. Saginaw River/Saginaw Bay
9. Collingwood Harbor
10. Penetang Bay to Sturgeon Bay
11. Spanish River Mouth

LAKE ONTARIO

12. Buffalo River
13. Eighteen Mile Creek
14. Rochester Embayment
15. Oswego River
16. Bay of Quinte
17. Port Hope
18. Toronto Waterfront
19. Hamilton Harbor

**CONNECTING
CHANNELS**

20. St. Mary's River
21. St. Clair River
22. Detroit River
23. Niagara River
24. St. Lawrence River

LAKE ERIE

25. Clinton River
26. Rouge River
27. Raisin River
28. Maumee River
29. Black River
30. Cuyahoga River
31. Ashtabula River
32. Wheatley Harbor

LAKE MICHIGAN

33. Manistique River
34. Menominee River
35. Fox River/
    Southern Green Bay
36. Sheboygan River
37. Milwaukee Estuary
38. Waukegan Harbor
39. Grand Calumet River/
    Indiana Harbor Canal
40. Kalamazoo River
41. Muskegon Lake
42. White Lake

Figure 14.2 A few of the many activities in drainage basins that can affect the quality of surface water. In natural drainage basins, flowing water (1) slowly dissolves and erodes rock, and wetlands (2) allow sediment to settle from water. In urban-industrial basins, municipal sewage treatment plants (3) fail to remove all the wastes added during use; storm sewer water (4) contains wastes that are washed from city streets and lots; industrial wastewater (5) contains a wide array of different water pollutants; acid water flows from mines and strip-mined land (6); heated water flows from power plants (7); industrial gases are washed out of the air forming acid rain (8); the transportation of oil (9) results in spills, including many minor spills during unloading; and faulty landfill sites (10) contaminate groundwater. In agricultural drainage basins, crop spraying (11) adds pesticides and herbicides to rain and runoff; fertilizers that are applied improperly (12) are dissolved in runoff; and animal wastes (13) are washed from farmland.

Nonpoint sources such as city streets and lawns contribute plant nutrients, pesticides, and road salt used for winter ice-control.

Water pollution problems in some metropolitan areas arise from inadequate sewer systems. An examination of basic sewer-system designs gives us an understanding of both their function and limitations. In cities, buildings and paved areas render a large part of the earth's surface impermeable to rainwater and snowmelt, thereby increasing the volume of runoff. To prevent flooding, large **storm-sewer** pipes are placed under city streets to collect runoff and channel it to the nearest river, lake, or ocean (fig. 14.3). During rainstorms, pollutants on the urban landscape (e.g., lawn fertilizers and pet droppings) are washed off the surface into the storm drain and are carried by storm-sewer pipes directly into nearby surface waterways.

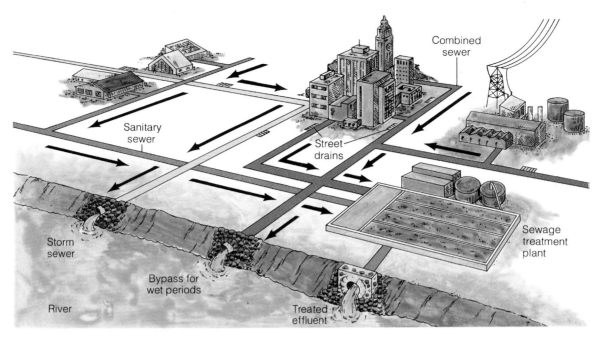

Figure 14.3 A city sewer system, showing both separated and combined sewer systems.

Most urbanized areas also have a second network of smaller-diameter pipes that functions as the **sanitary sewer.** This network carries wastewater from homes and commercial establishments to sewage treatment plants, where it is treated and discharged into nearby surface waters. Where both storm-sewer and sanitary sewer pipes service the same area, the two networks constitute a **separated sewer system.**

In some cities or parts of cities, a single, large-diameter pipe transports a combination of runoff and wastewater to a sewage treatment plant. This is a **combined sewer system.** In areas serviced by combined sewers, the treatment plant receives only wastewater during dry weather. When it rains, however, runoff is added to wastewater, and the volume of water that flows through the combined sewer often exceeds—by as much as one hundred times—the volume of water that can be adequately processed by the treatment plant. As a result, only a small fraction of the wastewater is treated. The overflow, which contains raw sewage, bypasses the treatment plant and is discharged directly into a surface water-way. Every time this happens, the health of the city's residents and their downstream neighbors is threatened, and the aquatic organisms in the receiving surface water are contaminated or killed.

To meet water-quality standards, large metropolitan areas with combined sewer systems face major economic difficulties. Restructuring an existing combined sewer system to create separate sanitary and storm-sewer systems would be very costly. In some cities with combined systems, officials choose to treat stormwater runoff. Chicago and Milwaukee, for example, are building huge underground caverns to store storm-sewer overflows. When dry weather returns, the stored water is pumped to the sewage treatment plant, treated, and then discharged.

Types of Water Pollutants

Waterborne diseases such as dysentery, hepatitis, and cholera have afflicted humans for millennia, but it was not until the early 1970s that Americans first began to realize that other types of water pollutants were destroying aquatic ecosystems. Some types of water pollutants and their potential hazards for people are obvious, for example, medical wastes, including used syringes that have washed up on some Atlantic and Great Lakes beaches. The effects of pollutants on aquatic ecosystems are often more subtle. In this section, we describe seven basic types of water pollutants (infectious agents, or pathogens; oxygen-demanding wastes; aquatic plant nutrients; sediments; toxic chemicals; oil spills; and heat) and some of the adverse effects of each on people and aquatic ecosystems. We also look briefly at strategies for minimizing the effects of each type of pollutant. Later in this chapter, we examine some of the effects of these pollutants on freshwater and salt-water ecosystems.

Pathogens

Historically, the first recognized water pollutants were **pathogens,** that is, disease-producing organisms. Pathogens, which include viruses, bacteria, protozoa (unicellular animals), and parasitic worms, cause such diseases as dysentery, typhoid fever, and cholera. Table 14.1 shows some important characteristics and symptoms of the more common waterborne diseases in humans.

Pathogens enter water mainly through the feces and urine of infected people and animals. The sources of body wastes may be (1) seepage from malfunctioning cesspools and septic tanks, (2) untreated sewage when sewer systems

Table 14.1 Some common waterborne diseases transmitted through drinking water and food

Disease	Type of Organism	Symptoms and Comments
cholera	bacterium	severe vomiting, diarrhea, and dehydration; often fatal if untreated; primary cases waterborne; secondary cases carried by contact with food and flies
typhoid fever	bacterium	severe vomiting, diarrhea, inflamed intestine, enlarged spleen; often fatal if untreated; primarily transmitted by water and food
bacterial dysentery	several species of bacteria	diarrhea; rarely fatal; transmitted through water contaminated with fecal material or by direct contact through milk, food, and flies
infectious hepatitis	virus	yellow jaundiced skin, enlarged liver, vomiting and abdominal pain; often permanent liver damage; transmitted through water and food, including shellfish foods
amoebic dysentery	protozoan	diarrhea, possibly prolonged; transmitted through food, including shellfish
giardia	protozoan	diarrhea, abdominal cramps, fatigue
schistosomiasis	parasitic worm	abdominal pain, anemia, chronic fatigue

become overloaded or treatment plants malfunction, (3) waste discharges from boats and ships, (4) untreated discharges from meat-processing plants, or (5) directly from swimmers, hikers, and so forth. In many less-developed nations, it is common to discharge human wastes directly into waterways without prior treatment.

Swimming in contaminated water frequently leads to an increased incidence of skin, ear, nose, and upper-respiratory-tract infections. Infection can be acquired indirectly by eating raw shellfish or fish harvested from contaminated water. For example, in the first five months of 1991, Peru reported 200,000 cases of cholera to the World Health Organization. The cholera epidemic virtually shut down the Peruvian fishing industry, because shellfish were identified as the source of contamination, and people dared not buy any seafood.

Because detecting the presence of specific pathogens is a costly process, microbiologists usually test water for a more readily identifiable group of microorganisms: **coliform bacteria.** Since these organisms are normally present in the intestinal tracts of humans and animals, large numbers of coliform bacteria in a water sample indicate recent contamination by untreated feces. If coliform bacteria are present in a sample, infectious pathogens are also likely to be present. In addition, coliform bacteria can act as pathogens in certain cases. Hence, when coliform bacteria exceed one organism per 100 milliliters of drinking water, a municipality must either chlorinate more heavily or seek an alternate water source. This same test is used to determine whether groundwater drawn from private wells is free from fecal contamination and to protect swimmers from contracting waterborne diseases. The Environmental Protection Agency (EPA) has set an upper limit of 200 coliform bacteria per 100 milliliters of water for recreational waters. If that limit is exceeded, the contaminated area and nearby beaches are usually closed.

Today the spread of waterborne pathogens can be almost completely controlled by disinfection of public drinking water supplies and sewage treatment plant effluent. Disinfecting water with chlorine, a process known as **chlorination,** or with ozone (O_3) either kills or inactivates most pathogens.

The question of whether or not to disinfect the effluent from sewage treatment plants is still actively debated among public health scientists. In the United States, most sewage treatment plants add chlorine as the final step in the treatment process before releasing wastewater into nearby surface waterways. But adding chlorine to water that still contains some organic material (such as sewage treatment plant effluent) leads to the formation of small amounts of chloroorganic compounds (for example, chloroform), some of which are known carcinogens. The impact of these chemicals on aquatic ecosystems has not been fully researched, but is thought to be minimal. In most European nations, chlorination of wastewater discharges is omitted for economic reasons and because waterborne diseases have not been attributed to discharges from sewage treatment plants. Only when direct evidence exists (e.g., a case of typhoid fever) that an outbreak of a waterborne disease is possible are European wastewater discharges chlorinated.

Although people in more-developed nations take the protection provided by sewage treatment facilities for granted, the potential for waterborne epidemics is always present. Waterborne epidemics are of special concern when a treatment plant malfunctions, when raw sewage is diverted around a treatment plant, or when water supplies are contaminated by floodwaters. During such episodes, people are urged to drink bottled water or to boil water used for drinking and food preparation. In less-developed nations, hundreds of millions of people live with the constant threat of infection by waterborne pathogens. One such disease, schistosomiasis, is considered in the issue on pages 288–289.

Oxygen-Demanding Wastes

Aquatic organisms, like terrestrial organisms, require oxygen for cellular respiration. One of the most important limiting factors for aquatic animals (e.g., fish, shellfish, and insects) is dissolved oxygen.* In surface waters, dissolved oxygen levels vary considerably, depending upon factors that govern the transfer of oxygen between the atmosphere and waterways.

*We know that solids dissolve in water (e.g., sugar in coffee), but it may be less obvious that gases also dissolve in water. A familiar example is carbon dioxide (CO_2) dissolved in a carbonated beverage. When we pop open a can of cola, tiny bubbles of CO_2 escape, giving the drink a fizzy taste.

Schistosomiasis, Scourge of Less-Developed Nations

Several forms of **schistosomiasis,** a waterborne disease that is practically unknown in the United States, infect an estimated 200 million people. Sometimes called *bilharzia,* this disease is particularly prevalent in Africa, Asia, and some of the West Indian Islands. In Egypt, schistosomiasis is the primary health problem. The disease is spread by a parasitic worm, which requires a certain species of snail as an intermediate host (fig. 14.4). Irrigation canals that are filled year-round provide a suitable habitat for host snails. Thus, construction of large hydroelectric dams and irrigation projects such as Egypt's Aswan Dam, has promoted the spread of schistosomiasis.

The disease spreads readily. A snail infected with a single schistosomiasis sporocyst (parasitic worm) can produce up to 200,000 free-swimming larvae, the form that infects people. Once infected, the digestive tracts of children and adults serve as hosts for the adult form of the parasitic worm. The adult female subsequently lays eggs in blood vessels, and some of them are excreted in urine and feces. If these eggs enter surface water, they hatch and infect snails, and the infection cycle continues. Eggs remaining in the body accumulate in the blood vessels, where they interfere with blood flow. Restricted blood flow to the kidneys, lungs, liver, and other vital organs

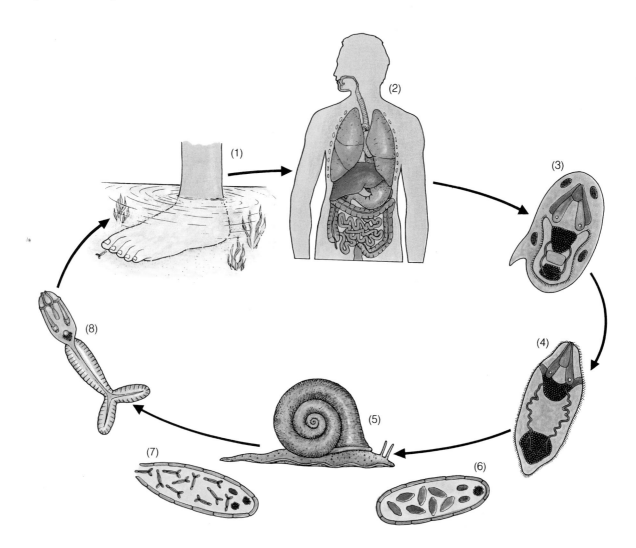

Figure 14.4   The life history of Schistosoma mansoni. (The sketches are drawn to different scales.) (1) Cercaria penetrate skin in water. (2) Adult worms live in blood vessels of intestines. (3) Eggshell containing developing miracidium. (4) Miracidium hatches in the water. It will enter a snail if it encounters one. (5) Inside a snail, a mother sporocyst (6) forms. It produces many daughter sporocysts. Each one (7) produces many cercaria larvae. (8) Cercaria larvae break out of snail's body and enter the water.

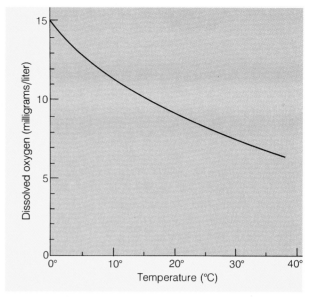

Figure 14.6 The amount of dissolved oxygen that water can hold at saturation declines as water temperature rises.

eventually results in their destruction. Symptoms of schistosome infection include dysentery, anemia, general weakness, and reduced resistance to other pathogens. In severe cases, brain damage and even death result after a number of years.

Although drugs are available to treat the disease, they are not widely used because of severe side effects. On a worldwide basis in 1987, the search for a better cure saw only $8 million directed toward this goal. We quickly realize how small this research effort is when we compare it to the $1.4 billion spent in the same year by the United States government on cancer research—a disease that affects ten million people.

People in less-developed nations are afflicted by many other waterborne diseases such as malaria (spread by the *Anopheles* mosquito), and onchocerciasis, or river blindness (spread by the black fly). Malaria is a life-threatening disease that is estimated to infect 100 million people worldwide, of whom about one million die per year. About 1,000 cases of malaria are reported in the United States each year. River blindness infections occur primarily in equatorial Africa (fig. 14.5), with Nigeria and Zaire having the most victims. The World Health Organization reports 17 million people infected with the disease, of whom about 326,000 are blind. In these areas, one sometimes sees chains of hand-holding blind people being led by an unimpaired person so they can meet their essential human needs.

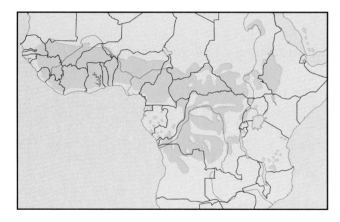

Figure 14.5 Geographical distribution of onchocerciasis in Africa.

Source: World Health Organization: WHO Expert Committee on Onchocerciasis.

Oxygen enters an aquatic system through the air-water interface and by the photosynthetic activities of aquatic plants (chapter 3). Thus the amount of dissolved oxygen in an aquatic ecosystem depends on the rate at which these processes occur. For example, the turbulence in waterfalls, rapids, and dam spillways, as well as wave activity in open water, increases the rate of oxygen transfer from air to water (unless the water is already saturated with oxygen). The profile of a waterway also affects oxygen transfer: A wide, shallow section of a river has a larger surface area for oxygen transfer than does a narrow, deep segment. Also, the amount of oxygen produced per unit area by photosynthesis is directly related to the density of aquatic plants.

Dissolved oxygen is removed from water primarily through cellular respiration by decomposers; much smaller amounts are removed by zooplankton, shellfish, and fish. Temperature influences the amount of dissolved oxygen that water retains because oxygen is less soluble in warm water (fig. 14.6). Furthermore, warm water enhances decomposer activity and thus increases the rate at which they remove oxygen through cellular respiration. Another factor that affects the level of dissolved oxygen is the amount of organic waste available to decomposers.

In most natural aquatic ecosystems, the quantity of organic waste (detritus) is small; therefore, the amount of dissolved oxygen that decomposers use is also small. Consequently, the concentration of dissolved oxygen usually remains relatively constant at more than 5 milligrams per liter—the minimum level considered essential for the survival of the most sensitive fish, such as trout. When saturated, natural waters typically have between 10 and 14 milligrams per liter of dissolved oxygen.

Industrial and municipal wastewater, however, typically contains high concentrations of organic wastes. This spurs the growth of decomposer populations, which consume large quantities of dissolved oxygen. Most industrial and municipal organic wastes are short-term pollutants; they are almost completely decomposed within several weeks after entering surface waters. In cold water, however, the rate of decomposition slows considerably.

The amount of dissolved oxygen that decomposers require to break down organic materials in a given volume of water is called the **biochemical oxygen demand (BOD).** Human wastes are a major source of BOD. Sewage-laden wastewater that enters a sanitary sewer typically has a BOD level of 250 milligrams per liter. However, that water is initially likely to contain only about 8 milligrams per liter of dissolved oxygen, so all of its dissolved oxygen is quickly taken up by the microbial decomposition of sewage. In fact, decomposition of the daily wastes of just one person requires all the dissolved oxygen in 9,000 liters (2,200 gallons) of water.

When water is used for processing organic materials such as vegetables, fruits, paper, meat, and dairy products, the BOD level in the wastewater is substantial. Some concentrated industrial wastewater has BOD levels that exceed by one thousand times the BOD levels in sewage. Other examples of high-level BOD wastes include runoff from livestock feedlots and organic sediments dredged from polluted harbors and canals.

When effluents with high levels of BOD are continually released into a stream or river, dissolved oxygen levels downstream respond in a characteristic pattern, called an **oxygen sag curve** (fig. 14.7). At the point of discharge, bacteria begin to consume the organic material. Bacterial populations grow in direct proportion to the concentration of organic matter, and they remove dissolved oxygen faster than it is naturally replenished. Thus, the dissolved oxygen content declines, producing a characteristic sag in the curve that describes its concentration. As the organic waste moves downstream, more of it is consumed by decomposers; thus, the BOD level declines. The dissolved oxygen deficit caused by the decomposers is slowly replaced by oxygen transfer from the atmosphere to the water and by photosynthesis in aquatic plants. Eventually, the rate of oxygen replacement exceeds the rate of removal, and dissolved oxygen returns to its original level.

Discharges of organic waste have their greatest impact on aquatic life during warm summer months because stream flow usually is relatively low at that time of year, and wastes are less diluted. Furthermore, warm water has less dissolved oxygen, and decomposers have much higher metabolic rates at elevated temperatures.

As we would predict from our discussion of tolerance limits in chapter 4, changes in an aquatic environment are accompanied by changes in the aquatic species that live there. The responses of aquatic organisms in rivers to the release of moderate amounts of organic waste have been well-documented (fig. 14.7). Upstream from a discharge point, a river can support a wide variety of fish, shellfish, and other organisms. Stretches of the river where dissolved oxygen levels are low, however, support only species of rough fish (such as carp, gar, and catfish) that can tolerate low dissolved oxygen levels. At some greater distance downstream, a more diverse and desirable community of fish can live. Sometimes a single BOD discharge may be so large that decomposers flourish to the point that they remove all the dissolved oxygen from the water. In other cases, a series of closely spaced point sources along the course of a river may totally deplete the entire downstream stretch of its dissolved oxygen.

Such a drastic change from an **aerobic** (with oxygen) **environment** to an **anaerobic** (without oxygen) **environment,** should lead us to expect a major change in species composition. In fact, **aerobic decomposers** (those that require oxygen) are replaced by **anaerobic decomposers** (those that require the *absence* of oxygen because it is toxic to them; chapter 3). The end products of aerobic and anaerobic bacterial decay differ markedly. The decay products of aerobic decomposers are mainly carbon dioxide (CO_2), water (H_2O), nitrate (NO_3^-), and sulfate (SO_4^{2-}), which are not usually harmful. On the other hand, some of the decay products formed by anaerobic bacteria, such as methane (CH_4), ammonia (NH_3), and hydrogen sulfide (H_2S), are potentially dangerous to humans and other species. Methane is potentially explosive when mixed with oxygen, and hydrogen sulfide, recognized by its rotten-egg odor, is a toxic gas at low concentrations. Where anaerobic conditions develop in surface waters, they become a putrid, turbid, decaying mess; bubbles of methane rise to the surface, accompanied by the odor of hydrogen sulfide. Under these conditions, even pollution-tolerant sludge worms can not survive, and the only living organisms are anaerobic bacteria (figure 14.7).

To reduce the impact of oxygen-demanding wastes, wastewater treatment facilities are designed to substantially reduce the BOD level of treated wastewater. In fact, the principal design criteria for wastewater treatment facilities is the reduction of the BOD level in the water prior to its release. Special Topic 14.1 on page 292 gives a brief overview of how a sewage treatment plant works.

Plant Nutrients and Cultural Eutrophication

Aquatic ecosystems support a wide variety of plant species, including rooted emergents (e.g., cattails and bullrushes) and free-floating and attached algae. Like terrestrial plants, aquatic plants require about fifteen essential nutrients to support their growth and reproduction (chapter 3). In most aquatic ecosystems, the two nutrients that usually limit plant growth are phosphorus and nitrogen (chapter 4).

Lakes and reservoirs with low rates of nutrient cycling (low fertility) are **oligotrophic ecosystems,** whereas those with high rates of nutrient cycling (high fertility) are **eutrophic ecosystems** (chapter 3). Because the level of nutrient cycling varies considerably from one lake (or reservoir) to another, aquatic plant productivity also varies greatly. Table 14.2 contrasts the general characteristics of oligotrophic and eutrophic lakes and reservoirs.

Two principal factors control nutrient supply: (1) the fertility of soils in the drainage basin, and (2) the depth of the lake or reservoir. Lakes located in basins with relatively infertile sandy or rocky soils tend to be oligotrophic; those located in basins with fertile soils tend to be eutrophic. The latter

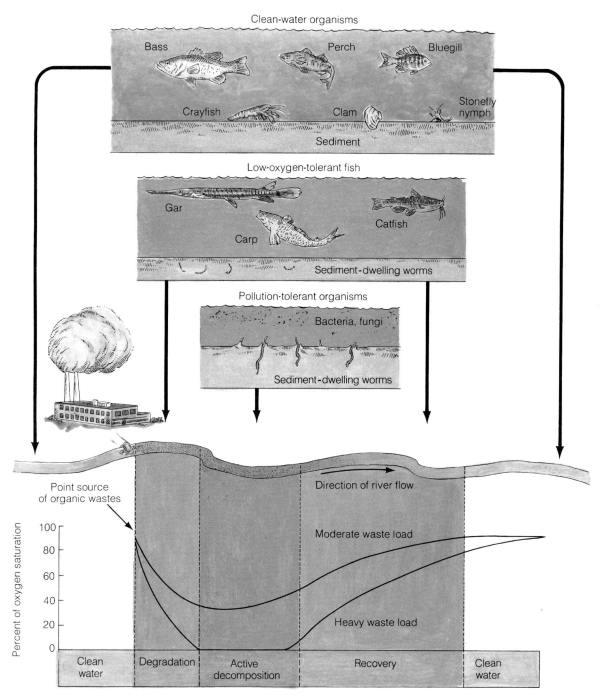

Figure 14.7 Dumping organic wastes in a river causes dissolved oxygen levels to fall. The level of dissolved oxygen affects the number and types of organisms that inhabit various sections of the river.

Table 14.2 Characteristics of lakes and reservoirs with low and high productivity		
Characteristic	**Oligotrophic**	**Eutrophic**
nutrient cycling	low	high
productivity (total biomass)	low	high
diversity of species	high*	low
relative numbers of undesirable fish	low	high
water quality	high	low

*Lakes that are extremely nonproductive (e.g., high mountain lakes) will have low species diversity.

experience luxuriant growth of rooted aquatic plants and algal blooms—conditions associated with high fertility. Although nutrients tend to settle out of the water into bottom sediments, wave action can resuspend them. Thus, shallow lakes and reservoirs tend to internally recycle nitrogen and phosphorus compounds more efficiently than do deep lakes, and consequently they tend to be more productive.

Because lake and reservoir basins slowly fill with nutrient-laden sediment (chapter 13), they gradually become shallower and recycle nutrients more efficiently, so that they produce more and more plant biomass. This natural aging process

Wastewater Treatment

As water flows through an urban-industrial water system, it is pumped from a source, treated, distributed, used, collected, treated again, and then discharged into the nearest appropriate surface waterway (river, lake, bay, or ocean). The drawing below illustrates this urban-industrial water-use cycle. Treatment of wastewater is designed to protect the quality of water in aquatic ecosystems. These waters not only serve as a habitat for aquatic organisms, they may be used for recreation and as sources of drinking water. We focus on sewage treatment because (1) most urban and suburban dwellers contribute wastes to such plants and pay significant fees to operate them, and (2) such plants provide the principal means of improving the quality of point-source discharges. In some instances, additional treatment processes are required to remove other contaminants from wastewater at considerably greater expense.

Activated sludge treatment is the most common technique used to treat sewage and industrial organic wastes. It combines biological (aerobic decomposition) and mechanical processes to remove and break down waterborne wastes. A simplified diagram of an activated sludge treatment plant is shown on page 293. The sewage flows by gravity or is pumped into the treatment plant. As it enters, sewage typically contains 99.9% water

and only 0.1% impurities. In the first step, the sewage passes through a screen that removes such debris as rags, sticks, and cans. It then flows into a grit tank, where dense materials, such as sand and pebbles, quickly settle to the bottom. This procedure protects pumps and other mechanical equipment from abrasion. Materials removed in these first two steps are usually hauled to a sanitary landfill (chapter 20).

Next, the sewage flows into the primary settling tank where it is held for about two hours. During this time, about 45% of the organic suspended solids settle to the bottom of the tank. This gravitational method of freeing wastewater of its suspended solids is called *clarification*. The settled solids, called *sludge,* are then pumped to sludge-handling equipment. The treatment steps described to this point are referred to as *primary treatment* and are largely physical processes.

Following clarification, sewage enters an aeration tank, where it is held for six to eight hours. Compressed air is pumped continually through the sewage to provide oxygen for aerobic bacteria, which decompose suspended and dissolved organic wastes. Afterwards, the wastewater and bacteria flow from the aeration tank into a final clarifying tank, where bacteria form clumps, called *floc.* Floc, along with most other

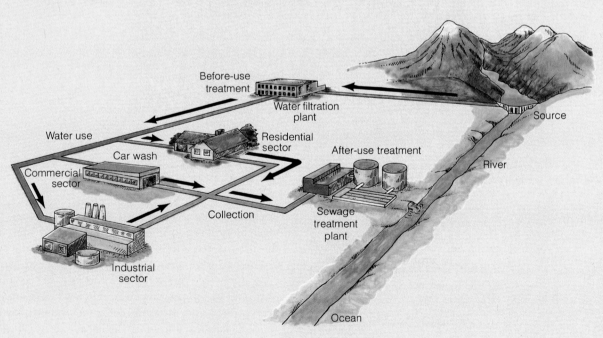

The water-use cycle in an urban area.

suspended material, settles out in the final clarifying, or settling, tank. Some of the floc is recycled back into the aeration tank to provide seed bacteria to maintain the bacterial colony there. Because recycled bacteria are already well adjusted to conditions in the aeration tank, they remove organic wastes efficiently. Unrecycled floc is pumped to sludge-handling equipment. Water that leaves the final settling tank is quite clear and contains only about 10% of the BOD that was in the incoming sewage. These steps beyond primary treatment are largely biological and constitute *secondary treatment*.

After secondary treatment, some 25-75% of the phosphorus, 90% of the nitrogen, and about 10% of the organic materials remain in the treated effluent. If these substances aggravate local water pollution problems, wastewater treatment plants are often required to add advanced steps called *tertiary treatment*. For example, cultural eutrophication is a problem in the Great Lakes. Thus, the EPA requires sewage treatment plants that discharge their effluents into the Great Lakes drainage basin to remove, on average, 90% of the phosphorus from the final effluent. Phosphorus removal is accomplished by adding chemicals that precipitate phosphorus during the final clarifying phase of secondary treatment.

Tertiary treatment for nitrogen removal from sewage effluent is not usually required. Excess nitrogen is sometimes removed, however, when too much is present in the form of ammonia (NH_3), a form toxic to fish, or if wastewater is discharged into a body of water where nitrogen is a limiting factor (chapter 4). For example, to preserve Lake Tahoe's world-renowned, pristine appearance, local citizens authorized costly nitrogen-removing steps at their wastewater treatment facility.

The final step in sewage treatment before discharge is chlorination. Chlorine is added to kill pathogenic organisms that may have survived earlier treatment steps. Although this step is considered necessary to prevent the spread of waterborne diseases, as noted elsewhere in this chapter, scientists have discovered that chlorination also results in the formation of small amounts of chloroorganic compounds, some of which are known carcinogens.

Sewage treatment plants, like all systems, must adhere to the law of conservation of matter (chapter 3). Although clean water flows out of a facility, we have to account for all the fecal matter, table scraps, and other dissolved and suspended

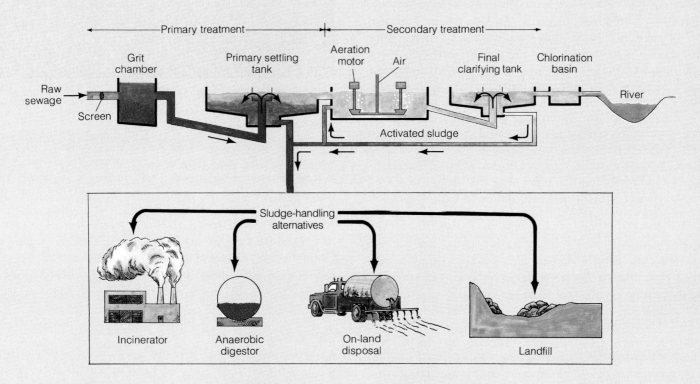

An activated sludge treatment plant for sewage. Such a plant removes approximately 90% of the organic material that enters it in the raw sewage.

materials that entered the facility. Wastes are not destroyed by the treatment process; they remain in the form of sludge.

What can be done with the large quantity of sludge that accumulates at a sewage treatment plant? Some communities incinerate sludge and dispose of the ash in a nearby landfill. Some treatment plants dispose of sludge through *anaerobic digestion,* in which anaerobic bacteria convert the organic components of sludge into methane and carbon dioxide. Methane is used as an energy source, often to run electric generators at the sewage treatment plant.

Because sewage sludge is relatively rich in organic matter and plant nutrients, some treatment plants apply it directly to agricultural land. A few plants manufacture fertilizer from sludge. For example, early in 1992, the Massachusetts Water Resources Authority put into operation a new $87 million sludge-pelletizing plant. The pelletized sludge is marketed as fertilizer. This facility is a key component of the $6.1 billion Boston Harbor cleanup project to upgrade sewage treatment for the Boston metropolitan area. For years Boston has been dumping untreated sewage into the harbor at the daily rate of 1.8 to 2.0 million liters (450,000 to 500,000 gallons).

is known as **eutrophication** and is a consequence of physical, chemical, and biological changes that occur in the body of water. When human activities accelerate the eutrophication process, we use the term **cultural eutrophication.** Runoff from agricultural and urban areas often contains high levels of phosphorus, nitrogen, and sediments, which contribute to cultural eutrophication.

Cultural eutrophication of freshwater resources is one of the most difficult water-quality problems facing us today. An estimated 94% of New York's 4,000 lakes are adversely impacted by cultural eutrophication. Dense growths of rooted aquatic plants hinder swimming and boating and may reduce shoreline property values (fig. 14.8). Also, some species of algae attach to rocks along shorelines, giving them an uninviting slimy-green appearance. In open waters, algae populations explode, causing algae blooms that give water the look of pea-green soup. Under nutrient-rich conditions, blue-green algae replace green algae and other algal forms.* Blue-green algae release foul-smelling and unpleasant-tasting substances that degrade the water's recreational value and reduce its quality as drinking water. Furthermore, the cost of removing blue-green algae and the odorous substances they release raises the cost of water supplied to municipalities and industries. Most blue-green algae are not consumed by members of the next higher trophic level, zooplankton. Thus, much of the production in highly eutrophic lakes is consumed by decomposers and detritus feeders (chapter 3). Fish populations in highly eutrophic systems tend to be dominated by trash fish such as carp.

During the summer months, most lakes and reservoirs in temperate latitudes stratify into two layers: a top layer of warm water, the epilimnion, and a bottom layer of cold water, the hypolimnion. Lakes and reservoirs undergo annual cycles that

Figure 14.8   One of the effects of cultural eutrophication is the dense growth of algae and aquatic weeds in shallow areas of lakes and reservoirs. © Bob Korth/ University of Wisconsin Extension

include periods when mixing occurs throughout the system and periods when mixing of the two layers is inhibited. The physical reasons for this are described in Special Topic 14.2. When large amounts of organic matter in the form of algae and other plant debris are produced, more organic material descends into the hypolimnion. Microbial decomposition of this additional organic material leads to greater oxygen depletion in the bottom layer. If oxygen levels drop below 5 milligrams per liter, cold-water species of fish such as trout, cisco, and whitefish die from oxygen starvation.

Several sources of nutrients contribute to cultural eutrophication. Domestic sewage is a significant point source. Approximately 50% of the phosphorus in sewage comes from phosphorus-based detergents (in areas where such detergents have not been banned). A modern treatment plant may remove as little as 10% of the phosphorus and 30% of the nitrogen in sewage. Consequently, it discharges water with

*Biologists now classify blue-green algae as a type of bacteria called *cyanobacteria.*

Mixing Cycles in Lakes and Reservoirs

*I*n temperate latitudes, lakes and reservoirs undergo an annual mixing cycle that replenishes dissolved oxygen. Depending on their relative size and depth, lakes tend to thermally stratify into two layers during the summer. (See diagram below.) Under bright summer sunshine, surface water warms and becomes less dense than the cold water below. The result is a stable layer of warmer, lighter water that floats on a layer of cooler, denser water. Little vertical mixing occurs between the two layers. The upper layer of a stratified lake is called the *epilimnion,* the lower layer is the *hypolimnion,* and the narrow transition zone between the two is known as the *thermocline.*

When a lake is stratified, only the epilimnion is replenished with dissolved oxygen by transfer through the air-water interface and by photosynthesis. In the hypolimnion, dissolved oxygen is removed by cellular respiration of decomposers that break down the detritus that settles into the hypolimnion from the epilimnion. Without input of oxygen-rich water from the epilimnion, dissolved oxygen levels in the hypolimnion decline. If stratification were to persist, dissolved oxygen levels would decline to the point that cold-water species such as trout and whitefish would die. But these fish usually do not die. How, then, is the dissolved oxygen supply of the hypolimnion replenished?

Lakes do not remain stratified year-round. In autumn, the surface water loses heat to the cooler atmosphere. Eventually,

the temperatures and therefore the densities of the epilimnion and hypolimnion become equal. Assisted by the force of wind, water mixes from the top to the bottom, and the mixing process replenishes the supply of dissolved oxygen throughout the lake. This special event in the annual cycle of a lake is called the *fall turnover.* Refer to the top diagram on page 296.

During winter months in high and middle latitudes, lakes are covered with ice, and the water temperature varies from 0° C (32° F) just below the ice to near 4° C (39° F) at the bottom. Ice is a barrier that prevents oxygen exchange between the water and the atmosphere. With the arrival of spring, the ice melts, and surface waters warm. Eventually, the temperature of the surface waters reaches 4° C (39° F), the temperature at which water is at its maximum density (see graph on page 296). The warmer, denser water sinks and causes mixing throughout the lake. This mixing is called *spring turnover.* Thus, twice a year, in fall and spring, surface and bottom water recirculate in temperate lakes. In contrast, a tropical or subtropical lake may experience only one turnover, during the winter when the surface of the lake cools.

Lake turnovers are essential to replenish the dissolved oxygen supply of bottom waters and permit the survival of cold-water species. Turnover periods also play an important role in recycling nutrients, especially nitrogen and phosphorus, from bottom sediments to the overlying water, where they become available to plants, especially algae.

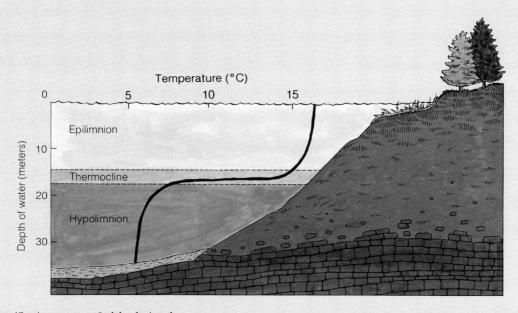

The thermal stratification pattern of a lake during the summer.

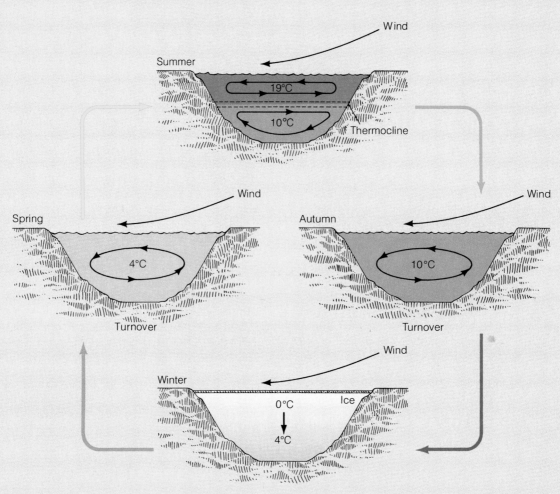

The seasonal mixing patterns of a lake or reservoir in temperate regions is determined by its temperature profile. Deep lakes undergo complete mixing in the spring and the fall; that turnover replenishes the dissolved oxygen of their deeper waters.

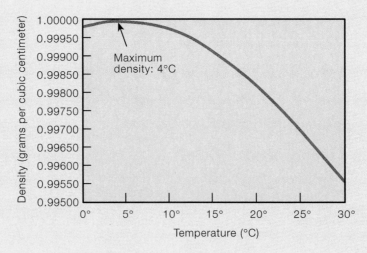

The effect of temperature on the density of water; maximum density is reached at 4° C.

Figure 14.9   Runoff from feedlots such as this one in Texas can be an important source of nutrients and oxygen-demanding substances for waterways.
© Junebug Clark/Photo Researchers, Inc.

phosphorus levels approximately one hundred times the concentration required to stimulate eutrophication. In some regions where cultural eutrophication is already a problem, the EPA may therefore require sewage treatment facilities to provide advanced treatment. With additional steps, a treatment plant can remove 90% or more of the phosphorus and thus reduce cultural eutrophication (see Special Topic 14.1 on pages 292-294).

Urban runoff is another major source of plant nutrients. Stormwater runoff from urban areas often contains levels of both nitrogen and phosphorus high enough to stimulate cultural eutrophication. The nutrients in runoff originate from lawn fertilizers (especially the portion that falls directly on sidewalks and streets), animal fecal material, dust, and leachate from leaves. Because of the nutrients and potential pathogens in runoff, a few cities (e.g., Chicago and Milwaukee) have been required to treat their stormwater runoff to remove nutrients prior to discharge.

Industrial sources of nutrients vary greatly. Paper mills produce enormous quantities of oxygen-demanding substances, but they discharge low concentrations of plant nutrients. In contrast, industrial facilities that have large surfaces to keep clean, such as creameries and car washes, often use large quantities of phosphorus-containing cleaning agents. Phosphate mining is another major source of nutrients, particularly in Florida, where most U.S. phosphate deposits are located.

Nutrient runoff from properly managed farmland is only slightly greater than that from natural areas. However, many farmers do not adequately follow soil conservation practices and often apply excessive amounts of fertilizers to ensure maximum crop yields. Thus, the quantities of nutrients washed from farmland are often detrimental. In northern states, such as Minnesota, Wisconsin, and New York, the practice of spreading livestock manure over frozen ground in winter adds nutrients to spring runoff. Feedlots that drain directly into streams are another important source of plant nutrients as well as BOD (fig. 14.9).

To minimize cultural eutrophication in lakes and reservoirs, nutrients must be managed within each drainage basin. This involves determining the relative size of each source, both point and nonpoint, and then implementing programs that will reduce the amount of nutrients that enter aquatic ecosystems. Even then, some lakes and reservoirs located in drainage basins with highly fertile soils may experience only minor improvements in water quality. Nonpoint sources of plant nutrients remain largely uncontrolled. Overall, in the United States in 1990, nonpoint sources were responsible for an estimated 79% of the nitrogen and 74% of the phosphorus in waterways.

Pollution by Sediments

When raindrops strike bare soil, they dislodge soil particles, temporarily suspending them in water. The finer particles remain suspended, and runoff carries them into streams, lakes, or the ocean, where they are deposited as **sediment.** Sediments may eventually fill lake or reservoir basins, harbors,

Figure 14.10   A floating dredge maintains the navigability of a ship channel by removing accumulated sediments from the bottom.
© Fred Whitehead/Earth Scenes

and river channels. As reservoirs behind dams fill with sediment, the capacity to generate electricity diminishes greatly, and the reservoirs are less suitable for recreation. Sediments also impede navigation (fig. 14.10), cover bottom-dwelling organisms, eliminate valuable fish-spawning grounds, and reduce the light penetration necessary for photosynthesis in aquatic plants. In addition, soils that are eroded from farmlands carry adhered nutrients (nitrogen and phosphorus compounds). Thus, most lakes and reservoirs that receive heavy sediment loads are usually eutrophic; they are shallow and experience high rates of nutrient cycling.

Rivers have always transported enormous quantities of sediment, but their sediment loads today are greater than ever. Soils that are laid bare by crop cultivation, timber cutting, strip mining, overgrazing, road building, and other construction activities are subject to high rates of erosion. In the United States, perhaps 85% of all soil erosion originates from cropland. Although small in area, construction sites are notorious for high rates of erosion. Estimates are that construction sites (fig. 14.11) contribute two thousand times more sedi-

ment per unit area than undisturbed forest land, two hundred times more than undisturbed grassland, and ten times more sediment than cultivated land. Drainage basins where strip mining has occurred also experience erosion rates that are comparable to those of cropland. Although soil losses as high as 1 metric ton per hectare (3 tons per acre) per year are viewed as acceptable for agriculture, such erosion rates are too high to maintain surface water quality.

Controlling erosion is the key to minimizing problems with sediments. Soil conservation techniques on cropland were discussed in chapter 8. Other strategies include (1) minimizing the time during which construction and mining sites are left unprotected by vegetative cover, (2) reducing the size of active construction areas as much as possible, and (3) using sediment-retention techniques such as basins and plastic fences (fig. 14.12). Constructing retention basins and directing runoff through swamps and marshes slows water and thereby allows sediments to settle out. Although billions of dollars have been spent on erosion control since the Dust Bowl days of the 1930s, soil erosion is still one of our most pervasive

Figure 14.11 This street is covered with sediment washed from a nearby unprotected construction site. Undoubtedly, a considerable amount of sediment poured into the storm sewers and into surface waterways.

Wisconsin Department of Natural Resources/Photo by Carol Holden

Figure 14.12 A sediment-retention basin captures sediment washed from construction. After construction is complete and stabilizing vegetation is planted, the basin is dredged and used as a pond.

Wisconsin Department of Natural Resources/Photo by Carol Holden

environmental problems. It affects both the productivity of agricultural systems and the quality of aquatic ecosystems receiving sediments.

Toxic Chemicals

Many industrial chemicals (e.g., ammonia, cyanides, sulfides, fluorides, and strong acids and alkalis) pose a direct threat to both humans and aquatic life. Most of these chemicals are used in concentrated forms that are directly toxic to organisms but indispensable to industry. For example, compounds containing cyanide, such as potassium cyanide (KCN), are essential for electroplating everything from electronic components to plumbing fixtures. Other toxic chemicals, such as phenols, are used in the manufacture of plastics. Accidental discharge of these highly toxic wastes may result in massive die-offs of aquatic organisms. Dumping these chemicals into sewage treatment plants can eliminate the microbial cultures that are essential in sewage treatment. Thus, every industry must have a means of on-site treatment and an emergency response plan to deal with a spill of the toxic chemicals used in their manufacturing processes.

People and aquatic organisms can be exposed to toxic chemicals directly or through food webs (chapter 12). Recall that some toxic substances, for example DDT and PCBs, are persistent and accumulate in food webs. Bioaccumulation is especially pronounced in aquatic food webs because they usually consist of four to six trophic levels; in contrast, terrestrial food webs usually have only two or three trophic levels. Perhaps the most dramatic example of food-web accumulation of a toxin occurred in Minamata, Japan in the 1950s. Methyl mercury discharged from a chemical plant into nearby waters bioaccumulated in fish and shellfish. People consuming these

contaminated food sources suffered central nervous system disorders—uncontrollable trembling, loss of coordination, and paralysis. Forty-six people died and another 3,500 experienced debilitating symptoms. For some, the damage was permanent, and they still suffer from their impairments today.

Chlorinated hydrocarbons (compounds composed of carbon, chlorine, and hydrogen) constitute a family of bioaccumulating chemicals that has caused major problems. Some members of this family, such as DDT, aldrin, and dieldrin are pesticides. Although their use in the United States has largely been banned, they are still heavily used in less-developed countries. Their environmental impact is described further in chapter 9. Another subgroup, called PCBs (polychlorinated biphenyls), was used as insulating materials in electric capacitors and transformers, in hydraulic fluids, as heat-exchanging fluids, and in the manufacture of some plastics and a few specialty paper products. Manufacture of PCBs was discontinued in the United States in 1977, but the problems associated with them have not disappeared. The significance of the bioaccumulation of chlorinated hydrocarbons is described in chapter 12.

Are we now unknowingly repeating past errors and introducing new chemicals into the environment that will bioaccumulate? Before we answer this question, we must put past problems in perspective. Scientists have been aware of bioaccumulation only for about two decades. Present problems are largely due to chemicals that were in general use before we recognized the significance of bioaccumulation. Banning of bioaccumulating chemicals has proved to be effective in protecting humans and the environment. For example, levels of DDT and PCBs have decreased in aquatic ecosystems since their use was banned. Scientists now know how to test chemicals for their potential to bioaccumulate,

Figure 14.13   Gold mine waste piles (tailings) in Cripple Creek, Colorado. It is easy to visualize how runoff readily washes these exposed mine wastes into nearby waterways.
© Leonard Lee Rue III/Photo Researchers, Inc.

and all new chemicals must be extensively tested and approved by government agencies such as the EPA before they can be marketed. Today we have a much better early warning system; scientists employ sophisticated instrumentation to detect pollutants in our food and the environment. Furthermore, government agencies and industries now perform much more extensive testing on samples of soil, plants, air, water, wildlife, and food than they did twenty years ago. While these measures do not completely protect us from mistakes such as those made with PCBs and DDT, problems of the magnitude generated in the past are less likely to occur again.

In addition to the indirect threat of bioaccumulation, some chemicals are directly toxic. One source of such chemicals is mine tailings, the waste generated during mineral or rock extraction and processing (chapter 17). Precipitation that seeps through coal-mine waste is often extremely acidic because sulfur-bearing minerals in the tailings react with water and oxygen from the air to form sulfuric acid. Two species of sulfur bacteria, *Thiobacillus ferrooxidans* and *Thiobacillus thiooxidans,* greatly increase the rate at which sulfuric acid is formed. Numerous streams in the Appalachian coal mining areas of Pennsylvania, West Virginia, Ohio, and Kentucky have been adversely affected by acid mine drainage. These streams are clear, but the only life they support is acid-loving sulfur bacteria.

Tailings generated by mining metallic ores can also lead to water pollution by soluble metals (often called **heavy metals**) such as lead (Pb), cadmium (Cd), mercury (Hg), and chromium (Cr). Heavy metals are toxic to aquatic life; they coat gill surfaces, and the animals suffocate. Colorado has a number of previously pristine mountain rivers with such problems (fig. 14.13). Little can be done to solve the problem at some sites because the accumulated volume of tailings is so vast that corrective measures are much too costly.

Although radioactive substances may accumulate in food webs, our exposure is usually more direct. In a few regions of the United States, well water, generally from deep aquifers, is naturally contaminated with radium, a radioactive element that increases the risk of cancer. If radium levels are too high, a municipality must use special treatment methods to remove it prior to distribution, or find an alternative source of drinking water. In the United States, the major pollution problems from radioactive substances occur in surface and groundwater around nuclear weapons production sites. Although nuclear power plants, universities, hospitals, and some industries use radioactive materials, their discharge and disposal is strictly regulated by the Nuclear Regulatory Commission. We discuss the problems of radioactive waste disposal in more detail in chapter 20.

Petroleum Spills

Because modern society cannot get along without oil, oil pollution is an ever-present threat to surface water quality, especially where oil is extracted or transported. In the past decade, several dozen major oil spills have made international news. The worst spill in U.S. history (fig. 14.14) occurred on March 24, 1989, when the submerged rocks of Bligh Reef in Prince William Sound, Alaska, ripped holes along a 200-meter (650-foot) section of the *Exxon Valdez.* The vessel had left the normal shipping lane to avoid icebergs that had calved off nearby glaciers. Nearly 10% of its load, some 42 million liters (11 million gallons) of crude oil spilled into Prince William Sound near the port of Valdez. The spill eventually fouled 1,900 kilometers (1,200 miles) of shoreline with an oily goo.

The oil slick's toll on wildlife in this remote region will never be known, but the tally included at least 36,000 sea birds and 1,011 sea otters (12-17% of the local population). The region's $100 billion per year commercial and sport fishing industry was jeopardized. To allay consumer fears about tainted fish from Alaska, an extensive system was set up to certify the quality of fish that were marketed.

Experts from Alaska, the federal government, and the Exxon Corporation struggled to reduce the spill's impact. Volunteers and 11,000 workers hired by Exxon used every technique imaginable to clean up the spill. Oil skimmers were used at sea to recover as much of the oil slick as possible. Volunteers washed living, oil-soaked birds with detergents and provided warm shelter and food to keep them alive (fig. 14.15). Exxon spent $30,000 per bird and $80,000 per otter that were eventually released back into the wild. Attempts to clean up blackened, slippery shorelines ranged from steam cleaning beaches to wiping individual rocks with absorbent toweling. (Follow-up studies on steam cleaning showed that the clean up effort actually killed more animal life under the rocky shores than the oil killed on shores that were left untreated.) In total, Exxon paid $2 billion for cleanup activities, $900 million in settlement charges, and $125 million in criminal penalties.

One notable benefit from the Alaskan oil spill is that scientists found a more effective way to clean oil-soaked beaches. In a $10 million experiment, they treated 110 kilometers (70 miles) of oil-coated Alaskan beaches with a fertil-

Figure 14.14 The largest oil spill in the history of the United States occurred when the supertanker *Exxon Valdez* struck a submerged rock ledge while navigating through Prince William Sound, Alaska.
© Randy Brandon/Peter Arnold, Inc.

Figure 14.15 Volunteers washing oil-coated duck rescued from an oil spill.
© Greg Vaughn/Tom Stack & Associates

izer called *Inipol.* This product contains nitrogen and phosphorus, which stimulate the activity of naturally occurring oil-consuming bacteria. Within fifteen days, treated beaches were less contaminated than untreated beaches. Elevated numbers of bacteria were still active five months after treatment.

Oil-consuming bacteria have since been used on the world's largest oil spill in the Persian Gulf. That oil spill during the Persian Gulf War of 1991 involved approximately forty times as much oil as the Alaskan spill. This oil has already biodegraded more completely than the spill of heavy Alaskan crude because of higher air temperatures and because the oil itself is composed of a larger proportion of biodegradable molecules.

Spilled oil is naturally dispersed via several routes. The insoluble fraction, which is by far the largest, is lighter than water and gradually spreads and thins to form an ever-widening oil slick on the water's surface. The most volatile (and toxic) components in crude oil (typically 30%) begin to vaporize immediately. Where there is wave action, the oil is whipped into an oil-water emulsion called *mousse,* which is 50% to 70% water. Wave action and currents also disperse the mousse throughout the upper layer of the water, where bacteria break down most of the oil's components. Oil that is not rapidly degraded by bacterial action collects to form floating tar balls varying in size from marbles to baseballs, which are unsightly but relatively nontoxic. Crude oil emulsions can collect sediment, and if it becomes denser than water, it sinks, coating and killing bottom-dwelling organisms. If a spill occurs near shore, waves transport oil and tar balls to beaches and tidal-zone wetlands.

The ecological consequences of oil spills are not easy to assess. Near-shore oil spills that do not disperse rapidly are particularly damaging because they affect many vulnerable marine organisms. Layers of oil coat birds (e.g., cormorants) and aquatic furry animals (e.g., sea otters), destroying their natural insulation and buoyancy. Most victims die of exposure or drown. Evidence also suggests that environmental stresses such as low temperatures act synergistically with oil (chapter 4). In several oil-spill incidents, even light coatings of oil have caused the death of seabirds. The effects of oil on the water surface generally last for a few days to a few weeks (a few months in cold climates). However, oil that sinks to the ocean floor is thought to have longer-term impact on marine ecosystems. Initially, pollutant levels from a spill are lethal to shellfish. As pollutant levels decline, shellfish begin to reestablish in the area, but they acquire an offensive taste and smell. Thus, commercial shellfisheries cannot market their catch for several seasons.

The long-term effects of oil spills are difficult to assess, but the major impacts (heavy oil coatings and bird and animal kills) are gone within one year on wave-exposed beaches. Protected beaches may have visible oil or oil buried under sediments for several years. The most vulnerable areas are the intertidal wetlands, aquatic ecosystems that experts recommend not be cleaned because cleanup crews would do more damage than good.

How do we prevent the economic, social, and environmental tragedies that accompany major oil spills? Two approaches are used: spill prevention and better cleanup methods. A successful spill-prevention program must focus on every aspect of handling oil, from the drilling rig to its ultimate destination. Large bulk shipments of oil, of course, present the greatest potential risk; thus, safe navigation practices are critical for the giant supertankers that move so much of the world's oil. Well-trained crews, stringent drug and alcohol abuse policies, and sophisticated navigation equipment lessen the chances of an accident.

The use of double-hulled ships has been proposed to minimize the number of spills associated with oil transport. (A double-hulled ship is a ship within a ship and costs about 15% to 25% more to build. Amoco expects to put the first one in service in 1993.) The U.S. Coast Guard estimates that double-hulled ships would eliminate 95% of all oil spills, but they probably would not prevent spills in the case of major accidents, such as the one involving the *Exxon Valdez.*

Quick response to oil spills is also essential to minimize their detrimental effects. Thus, communities must have emergency response plans designed to deal with a major accident. Personnel and equipment must be immediately available to contain a spill. (This was not the case for the *Exxon Valdez* spill.) There is no substitute for spill prevention when it comes to cutting the environmental costs associated with spills of petroleum products. But no matter how many precautions are taken to prepare for an accidental oil spill, experts agree that there is no way to prevent damage from a spill as large as that from the *Exxon Valdez.*

Thermal Pollution

Thermal pollution, the raising of water temperature to levels that harm aquatic organisms, is primarily associated with large-scale generation of electricity. Power plants produce tremendous amounts of waste heat and require enormous quantities of water for cooling. This water is usually drawn from the ocean (bays and inlets), major rivers, or large lakes. Heated discharge water affects a feather-shaped mixing zone called a **thermal plume** emerging from the discharge pipe. The thermal plume is confined to the top 1 or 2 meters (3 to 6 feet) of water.

Aquatic organisms such as fish, oysters, and clams, whose body temperatures are nearly the same as that of their surroundings, are particularly vulnerable to thermal pollution. When the water temperature rises, their body temperatures also rise, causing an increase in their respiration rates, which, in turn, increases their oxygen demand. As noted earlier, water contains less dissolved oxygen at higher temperatures, so those animals suffer oxygen stress. Two approaches are used to minimize thermal pollution problems. One strategy is to dissipate waste heat into the atmosphere by using cooling towers or extensive cooling ponds (fig. 14.16). The second approach is to find some use for the waste heat. A few power plants send their warm discharge water to a nearby industry that can use it in some process; such facilities are called *cogeneration plants* (chapter 19).

Maintenance and Rehabilitation of Aquatic Ecosystems

We have examined specific types of water pollutants, their effects, and some of the principal means used to control them. At first glance, we might think that water pollution problems can be solved simply by eliminating the sources of pollution, but many economic and environmental factors can hamper efforts to improve water quality. In the next two sections, we examine some of the strategies used to combat water pollutants in surface water and groundwater.

Earlier in this chapter, we pointed out that water pollution problems are best addressed on the basis of the drainage basin involved. Each drainage basin has a unique set of water-quality problems. To gain insight into the strategies for solving water pollution problems, we discuss the approach being taken in the Great Lakes region.

As noted at the beginning of this chapter, the United States and Canada work cooperatively to clean up the Great Lakes through the International Joint Commission (IJC). In 1987, the IJC identified forty-two **areas of concern (AOC),** around the Great Lakes. In each AOC, local, state, or provincial scientists, policymakers, and concerned citizens cooperate to develop a **remedial action plan** that addresses the specific problems of their AOC. Each plan is submitted to the International Joint Commission for review and comment. Each remedial action plan involves a three-stage process: In stage 1, pollution problems are defined; in stage 2, remedial and regulatory measures are selected; and in stage 3, beneficial uses of the AOC are restored. Most of the areas of concern have completed their stage 1 reports and are into stage 2.

(a)

(b)

Figure 14.16 The two principal means of dissipating waste heat at an electric-power generating plant are (*a*) evaporative cooling towers such as these at the Three Mile Island nuclear power plant near Middletown, Pennsylvania, and (*b*) a cooling canal system such as this one at Turkey Point nuclear power plant, Miami, Florida. (*a*) © Breck P. Kent/Earth Scenes (*b*) © John J. Bangma/Photo Researchers, Inc.

Some AOCs focus primarily on toxic substances. One project on the Clinton River north of Detroit, which empties into Lake St. Clair, involves the cleanup of a former waste-incineration site. The facility is heavily contaminated with benzene (a carcinogen), lead (a heavy metal), methylene chloride (a solvent), and other toxic substances. Approximately 850 different industries used the facility in the 1960s and 1970s. About 535 companies have agreed to share in the

$22 million cleanup costs; the rest of them are being pursued by EPA officials for payment. The United States EPA has agreed to use funds from its Superfund program to clean up this hazardous-waste site and a number of others in the watershed (chapter 20). Under the Superfund program, initial funding for cleanup of a hazardous-waste site is provided by the United States government, but it is empowered to recover cleanup costs from parties responsible for dumping the hazardous wastes. Other AOCs that focus on toxic materials include the Waukegan, Illinois, harbor (PCB-contaminated sediments), at an estimated cost of $21 million, and Duluth, Minnesota, which is dealing with the toxic residues from a former steel mill at an estimated cost of $3.1 million.

Some AOCs focus primarily on cultural eutrophication. These regions are conducting studies to determine the quantity of nutrients that must be removed to reverse cultural eutrophication. They are also developing plans on how to distribute the cost of nutrient removal fairly among the industrial and municipal treatment plants that discharge them. More stringent discharge limits may be required for industrial and municipal treatment plants to meet goals to reduce cultural eutrophication.

Figure 14.17 shows some of the pollutants that find their way into storm sewers and thus contribute to water-quality problems in urban regions. The following list suggests how we can help keep pollutants out of storm sewers and thus out of our waterways.

1. Keep soil, leaves, and grass clippings off driveways, sidewalks, and streets.
2. Do not pour waste oil, antifreeze, paints, pesticides, or other toxic or hazardous materials down the storm sewer.
3. Compost yard wastes and use them on gardens or shrubbery.
4. Dispose of pet wastes in the sanitary sewer (toilet), or bury them.

Efforts to combat cultural eutrophication also involve solving combined-sewer overflow problems, the cost of which is estimated to run into the billions of dollars. The cost for separating sewers in the Detroit metropolitan area alone is estimated to be $2.65 billion. In some AOCs, stormwater retention basins are being built to reduce the nutrient load. Toronto estimates that it will spend up to $2 billion to solve urban runoff problems associated with streams that drain into Lake Ontario.

The Great Lakes region is not the only place where major efforts are underway to improve water quality. Chesapeake Bay has also suffered a significant deterioration in water quality during the past several decades. This deterioration is discussed in "Reducing Nonpoint-Source Pollution in Chesapeake Bay" on page 305.

Labels visible in figure:

Leaves

Grass clippings

Auto exhaust
Motror oil
Lubricants
Gasoline
Tire wear

Pesticides,
Pet wastes

Phosphorus, nitrogen, bacteria, BOD

Bacteria, BOD

Toxic materials

Lead, cadmium, zinc, hydrocarbons

Paint

Lead

Eroded soil

Sediments, BOD

Metal corrosion

Fertilizer

Phosphorus, nitrogen

Zinc, copper, chromium

BOD, bacteria

Street litter

To nearby surface water

With thousands of storm-sewer inlets around town, stormwater is a major contributor to water pollution in urban areas. Although each storm-sewer inlet contributes only a small number of pollutants, when added together, pollution concentrations often exceed the limits established for industries and wastewater treatment plants. If the pollutants entering each of these inlets can be reduced, so will the pollution in area streams and bays.

Figure 14.17   Many of our activities in urban areas lead to contaminated storm-sewer runoff, which flows directly into surface waters.

Source: Adapted from Steve Bennett, University of Wisconsin Extension.

Reducing Nonpoint-Source Pollution in Chesapeake Bay

*I*f you travel in the Washington, D.C. area, including Maryland and Eastern Virginia, you are bound to see "Save the Bay" bumper stickers. They refer to the Chesapeake Bay, a 320-kilometer(200-mile)-long, shallow, fragile estuary whose drainage basin extends from the Finger Lakes region of New York to the coastal plain of Maryland (fig. 14.18). The bay still supports several thousand commercial fishermen who harvest half of the blue crabs and one-fifth of the oysters consumed each year in the United States. More than two million people hunt and fish for recreation on this famous mid-Atlantic estuary. However, in the last twenty-five years, the bay has lost 80-90% of the underwater grass beds that provide habitat for fish and food for birds. Many species of fish, including striped bass, shad, yellow perch, alewife, and blueback herring, have experienced steep declines in population. The oyster population is estimated to be only 1% of what it was a century ago because of pollution, disease, and overharvesting. Some of the losses of the fishery of Chesapeake Bay, however, have been partially offset by increased catches of saltwater fish, such as bluefish and menhaden, that migrate and feed in the bay.

What has caused so much of this shallow productive bay to become barren? Although overfishing has contributed to population declines, pollution from runoff appears to be the major culprit. The Chesapeake Bay is responding to changes in its 166,000-square-kilometer (64,000-square-mile) drainage basin. Nonpoint-source runoff feeds into the bay from fifty rivers. Through the years, that runoff has become more nutrient-rich and sediment-laden. Table 14.3 shows the importance of various sources of nitrogen and phosphorus from the Susquehanna River, the major contributing river at the north end of Chesapeake Bay. Excessive amounts of phosphorus and nitrogen have stimulated considerable growth of free-floating microscopic algae that shade the underwater grass beds. Sediment-laden water has had a similar effect. The overabundance of algae has also led to reduced levels of dissolved oxygen, which stressed or killed fish and shellfish.

Scientists believe it is possible to reverse these negative impacts on Chesapeake Bay. The plans for rehabilitating Chesapeake Bay appear straightforward, but implementation of such a multistate effort is a tremendous challenge. Changes in land use are required in both the agricultural and the urban-industrial sectors.

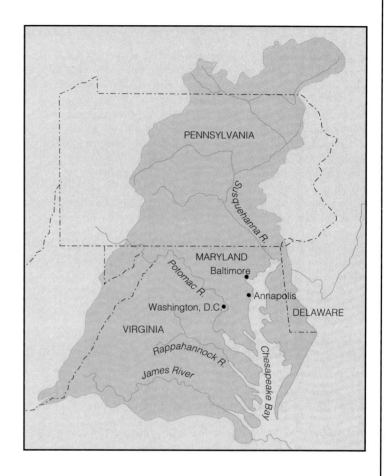

Figure 14.18   The Chesapeake drainage basin.

Table 14.3	Sources of nitrogen and phosphorus in the Susquehanna River	
Source	**Nitrogen (%)**	**Phosphorus (%)**
cropland	85	60
wastewater treatment plants	9	16
industry	1	7
other	5	17

Source: *Bay Fact Sheet*, Department of Agricultural Engineering, Pennsylvania State University.

Farmers, primarily in Pennsylvania, Maryland, and Virginia, will have to manage fertilizers much more carefully than in the past, which will require an extensive educational program. One strategy is to show farmers via demonstration plots throughout the basin that (1) heavy application of manure and/or commercial fertilizer does not always increase crop yields, (2) the highest profits come from producing the highest crop yield with the lowest costs; striving for maximum crop yield usually does not yield the greatest profit, (3) proper management of manure and commercial fertilizers saves money, and (4) testing soil and manure for nutrients is essential to sound management decisions. In addition, efforts are underway to reduce soil erosion on agricultural lands. Various state agencies within the drainage basin have established programs to help farmers meet these goals.

Urban areas also have to develop programs to reduce nonpoint-source pollutants in urban runoff. Nutrient-rich stormwater runoff from cities may have to be treated to reduce nutrient and sediment loads prior to its release into the estuary. Cities with combined sewer systems (for example Richmond, Virginia, and Washington, D.C.) must develop the means to deal with sewer overflows during heavy rains. Other remedial actions under consideration include (1) adoption and enforcement of erosion-control ordinances, (2) requiring stormwater management controls in new developments, (3) keeping streets cleaner, especially in spring and fall, and (4) developing programs to handle yard wastes.

Another major objective is the preservation and restoration of forests and grasslands. Landscapes with complete vegetative cover yield fewer sediments and nutrients to runoff than disturbed landscapes. Widespread conversion of natural areas to cropland and urban development has contributed significantly to nonpoint-source runoff problems. Thus, restoration of natural areas of forests and grasslands to intercept nutrients and sediments is a major goal. Projects are also underway in the drainage basin to stabilize streambanks, shorelines, and nontidal wetlands with buffer zones of natural plant communities.

In summary, the rehabilitation of Chesapeake Bay requires the cooperation of nearly everyone who lives in its drainage basin. Furthermore, rehabilitation depends as much on the control of nonpoint sources of pollutants from its vast drainage basin as it does on the traditional control of point sources of pollutants.

Groundwater Pollution

The same drainage-basin activities that pollute surface waters can also contaminate groundwater. Poorly designed landfills, agricultural activities, septic tanks, industrial waste lagoons, underground storage tanks (fig. 14.19), and petroleum and natural gas production can all lead to groundwater pollution (table 14.4). Groundwater pollution can be a serious problem in view of our dependence on this resource. As noted in chapter 13, about 22% of the U.S. freshwater supply is drawn from the ground, and many states rely on groundwater for more than 60% of their supply. Furthermore, as we also noted in chapter 13, groundwater is particularly vulnerable to contamination because of its slow and nonturbulent motion.

Groundwater Pollutants

As water seeps into the ground, its suspended and dissolved substances interact with soil and rock. Soils naturally filter suspended solids, bacteria, and some viruses, and retain them in the uppermost soil layers. Some chemicals (e.g., phosphate, many pesticides, and heavy metals) chemically bind to the surface of soil particles, which limits their movement. The uppermost layers of soil also host a wide array of soil organisms including bacteria, fungi, and earthworms. These organisms actively decompose most organic materials, such as dead plant and animal matter, as well as many pesticides.

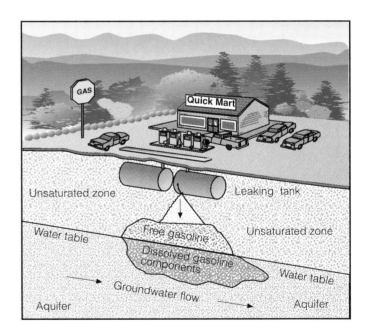

Figure 14.19 Petroleum liquids such as gasoline are less dense than water and therefore "float" on the water table. Enough gasoline dissolves in the groundwater, however, to contaminate it.

Table 14.4 Relatively well-documented organic contaminant plumes in sand–gravel aquifers[a]

Site location and plume map 0 5 km Flow →	Presumed sources	Predominant contaminants	Plume volume (liters)	Contaminant volume dissolved in plume (liters or 55-gal drums)
Ocean City, NJ	chemical plant	TCE TCA PER	5,700,000,000	15,000 (72 drums)
Mountain View, CA	electronics plants	TCE TCA	6,000,000,000	9800 (47 drums)
Cape Cod, MA	sewage infiltration beds	TCE PER Detergents	40,000,000,000	1500 (7 drums)
Traverse City, MI	aviation fuel storage	Toluene Xylene Benzene	400,000,000	1000 (5 drums)
Gloucester, ON, Canada	special waste landfill	1, 4 Dioxane Freon 113 DEE, THF	102,000,000	190 (0.9 drum)
San Jose, CA	electronics plant	TCA Freon 113 1, 1 DCE	5,000,000,000	130 (0.6 drum)
Denver, CO	trainyard, airport	TCE TCA DBCP	4,500,000,000	80 (0.4 drum)

[a]TCE = trichloroethylene; TCA = 1, 1, 1 trichloroethane; PER = per-, i.e., tetrachloroethylene; 1, 1DCE = 1, 1 dichloroethylene; DEE = diethyl ether; THF = tetrahydrofuran; DBCP = dibromochloropropane.

Adapted with permission from J. M. Thomas, D. M. MacKay, and J. A. Cherry, "Ground Water Contamination: Pump-and-Treat Remediation," *Environmental Science and Technology*, Vol. 23, No. 6, 1989, p. 631. Copyright © 1989 American Chemical Society.

Infiltration through soil is so effective a purifying mechanism that septic tanks are sited to take advantage of soil's filtration capacity (fig. 14.20). In some areas, industrial and municipal wastewater is sprayed on the ground surface so that percolation through the soil will purify it. Spray irrigation and ponding both employ infiltration to purify wastewater and recharge groundwater supplies at the same time (chapter 13). In the United States, approximately six hundred communities, mostly in arid and semiarid areas, withdraw groundwater from sources augmented by municipal effluent that has seeped through the zone of aeration to the zone of saturation.

Purification by filtration is impaired by (1) shallow water tables, (2) soils with low permeability, (3) extremely permeable soils, and (4) near-surface bedrock (especially if fractured). Where these conditions exist, wastewater is not in contact with the soil long enough for adequate purification. For example, water passes through sandy soils to the water table so quickly that effective filtration does not occur. If the water table is too close to the soil surface, the travel distance is too short for effective filtration. Highly fractured dolomite underlies the Yucatan Peninsula of Mexico; thus, pollutants from cesspools and accumulated animal wastes seep into cracks, which act as conduits that transport contaminants to drinking water wells (fig. 14.21). Contaminated drinking water has caused infectious hepatitis to reach epidemic proportions in the Yucatan, and dysentery now ranks as the leading cause of death in the region.

Choosing appropriate soil conditions as sites for septic disposal systems is important because about 4 trillion liters (1 trillion gallons) of wastewater per year are discharged directly into the ground from domestic sources. In the United States, approximately 20 million single-family housing units (40–50 million persons) use onsite wastewater disposal. Such systems are the most frequently cited sources of bacterial contamination of groundwater, especially the contamination of private wells. (Approximately one-half of the 520 incidents—111,000 people—of waterborne diseases reported in the United States between 1971 and 1985 were attributed to groundwater contamination.) The problem is particularly acute in heavily settled suburban areas served by individual septic disposal systems. The high density of private wells and septic tanks greatly increases the chances that people will drink contaminated groundwater.

In areas where the soils are not suitable for adequate filtration and a septic system is needed, some states require a mound system (fig. 14.22). These systems are constructed by hauling suitable soil to the site and shaping it into a mound equipped with a perforated-pipe drainage system. Wastewater

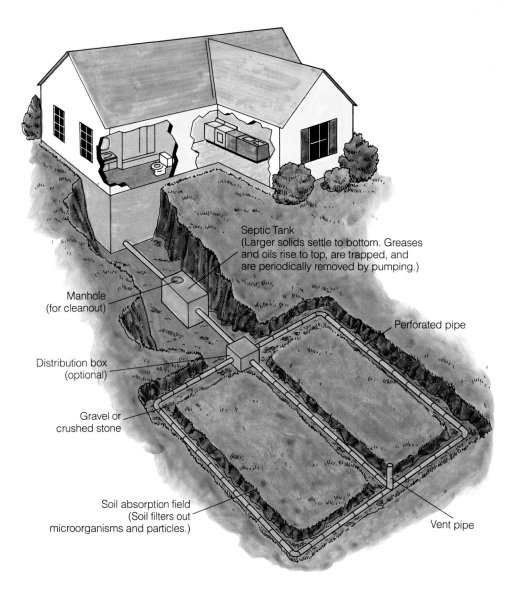

Figure 14.20   Cross section of a septic tank drainage field system, which is used for domestic waste disposal in rural and suburban areas.

Septic Tank
(Larger solids settle to bottom. Greases
and oils rise to top, are trapped, and
are periodically removed by pumping.)

Manhole
(for cleanout)

Distribution box
(optional)

Gravel or
crushed stone

Soil absorption field
(Soil filters out
microorganisms and particles.)

Perforated pipe

Vent pipe

Figure 14.21   Where soils are thin and fractured bedrock is close to the surface, downward seeping water is not adequately filtered.

is periodically pumped from a large underground holding tank into the drainage system.

Groundwater contamination usually occurs in shallow, unconfined aquifers (chapter 13), where the contamination source is nearby (fig. 14.23). Confined aquifers are often recharged tens or even hundreds of kilometers away from the point of water withdrawal; it can take decades for water (and pollutants) to travel from the point of recharge to wells that tap a deep, confined aquifer. In some cases, contamination of near-surface aquifers may force individuals or communities to resort to costly drilling in order to tap a deep, confined dwelling. In some regions, however, this option may not be available because there is only one aquifer or the deep aquifer(s) is (are) naturally contaminated with brine.

Some water pollutants are not retained by soil particles; they percolate directly into the groundwater. Examples of such contaminants include nitrate (NO_3^-), road salt (sodium chloride), petroleum products (gasoline and diesel fuel), the

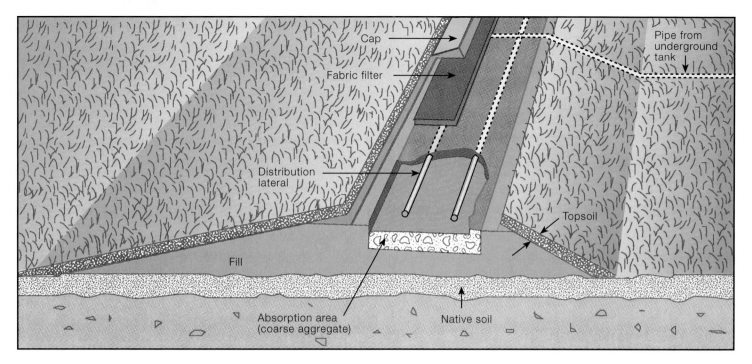

Figure 14.22 Mound septic disposal systems are used in rural areas for new homes or to retrofit failing septic tank systems located in unfavorable soils.

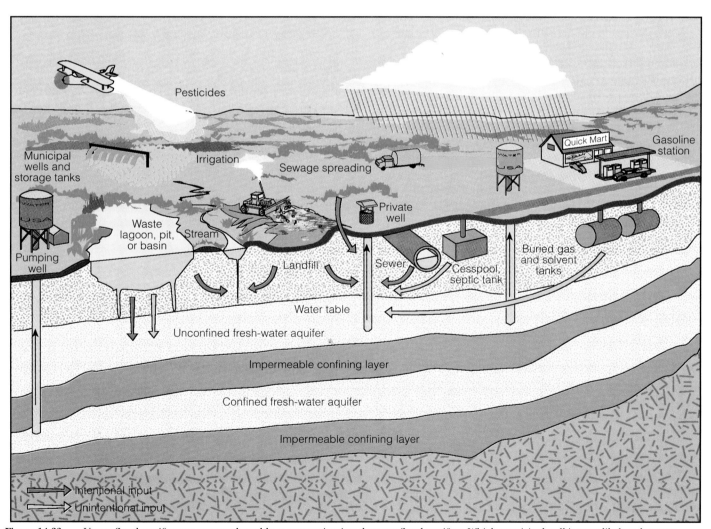

Figure 14.23 Unconfined aquifers are more vulnerable to contamination than confined aquifers. Which municipal well is most likely to become contaminated?

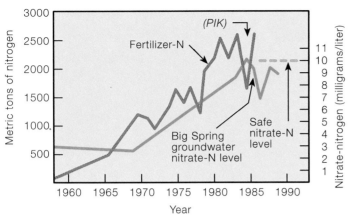

Figure 14.24 The amount of nitrogen-containing fertilizers used in Clayton County, Iowa's Big "Spring" Basin has increased dramatically over the past two decades as have the nitrate levels in the groundwater. The PIK on the diagram means Program In Kind, a government program that paid farmers for allowing land to lie fallow.

Source: Data from *Groundwater Monitoring in the Big Spring Basin 1984–1987: A Summary Review,* Iowa Department of Natural Resources, 1989.

herbicide atrazine, the cleaning solvent trichloroethylene (TCE), and the dry-cleaning chemical perchloroethylene.

Nitrate is a particularly troublesome substance. It seeps into the ground with water from fertilized fields and lawns, feedlots, and septic tanks. It is highly soluble in water and is not adsorbed by soil particles (chapter 3). Nitrate is a human health hazard when it gets into well water used for drinking. Its consumption is linked to methemoglobinemia, commonly called **blue-baby disease,** an often fatal condition in newborn infants that results from oxygen depletion in the blood. In the Central Valley of California, groundwater concentrations of nitrate (from heavy fertilizer application) have reached levels that are considered to be a public health hazard, and physicians recommend that only bottled water be used for infants' formulas. The corn-producing areas of the Midwest are also experiencing a rise in nitrate levels caused primarily by excessive use of fertilizer (fig. 14.24). Nitrate buildup in groundwater is occurring there even in regions where the soils are 20 meters (65 feet) or more thick.

Leaking Underground Storage Tanks

Groundwater contamination from leaking underground storage tanks (LUSTs) was first recognized in the 1980s as a serious source of groundwater contamination. The problem is particularly acute for petroleum products (gasoline and diesel fuel) stored in underground steel tanks at service stations, farms, and many industries and businesses. The tanks and the pipe networks used to fill or empty them corrode and develop slow leaks that often go undetected for years. The EPA estimates that 25% of all systems installed prior to 1980 are leaking. Some states are finding the failure rate much higher. For example, in Wisconsin alone in 1989, 1,800 cases of leaking underground storage tanks were under investigation.

In most cases, leaks from underground storage tanks travel undetected through the soil to the water table (fig. 14.19). Permeable soils and ample rainfall accelerate the process. Since the fuels are less dense and are insoluble in water, they float on the water table. Enough of the fuel, however, mixes with the groundwater to make it unfit for use. In a typical situation, the underground plume of contamination travels between 10 and 100 meters (32 and 320 feet) per year away from the site of entry.

The LUST problem was so serious that the EPA quickly initiated programs to control the threat to subsurface water supplies. A timetable was set for installation of leak-detection systems around buried fuel tanks. Tanks installed prior to 1965 had to have such systems in place by 1989. By December 1993, all underground tanks must be equipped with leak detectors. Owners of old tanks have usually chosen to dig up their tanks and replace them with new tanks (usually fiberglass).

One troublesome aspect of groundwater pollution by petroleum and solvents is that a small spill can contaminate a huge volume of groundwater. For example, at Denver, Colorado, 80 liters (20 gallons) of several organic solvents contaminated 4.5 trillion liters (1.1 trillion gallons) of groundwater. The contaminated area is about 5 kilometers (3 miles) in length.

Groundwater Cleanup and Protection

Groundwater contamination is extremely difficult and costly to clean up because both the water and the aquifer material (sediment or rock) must be cleaned. Petroleum leaks are usually cleaned up by drilling wells into the contaminated area and skimming the petroleum off the top of the water table with pumps. In addition, contaminated groundwater is pumped out and aerated to rid it of the dissolved fuel. In some cases, more elaborate water treatment (for example, carbon filtration) is required to achieve the desired water quality before water is returned to the environment. Other techniques to remedy soil-groundwater contamination problems include air stripping (compressed air is injected underground to remove volatile contaminants), biodegradation (bacteria are injected underground along with a supply of oxygen and nutrients to speed biodegradation), soil aeration (contaminated soil is removed and exposed to the open air to rid it of pollutants), and incineration (the contaminated soil is passed through an incinerator to burn combustible contaminates).

Some groundwater contamination may be virtually impossible to clean up because the aquifer materials are difficult to treat or the contamination has penetrated deep into the ground. Where cleanup is possible, the projects usually require tens to hundreds of thousands of dollars, even on small sites such as gasoline stations, and the remediation process typically requires several years to complete.

Most people agree that groundwater is a valuable resource and must be protected, but they often disagree about how it should be protected and who should pay for protective measures. Conflicts arise on such issues as whether to restrict the kinds of fertilizers and pesticides farmers can use and their rate and frequency of application. Water managers still argue about what constitutes an adequate groundwater monitoring program to assure that groundwater pollution is not occurring.

Water managers now recognize that in some regions, certain practices must be curtailed in order to protect groundwater quality. For example, persistent pesticides must not be used on groundwater recharge areas where soils do not retain pesticides. Although pesticides are vital to agriculture, some of them—especially water-soluble types—are now being found in groundwater, particularly in irrigated areas. In recent years, the insecticide aldicarb has been found in groundwater in Florida, Long Island, and in potato-growing areas in the upper Midwest. Other potentially dangerous practices include the heavy use of commercial fertilizers and manure, and the application of septic tank wastes to land, all of which can lead to nitrate contamination. We saw earlier (fig. 14.24) how groundwater nitrate levels have climbed in Clayton County, Iowa, over the past two decades as fertilizer use has increased in the region. That same study found the herbicide atrazine in 98% of the groundwater samples collected from the region between 1981 and 1987. Groundwater contamination from these sources can be reduced by lighter, less frequent application of commercial fertilizers, pesticides, and septic wastes. Many landowners are reluctant, however, to switch to such practices because they are more time-consuming, may cut crop yields, and are more expensive. Furthermore, assuring compliance with regulations regarding pesticide and fertilizer applications often presents a bureaucratic nightmare.

Contamination of groundwater reservoirs by toxic chemicals is another major problem. Spills from transportation mishaps, accidental discharges from plant equipment, and leakage from storage tanks occur every day. The chemical industry, transportation companies, and state and local governments have done much to prevent spills and to clean them up once they occur, but costs are high. The chemical industry has focused much attention on determining the type of equipment and procedures that are necessary for emergency response in the event of a spill. In fact, chemical plants are required by the 1986 amendments to the Comprehensive Environmental Response, Compensation, and Liability Act (Superfund) to report the amounts and types of hazardous chemicals they have on hand to the EPA. Furthermore, this law requires industries to help communities maintain emergency response units specifically equipped and trained to deal with spills of hazardous chemicals. The goal of these response teams is to reduce exposure of the public and the environment by minimizing promptly the size and duration of spills.

Legislative Responses to Water Pollution

The first federal action to improve water quality took place in 1899 with passage of the Rivers and Harbors Act, better known as the Refuse Act. This law prohibits the dumping of trash into rivers and harbors, but it had no provision against the discharge of sewage into navigable waterways. Unfortunately, the law was soon forgotten, and it was not used to prosecute violators until the 1970s.

The cornerstone of federal water-quality legislation is the 1972 Water Pollution Control Act and its subsequent amendments. This act, usually called the Clean Water Act, is one of the most ecologically enlightened and powerful pieces of environmental legislation ever passed in the United States. For example, it recognizes that some wastes concentrate in food webs; hence, it establishes procedures to control the release of bioaccumulating chemicals. It also recognizes that in some drainage basins, nonpoint-source pollutants pose as great a threat to water quality as point-source pollutants. The goal of the Clean Water Act was to make U.S. surface waters "fishable and swimmable." The act is responsible for marked improvement in the water quality of our lakes, rivers, and streams. There are still problem areas, however, such as Boston Harbor, where the goals of the Clean Water Act have not been met, and many other waterways still are not fishable and swimmable because of both point and nonpoint-source pollution.

Another example of important water-quality legislation is the Safe Drinking Water Act of 1974. One of its principal provisions sets maximum concentration levels for specific substances, so that they have "no known or anticipated adverse effects on the health of persons" and which "allow a margin of safety." The EPA was granted authority to establish and enforce these standards. Initially, the EPA set standards for only 17 chemicals (8 heavy metals, 6 pesticides, trihalomethanes, nitrate, and fluoride) and radioactivity. Amendments to the act in 1986 require the EPA to set maximum levels for 83 contaminants by 1991 and to set standards for another 25 contaminants every three years beginning in 1990. The amendments also call for water utilities that use surface water to install filtration systems that meet specific disinfection requirements for control of *giardia* (a protozoan prevalent in the United States) and enteric (intestinal) viruses. (See table 14.1 for symptoms of giardia.)

Conclusions

The last two decades have brought many improvements in water quality. Some rivers that had no fish in the 1970s and 1980s are again crowded with fishermen. Beaches that were closed have been reopened. Also levels of bioaccumulating chemicals in fish have declined in many areas over the past decade. Some difficult problems, however, persist. In many regions, improvements in surface water quality will come about only through control of nonpoint-source pollutants, sources that are difficult and costly to control. Financing such projects remains one of the major challenges of the 1990s.

The other major challenge is developing measures to slow groundwater contamination. As with surface waters, groundwater contaminants from nonpoint sources present the greatest challenge. To protect groundwater in the future, application of fertilizers and certain pesticides will probably have to be curtailed on a regional basis. Policies for sound water-quality management should consider surface water and groundwater as a single resource. To ensure the quality of future water supplies, each region of the country must determine its own optimal conditions for the use of surface and groundwater, and decide how to control effluents in ways that optimize user requirements without depleting or contaminating water resources.

Less-developed nations face enormous water-quality problems. They must first address the human suffering and death caused by waterborne pathogens. They also need to reduce other major pollutants such as plant nutrients, oxygen-demanding waste, and toxic chemicals such as pesticides. Financing such projects will be nearly impossible without significant monetary assistance from the more-developed nations.

Key Terms

point sources	anaerobic environment
nonpoint sources	aerobic decomposers
hard water	anaerobic decomposers
storm sewer	oligotrophic ecosystems
sanitary sewer	eutrophic ecosystems
separated sewer system	eutrophication
combined sewer system	cultural eutrophication
pathogens	sediment
coliform bacteria	heavy metals
chlorination	thermal pollution
schistosomiasis	thermal plume
biochemical oxygen demand (BOD)	areas of concern (AOC)
oxygen sag curve	remedial action plan
aerobic environment	blue-baby disease

Summary Statements

A drainage basin is a geographical area drained by a river and its tributaries. A variety of natural and human activities within a drainage basin govern the quality of surface and groundwater.

Until recently, waterborne diseases were the primary concern with regard to water pollution. Chlorination of water supplies has alleviated that danger in the United States and some other more-developed nations, but waterborne diseases still afflict hundreds of millions of people in the less-developed nations.

Decomposition of organic wastes in water by bacteria and fungi removes dissolved oxygen from the water. The concentration of organic material in water is measured by its biochemical oxygen demand (BOD). Surface waters that have reduced levels of dissolved oxygen are inhabited by pollution-tolerant organisms, most of which we consider undesirable.

Increased levels of phosphorus and nitrogen (plant nutrients) caused by human activity lead to cultural eutrophication. Eutrophic bodies of water have low water quality, making them less able to support desirable aquatic life and less suitable for municipal and industrial use.

Sediments fill reservoirs, lakes, harbors, and navigation channels. Suspended sediments impede photosynthesis and transport nutrients into surface waters; as sediments settle out, they cover bottom-dwelling organisms and destroy spawning areas.

Nonpoint-source pollutants often contribute to water-quality problems in a drainage basin. Control of nonpoint sources is more difficult and costlier than control of point sources. In many areas, these sources must be controlled before local water quality will improve.

Many chemicals are directly toxic to aquatic life; others become more toxic through bioaccumulation. Bioaccumulating chemicals can threaten the health of humans and animals that feed on fish and shellfish harvested from contaminated aquatic ecosystems. Chemicals used in the environment (e.g., pesticides) are more stringently screened and monitored today compared to a decade ago; some still cause problems.

Oil spills kill most organisms by coating them with oil. The long-term effects of oil spills are not thought to be serious, except for intertidal wetlands. Oil-spill prevention and immediate containment when a spill does occur are the best strategies for preventing the tragic results of oil spills.

Heated water discharged into the environment may cause oxygen stress in aquatic organisms. Therefore, it is cooled by either cooling towers or holding ponds prior to discharge.

The quality of groundwater is controlled mainly by purification processes—biodegradation by soil organisms, filtering, and adsorption—that occur as water infiltrates the soil. Areas with sandy soils or thin soils over fractured bedrock are especially susceptible to groundwater contamination.

Defective or improperly sited septic-tank systems in suburban and rural areas endanger groundwater quality; so do faulty landfills, wastewater ponds, and fuel storage tanks. To protect groundwater quality in some areas, it may be necessary to restrict such practices as heavy fertilization of fields and the use of certain chemicals in recharge areas.

Questions for Review

1. What is a drainage basin? What natural and human activities within a drainage basin affect the quality of surface and groundwater?

2. Distinguish between point and nonpoint sources of water pollution. Provide several examples of each.

3. How does a separated sewer system differ from a combined sewer system? Which system is preferred where the climate is rainy?

4. What are pathogenic organisms? How and why does the incidence of waterborne diseases differ between less-developed and more-developed nations?

5. Explain how the dissolved oxygen concentration varies with a water body's (a) temperature, (b) turbulence, (c) surface area, and (d) content of organic matter.

6. Identify several sources of oxygen-demanding wastes. Explain why the dissolved oxygen concentration decreases (sags) downstream from a point source of organic waste.

7. Distinguish between aerobic and anaerobic decomposition. Contrast the by-products of these two processes. Describe the appearance of a pond that is undergoing anaerobic decomposition.

8. In what sense does a river or stream cleanse itself?

9. Distinguish between eutrophication and cultural eutrophication. Describe the changes that take place as a lake undergoes eutrophication. What are the major sources of nutrients that cause cultural eutrophication?

10. How might an algal bloom affect the cold-water species of fish living in a deep lake or reservoir?

11. Explain why the species of organisms that inhabit an aquatic ecosystem change as the water quality changes.

12. What activities accelerate soil erosion? What is the impact of soil erosion on surface water quality?

13. What are heavy metals and why do they persist in the environment? Identify some sources of heavy metals that end up in waterways.

14. What properties make oil lethal to aquatic birds and furry animals?

15. How might a large oil spill affect a region's economy? Would you expect these effects to be permanent? Explain your response.

16. What is meant by thermal pollution, and how can it be prevented?

17. In what ways are problems with the quality of surface water linked to problems with groundwater quality?

18. Why is high nitrate concentration in groundwater a cause for concern?

19. Water that seeps into the ground is filtered (and hence cleansed) to some extent by interacting with soil particles and other sediments. What properties of the subsurface govern the degree of filtration?

20. What steps can be taken to safeguard groundwater quality in agricultural and urban-industrial areas?

Projects

1. Make a list of nonpoint-source pollutants that are causing surface water-quality problems in your area. Propose a strategy for controlling them. Do the same for groundwater contamination problems.

2. Determine to what extent the sanitary sewers and storm drains are separated in your area. What water-quality problems, if any, develop in your area during periods of heavy runoff? Are there proposals to improve stormwater quality in your area?

3. Arrange for a tour of your local sewage treatment plant. Find out how efficiently suspended solids, biochemical oxygen-demanding substances (BOD), and phosphorus and nitrogen compounds are removed from sewage. Does your plant meet the performance requirements imposed by your state or the EPA?

Selected Readings

Alm, A. L. "Nonpoint Sources of Water Pollution." *Environmental Science and Technology* 24(1990):967. Editorial on some new directions for dealing with nonpoint-source pollution problems.

Cairns, J., Jr., and D. I. Mount. "Aquatic Toxicology." *Environmental Science and Technology* 24(1990):154-61. A review of the fate and effects of toxic agents in aquatic ecosystems.

Clark, R. M., C. A. Fronk, and B. W. Lykins, Jr. "Removing Organic Contaminants from Groundwater." *Environmental Science and Technology* 22(1988):1126-30. A cost and performance evaluation of methods for removing contaminants from groundwater.

Environmental Protection Agency. "Saving the Nation's Great Water Bodies." *EPA Journal* 16(Nov/Dec 1990). Summaries of eighteen water-related remedial actions programs across the United States.

Hodgson, B. "Can the Wilderness Heal?" *National Geographic* 177(1990):3-43. An account of the Alaskan oil spill shortly after the event.

Hren, J., et al. "Regional Water Quality." *Environmental Science and Technology* 24(1990):1122-27. An evaluation of data for assessing conditions and trends in water quality.

Kelso, D. D., and M. Kendziorek. "Alaska's Response to the *Exxon Valdez* Oil Spill." *Environmental Science and Technology* 25(1991):16-23. A description of the extent and cleanup of the largest oil spill in U.S. history.

Maki, A. W. "The *Exxon Valdez* Oil Spill: Initial Environmental Impact Assessment." *Environmental Science and Technology* 25(1991):24-29. An evaluation of the effects of oil on various habitats.

Marshall, E. "Valdez: The Predicted Oil Spill." *Science* 244(1989):20-21. A description of the responses to the Alaskan oil spill.

Sun, M. "Groundwater Ills: Many Diagnoses, Few Remedies." *Science* 232(1986):1490-92. A review of some of the groundwater pollution problems detected in the United States and the frustrations of trying to prevent future incidents.

Sun, M. "Mud-Slinging over Sewage Technology." *Science* 246(1989):440-43. A discussion of some of the issues that face Boston and San Diego as they pursue clean-water goals.

Thomas, J. M., D. M. Mackay, and J. A. Cherry. "Groundwater Contamination: Pump-and-Treat Remediation." *Environmental Science and Technology* 23(1989):630-36. A review of methods used to solve groundwater pollution problems that are based on extraction of groundwater.

Ward, C. H. "In Situ Biorestoration of Organic Contaminants in the Subsurface." *Environmental Science and Technology* 23(1989):760-66. An evaluation of remediation methods that treat groundwater contamination problems in place.

World Health Organization. *WHO Expert Committee on Onchocerciasis.* Technical Report Series 752. Geneva: World Health Organization, 1987. Report on worldwide incidence of onchocerciasis and strategies to combat the disease.

Chapter 15

The Atmosphere
Weather and Climate

The earth's atmosphere is a thin envelope of gases, clouds, and tiny particles that surrounds the globe.
© Carl R. Sams II/Peter Arnold, Inc.

*I*n May of 1991, a typhoon (hurricane) with sustained winds in excess of 240 kilometers (150 miles) per hour drove a surge of ocean water more than 5-meters (16-feet) high over the low-lying deltas and floodplains of Bangladesh. The storm surge traveled well inland and combined with heavy rainfall to cause severe flooding that claimed the lives of more than 125,000 people and left 10 million homeless. Weather extremes such as this remind us of the awesome power of intense storms and our vulnerability to them. In fact, the vagaries of weather affect virtually all facets of life to some extent. Tranquil, pleasant weather allows us to enjoy outdoor activities, but stormy weather typically brings mixed blessings: Rain washes out our picnic, but also benefits crops. The impact of weather extremes may range from mere inconvenience to a disaster that is costly in human lives and property (table 15.1).

The role of weather in our day-to-day activities and the potential societal and economic impacts of severe weather are sufficient incentive for us to learn about the atmosphere and weather. But there are other reasons that have to do with some major environmental issues: (1) weather affects and is affected by air quality; (2) winds spread acid rain precursors thousands of kilometers from their sources; (3) weather influences supplies of energy, food, and fresh water; (4) air pollution threatens the atmosphere's protective ozone layer; and, (5) rising levels of atmospheric carbon dioxide and other gases may be changing the climate. Before considering those problems and what can be done to solve them, however, we must learn something about the basic properties of the atmosphere.

The Atmosphere

We live at the bottom of a very thin ocean of air known as the **atmosphere.** In fact, 99% of the atmosphere's mass is confined to a layer with a thickness of only about 0.5% of the earth's radius. Relative to the rest of the planet, the atmosphere is like the skin on an apple. It is thus a very limited resource that is nonetheless essential for life and the orderly functioning of the environment.

Composition of the Atmosphere

The earth's atmosphere is the product of a gradual evolutionary process that began at the planet's formation perhaps 4.5 billion years ago. As we will see, the atmosphere's composition continues to change as a consequence of air pollution. The atmosphere is a mixture of many different gases and tiny suspended particles. Because the lower atmosphere is in continual motion, the principal atmospheric gases occur almost everywhere in about the same proportions up to an altitude of about 80 kilometers (50 miles). Hence we may travel anywhere over the earth's surface and confidently breathe essentially the same type of air.

If we exclude water vapor, nitrogen (N_2) constitutes 78.08% by volume of the lower atmosphere (below 80 kilometers), and oxygen (O_2) makes up 20.95%. The next most

Table 15.1	Approximate annual losses to selected weather hazards in the United States	
Hazard	**Fatalities**	**Cost (in millions)**
flood	163	$3,175[a]
hurricane	33	796[b]
tornado	98	300
lightning	97	200
hail		750
drought		800

From: W. E. Riebsame et al., "The Social Burden of Weather and Climate Hazards," *Bulletin of the American Meteorological Society,* Vol. 67, No. 11, p. 1379, November 1986. Used by permission.
[a]In 1985 dollars
[b]In 1982 dollars

Table 15.2	Relative proportions of gases composing dry air in the lower atmosphere (below 80 km)	
Gas	**Percent by Volume**	**Parts per Million (ppm)**
nitrogen	78.08	780,840.0
oxygen	20.95	209,460.0
argon	0.93	9,340.0
carbon dioxide	0.035	350.0
neon	0.0018	18.0
helium	0.00052	5.2
methane	0.00014	1.4
krypton	0.00010	1.0
nitrous oxide	0.00005	0.5
hydrogen	0.00005	0.5
ozone	0.000007	0.07
xenon	0.000009	0.09

abundant gases are argon (Ar, 0.93%) and carbon dioxide (CO_2, 0.035%). The atmosphere also contains trace quantities of helium (He), methane (CH_4), hydrogen (H_2), ozone (O_3), water vapor (H_2O), and several other gases (table 15.2). The volume of some trace gases (e.g., carbon dioxide and water vapor) varies from one place to another and with time.

The atmosphere also contains minute liquid or solid particles, which collectively are called **aerosols.** Most aerosols occur in the lower atmosphere near their source, the earth's surface. They originate naturally from forest fires and wind erosion of soil, as sea-salt crystals from ocean spray, and in volcanic eruptions, as well as from industrial smokestacks and agricultural activities.

Although it may be tempting to dismiss as unimportant those substances that make up only a tiny fraction of the atmosphere, the significance of an atmospheric gas or aerosol is not necessarily related to its relative abundance. For example, water vapor, carbon dioxide, and ozone occur in minute concentrations, yet they are essential for life. Most water vapor occurs within a few kilometers of the earth's surface and even in the most humid tropical regions of the world, air is less than 4% water vapor. But without water vapor, no rain or snow would fall to replenish soil moisture, rivers, lakes, and seas (chapter 13). Humans and all other organisms would perish without life-sustaining water. Carbon dioxide (CO_2) is essential for photosynthesis. Ozone (O_3), which forms in the upper atmosphere, shields all living things from exposure to potentially lethal levels of solar ultraviolet radiation (chapter 3). The aerosol content of the atmosphere is also relatively small, yet those suspended particles participate in several important processes. Some aerosols function as nuclei for the development of clouds and precipitation (chapter 13), while others influence the air temperature by interacting with sunlight.

Air Pressure

As noted in chapter 13, we can think of **air pressure** as the weight of the atmosphere over some unit area. Gravity, the force that holds everything on the earth's surface, compresses the atmosphere, so that air pressure drops very rapidly with increasing altitude. At sea level, the average air pressure is about 1.0 kilogram per square centimeter (14.7 pounds per square inch). At an altitude of 5.5 kilometers (3.4 miles), air pressure drops to half its sea-level value, and at 32 kilometers (20 miles), the air pressure is only about 1% of that at sea level.

As air pressure decreases with altitude, so too does air density (mass per unit volume). That is, air rapidly thins with altitude, so that there is less and less oxygen (and other gases) per unit volume of air. As noted in chapter 4, visitors to mountainous regions adjust (acclimatize) to relatively low oxygen levels. However, most people cannot adjust to the low oxygen concentrations at altitudes greater than about 5,500 meters (18,000 feet).

Viewers of televised weather reports are also well aware that air pressure varies from one place to another. Furthermore, at any locality, air pressure fluctuates from day to day and even from one hour to the next. These spatial and temporal changes in air pressure occur independently of altitude (or elevation of the land surface). Although such changes in air pressure are small, they may be accompanied by significant changes in weather. As a general rule, the weather tends to be fair when the air pressure is relatively high and rising, and stormy when air pressure is relatively low and falling. Hence, on weather maps (e.g., fig. 1.4), regions under the influence of high pressure (symbolized by "H" or "HIGH") usually experience fair weather, and regions under low pressure (symbolized by "L" or "LOW") are stormy. High- and low-pressure systems travel over the face of the globe and bring about changes in the weather.

Subdivisions of the Atmosphere

For convenience of study, atmospheric scientists usually subdivide the atmosphere into four concentric layers on the basis of its vertical temperature profile (fig. 15.1): troposphere, stratosphere, mesosphere, and thermosphere. Virtually all

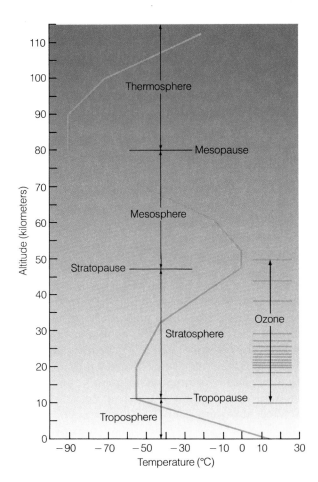

Figure 15.1   The atmosphere is subdivided vertically into zones that are based on the profile of air temperature. The ozone concentration within the stratosphere is greatest in the areas where the horizontal lines are closest together.

weather is confined to the lowest layer, the **troposphere,** which extends from the earth's surface to an altitude that ranges between 16 kilometers (10 miles) at the equator down to 8 kilometers (5 miles) at the poles. Temperatures within the troposphere normally (but not always) drop with altitude, so that air temperatures are usually lower on mountain tops than in surrounding lowlands.

Above the troposphere is the **stratosphere,** which extends up to an average altitude of about 50 kilometers (30 miles). It features isothermal (constant temperature) conditions in its lower portion and a gradual rise in temperature at higher altitudes. The earth's protective ozone shield is located within the stratosphere. Pilots of jet aircraft prefer to fly in the stratosphere because that layer is above most weather and hence offers excellent visibility and generally smooth flying conditions.

From the perspective of environmental quality, we are most concerned with the impact of human activity on the troposphere and stratosphere. As we will see in chapter 16, most air pollutants originate in and are confined to the troposphere. However, as we discuss later in this chapter, some pollutants that escape the troposphere and move into the stratosphere threaten to erode the ozone shield. Human activities appear to have had little direct impact so far on the outermost atmospheric layers, the mesosphere and thermosphere.

The Dynamism of the Atmosphere

Regardless of where we live on the planet, we are well aware from personal experience that weather is dynamic; that is, it varies from place to place and with time. Hence, we define **weather** as the state of the atmosphere at some place and time, described in terms of such variables as temperature, cloudiness, precipitation, and wind. **Climate** is defined as weather at a particular location, averaged over some period of time plus extremes in weather that occurred during that same period. Climate is the ultimate environmental control that determines, for example, where particular crops can be cultivated, the long-term freshwater supply, and average heating and cooling requirements for homes.

Weather is a composite of a wide variety of phenomena—spectacular lightning displays, devastating tornadoes, blizzards, and gentle spring breezes, to mention only a few. Although these phenomena are very different, each one manifests the dynamic nature of the atmosphere. The ceaseless flow of energy from the sun to the earth ultimately maintains that dynamism. We now examine the characteristics of solar energy input.

Solar Energy Input

More than 99% of the energy involved in weather phenomena comes from the sun. The sun's energy travels through space at the astonishing speed of 300,000 kilometers (186,000 miles) per second in the form of oscillating waves (fig. 15.2). These waves have both electrical and magnetic properties and constitute **electromagnetic radiation.** Familiar types of electromagnetic radiation include X rays, visible light, microwaves, and radio waves. The **electromagnetic spectrum** is shown in figure 15.3. The various forms of electromagnetic radiation are distinguished on the basis of

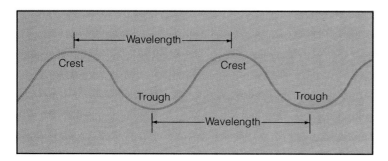

Figure 15.2   The wavelength of an electromagnetic wave is the distance between successive crests or, equivalently, between successive troughs.

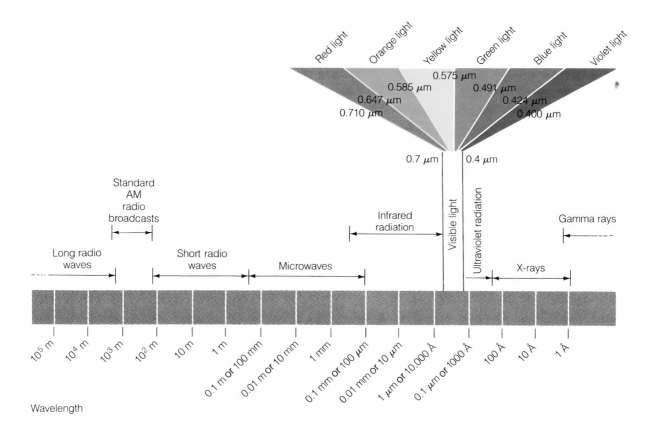

Figure 15.3   The electromagnetic spectrum. The various types of electromagnetic radiation are arranged according to wavelength.

Electromagnetic Radiation and How We Use It

Over the past century, scientists and engineers have harnessed electromagnetic radiation in many ways to improve the quality of life. Today, various forms of electromagnetic radiation are used to diagnose and treat disease, cook and preserve food, and transmit sound and visual images. But all technologies have their trade-offs. Overexposure to some forms of electromagnetic radiation can be lethal. Furthermore, some radioactive materials emit dangerous levels of electromagnetic radiation. Before we can better evaluate the role of electromagnetic radiation in our lives, we need to know more about its properties.

Electromagnetic radiation travels in the form of waves that are usually described in terms of wavelength or frequency. *Wavelength* is the distance between successive wave crests (or equivalently, wave troughs), as shown in figure 15.2. *Wave frequency* is defined as the number of crests (or troughs) that pass a given point in a specified period of time, usually 1 second. The passage of one complete wave is called a cycle, and a frequency of one cycle per second is defined as 1 hertz (Hz). Wave frequency is inversely proportional to wavelength; that is, the higher the frequency, the shorter the wavelength. Radio waves have frequencies of millions of hertz and wavelengths of up to hundreds of kilometers. Gamma rays, in contrast, have frequencies as high as 10^{24} (a trillion trillion) Hz and wavelengths as short as 10^{-14} (a hundred-trillionth) meter.

Electromagnetic waves can travel through space as well as through gases, liquids, and solids. In a vacuum, all electromagnetic waves travel at their maximum speed: 300,000 kilometers (186,000 miles) per second. All forms of electromagnetic radiation slow down when passing through materials, their speed depending on wavelength and the type of material. As electromagnetic radiation passes from one medium to another, it may be reflected or refracted (that is, bent) at the interface. That happens, for example, when solar radiation strikes the ocean surface: Some is reflected, and some bends when it penetrates the water. Electromagnetic radiation may also be absorbed, that is, converted to heat.

Although the electromagnetic spectrum is continuous, it is convenient to assign different names to different segments of it, because we detect, measure, generate, and use different segments in different ways. The types of electromagnetic radiation do not begin or end at precisely marked points along the spectrum. For example, red light shades into invisible infrared radiation (*infrared,* meaning "below red") on the frequency scale. At the other end of the visible portion of the electromagnetic spectrum, violet light shades into invisible ultraviolet radiation (*ultraviolet,* meaning "beyond violet").

We now briefly consider each segment of the electromagnetic spectrum, beginning at the high-energy, high-frequency, and short-wavelength end.

Beyond visible light on the electromagnetic spectrum, in order of increasing frequencies, increasing energy levels, and decreasing wavelengths, are *ultraviolet (UV) radiation, X rays,* and *gamma rays.* All three types of radiation occur naturally, and all can be produced artificially. All have medical uses: UV is a potent germicide; X rays are a powerful diagnostic tool; and both X rays and gamma radiation are used in treating cancer patients.

---

energy level, wavelength (distance between successive crests or troughs), and frequency (number of waves passing a given point in a specified period of time). Some key characteristics of radiation and how we use it are summarized in Special Topic 15.1.

Of the enormous amount of energy that is continuously radiated by the sun, only about one two-billionth of the total is intercepted by the planet. The amount is so small because of the great distance of the earth from the sun and the relatively small dimensions of the planet. As the earth moves in space, that radiant energy is distributed over its surface. The earth's rotation accounts for day-to-night variations in the amount of energy received at a given place. Seasons arise from the fact that the earth's equatorial plane is inclined at 23.5° to its orbital plane. As shown in figure 15.4, at that angle, the most intense solar beam is focused on the southern hemisphere for one-half of the year and on the northern hemisphere during the other half.

Radiation Interactions

As solar radiation travels through the atmosphere, it interacts with the atmospheric gases and aerosols. Some solar radiation is absorbed by oxygen, ozone, water vapor, clouds, and aerosols. Through **absorption,** solar energy is converted to heat, and the air is warmed to some extent. Some solar energy undergoes **reflection;** that is, solar energy is thrown back upon striking a surface such as the top of a cloud. Reflected solar radiation merely changes direction; it is not converted to heat. Another portion of incoming solar radiation is subject to **scattering;** that is, it is dispersed in all directions. (In fact, it is the scattering of the blue portion of visible sunlight by nitrogen and oxygen molecules that gives the daytime sky its blue color.) The portion of solar radiation that is not reflected or scattered back to space or absorbed by gases and aerosols reaches the earth's surface.

These three highly energetic forms of electromagnetic radiation are dangerous as well as useful. Ultraviolet radiation causes irreparable damage to the light-sensitive cells of the eye. One can be permanently blinded by looking at the sun (say, during a partial solar eclipse), unless a filter is used to block out ultraviolet radiation. Also, overexposure to UV radiation, X rays, or gamma rays can cause sterilization, cancer, or mutations. Fetal tissue is particularly sensitive to these forms of radiation. Some radioactive materials emit dangerous levels of X rays and gamma rays, as do such devices as X-ray machines. Overexposure to high-energy electromagnetic radiation, whatever its origin, must be avoided.

Fortunately for us, the atmosphere blocks out most incoming UV radiation and virtually all X rays and gamma radiation. Without that protective shield, all life on earth would be destroyed quickly. Although all nations have safety programs to shield their citizens from the harmful effects of radioactive materials, some are considerably more diligent than others. Furthermore, no nation has a permanent repository for its radioactive wastes. We have more to say on the disposal of radioactive wastes in chapter 20.

At lower frequencies, UV radiation shades into the highest frequency and shortest wavelength *visible radiation,* which is violet light. Wavelengths of visible light are in the range of approximately 0.40 micrometers at the violet end to about 0.70 micrometers at the red end.* Visible light is essential for many activities of plants and animals. In plants, light provides the energy needed for photosynthesis; day length also coordinates the opening of buds and flowers and the dropping of leaves. In animals, day length regulates the timing of reproduction, hibernation, and migration.

Between visible light and microwaves is *infrared radiation (IR).* IR is not sufficiently energetic to be visible, but we can feel the heat it generates if it is intense—as it is, for example, when emitted from a hot stove. Actually, small amounts of infrared radiation are emitted by all objects, including you and this book. Absorption of IR by certain atmospheric gases is responsible for significant warming of the earth's surface (the greenhouse effect).

Next comes the *microwave* portion of the electromagnetic spectrum, which includes wavelengths that range from about 0.1 millimeter to 300 millimeters. Some microwave frequencies are used for radio communication, for microwave cooking, and for tracking weather systems (radar).

At the low-energy (low-frequency, long-wavelength) end of the electromagnetic spectrum are *radio waves.* Their wavelengths range from a fraction of a meter up to hundreds of kilometers, and their frequencies can extend to a billion Hz. FM (frequency modulation) radio waves, for example, span 88 million to 108 million Hz; hence the familiar 88 and 108 at opposite ends of the FM radio dial. Exposure to low levels of microwaves and radio waves appears to pose no significant health hazard, although some controversy persists.

*A micrometer is one-millionth of a meter.

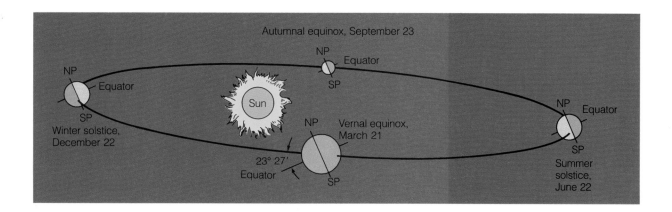

Figure 15.4   The seasons are the result of the inclination of the earth's rotational axis. The earth rotates on its axis once every 24 hours and revolves about the sun once every 365.25 days. Note that the eccentricity of the orbit is greatly exaggerated in this figure.

Table 15.3	Reflectivity (albedo) of some common surface types for visible solar radiation
Surface	**Albedo (% reflected)**
grass	16–26
deciduous forest	15–20
coniferous forest	5–15
crops	15–25
tundra	15–20
desert	25–30
blacktopped road	5–10
sea ice	30–40
fresh snow	75–95
old snow	40–70
glacier ice	20–40
water (high sun)	3–10
water (low sun)	10–100

Figure 15.5 The oceans have a much lower albedo for solar radiation than does land. Note in this satellite image of earth that the oceans appear much darker than the continents.
NASA

Table 15.4	Global solar radiation budget
scattered and reflected to space	31%
absorbed by the atmosphere	23%
absorbed at the earth's surface	46%
total	100%

Solar radiation striking the earth's surface is either absorbed (that is, converted to heat) or reflected. The fraction (or percentage) of incident solar radiation that is reflected by some surface is known as the **albedo** of that surface. As a rule, surfaces that have a high albedo appear light-colored, and surfaces with a low albedo are perceived as dark. For example, skiers who have been sunburned on the slopes are well aware that snow is highly reflective. Fresh snow typically has an albedo of between 75% and 95%; that is, 75% to 95% of the solar radiation that strikes the snow is reflected. In contrast, the albedo of a dark surface, such as a blacktopped road or a green forest, may be as low as 5%. Thus, it is obvious why light-colored clothing is usually more comfortable than dark-colored clothing during sunny, hot weather. The albedos of some common surface types are listed in table 15.3.

One of the essential interactions between solar radiation and the atmosphere produces stratospheric ozone. Recent evidence demonstrates that this protective shield is thinning. For a discussion of the causes of this erosion, see the issue on page 321.

Solar Radiation Budget

Satellite measurements indicate that approximately 31% of the total incoming solar radiation is reflected or scattered by the earth-atmosphere system (the earth's surface and the atmosphere considered together), and is lost to space. That percentage of reflected and scattered light (seen as *earthshine* by American astronauts on the moon) is known as the **planetary albedo.** The remaining 69% of solar radiation intercepted by the earth-atmosphere system is absorbed (converted to heat) and is ultimately involved in the functioning of the environment.

Of the total amount of solar radiation intercepted by the planet, only 23% is absorbed directly by the atmosphere; in other words, the atmosphere is relatively transparent to solar radiation. The remaining 46% of the solar radiation is ab-

sorbed by the earth's surface—most of it by ocean waters, which have an average albedo of only about 8% and cover nearly three-quarters of the globe. The oceans' relatively low reflectivity (and hence, high absorption of solar radiation) is evident in the satellite image of the earth in figure 15.5. Note how the oceans appear darker (less reflective) than the adjacent continental land masses.

It follows from the global distribution of solar radiation, summarized in table 15.4, that the earth's surface is the principal locus of solar heating. The earth's surface, in turn, continuously transfers heat back to the atmosphere, which eventually radiates it off to space. Hence, the earth's surface is also the main source of heat for the lower atmosphere, which is evident in the usual vertical temperature profile of the troposphere (fig. 15.1). Normally, air in the troposphere is warmest close to the earth's surface and becomes cooler with altitude, that is, away from the main source of heat.

Infrared Response and the Greenhouse Effect

If the planet continually absorbed solar radiation without any compensating flow of heat off into space, the global surface

The Thinning Ozone Shield

Threats to the planet's ozone shield and the potential impact of its reduction on human health are among today's major environmental concerns. Certain chemicals that persist in the atmosphere erode stratospheric ozone, allowing more intense solar ultraviolet radiation to reach the earth's surface. More intense UV radiation is likely to raise the incidence of skin cancer and cataracts of the eye (chapter 3). In late 1991, a U.N. panel concluded that a 10% reduction in stratospheric ozone (possible by the close of the century) could cause 300,000 additional cases of skin cancer and 1.6 million additional cases of cataracts annually worldwide. Also, more intense UV radiation is likely to adversely affect crops, marine plankton, and the functioning of food webs.

Ozone occurs within the stratosphere primarily between altitudes of 15 and 35 kilometers (9 and 22 miles). (As we will see in chapter 16, ozone is also present in the troposphere as a component of photochemical smog and is a serious air pollutant.) Ultraviolet radiation, which is a small fraction of incoming solar radiation, spans a band of wavelengths, and different parts of that band are absorbed by oxygen, nitrogen, and ozone. Within the stratosphere, two sets of UV-absorbing chemical reactions take place. One set generates ozone, and the other destroys it. The net result of these opposing chemical reactions is a minute reservoir of ozone that peaks at only about 10 ppm within the middle stratosphere. These absorption processes filter out much of the incoming solar UV radiation and shield organisms from exposure to potentially lethal intensities of UV radiation; hence the term **ozone shield.**

A portion of the UV radiation that reaches the earth's surface is responsible for sunburn and causes or contributes to skin cancer. This biologically effective radiation, called **UVB,** spans the wavelength band from 0.28 to 0.32 micrometers.

As a rule of thumb, every 1% decline in stratospheric ozone concentration translates to a 2% increase in the intensity of UVB that passes through the ozone shield. The amount of increased UVB that actually reaches the earth's surface hinges on atmospheric conditions such as cloudiness and dustiness. Various studies suggest that a 2.5% thinning of the ozone shield could boost the rate of human skin cancer by 10%. For more information on the health hazards of overexposure to UVB, refer back to Special Topic 3.1.

The most serious threat to the ozone shield is from a group of chemicals, collectively known as CFCs (for chlorofluorocarbons), which were first developed in 1931. Most of us use several types of CFCs, perhaps unknowingly, in our daily lives (table 15.5). CFCs are still widely used as chilling (heat-transfer) agents in refrigerators and air conditioners, for cleaning electronic circuit boards, and in the manufacture of foams used for insulation. Their use in common household aerosol sprays such as deodorants, hairsprays, and furniture polish was banned from the United States and Canada in the late 1970s. Less significant ozone-depleting chemicals include halons (used in fire extinguishers) and the industrial chemicals carbon tetrachloride and methyl chloroform.

F. S. Rowland and M. J. Molina of the University of California at Irvine first warned of the danger of CFCs to the ozone shield in 1974. They proposed the following disturbing scenario. Certain CFCs are inert (chemically nonreactive) in the troposphere, where they have been accumulating since production began. The inert CFCs gradually migrate upward into the stratosphere. At altitudes above about 25 kilometers (15 miles), intense UV radiation breaks down the CFCs, causing them to release chlorine, a gas that readily reacts with and destroys ozone. Through a chain reaction, each chlorine atom destroys perhaps 100,000 ozone molecules.

Although almost twenty years have passed since scientists first warned of threats to stratospheric ozone, evidence that the ozone shield is thinning appeared only recently. The first indications of a thinning ozone shield came from the most remote and desolate region of the globe: Antarctica.

For about six weeks, during the southern hemisphere spring (mainly in September and October), the ozone layer in the Antarctic stratosphere (mostly at altitudes between 15 and 23 kilometers) thins drastically. Every November the ozone level recovers. Although the British Antarctic Survey team first reported this phenomenon in 1985, massive ozone depletion had been measured during the prior eight Antarctic springs, but was erroneously dismissed as the product of instrument error. The area of Antarctic ozone depletion, dubbed an **ozone hole,** is about the size of the continental United States (fig. 15.6).

Satellite measurements indicate that the Antarctic ozone hole deepened from the late 1970s into the 1980s. This prompted speculation of a possible link to CFCs and led to an intensive field investigation by specially instrumented aircraft during the Antarctic spring of 1987. That investigation not only measured a record ozone loss of 50%, but also detected exceptionally high concentrations of chlorine monoxide (ClO),

Table 15.5	The most commonly used chlorofluorocarbons (CFCs)		
Type	**Formula**	**Primary Use**	**Residence Time[a]**
CFC-11	CCl_3F	foam-blowing agent; aerosol propellant	76 yr
CFC-12	CCl_2F_2	refrigeration and air conditioning	139 yr
CFC-113	$C_2Cl_3F_3$	solvent for cleaning electronic microcircuits	92 yr

[a]Time required for 63% of the CFC to be washed from the atmosphere.

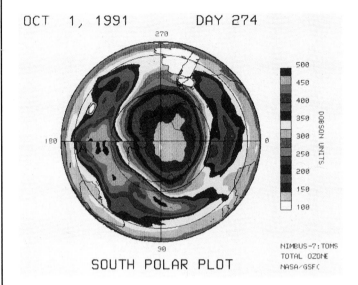

OCT 1, 1991   DAY 274

270

180   0

90

SOUTH POLAR PLOT

500
450
400
350
300
250
200
150
100

DOBSON UNITS

NIMBUS-7:TOMS
TOTAL OZONE
NASA/GSFC

Figure 15.6 Map of total ozone in the southern hemisphere on 1 October 1991, illustrating the Antarctic ozone hole. This image was produced at NASA's Goddard Space Flight Center using data obtained by the TOMS (Total Ozone Mapping Spectrometer) on board the Nimbus-7 satellite. The ozone hole is portrayed in pink and violet colors and is surrounded by a ring of relatively high total ozone (yellow, green, and brown) at middle latitudes. As illustrated in the color bar at the right, ozone concentrations are expressed in Dobson Units. One Dobson Unit is a hundredth of a millimeter and refers to the depth of ozone produced if all the ozone in a column of the atmosphere were brought to sea-level temperature and pressure. NASA

a product of chemical reactions known to destroy ozone. Record or near-record ozone depletion also occurred in the Antarctic stratosphere during the springs of 1989, 1990, 1991, and 1992.

What causes the Antarctic ozone hole, and why does it fill in by November? During the long, dark Antarctic winter, extreme radiational cooling causes temperatures in the stratosphere to plunge below −85° C (−121° F). At such frigid temperatures, what little water vapor there is in the stratosphere deposits as ice-crystal clouds. The formation of those ice crystals is key to ozone depletion; they provide surfaces for reactions whereby chlorine compounds that are inert toward ozone are converted to active forms that destroy ozone. Once the sun reappears in spring, solar radiation supplies the energy that causes active forms of chlorine to begin destroying ozone.

Ozone depletion takes place while the Antarctic atmosphere is essentially cut off from the rest of the planet's atmospheric circulation by the **circumpolar vortex,** a band of strong winds that encircles the margins of the Antarctic continent. A month or so into spring, however, the circumpolar vortex begins to weaken and allows warmer, ozone-rich air from lower latitudes to invade the Antarctic stratosphere. Ice-crystal clouds vaporize, and the stratospheric ozone concentration returns to normal levels; that is, the ozone hole fills in.

Scientists investigating stratospheric chemistry in the Arctic in early 1989 discovered ozone-destroying chlorine compounds and a slight thinning of ozone. An Arctic ozone hole comparable in magnitude to that of Antarctica is considered unlikely for two reasons: (1) in winter, the Arctic

and air temperatures would rise steadily until the oceans boiled. In reality, however, little change occurs in global air temperature from year to year because the planet continually emits heat to space in the form of infrared radiation (IR). A global radiative equilibrium prevails such that the input (to earth) of solar radiation equals the output (to space) of infrared radiation. But why does the planet radiate in the infrared portion of the electromagnetic spectrum?

The higher the surface temperature of a radiating object (such as a hot stove, the sun, or the earth-atmosphere system), the shorter are the wavelengths of its most intense emitted radiation. The sun, which radiates at about 6,000° C (11,000° F), emits relatively short electromagnetic waves. It emits a band of radiation (mostly between 0.25 and 2.5 micrometers) that is most intense at a wavelength of about 0.5 micrometer (in the green portion of visible light).

The earth's surface, radiating at an average temperature of approximately 15° C (59° F), emits longer-wavelength radiation. As figure 15.7 shows, the earth's surface emits a band

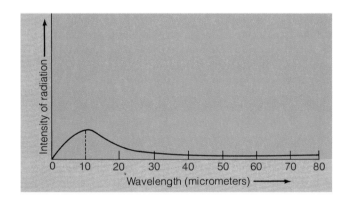

Figure 15.7 The intensity of terrestrial radiation as a function of wavelength. Peak energy intensity occurs in the form of waves that are about 10 micrometers long in the infrared region of the electromagnetic spectrum.
From H. R. Byers, *Meteorology.* Copyright © 1959 McGraw-Hill, Inc. Reproduced with permission of McGraw–Hill, Inc.

stratosphere averages about 10 C° (18 F°) warmer than the Antarctic stratosphere, making the development of ice-crystal clouds unlikely;* (2) the circumpolar vortex that surrounds the Arctic weakens earlier than its Antarctic counterpart. An exceptionally cold Arctic winter coupled with an unusually persistent circumpolar vortex could, however, translate into considerable ozone depletion in an area not very far from population centers.

Discovery of the Antarctic ozone hole and speculation about a possible link to CFCs spurred questions about trends in ozone levels over North America. This led to a reanalysis of global ozone measurements made by ground-based and satellite instruments between 1969 and 1986. In March 1988, NASA's Ozone Trends Panel reported that average ozone levels were declining by about 1-3% per decade in the latitude belt roughly bounded by New Orleans and Seattle. An update of this study released in April 1991 by the EPA reports ozone losses in the same latitude belt at about twice the previous rate: 4-5% per decade. The EPA also noted that a thinner ozone shield persists into spring when more people are outdoors and exposed to more solar radiation. While many scientists believe that CFCs and other ozone-destroying chemicals are responsible for thinning the ozone shield over midlatitudes, a precise cause-effect relationship is yet to be demonstrated.

The first international response to threats to the ozone shield was the so-called Montreal Protocol of September, 1987. Under United Nations auspices, representatives of twenty-three nations meeting in Montreal negotiated a plan to cut by 50% (from 1986 levels) the global production and consumption of

CFCs by June 1998. By June of 1990, however, ominous signs of ozone depletion in midlatitudes and the Arctic prompted a revision of the Montreal Protocol, calling for a complete phase-out of CFC production and use by the year 2000 for more-developed nations and by 2010 for less-developed nations. All this spurred the chemical industry to search for environmentally safe, nontoxic alternatives to CFCs.

In early 1992, the U. S. Senate passed a resolution calling for a complete phase-out of CFC production by 1995. This change in timetable was in response to the late 1991 discovery of surprisingly high levels of ozone-destroying chlorine monoxide (ClO) in the stratosphere of middle and high latitudes of the northern hemisphere. This discovery was made by high-altitude NASA aircraft surveys. As of this writing, CFC emission rates have dropped by 50% since 1988.

Even if production and use of CFCs and other ozone-depleting chemicals cease within the next several years, the atmospheric content of ozone-destroying chlorine will continue to grow because of the long residence times of CFCs (see table 15.5). CFCs may remain for a century or more before being washed from the atmosphere. Consequently, stratospheric chlorine concentration, now at 3.4 ppb, may top out at 4.1 ppb by the turn of the century.

*Some scientists believe, however, that stratospheric sulfate particles (such as those emitted by a volcanic eruption) may play a similar role as ice particles in activating ozone-destroying chlorine compounds.

of infrared radiation (mostly between 4 and 24 micrometers) that has peak intensity at a wavelength of about 10 micrometers. Hence, the earth-atmosphere system responds to solar radiational heating by emitting infrared radiation.

Because solar radiation and the earth's radiation peak in different portions of the electromagnetic spectrum, they have different properties, and their interactions with the atmosphere also differ. We noted earlier that the atmosphere absorbs only about 23% of the solar radiation intercepted by the planet, but certain gases in the atmosphere absorb a larger percentage of the infrared radiation emitted by the earth's surface. As a result, the atmosphere is heated, and some of this heat is reradiated back toward the earth's surface (fig. 15.8). Absorption and reradiation slow the escape of heat into space and elevate the average temperature of the lower atmosphere and the surface of the earth. While the earth's surface radiates at an average temperature of about 15° C (59° F), the planet viewed from space radiates at about −18° C (0° F). Hence, thanks to internal reradiation of IR, the earth's

average surface temperature is some 33 C° (59 F°) higher than it would be otherwise. Clearly, the earth would be a most inhospitable place without this warming.

Atmospheric gases that absorb large amounts of infrared radiation are primarily water vapor and, to a lesser extent, carbon dioxide, ozone, methane, nitrous oxide, and CFCs. The percentage of infrared radiation that these gases absorb varies with wavelength; it is very low in the wavelength bands around 8 and 11 micrometers (which include the wavelength of peak IR intensity emitted by the earth). Much of the planet's heat eventually escapes to space through those so-called **atmospheric windows.**

Like the earth's atmosphere, window glass is relatively transparent to visible solar radiation but slows the transmission of infrared radiation. Greenhouses are designed to take advantage of this property of glass; they are constructed almost entirely of glass panes (fig. 15.9). Sunlight readily passes through the glass and is absorbed (i.e., converted to heat) within the greenhouse. Some of the heat is radiated as IR,

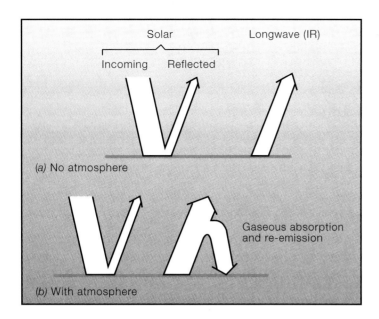

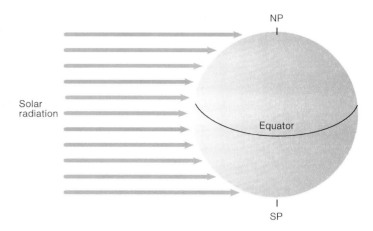

Figure 15.10 Solar radiation strikes tropical latitudes more directly than higher latitudes. Hence, the intensity of solar radiation also varies with latitude.

Figure 15.8 Schematic illustration of the greenhouse effect. (*a*) With no atmosphere, infrared radiation escapes readily to space. (*b*) With the atmosphere, certain gases (mainly water vapor) absorb and reradiate infrared radiation both downward and upward. The net effect is a warming of the earth's surface.

Figure 15.9 The glass of a greenhouse behaves in an analogous manner to certain gases in the atmosphere that absorb large amounts of infrared radiation. Greenhouse gases include water vapor and carbon dioxide.

which is absorbed and not transmitted by the glass. The behavior of IR-absorbing atmospheric gases, which have radiative properties similar to those of glass, is often referred to as the **greenhouse effect,** and the gases contributing to the effect are known as **greenhouse gases.**

To illustrate the greenhouse effect in the atmosphere, we compare the typical summer weather of the American Southwest with that of the Southeast. Both areas are at nearly the same latitude and therefore receive essentially the same intensity of solar radiation. Consequently, both areas commonly experience afternoon temperatures that top 30° C (86° F). At night, however, air temperatures in the two regions often differ markedly. In the Southwest, the air is relatively dry, so

that infrared radiation readily escapes to space; thus, the surface air temperatures may fall below 15° C (59° F) by dawn. In the Southeast, the air is more humid and hence absorbs more infrared radiation. Because a portion of that heat is reradiated back toward the earth's surface, the surface air temperature may fall only to about 25° C (or 77° F) by sunrise.

Furthermore, since clouds are composed of water droplets and ice crystals that also absorb IR, they too produce a greenhouse effect. Hence, as we know from experience, all other factors being equal, nights are usually colder when the sky is clear than when it is cloud-covered.

We have more to say about the greenhouse effect later in this chapter. There we are concerned about the potential effect on global climate of a rising trend in atmospheric carbon dioxide and other greenhouse gases.

Atmospheric Circulation

So far, our discussion has focused on the flow of solar radiation and terrestrial IR radiation within the earth-atmosphere system. We have seen that absorption of solar radiation causes heating, and emission of infrared radiation causes cooling. We refer to these processes as **radiational heating** and **radiational cooling,** respectively. In these processes, radiational energy is neither created nor destroyed; radiational heating and cooling follow the **law of energy conservation** (described in chapter 3). However, the rates of radiational heating and cooling are not uniform everywhere within the earth-atmosphere system, and this is the basic reason why the atmosphere circulates and why weather is variable.

Consider the following illustration of the link between imbalances in radiational heating/cooling and atmospheric circulation. Because the planet is nearly spherical, parallel beams of incoming solar radiation strike tropical latitudes more directly than higher latitudes (fig. 15.10). Hence, at higher latitudes, solar radiation is less intense per unit of surface area. The output of infrared radiation per unit of surface area, on the other hand, varies much less with latitude

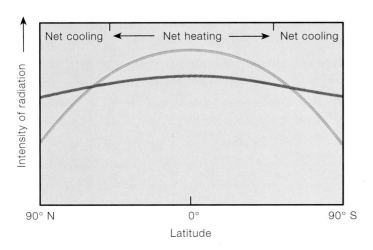

Figure 15.11 Variation by latitude of absorbed solar radiation (blue line) and outgoing infrared radiation (purple line) over the course of a year.

(fig. 15.11). Consequently, over the course of a year at middle and high latitudes (poleward of about 30° latitude), the rate of infrared cooling exceeds the rate of solar heating. In tropical latitudes, the reverse is true: The rate of warming by the absorption of solar radiation is greater than the rate of infrared cooling.

These imbalances in radiational heating and cooling imply that higher latitudes experience net cooling, while tropical latitudes experience net warming. In reality, lower latitudes do not become increasingly warmer relative to higher latitudes, because the excess heat in the tropics is transported to middle and higher latitudes. In the northern hemisphere, this **poleward heat transport** is brought about primarily by air mass exchange: Warm air masses that form in lower latitudes flow northward and replace cold air masses that flow southward (fig. 15.12). An **air mass** is a huge volume of air covering thousands of square kilometers that is relatively uniform in

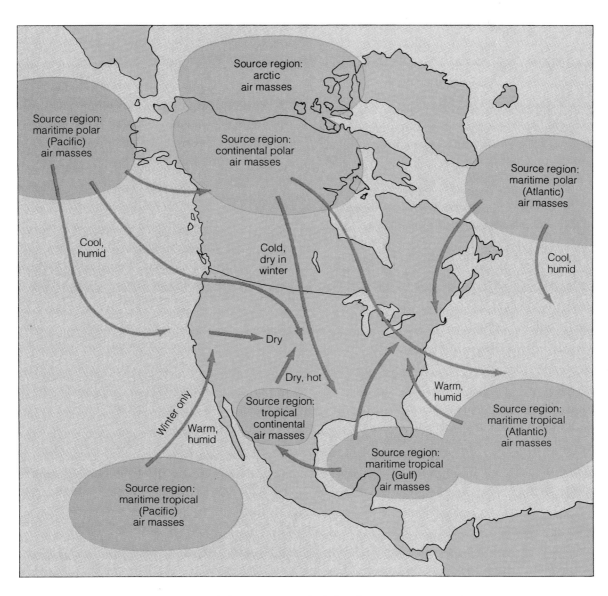

Figure 15.12 Source regions for the air masses that regularly stream across North America.

After Haynes, U.S. Department of Commerce. Reprinted by permission of Arthur N. Strahler.

temperature (warm or cold) and humidity (humid or dry). Storms and ocean currents play secondary roles in poleward heat transport. Hence, atmospheric circulation serves to counter imbalances in radiational heating and cooling.

On a global scale over the course of a year, imbalances in radiational heating and cooling also occur between the earth's surface and the overlying troposphere. At the earth's surface, the rate of radiational heating (due to absorption of solar radiation) is greater than the rate of radiational cooling (due to emission of infrared radiation). On the other hand, within the troposphere, the rate of radiational cooling exceeds the rate of radiational heating. This imbalance in radiational heating and cooling implies that the earth's surface experiences net warming, and the troposphere experiences net cooling. Actually, the earth's surface does become increasingly warm relative to the overlying troposphere. This is because excess heat at the earth's surface is transported into the troposphere via circulation of air.

Convection and water play key roles in the vertical transport of heat from the earth's surface to the troposphere. Recall our discussion of convection currents in chapter 13. The sun heats the earth's surface, and some of that heat is conducted to the overlying air. Heated air rises, undergoes expansional cooling, sinks back to the earth's surface, is heated again, and the process is repeated. Some of the solar radiation absorbed at the earth's surface is also used to vaporize water from the soil, lakes, rivers, and seas. As we saw in chapter 13, if convection currents reach high enough into the atmosphere, the relative humidity of rising air approaches 100%, and condensation (or deposition) takes place, that is, clouds form. When water changes phase from a vapor to a liquid (or solid), its molecules become less active, and heat is released to the environment. Hence, heat is transported from the earth's surface into the atmosphere via (1) convection and (2) phase changes of water. Of the two vertical heat-transport processes, phase changes of water are much more important.

Climate and Climatic Variability

As we saw earlier (chapters 4, 5, and 8), climate is a primary regulator (limiting factor) for organisms. Thus, climate is the principal factor that controls the geographical distribution of the various worldwide ecosystems, and changes in climate can have profound effects on the environment—and thus on humankind. Before we can assess the possible consequences of such variations, however, we must have some understanding of the basics of climate.

Some Principles of Climate

The globe is a mosaic of many climates. A variety of controls interact to shape the climate of any particular region. Controls that exert a regular and more or less predictable influence on the climate are latitude, proximity to large bodies of water, and topography.

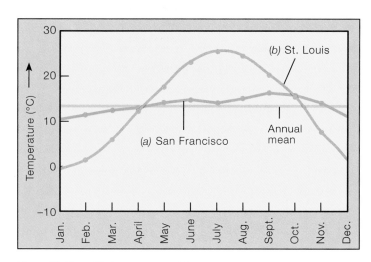

Figure 15.13 Variation in monthly mean air temperature at (a) San Francisco and (b) St. Louis. Although both cities are at about the same latitude, St. Louis experiences much greater seasonal contrast. This is because San Francisco is downwind of the Pacific Ocean.

Latitude is the chief climatic control because, as we saw earlier, it determines the intensity of solar radiation that strikes the earth's surface. Air temperatures over land areas that are downwind of large bodies of water, such as seas and lakes, exhibit less seasonal variability and fewer extremes than inland locations (fig. 15.13). This is because water temperature changes relatively little with the seasons, so that bodies of water dampen the temperature fluctuations of air masses that travel over them. Topography is important because it affects air temperature and the type and amount of precipitation that falls. Air temperatures usually drop and precipitation usually increases with land elevation. Recall our discussion of orographic precipitation in chapter 13. Also, mountain barriers such as the Rocky and Himalayan Mountains deflect the prevailing atmospheric circulation.

In addition to the fixed controls of climate, the variable atmospheric circulation strongly influences the climate of particular regions. The prevailing circulation determines the types of air masses that regularly develop over a region or invade it from other areas. For example, in some regions of the world, such as India, Southeast Asia, and North Africa, the prevailing atmospheric circulation imposes a distinct seasonality to rainfall. In those areas of monsoon circulation, winters are generally dry, and summers are wet. It is useful to elaborate on this climatic control, particularly as atmospheric circulation affects middle and high latitudes.

The moisture and temperature characteristics of an air mass are determined by the surface over which the air mass develops and travels. Air that remains for a long period of time over cold, snow-covered ground (such as northern Canada in winter) becomes cold and dry. On the other hand, an air mass that forms over a warm, wet surface (such as the Gulf of Mexico) becomes warm and humid. As noted earlier, in poleward heat transport, cold air masses that develop in the north push southward, while the warm air masses that develop to the south stream northward.

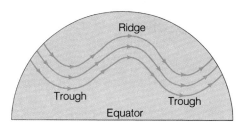

Figure 15.14 At middle and higher latitudes, the westerly winds encircle the globe in a wavelike pattern of ridges and troughs. To the east of a trough (and west of a ridge), warm air moves northeastward. To the west of a trough (and east of a ridge), cold air moves southeastward.

As air masses move from one place to another, their temperature and humidity change to some extent. In the winter, a cold air mass loses some of its punch as it plunges southeastward from Canada into the Central United States. Cold air masses tend to modify much more rapidly than warm air masses. Thus, a heat wave may spread from the Gulf Coast into southern Canada with about the same intensity, but a cold wave usually moderates considerably by the time it reaches the deep South. Also, the temperature of air masses changes with the seasons, warming during the summer and cooling during the winter.

The particular type of air mass that flows into a middle- or high-latitude region is determined by the pattern of west-to-east winds in the middle troposphere. These winds are known as the global **westerlies,** and their role in governing the movement of air masses (and storms) has earned them the name of *steering winds*. Aloft, the westerlies girdle the middle latitudes in wavelike patterns of troughs and ridges (fig. 15.14). Thus, the westerlies actually consist of a north-south airflow that is superimposed on an overall west-to-east wind. In regions where the westerlies have a component that blows from north to south, cold air masses are steered southward; where the westerlies have a component directed from south to north, warm air masses are funneled northward. Hence, the type of air mass that invades a region depends on the location of that region with respect to the westerly wave pattern.

A particular wave pattern in the westerlies tends to persist for anywhere from several days to weeks or longer. Then, abruptly (usually in a single day or less), the wave pattern shifts. This shift is usually accompanied by a change in the regional distribution of air masses; areas that have been cold and dry may suddenly turn mild and stormy, while areas that were warm and moist may become cool and dry. Thus, the weather consists of a sequence of episodes that last for varying periods of time and are punctuated by abrupt periods of change.

Different wave patterns of the westerlies prevail at different times of the year. Hence, in many regions different air masses dominate, depending on the season. In portions of the Midwest, cold and dry air from Canada is the most frequent type of air mass in January, while warm, humid air from the Gulf of Mexico prevails in July. A change in climate, then, means a change in prevailing atmospheric circulation patterns and, at least in some regions, a change in the frequency of various types of air masses.

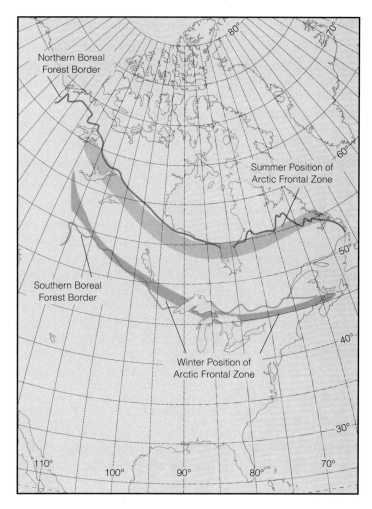

Figure 15.15 The northern edge of Canada's boreal forest closely corresponds to the mean summer position of the leading edge of arctic air and the southern border of the forest approximates the mean winter position of the leading edge of arctic air.

The influence of atmospheric circulation patterns and air-mass frequency on climate is evident in the geographic distribution of certain ecosystems. For example, the region of Canada that is dominated by cold, dry Arctic air closely corresponds with the location of the boreal forest. As shown in figure 15.15, the southern boundary of the boreal forest coincides with the average position of the leading edge of Arctic air during the winter, and its northern border closely matches the average position of the leading edge of Arctic air during summer.

Similarly, the location of the North American grasslands can be explained in terms of air-mass frequency. Grasslands occupy a wedge-shaped geographical region on the western plains where the weather is dominated by mild, dry air masses. These air masses originate over the Pacific ocean, but lose much of their moisture on the windward slopes of the western mountains (chapter 13). The air masses descend onto the western plains and are compressionally warmed. Hence, climatic variations that alter the frequency of occurrence of air masses may also change the boundaries of the boreal forest, grasslands, and other ecosystems.

Average Climate and Climatic Anomalies

Like weather, climate is inherently variable. Hence, the term *normal* is potentially misleading when applied to climate. Some people equate *normal* to *average,* but actually the **normal climate** encompasses the total variation in the climatic record, that is, averages plus extremes. An exceptionally cold winter may not be really abnormal for a region if it falls within the expected range of variability of winter temperature based on the past climatic record. Thus, an unusually cold winter would be an expected albeit infrequent and extreme event.

Through international agreement by climatologists, average values for temperature and precipitation (and other elements of climate) for various locations are computed from measurements compiled over a thirty-year period. (Extreme values are usually drawn from the entire period of record.) Current climatic averages are derived from the period 1961-1990. The assumption is that averages of temperature and precipitation recorded during that period constitute a reasonable guide for future weather expectations.

Problems arise, however, when we assume that the weather of this or any thirty-year period actually represents the normal climate. First, climate can change significantly in periods much shorter than thirty years. In addition, a thirty-year period provides a very restricted view of the climatic record and the variability that climate can exhibit. The weather during this century, for example, has been unusually mild when compared to the climatic record of the last several hundred years (fig. 15.16).

It is useful to compare weather over some specified period with the climatic record. Suppose, for example, that we compare last January's average temperature to the long-term (thirty-year) average January temperature. Typically, we compute a departure from the long-term average, known as a **climatic anomaly;** that is, we will find that the recent January was colder or warmer than the thirty-year average.

If climatic anomalies are computed for the same period of time at numerous sites across the nation, they form a spatial pattern of anomalies. Over an area as large as the United States or Canada, climatic anomalies are always geographically nonuniform in both magnitude and direction (fig. 15.17). That is, for example, some areas experience anomalously high temperatures, while others have anomalously low temperatures. This also means that it is highly unlikely that a weather extreme such as drought or severe cold would grip the entire continent at the same time.

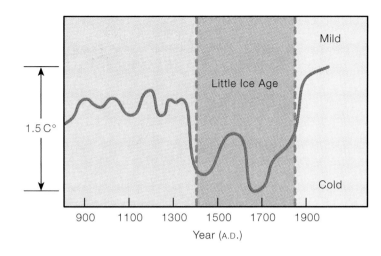

Figure 15.16   Generalized curve of air temperatures in eastern Europe over the past thousand years based on a winter severity index.

The nonuniformity of climatic anomalies—in both magnitude and direction—is linked to the westerly wind pattern aloft. Suppose that a certain westerly wave pattern causes abnormally cold conditions in the eastern United States, exceptionally mild weather in the western United States, and near-average temperatures in between. Now suppose that this pattern persists through the course of a winter, as it did during most of the notorious winter of 1976-1977 (fig. 15.18). During such winters, the temperature anomaly pattern across the country mirrors the characteristics of the dominating westerly wave pattern: Temperatures are lower than the long-term average in the east, higher than the long-term average in the west, and near average elsewhere.

From an agricultural perspective, the geographic nonuniformity of climatic anomalies may be advantageous because some compensation may occur. As we saw in chapter 8, good growing weather and consequent high crop yields in one area may compensate to some extent for poor growing weather and low yields elsewhere. This offsetting effect is known as **agroclimatic compensation.** For example, corn yields may be down in Iowa and Kansas, perhaps because of a wet spring that delayed planting, but the loss may be offset by higher corn yields in the eastern corn belt states of Indiana and Ohio, where growing weather was better. Of course, the better growing weather may well affect a region where soils are less fertile or no crops are grown, so that agroclimatic compensation cannot always be relied upon to ensure an adequate harvest.

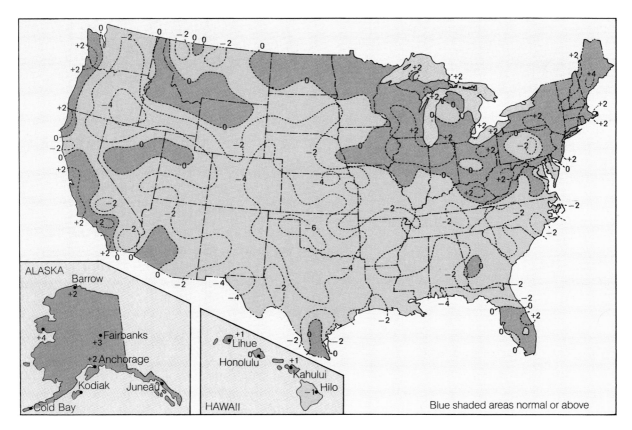

Figure 15.17   Monthly anomaly pattern (departures from long-term averages) of temperature (°F) for the United States in June, 1983.
Source: Data from NOAA.

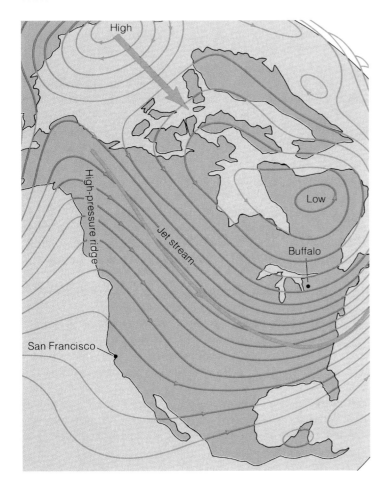

Figure 15.18   The westerly wind pattern that prevailed during the winter of 1976-1977, bringing record cold to the East and record drought and heat to the West.
Source: Data from NOAA.

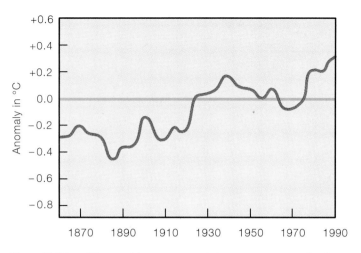

Figure 15.19 The trend in mean annual air temperature over land areas of the northern hemisphere. Temperature is expressed in terms of departure from the 1951-1980 mean.

Northern Hemisphere Temperature Trend

Geographic nonuniformity also characterizes trends in climate. For example, consider the variation in the average annual temperature of the northern hemisphere for the period 1880-1988 (fig. 15.19). The total temperature fluctuation amounts to only ±0.6 C° (±1.1 F°) about the average for the entire period of record. Note that a general warming trend began in the 1890s and continued into the late 1930s. Then, an oscillatory cooling trend set in and ended by the late 1970s. Since then, the temperature trend has been upward; in fact, the 1980s were the warmest decade of record.* Because of the geographical nonuniformity of climatic anomalies, however, the hemispheric temperature trend is not representative of all locations within the hemisphere. Over the same period, some locations experienced cooling trends, while others experienced warming trends, regardless of the direction of the hemispheric temperature trend. In the United States, for example, while the cooling trend of the 1960s and 1970s was quite marked in the southeastern states, a warming trend characterized the same period in the Pacific Northwest.

Just as it is misleading to apply the direction of hemispheric climatic trends to all regions, it is also misleading to assume that the magnitude of climatic change is the same everywhere. In fact, a small change in average hemispheric temperature may translate into a much larger change in certain regions. For example, the slight fall in average hemispheric air temperature in the decades after 1938 was accompanied by a considerably greater drop in temperature in polar regions. Thus, it follows that in any meaningful assessment of the potential impact of hemispheric climatic trends on ecosys-

tems, it is unwise to underestimate the regional response to small changes in average hemispheric temperature. Locally, hemispheric trends may be amplified significantly or even reversed, to the point of disrupting physical processes and the activities of plants and animals.

Responses of Ecosystems to Climatic Change

As a general principle, we know that plants and animals respond to changes in climate (chapter 4). A major climatic shift may exceed the tolerance limits of certain organisms, and the kinds of species that live in the region will change. Organisms vary greatly in their ability to disperse and become established. Hence, a changing climate may disrupt population interactions for many years; some organisms may succumb, while others may migrate into or out of a region. Recall from earlier discussions in chapters 4, 9, and 10 that serious disruptions often follow new interactions among populations. Centuries may pass before a climax ecosystem can become established that is adapted to the new climate if, indeed, it ever does.

Extreme examples of changes in ecosystem distribution come from the Ice Age. (Some scientists suggest that the forests of North America are still not in equilibrium thousands of years after glacial retreat.) Global cooling about 25,000 years ago triggered the final advance of a huge glacial ice sheet that spread over most of Canada and, by 18,000 years ago, engulfed the northern tier states of the United States (fig. 15.20). In nonglaciated areas, south of the ice sheet, climatic shifts disrupted ecosystems. For example, tundra developed along a narrow belt at the edge of the ice sheet, and prairies were replaced by coniferous forests. In the American Southwest, moist conditions erased arid ecosystems; in the East, deciduous forest species retreated to tiny refuges. Today's pattern of ecosystem distribution reflects merely the current transient stage of species migration following glacial retreat.

In historical times, abrupt changes in climate have also had a disruptive effect on agriculture. For example, severe drought may have forced the Anasazi Indians of Mesa Verde to abandon their southwestern Colorado cliff dwellings about 1300 A.D. (fig. 15.21). More recently, during the Dust Bowl era of the 1930s, farms were abandoned by the score on the western Great Plains.

Agriculture that is most vulnerable to climatic change is located in areas where the climate is already marginal for food production; that is, just enough rain falls, or the growing season is barely long enough for crops to mature. Thus, even a small unfavorable change in these critical parameters can spell disaster. Apparently, this is what happened to the early European settlements in Greenland, described in Special Topic 15.2.

Today there is increasing concern over the potential for climatic change due to rising levels of greenhouse gases. For an update, see the issue on page 333.

*NASA reports that the global mean temperature in 1991 was second to 1990 as the highest since reliable records began in 1880, and that the eight warmest years on record occurred during the past twelve years (1980-1991).

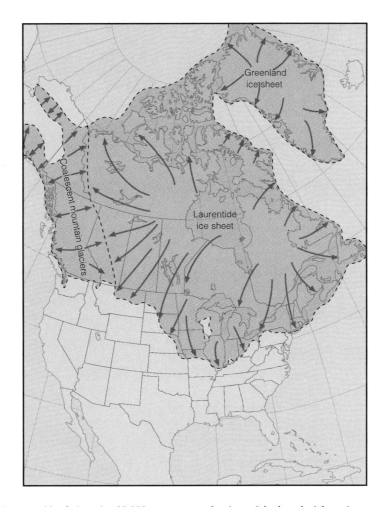

Figure 15.20 The extent of glaciation over North America 18,000 years ago, at the time of the last glacial maximum.

Figure 15.21 Prolonged drought probably forced the Anasazi Indians to abandon their cliff dwellings (located in southwest Colorado) about A.D. 1300.

Climatic Change and the Norse Greenland Tragedy

The late ninth century saw the beginning of a long period of unusually mild conditions throughout much of the northern hemisphere. A relatively mild climate enabled Viking explorers to probe the northerly reaches of the Atlantic Ocean. Previously, severe cold and extensive drift ice had proven insurmountable obstacles to European navigators. By 930 A.D., the Vikings established the first permanent settlement in Iceland, some 970 kilometers (600 miles) west of Norway and just south of the Arctic Circle.

Among Iceland's early inhabitants was Eric the Red, a troublesome individual whose exploits eventually caused his banishment from Iceland in 982 A.D. He sailed west and discovered a new land, which he named Greenland. Then as now, much of Greenland was buried under a massive glacial ice sheet. The only habitable lands were small patches scattered along the coast, hemmed in by the sea and ice sheet and separated by treacherous mountain ridges and deep fjords. Some historians speculate that Eric called the new land Greenland to entice others to follow him. At the time, however, the climate was so mild that some sheltered valleys were probably greener than they are today.

In such a place, on Greenland's southwest shore, Eric founded the first of two Norse settlements. Although never prosperous, Norse colonization of Greenland persisted for nearly five centuries. The Norse subsistence economy was primarily agrarian and dependent on cattle, sheep, and goats. In addition, the Norse hunted migratory harp seals in spring and caribou in autumn. They even embarked on long, dangerous hunting expeditions northward along Greenland's west coast in search of walrus and polar bear, whose valuable tusks and hides they traded to Europeans for durable goods.

Initially, the colonists fared relatively well; the population of the Eastern Settlement climbed to an estimated 4,000-5,000, and that of the Western Settlement peaked at perhaps 1,000-1,500. By about 1350 A.D., however, the Western Settlement was vacant, apparently the victim of some sudden calamity. The larger Eastern Settlement also succumbed, but more gradually, so that by 1500 A.D., Norse society in Greenland had been erased. What happened to the Norse settlements in Greenland can only be inferred from their graves, the ruins of their homes and barns, and a few chronicles. There were no survivors. In 1921, an expedition from Denmark examined the Norse remains and found evidence that the settlers had suffered a painful annihilation. Grazing land was buried under advancing lobes of glacial ice, and most farmland was made useless by permafrost. Near the end, the descendants of robust and hardy people were ravaged by famine; they were crippled, dwarflike, and diseased.

Many explanations have been proposed for the extinction of Norse society in Greenland, and there were probably many contributing factors. Climatic change and the inability (or unwillingness) of the Norse to adapt to an increasingly stressful climate, along with unreliable food sources, however, appear to have been the major causes of disaster. By the 1300s, the climate was cooling rapidly, heralding the Little Ice Age. Drift ice expanded over the North Atlantic, hampering and eventually halting navigation between Greenland and Iceland. All contact between the Norse settlements and the outside world ended shortly after 1400 A.D.

Deteriorating climate in Greenland meant wetter summers with poor haying conditions that caused major livestock losses. Snowier and longer winters probably took an even greater toll on livestock—especially among newborn animals. Climatic change also disrupted the harp seal migration and decimated caribou herds, further reducing food sources. About the same time, the Norse had to compete for shrinking food sources with the Inuit people who had migrated southward along Greenland's west coast. Competition for food evidently led to hostilities between the two groups.

Unable to survive the climatic stress, the Norse succumbed to famine; yet the neighboring Inuit people survived. The Inuits probably succeeded because their hunting skills and techniques were more suited to the hostile climate. The Norse refused to adopt the hunting practices of the Inuits and clung tenaciously to their traditional agrarian subsistence methods, which were no longer suited to the new climate.

The Norse tragedy in Greenland may be one of only a very few historical examples of the extinction of a European society in North America. The lesson is that climate can change rapidly—sometimes with serious, even disastrous, consequences. Nowhere are people more vulnerable to climatic shifts than in regions where the climate is already marginal for their survival—where barely enough rain falls to sustain crops and livestock or where mean temperatures are so low and the growing season so short that only a few hardy crops can be cultivated successfully. In such regions, even a small change in climate can make agriculture impossible. If inhabitants cannot locate and utilize new food sources, if they cannot migrate to more hospitable lands, or if humanitarian aid is not provided, their fate may be similar to that of the Greenland Norse.

Greenhouse Gases and Global Warming

*I*n recent years, many scientists have voiced concern about the possible climatic ramifications of steadily rising concentrations of atmospheric carbon dioxide (CO_2) and other greenhouse gases. Higher levels of these gases are likely to enhance the greenhouse effect—that is, increase absorption and reradiation of infrared radiation and consequently warm the lower atmosphere. Numerical models of the earth-atmosphere system project that global warming of the order of 2 C° to 5 C° (4 F° to 9 F°) would accompany a doubling of carbon dioxide concentration—possible by the middle of the next century. As we saw in chapter 3, the agricultural and socioeconomic implications of such a climatic change are likely to be far-reaching and disruptive.

Although G. S. Callendar, a British engineer, first reported an upward trend in atmospheric carbon dioxide in 1939, the increase probably began during the Industrial Revolution. Systematic measurements of atmospheric carbon dioxide levels began in 1957 at the Mauna Loa Observatory in Hawaii under the direction of Charles D. Keeling of the Scripps Institution of Oceanography. That record is shown in figure 15.22. At 3,400 meters (10,200 feet) above sea level in the middle of the Pacific Ocean, Mauna Loa Observatory is far enough away from major industrial sources of air pollution that carbon dioxide levels monitored there are considered representative of at least the northern hemisphere. Also beginning in 1957, atmospheric carbon dioxide has been monitored at the South Pole station of the U.S. Antarctic Program. Interestingly, the South Pole carbon dioxide trend closely parallels that of Mauna Loa.

The Mauna Loa and South Pole records show an annual cycle due to seasonal changes in northern hemisphere vegetation; carbon dioxide levels fall during the growing season (when photosynthetic removal exceeds cellular respiration), reaching a minimum in October, and recover in winter (when cellular respiration exceeds photosynthesis), reaching a maximum in May. This annual cycle is superimposed on a nearly exponential rate of growth: The annual rate of increase of carbon dioxide rose from about 0.7 ppm in 1958 to 1.0-1.5 ppm during the 1980s. The average carbon dioxide concentration increased from 315 ppm in 1958 to 353 ppm in 1990. Sketchy data suggest that atmospheric carbon dioxide may have climbed by 24% since its estimated preindustrial level of 280 ppm.

As noted in chapter 3, the concentration of atmospheric carbon dioxide is rising primarily because of the burning of fossil fuels (coal and oil, especially), which produces carbon dioxide as a by-product. The clearing of tropical forests also contributes to higher atmospheric levels of carbon dioxide via burning, decay of wood residue, and reduced photosynthetic removal of carbon dioxide from the atmosphere.

Warming induced by carbon dioxide will probably be enhanced by rising levels of certain other greenhouse gases whose atmospheric concentrations are so small that they are

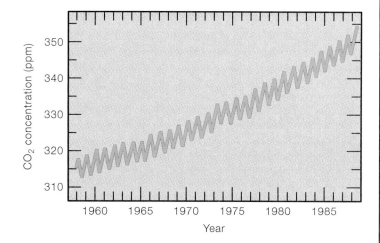

Figure 15.22   Upward trend in atmospheric carbon dioxide concentration as measured at Mauna Loa Observatory, Hawaii. The annual photosynthesis/respiration cycle is responsible for the annual fluctuation in CO_2.

Source: *Geophysical Monitoring for Climate Change*, NOAA.

expressed in parts per billion (ppb). These gases include methane (CH_4) at 1,720 ppb, nitrous oxide (N_2O) at 310 ppb, and chlorofluorocarbons (CFCs) at less than 0.5 ppb. The climatic significance of these trace gases lies in their strong absorption of infrared radiation in the atmospheric windows, especially at wavelengths of 7-13 micrometers. (Recall from earlier in this chapter that within atmospheric windows, IR absorption by water vapor, carbon dioxide, and ozone is minimal.) Trace-gas absorption is directly proportional to concentration, so that doubling the concentration of a gas doubles the absorption. The combined climatic impact of rising levels of these greenhouse gases could match that of increasing carbon dioxide (fig. 15.23).

The concentration of methane in the atmosphere is increasing by about 1% per year. Recent studies of air bubbles trapped in ancient glacial ice cores suggest that the methane increase began more than a century ago and has doubled since then. Nitrous oxide is increasing at about 0.3% per year and, based on ice-core analysis, is now about 9% higher than it was at the turn of the century. CFCs are currently the fastest growing of the greenhouse trace gases. The two most common CFCs (CFC-11 and CFC-12) are increasing at about 4% per year.

The methane increase is probably due mostly to biological decay in wetlands and bacteria in the digestive systems of ruminants (cud-chewing animals) such as cattle and sheep. Termites, rice paddies, and landfills are additional sources of methane. Increases in nitrous oxide are probably linked to industrial and agricultural activity. Recall from our earlier discussion regarding the thinning of the ozone shield that there are several sources of CFCs.

If all other climate controls remain the same, then rising concentrations of atmospheric carbon dioxide and other greenhouse gases could very well cause global warming. To what extent all other controls (e.g., solar energy output, volcanic eruptions) remain the same, however, is open to question. Furthermore, although weather forecasters now have many sophisticated tools, including satellites and computers, long-range weather forecasters still cannot predict the weather for more than about 6 to 10 days in advance. Difficulties stem from the margin of error that seems to be inherent in the worldwide weather monitoring network and from the lack of observational data from vast areas of the ocean. A more imposing challenge is the difficulty of properly modeling the complex behavior of climate. Models used to predict the climatic future vary in sophistication and in their ability to simulate the climatic influence of ocean currents, cloud cover, and atmospheric water vapor. Furthermore, the current generation of climatic models does not do well in forecasting the regional response to global or hemispheric climatic change.

The climatic future will remain a mystery until we have more accurate numerical models of the atmosphere, denser weather-observation networks, and a better understanding of how climatic controls interact. The climatic future, like the climatic past, will be shaped by many interacting factors. At present, we are confident only that the climate will change (as it has in the past); precisely how it will change we do not know.

In spite of the current scientific uncertainty, some experts argue that so much is at stake that action must be taken now to head off enhanced greenhouse warming. They call for (1) at least a 50% reduction in global fossil-fuel consumption, (2) greater reliance on nonfossil fuels (e.g., solar power), (3) higher energy efficiencies (e.g., more vehicle miles per gallon), (4) massive reforestation and a halt to deforestation, and (5) phasing out of CFCs. They hasten to point out that these actions are advisable even if greenhouse warming fails to materialize because they will help alleviate other serious environmental problems. For example, phasing down fossil fuels will also reduce the problem of acid deposition (chapter 16). Similarly, eliminating CFCs will lessen the threat to the ozone shield.

The United States is the world's largest single contributor of greenhouse gases. With only about 5% of the world's

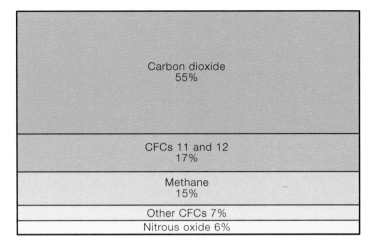

Figure 15.23   The contribution of various greenhouse gases to potential global warming.
Source: Data from Houghton, et al., *Climate Change,* IPPC 1990, Cambridge University Press.

population, the United States could account for 20% of possible greenhouse warming. But as of this writing, the Bush administration has refused to join other industrialized nations in pledging to cut carbon dioxide emissions. The Bush administration argues that in view of scientific uncertainties surrounding possible global warming, major cuts in emissions of CO_2 and other greenhouse gases would impose an excessive financial burden on industry. At the 1992 Earth Summit in Rio de Janeiro, the Bush Administration refused to sign a global treaty that restricted greenhouse gas emissions until provisions mandating compliance were stricken from the document.

A special panel of the National Academies of Science and Engineering has taken issue with Bush's economic assessment. The panel claims that the United States could cut emissions of greenhouse gases by 10-40% at relatively little cost, primarily through greater energy efficiency and conservation. The panel called for a national energy policy that includes regulations and price incentives designed to spur adoption of energy-efficient technology (that in time would pay for itself in energy savings). The panel also recommended more research and development of solar power and safer nuclear power plants. For more details on proposed strategies to cut U.S. emissions of greenhouse gases, see table 15.6.

Table 15.6   Selected strategies to lower U.S. emissions of greenhouse gases

Measures That Affect All Sectors

Promising policy options:

carbon emission tax

marketable permits for carbon emissions

RD&D for noncarbon energy sources and energy-efficient technologies

Buildings

Technical options

Better insulation, windows, and improved construction methods to lower heating and cooling needs

more efficient furnaces, air conditioners, and appliances

improved lighting, especially in commercial buildings

Promising policy options

efficiency standards such as tighter appliance standards and building energy codes for new construction

"demand-side management" programs by electric and gas utilities to facilitate improvements in existing buildings

information programs such as energy audits and home energy ratings as a requirement for mortgages

Transportation

Technical options

fuel efficiency improvements

increased use of mass transit, van-pools, and carpools

more efficient vehicle operation through measures to reduce congestion, reinstate 55-mph speed limit

Promising policy options

tighter fuel efficiency standards for cars and trucks

gasoline tax

transportation control measures (e.g., high-occupancy vehicle lanes and parking management), better mass transit, and improved urban design

Manufacturing

Technical options

continuing process improvements—especially for iron and steel, chemicals, paper, and petroleum

increased cogeneration of electricity and heat

improved electric motors

fuel switching from coal to gas

Promising policy options

carbon emission tax

joint industry/utility programs for "demand-side management"

marketable carbon emission permits

Electricity generation

Technical options

increased use of nonfossil sources (e.g., hydro, wind, solar, nuclear)

co-firing existing coal-fired plants with natural gas

retiring existing fossil-fuel-fired plants after 40 years

Promising policy options

marketable carbon emission permits

carbon emission tax

efficiency standards, carbon emission limits, and/or moratorium on new coal-fired powerplants

RD&D for noncarbon energy sources

Forestry

Technical options

increasing productivity of forests

planting trees on marginal croplands, other nonforested rural lands, and in urban areas

growing short-rotation woody crops for use as biomass fuels

Promising policy options

increase existing Federal forestry assistance programs

expand Conservation Reserve Program

tax incentives for forest management, tree planting

Food

Technical options

decrease livestock methane emissions

increase efficiency of nitrogen fertilizer use to reduce nitrous oxide emissions

take marginal croplands out of production and rotate crops

Promising policy options

increase research funding to evaluate efficacy of methane-reducing practices

require farmers to adopt environmentally sound agricultural practices as a prerequisite for receiving Federal price and income supports

continue to remove disincentives to crop rotations

Source: "Changing by Degrees: Steps to Reduce Greenhouse Gasses," *Office of Technology Assessment Report Brief*, February 1991, U.S. Congressional Office of Technology Assessment.

Conclusions

This chapter has considered some of the basic properties of the atmosphere: its composition, pressure and density, temperature profile, and dynamism. We described how solar energy drives the circulation of the atmosphere. We also considered two major environmental issues: threats to the ozone shield by CFCs and the possibility of global warming through an enhanced greenhouse effect.

We learned that weather is far from a capricious act of nature; it is a response to unequal radiational heating and cooling within the earth-atmosphere system. Weather conditions averaged over long periods of time plus extremes in weather behavior constitute the climate. The climate in middle latitudes is strongly influenced by patterns that characterize the prevailing westerly winds. When prevailing circulation patterns change, the climate also changes. Since the well-being of ecosystems depends on climate, a climatic change could cause serious disruptions.

Our examination of the basic properties of the atmosphere provides the necessary background for an analysis of problems related to air quality. In the next chapter, we focus on the sources of air pollutants, the impact of polluted air on humans and other organisms, atmospheric conditions that govern the dispersal of pollutants, and the strategies being used to manage air quality.

Key Terms

atmosphere	UVB
aerosols	ozone hole
air pressure	circumpolar vortex
troposphere	atmospheric windows
stratosphere	greenhouse effect
weather	greenhouse gases
climate	radiational heating
electromagnetic radiation	radiational cooling
electromagnetic spectrum	law of energy conservation
absorption	poleward heat transport
reflection	air mass
scattering	westerlies
albedo	normal climate
planetary albedo	climatic anomaly
ozone shield	agroclimatic compensation

Summary Statements

The atmosphere is a limited resource that is composed of a mixture of gases (mostly nitrogen and oxygen) and suspended aerosols.

The environmental significance of an atmospheric gas or aerosol is not necessarily related to its relative concentration.

Air pressure is the weight of the atmosphere per unit area. It drops rapidly with altitude and, at the earth's surface, exhibits relatively small spatial and temporal fluctuations that are accompanied by significant changes in the weather.

The atmosphere is subdivided into four concentric layers on the basis of the vertical profile of temperature. The troposphere is the site of most weather. Both the troposphere and the stratosphere are sites of air pollution problems.

Weather is defined as the state of the atmosphere at a specified place and time; climate refers to average weather plus weather extremes.

The energy that drives the atmosphere originates in the sun and travels through space in the form of electromagnetic radiation.

Solar radiation that is not absorbed by the atmosphere or reflected or scattered by it to space reaches the earth's surface, where it is either reflected or absorbed, depending on the surface albedo.

Ozone formation within the stratosphere protects life on earth by filtering out harmful intensities of solar ultraviolet radiation. The principal threat to the ozone shield is a group of chemicals known as chlorofluorocarbons (CFCs). Thinning of the ozone shield is likely to lead to a greater incidence of human skin cancers and cataracts of the eye.

The earth-atmosphere system emits primarily infrared radiation (IR). In the greenhouse effect, water vapor and, to a lesser extent, carbon dioxide and other atmospheric gases absorb and reradiate IR, warming the lower atmosphere.

Differences in rates of radiational heating and cooling cause a heat imbalance between tropical and higher latitudes. Excess heat is transported poleward from the tropics via circulation of the atmosphere and the oceans.

Differences in rates of radiational heating and cooling also cause a heat imbalance between the earth's surface and the troposphere. Excess heat at the earth's surface is transported into the troposphere via convection currents and phase changes of water.

The climate of a region is shaped by its latitude, proximity to large bodies of water, and land elevation, as well as by the prevailing pattern of atmospheric circulation.

The wavelike pattern of tropospheric westerly winds that encircle the middle and high latitudes is responsible for the regional distribution of air masses and the episodic behavior of weather.

Normal climate encompasses both averages and extremes. Both anomalies and trends in climate are geographically nonuniform in both magnitude and direction. Locally, hemispheric or global trends in climate may be significantly amplified and even reversed, thereby disrupting ecosystems.

Fossil-fuel combustion and deforestation are raising atmospheric carbon dioxide levels. Uncompensated, this trend is likely to enhance the greenhouse effect. Furthermore, carbon-dioxide-induced warming may be added to by rising levels of other greenhouse gases, including methane, nitrous oxide, and CFCs, all of which absorb IR within atmospheric windows.

Change is an inherent characteristic of climate. The climatic future, like the climatic past, is likely to be shaped by the interaction of many climatic controls.

Questions for Review

1. In what sense is the atmosphere a limited resource?
2. Present several examples of how *minor* atmospheric gases or aerosols play important roles in the functioning of the environment.
3. How does the troposphere differ from the stratosphere? Have you ever been in the stratosphere?
4. Distinguish between weather and climate. Provide some examples of how weather and climate affect your everyday life and the environment.
5. On a winter day, is the air temperature more likely to be higher if the ground is snow-covered or if it is bare? Explain your answer.
6. Distinguish between radiational heating and radiational cooling.
7. What is meant by the ozone shield? How might CFCs affect the ozone shield, and what might that mean for human health?
8. What is the environmental significance of the deepening Antarctic ozone hole?
9. Describe in your own words the greenhouse effect. All other factors being equal, why is a clear night usually colder than a cloudy night?
10. Explain why and how the atmosphere is heated from below.
11. For the entire globe, why must incoming solar radiation balance outgoing infrared radiation? What would happen if this equilibrium did not prevail?
12. What is meant by (a) normal climate, (b) climatic anomalies, and (c) climatic controls?
13. Describe how climate governs the geographical distribution of the boreal forest and the prairies of North America.
14. How do we know that climate is variable?
15. Explain what is meant by the geographic nonuniformity of climatic anomalies. What are the agricultural implications of this characteristic of climatic behavior?
16. What human activities are contributing to the upward trend in atmospheric carbon dioxide concentration? What might this trend mean for our future climate?
17. List some of the positive and negative impacts of an enhanced greenhouse effect.
18. What—if anything—can we do now to head off an enhanced greenhouse effect?
19. CFCs are implicated in both the thinning of the ozone shield and the enhanced greenhouse effect. Please explain.
20. List some factors other than IR-absorbing gases that might contribute to future global climatic change.

Projects

1. List some things that you can do to reduce your contribution to greenhouse gases.
2. Speculate on the economic impact on your region should an enhanced greenhouse effect cause a trend toward warmer and drier summers. Consider both the benefits and drawbacks.
3. Find out how weather data is gathered in your community. Locate the station record and analyze it for trends. (Try the government documents section of your library.) How does your local temperature trend compare in direction and magnitude with the northern hemispheric temperature trend?

Selected Readings

Ausubel, J. H. "A Second Look at the Impacts of Climate Change." *American Scientist* 79 (1991):210-21. A critical look at several widely held assumptions regarding an enhanced greenhouse effect and climatic change.

Bryson, R. A., and T. J. Murray. *Climates of Hunger.* Madison, Wis.: Univ. of Wisconsin Press, 1977. Describes how past variations in climate have affected people living in agriculturally marginal regions of the world.

Houghton, J. T., G. J. Jenkins, and J. J. Ephraums, eds. *Climate Change, The IPCC Scientific Assessment.* Cambridge: Cambridge Univ. Press, 1990. Provides a comprehensive review of what is currently understood about the nature and impact of climatic change.

Kellogg, W. W. "Response to Skeptics of Global Warming." *Bulletin of the American Meteorological Society* 72, no. 4 (1991):499-511. Reviews the arguments for and against greenhouse warming.

Moran, J. M., and M. D. Morgan. *Meteorology: The Atmosphere and the Science of Weather.* 3d ed. New York: Macmillan, 1991. Covers in detail the principles governing weather and climate.

Riebsame, W. E., et al. "The Social Burden of Weather and Climate Hazards." *Bulletin of the American Meteorological Society* 67 (1986):1378-88. Includes a summary of trends in mortality and property damage caused by weather extremes.

Schneider, S. H. "Climate Modeling." *Scientific American* 256 (May 1987):72-80. Presents a critique of climate modeling as applied to projected trends in atmospheric carbon dioxide.

Schneider, S. H. *Global Warming.* San Francisco: Sierra Club Books, 1989. Discusses prospects for an enhanced greenhouse effect, the potential impacts, and what can be done.

Stolarski, R. S. "The Antarctic Ozone Hole." *Scientific American* 258 (Jan. 1988):30-36. Reviews the thinning of stratospheric ozone during the Antarctic spring and possible causes for the recent acceleration of that thinning.

Toon, O. B., and R. P. Turco. "Polar Stratospheric Clouds and Ozone Depletion." *Scientific American* 264 (June 1991):68-74. Describes the mechanisms involved in ozone destruction in the Antarctic.

White, R. M. "Greenhouse Policy and Climate Uncertainty." *Bulletin of the American Meteorological Society* 70 (1989):1123-27. Provides a concise summary of problems encountered when public policy-making decisions are based on uncertain climate forecasts.

Air Quality

Industrial and other human activities threaten the quality of air.
© Gary Milburn/Tom Stack & Associates

*P*erhaps more than any other American city, Los Angeles is known for its air-quality problems. Ultimately, those problems stem from the city's love affair with the automobile, along with regional climate and topography that trap the exhaust of congested urban traffic. Bright sunshine triggers a complex set of chemical reactions that convert motor vehicle exhaust and industrial emissions to an unhealthy mixture of pollutants known as *photochemical smog* (fig. 16.1).

Although we usually think of air pollution as an undesirable by-product of modern industrialism, it is at least as old as civilization. The first episode of air pollution probably occurred when some of the first humans made a fire in a poorly ventilated cave. Reference to polluted air appears as early as Genesis 19:28: "Abraham beheld the smoke of the country go up as the smoke of a furnace." Hippocrates noted the pollution of city air in about 400 B.C., and in 1170, Maimonides, referring to Rome, wrote, "The relation between city air and country air may be compared to the relation between grossly contaminated, filthy air, and its clear, lucid counterpart."

The Industrial Revolution was the single greatest contributor to air pollution as a chronic problem in Europe and North America. In the United States, in post-Civil War days, cities swelled with new industries and immigrants to work them. By the turn of the century, the urban environment was increasingly fouled by the fumes of foundries and steel mills. In those days, a city took pride in its smokestacks (see fig. 16.2), which were considered a sign of a prosperous economy. Efforts to regulate air quality were meager, and little was known about the effects of air pollution on human health. In an attempt to placate the wheezing and coughing citizenry, some physicians even argued that polluted air had medicinal value.

Even today, concern over polluted air does not stem from disenchantment with the fruits of industrialism, but many people are troubled by reports that polluted air adversely affects our health, agricultural productivity, and the weather. In this chapter, we examine several issues related to these concerns: (1) how polluted air affects the well-being of living things; (2) the problem of acid deposition; (3) how weather conditions influence levels of air pollution; and (4) how air quality is managed. First, however, we consider the general nature of air pollutants.

Air Pollutants: Types and Sources

An **air pollutant** is an airborne gas or aerosol that occurs in concentrations that threaten the well-being of organisms or disrupt the natural processes upon which they depend. Many of these substances are natural constituents of the atmosphere. For example, sulfur dioxide (SO_2) and carbon monoxide

Figure 16.1   A layer of smog is clearly visible in this photograph taken from an aircraft approaching Los Angeles International Airport. Smog is a chronic problem in the Los Angeles basin because of a combination of weather conditions, topography, and considerable motor vehicle traffic.

(CO) are normal components of the atmosphere that become pollutants when their concentrations approach or exceed the tolerance limits of organisms (chapter 4). On the other hand, some air pollutants do not occur naturally in the atmosphere and may be hazardous in any concentration. Asbestos fibers, which are known carcinogens (chapter 12), fall into this category.

Air pollutants are products of both natural events and human activities. Natural sources of air pollutants include forest fires, pollen dispersal, wind erosion of soil, decay of organic matter, and volcanic eruptions. The single most important source of atmospheric pollutants attributed to human activity is the motor vehicle. According to the U.S. Environmental Protection Agency (EPA), each year transportation vehicles emit about 57 million metric tons (64 million tons) of the major air pollutants. Many industrial sources contribute as well. Unless their emissions are controlled, pulp and paper mills, iron and steel mills, oil refineries, smelters, and chemical plants can be prodigious producers of air pollutants. Additional sources of pollutants include fuel combustion for space heating and the generation of electricity, refuse burning, and various agricultural activities, such as crop dusting and cultivation. In the United States, almost 128 million metric tons (143 million tons) of the chief air contaminants are emitted into the atmosphere each year as a result of human activities—almost 0.5 metric ton per person (table 16.1).

Some substances are potentially harmful immediately upon emission into the atmosphere and are designated **primary air pollutants** (e.g., carbon monoxide from automobile exhaust). Within the atmosphere, chemical reactions involving primary air pollutants, both gases and aerosols, produce **secondary air pollutants** (e.g., photochemical smog). The environmental impact of certain primary pollutants is less severe than that of the secondary pollutants they form. In Special Topic 16.1 on pages 342-343, we survey the major air pollutants, noting their natural cycling within the environment and the contributions of human activities.

Air Pollution and Human Health

The potential impact of air pollution on human health is the chief reason for our concern. In this section, we examine how humans respond to different types of air pollutants.

The initial impact of air pollutants is primarily on the **respiratory system** (fig. 16.3). Air pollutants enter the respiratory system along with air through either the oral or nasal cavities, which join to form the pharynx (throat). Air then travels down the pharynx through the larynx (voice box) and then through the trachea (windpipe), which splits into two bronchi, each of which leads to a lung. In the lungs, each bronchus splits into successively smaller tubes called **bronchioles,** which terminate in tiny pouches known as **air sacs** (*alveoli*). Oxygen then moves through the walls of the air sacs

Figure 16.2   Industrial smokestacks in Pittsburgh in 1906.
Carnegie Library of Pittsburgh

Table 16.1   Estimated U.S. emissions of principal air pollutants in 1987 (in millions of metric tons)

Source	CO	SO$_x$	VOCs[a]	TSP[b]	NO$_x$	Total
transportation	40.7	0.9	6.0	1.4	8.4	57.4
fuel combustion	7.2	16.4	2.3	1.8	10.3	38.0
industrial processes	4.7	3.1	8.3	2.5	0.6	19.2
solid waste disposal	1.7	—	0.6	0.3	0.1	2.7
others	7.1	—	2.4	1.0	0.1	10.6
totals	61.4	20.4	19.6	7.0	19.5	127.9

Source: *National Air Pollutant Emission Estimates, 1940–1987*, U.S. Environmental Protection Agency, March, 1989.

[a]Volatile organic compounds (hydrocarbons)
[b]Total suspended particulates

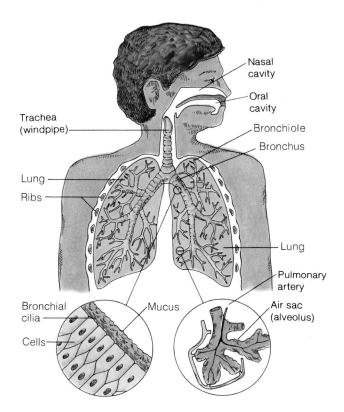

Figure 16.3   The human respiratory system.

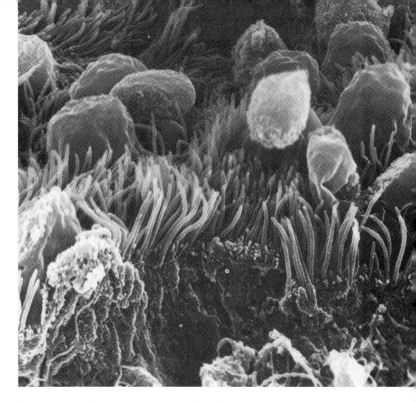

Figure 16.4   Photomicrograph of cilia within a human respiratory system. The coordinated flailing motion of these tiny, hairlike structures transports pollutant-laden mucus into the mouth. There, mucus is either swallowed or expelled.
© David M. Phillips/The Science Population Council/Science Source/Photo Researchers, Inc.

into capillary beds where it is captured by hemoglobin molecules contained within red blood cells. To ensure an adequate uptake of oxygen, our lungs contain over 300 million air sacs, having a combined surface area of about 700 square meters (7,500 square feet), which is nearly 1.5 times the area of a basketball court.

Human Defense Mechanisms

Because humans and natural air pollutants have existed together for millennia, we have evolved defense mechanisms against some air pollutants. For example, nasal hairs filter out larger particulates. Furthermore, the entire respiratory passage from nose to bronchioles is coated by a layer of sticky mucus produced by specialized cells in the lining of the respiratory tract. Mucus captures smaller particles not trapped by nasal hairs and may also absorb some gaseous pollutants. Mucus, with its load of pollutants, is transported to the mouth by the continuous, wavelike beating of millions of **cilia,** tiny hairlike structures that are extensions of other specialized cells that line much of the respiratory tract (fig. 16.4). There are about two hundred cilia per cell, and each one beats about 10-20 times per second. These activities move mucus about one centimeter per minute. Upon reaching the mouth, the pollutant-bearing mucus is either swallowed or expelled.

The smallest particles may eventually reach the air sacs, which have no mucus-producing or ciliated cells. Alveoli, however, are not without their own specialized defenses. Within the alveoli are free-living cells known as **macrophages** (fig. 16.5). These cells, which are part of our immune system,

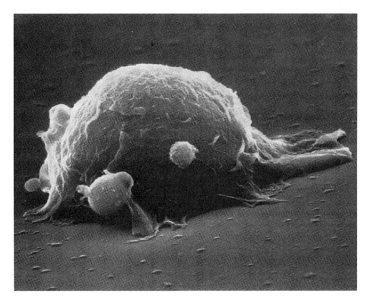

Figure 16.5   A lung macrophage surrounded by particles of fly ash, viewed under a scanning electron microscope. Each of these free-living cells is capable of engulfing many particles of fly ash or other pollutants.
Courtesy of the Lawrence Berkeley Laboratory, University of California, Berkeley

engulf and digest foreign particles such as bacteria, viruses, and pollen grains. Macrophages also possess powerful toxins that kill bacteria even if the cells cannot digest them. Eventually macrophages succumb to the very poisons that they produce to kill invaders. Many particle-laden, dead macrophages

Principal Air Pollutants

Here we summarize the sources, cycling patterns, and some of the impacts of the major air pollutants.

Oxides of Carbon

Through cellular respiration, organisms release carbon dioxide (CO_2) to the atmosphere, and through photosynthesis, plants take up carbon dioxide. Other natural sources of CO_2 include forest and brush fires and volcanic activity. Combustion of fossil fuels for electric power generation and space heating releases CO_2 into the atmosphere. About 55% of it remains in the atmosphere; most of the remainder dissolves in ocean water. Today the concentration of CO_2 in the atmosphere is about 350 parts per million (ppm) and is rising at a rate of about 16 ppm per decade. As we saw in chapters 3 and 15, this upward trend in atmospheric CO_2 may affect the global climate.

By far the most important natural source of atmospheric carbon monoxide (CO) is the combination of oxygen with methane and other volatile organic compounds. Carbon monoxide is removed from the atmosphere by the activity of certain soil microorganisms and by chemical reactions that convert CO to CO_2. The net result is a harmless northern hemisphere average concentration of less than 0.15 ppm.

In more-developed nations of the northern hemisphere, the principal source of CO derived from human activity is incomplete combustion of fossil fuels, especially by motor vehicles. Recently, scientists discovered that the major source of CO in the southern hemisphere and the tropics is the burning of forests and savannas to clear land. Instrument measurements from aircraft and a space shuttle mission indicate that this contribution may compare in magnitude to global fossil fuel combustion. Carbon monoxide is a colorless, odorless, and tasteless gas that defies direct human detection. It is an asphyxiating agent that in sufficiently high concentrations causes death.

Volatile Organic Compounds

Volatile organic compounds, commonly called *hydrocarbons*, include a wide variety of chemicals that contain only carbon (C) and hydrogen (H). Of the hydrocarbons that occur naturally in the atmosphere, methane (CH_4) has the highest concentration (1.68 ppm, on average). Methane is produced when organic material decays anaerobically (in the absence of oxygen), for example, in swamps, rice paddies, and the stomach of cattle. At normal background levels, methane is nonreactive; that is, it does not chemically interact with other substances and is not harmful. However, methane is the chief component of natural gas, and air with a methane concentration of 5% or greater is explosive. Also, as we saw in chapter 15, some scientists are concerned about the potential effect of methane on global climate: It is a greenhouse gas, and its mean global concentration is rising.

All vegetation emits various hydrocarbons. Among them are terpenes, which occur in concentrations of less than 0.1 ppm. They are chemically reactive and responsible for the aromas of pine, sandalwood, and eucalyptus trees. Recent studies suggest that such natural hydrocarbons (other than methane) may play an important role in the formation of photochemical smog.

Reactive hydrocarbons are also emitted during the incomplete combustion of gasoline by motor vehicles. This source is responsible for hundreds of different hydrocarbons. Because gasoline is very volatile, some hydrocarbons (perhaps as much as 15% of the total in some cities) also escape to the air during gasoline delivery and refueling operations at service stations. Hydrocarbons are also emitted by solvents used in a variety of industrial, commercial, and home processes, from painting and cleaning heavy equipment to cleaning paintbrushes. Chemical manufacturing and petroleum refining also emit hydrocarbons.

Although our understanding of the natural cycling of hydrocarbons in the atmosphere is still incomplete, we do know that the typically low concentrations of most hydrocarbons found in city air appear to pose little direct environmental threat. However, serious health hazards arise from the products of complex chemical reactions involving hydrocarbons and other air pollutants (particularly nitrogen oxides) in the presence of sunlight. These reactions produce *photochemical smog*. Some hydrocarbons, such as benzene (C_6H_6), an industrial solvent, and benzopyrene, a product of fossil fuel and tobacco combustion, are also carcinogenic.

Oxides of Nitrogen

The action of soil bacteria is responsible for most of the nitric oxide (NO) that is produced naturally and released to the atmosphere. Within the atmosphere, NO combines readily with oxygen to form nitrogen dioxide (NO_2). Together, these two oxides of nitrogen are usually referred to as NO_x.

Although human activities contribute only about 10% of the atmosphere's total load of NO_x, our contributions tend to be much more concentrated than the natural atmospheric average. NO_x forms when high combustion temperatures, such as those inside an automobile engine, cause nitrogen (N_2) and oxygen (O_2) in the air to combine. In addition, NO_x forms when nitrogen compounds in a fuel, such as coal, are oxidized (combine with O_2). For both modes of formation, NO is generated initially, and when it is vented and cooled, some of the NO converts to NO_2. About half of NO_x is emitted from stationary sources (power plants, primarily), and the other half comes from motor vehicles.

Nitrogen dioxide (NO_2) is a much more serious air pollutant than its precursor NO; the toxicity of NO_2 is about four times that of NO. Nitrogen dioxide at high levels is believed to contribute to heart, lung, liver, and kidney damage, and it is linked to the incidence of bronchitis and pneumonia. Moreover, because nitrogen dioxide occurs as a brownish haze, it reduces visibility. Oxides of nitrogen are precursors of photochemical smog. In addition, NO_2 combines with moisture in the air to form nitric acid (HNO_3), a corrosive substance and acid-rain precursor.

Compounds of Sulfur

Sulfur enters the atmosphere naturally as sulfur dioxide (SO_2) from volcanic eruptions, as sulfate (SO_4^{2-}) containing particles from sea spray, and as hydrogen sulfide (H_2S) produced when organic matter decays anaerobically. These sulfur compounds are washed from the air by precipitation and are taken up by soil, vegetation, and surface waters.

Human activities also release sulfur compounds into the atmosphere, at about one-third the rate of emission by natural sources. Most of our contribution comes from fossil fuels (chiefly coal and oil) that contain sulfur as an impurity and emit sulfur dioxide when burned. The principal U.S. source of sulfur dioxide is coal burning for electric utilities (65.7%). Industrial processes contribute 16.4%, transportation 4.4%, and nonutility fuel combustion from stationary sources, 13.5%.

Within the atmosphere, sulfur dioxide is converted to sulfur trioxide (SO_3) and sulfate-containing aerosols. Sulfate aerosols restrict visibility and in the presence of water form droplets of sulfuric acid (H_2SO_4), a highly corrosive substance and acid-rain precursor. Both SO_2 and SO_3 irritate respiratory passages and can aggravate asthma, emphysema, and bronchitis. Sulfate particles and sulfuric acid droplets are thought to increase human vulnerability to respiratory infection.

Certain industrial activities, including paper and pulp processing, emit hydrogen sulfide (H_2S) as well as a family of organic, sulfur-containing gases called *mercaptans*. Even at extremely low concentrations, these compounds are foul-smelling. Hydrogen sulfide tarnishes silverware and copper facings and blackens lead-based paints.

Photochemical Smog

When vehicular traffic is heavy (during rush hours, for example), photochemical smog is likely to form. Oxides of nitrogen in motor vehicle exhaust and hydrocarbons (from various anthropogenic and biogenic sources) react in the presence of sunlight to produce a noxious, hazy mixture of suspended particles and gases. Products include ozone (O_3), formaldehyde (CH_2O), ketones, and PAN (peroxyacetyl nitrates), substances that irritate the eyes and damage the respiratory system. Photochemical smog is most common in urban areas, but winds can transport auto exhaust into suburban and rural areas, where the sun's rays also trigger smog development.

While levels of ozone at the earth's surface average only about 0.02 ppm, ozone concentrations may exceed 0.5 ppm in thick photochemical smog. Exposure to these relatively high concentrations for an hour or two can cause healthy people to experience coughing, painful breathing, and temporary loss of some lung function when they engage in vigorous physical activity. Some medical experts fear that chronic exposure to ozone may cause structural changes in the lung, perhaps leading to lung disease. It also degrades rubber and fabrics, retards tree growth, and damages some crops.

Suspended Particulates

So far, our survey of the major air pollutants has focused primarily on gases. Now we turn our attention to the multitude of tiny solid particles and liquid droplets that are suspended in the atmosphere. Sea-salt spray, soil erosion, volcanic activity, and various industrial emissions account for about one-half of the atmosphere's total aerosol load. The other half is largely the consequence of atmospheric reactions among various gases.

Perhaps the most common particulates are dust and soot. Most dust is produced when wind erodes soil; agricultural activity often accelerates such erosion. Soot, tiny solid particles of carbon, is emitted during the incomplete combustion of coal, oil, and refuse. In urban-industrial air, suspended particulates may include a wide variety of materials, depending on the specific types of manufacturing. Urban-industrial particulates usually include a diverse array of trace metals such as lead, nickel, iron, zinc, copper, magnesium, and cadmium. These particulates pose a significant health hazard because their small size allows them to penetrate the defense mechanisms of the upper respiratory tract and enter the lungs. In addition, air may contain asbestos fibers, pesticides, and fertilizer dust.

Air normally also contains fungal spores and pollen. Disturbance of the land by wildfires, farming, or construction promotes the abundant growth of ragweed and other weeds whose pollen evokes allergic reactions (e.g., hay fever) in roughly one out of every twenty people.

reach the lower bronchioles, where they are caught up in mucus and transported by cilia to the mouth.

Pollutants may also induce defensive responses that force fully expel them from the respiratory system: the sneeze and cough reflexes. When nasal passageways are irritated, a message is sent to the brain that initiates the sneeze reflex. In much the same way, the cough reflex is a response to an irritation of the trachea or other passageways of the lower respiratory tract. The sneeze and cough reflexes are strong enough to expel air at velocities that can approach 160 kilometers (100 miles) per hour.

Health Effects of Gases and Particulates

Our natural defenses cannot protect us from all the effects of air pollutants. For example, the ability of mucus to capture gases is quite limited; thus, gaseous pollutants usually penetrate deeply into the respiratory tract. Some gaseous pollutants act as **asphyxiating agents;** people who are exposed to high enough concentrations of these gases die of oxygen starvation. Other gases irritate the respiratory lining.

Carbon monoxide (CO) is an invisible, odorless, tasteless asphyxiant. Because it bonds with hemoglobin molecules two hundred times more readily than oxygen does, it prevents them from picking up oxygen in the air sacs. Hence, as increasing concentrations of CO are inhaled, the quantity of life-sustaining oxygen that the bloodstream transports from the lungs diminishes. If CO ties up more than 50-80% of hemoglobin molecules, a person dies of oxygen starvation.

Hydrogen sulfide (H_2S), produced by anaerobic decomposition of organic matter, is also an asphyxiating gas, but unlike carbon monoxide, it does not displace oxygen in the blood. Rather, the bloodstream transports hydrogen sulfide to the brain where, at sufficiently high concentrations, it impairs the part of the brain that controls the chest movements essential for normal breathing. At 1,000 ppm, death is essentially instantaneous. Although high concentrations can develop in closed spaces with limited ventilation, such as sewers and mines, hydrogen sulfide is relatively harmless in the very low concentrations that we normally encounter, although we may find its rotten-egg odor unpleasant.

Gases that act mainly as irritants of the respiratory tract include ozone (O_3), sulfur dioxide (SO_2), sulfur trioxide (SO_3), and nitrogen dioxide (NO_2). Although each one causes somewhat different reactions, they generally cause persistent coughing and heavy secretion of mucus.

Particulates that penetrate the natural defenses of the respiratory system create the most serious problems when exposure occurs over an extended period of time. As we saw in chapter 12, chronic exposure to relatively high levels of dust in occupations such as mining, metal grinding, and the manufacturing of abrasives, can eventually cause lung diseases such as black lung, silicosis, asbestosis, and byssinosis. Depending on the type of particulate and the concentrations inhaled, the lungs may suffer irritation, allergic reactions, or

scarring of tissues, which can become the basis for tumor development. Victims usually experience coughing and shortness of breath and may eventually develop pneumonia, chronic bronchitis, emphysema, or lung cancer.

As noted earlier, mucus has a limited capacity to absorb gases and particulates, and other respiratory defenses are often impaired by various inhaled agents as well. For example, some gases and particulates reduce the activity of the cilia-mucus transport system, thereby increasing the residence time of pollutants within the respiratory tract. Such impairment may increase the likelihood of cancer if carcinogenic agents are retained in the upper respiratory system. Likewise, pollutant damage to macrophages may play a role in chronic lung diseases such as emphysema and silicosis.

Some people argue that no definite links have been established between air pollution and respiratory diseases such as emphysema and chronic bronchitis. But symptoms of bronchitis and emphysema are similar to those attributed to chronic inhalation of air pollutants, which suggests that these diseases may at least be aggravated by polluted air. In addition, the incidence of respiratory diseases is greatest in regions that report the highest levels of air pollution.

Still, direct evidence that unequivocally links elevated atmospheric levels of a specific air pollutant to a particular respiratory disease is rare. Recall our discussion in chapter 12 regarding the difficulties of determining the toxicity of chemicals, including air pollutants. It is extremely difficult to prove the existence of such a connection, because the urban-industrial atmosphere contains numerous gases and particulates whose relative concentrations continually fluctuate and whose interactions may be complex. Also, age, level of physical activity, and general health, as well as the duration and frequency of exposure, affect a person's response to polluted air.

We usually associate air pollution with outdoor air or occupational exposure. Recent years, however, have brought increasing concern with the quality of air that we breath in our homes (or other buildings). After all, most of us spend more than half of each day at home. In fact, some scientists argue that health risks caused by indoor air pollution may be of the same magnitude as the risks of occupational exposure to chemicals or radiation. For more on this topic, refer to the issue on page 345.

Impact of Air Pollution on Plants

In North America, ozone (O_3) appears to cause the greatest amount of damage to vegetation. Ozone is a component of photochemical smog that forms within and downwind of urban areas. Exposure to ozone reduces photosynthesis and subsequent growth in many species, including soybeans, winter wheat, red clover, sugar maple, eastern white pine, and northern red oak. In fact, for four important cash crops—corn, soybeans, wheat, and peanuts—the annual loss due to ozone damage is estimated at $1.9-$4.5 billion.

Indoor Air Quality

*I*ndoor air pollutant sources include building materials (especially those that contain formaldehyde resins or asbestos fibers), unvented cooking or heating appliances, smoking materials, cleaning compounds, and personal-care products. Table 16.2 is a list of indoor air pollutants, their sources, typical concentrations, and how their concentrations compare to those of outdoor air pollutants.

Some efforts to conserve energy have compounded the problem of indoor air quality. Tighter, better-insulated homes reduce the amount of air that is exchanged between indoors and outdoors; thus, the levels of indoor air pollutants may climb to unhealthy concentrations. Also, the use of wood, coal, and kerosene-burning heaters may expose people to toxic and carcinogenic gases and airborne particles.

Potentially the most serious indoor air pollution problem involves radon. Like carbon monoxide, radon is a gas that defies direct human detection, because it is invisible, odorless, and tasteless. Radon is a radioactive decay product of radium-226, a natural trace component of soils and bedrock in many parts of the United States. (We discuss radioactivity in detail in chapter 20.) Radon escapes from building materials (such as brick or stone) and may enter homes through cracks in basement walls or, to a lesser extent, via well water. The problem with radon stems not from the gas itself but from the radioactive elements that are produced as radon decays.

When inhaled, most radon atoms are exhaled almost immediately. However, if radon decays in the lungs, its radioactive decay products are left behind on tracheal-bronchial surfaces. There they emit damaging alpha particles (each consisting of two protons and two neutrons) that penetrate nearby cells and may trigger lung cancer. In 1988, the Committee on Biological Effects of Ionizing Radiation of the National Academy of Sciences estimated that as many as 13,000 radon-related lung cancer deaths may occur annually in the United States. That estimate spurred considerable controversy, and today some scientists believe that radon-related deaths are considerably under 13,000 per year.

Ordinarily, radon levels are too low to pose a serious health hazard, but with the restricted air exchange of energy-efficient homes and in regions where the soil or bedrock is enriched in radium, radon levels (especially in the basements of homes) may climb to two hundred times the background (average) level. At that concentration, the risk of lung cancer is about the same as smoking 2.5-10 packs of cigarettes a day. Obviously, cigarette smoking and exposure to high radon levels are a lethal combination.

In a random national survey, researchers at the Lawrence Berkeley Laboratory found that the average annual concentration of radon in the living spaces of homes was 1.5 picocuries (pCi) per liter (L) of air. (A picocurie is a measure of radioactivity.) This compares to the EPA recommended threshold of 4 pCi/L, above which some corrective action is recommended. Researchers further estimated that 7% of American single-family homes had an average annual value above the EPA threshold and 1-3% exceeded 8 pCi/L. (There are about 65 million houses in the United States.)

What can be done about radon in the home? The first step is to determine whether a problem exists, through long-term (several months to a year) monitoring using a radon-detection kit available at most hardware stores. The typical cost of testing is $10 to $30. A standard radon detector consists of a small box containing a strip of special polymer film that records tracks made by alpha particles. Detectors should be placed in one or more living spaces—not in the basement where radon levels are typically twice what they are in the rest of the house. At the end of the monitoring period, detectors are sent back to the store (or lab) for a reading. Readings greater than 4 pCi/L should be followed by additional testing. If the level is confirmed, remedial action is recommended.

If there is a radon problem in the home, the solution is improved ventilation, particularly of the radon-laden air in the basement. (This may well reduce heating efficiency.) In addition to improved ventilation, other means of reducing radon include covering exposed earth and sealing cracks in basement walls and floors.

The goal of the Indoor Radon Abatement Act of 1988 is to achieve indoor radon levels comparable to average outdoor levels nationwide. In some regions, such a goal is not yet technically possible. In most cases, however, a reduction to the 4 pCi/L threshold should cost $500 to $1500 per house.

Ozone also appears to be a more significant threat to the health of North American forests than either acid deposition (discussed later in this chapter) or particulates (e.g., heavy metals). The impact of moderate levels of air pollutants on trees varies with species but may include impaired metabolism, predisposition to damage by insects, or increased susceptibility to disease. The problem of ozone damage is particularly acute in the San Bernardino National Forest in southern California. Exceptionally high concentrations of windborne ozone (up to 300 to 400 ppb) originating in Los Angeles smog some 125 kilometers (80 miles) upwind of the forest are destroying hundreds of thousands

Table 16.2 Indoor air pollutants

Pollutant	Major Emission Sources	Typical Indoor Concentrations	Indoor/Outdoor Concentration Ratio
Origin: predominantly outdoors			
sulfur oxides (gases, particles)	fuel combustion, smelters	0–15 μg/m^3	$<$ 1
ozone	photochemical reactions	0–10 ppb	$\ll$ 1
pollens	trees, grass, weeds, plants	L.V.*	$<$ 1
lead, manganese	automobiles	L.V.	$<$ 1
calcium, chlorine, silicon, cadmium	suspension of soils, industrial emissions	N.A.†	$<$ 1
organic substances	petrochemical solvents, natural sources, vaporization of unburned fuels	N.A.	$<$ 1
Origin: indoors or outdoors			
nitric oxide, nitrogen dioxide	fuel burning	10–120 μg/m^3‡ 200–700 μg/m^3§	$\gg$ 1
carbon monoxide	fuel burning	5–50 ppm	$\gg$ 1
carbon dioxide	metabolic activity, combustion	2000–3000 ppm	$\gg$ 1
particles	resuspension, condensation of vapors, combustion products	10–1000 μg/m^3	1
water vapor	biological activity, combustion evaporation	N.A.	$>$ 1
organic substances	volatilization, combustion, paint, metabolic action, pesticides	N.A.	$\gg$ 1
spores	fungi, molds	N.A.	$>$ 1
Origin: predominantly indoors			
radon	building construction materials (concrete, stone), water	0.01–4 pCi/liter	$\gg$ 1
formaldehyde	particleboard, insulation, furnishings, tobacco smoke	0.01–0.5 ppm	$>$1
asbestos, mineral, and synthetic fibers	fire retardant materials, insulation	0–1 fiber/ml	1
organic substances	adhesives, solvents, cooking, cosmetics	L.V.	$>$ 1
ammonia	metabolic activity, cleaning products	N.A.	$>$ 1
polycyclic hydrocarbons, arsenic, nicotine, acrolein, and so forth	tobacco smoke	L.V.	$\gg$ 1
mercury	fungicides, paints, spills in dental-care facilities or labs, thermometer breakage	L.V.	$>$ 1
aerosols	consumer products	N.A.	$\gg$ 1
microorganisms	people, animals, plants	L.V.	$>$ 1
allergens	house dust, animal dander, insect parts	L.V.	$\gg$ 1

From J. D. Spengler and K. Sexton, "Indoor Air Pollution: A Public Health Perspective," *Science,* Vol. 221, pp. 9–17, July 1, 1983. Copyright 1983 by the AAAS. Used by permission.

*L.V., limited and variable (limited measurements, high variation).
†N.A., not applicable.
‡Annual average.
§One-hour average in homes with gas stoves, during cooking.
$<$ Less than
$>$ Greater than
$\gg$ Much greater than

Figure 16.6 The forests of the San Bernardino Mountains located downwind of the Los Angeles metropolitan area show the devastating effects of wind-borne ozone.
© John D. Cunningham/Visuals Unlimited

of trees (fig. 16.6). Needles turn yellow and eventually fall off. Also, the trees are made vulnerable to infestation by bark beetles or microbial root infection, either of which may be lethal.*

Sulfur dioxide (SO_2) can also harm vegetation. This pollutant is emitted in the effluent of iron and copper smelters; such emissions have virtually eliminated vegetation in large areas adjacent to smelters. A white pine showing the effects of sulfur dioxide poisoning and a healthy tree of the same species appear in figure 16.7.

Most commonly, air pollutants damage the leaves of plants, but leaves also possess defense mechanisms against air pollutants. The natural waxy layer that covers leaves and protects them from severe water vapor loss also significantly limits direct exposure to air pollutants. However, the tiny pores (stomates) on the surfaces of leaves that allow carbon dioxide to enter for photosynthesis and water vapor to escape (transpiration), also provide access for gaseous pollutants. Although stomates close when leaves transpire more water than is taken up by plant roots (resulting in wilting), enough water is usually available so that pores remain open. Under these circum-

*For more information on the impact of air contaminants on North American forests, see the article by Smith listed at the end of the chapter.

stances, gaseous pollutants may enter through the pores along with carbon dioxide.

When pollutants such as sulfur dioxide and ozone enter leaves, they dissolve in the water that adheres to cell-wall surfaces. At that point, a variety of complex and poorly understood chemical reactions occurs. The first sign that a plant has suffered air-pollution damage is the appearance of **chlorosis,** a yellowing of leaves that results from chlorophyll loss (fig. 16.8). Without chlorophyll, photosynthesis cannot occur, and without a continual supply of newly synthesized sugars, insufficient energy is available for tissue growth and repair. Frequently, air-pollution damage makes plants more vulnerable to other stresses such as drought and pathogens. Thus, plants may eventually succumb to some combination of these stresses.

Particulates are usually too large to enter leaf pores, so they produce fewer adverse effects. In some cases, however, particulates coat the surfaces of leaves, reducing the amount of sunlight available for photosynthesis.

Tolerance of vegetation to air pollution varies considerably among species and the pollutants that they encounter. Scientists do not yet know why certain plant species are more sensitive than others to gaseous pollutants, but in a few cases they have proposed explanations. For example, in varieties of

(a)

(b)

Figure 16.7   (a) A 10-year-old white pine severely damaged by relatively low concentrations of sulfur dioxide possibly mixed with ozone. (b) A healthy 6-year-old white pine from the same area grown in filtered air.
U.S. Department of Agriculture

onions that are resistant to ozone, the leaves are sensitive to ozone and leaf pores close in its presence, thereby protecting the interior of the leaf.

We close our discussion of the impact of air pollution by considering the effects of acid deposition on ecosystems. Refer to the issue on page 349 for more on this topic.

Air-Pollution Episodes

On the morning of 26 October 1948, a fog blanket that reeked of pungent sulfur dioxide fumes spread over the town of Donora in Pennsylvania's Monongahela Valley. Before the fog lifted five days later, almost half of the areas's 14,000 inhabitants had fallen ill, and twenty had died. That killer fog resulted from a combination of mountainous topography and weather conditions that trapped and concentrated deadly effluents from the community's steel mill, zinc smelter, and sulfuric acid plant.

Air pollutants are especially dangerous when atmospheric conditions reduce their rate of dilution. Upon emission into the atmosphere, air-pollutant concentrations begin to decline.

Figure 16.8   Sulfur dioxide can destroy the chlorophyll-bearing cells of leaves. The affected leaves develop white areas between the veins. The leaf on the right is healthy.
U.S. Department of Agriculture

Acid Deposition

As we saw in chapter 13, the atmospheric subcycle of the global water cycle purifies water. That is, when water vaporizes, dissolved and suspended substances are left behind. As raindrops and snowflakes fall from clouds to the earth's surface, however, they wash pollutants from the air, and the chemistry of the precipitation changes.

Rain and snow are normally slightly acidic because they dissolve some atmospheric carbon dioxide, producing a weak acidic solution. Where air is polluted with oxides of sulfur and oxides of nitrogen, these gases interact with moisture in the atmosphere to produce tiny droplets of sulfuric acid (H_2SO_4) and nitric acid (HNO_3). These substances dissolve in precipitation and increase its acidity. Precipitation that falls through such contaminated air may become two hundred times more acidic than normal. Furthermore, in the absence of precipitation, sulfuric acid droplets convert to acidic particles that reduce visibility and may cause human health problems when inhaled. Eventually, acidic particles settle to the ground as dry deposition. The combination of acid precipitation and dry deposition is often referred to as **acid deposition.**

The range of acidity and alkalinity, called the **pH scale,** is shown in figure 16.9, which compares the normal acidity of rainwater with the pH values of some other familiar substances. The normal pH of rainwater is 5.6; rain that is more acidic than normal is called **acid rain.** Note that the pH scale is logarithmic; that is, each unit increment corresponds to a tenfold change in acidity. Hence, a drop in pH from 5.6 to 3.6 represents a hundredfold (10×10) increase in acidity.

One of the first alarms regarding acid rain was sounded by Gene E. Likens, past director of the Institute for Ecosystem Studies of the New York Botanical Garden. Likens and his colleagues reported an increase in the acidity of rainfall over the eastern United States between 1955 and 1973.* Their findings were later confirmed and updated by measurements made by the National Atmospheric Deposition Program in the United States and by the Canadian Network for Sampling Precipitation (fig. 16.10). In general, the mean annual pH of precipitation is under 5 east of the Mississippi and over 5 west of the Mississippi. Rain and snow is most acidic in the northeastern United States and adjacent portions of Canada, where mean annual pH in 1991 ranged from 3.0 to 5.5, and individual storms produced rainfall with pH values as low as 2 to 3.

Acid precipitation is attributed to gaseous precursors emitted as by-products of fuel combustion for electric power, industry, and motor vehicles. Coal burning for electric power generation is the principal source of sulfur oxides, while high-temperature industrial processes and internal combustion engines (in motor vehicles) produce nitrogen oxides.

Where soils are thin and the bedrock composition is noncarbonate (neither limestone nor dolomite), acid rains and snowmelt are not usually neutralized. Figure 16.11 shows acid-sensitive areas of North America. Thus, acidic drainage can

*See the article by Likens et al., listed at the end of the chapter.

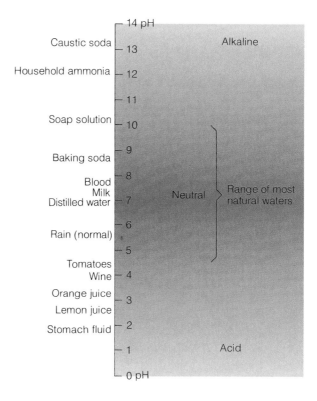

Figure 16.9   The pH scale, or scale of acidity and alkalinity, and the pH of some common substances. It is a logarithmic scale, which means that an increase or decrease of 1 on the scale represents a tenfold decrease or increase in acidity.

lower the pH of lakes and streams. Spring snowmelt is a potentially serious problem because of the sudden pulse of acidic meltwater entering waterways. A 1991 EPA survey of 1,180 lakes and 4,670 streams in acid-sensitive regions of the United States found that of the waterways with excessive acidity, acid deposition was the principal cause in 75% of the lakes and 47% of the streams. Acid mine drainage (chapter 17) was responsible in 26% of the acidified streams, and natural organic sources were implicated in 25% of the lakes and streams with below normal pH.

Excessively acidic lake or stream water disrupts the reproductive cycles of fish. Acid rains also leach metals (such as aluminum) from the soil, washing them into lakes and streams where they may cause direct harm to fish or to the aquatic organisms that form the base of food webs for fish. Increased acidity has caused the decline or elimination of fish populations in some lakes and streams in Norway, Sweden, eastern Canada, and the northeastern United States.

As part of the EPA's National Acid Precipitation Assessment Program, scientists artificially acidified a portion of a small lake in northern Wisconsin to determine the impact on the lake's aquatic life under controlled conditions. The 18-hectare (45-acre) Little Rock Lake is typical of the thousands of small,

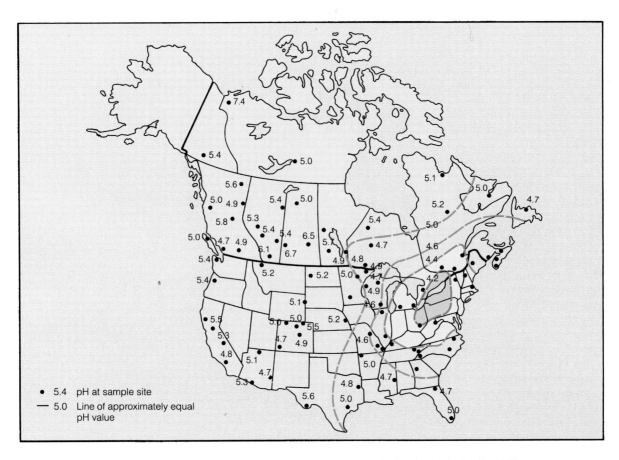

Figure 16.10   The acidity of precipitation over North America. Figures are average annual pH of precipitation for 1982.
Sources: From data developed by the National Atmospheric Deposition Program in the United States, and the Canadian Network for Sampling of Precipitation.

acid-sensitive lakes that dot northern Wisconsin, Minnesota, and Upper Michigan. Most of these lakes are fed by precipitation and receive no input from streams or groundwater. Hence, they are particularly vulnerable to acid deposition. Scientists anticipate that the artificial acidification experiment will enable them to predict the likely long-term response of similar lakes to acid deposition.

Little Rock Lake is shaped like an hourglass: Two basins of about equal area are joined by a narrows. Following an intensive field study of the lake's properties and species composition, scientists installed a 71-meter (230-foot) plastic barrier to separate the two basins at the narrows (fig. 16.12). Between 1985 and 1990, about 1,500 liters (400 gallons) of sulfuric acid were gradually added to one of the lake basins. In response, the pH of the acidified basin declined from its initial value of 6.1 to 5.6 in 1985, 5.2 in 1987, 4.9 in 1989, and 4.7 in 1990. The other lake basin was not disturbed and served as the control for comparison with the acidified basin (chapter 1).

What were the effects of acidification? Researchers found that an algal mat up to a meter thick now covers the bottom, smothering nesting sites for sport fish. Populations of rock and largemouth bass appear doomed; reproductive activity has ceased and all juvenile bass have been eliminated. In addition, many species of zooplankton near the base of the aquatic food web have disappeared. Furthermore, increased acidity facilitates a change in mercury to a form (methyl mercury) that is readily taken up by fish. Most of the mercury in the lake is scavenged from the atmosphere by rain and snow. Consequently, the fish have elevated levels of mercury (chapter 12). As of this writing, the acidified lake basin is very slowly recovering without human intervention.

In addition to threatening aquatic ecosystems, acid deposition may contribute to the dieback of coniferous forests in the Appalachian Mountains from North Carolina to New England (chapter 4). However, acid deposition probably plays a secondary role to ozone in forest decline.

Another costly impact of acid precipitation is accelerated weathering of building materials, especially limestone, marble, and concrete (fig. 16.13). Metals, too, corrode faster than normal when exposed to acidic moisture.

Winds aloft can transport oxides of sulfur and nitrogen many thousands of kilometers from tall stack sources, so acid rain is becoming a global problem. Acid rains have been reported from such isolated localities as the Hawaiian Islands and the central Indian Ocean. Long-range transport of acid-rain precursors has even strained the traditionally amiable relationship between the United States and Canada.

Figure 16.11 Areas in North America where lake waters are particularly susceptible to acidification. These areas lack natural buffers (neutralizers) in the soil or bedrock.

Figure 16.12 Little Rock Lake in northern Wisconsin is the subject of an ongoing experiment to determine the effects of acidified waters on aquatic organisms. The basin to the left of the barrier has been artificially acidified, while the basin to the right is unaltered and serves as a standard (control) for comparison with the acidified basin.

Figure 16.13 Acid rain damage to a statue on the exterior of the Field Museum in Chicago, Illinois.
© Gary Milburn/Tom Stack & Associates

Acid rains threaten Canada's primary industries: lumber and fisheries. An estimated 300,000 lakes in eastern Canada are susceptible to acidic deposition; the number in the eastern United States is 11,000. In 1985, the United States exported more than three times as much sulfur dioxide (SO_2), the chief acid rain precursor, to Canada than Canada exported to the United States. Perhaps half of the acidic rainfall in Canada can be traced to industrial emissions from Ohio Valley coal-fired power plants and industries. Canadian government officials have pressed their Washington counterparts to enact stricter controls on polluting industries and power plants in order to ease the problem, which some Canadians feel will become severe within the next decade. In partial response to this international issue, the 1990 amendments to the Clean Air Act call for a sharp reduction in emissions of sulfur dioxide.

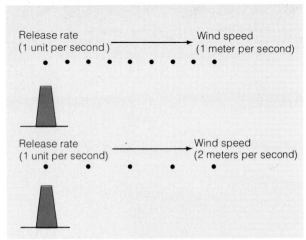

Figure 16.14 The effect of wind speed on air-pollution concentrations. A doubling of wind speed from 1 meter per second to 2 meters per second increases the spacing between puffs of smoke by a factor of 2, thereby reducing concentrations by one-half.

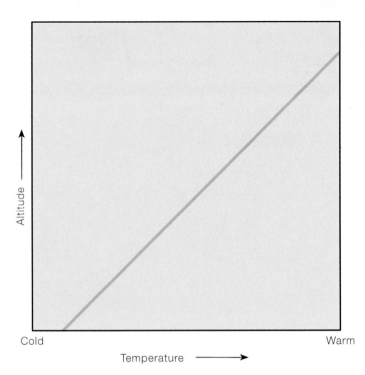

Figure 16.15 An air layer characterized by a temperature inversion features an increase in temperature with altitude.

The rate of decline depends partially on the extent to which polluted air mixes with cleaner air. The more thorough the mixing, the more rapid is the dilution. When conditions in the atmosphere favor rapid dilution, the impact of air pollutants is often minor. On other occasions, called **air-pollution episodes,** conditions in the atmosphere minimize dilution, and the impact can be serious, especially on human health.

Weather Conditions

Intuitively, we know that air is likely to mix more vigorously on a windy day than on a calm one. As a general rule, a doubling of wind speed cuts the concentration of air pollutants in half (fig. 16.14). Ironically, dilution of air pollutants by wind is particularly impeded in urban areas—the very places where most of the contaminants are generated. In a city, the canyonlike topography of tall buildings and narrow streets creates a rough surface that slows the wind; average near-surface wind speeds may be 25% lower in a city than in the surrounding countryside. Furthermore, regions of great topographical relief often have air-pollution problems partly because hills and mountain ranges block horizontal winds that could disperse pollutants.

In addition to weak or calm winds, another weather condition that contributes to the development of an air-pollution episode is a persistent temperature inversion. In an air layer characterized by a **temperature inversion,** air temperature increases with altitude; that is, warmer air overlies cooler air (fig. 16.15). Note that a temperature inversion is the *inverse* of the usual temperature profile within the troposphere (refer to fig. 15.1). Because warm air is less dense than cool air, warm air over cool air is an extremely stable stratification that strongly inhibits mixing and dilution of pollutants. A temperature inversion can form by subsidence of air or extreme radiational cooling. The inversion may form aloft or near the earth's surface.

A **subsidence temperature inversion** forms a lid over a broad area, often encompassing several states at once. It develops during a period of fair weather when the atmospheric circulation pattern causes a high-pressure system to stall. A high-pressure system brings fair weather (chapter 15) and is also characterized by subsiding air that warms by compression (chapter 13). Subsiding, warm air is prevented from reaching the earth's surface by the **mixing layer,** in which air is mixed by convection (fig. 16.16). The air temperature within the mixing layer declines with altitude, but air just above the mixing layer, having been warmed by compression, is warmer than air at the top of the mixing layer. A temperature inversion thus separates the mixing layer from the compressionally-warmed air above. Under these conditions, air pollutants are distributed throughout the mixing layer, but no higher than the temperature inversion. Pollutants are thus confined to a relatively small volume of air, and continual emissions will elevate concentrations.

A **radiational temperature inversion** is perhaps more common and often more localized than a subsidence temperature inversion. Under clear night skies and light winds, radiational cooling (chapter 15) chills the ground, which in turn chills the air in contact with it. Consequently, a low-level temperature inversion develops. Such temperature inversions usually disappear within a few hours after sunrise because the sun heats the ground, which heats the overlying air and eventually reestablishes the usual temperature profile (fig. 16.17). In the winter, when the sun's rays are weak and the ground may be covered by a highly reflective layer of snow, a radiational temperature inversion may persist for days.

We can sometimes estimate how well mixing (and hence, dilution) is taking place by observing the behavior of a plume of smoke belching from a stack. If the plume undulates down-

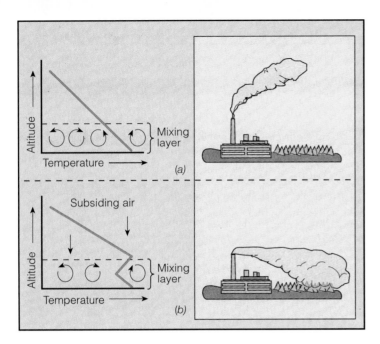

Figure 16.16 A temperature inversion can develop aloft through subsidence of air. A temperature profile prior to subsidence (*a*) is compared with the profile during subsidence (*b*). The temperature inversion acts as a lid over the lower atmosphere, trapping air pollutants.

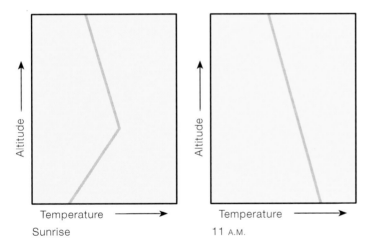

Figure 16.17 During the morning, the sun heats the earth's surface, which in turn heats the overlying air. Hence, several hours after sunrise, a radiational temperature inversion is replaced by a more usual temperature profile, in which temperature drops with altitude.

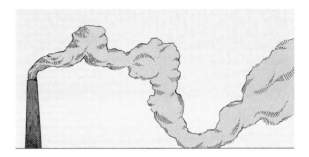

Figure 16.18 If the air temperature drops rapidly with altitude, a smoke plume undulates downwind as it readily mixes with the ambient air.

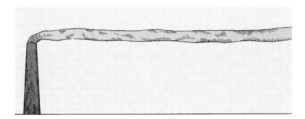

Figure 16.19 In an air layer characterized by a temperature inversion, a smoke plume forms a thin ribbon that slowly drifts downwind; mixing and dilution are minimal.

wind, as shown in figure 16.18, then polluted air is mixing readily with the surrounding cleaner air, thereby facilitating dilution. The net effect is improved air quality—except where the plume may loop to the ground. On the other hand, a plume of smoke that flattens and spreads slowly downwind, as shown in figure 16.19, indicates a temperature inversion and minimal dilution.

Not only do weather conditions affect air quality, but air pollution also influences weather and climate. In chapter 15, we considered the potential impact of rising levels of carbon dioxide and other greenhouse gases on global temperature.

Air pollution also affects the amount of cloudiness, and the quantity and quality of precipitation, especially downwind from large urban-industrial areas. For more on this problem, refer to Special Topic 16.2.

Air-Pollution Potential

Weather conditions that reduce dilution and hence favor air-pollution episodes occur with varying frequency in different places and at different seasons of the year. Areas with particularly high potential for air pollution include southern and coastal California, parts of the Rocky Mountain states, and the mountainous portions of the mid-Atlantic states. Temperature inversions are quite common in deep mountain valleys and other lowlands. At night, cold dense air drains down hillslopes and settles in low areas, so that by morning a temperature inversion is firmly established.

In general, in much of the West, air quality is lowest during winter, while in the East, air quality is lowest during autumn. In southern California, the potential for air pollution is highest during summer. Those seasonal changes in air quality are caused primarily by the usual seasonal shifts in atmospheric circulation patterns.

Los Angeles is particularly susceptible to air-pollution episodes because of frequent temperature inversions, topographic setting, and high concentration of pollutant sources (9 millon motor vehicles). Figure 16.20 shows the air circulation and topographic features that influence air quality in that city. Weather in the Los Angeles area, like the weather throughout much of California, is strongly influenced by a huge high-pressure system that persists over the subtropical Pacific Ocean. This high is responsible not only for California's

Air Pollution and Urban Climate

The climate of large metropolitan areas is somewhat different from that of adjacent rural areas, partly due to human modification of the environment. Cities are warmer than their surroundings, and cloudiness and precipitation are more frequent downwind of a city.

The average annual air temperature of a city is usually only slightly higher than that of the surrounding countryside, but on some days the temperature contrast may be 10 C° (18 F°) or more. Consequently, snow melts faster, and flowers bloom earlier in a city. This climatic effect is known as the *urban heat island.*

Several factors contribute to the development of an urban heat island. One is the relatively high concentration of heat sources in cities (e.g., people, cars, industry, air conditioners, and furnaces). Because all heat from every source eventually escapes to the atmosphere, the air of a large city receives a considerable input of waste heat. A second contributing factor is the material composition of cities. Concrete, asphalt, and brick conduct heat more readily than do the soil and vegetative cover of rural areas. The heat loss at night by infrared radiation to the atmosphere and space is thus partially compensated for in cities by a release of heat from buildings and streets.

Moisture availability is a third factor involved in the temperature contrast between city and country. Urban drainage systems (chapter 14) efficiently remove most runoff from rain and snowmelt, so that there is relatively little standing water in a city. Hence, during the day, less incoming solar radiation is used for evaporation, and more is used to heat the ground and the overlying air. On the other hand, the relatively moist surfaces of the countryside (lakes, streams, soil, vegetation) increase the fraction of solar radiation that is used to vaporize water. Hence, less heat is available to warm the ground and the overlying air.

Certain air pollutants usually found in urban air, including a variety of dust particles and acid droplets, can influence the development of clouds and precipitation within and downwind from a city. These pollutants, many of which are hygroscopic, serve as nuclei for cloud droplets and thus accelerate condensation (chapter 13). Also, the relative warmth of large urban areas spurs rising air currents and cloud formation.

The influence of urban air pollution on condensation and precipitation is illustrated by typical climatic contrasts between urban and rural areas. Winter fogs occur about twice as frequently in cities as in the surrounding countryside. Downwind from cities, rainfall may be enhanced by 5-10%. The greater contrasts tend to occur on weekdays, when urban-industrial activity is at its peak, suggesting that increased precipitation is at least partially due to urban-industrial air pollution.

Data from the *Metropolitan Meteorological Experiment (METROMEX)* indicate significantly more precipitation enhancement downwind of St. Louis. METROMEX scientists analyzed weather observations during an intensive field study over five years (1971-1975) and concluded that summer rainfall was up to 30% greater downwind of St. Louis than upwind of the city. This rainfall anomaly was attributed to the combined effect of heat and cloud condensation nuclei from urban sources.

Because precipitation, fog, and cloudiness in urban areas often adversely affect both surface and air transportation, any artificial increase in these conditions is potentially troublesome. Reduced visibility, for example, slows surface traffic, curtails air travel, and contributes to auto accidents. In the last two decades, reports show significant improvement in visibility in some urban areas, apparently due to the enforcement of stricter air-quality standards.

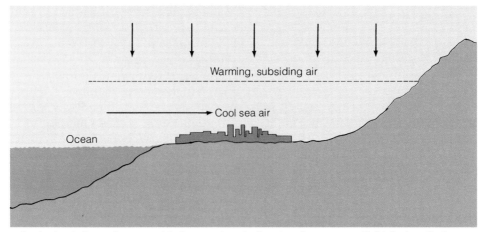

Figure 16.20   The atmospheric circulation and topographic features that give Los Angeles an unusually high air-pollution potential.

famous fair weather, but also for air that gently subsides over the city. Subsiding air is prevented from reaching the ground, however, by a shallow onshore flow of cool marine air from over the adjacent ocean waters. Compression causes the subsiding air aloft to be warmer than the underlying marine air layer. Consequently, a subsidence temperature inversion typically develops at about 700 meters (2,300 feet) over Los Angeles, and convection (mixing and dilution of pollutants) is confined to the shallow marine layer. This weather pattern contributes to a relatively high air pollution potential and occurs on perhaps two out of every three days of the year.

The exceptionally high incidence of temperature inversions over Los Angeles is aggravated by topographic barriers. The city is situated in a basin that is rimmed on three sides by mountains. Cool breezes that sweep inland from the ocean are unable to flush pollutants out of the city. Mountains and a temperature inversion aloft thus encase the city in its own fumes, within which a complex photochemistry takes place that produces photochemical smog.

Natural Cleansing Processes

Conditions that favor accumulation and concentration of pollutants in the air are countered to some extent by natural removal (cleansing) mechanisms. Some particulates are removed from the air when they strike and adhere to the surfaces of buildings and other structures, a process called **impaction.** Particulates with radii greater than 0.1 micrometer are also subject to **gravitational settling.** Heavier (and larger) particulates settle more rapidly than smaller ones, so that larger particulates tend to settle out near their sources, whereas smaller ones may be carried for many kilometers and to great altitudes before they finally settle to the earth's surface.

The most effective natural pollution-removal mechanism is **scavenging** by rain and snow. From personal experience, we know that the air is fresher just after a rain shower. In fact, in regions that experience moderate precipitation, scavenging accomplishes as much as 90% of particulate removal. Although gaseous pollutants are somewhat less susceptible to scavenging than particulates, they dissolve to some extent in raindrops and cloud droplets. While scavenging improves air quality, it degrades the quality of rain and snow—sometimes to the point of polluting surface water and harming aquatic life, as discussed earlier.

Air-Quality Management

The adverse effects of air pollution have spurred a national effort to improve air quality. Today, in most areas of the U.S., people breathe air that is significantly healthier for them than it was only a decade ago. Much of this improvement stems from strong pollution-control legislation and reduction of motor vehicle and industrial emissions. Serious problems remain, however, and further advances in air quality will be very costly. We begin our discussion of the challenges of air-quality management with a description of air-quality legislation and air-quality standards.

Air-Quality Legislation

After the Industrial Revolution, state and local governments took some steps to control air quality, but those steps were generally limited to ordinances that regulated smokey nuisances caused by coal or refuse burning. By the late 1940s, however, smog-plagued Los Angeles began to make progress in controlling air pollutants generated by motor vehicles. Since then, California has led the way in enacting strict legislation to control automobile emissions.

Federal law, more or less, is modeled on the pioneering example of California. Although federal air-quality laws enacted during the 1950s and 1960s had good intentions, they actually were weak and accomplished very little. In the late 1960s, however, bolstered by a wave of public concern about environmental quality, Congress developed a new set of strict and comprehensive amendments to the 1963 Clean Air Act. The goal of those amendments, signed into law by President Nixon on 31 December 1970, was to achieve the national air quality necessary "to protect the public health . . . with an adequate margin of safety." The distinctive feature of this landmark legislation is strictness; it set a schedule for compliance and was the first to mandate stiff fines for violators.

The 1970 law required the EPA to develop uniform air-quality standards for the ambient (outdoor) air. The individual states drew up **state implementation plans (SIPs)** and enforced air-quality standards for stationary sources (e.g., incinerators and power plants). The EPA must approve all SIPs. Congress specified standards for motor vehicle emissions and assigned responsibility for enforcement to the federal government.

The 1970 law failed to achieve all its goals, however. By the early 1980s, it was apparent that significant revisions of the law were needed. Especially troublesome were (1) slow progress in dealing with hazardous (toxic) air pollutants and (2) the impact of a rapid surge in vehicular traffic in already congested urban areas. Although car and truck emissions were down per vehicle, more and more were on the road. In fact, the number of vehicle-miles traveled is growing much faster than the U.S. population, doubling from 1 trillion in 1970 to 2 trillion in 1990.

On 15 November 1990, President Bush signed into law sweeping revisions of the Clean Air Act. When fully implemented by 2005, this law will force deep cuts in motor vehicle emissions, acid precipitation precursors, and hazardous air pollutants. The 1990 law includes the following provisions:

1. Cut auto emissions of hydrocarbons by 30% and nitrogen oxides by 60% by the 1994 model year.
2. Set tough standards for cleaner-burning fuels in the nation's nine cities with the worst ozone problems.

Table 16.3 National ambient air-quality standards for criteria pollutants

Pollutant	Averaging Time*	Primary Standard	Secondary Standard
total suspended particulates	1 Year	75	60
(micrograms per cubic meter)†	24 Hours	260‡	150
sulfur oxides (ppm)	1 Year	0.03	
	24 Hours	0.14‡	—
	3 Hours	—	0.5
carbon monoxide (ppm)	8 Hours	9‡	9
	1 Hour	35‡	35
nitrogen dioxide (ppm)	1 Year	0.053	0.053
ozone (ppm)‡	1 Hour	0.12	0.12
lead (micrograms per cubic meter)†	3 Months	1.5	—

Source: U.S. Environmental Protection Agency.
*Averaging time is the time period over which concentrations are measured and averaged.
†A microgram is one-millionth of a gram.
‡Concentration not to be exceeded more than once (on separate days) per year.

3. Require introduction of 150,000 *clean-fuel* autos in California beginning in 1996.
4. Reduce annual emissions of SO_2 to 40% of 1980 levels.
5. Identify and set standards for 189 hazardous (toxic) pollutants over a ten-year period.
6. Phase out CFCs and other stratospheric ozone-destroying compounds.

Air-Quality Standards

To date, national ambient air-quality standards have been established for six **criteria air pollutants** (table 16.3). Standards are of two types: primary and secondary. **Primary air-quality standards** are defined as the maximum exposure levels that can be tolerated by human beings without ill effects. **Secondary-air quality standards** are defined as the maximum levels of air pollutants that are allowable to minimize the impact on materials, crops, visibility, personal comfort, and climate. Actually, primary and secondary standards are identical in most cases, and, for total suspended particulates (TSP), the secondary standard is more stringent than the primary standard. Emission standards for stationary sources (e.g., power plants) and mobile sources (e.g., automobiles) are set to ensure, theoretically at least, that once pollutants are emitted, ambient air-quality standards are not exceeded.

Control standards for the six criteria air pollutants are derived from scientific studies indicating that ill effects are likely only after concentrations exceed a specific *threshold* value (hence, the term *criteria*). Certain contaminants, however, called **hazardous air pollutants,** pose a human health hazard even at extremely low concentrations. Thus, the 1970 amendments also required the EPA to propose emission standards for hazardous air pollutants.

Progress in regulating hazardous (toxic) air pollutants has been slow, however, because of economic pressures and scientific uncertainties. To date, the EPA has designated as hazardous only eight air pollutants: asbestos, beryllium, benzene, mercury, vinyl chloride, radionuclides, inorganic arsenic, and coke-oven emissions. Many potentially hazardous air pol-

lutants are trace by-products of important industries, including steel-making, petroleum refining, and chemical manufacturing. Thus, strict adherence to the law would require those industries to eliminate *all* emissions, at tremendous cost. Hence, the EPA has opted to list as hazardous only those substances for which the scientific evidence is "almost irrefutable." As noted earlier, however, provisions of the 1990 Clean Air Act are expected to spur progress in control of an additional 189 hazardous air pollutants.

Air-Pollution Abatement

Although efforts to control air and water quality are analogous in purpose, abatement of air pollution is generally more challenging because air is more mobile than water. Hence, although it is convenient and appropriate to manage water quality in the context of a drainage basin or watershed (chapter 14), it is almost impossible to define an *airshed*. Winds and the pollutants they transport are not restricted to fixed geographic boundaries as are the tributaries of a river. For example, the air pollution plume of Chicago may extend over Wisconsin on one day and over Indiana or Michigan on the next, depending on the wind direction. In short, we cannot define an airshed on which to focus our pollution-abatement efforts.

The only feasible approach to the problem of air pollution is to control emissions from individual point sources—that is, from exhaust pipes, chimneys, and smokestacks. Automobile and other industries have adopted a wide variety of technologies in efforts to reduce emissions of air pollutants. Those technologies are tailored to the specific type of industry and its major emissions.

Motor Vehicle Emissions Control

The internal combustion engine that powers virtually all motor vehicles is by far the chief source of air pollutants. The engine emits several types of contaminants (fig. 16.21).

Carbon monoxide, oxides of nitrogen, and hydrocarbons are important constituents of motor vehicle exhaust. Hydrocarbons also escape from the crankcase and carburetor and evaporate from the fuel tank. Lead is emitted in exhaust when leaded gasoline is burned.

A number of techniques exist to control the various pollutants emitted by motor vehicles. The **positive crankcase ventilation (PCV)** system reduces hydrocarbon emissions from crankcases by channeling crankcase blow-by gases back to the engine, where they are burned. The PCV system first appeared on autos during the 1963 model year. The escape of hydrocarbons from the carburetor and fuel system was reduced by modifications introduced in 1971 models (1970 in California). More recently, however, electronic feedback controls and fuel-injection systems have replaced carburetors. A microcomputer precisely regulates the amount of fuel needed, so that combustion is more efficient and hence less polluting.

Exhaust emissions, however, are more difficult to control, since concentrations of exhaust gases depend on many variables that relate to engine performance. The single most important emission-control strategy is the **catalytic converter** (fig. 16.22). In the so-called three-way catalytic converter, the catalyst causes nitric oxide (NO) to oxidize carbon monoxide (CO) and hydrocarbons, thereby releasing molecular nitrogen (N_2), carbon dioxide (CO_2), and water vapor (H_2O). Over the life of a typical vehicle, three-way catalytic converters cut emissions of hydrocarbons by an average 87%, carbon monoxide by 85%, and nitrogen oxides by 62%. That device works only with unleaded gasoline and, thus, today lead has been virtually eliminated from gasoline sold in the United States.

The traditional strategy for controlling automobile emissions has been to modify the internal combustion engine or its exhaust system. Today, however, other options are being explored. One alternative is to change the fuel so that it burns more cleanly.

Methanol (wood alcohol) is currently a candidate for replacing some gasoline. Mixtures of 85% methanol and 15% unleaded gasoline have been successful. Methanol burns more thoroughly than gasoline and cuts hydrocarbon and nitrogen oxide emissions, thus alleviating the photochemical smog problem. However, methanol does little to cut carbon monoxide emissions, is more corrosive than gasoline, and produces only half the energy of gasoline. The last point means that a motor vehicle goes only half as far on a tank of methanol. Although incomplete combustion of methanol produces formaldehyde, a carcinogen, so does gasoline. On balance, the EPA is convinced that methanol is superior to gasoline from an air-quality perspective. Indeed, persistent problems with smog have spurred California's plan to substitute methanol for about a third of gasoline consumption in air-quality problem areas by the end of the 1990s.

Where carbon monoxide (CO) is the main air-quality problem, different blends of gasoline may be required. In

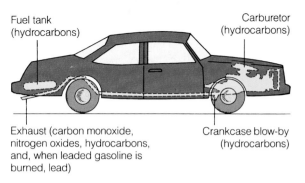

Figure 16.21 Principal sources of air-pollutant emissions in the automobile.

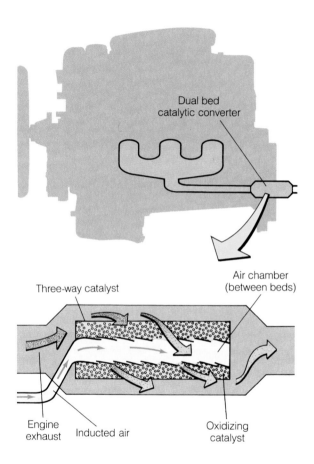

Figure 16.22 A catalytic converter chemically changes nitric oxide, hydrocarbons, and carbon monoxide in automobile exhaust into carbon dioxide, water vapor, and nitrogen.

fact, the 1990 amendments to the Clean Air Act mandate that motor vehicles in the forty-one U.S. cities with the highest CO levels burn *oxygenated fuels*, that is, gasoline blends with a minimum oxygen content of 2.7%. This requirement is for winter, when atmospheric circulation patterns favor peak CO concentrations. For several years, many western cities (e.g., Denver, Albuquerque, Las Vegas) have required motorists to burn such fuels during winter. Oxygenated fuel consists of either ethanol (grain alcohol) or MTBE (methyl-tertiary-butyl-ether) blends.

Figure 16.23 Air pollutant emissions, fuel consumption, and sometimes even travel times are reduced when commuters use mass transit. This roadway in downtown Boise, Idaho, has a special lane for buses only.
© David R. Frazier Photolibrary

Figure 16.24 Wisconsin's Department of Natural Resources operates this vehicle inspection station in Milwaukee. Automobiles are monitored for emission of major air pollutants.

According to the General Accounting Office, such blends reduce CO emissions by up to 20%.

Another option is the phasing out of the internal combustion engine entirely in favor of steam- or battery-powered engines. These engines do not eliminate air-pollution, but they significantly reduce emissions. Ideally, the electricity for charging battery-powered vehicles would be derived from sources other than coal in order to maximize the air-quality benefits. California law now mandates increasing reliance on "zero emission" motor vehicles, generally taken to mean battery-powered autos. These motor vehicles must make up 2% (about 40,000) of new cars sold in the state by 1998, 10% (or 200,000) by 2003, and 17% by 2010. This is part of California's strategy to meet federal ambient air-quality standards.

Nontraditional engines require more refinement, however, before they are likely to gain widespread acceptance. At present, they fail to match the internal combustion engine in durability, performance, or cost. The range of electric cars is limited to about 80-325 kilometers (50-200 miles) before recharging. Also, after about 32,000 kilometers (20,000 miles), the batteries must be replaced at considerable expense (about $1,500). On the positive side, besides lowering pollutant emissions, electric vehicles are less expensive to operate; fuel (electricity) is about one-third the price of gasoline per kilometer.

By law, the automobile industry is responsible for reducing emissions from motor vehicles. Each of us can also contribute to improved air quality by voluntarily restricting our use of motor vehicles and keeping them well-tuned. Municipalities with extreme air-pollution problems might decide to enforce restrictions on motor vehicle use, for example, by encouraging car pools, establishing bus lanes (fig. 16.23), restricting downtown parking, and expanding mass transit. In California, for example, the South Coast Air Quality Management District, in an effort to boost ridership to an average 1.5 passengers per auto, requires all firms with more than one hundred employees to provide car or van-pooling options. Some employers offer incentives in the form of subsidized vans or free parking for car and van poolers.

Regions that have failed to achieve ambient air-quality standards, so-called **nonattainment areas,** have enacted mandatory vehicle inspection and maintenance programs (fig. 16.24). Such programs force repair of cars needing maintenance or adjustment to reduce emissions. Currently, seventy U.S. cities have annual vehicle inspections, but this number should soon rise by another forty cities (mostly in the Northeast) in response to mandates of the 1990 amendments to the Clean Air Act.

Industrial Control Technologies

A popular industrial strategy employed to improve local air quality is to construct taller smokestacks (fig. 16.25), which reduce local concentrations of ground-level air pollutants by emitting pollutants into stronger winds aloft. (Wind speed generally increases with altitude.) Taller stacks do not improve regional air quality, however, because they do not reduce the total quantity of pollutants emitted. Also, tall stacks in the Midwest may only exacer-

Figure 16.25   Construction of tall stacks is one strategy used by industry to reduce local ground-level concentrations of air pollutants.

bate the acid deposition problem downwind to the east. Furthermore, local climatic or topographic conditions may negate any advantage of tall stacks for improving local air quality. For example, even the world's tallest chimney would not penetrate the temperature inversion over the Los Angeles basin. Thus, other control strategies are necessary that are directed at point sources (stacks) and designed to control emissions of both particulates and gases.

Three methods of removing particulates from stack emissions are currently in wide use: electrostatic precipitation, filtering, and gravitational settling. In an **electrostatic precipitator** (fig. 16.26), effluent particulates are electrically charged and then attracted to plates having the opposite charge. At maximum efficiency, electrostatic precipitators can remove up to 99% of the fly-ash emissions from the stacks of coal-fired electric power plants. Fly-ash is a solid by-product of coal burning.

A particle-laden airstream may be forced through a series of **filters,** most commonly composed of fiberglass. Filters are porous enough to permit flue gases to pass through while trapping particulates, and they may take the form of cylindrical bags, like the one shown in figure 16.27. For coal-fired electric power plants, these so-called bag houses are common alternatives to electrostatic precipitators. Bag houses match precipitators in efficiency, reliability, and overall cost, and they are more effective in collecting fine particulates, especially from burning of low-sulfur coal. Larger particulates can be separated from the effluent stream by a **cyclone collector,**

which induces gravitational settling of heavier particles. By using these techniques, some industries have lowered their particulate emissions by as much as 90%.

Like our own respiratory systems, industries have more difficulty removing gases than particulates. Much effort focuses on the removal of sulfur dioxide (SO_2) because of its role as an acid rain precursor. Some industries—especially coal-fired electric utilities—have made much progress recently in reducing SO_2 emissions. They have done so in response to the EPA's stringent New Source Performance Standards (NSPS), which limit sulfur dioxide emissions from new facilities and from old facilities that are undergoing major modifications. Many technologies for controlling sulfur dioxide emissions are either currently in use or in the testing and developing phase. Today, flue-gas desulfurization (FGD), more commonly called **scrubbing,** is the most common method of controlling sulfur dioxide emissions.

In one of the most effective FGD systems currently in use, effluents are channeled through a slurry of water and ground limestone ($CaCO_3$). In that process, the calcium in the limestone combines chemically with the sulfur to produce calcium sulfate ($CaSO_4$), which is subsequently collected and disposed of. Up to 90% of the sulfur dioxide in flue gases can be removed this way.

Coal-fired electric power utilities also control sulfur dioxide emissions by burning low-sulfur coal (chapter 17) or by physically cleaning coal. Low-sulfur coal (typically about 0.6% sulfur) mined in the West currently supplies almost 40% of

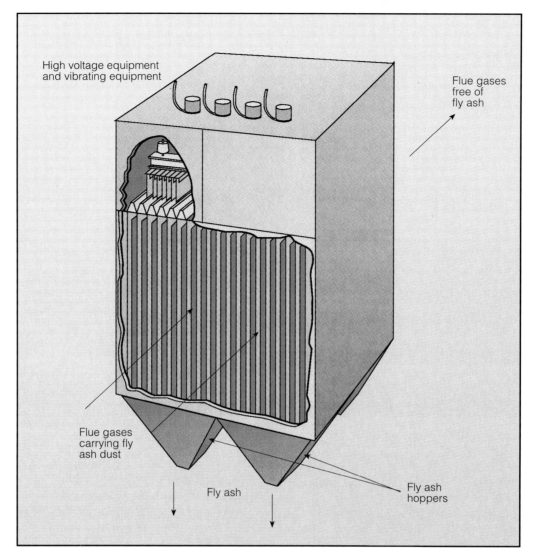

Figure 16.26 An electrostatic precipitator removes particulates from flue gases by inducing opposite electrical charges on the particulates and the collection plates.

the coal used by electric utilities. Transportation costs are so high, however, that potential users in the East may find western coal economically unattractive. Those users find it cheaper to burn high-sulfur coal (typically 2-3.5% sulfur) mined in northern Appalachia and the lower Midwest.

Between 40% and 90% of the inorganic sulfur (iron sulfide) contained in high-sulfur coal can be removed by mechanical means (washing) before it is burned. (Some residual organic sulfur is chemically bound in the coal and cannot be removed by any standard technique prior to burning.) Thus, many power plants meet the emission standards for sulfur dioxide by combining washing of coal with scrubbing of flue gases.

Some industries also have an ongoing problem in reducing emissions of oxides of nitrogen (NO_x), although several promising industrial control strategies for NO_x are currently under study. Control strategies attempt either to prevent oxidation (the combining of N_2 with O_2) in the first place, or to reconvert NO_x back to harmless molecular nitrogen (N_2).

Although today's industrial air-pollution control technologies can clean up stack effluents to some extent, they are not a panacea for the problem of industrial emissions. Many difficulties remain. For example, the particulates that cannot be readily removed from an effluent are often the very small ones (2.5 micrometers or less in diameter), which are precisely the ones that pose the greatest health hazard, since they can evade the defenses of the respiratory tract and penetrate deeply into the lungs. Furthermore, control devices operate at maximum efficiency only if they are properly maintained. Many devices quickly lose efficiency after they come into contact with the corrosive substances they control, which damage or destroy them.

Ironically, adequate programs to control air quality can create other environmental problems. After all, air quality controls must obey the law of conservation of matter (chapter 3); that is, all air contaminants removed from an effluent stream must be accounted for. Ideally they are either recycled (if they have economic value) or disposed of in an environ-

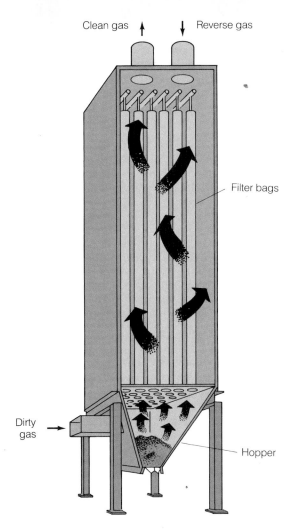

Clean gas   Reverse gas

Filter bags

Dirty gas

Hopper

Figure 16.27 A typical industrial baghouse compartment for removal of particulates. Each bag is up to 11 meters (35 feet) in height and 300 millimeters (12 inches) in diameter. Accumulated dust is dislodged by some of the filtered gas that is channeled back into the baghouse ("reverse gas"). Dislodged particulates accumulate in the hopper.

Table 16.4 Trends in total emissions of criteria air pollutants, 1980–1987 (in millions of metric tons, except for lead)

Criteria Pollutant	1980	1987	Percent Change
carbon monoxide	77.0	61.4	−20.3
sulfur oxides	23.4	20.4	−12.8
VOCs[a]	22.3	19.6	−12.1
TSP[b]	8.5	7.0	−17.6
nitrogen oxides	20.4	19.5	−4.4
lead (thousands of metric tons)	70.6	8.1	−88.5

Source: *National Air Pollutant Emission Estimates, 1940–1987*, U.S. Environmental Protection Agency, March 1989.
[a]Volatile Organic Compounds (hydrocarbons)
[b]Total Suspended Particulates

carbon monoxide was down 32%, ozone 13%, sulfur dioxide 38%, particulates 21%, nitrogen dioxide 12%, and lead 88%.

Probably the chief reason for the upswing in national ambient air quality is control of motor vehicle emissions. The 1970 amendments to the Clean Air Act required auto manufacturers to follow a strict schedule for reduction of hydrocarbons, carbon monoxide, and nitrogen oxide emissions. Although manufacturers requested and were granted numerous delays in complying with the schedule, they still achieved substantial reductions, especially for carbon monoxide and hydrocarbons. Phasing out the production and use of leaded gasoline reduced substantially the amount of lead particles in the air, so that average lead levels in the blood of U.S. residents have declined by more than a third.

On the other hand, major air-pollution problems persist in many urban areas. As of 1991, an estimated forty-one cities (combined population: 22 million) still failed to meet the primary standard for carbon monoxide (fig. 16.28). Typically, these are congested cities, where exhaust from heavy rush-hour traffic is trapped in the canyonlike topography of tall buildings and narrow streets. In 1991, ninety-six U.S. cities, home to nearly half the population, did not meet the federal standard for ozone, the principal component of photochemical smog (fig. 16.29), in spite of significant cuts in emissions of the volatile organic compounds (hydrocarbons) that are ozone precursors. (A complication in the urban ozone problem is the recent discovery that trees emit hydrocarbons that contribute significantly to smog formation.) Also in 1991, seventy-two cities exceeded the particulate standard and several cities failed to meet the primary standard for sulfur dioxide. Such persistent problems led to enactment of the 1990 amendments to the Clean Air Act and a renewed effort to substantially improve air quality.

Economics of Air-Quality Control

What does air pollution cost our society? The corrosion of buildings and the damage to crops are obvious effects whose cost can be estimated, but much of the cost of air pollution is

mentally safe manner (chapter 20). The amount of material that must be disposed is enormous. For example, 10% of the volume of coal burned at a power plant ends up as fly ash. Sometimes, however, extracted air pollutants are dumped illegally into wetlands or onto other sites where they threaten groundwater quality. Thus, an effective program to control air quality must require both adequate control strategies and reuse or proper disposition of collected air pollutants.

Trends in Air Quality

The effectiveness of air-quality legislation and the response of industry in abating the nation's air-pollution problems are becoming apparent, although serious problems remain. For the nation as a whole, recent trends show significant declines in emissions of all six criteria air pollutants (table 16.4), so that the national average ambient air concentrations of those pollutants are decreasing. In the ten-year period 1978–1987,

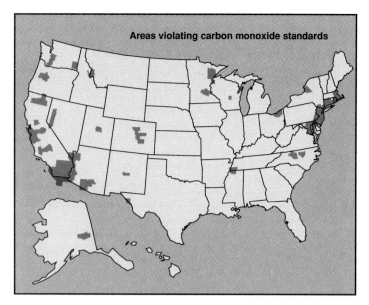

Figure 16.28   As of early 1991, a total of forty-one urban areas failed to meet the national air-quality standard for carbon monoxide.
Source: Office of Air and Radiation, Environmental Protection Agency.

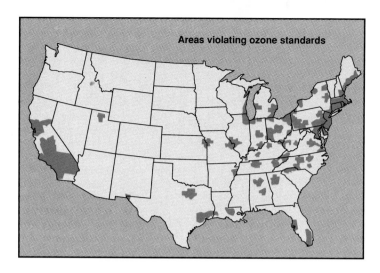

Figure 16.29   As of early 1991, a total of ninety-six urban areas failed to meet the national ambient-air standard for ozone.
Source: Office of Air and Radiation, Environmental Protection Agency.

attributed to elusive, less obvious, and less direct effects. For example, while most of us would agree that esthetic losses result from polluted air, it would be virtually impossible to estimate the dollar value of such losses.

The indirect effects of air pollution are often linked with other factors that contribute to those effects—especially in the case of human health. For example, although high levels of air pollution may contribute to emphysema and other lung disorders, factors such as smoking, age, diet, and general health also contribute to the incidence and severity of those illnesses. Thus, we cannot assign a precise dollar value to each contributing factor when we attempt to account for the costs of hospitalization and treatment, and loss of productivity due to missed work. Such problems of cost assessment arise whenever the link between pollution and environmental damage is real but not measurable, which is a common circumstance.

The costs of complying with federal air-quality standards are great and will continue to be so. A great deal of money will be required to develop new abatement techniques, and industries will continue to spend increasing amounts on the purchase and maintenance of emission-control devices. The U.S. Commerce Department reports that in 1988, the chemical industry spent $450 million on air-pollution abatement; this is about 36% of their total spending for all forms of pollution control and 2.5% of total capital spending. The EPA and the President's Council of Economic Advisors estimate that requirements of the 1990 amendments to the Clean Air Act will cost about $12 billion per year by 1995 and $25 billion per year when fully implemented in 2005. (To put these figures into perspective, however, that works out to about 24 cents per person per day.)

Damage and abatement account for most of the economic impact of air pollution, but a significant sum also goes for monitoring air quality in order to document progress. However, monitoring can also save money (as well as lives) indirectly by warning the public of potential episodes of air pollution. For example, city health departments issue air-pollution alerts when ozone levels approach hazardous concentrations, and they advise people with respiratory problems to stay indoors and to avoid strenuous activity. Thus, expensive monitoring procedures result in savings associated with human health.

At least through the early 1990s, the White House Council on Environmental Quality has maintained that public and private funds spent on air-quality control are exceeded by the value of benefits that result from reduced air-pollution damage. Some businesses and industrial leaders, however, have expressed serious doubt about whether that positive cost-benefit ratio will continue through the remainder of the decade.

Some industries claim that the costs of pollution control in general are so high that full compliance would force them out of business. Indeed, some industrial plants have had to close their doors because they were unable to afford control costs. Another objection to enforcement is that the appropriate technology is not yet available to meet certain standards. Industries that take this position have gone to court to battle with environmentalists in an attempt to weaken or avoid regulations.

Some environmentalists assign full responsibility to industry and big business for what they consider to be the less-than-rapid pace of our progress toward improved air quality. But the situation is not that simple. In industrial societies, a nation's economic well-being is tied to the well-being of its key industries. We depend on business and industry for employment and for goods and services that enhance the quality of life. A growing economy, however, means more resource utilization and more air pollution. Still, we are all entitled to breathe air that will not impair our health or shorten our life expectancy. Thus, our major problem is to improve air quality while maintaining a healthy and sustainable economy.

Typically, we must resolve questions of policy in the political arena, because the interests of the parties involved are often in fundamental conflict. Environmentalists may argue convincingly for the health benefits of cleaner air, but industries must also maintain their economic viability. For example, it is unlikely that an industry would voluntarily install pollution-control equipment, because the cost of doing so would probably put them at a competitive disadvantage, especially with foreign producers who may face less stringent pollution-control requirements. Furthermore, few consumers will purchase a more expensive, but otherwise equal, product just because it was manufactured by a nonpolluting industry. Hence, the courts and the legislature are the routes through which industries and environmentalists seek to resolve conflicts.

Conclusions

There is little doubt that polluted air has adverse effects on the health and well-being of all living organisms. Humans (and other organisms) have certain defense mechanisms that provide some protection from air pollution. These defenses are limited, however, and we are well-advised to continue efforts to reduce air-pollutant emissions. In this chapter, we have seen how a combination of weather and topographic conditions can combine to trap air pollutants and give rise to health-threatening situations.

Nationwide, strict federal regulations have spurred a significant improvement in ambient air quality as measured by recent trends in the six criteria air pollutants. However, many urban areas still suffer from photochemical smog and elevated levels of carbon monoxide. We have made only slow progress in cutting emissions of hazardous air pollutants. We also face monumental global environmental changes caused by thinning of the stratospheric ozone shield, acid deposition, and a possible enhancement of the greenhouse effect.

The cost of air-pollution abatement will continue to be a significant burden for industries and consumers alike. The scientific validity of several existing standards has been challenged. We must also overcome many major technical obstacles before we can control certain hazardous air pollutants. In view of such difficulties, we may very well decide that we must learn to live with tolerable levels of air pollution—that is, concentrations that pose some risk to public health rather than no risk at all.

Key Terms

air pollutant	air sacs
primary air pollutants	cilia
secondary air pollutants	macrophages
respiratory system	asphyxiating agents
bronchioles	chlorosis

acid deposition	primary air-quality standards
pH scale	secondary air-quality standards
acid rain	standards
air-pollution episodes	hazardous air pollutants
temperature inversion	positive crankcase ventilation (PCV)
subsidence temperature inversion	
mixing layer	catalytic converter
radiational temperature inversion	nonattainment areas
impaction	electrostatic precipitator
gravitational settling	filters
scavenging	cyclone collector
state implementation plans (SIPs)	scrubbing
criteria air pollutants	

Summary Statements

Most potential air pollutants are natural components of the atmosphere; they become pollutants when their concentrations reach levels that threaten the well-being of organisms (including humans) or the life-support systems upon which they depend.

The human respiratory system is armed with mechanisms that defend against air pollutants. Those mechanisms include nasal hairs, a cilia-mucus transport system, macrophages, and the sneeze and cough reflexes.

In sufficiently high concentrations, some air pollutants are asphyxiating agents, and others irritate the respiratory tract.

Radon is a gaseous decay product of radium-226. Within the respiratory system, radon decays to radioactive elements that emit alpha particles, which can lead to lung cancer.

Pollutant damage to vegetation occurs primarily when certain gases enter leaf pores. The tolerance of vegetation to air pollution varies considerably with the plant species and type of pollutant.

Rain and snow are normally slightly acidic because they dissolve some atmospheric carbon dioxide. In regions where air is polluted by oxides of sulfur or nitrogen, precipitation becomes strongly acidic and may threaten aquatic life and corrode structures. The greatest threat is to aquatic ecosystems in regions where soils are thin and the bedrock is noncarbonate.

An air-pollution episode is most likely to develop when winds are weak and the air is characterized by a temperature inversion. Weak winds slow the rate of dilution of air pollutants, and strong winds enhance dilution. A temperature inversion consists of an extremely stable stratification of warmer, lighter air over cooler, denser air.

Regions of great topographic relief often have a relatively high air-pollution potential because hills and mountain ranges can block horizontal winds that would disperse polluted air. Also, mountain valleys are favorable sites for the development of temperature inversions.

Atmospheric conditions that reduce dispersal of pollutants are countered to some extent by natural cleansing processes, including impaction, gravitational settling, and scavenging by rain or snow. Of these, scavenging is the most effective (within the troposphere).

The 1970 Clean Air Act Amendments set rigid standards for ambient air quality and for emissions from motor vehicles and stationary sources (smokestacks). However, that law failed to achieve all its goals. Sweeping revisions to the Clean Air Act in 1990 promise to revitalize efforts to improve air quality.

Primary standards for air quality are based on potential human health effects, while secondary standards are designed to minimize the impact of air pollution on crops, visibility, climate, materials, and personal comfort.

Control standards for criteria air pollutants are based on the assumption that ill effects are likely if concentrations exceed some threshold level. Hazardous air pollutants, on the other hand, pose a health risk even at extremely low concentrations.

Control of automobile emissions so far has primarily involved modification of the internal combustion engine and exhaust system. Other options for reducing emissions include burning methanol as a fuel, phasing out the internal combustion engine, encouraging car pools and mass transit systems, and establishing bus lanes.

Electrostatic precipitators, filters, and cyclone collectors can reduce industrial particulate emissions by up to 90%. However, the particulates that cannot be removed are often the very small ones that pose the greatest health hazard. Scrubbers effectively remove water-soluble gases, including sulfur oxides, from industrial effluents.

An overall upswing in national ambient air quality is indicated by recent trends in the average levels of all six criteria air pollutants. Probably the chief reason for this improvement is control of motor vehicle emissions. Yet air pollution remains a serious problem in many of our highly-congested U.S. cities.

The total monetary cost of air pollution is difficult to assess. Costs stem from pollution damage, control technologies, monitoring, and cleanup. Those costs must be weighed against the benefits that accrue from cleaner air. So far, the benefits outweigh the costs.

Questions for Review

1. Distinguish between primary air pollutants and secondary air pollutants.
2. Speculate on the major sources and types of air pollutants in nonindustrialized nations.
3. Describe the various mechanisms that protect the human respiratory system from the intrusion of air pollutants.
4. How does the size of particulates relate to their potential as a health threat?
5. Why is it difficult for scientists to isolate specific cause-effect relationships between air pollution and human health?
6. What can be done to lessen the radon hazard?
7. Identify the principal acid rain precursors. What are the major sources of these gases?
8. How do taller smokestacks contribute to the acid deposition problem?
9. What factors govern the vulnerability of an aquatic ecosystem to acid precipitation? What is the significance of spring snowmelt in the acidification of streams and lakes?
10. Describe how wind speed and temperature inversions influence the dispersal and dilution of air pollutants.
11. Distinguish between a subsidence temperature inversion and a radiational temperature inversion. Which one tends to persist longer and affect a broader area?
12. Explain why Los Angeles has a relatively high air-pollution potential.

13. Compare and contrast the natural cleansing processes in the troposphere with those in the stratosphere.
14. Does current federal legislation to control air quality adequately address the problem of interstate transport of air pollutants? Support your response.
15. Distinguish between primary and secondary air-quality standards. Emission standards for stationary and mobile sources are based on what fundamental assumption?
16. Identify and describe briefly the various techniques and devices that industry uses to control pollutant emissions. Comment on the effectiveness of each.
17. How might effective air-pollution abatement technology actually aggravate water pollution problems?
18. Identify and evaluate strategies for reducing our use of motor vehicles in the interest of better air quality.
19. Comment on the notion that air pollution is an inevitable consequence of our way of life.
20. Why is it impossible to design measures to control air quality in the context of an airshed?

Projects

1. The types of pollutants that foul the air of a particular area depend on the specific kinds of industrial, domestic, and agricultural activities that take place there. Prepare a list of sources and types of air pollutants in your community. Is information available on the amounts of air pollutants emitted by individual sources?
2. Maintain a daily log recording the appearance of local industrial stack plumes. Determine whether the shape and behavior of the plumes indicates good mixing or restricted mixing. Taking daily photographs of the plumes may aid your analysis.
3. Collect a sample of snow, melt it, and filter the meltwater. Examine the residue on the filter paper under a microscope. Describe what you see, and speculate on its origins. Compare the appearance of samples taken from different locations in your community.

Selected Readings

Hahn, J. "Environmental Effects of the Kuwaiti Oil Field Fires." *Environmental Science & Technology* 25, no. 9(1991): 1531-32. Concludes that smoke from burning oil wells had a significant impact on regional climate.

Hall, J. V., et al. "Valuing the Health Benefits of Clean Air." *Science* 255 (1992):812-17. Assesses the health effects of elevated levels of ozone and particulates in Southern California.

Krupnick, A. J., and P. R. Portney. "Controlling Urban Air Pollution: A Benefit-Cost Assessment." *Science* 252 (1991):522-28. Concludes that the costs of new controls to cut ozone levels for the nation as a whole and for Los Angeles exceed benefits.

Likens, G. E., R. F. Wright, J. H. Galloway, and T. J. Butler. "Acid Rain." *Scientific American* 241 (Oct. 1979):43-51. Describes trends in acid deposition over North America and Western Europe.

Meyer, W. B. "Urban Heat Island and Urban Health: Early American Perspectives." *Professional Geographer* 43, no. 1 (1991):38-48. A historical perspective on the urban heat island.

Nero, A. V., Jr. "Controlling Indoor Air Pollution." *Scientific American* 258 (May 1988):42-48. Includes an excellent summary of the radon hazard.

Newell, R. E., H. G. Reichle, Jr., and W. Seiler. "Carbon Monoxide and the Burning Earth." *Scientific American* 261 (Oct. 1989):82-88. Relates the discovery of a major source of CO in tropical latitudes.

Renner, M. "Rethinking the Role of the Automobile." *Worldwatch Paper* 84 (1988):1-70. Assesses the environmental impact of reliance on autos and includes a section on emissions reduction.

Rice, F. "Do You Work in a Sick Building?" *Fortune* 122, no. 1 (1990):86-88. Discusses indoor air pollution, especially in the office building.

Schindler, D. W. "Effects of Acid Rain on Freshwater Ecosystems." *Science* 239(1988):149-56. Summarizes what is currently understood about the impact of acid precipitation.

Shaw, R. W. "Air Pollution by Particles." *Scientific American* 257 (Aug. 1987):96-103. Discusses the role of acidic sulfate particles in the development of haze.

Smith, W. H. "Air Pollution and Forest Damage." *Chemical & Engineering News* 69, no. 45 (1991):30-43. Summarizes the impact of regional air pollutants on the health of forests.

U.S. Environmental Protection Agency. *EPA Journal* 17, no. 1 (1991). An entire issue devoted to provisions and implementation of the 1990 Clean Air Act.

Chapter *17*

Rock and Mineral Resources

Our highly technological society relies on the
properties of numerous rocks, minerals, and
fuels that are mined from the ground. A by-
product is environmental disruption.
© Doug Sherman

The late 1980s and early 1990s saw a renaissance of mining activity in Alaska, including the more populous southeast portion of the state. In 1989, a silver mine opened on Admiralty Island in the Tongass National Forest and soon became the nation's top producer of that precious metal. The Red Dog Mine near Kotzebue, in full operation since 1990, is expected soon to become the world's biggest producer of zinc. Plans are underway to reopen the A-J gold mine near downtown Juneau and to develop the Quartz Hill molybdenum deposit located near a wilderness area outside Ketchikan. The A-J mine could become the leading gold producer in the western hemisphere, and the Quartz Hill deposit contains about 10% of the world's known reserve of molybdenum, an important metal that imparts strength and corrosion-resistance to steel and iron alloys.

While national and regional economies will benefit from renewed mining activity in Alaska, all this is coming at a time of heightened concern about environmental quality. Opponents of mining argue that even the most environmentally sensitive mining operation will permanently scar the land, degrade water quality, and convert natural habitats and recreational lands into dumps for mine wastes. On the other hand, proponents of mining argue that the nation must develop its domestic resources in order to meet future demand and reduce dependency on foreign sources and thus help improve our trade deficit—currently at more than $100 billion. As with other controversies involving the environment and the economy, no ready resolution is on the horizon.

In the early days, discoveries of new resources were more than adequate to keep pace with the mineral and fuel needs of a growing population. Today, however, new discoveries are rarer, and as demand for minerals and fuels continues to climb, we are reminded of the finite nature of those resources. In this chapter, we survey the earth's rock, mineral, and fuel resources, the methods by which they are extracted from the ground, and the impact of that extraction on environmental quality. We also examine the problem of managing mineral resources in the face of growing demand.

Generation of Rock, Mineral, and Fuel Resources

Earth provides us with the energy and mineral resources essential (and sometimes not so essential) for our way of life. Resources include:

1. Fossil fuels (coal, oil, and natural gas) and uranium, which provide us with energy
2. Metals (e.g., iron, aluminum, and copper), which are used in the manufacture of durable goods such as cars and appliances
3. Materials used in industrial and agricultural processes, such as sand and gravel for construction and phosphate rock for fertilizer
4. Gemstones (e.g., diamonds and emeralds) and native metals (e.g., gold) used for investment, industrial purposes, and personal adornment

Our appetite for these resources is huge. Although we in the United States account for less than 5% of the world's population, we consume a disproportionately large percentage of the world's annual production of many critical resources. Each year we consume about 26% of total petroleum production, 27% of the silver, and more than 21% of all copper and lead. The total rock, mineral, and fuel used each year in the United States amounts to about 15 metric tons (13.5 tons) per person. No doubt, as some less-developed nations continue to industrialize, world demand for these resources will rise, and competition for the available supply will increase.

We begin our analysis by examining how rock, mineral, and fuel resources form and thus gain an appreciation of their finite nature.

Rocks and Minerals

From studies of earthquake vibrations, geologists have determined that the earth's interior is divided into three main layers: the crust, mantle, and core (fig. 17.1). The thin outermost skin of the planet, called the **crust,** is the source of virtually all our metals, nonmetallic minerals, rocks, and fuels. The thickness of the crust ranges from only about 6 kilometers (4 miles) under the oceans to 70 kilometers (45 miles) in some high mountain belts.

Two sets of geological processes act on the crust to generate rock, mineral, and fuel resources: internal processes and surface processes. **Internal geological processes** include volcanic eruptions and fracturing and bending (folding) of rock under great pressure—processes that are sustained by energy that originates in the earth's interior. **Surface geological processes** involve interactions between the surface of the earth and running water, glaciers, wind, and living organisms—all of which are ultimately sustained by solar energy.

Internal and surface geological processes lead to the formation of a multitude of rock types. Some rocks are economically important without further processing; others are important because valuable minerals can be extracted from them. A **mineral** is a naturally occurring solid that is characterized by an orderly internal arrangement of atoms (the basic structural units of all matter). The atoms in a specific mineral are arranged in a distinctive three-dimensional array, and that geometry, together with the mineral's chemical composition, determine its chemical and physical properties. A few rocks are composed of a single mineral. Rock salt, for example, consists entirely of the mineral halite (sodium chloride, NaCl); most rocks, however, are aggregates of many minerals.

Rocks are classified into three families—igneous, sedimentary, or metamorphic—according to the general way in which they formed. **Igneous rocks** form when magma cools and solidifies. **Magma** is molten rock that originates in the upper mantle and slowly invades the crust. Magma may remain within the crust and solidify slowly to form coarse-grained rock, or it may spew forth as **lava** through volcanoes or fractures at the earth's surface and solidify rapidly to form fine-grained or glassy rock. Granite is a particularly durable,

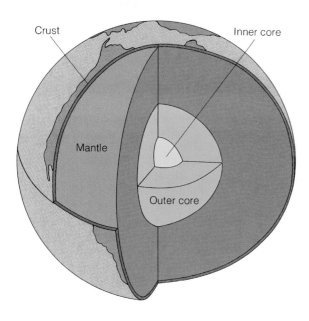

Figure 17.1   The internal structure of the earth. All rock, mineral, and fuel resources are extracted from the crust—the thin, outermost layer.

Figure 17.2   During the Ice Age, glacial erosion removed overlying sedimentary rock layers and exposed this ancient metamorphic rock mass.

coarse-grained, igneous rock that is used as monument stone and building material. Rhyolite is a fine-grained igneous rock that is crushed and used as roadbed material for both highways and railroads.

Sedimentary rocks are composed of compacted and cemented particles of rock and/or partially decomposed remains of plants or animals. A common sedimentary rock-forming process begins when rain and wind gradually break down solid rock into small fragments, called **sediments.** Fragmentation of rock by physical and chemical processes is known as **weathering.** Sediments are washed into rivers that transport them to the sea, where they settle to the sea floor. The removal and transport of sediments is known as **erosion.** Over long periods of time, seafloor sediments are gradually compacted by their own weight as they accumulate and are subsequently cemented together into solid rock by migrating fluids. The product is sandstone, a common sedimentary rock that is sometimes used as a building material. Another widespread and economically important sedimentary rock is limestone, which is composed primarily of the calcium carbonate ($CaCO_3$) shells and skeletal remains of marine organisms. Limestone is used in making portland cement, in steel furnaces, and in scrubbers for air-quality control (chapter 16).

Like many sedimentary rocks, **metamorphic rocks** are derived from other rocks. A rock is metamorphosed, or changed in form, when it is exposed to high pressures, intense heat, and chemically active fluids—conditions that develop deep within mountain belts. During metamorphism, the mineral components of rock sometimes become aligned in a banded structure that may facilitate the cleaving of the rock into plate-like slabs. Slate, which is used as a roofing or flooring material, is a good example. In other rocks, metamorphism causes

the constituent particles to recrystallize into fewer but larger crystals. For example, metamorphism of limestone produces marble, a coarse-grained rock that is valued by sculptors and builders.

Although the earth's crust is composed of rocks that belong to all three rock families, igneous rocks are by far the most abundant, making up about 95% of the crust. Sedimentary rocks, however, are the most conspicuous because they form a relatively thin veneer over nearly 75% of the earth's surface. In some areas, weathering and erosion have exposed coarse-grained igneous and metamorphic rocks that were formerly located deep in the crust (fig. 17.2). In other regions, volcanic activity has covered, or is in the process of covering, the surface with new igneous rocks (fig. 17.3). In most land areas, bedrock is hidden beneath soil and vegetation. Although bedrock is usually prominently exposed in mountainous regions, elsewhere geologists often must search for scattered outcrops or drill through surface soil and sediments to determine the composition of the local bedrock.

Most valuable resources are concentrated in specific areas within the upper portion of the crust. If those resources were disseminated uniformly throughout the crust, their concentrations would be so low that extraction would be neither technically nor economically feasible. As table 17.1 shows, only four economically important elements account for more than 1% of the total weight of the crust: aluminum (8.00%), iron (5.80%), magnesium (2.77%), and potassium (1.68%). Internal and surface geological processes, however, operating over millions of years, have selectively concentrated ingredients of the earth's crust into ores. An **ore** is a deposit that is sufficiently concentrated to be profitably mined. For example, copper constitutes only 0.0058% of the crust by weight, but in areas where that concentration has been enriched at least 80-100 times, copper can be extracted profitably.

The principal crustal enrichment processes are listed below.

1. Precipitation from hot water (hydrothermal) solutions that circulate through permeable rock. The origins of the solutions are thought to be magmas deep underground. The world's most significant deposits of gold, copper, lead, and some other metals were formed in this way.
2. Separation of crystallizing minerals within masses of magma. This process is responsible for important deposits of chromium, titanium, and nickel.
3. Precipitation from lake water or sea water. When water evaporates, dissolved substances are left behind; that is, they precipitate from solution. In this way, lake water may yield borax, and sea water yields gypsum, halite, manganese, and phosphorus minerals.
4. Weathering and erosion. These processes concentrate resources either by adding desired mineral matter or by removing unwanted host rock or sediments. For example, weathering of certain igneous rocks produces bauxite, the source of aluminum. Gold, diamonds, and platinum are weathered out of their host rock and become mixed with sand and gravel in streambeds. Such sedimentary deposits of heavy minerals are called **placer deposits.** Economically important sand and gravel deposits occur in areas where erosion by wind or water removes the finer sediments, thereby concentrating the larger sediments, for example in beach and stream channel deposits.

Geological processes that concentrate deposits of important resources have occurred in different regions at different times over hundreds of millions of years. Consequently,

Figure 17.3 New bedrock forms when lava flows through fractures in the Earth's crust, spreads over the surface, and rapidly cools and solidifies. These Hawaiian lava flows took place in historical time and are now solid rock.

significant deposits occur in some areas and not in others. Figure 17.4 maps locations in the United States where past geological events have favored the development of metallic ores (e.g., iron and copper). Note that the most concentrated deposits are in the West and Alaska.

Fossil Fuels

When we burn coal, oil, or natural gas for fuel, we are liberating solar energy that was locked in the remains of ancient organisms (chapter 3). Hence, these resources are known as **fossil fuels.** Coal occurs in the crust of the earth as distinct layers, or seams, and oil and natural gas are trapped in the pore spaces of permeable sedimentary rock that is sandwiched between less permeable layers of rock. As with all other crustal resources, a special set of circumstances was necessary to generate those important energy sources.

Most of today's coal deposits developed through the anaerobic decay of massive amounts of vegetative material 250 to 350 million years ago. During that period, giant tree

Table 17.1	The relative abundance of economically important elements in the earth's crust	
Element	**Chemical Symbol**	**Crustal Abundance (percent by weight)**
aluminum	Al	8.00
iron	Fe	5.80
magnesium	Mg	2.77
potassium	K	1.68
titanium	Ti	0.86
hydrogen	H	0.14
phosphorus	P	0.101
manganese	Mn	0.100
fluorine	F	0.0460
sulfur	S	0.030
chlorine	Cl	0.019
vanadium	V	0.017
chromium	Cr	0.0096
zinc	Zn	0.0082
nickel	Ni	0.0072
copper	Cu	0.0058
cobalt	Co	0.0028
lead	Pb	0.00010
boron	B	0.0007
beryllium	Be	0.00020
arsenic	As	0.00020
tin	Sn	0.00015
molybdenum	Mo	0.00012
uranium	U	0.00016
tungsten	W	0.00010
silver	Ag	0.000008
mercury	Hg	0.000002
platinum	Pt	0.0000005
gold	Au	0.0000002

ferns and so-called scale trees (gymnosperms) thrived in swampy terrain, similar to the Great Dismal Swamp that now straddles the North Carolina-Virginia border. Over many centuries, as tree ferns and other plants died, a thick vegetative mat accumulated in the swampy waters. Anaerobic bacteria partially decomposed the matted plant remains and concentrated carbon (just as they do in lakes and bogs today). The mass of carbon-rich, partially decomposed plant remains, called *peat*, was further compacted under the weight of more sediment and dead plants. Increasing heat and pressure eventually transformed some of the peat into *lignite* (brown coal), and then into *bituminous coal* (soft coal). In regions where heat and pressure were particularly intense, bituminous coal was metamorphosed to *anthracite coal* (hard coal).

The stages in the sequence from lignite to anthracite are called **ranks of coal.** As table 17.2 shows, the higher the rank (from lignite to bituminous to anthracite), the greater the concentration of carbon and, in general, the more heat liberated per unit mass during combustion. Furthermore, the higher the rank of coal, the smaller the quantity of ash and volatile matter released during combustion. Thus anthracite has the lowest potential for air pollution. Anthracite, however, is in short supply, and most of it is used by the metallurgical industry.

Bituminous coal (which constitutes about half the world's fossil-fuel reserves) is the rank of coal primarily used in electric power plants and in some industrial processes. Unfortunately, bituminous coal typically contains significant concentrations of sulfur and, during combustion, yields oxides of sulfur (chapter 16). Western coal, which is mostly either lignite or on the border between lignite and bituminous, is relatively low in sulfur (0.6% is typical). By contrast, the bulk of coal (mostly bituminous) from the Appalachians and central states (Illinois, Indiana, Ohio, and western Kentucky) contains 2% to 3.5% sulfur. Recall from chapter 16 that the 1990 amendments to the Clean Air Act require coal-fired electric power plants to slash sulfur oxide emissions by the turn of the century. Many electric utilities were already slowly moving in the direction of this mandate by switching to low-sulfur coal from the Western states. According to the Energy Information Administration, 30% of U.S. coal production in 1980 came from west of the Mississippi. This

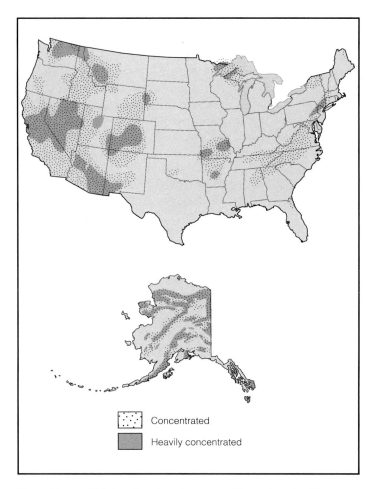

Figure 17.4 U.S. locations favorable to metallic ore deposits. No significant deposits are known from Hawaii.

Sources: Data from U.S. Geological Survey and *Surface Mining of Non-Coal Materials,* National Research Council, 1979.

Table 17.2 Composition and heat potential of coal by rank and subrank

Rank	Percent Carbon	Percent Volatiles[a]	Heating Value (BTU/lb)
lignite	37.8	18.8	7400
bituminous			
subbituminous	42.4	34.2	9720
low bituminous	47.0	41.4	12880
medium bituminous	54.2	40.8	13880
high bituminous	64.6	32.2	15160
super bituminous	75.0–83.4	11.6–22.0	15420
anthracite			
subanthracite	83.8	10.2	
anthracite	95.6	1.2	14400

From *At the Crossroads: The Mineral Problems of the United States* by E. N. Cameron. Copyright © 1986 John Wiley & Sons, Inc. Reprinted by permission of John Wiley & Sons, Inc.

[a]Volatile matter includes carbon, hydrogen, oxygen, nitrogen, and sulfur.

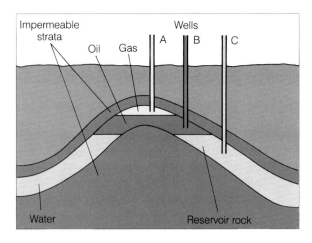

Figure 17.5 Layers of folded rock form a trap for oil and natural gas. Well A yields natural gas, well B yields oil, and well C yields groundwater. Copyright © 1976 by Arthur N. Strahler. Used by permission.

figure climbed to about 37% by 1987 and is expected to reach 39% by 2000 (assuming moderate growth in national energy demand).

Coal seams are often sandwiched between other sedimentary rock layers and occur in varying thicknesses and depths below the surface. West of the Mississippi, coal seams are very thick (3 to 60 meters, typically) and generally lie relatively close to the surface. Eastern coal seams are thinner (1 to 3 meters), and many occur at great depths. As we will see later, the depth of the coal seam dictates the method of mining employed to extract it.

Like coal, oil and natural gas (the chief forms of petroleum) are organic in origin. However, the process by which the original organic material (which included both plant and animal remains) was converted to petroleum is extremely complex and poorly understood. Both are composed of hydrocarbons, molecules that contain only hydrogen and carbon. Oil is a mixture of thousands of different hydrocarbon molecules and other organic molecules. Natural gas is a mixture of gases that consists primarily of methane (up to 99% by volume), plus small quantities of ethane (C_2H_6), propane (C_3H_8), and butane (C_4H_{10}).

Petroleum usually originates within subsurface layers of shale, but under pressure, petroleum is squeezed into adjacent layers of permeable sandstone or limestone and partially displaces the groundwater that occupies the pore spaces in those rocks. An overlying layer of less permeable rock traps the petroleum in the so-called **reservoir rock.** A reservoir rock may contain separate layers of natural gas, oil, and water (fig. 17.5). A cubic meter of reservoir rock may contain as much as 100 liters (26.4 gallons) of oil and 50 liters (13.2 gallons) of water.

The air-pollution potential of burning or processing oil is much lower than that of burning coal; that is, it yields less carbon dioxide (CO_2), oxides of sulfur (SO_x), and particulates per unit of energy generated. The sulfur content of oil varies, depending on the source locality, but it is usually less than 2%. Natural gas is an even more desirable fuel than oil in terms of air-quality, because it contains only a trace amount of sulfur that is easily removed prior to combustion.

Nuclear Fuels

Nuclear fuels, primarily uranium, are used to power nuclear reactors that generate electricity (chapter 18). In the United States, the primary source of uranium ore is certain sandstone strata located in the Colorado Plateau, Wyoming basins, and along the Texas coastal plain. In those areas, uranium ore occupies pore spaces within the sandstones. Apparently, the uranium was weathered out of granite and then precipitated from groundwater that circulated through ancient sandstones. In other parts of the world, uranium ores occur as fillings within rock fractures. The most important domestic ores occur in grades that range from 0.08% to 0.30% uranium minerals.

The Rock Cycle

Geological processes may alter rock from any of the three rock families (igneous, sedimentary, or metamorphic) and eventually transform it into rock of another family, as shown in figure 17.6. Hence, the so-called **rock cycle** means that rocks and their component minerals and fuels are continually regenerated. For example, weathering processes fragment an exposed igneous rock mass to sediments, which are subsequently eroded and redeposited in a low-lying basin. Then the accumulated sediments are gradually transformed to sedimentary rock. The mounting weight of accumulating sediments forces the sedimentary rock to greater depths within the crust. Temperature and confining pressure increase with depth, so that eventually the rock is metamorphosed. At some depth in the subsurface, temperatures may be so extreme that the metamorphic rock melts into magma. The magma may subsequently well upward, cool, and solidify to produce igneous rock, thereby completing the rock cycle.

Transformations that are involved in the rock cycle, however, are extremely slow. Typically, the regeneration of rocks and minerals takes millions of years. Hence, in the time frame of a human life or even of civilization, the rate of regeneration of crustal resources is so slow that their supply for all practical purposes is fixed and finite. Crustal materials are therefore considered to be **nonrenewable resources.**

The Mining of Rock and Mineral Resources

Now that we have outlined the origin and some of the basic characteristics of rock, mineral, and fuel resources, we turn to the methods of extracting them from the earth's crust. Our primary concern here is with the environmental impact of surface and subsurface mining. Unless precautions are taken, mining can permanently disrupt the landscape and destroy natural habitats, contaminate surface and ground water, and pollute the air.

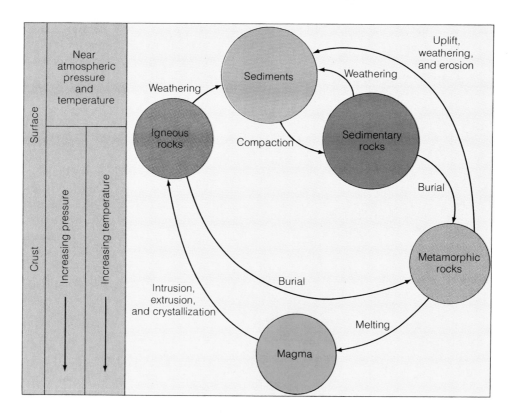

Figure 17.6 Rocks in each of the three rock families are transformed by geological processes involved in the rock cycle. The rock cycle implies a very slow generation of rocks, minerals, and fuels.

In contrast to many other land uses, mining directly affects a relatively small area. The U.S. Bureau of Mines reports that during the half-century between 1930 and 1980, resource extraction and processing disturbed only about 0.25% of U.S. land, some 2.3 million hectares (5.7 million acres). Approximately 70% of that land was devoted to actual excavation; the remainder was used for waste disposal. Almost half the land was used for the mining of bituminous coal, but even in the coal-mining states of Kentucky, Pennsylvania, and West Virginia, only about 2% of the total land area has been mined.

Although it affects a relatively small area, mining is a very conspicuous form of land use because of its potentially severe and long-lasting disruption of the environment. The environmental impact of mining depends upon the (1) specific mining method, (2) physical and chemical properties of the resource and its waste by-products, (3) habitat conditions of the mine site, (4) use of available technology to reduce adverse impacts, and (5) enforcement of environmental safeguards by local, state, and federal agencies.

Surface Mining Methods

Surface mining is used to extract about 80% of the total tonnage of minerals produced each year in the United States and more than 60% of the nation's annual coal output. **Surface mining** is particularly disruptive of the landscape because it requires the complete removal of **overburden,** that is, the vegetation, soil, rock, and anything else covering the deposit.

Even in regions where underground mining techniques are feasible, mine operators often favor surface mining because it permits more complete extraction of the deposit and poses fewer safety hazards. Surface mining can remove 80% or more of the deposit (whereas subsurface mining may remove less than 50%). The danger of collapse or explosion is relatively low in surface mining, and it is less expensive than underground mining. As a rule, underground mining may be ten or more times more expensive per metric ton extracted than surface mining.

Several techniques of surface mining are practiced depending on the type of deposit. Small **sand and gravel pits** (fig. 17.7) dot the landscape, while **quarries** (fig. 17.8) are excavated to remove limestone, granite, marble, and other rock. Most of the more than 6,000 sand and gravel pits and rock quarries nationwide are located near metropolitan areas where the resource is mainly used in construction.

In parts of the western United States and in the Mesabi Range near Lake Superior, metallic ores (mostly copper and iron) are removed from **open pit mines,** huge excavations that are dug to great depths (fig. 17.9). The open-pit copper mine in Bingham Canyon, Utah, the world's largest excavation, is 4 kilometers (2.5 miles) in diameter and 770 meters (2,500 feet) deep.

Another surface mining technique is **dredging,** which is performed with chain buckets and draglines. Dredging is used to recover streambed sand and gravel deposits or to mine placer deposits for gold, titanium, and a few other metals.

Figure 17.7   This sand and gravel pit provides the matrix material for asphalt. The sand and gravel were deposited here about 12,000 years ago by the meltwaters of massive glaciers.

Figure 17.9   Open-pit mines are used to extract metallic ores. They are huge excavations that require the use of enormous earth-moving equipment to remove the overburden.
© David R. Frazier Photolibrary

Figure 17.8   This rock quarry supplies crushed stone for use in construction of highway and railroad beds.

Figure 17.10   A huge power shovel removes coal in a strip mine near Central City, Kentucky.
© S. D. Halperin/Earth Scenes

Strip mining is the method that disrupts the landscape most extensively, primarily because it affects a much larger surface area for the resource that is recovered than other surface techniques. In strip mining, huge power shovels or stripping wheels literally chew up the land in gulps of 10–100 cubic meters (fig. 17.10). Strip mining is used to extract coal from seams that average 2 meters (6 feet) thick and occur within about 30 meters (100 feet) of the surface. It is also used to recover phosphate rock and gypsum deposits.

Two methods of strip mining are used: area and contour. **Area strip mining** occurs in flat or gently rolling terrain. It is used in the central and western coal fields, and for extracting phosphate deposits in Florida. Area strip mining involves the stripping away of overburden to form a series of parallel trenches. When coal or phosphate extraction from one trench is completed, it is filled in with the overburden removed from the adjacent trench. Unless it includes regrading, this mining method leaves behind a rugged topography that resembles a washboard (fig. 17.11).

In hilly or mountainous terrain, such as the Appalachian Mountains, near-surface coal seams are mined by **contour strip mining,** which requires cutting a shelf, or bench, into the steep flanks of a mountain. One shelf is cut for each coal seam, and the overburden is dumped downhill from each successively higher terrace onto the one below. In areas where the overburden is too thick to be removed economically, the coal may be extracted by augers, huge drills that burrow up to 70 meters (225 feet) horizontally into a coal seam.

Most strip-mined land is in the Appalachian coal fields, where more than 1.6 million hectares (4 million acres) have been disturbed since mining began. In the future, increasing demand for low-sulfur coal is likely to make the western coal fields the principal site of strip mining. These fields contain an estimated 75% of the nation's surface mineable coal.

Figure 17.11   If not reclaimed, area strip-mined land resembles a washboard and is prone to severe erosion.
© Science VU/Visuals Unlimited

Subsurface Mining Methods

When a mineral, rock, or fuel deposit occurs at so great a depth that the cost of removing the overburden is prohibitive, the resource is extracted by **subsurface mining.** Obviously, in the case of fluid resources (oil or natural gas, for example), only subsurface mining is technically feasible. To extract deep deposits of metallic ore, coal, and even marble, mine operators excavate a system of subsurface shafts, tunnels, and rooms by drilling and blasting the rock. The world's deepest mine shafts (gold mines of South Africa) extend to nearly 4,000 meters (13,000 feet) below the surface. In most cases, a portion of the deposit must be left behind to support the mine roofs. In the room-and-pillar system of subsurface coal mining, as much as half the coal is left in place to serve as supporting pillars, although the pillars are commonly *robbed* in the end.

Certain soluble minerals (such as potash and salt) are removed from the subsurface by **solution mining.** Water is pumped down an injection well to the deposit, where it dissolves the minerals. The solution is then brought back to the surface through extraction wells. One of the major problems with solution mining is its potential for groundwater contamination.

Oil and natural gas are extracted by wells that are drilled into reservoir rock to depths as great as 10,000 meters (32,500 feet). Within the reservoir rock, oil is usually under pressure, which is caused by water pushing up from below or from a cap of natural gas. If the overlying rock is fractured, the oil may rise to the surface, forming a natural oil seep. Although the pressure in the oil reservoir is normally advantageous for oil extraction, precautions must be taken during exploratory drilling to prevent oil from bursting to the surface as a gusher or to prevent a gas blowout. After the natural pressure is relieved, pumps extract the petroleum.

Even with pumping, typically less than half the oil is recovered; the rest remains behind in the pore spaces, clinging to grains of sediment. A second step, called **secondary recovery,** removes a portion of the remaining oil. In that procedure, steam, water, or natural gas is pumped down the well and into the reservoir rock to drive more oil toward the production wells. Secondary recovery in Alaska's Prudhoe Bay oil fields is expected to increase the oil yield by more than 1 billion barrels over the estimated 8 billion barrels that are recoverable with pumping and natural pressure. However, secondary recovery requires a much greater investment of energy for a lower rate of return.

Although secondary recovery has been somewhat successful in boosting production, an estimated 300 billion barrels of domestic oil lie beyond the reach of traditional recovery techniques. (In comparison, the United States so far has withdrawn about 150 billion barrels of its oil reserves.) It is not surprising, then, that considerable research is now directed at methods of enhancing oil recovery from both producing and abandoned wells.

Although oil rigs usually have a minimal impact on the landscape, the danger of well blowouts and oil spills always exists. The potential hazard is particularly great in offshore drilling operations, where oil spills are difficult to contain and threaten marine life and beaches. Blowouts and oil spills are also especially damaging in the frigid climate of the Arctic, where the breakdown of oil by microorganisms occurs very slowly (chapter 14).

Mine Waste

A major environmental problem associated with mining is the disposal of mine waste: the overburden plus **tailings** (residue of ore processing). Frequently, the desired mineral represents only a tiny fraction of a huge rock mass. Miners must therefore extract enormous quantities of rock from the ground and then separate the desired mineral from its host rock. The waste produced can be enormous. For example, the average grade of copper mined today is only about 0.6%. This means that every 1,000 metric tons of copper-bearing rock (ore) mined yields an average of only 6 metric tons of copper, leaving 994 metric tons of waste rock. In some cases, the waste produced is toxic or excessively acidic or alkaline. Furthermore, mechanical processes, such as blasting, crushing, and sorting, that are required to separate the target mineral from the host rock typically require huge inputs of energy and water.

Nonmetallic minerals (e.g., calcite, gypsum, and halite) typically are useful in their natural state, as are certain metals, such as gold and silver, which occur as elements. However, most metals are united chemically with other elements, such as sulfur. For example, lead occurs as the mineral galena (PbS), a chemical combination of lead and sulfur, and zinc occurs as the mineral sphalerite (ZnS), a chemical combination of zinc and sulfur. To liberate the desired metallic elements, the mineral must be broken down chemically through smelting or refining. Those processes usually require a major input of energy and produce solid, liquid, and gaseous waste products that must be disposed of in a manner that does not adversely impact the environment (chapter 20). Recall from chapter 16 that smelting of metallic ores is a source of air pollution, especially sulfur dioxide (SO_2), which harms the human respiratory system and is an acid rain precursor.

Surface mining accounts for nearly 99% of all mine waste, and coal mining produces more waste than any other type of mining. Waste produced by the surface mining of metallic minerals, on the other hand, amounts to only about 5% of coal-mine spoils. Overall, in the United States, an estimated 1.3 billion metric tons of mine waste, or approximately 5.2 metric tons per person, are generated each year. Most of those spoils are left at the mine sites. Since the spoils lack nutrients and are excessively stony, those waste piles inhibit the growth of most types of stabilizing vegetation. Thus rapid erosion and dangerous sliding can occur. In Wales, in 1966, a 125-meter (400-foot) pile of coal-mine wastes suddenly slid down into the town of Aberfan, smashing buildings, including a school house, and killing 144 people, most of whom were children.

As noted in chapter 14, mining can seriously degrade water quality. According to a 1990 report by the U.S. Bureau of Mines, mining activity in the United States is reducing water quality along more than 19,300 kilometers (12,000 miles) of rivers and streams and more than 730 square kilometers (280 square miles) of lakes and reservoirs. Perhaps one-third of the contamination is due to **acid mine drainage (AMD).** Seepage of rainwater and snowmelt through mines or mine waste piles containing sulfur minerals (e.g., iron sulfide [FeS_2]) produces sulfuric acid. Acid mine drainage that enters waterways lowers the pH and threatens aquatic life. In the East, coal mining is the principal source of AMD; in the West, it is metallic ore mines.

Environmental regulations designed to curb acid mine drainage have had some success. The Bureau of Mines reports that over the past twenty years, adverse impacts from AMD have declined by about one-third. This reduction is primarily the consequence of (1) chemical neutralization prior to discharge of mine water, (2) reclamation of abandoned mines, and (3) impoundment of drainage water in collection basins. Today, acid mine drainage is primarily a problem arising from mines that were closed and abandoned before current regulations were enacted.

Besides producing acidic drainage, mines and tailings may be sources of heavy metals and other sediments. Heavy metals that wash into waterways may be toxic to organisms even at relatively low concentrations. For example, mine drainage waters containing arsenic, lead, zinc, cadmium, and other heavy metals have contaminated a 200-kilometer (124-mile) stretch of Silver Bow Creek and the Clark Fort River in Montana. Sediments also turn waterways turbid and reduce photosynthesis. One solution to these problems may be to divert mine drainage to impoundment basins. Furthermore, dust lifted from mine dumps by the wind reduces air quality, and tailings foul both natural and human habitats.

Although subsurface mining produces considerably less waste than surface mining, spoils are heaped on the ground at mine sites, where they can cause the same problems that attend surface mining spoils—acid runoff, landslides, and air pollution.

Ground Subsidence

In subsurface mining, materials are extracted without being replaced, so that mine collapse and consequent **ground subsidence** are constant hazards. Those events in turn, can have serious and costly effects on the landscape, such as damage to buildings, disruption of surface and subsurface drainage, and disturbance of natural habitats. (Recall from chapter 13 that ground subsidence is also a potential consequence of groundwater mining.) Mine collapse is most likely to occur in the more than 90,000 abandoned coal mines in the United States. Those mines collapse because pillars of coal left behind to support the ceilings of subsurface caverns and tunnels are weak and prone to failure. To prevent collapse, mine operators can inject a mud slurry into an abandoned mine. The water drains off leaving behind a solid material that shores up mine ceilings.

Ground subsidence caused by withdrawal of oil and natural gas can be especially costly when it occurs beneath urban areas. One particularly dramatic example occurred in the Wilmington and Signal Hill oil fields in the Los Angeles-Long Beach harbor area. Large-scale oil withdrawal beginning in 1928 triggered severe ground subsidence that in some localities totaled as much as 9 meters (29 feet). Elaborate flood-control measures were necessary, including construction of levees and seawalls to hold back the Pacific Ocean. Subsidence was not halted until 1968, after the subsurface reservoirs had been recharged through injection of sea water. By then, however, power plants, railroad terminals, docks, and most of the water and sewer systems of the city of Wilmington had to be rebuilt.

Land Reclamation

The long-term environmental impact of surface-mined lands can be mitigated to some degree by land rehabilitation or reclamation. **Land reclamation** encompasses activities that accelerate ecological succession on land disturbed by mining (chapter 4). Generally, those activities include contouring (shaping) mine spoils and overburden to minimize erosion, applying topsoil and fertilizer, and planting and maintaining vegetation to match the species assemblage that is native to the area.

The specific reclamation strategy hinges on the physical and chemical properties of the spoils, the type of ecosystem affected by mining, and climate. Vegetation that is planted on acidic waste heaps may not be appropriate on spoils that have a high salt content. In a desert ecosystem, the use of fertilizer should be minimal because it favors less desirable early successional plant species rather than more desirable late successional species. The amount and reliability of rainfall affects the germination of seeds and the establishment of seedlings.

Reclamation activities are most effective and economical when they are an integral component of the mining operation rather than as remedial action following mining. For example, contour strip mining is not nearly so destructive of the landscape when miners employ what is known as the *haul-back method*. With that technique, trucks haul spoils back along the contour terrace to fill in and restore the original slope of the land rather than dumping the spoils downhill. That practice virtually eliminates the danger of landslides and allows considerably more land to remain undisturbed than does conventional contour strip mining. It also lowers the cost of other reclamation measures.

Reclamation of surface coal-mined lands was given a major boost nationwide on 3 August 1977 with the long-awaited enactment of the Surface Mining Control and Reclamation Act (SMCRA). This is the first federal law to regulate strip mining nationwide, and sets standards for leasing, mining, and reclamation. SMCRA requires that strip-mined land be restored to *a condition capable of supporting the uses it was capable*

Figure 17.12 Containment lagoons such as this one are designed to capture runoff from mine waste and thereby protect the quality of surface waterways.
© David R. Frazier Photolibrary

of supporting prior to any mining. Mine operators must demonstrate that they can reclaim the land before they are granted a permit to mine. To minimize erosion and contamination, they must store soil layers. After mining, the company must regrade the land to approximately its original contour, and must replace and reseed topsoil. Any toxic waste must be buried beneath the plant root zone. In areas where the mined land was originally agricultural, the restored land must be as productive as it was prior to mining. The law also requires that landowners give written consent for mining in those places where mineral rights are held by someone other than the landowner. Also, drainage waters must be impounded and treated (fig. 17.12). Finally, the law encourages miners to use the haul-back method of contour strip mining.

In a 1982 assessment of the reclamation situation, the Bureau of Mines concluded that 75% of United States land utilized for bituminous coal mining between 1930 and 1980 had been reclaimed. And overall, for 1930 through 1980, 47% of the land used for all mineral extraction and processing had been reclaimed. However, only 8% of the land mined for metals was reclaimed, and only 27% of the land mined for nonmetallic minerals was reclaimed. The slow progress in reclaiming land that is mined for nonfuel resources is attributed to several factors. (1) Over the years, public pressure has been greater for regulation of strip mining for coal. Hence, no federal laws specifically address reclamation of other mined lands, and state laws on this subject are generally weak. (2) Large-scale metal and mineral mining usually takes place in remote and sparsely populated areas, so the environmental impact is not as visible. (3) Open pit mines and large quarries disturb relatively small surface areas, considering the enormous quantity of ore that is removed.

In the coal fields of the eastern and central United States, reclamation has been underway for more than four decades. However, in the western coal fields, reclamation has been slower, primarily because the region gets less rainfall and research on revegetation has been underway only in recent decades. Rainfall is particularly marginal in rangeland, mak-

Mining versus Water: A Case Study

In 1966, the Navajo and Hopi Indians granted the Peabody Coal Company a thirty-five-year lease to develop a coal mine on land that straddles the Navajo and Hopi Reservations in northeast Arizona. The reservations are home to 200,000 Navajos and 15,000 Hopi. As of 1990, the Black Mesa-Kayenta Mine was the largest operating coal mine in the United States. Crushed coal is mixed with water to form a slurry that is transported by pipeline about 440 kilometers (275 miles) to Southern California Edison's Mojave power plant in Laughlin, Nevada.

Through the years, the Navajo and Hopi have benefited economically from their agreement with Peabody Coal. The mine employs more than 900 Navajo and about 25 Hopi. Also, Peabody pays the tribes millions of dollars in royalties each year. In a land of few other economic opportunities, the contribution of the Peabody Coal Company to the tribal economy is highly visible and significant.

Recent events, however, have strained relations between the tribes and the Peabody Coal Company. Peabody proposes to more than triple the mined area (currently encompassing about 2,600 hectares, or 6,500 acres) and greatly increase coal output, primarily to supply overseas customers. Tribal leaders have raised concerns that a larger mine would place excessive

demands on the Navajo Aquifer, the source of water for both the slurry pipeline and the Navajo and Hopi people. Currently, each hour the pipeline uses more than 450,000 liters (120,000 gallons) of water to transport 700 metric tons (640 tons) of coal. The same aquifer supplies water to wells, springs, and streams that for centuries have sustained agriculture and life in this arid land. A recent drought saw the drying up of many traditional water sources and serious water shortages heightened fears that further expansion of the Black Mesa-Kayenta mine would threaten the tribes' essential water supply.

Tribal leaders are well aware of Peabody's contribution to their economy, but they also want to protect the Navajo Aquifer. They point out that as a condition of the original mining lease, Peabody is entitled to subsurface water as long as the tribal water supply is not harmed. They propose that Peabody replace the pipeline with a railroad or drill down to another aquifer (under the Navajo Aquifer) that contains low-quality water suitable for the slurry pipeline. As of this writing, the conflict is unresolved. In July 1990, then Secretary of the Interior, Manuel Lujan, Jr., decided to allow Peabody to go on using the pipeline while a study is done to assess the impact of mining on the tribal water supply and of alternatives to the pipeline.

---

ing the reestablishment of grasses difficult (chapter 10). Also, rehabilitation of mine spoils in the mountainous West is a great challenge, because revegetation proceeds more slowly at high elevations: Temperatures are lower, growing seasons are shorter, soils are shallower, and fewer plant species are adapted to such habitat conditions.

Compliance with regulations for surface mining of coal is costly. Depending on the specific geological structure of the deposit, the cost that is added to surface-mined coal can range from $1 to $5 or more per metric ton. Thus some mine operators view the provisions of the 1977 law as being too restrictive and the regulatory costs too burdensome.

Mining often requires enormous quantities of water. In regions where water is already in short supply, conflicts can arise in allocating this vital resource between mining and other uses. For an example, refer to the issue discussed above.

Resource Management

Geological processes have provided us with finite quantities of rock, mineral, and fuel resources. When we extract those crustal resources, we also sacrifice environmental quality. If we are to ensure future generations an adequate supply of those resources and also maintain environmental quality, we

must develop sound management practices. The principal objectives of such a management effort are to

1. Conserve resources
2. Maximize exploration for new reserves
3. Minimize environmental disruption

In this section, we focus mostly on managing nonfuel resources; managing fossil and nuclear fuels is discussed in chapters 18 and 19.

Throughout this century, a gradual but significant shift took place in the federal government's regulation of crustal resources (table 17.3). In the early days, regulation was minimal. Mineral development was encouraged in the interest of settling frontier lands and spurring economic growth. Later years brought a shift toward stricter regulation at both the federal and state levels. Today's mining laws attempt to address the need for continued economic growth and national security, as well as environmental protection.

The Mineral Supply

Of the nearly three thousand different minerals that compose the rocks of the earth's crust, modern society relies on the properties of more than ninety metallic and nonmetallic minerals. Assessing supplies of those resources is clouded

Table 17.3   Important Federal laws related to mineral exploration and mining on public lands

1872	General Mining Law	Established the principles of discovery, right of possession, and patent provisions covering hard-rock minerals on public lands.
1920	Mineral Lands Leasing Act	Authorized the Secretary of the Interior to issue leases for mining of certain minerals on National Forest lands.
1947	Mineral Materials Act	Regulated mining of minerals on public lands through bidding, negotiated contracts, or free use.
1955	Multiple-Use Mining Act	Extended provisions of the 1947 Mineral Materials Act to sand and gravel deposits.
1970	Mining and Mineral Policy Act	Restated the continuing policy of the federal government to encourage private enterprise in the development of economically sound and stable domestic mining.
1976	Coal Leasing Amendments Act	Required a comprehensive land use plan prior to issuance of coal leases on National Forest lands.
1977	Surface Mining Control and Reclamation Act	Regulated surface coal mining; also restricted or prohibited surface coal mining operations on National Forest lands.
1980	National Materials and Minerals Policy, Research, and Development Act	Restated congressional policy to promote an adequate and stable supply of materials with appropriate attention to a long-term balance among resource production, healthy environment, and national resource conservation.
1987	Federal Onshore Oil and Gas Leasing Reform Act	Authorized the Forest Service to approve any surface-disturbing activity on a federal oil and gas lease. Secretary of the Interior cannot issue a lease on any National Forest land over the objection of the Secretary of Agriculture.

Source: From "An Analysis of the Minerals Situation in the United States: 1989–2040," *General Technical Report RM-179*, Rocky Mountain Forest and Range Experiment Station, USDA, 1989, pp. 3–4.

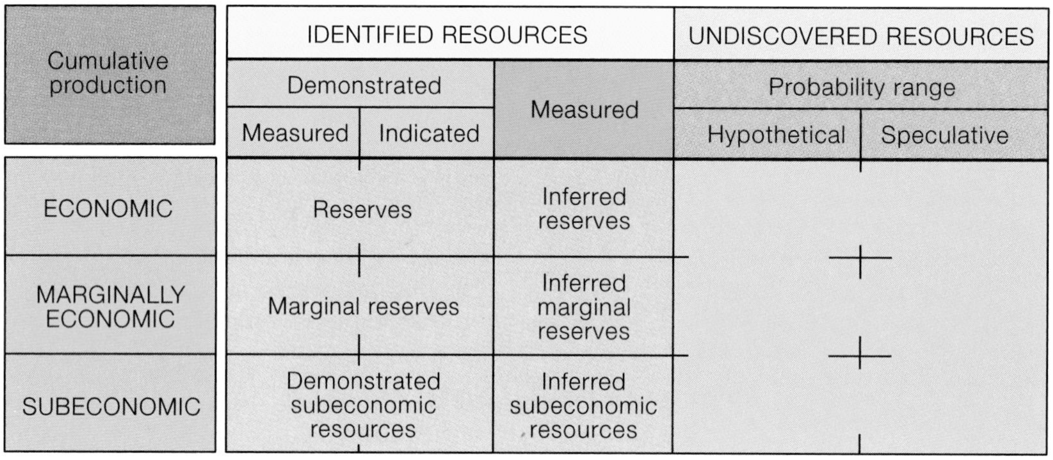

Cumulative production	IDENTIFIED RESOURCES			UNDISCOVERED RESOURCES	
	Demonstrated		Measured	Probability range	
	Measured	Indicated		Hypothetical	Speculative
ECONOMIC	Reserves		Inferred reserves		
MARGINALLY ECONOMIC	Marginal reserves		Inferred marginal reserves		
SUBECONOMIC	Demonstrated subeconomic resources		Inferred subeconomic resources		

Figure 17.13   Classification scheme for resources and reserves.

somewhat by a complex terminology. One important distinction is between minerals and ores. *Mineral* is a geological term for the components of rock, while *ore* is an economic term for a material that contains a mineral (or metal) that can be profitably mined. Furthermore, the United States Geological Survey and the Bureau of Mines use a resource classification system (fig. 17.13) that is based on geological information and the economics of mineral extraction.

The resource classification scheme distinguishes between *undiscovered* resources (hypothetical or speculative, based on geological theory) and resources that have been *identified* as to quantity and grade, based on actual field investigations. An identified mineral or fuel resource is labeled a **reserve** when it is either economically feasible, or marginally so, to recover it using present technology. Laws or regulations, however, may restrict the extraction of a part of any reserve or resource category. Resources that are subeconomic, hypothetical, or speculative are called **nonreserve resources.**

The Bureau of Mines defines a reserve base for specific mineral resources. A **reserve base** is that part of a resource that could be mined given current mining and production technology; it includes all economic and marginal reserves plus a portion of the demonstrated subeconomic resources. New discoveries, technological developments, and fluctuations

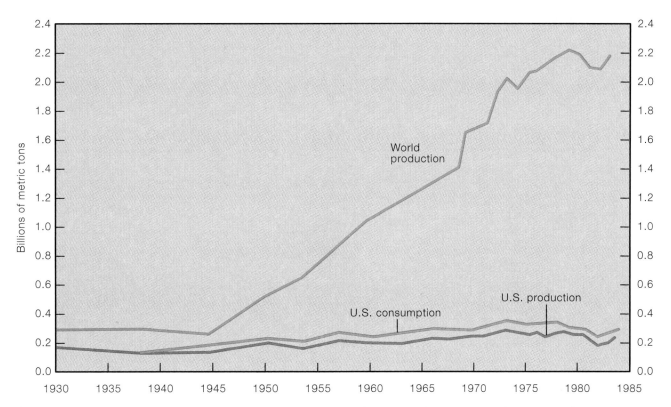

Figure 17.14 U.S. consumption versus production plus world production of eighteen nonfuel resources, 1930-1984: iron ore, bauxite, copper, lead, zinc, tungsten, chromium, nickel, molybdenum, manganese, tin, vanadium, fluorspar, phosphate, cement, gypsum, potash, and sulfur.
Source: Data from E. N. Cameron, *At the Crossroads: The Mineral Problems of the U.S.* Copyright © 1986 John Wiley & Sons, Inc.

in local, national, and global economies mean that the reserve base varies for most resources. The reserve base of a specific mineral increases if (1) changing economic conditions make previously uneconomical sources more attractive, (2) previously unknown sources are discovered, or (3) technological advances increase the accessibility of the mineral.

The domestic reserve base of some rocks and minerals is large enough to ensure an adequate supply far into the future. Examples include crushed stone, sand, and gravel, but even here there is some cause for concern. As pointed out earlier, most sand and gravel pits are located near urban areas. Urban development, zoning restrictions, and environmental regulations may force the closing of nearby pits and the opening of pits at greater distances from cities. The consequent increase in transportation costs may thus erode the reserve base. In fact, on the eastern seaboard, demand for sand and gravel is so great that in the future offshore deposits are likely to be mined.

Perhaps of greater importance is the reserve base of certain minerals that are essential for national security and the maintenance and continued growth of highly technological societies. Since about 1970, the demand for metallic minerals has actually declined. However, the Bureau of Mines expects this trend to reverse and the demand for metals (especially scarce and costly specialty metals used in the manufacture of certain alloys) to increase well into the next century. The actual strength of this demand will depend upon a number of

variables, including consumer demand for durable goods. Consumer demand, in turn, hinges on the rate of growth of (1) the population, (2) the gross national product, and (3) disposable income. In addition, world economic and political conditions and military events will influence mineral supply and consumer prices. Where will these resources come from? They may be supplied either domestically or from other nations.

Geological processes that are responsible for concentrating ore deposits occurred sporadically around the globe through geologic time. Some nations are favored geologically more than others, but none is entirely self-sufficient in metals, nonmetallic minerals, and fuels. Thus, countries must engage in international trade to acquire their needed resources. For example, 44% of oil, 14% of natural gas, and 11% of coal are traded internationally.

Since the late 1930s, worldwide production of eighteen critical nonfuel minerals has soared (fig. 17.14), but in the United States, domestic production has failed to keep pace with consumption. Consequently, the United States now imports more than half of twenty-four of the thirty-two nonfuel resources that are most essential for its economic viability. Figure 17.15 shows the twenty-five countries that supply the U.S. with strategic minerals. Although many of those nations are allies and politically stable, some are controlled by unstable or unfriendly governments, and the resource supply is vulnerable to disruption during regional conflicts.

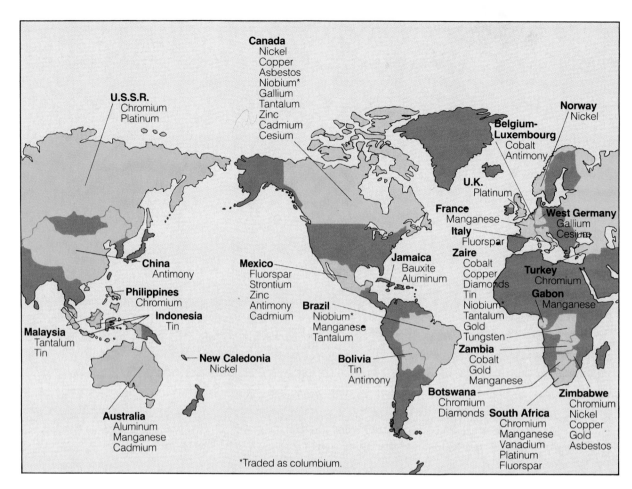

Figure 17.15 The United States depends on twenty-five other nations for more than half of its supply of twenty-four strategic materials.
Reprinted with permission from Lois R. Ember, "Many Forces Shaping the Strategic Minerals Policy," *Chemical and Engineering News*, May 11, 1981, 59(19), p. 12. Copyright © 1981 American Chemical Society.

The United States and other more-developed nations are increasingly dependent on less-developed nations for their mineral needs. In turn, the economies of some less-developed nations are becoming more dependent on mineral exports to industrialized nations. In fact, for about a dozen less-developed nations, minerals generate at least one-third of their total export revenue. This makes the economies of those nations particularly vulnerable to fluctuations in the international minerals market. Furthermore, less-developed nations tend to have weak (if any) environmental regulations. Hence, the considerable environmental cost of mineral mining and processing is borne primarily by the source nations rather than the user nations.

It is likely that the United States will continue to rely on foreign sources for many minerals of economic and strategic importance. For some metals, the nation has no choice, since mineable deposits do not occur within its borders. It also seems likely that increased demand for resources will spur greater domestic production, that is, more mining. On the plus side, this would mean an improved balance of trade with foreign nations and a healthier domestic mineral industry. On the negative side, this would also mean more threats to environmental quality. What happens in the future depends

to a large degree upon the cost of domestic exploration and mining versus import prices.

With respect to domestic reserves, some observers project that, at current rates of use and mine production, the United States has less than fifty years' supply of lead, zinc, copper, and iron. These estimates are conservative, however, since they do not account for possible new discoveries, changes in the economic climate, and technological innovations that could make feasible the extraction of lower-grade deposits (i.e., deposits that contain less of the desired material and more waste rock).

The market value of each resource ultimately determines not only the grade of the deposit that can be mined economically, but also whether a deposit will be mined at all. For example, a worldwide slump in mineral prices occurred during the recession of the early 1980s, and reduced demand for consumer goods, such as automobiles and appliances, meant less demand for raw materials. Consequently, many mines closed, which disrupted local and regional economies. In Canada's Yukon Territory, an estimated 25% of the population left because mines closed. Then, by the spring of 1982, plummeting copper prices forced the shutdown of fifteen of Arizona's twenty-eight copper mines and the loss of 3,200 jobs. Since then, the economic climate has improved, and as pointed out at the beginning of this chapter, some areas (e.g., Alaska) experienced a resurgence in mining activity by the late 1980s.

Although geological processes ultimately determine the world's supply of minerals, economic conditions control the actual supply at a given moment. Thus, as the demand for a mineral resource increases, its price rises, and it becomes more worthwhile to expend more energy and other resources to mine and process lower-grade deposits of that mineral. But the mining of low-grade minerals increases environmental degradation; it produces more air and water pollution and more waste heaps per unit of mineral extracted. Consequently, more resources are required to maintain environmental quality.

The principal obstacle to mining lower-grade deposits in the future is likely to be the cost of energy. In the past, fossil fuels were abundant and inexpensive. Thus, as technological advances made mining and processing more efficient, miners could extract leaner and deeper deposits than ever before, since they had the fuel resources to do so. For example, it was not feasible in 1880 to mine copper deposits containing less than 3% copper. The average cutoff grade declined to 2% by 1910, 1% by 1950, 0.72% in 1960, 0.59% in 1970, and leveled out in the mid-1970s. Since then, the cutoff grade has increased somewhat, largely due to higher fuel costs and worldwide recession in the 1980s.

Encouraged by past trends of ever-increasing reserves, some earth scientists dismiss the likelihood of a serious crisis in mineral supply. They point out that the earth's crust contains an enormous quantity of critical minerals in dilute concentrations, and they believe that, given the appropriate technological innovations, that supply will carry us thousands of years into the future. That optimism may be misplaced, however. Even though we have reason to feel confident that we will devise methods for mining exceptionally low-grade deposits, inexpensive fuel may run out before we can apply those innovations. Unless we develop new and extensive energy sources, the mining of very low-grade mineral deposits will be uneconomical because of high fuel costs (chapters 18 and 19).

Exploration

One strategy to help meet our future resource needs is to maximize our search for new ore deposits. Exploration, however, has always been a financially risky venture that requires a large capital investment and, traditionally, a free-wheeling spirit. Typically, of ten thousand sites where a deposit might exist according to our understanding of local geology, only one thousand merit detailed scientific study, only one hundred warrant costly drilling, trenching, or tunneling, and, typically, only one of those excavations will prove to be a productive mine.

Exploration efforts have been aided in recent years by the development of sophisticated instruments and new geological concepts. For example, remote sensing by orbiting satellites (e.g., LANDSAT) has proved valuable for selecting likely exploration sites. Also, the concept of *plate tectonics* has revolutionized geologists' ideas about how metallic ores are formed. For more on plate tectonics and mineral deposits, see Special Topic 17.1.

Efforts to expand exploration have been hindered by those who are concerned about the potential environmental degradation caused by extraction of crustal resources. As noted in chapter 10, in response to mounting pressure from these groups, the federal government has increasingly limited exploration and mining activities on federal (public) lands. Another problem is that only about half of the nation's land has been mapped geologically in sufficient detail to aid in mineral exploration. Also, at the federal level, management of rock and mineral resources is fragmented among many agencies. The Department of Energy, EPA, Bureau of Land Management (BLM), U.S. Geological Survey, Bureau of Mines, Forest Service, and others all have an interest in minerals policy. The lack of coordination among these agencies hinders orderly and timely exploration for mineral and fuel resources (chapter 2).

Seabed Resources

Although largely unexplored, the seabed is a potential source of mineral and fuel deposits. Because continental shelves are actually submerged extensions of the continents, they probably contain many of the same resources that are mined on land. Geologists already know that huge metallic deposits exist on the deep ocean floor.

An unfavorable economic climate has so far discouraged extensive mining of rocks and minerals on the continental shelf. Exceptions include Thailand, where offshore mining accounts for about one-half of that nation's annual tin production. (In fact, among nonfuel resources, only offshore tin supplies more than 1% of its world market.) In U.S. waters, some dredging of phosphate rock (the source of phosphorus for fertilizers) is taking place. Although deposits on the continental shelf are enormous, they are of lower grade than those on the land. Therefore, the industry still favors terrestrial mining of phosphate rock. In some shelf regions, wave action and ocean currents have concentrated heavy minerals in sediments, and those marine placer deposits are profitably mined on a limited scale.

Seabed **manganese nodules** have received considerable attention in recent decades because they contain metallic elements. Typically, nodules contain about 25% manganese (used in certain steel alloys) and, more important, small amounts of copper, cobalt, nickel, and molybdenum. Manganese nodules vary in size from tiny grains to masses that weigh hundreds of kilograms; most, however, are the size of potatoes. They carpet about one-quarter of the deep ocean bottom—mostly in international waters at depths of 4,000-5,000 meters (13,000-16,000 feet). The greatest potential for gathering manganese nodules exists in an east-west belt about 5,000 kilometers (3,000 miles) long on the floor of the tropical Pacific southeast of Hawaii.

The means for harvesting those billions of metric tons of nodules basically consists of a monster vacuum cleaner, which pulls nodules from the muds of the ocean bottom and delivers them to a ship. The process has been likened to sucking grains of sand through a straw to the top of a high-rise building. A second method employs a chain of buckets that sweeps the ocean floor. Although the technology for recovering nodules exists, economics do not yet favor mining.

In addition to manganese nodules, trillions of dollars' worth of cobalt-enriched crusts occur on the Pacific floor. These deposits contain several minerals and are generally more accessible than manganese nodules since they occur in much shallower water. The deposits most likely to be exploited occur in the western South Pacific near the Marshall Islands, a U.S. Trust Territory, but even there, mining is unlikely before the turn of the century.

Since 1958, the United Nations has been working to formulate an international policy concerning the exploitation of resources (including minerals and fisheries) in international waters. Progress has been slowed by conflicts among the more than 150 nations involved in negotiations; the principal philosophical divisions occur between less-developed and more-developed nations. Less-developed nations view ocean resources as the common heritage of all people, but they fear that the world's richest and most technologically sophisticated nations will reap the bulk of the harvest. While the United States and many other more-developed nations agree that ocean resources are the common heritage of all nations, they fear that too much power would be vested in the governments of less-developed nations if they were granted significant say in exploiting ocean resources.

In December 1982, representatives of 117 nations signed the United Nations' Law of the Sea Treaty. One provision of that pact establishes an International Seabed Authority (ISA) within the United Nations that would hold (or license) exclusive rights to deep-sea mining, with the revenue going to less-developed nations. The United States and forty-six other nations (including Belgium, West Germany, England, Italy, and Japan) rejected this plan, and consortia of more-developed Western nations, including the United States, have forged ahead independently with plans to begin mining manganese nodules when economic conditions are favorable.

In March of 1983, the United States (later joined by fifty other nations) defined its jurisdiction over ocean resources (including minerals, fuels, and fish) to extend 370 kilometers (200 nautical miles) offshore. This so-called Exclusive Economic Zone (EEZ) encompasses an area that is 1.7 times that of the total land of the United States and its territories. Within the EEZ, the federal government regulates all economic activity. The National Oceanic and Atmospheric Administration (NOAA) and the Interior Department share responsibility for managing seabed mineral resources.

Offshore petroleum deposits are currently being extracted in the Middle East, the North Sea, the Gulf of Mexico, and off the coasts of Nigeria, Angola, Indonesia, and Australia. Perhaps as much as one-third of U.S. oil and natural gas may lie untapped in offshore reservoirs. However, within a fifteen-month period in 1989-90, a series of three well-publicized oil tanker accidents produced major oil spills in coastal waters, causing untold damage to estuaries and marine ecosystems (chapter 14), and soured public opinion on new offshore drilling. In response, some Congressional leaders called for a total ban on drilling along the entire Atlantic and Pacific coasts as well as a portion of the Gulf Coast of Florida. President Bush initially agreed with the spirit of Congressional concern, and in mid-summer of 1990 he imposed a ten-year moratorium on offshore oil exploration in the North Atlantic, the Gulf of Mexico off southwest Florida, and the entire Pacific Coast except off Santa Barbara, California.

Predictably, the oil industry (which has already invested more than $80 billion in outer continental shelf oil leases and royalties) is greatly disturbed by these limitations on development of domestic petroleum reserves. Spokespeople for the industry point out that the plan is to pipe rather than ship the oil to shore from offshore drilling platforms, and that so far thousands of exploratory wells have been drilled in U.S. waters without a single blowout. Furthermore, they argue that restricting the development of domestic petroleum deposits will force greater dependence on foreign sources, create more tanker traffic in coastal waters, and therefore increase the likelihood of oil spills.

The debate on offshore petroleum development changed somewhat after the Persian Gulf War. In 1991, President Bush announced a national energy policy that emphasizes more exploration with the goal of fully developing the nation's energy resources. We have much more to say on this issue in the context of overall energy needs in chapters 18 and 19.

Conservation

Prospects for immediate recovery of substantial quantities of high-grade minerals from new discoveries on land and at sea are not very promising. Time and effort are needed to develop the technology and energy sources required to explore for and extract new reserves. Also, the more we mine, the greater the environmental degradation. **Resource conservation** is therefore all the more important. Resource conservation has the advantages of (1) reducing environmental damage, (2) cutting waste, (3) extending the supplies of vital minerals, and (4) saving energy.

Efforts are being made to use existing crustal resources more efficiently and to reduce the present rates of depletion. In the past, mine operators often bypassed low-grade deposits in the quest for more valuable grades. In fact, mines were frequently abandoned because it was not economically feasible to extract their low-grade deposits. Now, however, miners are encouraged to remove both high and low-grade deposits, stockpiling the low-grade materials until economics or technology makes processing feasible. In fact, some metals (e.g., copper) are now being recovered from old mine dumps at a significant energy savings (for copper, approximately 20%).

Plate Tectonics and Mineral Resources

The rigid uppermost portion of the earth's mantle plus the overlying crust constitutes the *lithosphere*. The lithosphere is broken into a dozen huge plates, each of which averages about 100 kilometers (62 miles) thick (see drawing below). Plates slowly drift (typically less than 20 centimeters per year) across the face of the globe. The continents are part of some plates so that as those plates move, the continents drift. Although the concept of *continental drift* was first formally proposed in 1912, it was not widely accepted until the 1960s as mounting geological evidence became convincing.

Continental drift explains such strange discoveries as glacial sediments in the Sahara Desert, fossil tropical plants in Greenland, and fossil coral reefs in Michigan. Such discoveries reflect climatic conditions millions of years ago when the continents were situated at different latitudes. Geological evidence suggests that about 225 million years ago, Eurasia, Africa, and the Americas were combined as a single huge continent, called Pangaea. Subsequently, Pangaea broke up, and its constituent land masses, the continents we know today, slowly drifted apart and eventually reached their present locations, as shown in the top drawing on page 384.

Many large-scale landscape features, including mountain ranges, volcanoes, and deep ocean trenches, are associated with geological processes that occur at boundaries between plates. Those same processes concentrate elements into mineable deposits (ores). In terms of relative plate movement, we can distinguish three basic types of plate boundaries: divergent, convergent, and parallel.

At *divergent plate boundaries* (see drawings on the bottom of page 384), adjacent plates drift apart, allowing magma to well up from the mantle below and fill the gap between plates. Magma cools and solidifies to form new crust. For example, the mid-Atlantic ridge, which divides the Atlantic Ocean into two roughly symmetrical basins, marks a plate boundary that has been diverging for perhaps 200 million years. Divergence at that boundary split Pangaea apart and opened the Atlantic Ocean basin.

At *convergent plate boundaries* (see drawing on page 385), one plate slides downward and under an adjacent plate, bending the crust into an oceanic trench. Within the so-called *subduction zone*, the downward-moving plate is forced into the mantle, a region of high temperatures and pressures, where the plate

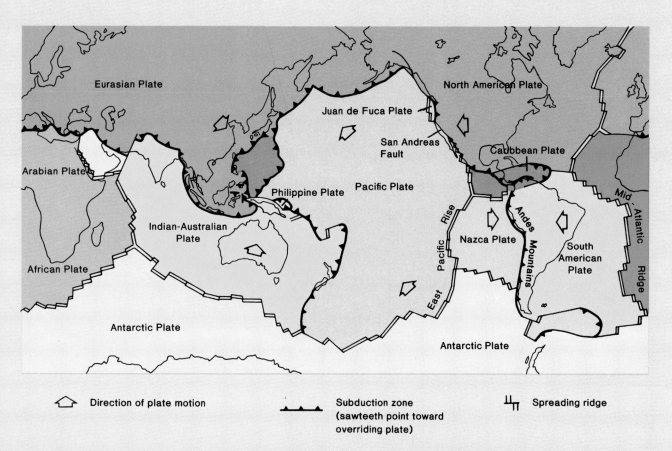

Direction of plate motion   Subduction zone (sawteeth point toward overriding plate)   Spreading ridge

The earth's lithosphere is divided into about a dozen shell-like plates that drift very slowly across the face of the globe.

melts. Hence, crust forms at divergent boundaries and is destroyed at convergent boundaries. Meanwhile the overriding plate is bent and fractured into mountain belts. The subduction zone is also a favorable site for volcanic eruptions and major earthquakes.

Some plates move parallel to one another. Although crust is neither created nor destroyed in those instances, the slipping of plates horizontally past one another can trigger earthquakes (chapter 21). The San Andreas fault of California, which was the site of the strong earthquake that destroyed much of San Francisco in 1906, marks such a boundary. There, the Pacific plate slides toward the northwest, past the North American plate.

Plate boundaries are zones of weakness that allow magma to invade the crust and, in some cases, feed volcanoes. Magma that reaches the earth's surface is called *lava*. Lava may form huge volcanoes or spread as successive flows over large areas of the earth's surface. For example, the Hawaiian Islands consist of overlapping volcanoes that tower up to 10,700 meters (35,000 feet) above the Pacific Ocean floor; in the Pacific Northwest, the Columbia Plateau is covered by many layers of solidified lava, which in some places are more than 1,400 meters (4,500 feet) thick.

Intrusion of magma into rock and extrusions of lava on the earth's surface may be associated with commercially important mineral deposits. For example, along divergent plate boundaries under the sea, lava that moves upward from the mantle interacts chemically with the sea water that enters fractures in the seafloor. Through a complex series of chemical reactions, massive deposits of metallic sulfide ore (copper and zinc deposits, for example) precipitate from *hydrothermal* (hot water) *solutions*. Many such deposits have been discovered along oceanic ridges, but deep-sea metallic deposits are unlikely to be exploited for many decades.

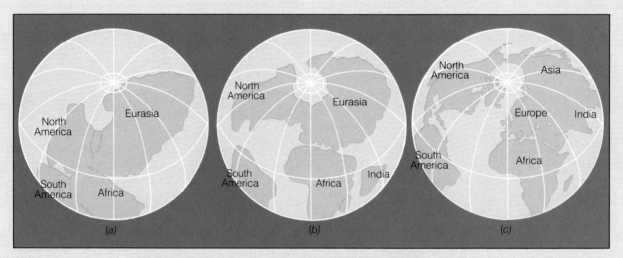

As plates move, continents drift: (*a*) 200 million years ago, (*b*) 65 million years ago, and (*c*) today.
From P. J. Wyllie, *The Way the Earth Works*. Copyright © 1976 John Wiley & Sons, Inc. Reprinted by permission of John Wiley & Sons, Inc.

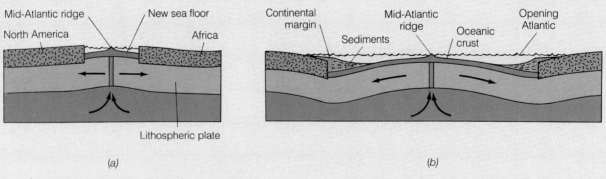

A divergent plate boundary. (*a*) As plates move apart, magma wells up from the mantle below and cools and solidifies to form new oceanic crust. (*b*) Lateral spreading of the seafloor opens the ocean basin.
From "Geosynclines, Mountains, and Continent-Building" by R. S. Dietz. Copyright 1972 by Scientific American, Inc. All rights reserved.

Metallic deposits may also occur at convergent plate boundaries. The rising temperatures and pressures in those zones cause descending plates to partially melt, which liberates metallic elements. The metal-rich fluid becomes concentrated in the magma, ascends into the crust, cools, and solidifies as metal-rich rock.

In areas where magma invades solid rock, the heat, pressure, and chemically active fluids associated with magma trigger chemical alterations when the magma and host rock come in contact. That process, known as *contact metamorphism*, is responsible for the formation of some important deposits of lead and silver. On a larger scale, the processes that occur within a subduction zone can elevate crustal temperatures and pressures over a broad area. That process, called *regional metamorphism*, has produced large ore deposits of asbestos, talc, and graphite.

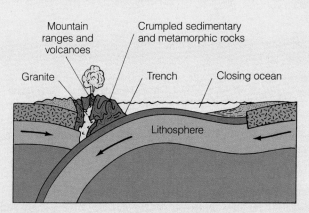

A convergent plate boundary. In areas where plates collide, one plate plunges under the other and into the mantle.

From "Geosynclines, Mountains, and Continent-Building" by R. S. Dietz. Copyright 1972 by Scientific American, Inc. All rights reserved.

Figure 17.16 Aluminum is recycled in significant quantities. This reverse vending machine at a shopping center provides cash for empty aluminum beverage cans.

Another conservation strategy is the substitution of renewable resources for essentially nonrenewable mineral resources, such as using wood instead of aluminum in construction. That practice also results in a significant energy savings. A study prepared under the auspices of the National Research Council concluded that it takes five times as much energy to make a piece of aluminum siding as it does to produce the same amount of wood siding.

One depletable resource may often be replaced by another that is more abundant, as when glass containers are substituted for tin-plated cans or plastic is used in place of copper or aluminum. Most drainage pipes are now made of plastic, whereas formerly they were copper or cast iron. In addition, research is underway to lengthen the life span of metals by developing more corrosion-resistant alloys.

Recycling metals is an essential component of a national resource conservation program. (Nonmetallic minerals are not readily recycled.) Although metals are not destroyed in manufacturing processes, they are widely dispersed as components of many different products. When those items lose their value, they are discarded, and their metallic components are lost unless we make the effort to recover them. Hence, recycling initially requires the recovery and separation of metallic components.

Some metals, such as aluminum and lead, are recycled in significant quantities. Aluminum is recovered at a rate that equals about 15% of all the aluminum consumed in the United States each year. The U.S. recovers more aluminum from recycled aluminum cans (fig. 17.16) than Africa produces from all sources. Recycled lead (used mostly in storage batteries) now accounts for about 65% of annual domestic consumption. American industry also recycles gold (42%), copper (22%), tin (21%), and silver (12%), but for other strategic materials, the rate of recycling is much lower. Typically, those materials, such as tungsten, are so widely dispersed in low concentrations that the energy required to collect and separate them virtually prohibits recycling.

Metal-recycling rates depend upon (1) the comparative price of reclamation versus new supplies, (2) how concentrated

Hard-Rock Mining on Federal Lands

*T*he General Mining Law of 1872 is a relic of a bygone era. Its purpose was to encourage development of the West by essentially giving away the energy and hard-rock (e.g., gold, silver, copper, iron) resources on federal (public) lands. Miners needed only to file a claim and sell the resources; they had no financial obligation to the government. The claim continued indefinitely as long as at least $100 worth of work was done each year. Miners could obtain legal title to the land with proof of valuable minerals and $500 in work performed. The law set a price of $12 per hectare or less. When resources were sold, the federal government received no royalties or leasing fees. Reflecting the prevailing spirit of the day, the 1872 law ignored mining's legacy of degraded landscapes and had no provision for land reclamation.

The 1872 law succeeded in luring thousands of prospectors to the West in search of fortunes. But times changed. The Mineral Lands Leasing Act of 1920 retained federal ownership of lands having deposits of energy resources (e.g., coal, oil, natural gas). Energy resource developers leased the land and had to pay royalties on profits. As the environmental movement gained strength in the early 1970s, laws were enacted that began to address the environmental impact of mining. In spite of these reforms, however, the 1872 law remains on the books.

Hard-rock miners are still paying 1872 prices for federal land; their activities are largely unregulated, no land reclamation is required, and the public coffers receive no income when valuable resources are sold. About $4 billion worth of hard-rock minerals are sold each year from federal lands. Today about 1.2 million mining claims covering 18 million hectares (44 million acres) of federal land are still outstanding under the 1872 law. A 1988 General Accounting Office report estimated that nearly 170,000 hectares (425,000 acres) of federal land in eleven states are unreclaimed. They further estimate that at least $284 million would be required to repair the environmental damage caused by mining on that land.

Major efforts are now underway to reform the 1872 Mining Law in order to address the issues of fair payment of royalties, environmental protection, and land reclamation. Although conservation groups praise such efforts, the mining industry defends the status quo. They point out that mineral exploration is financially risky and mining companies typically must spend hundreds of thousands of dollars to develop a claim prior to filing. Mining interests also argue that government regulation would interfere with the search for strategic minerals. In addition, they point to the economic benefits that mining brings to rural areas. As of this writing, the debate over mining reforms continues.

---

the metal is in the waste stream, and (3) the available recycling technology. Some scrap metal is favored over new metal produced from ores, because it is less expensive to recycle—partly because of the high cost of energy. This is especially the case for aluminum, which requires a considerable input of electricity when produced from bauxite, its source ore. An energy savings of about 90% is realized when scrap aluminum is used in place of aluminum from ore. We have more to say about recycling in chapter 20.

It is interesting to note that the hard-rock mining industry is essentially the only sector of the nation's economy that is not held accountable for much of its environmental impact. For more on this, see the issue discussed above.

at sea are not very promising, we will be forced to mine lower grades of minerals and potentially disrupt the environment ever more severely. The mining and processing of lower grades of minerals will also consume greater supplies of fossil fuels. Thus, to reduce conflicts among resource development, resource depletion, and environmental preservation, we must develop new techniques to discover, recover, and process crustal resources that will reduce environmental degradation. Of equal importance is our need to develop new, inexpensive, and abundant energy sources. Implementing any new technology takes time, making conservation of crustal resources a higher priority than ever before. Conservation has the added advantages of significantly reducing environmental degradation, and cutting energy demand.

Conclusions

Efforts to explore and develop the planet's rock, mineral, and fuel resources will probably continue to conflict with efforts to protect environmental quality. Since demand for these resources will continue to grow well into the next century, and prospects for immediate recovery of substantial quantities of high-grade minerals from new discoveries on land and

Key Terms

crust	lava
internal geological processes	sedimentary rocks
surface geological processes	sediments
mineral	weathering
igneous rocks	erosion
magma	metamorphic rocks

ore
placer deposits
fossil fuels
ranks of coal
reservoir rock
rock cycle
nonrenewable resources
surface mining
overburden
sand and gravel pits
quarries
open pit mines
dredging
strip mining
area strip mining

contour strip mining
subsurface mining
solution mining
secondary recovery
tailings
acid mine drainage (AMD)
ground subsidence
land reclamation
reserve
nonreserve resources
reserve base
manganese nodules
resource conservation
recycling

Summary Statements

More-developed nations depend heavily on a multitude of rock, mineral, and fuel resources that are extracted from the crust, the thin outermost skin of the earth.

A variety of geological processes have acted over millions of years to generate rock, mineral, and fuel resources. Those processes include folding and fracturing of rock, volcanic activity, weathering, and erosion.

Internal and surface geological processes lead to the formation of a variety of rock types. A rock is an aggregate of one or more minerals. A mineral is a solid that is characterized by an orderly internal arrangement of atoms. That internal structure plus the chemical composition of a mineral determine its physical and chemical properties.

Rocks are classified according to their mode of origin as igneous, sedimentary, or metamorphic. Igneous rocks are the most abundant, but sedimentary rocks are the most conspicuous.

Geological processes concentrate crustal materials into ores, that is, deposits that can be mined profitably. Enrichment processes include precipitation from hydrothermal solutions, separation of crystallizing minerals within magma, precipitation from lake water or sea water, and weathering and erosion.

Coal deposits are the consequence of anaerobic decomposition of plant matter in swampy terrain millions of years ago. Upon subjection to heat and pressure, layers of peat changed to lignite, then to bituminous coal, and in some regions, to anthracite coal. With increasing rank (from lignite to bituminous to anthracite), coal is cleaner burning and generally liberates more heat per unit of mass burned.

Oil and natural gas are derived from organic materials (both plants and animals) and occupy pore spaces in reservoir rock, that is, certain permeable sedimentary rocks.

Geological processes may alter rocks (and their mineral and fuel components) from one type to another as part of the rock cycle. The typically slow pace of those processes, however, essentially fixes the supply of rock, mineral, and fuel resources.

Surface mining removes vegetation, soil, sediments, rock, and anything else that overlies a mineral deposit. Surface mines include sand and gravel pits, rock quarries, open pit mines, and strip mines.

Deep ore and coal deposits are extracted from subsurface shafts and tunnels, which are blasted and drilled into bedrock. Collapse and consequent ground subsidence are hazards of subsurface mining.

Some minerals of economic value (ores) occur as minor components of large rock masses, and some target elements occur in chemical combination with other elements. Hence, energy and water are needed for physical separation, smelting, or refining of these minerals. Solid, liquid, and gaseous wastes are by-products of these processes.

Rain seeping through mine wastes triggers erosion, dangerous sliding, and pollution of drainageways.

Reclamation includes activities that accelerate ecological succession on land that has been disturbed by mining. Federal law requires reclamation of land strip-mined for coal.

The reserve base of many rocks and minerals is extensive enough to assure us of an adequate supply well into the future. For some strategically and technically important minerals, however, the situation is less favorable.

Economic conditions and environmental concerns do not currently favor extensive development of seabed resources.

The principal obstacle to mining low-grade deposits in the future may be high energy costs.

Conservation strategies include reducing depletion rates, substituting renewable resources for essentially nonrenewable ones, and recycling.

Questions for Review

1. Is a mineral always an ore? Explain your response.
2. Distinguish between internal and surface geological processes. How do they relate to the formation of rocks and minerals?
3. Identify the geological processes that have shaped and are shaping the landscape of your region.
4. Explain why rock, mineral, and fuel resources are not uniformly distributed around the globe.
5. Distinguish among the general geological conditions required for the formation of igneous, sedimentary, and metamorphic rocks.
6. Distinguish between rocks and minerals. What ultimately determines the physical and chemical properties of a specific mineral?
7. What is the rock cycle? What is the significance of the rock cycle for the global supply of rock and mineral resources?
8. Describe in general terms the sources of waste when resources are extracted from the ground.
9. Distinguish among the various types of surface mining. Why is strip mining generally considered to be the most environmentally disruptive mining method?
10. Suggest some uses for an abandoned sand and gravel pit or rock quarry.
11. How might surface and subsurface mining affect surface and groundwater quality?
12. Identify some of the environmental problems caused by mine wastes. Also, describe some measures that could reduce the adverse impact of those wastes.

13. Reclamation of mined lands involves what basic processes? Describe the provisions of the 1977 federal law that requires reclamation of strip-mined lands.
14. Distinguish between resources and reserves. What is meant by reserve base?
15. What are some of the obstacles to mining very low-grade mineral deposits?
16. Why must the United States import many of its needed minerals from other nations?
17. Describe prospects for recovering high-grade mineral deposits from the seafloor and for new discoveries on land.
18. Identify some conservation strategies that make more efficient use of resources and reduce the present rates of depletion.
19. How is the grade of a mineral deposit related to the environmental impact of mining?
20. List some of the advantages of recycling metals.

Projects

1. Arrange for a class field trip to a nearby quarry or sand and gravel pit. Find out what precautions are being taken to safeguard environmental quality.
2. Determine what minerals or fuels are mined in your state. What types of mining methods are used? How would you assess the environmental protection provisions of your state's mining laws?
3. Would land reclamation be more challenging in some areas of your state than in others? Elaborate on your response.

Selected Readings

Borgese, E. M. "The Law of the Sea." *Scientific American* 248 (March 1983):42–49. Reviews the international convention that regulates mining of resources in international waters.

Broadus, J. M. "Seabed Materials." *Science* 235 (1987):853–60. Reviews the present status of and future prospects for exploiting resources on the seafloor.

Cameron E. N. *At the Crossroads: The Mineral Problems of the United States.* New York: John Wiley & Sons, 1986. A comprehensive analysis of the U.S. mineral position and prospects for the future.

Dennen, W. H. *Mineral Resources.* New York: Taylor & Francis, 1989. A detailed study of the geology and mining of minerals; includes a section on economics, regulation, and trade.

Frosch, R. A., and N. E. Gallopoulos. "Strategies for Manufacturing." *Scientific American* 261 (Sept. 1989):144–52. Discusses issues related to metal recycling by industry.

Gillis, A. M. "Bringing Back the Land." *BioScience* 41 (1991):68–71. Reviews the challenges and progress in reclaiming surface-mined coal lands in Wyoming.

McGregor, B. A., and T. W. Offield. *The Exclusive Economic Zone: An Exciting New Frontier.* U.S. Geological Survey, Washington, D.C.: U.S. Government Printing Office, 1987. Describes the geology, mineral, and petroleum potential of the EEZ.

Plummer, C. C., and D. McGeary. *Physical Geology,* 5th ed. Dubuque, Iowa: Wm. C. Brown Publishers, Inc. 1991. An introductory textbook that describes in some detail the various forces that shape the earth's landscapes.

Powell, C. S. "Peering Inward." *Scientific American* 264 (June 1991):100–111. Describes how earth scientists are able to determine the structure of the earth's interior.

Throop, A. H. "Should the Pits Be Filled?" *Geotimes* 36 (Nov. 1991):20–22. Explains why reclamation of land strip-mined for coal is more feasible than reclamation of open pit metallic mines.

Young, J. E. "Mining the Earth," in *State of the World, 1992.* Worldwatch Institute, New York: W. W. Norton, 1992. Argues for a drastic change in the nation's mineral policy toward more efficient use of minerals and ending subsidies for mining.

Current Energy Supply and Demand

Chapter 18

An oil lake and burning oil wells in Kuwait, among the many environmental impacts of the 1991 Persian Gulf War.

© Peter Menzel/Stock Boston

The summer of 1990 brought the most dramatic event ever to affect world energy supplies: Iraq invaded and took over oil-rich Kuwait. The Iraqi invasion of Kuwait came at a time when the United States was importing about 40% of its oil and caused gasoline prices in the United States to rise by 25% to 30% within a matter of days. President Bush quickly issued orders to move 60,000 military personnel and equipment to the Middle East to protect other rich oil reserves, primarily those in Saudi Arabia, from hostile take-over by Iraq. The major industrialized nations joined together to condemn the takeover of Kuwait and demanded quick withdrawal by Iraq—a demand that was not met. The more-developed nations that rely on oil as the lifeblood of their modern economies were concerned because a single nation ruled by an unfriendly government now controlled nearly 20% of the world's oil reserves in Iraq and Kuwait.

Largely through prompting by the United States, the United Nations responded to the invasion by imposing a trade embargo. A blockade cut off Iraq's ability to sell most of its oil and to import goods needed to support some essential services. Troop buildup continued until 450,000 Americans were bivouacked in the Saudi Arabian desert. At the same time, Iraq held Americans and citizens of other countries hostage, placing them in strategic military positions throughout the country to deter possible attack. Four months later the hostages were released.

An impasse continued for five months. The United Nations demanded Iraq's withdrawal, but Iraq refused. Leaders of the major nations debated as to the best means of dealing with this act of aggression. Finally, the United Nations passed a resolution that gave Iraq's president, Saddam Hussein, until 15 January 1991 to withdraw from Kuwait. Hussein refused.

On 16 January 1991, President Bush on behalf of a coalition of nations ordered an attack on Iraq's extensive military presence in Kuwait. Only four days later, the Iraqi military was routed, and the remnants retreated to Iraq. Before leaving, however, they blew up and set fire to over five hundred oil wells (fig. 18.1). Oil gushing from these wells fueled fires that caused considerable regional air pollution. It took nine months and a considerable amount of money to put out all the fires. Release of pressure by the uncapped wells also did permanent damage to subsurface reservoirs, making oil extraction more expensive. The outcome of this conflict will influence the oil supplies and energy policies of nations throughout the world for decades to come.

The Persian Gulf War underscored the dependence of many more-developed nations on foreign sources of energy to maintain their affluent lifestyles. We have already seen how energy use is involved in many of today's pressing environmental problems. In this chapter, we examine in some detail our society's sources of energy and prospects for meeting our future energy demands.

Figure 18.1   In January 1991, prior to leaving Kuwait, the Iraqi military blew up or set fire to more than 500 oil wells.
© Wesley Bocxe/Photo Researchers, Inc.

Historical Perspective

The Iraqi invasion of Kuwait was not the first time the oil-importing nations (the United States, western Europe, and Japan) experienced trauma caused by disruption of oil supplies. Two earlier events had dramatically pointed out the tenuous nature of the world energy-supply network. The first was the Arab Oil Embargo in 1973. The second was political turmoil in Iran in 1979. When the Arab oil countries shut off all shipments of oil in 1973 to the United States to press for higher prices, the American public received its first clear signal that energy supplies were finite. Although only a small amount of oil was withheld from world markets (2.7 million barrels per day), a great deal of personal inconvenience and discomfort resulted. People had to wait in line for hours at local service stations, live in cooler homes during the winter, and adjust to a lower speed limit of 88 kilometers (55 miles) per hour. The Iranian turmoil also reduced the amount of oil on the world market, causing higher prices that contributed to worldwide recession.

Following each of these disruptions, the short-term energy-supply picture significantly improved. Nonetheless, the energy crises did spur some efforts to reduce energy consumption and improve conservation. Presidents Richard Nixon and Jimmy Carter attempted to put U.S. energy policies on firmer ground. In 1973, President Nixon set a national goal for the United States to be totally independent of energy supplies from foreign countries. During the Carter administration, federal funds were earmarked for research and development on energy conservation strategies and alternate energy sources such as solar power.

In the 1980s, however, a glut of oil and declining energy prices worldwide created a political climate, fostered by the Reagan Administration, that (1) reduced constraints on energy use, (2) cut support for research on alternative energy technologies, and (3) downgraded earlier efforts to develop a national energy policy. By 1990, energy conservation was largely ignored by the American public and policymakers. The roots of a future crisis remained alive, however, and bloomed with the invasion of Kuwait.

What were the causes of the energy crises of the past twenty years? For centuries, people have been inventing devices that save human labor, provide entertainment, make life more comfortable, and amplify human muscle power in their desire to improve the quality of life. That trend has led us to depend on a tremendous variety of energy-consuming machines without which modern life would be nearly inconceivable. Society would quickly come to a halt without automobiles, trucks, airplanes, trains, lighting, computers, and electric motors that power elevators, escalators and so on. Most of our entertainment depends on energy-using devices: televisions, stereos, motorcycles, boats, jet skis, ski lifts, and amusement parks, to name a few.

We are not always aware of all the energy involved in providing goods and services. Consider, for example, the hamburger. Long before we bite into a savory burger, energy was necessary to power the farmer's tractor, produce and apply fertilizer and pesticides, harvest the crop, dry the grain, transport and distribute the grain to cattle, transport and slaughter the cattle, chill the carcasses, grind the meat, transport the meat, make the packaging, freeze the burger, transport it, and finally fry it.

Throughout the last century, but primarily in the last four decades, human ingenuity has not only invented thousands of energy-dependent devices, but has also built and maintained an elaborate infrastructure to fuel them. Some of the major links in the support network include oil-drilling rigs, oil pipelines, supertankers, deep-water ports, oil refineries, petroleum barges, petroleum tank farms, eighteen-wheeled fuel transport trucks, underground storage tanks, service stations, natural gas pipelines, surface and underground coal mines, huge coal shovels, railroads, electric power plants, and power line grids.

As long as ample, inexpensive energy continues to flow, few people concern themselves with the vulnerability of energy sources. Today our society is totally dependent on three fossil fuels: coal, oil, and natural gas. We depend upon these nonrenewable resources for a wide array of energy services. Declines in world energy supply and the 1991 crisis in the Middle East leave little doubt that the world's energy picture is entering a period of dramatic change. In fact, many people are now viewing energy resources in a new light. For contrasting views on energy, see the issue on page 392.

Changes on the Horizon

Many factors contribute to changes in the infrastructure of our energy systems. Tighter energy supplies in the near future are likely to make energy more expensive, encouraging a switch to more energy-efficient technologies. In addition, 1990 amendments to the Clean Air Act (chapter 16) require dramatic cutbacks in the levels of air pollution, which means that we must replace some technologies for producing energy. Furthermore, many people would feel more secure if their energy supply was less subject to the policies of leaders in foreign countries. Finally, some people feel that citizens of more-developed nations have a moral obligation to help less-developed countries feed themselves and raise their standard of living—which will lead to a much greater consumption of energy resources by those countries.

As we will see, we need to move toward a different mix of energy sources. Thus, we have the opportunity to determine our future energy demand and how we meet it. Before most people will take any action to solve a problem, however, they must be convinced that a problem actually exists. As we consider what we should do about the future of energy resources, we need to consider many questions. What are our current energy resources? How long can existing energy resources continue to meet our demands? How is energy used within the various sectors of our economy? Can we sustain energy consumption at present rates over both the short and long terms? Do our current energy-consuming practices minimize energy consumption? What are the environmental and human consequences of using alternate energy resources? What are the political consequences of using various energy resources? What are sound national goals for providing energy services in the future?

The answers must consider not only energy supply and demand but also protection of the environment and public health, continued economic growth, equity among geographic regions and economic classes, and national security. Energy problems are not confined to the United States. Most of the other more-developed nations face the same challenges, and the less-developed nations, attempting to raise their standard of living, will want a larger piece of the world's energy pie.

Contrasting Views on Energy

Different people view the role of energy in our society differently. Some consider energy resources as *commodities* to be bought and sold. Others view energy as a *natural resource*, the use of which has implications (especially ecological) that range far beyond the buyer and seller. Still others view access to energy supplies as a *social necessity* because any curtailment in supply affects their life-style and well-being. Government officials responsible for national security view energy supplies as a *strategic resource*. Thus it is clear that no single shared concept prevails concerning the role of energy resources in our society.

When energy is viewed solely as a commodity, discussion focuses on policies to develop and market coal, oil, natural gas, and electricity. The commodity concept has dominated energy policies in the United States for most of this century, but its critics are growing in number. Many feel that the commodity view (in which supply and demand determine price) is too narrow. They point out that people or groups who hold this viewpoint ignore important long-term concerns, such as future access to dwindling supplies, dependence on foreign supplies, and poor accountability for pricing by regulated monopolies (utility companies).

When energy is viewed as a natural resource, we must consider the specific impacts of its use on ecosystem components—water, soil, climate, and plant and animal communities. Those who take this perspective consider the long-term implications of pollutants and resource availability, and classify energy resources as being polluting or nonpolluting, renewable or nonrenewable, and exhaustible or inexhaustible. Although they recognize that the final service provided by such resources is the main concern, they also recognize the various environmental impacts associated with the methods of providing energy services. For example, all other things being equal, persons with this viewpoint would choose a combination of weatherization and solar power rather than oil or natural gas to meet their space-heating needs; they would prefer using an essentially nonpolluting, renewable resource instead of a polluting, nonrenewable resource.

During most of this century, few authorities have challenged the view that energy is anything more than a commodity. After all, energy was relatively inexpensive, resources appeared large in comparison to the rates at which they were being used, and pollution caused by certain types of energy resources was largely ignored. However, during the 1970s, energy shortages and a growing concern about pollution brought the natural-resource viewpoint to the public's attention. Rapidly rising energy prices during this period and again in late 1990 and the resulting hardships on the poor helped to focus attention on the social necessity of energy. The energy shortages during 1973-74 also underscored the strategic importance of conserving energy resources both for industry (because the most energy-efficient companies are the most likely to survive) and for national security.

No doubt, the relative importance of these competing viewpoints on energy will change with specific circumstances, but the confrontation with Iraq over its invasion of Kuwait should strengthen the natural-resource viewpoint (fig. 18.1). Certainly, protecting access to Middle East oil was one major reason for going to war against Iraq. Nevertheless, widespread recognition of the variety of viewpoints helps focus the debate on energy policies.

(Recall from chapter 6 that 90% of the world's human population growth over the next century is expected to be in the less-developed countries.) Thus, we must begin our examination of energy problems from a worldwide perspective by identifying current trends in the use of energy and assessing the prospects for future energy supplies.

Energy Consumption Patterns Today

At the start of the Industrial Revolution, a few hundred million people contributed to the demand for commercial fuels—first coal, then oil, and then natural gas. Today approximately one billion people contribute to the energy demand of an industrialized world. They heat and cool an estimated one billion dwellings and drive 500 million motor vehicles. On a worldwide basis, however, the individual demand for energy is far from uniform.

An average citizen of a highly industrialized nation such as the United States or Canada consumes the energy equivalent of just over 40 barrels of oil per year, while the average Mexican uses the equivalent of less than 10 barrels of oil per year. Elsewhere, in poverty-stricken nations such as Ethiopia and Bangledesh, an individual typically uses the energy equivalent of 1 to 2 barrels of oil in the form of firewood, animal dung, or plant debris. In addition, they may use indirectly the energy equivalent of 1 to 2 barrels of commercial fuels in the form of various products and services.

Most people are not concerned with the specific source of the energy they consume; rather, they want assurance that the service energy provides will be available and convenient. Most people do not care if a car is powered by gasoline, diesel fuel, natural gas, or electricity; what they want is the freedom to move from place to place whenever the need arises. Most people who turn on a television or a lamp are not concerned with whether coal, nuclear, oil, hydropower, or

How Energy Resources Are Measured

The United States uses the nonmetric British thermal unit as the basic measure of the energy content of fuels. One *British thermal unit (BTU)* is defined as the amount of heat required to raise the temperature of 1 pound of water 1 Fahrenheit degree.

A much larger energy unit, the quad, is generally used in discussions of energy supply and demand in the United States and the world. One *quad* is equal to 1 quadrillion, or 1.0×10^{15} BTUs. Thus, burning approximately 172 million barrels of oil releases 1 quad of energy. (The United States consumed 82 quads of energy resources in 1990.) By expressing energy in units of BTUs and quads, we avoid having to make conversions between the different measures of the various energy resources. For example, natural gas is sold by the cubic foot (1,020 BTUs) or therm (100,000 BTUs); oil is sold by the barrel (42 gallons yields 5.8 million BTUs); coal is sold by the ton (approximately 22 million BTUs); and electricity is sold by the kilowatt-hour (2,312 BTUs). Students who want to become more familiar with converting energy units should consult Appendix A for conversion factors.

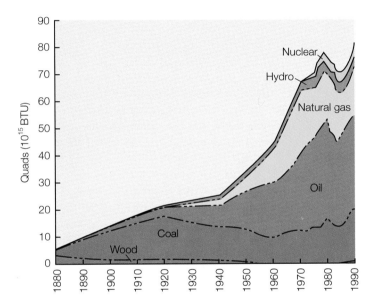

Figure 18.2   Energy use in the United States during the last century. Approximately 2.5 quads of biofuel energy are not accounted for in recent data. The amount of each type of energy used is the distance between the curve for that fuel and the curve immediately below it. Thus nuclear and hydroelectric energy each provided approximately 6 quads of energy in 1990, whereas oil provided 34 quads. The sharp downturns in 1979 were due to a combination of economic recession and stepped-up conservation. Sources: Data for 1880–1976: *Energy in Focus—Basic Data*, Federal Energy Administration, 1977; data for 1977–1990: *Monthly Energy Review*, U.S. Department of Energy.

Table 18.1	Percent of the world's energy consumption supplied by various sources in 1990 and at their peaks		
Source	**1990**	**Peak**	**Year**
coal	26%	70%	1920
oil	38%	40%	1972
natural gas	19%	19%	1990
nuclear, hydropower, and biofuels	17%	17%	1990

the percent each source supplied at its peak year. Special Topic 18.1 explains the units that are used to measure energy resources.

Today the 5.4 billion people of the world consume 0.71 quads of energy every day (the equivalent of 123 million barrels of oil). This rate of consumption is approximately fifteen times greater than it was a century ago. If wood fuel is excluded from the calculation, the past century has seen a sixty-fold increase in the rate of fossil fuel consumption. This dramatic rise in consumption is attributable, in roughly equal parts, to rapid population growth and soaring per capita energy demand. However, the demand per capita has eased somewhat since the Arab oil embargo of 1973 chiefly because of conservation.

Today's global economy consumes fossil fuels 100,000 times faster than they were formed by geological processes (chapter 17). All sectors of the world economy—residential-commercial (38.4%), industrial (37.5%), and transportation (24.1%)—help deplete this legacy of the geologic past. Just keeping the world's fleet of motor vehicles rolling consumes half of the world's daily oil production.

Energy is used in homes and commercial buildings for basically the same purposes: heating, air conditioning, lighting, ventilation, and running small electrical appliances. In the U.S., natural gas and oil are the two most important fuels used directly by the residential-commercial

natural gas, was used to generate the electricity; they just want the TV set or lamp to come on when they flick the switch.

Through most of the twentieth century, development and expansion of highly energy-dependent societies have been major endeavors, with remarkable success. Fossil fuels have served as the major source of energy, but the fraction of energy provided by each fuel type has changed significantly. Figure 18.2 shows the change in U.S. energy consumption patterns between 1880 and 1990. Table 18.1 gives the percent of the world's energy provided by various sources today and

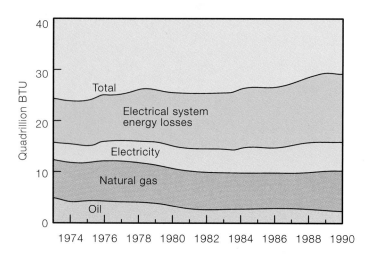

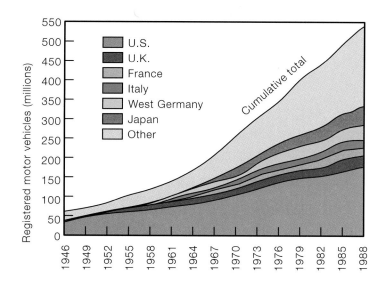

Figure 18.3 Sources of energy for the residential-commercial economic sectors of the United States. The total amount of energy that was consumed by all sectors in 1990 was 82 quads.

Source: *Monthly Energy Review,* December 1990, Energy Information Administration.

Figure 18.4 The number of motor vehicles on world roadways has increased by a factor of 12 since 1945. Most of the growth has been in industrialized countries. Eastern Europe and developing countries have the greatest potential for future growth.

sector (fig. 18.3).

Space heating and cooling are the single most important uses of energy in homes and businesses. Migration to the sunbelt and rising affluence have made air conditioning a major contributing factor to the recent rise in energy consumption in the United States. For example, in 1950, fewer than 1% of all U.S. homes had any type of air conditioning; today, over 50% are so equipped. Other appliances found in most U.S. homes include clothes dryers, self-defrosting refrigerators, television sets, water beds, videocassette recorders, stereos, and improved, but often excessive, lighting. Thus, given that the United States' population continues to increase and new energy-consuming devices are continually coming on the market, total energy consumption will not decline in the residential-commercial sector unless we learn to use energy more wisely—a topic we address in chapter 19.

The world's fleet of cars, trucks, buses, trains, planes, ships, and barges create another major demand for energy—24.1% of the total. The number of vehicles on the roads of the world has increased about 5% per year since 1970 (fig. 18.4). If the trend continues, the year 2030 will see one billion vehicles, twice the number on the highways in 1990. Most of this growth is projected for the eastern European countries and the less-developed nations. As a result, more nations will be importing oil, and the growing vehicular fleet will aggravate regional air pollution problems and increase the rate at which carbon dioxide (a greenhouse gas) is released to the atmosphere. Today, the U.S. fleet burns the equivalent of 6.4 million barrels of oil each day—mostly (74%) consumed by cars and light trucks. Sixty percent of the oil burned in the United States is consumed for transportation. The vast amount of oil used for transportation suggests that improvements in fuel efficiency for replacement vehicles should be a global goal.

The industrial sector of the world economy also consumes a large amount of energy—37.5% of the total, most of which is used to fire boilers. The chemical industry is the largest single energy consumer within the industrial sector, because it not only requires energy for manufacturing, but also uses petroleum as the raw material for many of its products (e.g., fibers, pharmaceuticals, plastics, and pesticides). Also, the steel-producing, petroleum-refining, paper, and ore-processing industries are energy-intensive. Industries can usually switch from one type of fuel to another, however, if they have an economic incentive. For industries, the most important fuel characteristics are cost and uninterrupted supply—and in the foreseeable future, coal is the fuel whose supply seems to be most secure. Some U.S. industries, however, may not be able to convert to coal because they are located in regions that do not meet federally mandated air-quality standards (chapter 16).

Total U.S. industrial demand for energy has actually declined by about 1% per year over the past two decades, despite an increase in production. Many older industrial processes used energy inefficiently. An increasingly competitive world economy has forced industries to cut costs by developing more energy-efficient processes. Consumers have also played a part by demanding appliances and cars that require less energy.

Nippon Steel Corporation of Japan has shown what can be accomplished when energy conservation becomes a central focus for redesigning an industrial process. By conserving energy in every step of the steel-making process, engineers were able to cut the energy required to produce a metric ton of steel by 25%. Such savings are not trivial; the energy saved is enough to keep a city of 2 million running! Similar efforts throughout Japan have made it the world's most fuel-efficient economy. Such improvements have contributed to Japan's rise as a world economic leader.

Nonrenewable Energy Resources

Supertankers in the Persian Gulf lined up for refilling, a railroad yard choked with coal cars, and a mountain of coal piled next to a giant electric utility are scenes that convey the dependence of the world economy on tremendous quantities of fossil fuels. Since the world consumes nonrenewable energy resources much faster than they are generated (chapter 17), we must ask how long we can expect these resources to last.

Nonrenewable energy resources are defined as resources that are not regenerated within a meaningful period of time. Fossil fuels are prime examples of nonrenewable resources. The world's dependence on fossil fuels is shown by the fact that oil, natural gas, and coal meet 88% of current world energy demand—321 quads (fig. 18.5).

To determine the nonrenewable energy resource supply, we must consider two factors: (1) the amount of available resources, and (2) our rate of consumption. With this information on supply and demand, we can calculate the expected lifetime of a resource. For example, a small automobile on the open highway consumes about 8 liters (2.1 gallons) of fuel per hour and has a fuel capacity of 40 liters (10.6 gallons). Thus, if we drive that car nonstop for 5 hours, it would consume a tank of fuel (40 liters divided by 8 liters per hour equals 5 hours). We could make similar calculations for fossil-fuel resources of the world or the United States, but the answer would not be as precise as our example implies. Several factors complicate the forecasting of how long energy supplies will last.

Although our present consumption rates of fossil fuels are known with a fair degree of accuracy, the total amount of recoverable fossil fuels (reserves) is more uncertain. Undiscovered energy resources exist, but estimates of their size are speculative. Some energy resources are low-grade, meaning that prices must rise before they can be developed. For these reasons, calculating how long our fossil-fuel reserves will last yields only rough approximations (some would say guesses). Furthermore, energy experts expect future patterns of consumption to change in response to changing prices and availability. In addition, population and economic growth will vary with time and country. Each of these factors introduces more error into projections of future supply. Nevertheless, these calculations can tell us which resources are likely to be depleted first, and they can indicate where we should make the greatest efforts to reduce consumption. Such analyses also suggest which mix of alternative energy sources is most likely to meet our future energy demands. We now examine the supply and demand picture of global and U.S. fossil-fuel resources: oil, natural gas, and coal.

Oil

In this century, crude oil has become one of the primary raw materials for the world's technological society, supplying 38% of its energy demand (table 18.1). Oil supplies essentially the same percentage of U.S. energy demand. Oil is the preferred fuel because its liquid form makes it especially versatile and

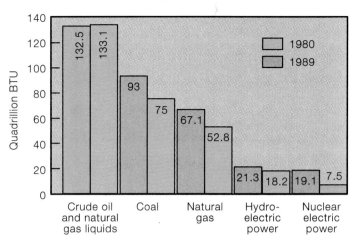

Figure 18.5 World energy consumption in 1980 and 1989 by source. In 1990, a total of 333 quads were consumed, whereas in 1980, 287 quads were used. Biofuel consumption is excluded.

Source: *Energy Facts 1989*, Energy Information Administration.

easy to handle. Petroleum products power our transportation system, fuel our farms and factories, and heat many of our homes and businesses. A petroleum-refining residue called *asphalt* is used to pave roads, and oil products are ingredients in more than three thousand commercial products (e.g., lubricants and paints), as well as being the raw materials for another three thousand petrochemicals (e.g., plastics and pesticides). In all, production of goods and services for an average family in the United States requires approximately 73 liters (0.45 barrels) of oil-derived products each week.

Energy experts put world oil reserves at 7,000 quads. At the present rate of consumption, 121 quads per year, this oil would last another fifty-eight years. Figure 18.6 gives a breakdown of oil reserves around the world and total production up to 1990, which reveals two glaring facts: (1) the enormity of known oil reserves in the Middle East, as shown in more detail in table 18.2, and (2) how little of the original oil deposits are left in the United States. Although the U.S. did not start to use oil until the beginning of the twentieth century, known and estimated domestic oil reserves are now about two-thirds gone. The United States, however, continues to consume oil at a near record rate. If 1990 consumption rates continue, and oil is imported at the same rate, domestic oil reserves will be used up in only sixteen years.

Figure 18.7 shows the historical pattern of oil production in the United States as well as the estimated size of oil reserves. Note that the curve exhibits a rate-of-use curve that is typical of any finite resource. The rate of production initially increases, reaches a maximum, and then gradually tails off. These data demonstrate that the United States has clearly passed the point of its maximum rate of oil production.

Some people, however, argue that the projected production curve for the United States should have a longer tail because of undiscovered reserves. If such production is to occur, however, oil must be discovered at a rate greater than today's rate of discovery. Several lines of evidence suggest that current oil discovery rates are not likely to increase. Both in the United States and worldwide, the rate of discovery has

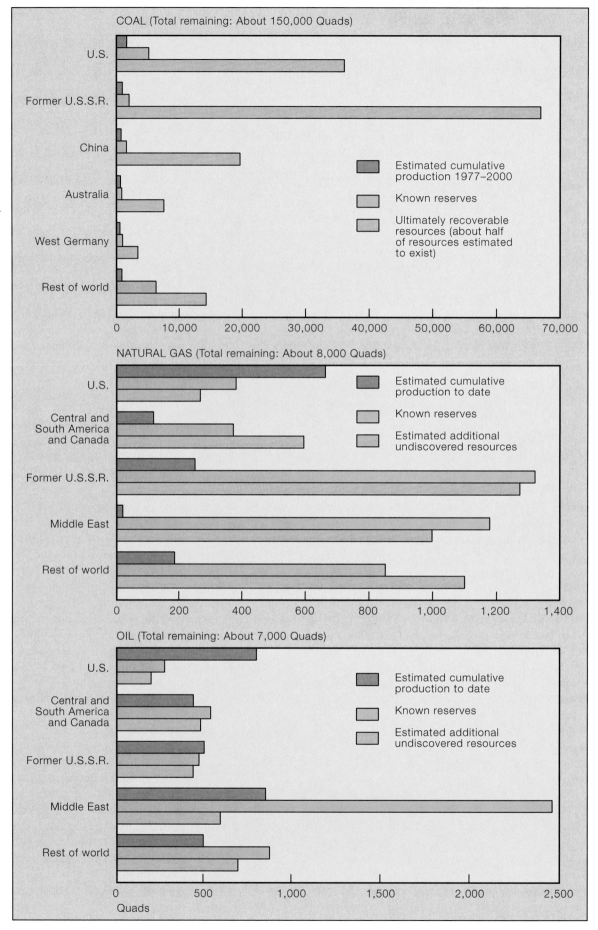

Figure 18.6 Estimates of known and undiscovered fossil fuel resources in the world. Note the size of the cumulative production (top line) relative to known reserves (middle line), especially for oil in the United States. Also note the differences in scales for the three diagrams.

From "Energy from Fossil Fuels" by William Fulkerson, et al. Copyright September 1990, Scientific American, Inc. All rights reserved.

Table 18.2	Proven oil reserves in the Organization of Oil Exporting Countries (OPEC) in 1989
Nation	**Billions of Barrels**
Saudi Arabia	255
Kuwait[a]	95
United Arab Emirates	98
Iran	94
Iraq	100
Other OPEC nations[b]	118

[a]Estimates are that Kuwait lost 10% of these reserves because of the Persian Gulf War.
[b]Includes Algeria, Ecuador, Gabon, Indonesia, Libya, Nigeria, Qatar, and Venezuela.

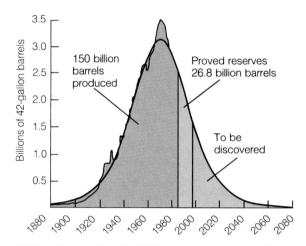

Figure 18.7 Actual (through 1990) and projected crude oil production rates from oil fields in the United States. Proven reserve estimate is from the Energy Information Administration.

declined during the past fifteen years despite improved methods of prospecting. Furthermore, only one new giant oil field has been discovered in the U.S. since 1965: Alaska's North Slope deposits. **Giant oil fields** are those containing more than 100 million barrels.

Giant oil fields, such as those discovered in the 1930s in East Texas and in Wilmington, California, still dominate current production in the United States. Petroleum geologists doubt that many new giant oil fields will be discovered, particularly in the United States, where extensive drilling has already been done. Furthermore, even if a new giant oil field is discovered, it would extend our oil supplies for only a few weeks to a few years at our current rate of consumption. The output of a giant oil field that produced 100 million barrels of oil over its lifetime would be used up in the United States in a week at the current rate of consumption (16 million barrels of oil per day.)

Five independent forecasts of future U.S. oil production all show production declining from about 14.6 quads per year in 1990 to 10.4 quads per year in the year 2000 under low price conditions ($12 to $18 per barrel). Only with higher prices will the relatively large low-grade oil resources

contribute significantly to future domestic oil production. Such low-grade deposits are accessible only by expensive secondary recovery methods such as steam injection (chapter 17). Even under a high-price scenario, domestic oil production is expected to decline.

There is no doubt that the United States will have to go on importing oil even with stringent conservation measures. Rising oil costs will only exacerbate the problems that oil-importing nations experience with balance of payments—the cost of imports versus the cost of exports. The United States will probably import 60%-70% of its oil by the year 2000, and that oil is likely to cost $100 to $200 billion per year—a cost so huge that it may seriously weaken the U.S. economy.

The issue of obtaining oil at low cost is not likely to fade away. Many experts feel that competition for oil will intensify in the future. They cite (1) the desire of the more-developed nations to maintain their energy-intensive life-style, (2) increasing worldwide competition for oil by less-developed nations, and (3) decreasing worldwide production within the next decade. These factors are likely to keep the oil-rich Middle East a center of continued political unrest, entangling many oil-dependent nations. Indeed, if every person in the world used as much oil as citizens of the United States, the total estimated oil resources throughout the world would be used up in only a decade!

Examining past consumption rates and future oil supplies makes it clear that the oil era is coming to a close. It was a period in which relative energy costs (the proportion of income spent for energy) were small. Easy access to oil also made products derived from it inexpensive. However, to maintain our current living standard over the coming decades, we need to make a determined transition to other sources of energy and use more effective methods of energy conservation, particularly with respect to oil.

Natural Gas

Natural gas accounted for 24% of the total U.S. energy demand in 1990. Forty percent of the total energy supplied by natural gas was used in residences and commercial establishments, primarily for space heating—a use that could not be converted easily to another energy source without great cost and increased air pollution. Industry and electric utilities consumed the remaining 60%. Industries, however, can convert to other energy sources, because their size can justify the financial investment in air-pollution control devices.

Natural gas has many environmental and economic advantages. After removal from subsurface reservoirs (chapter 17), it is easily processed and transported by pipeline, primarily as methane (CH_4). Of the three fossil fuels, natural gas burns cleanest, producing only carbon dioxide and water as end products. If dirtier fuels (coal, oil, or wood) were used exclusively in home heating systems in large urban areas, the resulting air-pollution would necessitate pollution-control devices on individual systems, pollutant removal from fuels, or both. Of all fossil fuels, natural gas produces the least amount of the greenhouse gas carbon dioxide per unit of

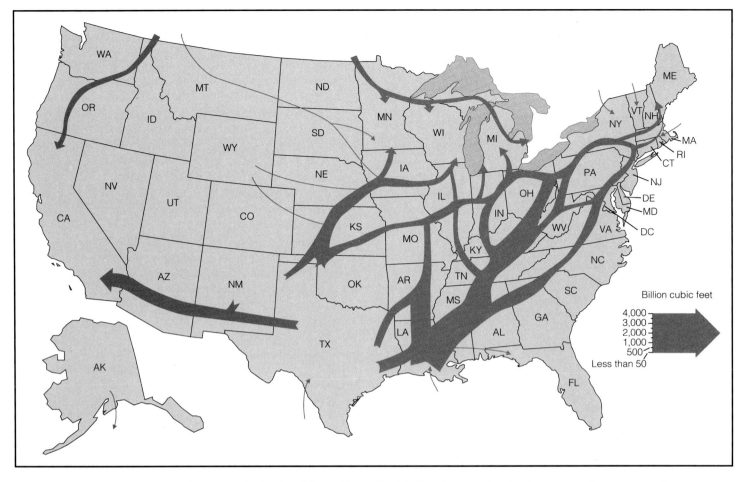

Figure 18.8   The origin and flow of natural gas in the United States. The width of the lines is proportional to the amount of gas transported.
Source: Energy Information Administration, DOE/EIA-0131 (88).

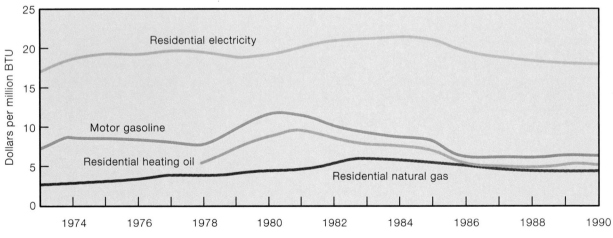

Figure 18.9   The relative cost of energy to U.S. consumers (constant 1982–1984 dollars).
Source: *Monthly Energy Review*, December 1990, Energy Information Administration.

energy it yields when burned. On a per-energy-unit basis, natural gas is also the safest fossil fuel in terms of the number of accidents and deaths that occur during extraction and use.

A pipeline system 400,000 kilometers (250,000 miles) long transports natural gas from reservoirs to consumers across the United States (fig. 18.8). Most of the natural gas used in the United States is produced in the lower forty-eight states; approximately 7% is imported from Canada. Natural gas has been available to U.S. consumers at exceptionally low prices, relative to the cost of other forms of energy (fig. 18.9), because the federal government has regulated its price. However, in 1989, President Bush signed into law the Natural Gas

Wellhead Decontrol Act, which provides for the gradual removal of all remaining price ceilings on natural gas. The Act is expected to spur development of new natural gas resources, thus increasing supply.

World natural gas reserves are estimated at 8,000 quads. At the present rate of consumption, 65 quads per year, natural gas would last another 125 years. Figure 18.6 gives a breakdown of natural gas reserves around the world and total production up to 1990. The Organization of Oil Exporting Countries (OPEC), most of which are in the Middle East, and the former USSR each have an estimated 38% of the world's natural gas reserves.

The Department of Energy estimates the size of technically recoverable natural gas reserves (including undiscovered resources) in the United States to be approximately 1,000 quads. Thus, if we use the 1990 natural-gas consumption rate of 17 quads per year, the discovered and undiscovered resources would last 50-60 years if we do not change our rate of consumption. Since natural gas can often be substituted for oil, however, its consumption rate is likely to increase as oil supplies decline and prices rise.

During the early 1980s, the outlook for the natural gas supply appeared to improve somewhat in the United States. Potential deep gas reserves were identified. New frontiers for gas exploration in Alaska, the offshore Atlantic, and geological formations in the West were thought to contain sizeable reserves. However, a 1989 report by the United States Department of the Interior put an end to this optimism. Many of the promising new exploration sites failed to meet expectations, and today experts are more conservative in their estimates of the potential size of U.S. gas reserves.

One bright spot is the amount of natural gas that can be extracted from coal beds. The Gas Research Institute has demonstrated that coal seams can be an important source of methane, the chief constituent of natural gas. Significant quantities of methane can be removed from the pore spaces in a coal bed prior to mining, and this procedure also makes underground coal mining safer. In 1990, the federal government took an important step to promote extraction of natural gas from coal beds by providing a tax-credit incentive of $0.80 per million BTUs on coal-bed gas—a significant incentive when one considers that, at the time, gas at the wellhead was selling for $1.20 per million BTUs.

Significant quantities of natural gas are located in Alaska's North Slope—an estimated 75 quads, but projections indicate that the price of natural gas would have to triple before it would be profitable to build a pipeline to transport the gas to the lower forty-eight states. Significant quantities of natural gas also are under Arctic Alaska National Wildlife Refuge. These reserves are not included in the above estimates because federal law currently prohibits their extraction (chapter 10).

The United States could extend natural gas supplies by importing natural gas in liquefied form from the Middle East. Those nations currently flare (burn-off) natural gas as a waste by-product of oil production because they have no local market for it. Cooling natural gas to -162° C (-259° F) converts it to a liquid that can be shipped by tanker. **Liquefied natural gas (LNG)** occupies only about 1/600 of its gaseous-state volume. Hence, each tanker can carry an enormous amount of LNG. At the tanker's destination, the LNG is regasified. LNG is extremely hazardous to handle, however, and elaborate precautions must be taken to prevent gas leaks and possible explosions during loading, transport, and unloading. Specialized tankers plus the liquefaction/gasification processes make LNG more expensive than natural gas supplied by pipelines. Today, there are only 65 LNG tankers, as compared to 2,600 oil tankers. Japan has developed technology to handle LNG and is the world's largest user of LNG.

Synthetic gas can be produced from coal and a variety of other sources, including oil shale, agricultural wastes, and municipal and some industrial wastes (chapter 20). We examine how synthetic gases are produced from coal and oil shale later in this chapter.

Coal

In 1990, the United States derived 23% of its energy from coal, compared with 93% at the turn of the century. In 1950, coal heated 35% of U.S. homes, but now less than 1% of American homes still use coal. Today, coal is used primarily to produce steam that drives turbines that generate electricity. Indeed, increased demand for electricity is the reason for a resurgence in the use of coal since 1970. Currently, about 765 million tons (700 million metric tons) of coal are burned each year to generate electricity in the United States. Coal is also an important chemical in metallurgy, the production of metals from their native ores.

Coal has a great competitive advantage in the generation of electricity, because it can be delivered at a very low cost per unit of energy. For example, in 1990, coal was one-half as expensive as natural gas and one-third as expensive as oil for generating the same amount of electricity. Even though capital costs for pollution control are higher in coal-fired power plants, most utilities find that their total costs are lowest when they burn coal.

The use of coal poses many environmental problems, however. Recall from chapter 16 that coal is a dirty fuel; its combustion produces large amounts of sulfur dioxide (SO_2) and particulates. The use of coal to generate electricity became an even more complex issue in the early 1980s, when sulfur dioxide was identified as one of the major precursors of acid rain. Furthermore, the insistence of the Canadian government that the United States do something to control its emissions of sulfur dioxide has made the issue an international one. As a result, 1990 amendments to the Clean Air Act aim for sharp reductions in sulfur dioxide emissions from sources in all states east of the Mississippi and in five states west of it. The Department of Energy estimates that as a consequence, $200 to $300 billion may be added to the cost of generating electricity from coal over the next thirty years. Each person would pay an additional $30 to $45 per year for electricity. Furthermore, utilities would generate more than twice as much solid waste, since plants that use fuel-gas

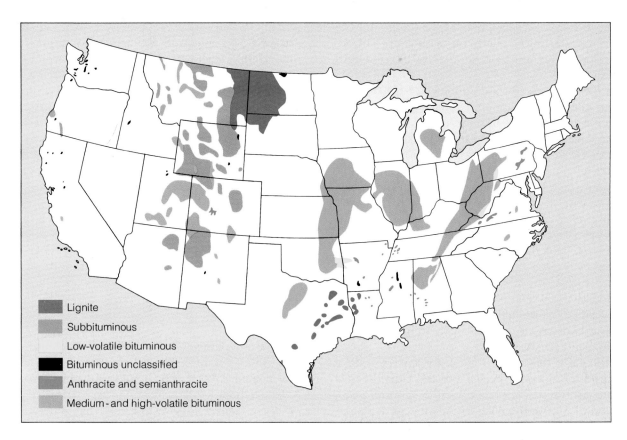

Figure 18.10   Locations and types of coal deposits in the lower forty-eight states. Alaska also has a considerable amount of coal.

Source: *Trace Elements in Coal,* Circular 499, Illinois Geological Survey, 1977.

desulfurization scrubbers (chapter 16) produce more solid waste than plants that do not. Holding those wastes in artificial ponds, which is the current method, is viewed as an unacceptable long-term method of disposal.

Burning coal degrades the environment in several other ways. Its combustion is a principal contributor to the buildup of atmospheric carbon dioxide, which may lead to global warming. Strip mining for coal accounts for about half of the U.S. land disturbed by mining. Seepage of water through coal mine wastes leads to the formation of sulfuric acid and to the acidification of lakes and streams that receive runoff from waste heaps. Furthermore, when compared to other fuels (including nuclear fuel) on an energy-equivalent basis, coal is the most dangerous in terms of number of deaths, illnesses, and injuries resulting from mining and handling.

Ironically, despite these major environmental and safety problems, many energy experts favor coal as the answer to the energy problems of the twenty-first century. Why? Global reserves of coal are by far the largest of the fossil fuels, estimated to be equivalent to 150,000 quads of energy. The present rate of world consumption of coal is only 96 quads per year; thus the supply could last for centuries. Figure 18.6 shows the locations of world coal reserves and the estimated total production between 1977 and the year 2000.

Coal reserves in the United States are the largest in the world, estimated by the Department of Energy to represent 28% of the world's recoverable reserves. Energy experts estimate that economically recoverable deposits in the former USSR exceed those of the United States (figure 18.6). Coal deposits are scattered across the lower forty-eight states (fig. 18.10). Three-quarters of the surface-mineable coal is west of the Mississippi, and virtually all of those deposits are composed of subbituminous coal and lignite (chapter 17). Approximately equal quantities of subsurface-mineable coal are found east and west of the Mississippi. The East also has small reserves of anthracite coal. The characteristics of eastern and western coals are summarized in table 18.3.

Some methods of transporting coal have a significant impact on the environment. Increased rail transport of western coal, for example, has caused many towns to ask that railroads be rerouted around them to avoid the noise and the interruption of local traffic patterns. Coal unit trains are nearly 2 kilometers (1.2 miles) long and take several minutes to pass a highway crossing. In the mid-Atlantic states, a number of canals and locks must be enlarged to handle the larger coal barges required to meet increased demand. Furthermore, truck transport, which is used primarily in the East, is the most costly means of transporting coal and increases dust, the risk of traffic accidents, and noise.

Coal-slurry pipelines are a relatively new method of transporting coal that avoids some of the negative impacts of the other methods. These pipelines pump a mixture of 50%

Table 18.3 Characteristics of eastern and western coal

Coal Type	Heating Value (BTU/pound)	Sulfur Content (percent)	Seam Thickness (meters)
Western coal			
lignite (brown coal)	6,000–7,500	0.6	3–60
sub-bituminous coal	9,000–11,000	0.6	—
Eastern coal			
bituminous coal	11,000–14,000	0.6–6.0 (2.0–3.5 ave.)	1–2.5
anthracite	13,500–15,000	0.5	—

ground-up coal and 50% water from coal mines to power plants or ports. The major incentive for building coal-slurry pipelines is that they can transport coal more cost-effectively in regions where rail lines need to be expanded or upgraded, or where the terrain limits access to rail lines. Also, fewer traffic deaths result when coal is transported by pipeline rather than by rail.

The two major impediments to coal-slurry pipelines are obtaining rights-of-way for building pipelines and water availability. For twenty years, the coal industry has tried unsuccessfully to gain the right of **eminent domain** (the right to condemn land and purchase it for public benefit) so it could build coal-slurry pipelines. Although water shortages are not expected to be a problem in eastern states, complex laws concerning water rights complicate water use in the water-short western states. Nevertheless, a 440-kilometer (275-mile) coal-slurry pipeline has been transporting 5 million metric tons of coal per year from the Black Mesa-Kayenta mine in Arizona, to Laughlin, Nevada, since 1970 (chapter 17).

Although oil and natural gas remain the preferred fuels for many purposes, U.S. and global reserves of these precious fuels are being rapidly depleted. Ironically, U.S. and global reserves of coal are immense. Hence, policy decisions regarding how and where coal will be extracted and how it will be used will significantly affect the U.S. and global economy and environment well into the future. One way to use coal is to convert it to cleaner and more versatile forms of fuel. In the next section, we consider the future role of such synfuels.

Synthetic Fuels (Synfuels)

Fuel substitutes for oil and natural gas, called **synthetic fuels** (or **synfuels**), can be produced from coal or oil shale, extracted from tar sands, or obtained by mining oil formations that have already been drained by oil wells. (A typical oil well recovers less than half of the oil that is present in the reservoir rock.)

Much of the developmental work for synfuel production is already complete. In the wake of the Arab oil embargo and skyrocketing fuel prices, legislators during the late 1970s felt that the government should explore commercial-scale synfuel plants to determine their economic and environmental feasibility. President Carter signed into law the Energy Security Act of 1980 and set up the U.S. Synthetic Fuels Corporation (SFC) to develop the technology necessary for commercial production of synfuels. That act committed $88 billion through 1992 for loans, established price guarantees (for products whose market price is less than their manufactured cost), and provided for joint-venture participation for projects that could not obtain private funding, with the SFC as a part owner.

The SFC began with major awards to fund development of coal-to-gas and coal-to-gasoline conversions. However, as energy supplies improved and prices subsequently dropped during the 1980s, the public and the federal government lost interest in synfuels. Thus, the massive effort by the government to develop a profitable synfuels industry was scrapped, and only limited funding for synfuel development was available in the early 1990s.

Industries, however, did realize some significant technological advances under these federally funded programs. Commercial-scale processes were developed that can produce fuels (synthetic natural gas, oil, and gasoline) that are essentially equivalent to those in use today. Consider some examples.

Since coal is the largest fossil-fuel resource in the United States, coal gasification could become an important source of energy. **Coal gasification** is the conversion of coal to natural gas. The technology is already well developed. For example, a plant at Beulah, North Dakota, has produced 140 million cubic feet per day of synthetic natural gas since 1985 and is now owned by a local utility, which purchased it from the federal government (SFC) in 1988.

Coal is gasified when crushed coal reacts with steam and either air or pure oxygen to form a mixture of carbon monoxide, hydrogen, and varying amounts of methane. If the gasification process uses air, the gas mixture produced is called **low-BTU gas** because it has a heating value of only about 100 BTUs per cubic foot, roughly one-tenth the heating value of natural gas. Injection of pure oxygen produces **medium-BTU gas,** which has a heating value of approximately 300 BTUs per cubic foot. Neither of those gas mixtures is economical to pipe long distances, but they are both well suited for industrial purposes in nearby boilers or as raw materials for local chemical industries.

The only synfuel that can be transported economically over long distances is **high-BTU gas,** also called **substitute natural gas.** Its production requires two additional steps. Substitute natural gas has the same heat content as natural gas, but it costs 20-40% more to produce.

The process used to manufacture substitute natural gas removes many of the pollutants (especially sulfur) from the product, but the removal processes produce both gaseous emissions and solid wastes that must be controlled. Substitute natural gas plants face air-pollution problems that are similar to those of an oil refinery, because they both produce tars and aromatic hydrocarbons (benzene-related compounds). Some of those hydrocarbons are known carcinogens, but existing technology can control emissions of these substances.

Coal gasification can also be accomplished by **in situ recovery.** In this method, part of a coal seam is burned by supplying oxygen through an injection well. After the coal is ignited, water is injected into the seam, converting the coal to a mixture of gases, which constitute medium BTU gas. It is also possible to manufacture substitute natural gas by this method. Initial cost estimates are $4.35 per million BTUs for the medium BTU gas and $6.25 per million BTUs for substitute natural gas. If these estimates are correct, the cost of producing substitute natural gas by this method must drop significantly before it can compete with natural gas. The 1990 price for natural gas supplied to residential customers was approximately $4.50 per million BTUs.

Despite its current economic disadvantages, underground (in situ) gasification technology has special appeal because it can withdraw energy from coal deposits that cannot be extracted by conventional techniques. The amount of coal estimated to be available for underground gasification constitutes 35,000 quads (1.6 trillion metric tons of coal)—nearly five times the amount that is mineable by conventional means. The technique also avoids some of the environmental problems associated with the use of coal, such as sulfur dioxide emissions and fly-ash disposal. However, a few problems can occur, such as ground subsidence and the possible disturbance of adjacent aquifers, which might lead to contamination or a change in the direction of groundwater flow.

Synthetic natural gas produced from coal can serve as an intermediate material in the production of methanol and other important liquid industrial chemicals. These products will probably become the major source of the basic chemicals used by the chemical industry to manufacture many modern products, such as synthetic fabrics. Furthermore, methanol can be used in proven, production-scale processes for the manufacture of gasoline. One such process, invented by Mobil Oil, is being used in New Zealand to produce 57,000 liters (15,000 gallons) of gasoline per day.

In addition to coal gasification, the manufacture of petroleum-like liquids from coal, called **coal liquefaction,** has received much research attention, because liquid fuels are better suited than gaseous fuels for use in motor vehicles. However, coal liquefaction is more complicated and costly than coal gasification. Amoco bought and redesigned one of the large-scale coal liquefaction plants financed by the federal government in the 1970s and 1980s. Each day, this Texas plant processes 13,600 metric tons (15,000 tons) of coal and 6.4 million cubic meters of natural gas into 67,000 barrels of regular gasoline and 17,000 barrels of easily liquefiable gases, propane and butane. Each metric ton of coal yields the equivalent of 6.1 barrels of petroleum liquids, of which 5.0 barrels are gasoline. The company figures the cost of a barrel at $35 if it includes a 10% return on investment. With further improvements in design, the company thinks it can cut its cost to $30 per barrel. If such costs are realized and the higher fuel prices predicted for the late 1990s become reality, more such plants will be producing synthetic fuels from coal within a decade.

Oil shale is neither oil nor shale, but a fine-grained rock called *marlstone,* which contains varying amounts of gray-to-brown organic material known as **kerogen.** When oil shale is heated to approximately 480° C (900° F), kerogen decomposes, forming combustible gases and petroleum-like liquids. Vast, high-grade deposits of oil shale known as the Green River formation, exist in the Colorado-Wyoming-Utah area. These deposits yield at least 106 liters (28 gallons) of premium-grade oil per metric ton of marlstone. Eventually, recoverable high-grade oil shale deposits in the United States could yield as much as 3,500 to 4,060 quads (600 to 700 billion barrels) of energy—roughly a one hundred-year supply of petroleum at present rates of consumption.

Lower-grade oil shale deposits also exist in a roughly triangular region that extends from Michigan to western Pennsylvania to Mississippi. However, these deposits contain less than 62 liters (17 gallons) of oil per metric ton. Estimates put their energy content at 17,000 quads (3 trillion barrels) of energy. Combining the high- and low-grade oil shale deposits makes oil shale second to coal in terms of the amount of energy resources in the United States.

Serious environmental considerations inhibit exploitation of oil shale. For example, every liter of oil produced requires 2-4 liters of water. Ironically, essentially all high-grade oil shale deposits are located in semiarid states. Thus water shortages in these areas would probably limit oil shale production to approximately 170 million liters (1 million barrels) per day, which is about 6% of the nation's 1990 rate of oil consumption.

A second and almost overwhelming obstacle to processing oil shale is disposal of huge volumes of waste. For every 170 liters (1 barrel) of oil produced, 1.4 metric tons of shale must be processed. Thus, to realize the probable maximum production of 170 million liters (1 million barrels) of oil per day, 1.4 million metric tons of shale must be processed, resulting in 1.3 million metric tons of waste per day. This residue is highly alkaline, low in nutrients, and, especially in semiarid climates, difficult to revegetate.

Given inadequate water resources and the difficulty of waste disposal, conventional oil shale mining operations are likely to remain small. Because in situ mining causes fewer environmental problems, research has focused on this technology. However, in situ mining reduces the percentage of oil recovered from the shale.

Figure 18.11 Tar sands processing facility in Alberta, Canada. Aerial view shows tar sand pile, which descends by gravity to belts that transport the sands to the extraction plant.
American Petroleum Institute

Imported oil remains less expensive than oil derived from oil-shale. Hence, large-scale, oil shale production will probably have to wait for a substantial rise in world oil prices or perhaps a change in political climate. As of this writing, it remains to be seen how much the political leadership will subsidize the future development of oil shale. Over the short term, oil shale will only make a minor contribution to our domestic oil supply, but over the long term, its potential is enormous.

Tar sands are another potential source of petroleum liquids. The U.S. Department of Energy estimates that some 30 billion barrels of oil (a 5-6 year supply at the 1990 rate of consumption) occur in deposits scattered across twenty-one states, with Utah having the largest amounts. Although several companies are pursuing ways to extract oil from tar sands, the most promising technique involves injecting steam to reduce the viscosity of oil and then collecting it at

recovery wells—a technique similar to the one used for enhanced oil recovery. Most deposits of tar sands in the United States are covered by a thick overburden, which requires recovery by in situ techniques, whereas in Alberta, Canada (fig. 18.11), surface deposits can be mined. Canada has more extensive deposits of tar sands than the United States. Its two major plants now produce a total of 300,000 barrels of oil per day. By the year 2000, Canada expects to obtain 40% of its oil from tar sands.

Other possible methods for enhancing the future supply of petroleum include peat gasification and enhanced oil recovery from depleted fields, some of which may still contain up to 90% of the original amount of oil. The amounts of petroleum-like fuels that will be produced by these various techniques in the future will depend greatly on world prices for competing energy sources.

Nuclear Fuels

Most nuclear fuels are used for the generation of electricity. About 16% of the world's electrical energy is derived from nuclear power. The equivalent of 350 1,000-megawatt plants were producing power in 1990. (A 1,000-megawatt plant is large enough to provide power for 500,000 homes.) Worldwide, plants under construction are expected to bring the total capacity to the equivalent of 400 1,000-megawatt plants by the mid 1990s.

Today, most nuclear power plants utilize **light-water nuclear reactors**—so named because ordinary water is used as a medium to transfer heat from the reactor core to the boilers. Light-water reactors require uranium-235 as fuel, an isotope of uranium that is relatively rare.* As a consequence, light-water reactors utilize only about 0.7% of the uranium in ores. If the light-water reactor continues as the technology of choice, then U.S. uranium reserves will last approximately one hundred years. This estimate assumes that the unused uranium-235 in nuclear fuels is not recovered through reprocessing. In this scenario, 99.3% of the uranium in nuclear fuel would not be used and would have to be stored as high-level radioactive waste (chapter 20).

An available technology that can improve the nuclear fuel potential of uranium by one hundred times, is the **fast breeder reactor.** This technology has passed engineering feasibility tests and is near the commercial development stage. Fast breeder reactors have tremendous appeal because they can extend uranium reserves by a factor of five hundred. Not only could breeder reactors utilize up to 60% of the uranium in the ore, but lower-grade uranium ores would also become economical to mine. The breeder reactor technology of the past required the reprocessing of used nuclear fuels and the isolation of weapons-grade materials—a prospect many countries, including the United States, are not willing to accept. New designs for breeder reactors, however, do not require weapons-grade materials.

*An isotope is any of two or more forms of an element having different atomic masses.

Renewable Energy Resources

Renewable energy resources are essentially inexhaustible. They include electricity generated by hydropower, solar thermal, ocean thermal, photovoltaics, wind, and biofuels. **Biofuels** include wood, municipal and agricultural waste, landfill and sewage gas, and methanol (wood alcohol) and ethanol (grain alcohol) produced from biological sources. In 1990, renewable resources contributed about 8.5% of the total energy consumed in the United States. Nearly one-half is from hydroelectric power facilities. Table 18.4 gives the specific amounts of renewable energy resources consumed.

Renewable energy sources are expected to contribute more significantly to energy supplies in the future. The principal forces behind these changes are the increasing cost of conventional energy sources, the decreasing cost of energy supplied by nonrenewable resources, and a higher level of concern for the environment. While renewable energy resources are generally thought to be more environmentally compatible, in fact significant environmental impacts may be associated with their production and use. Chapter 19 explores renewable energy technologies and their environmental impact in more detail.

Electricity

Electricity is indispensable in modern society. We expect it to be available at the flick of a switch. During power outages, serious problems can arise unless electrical systems are backed up with emergency generators. Life-saving equipment in hospitals will stop; frozen and refrigerated foods in homes, restaurants, and warehouses will spoil; stalled pumps will cause basements, tunnels, and sewage treatment plants to flood; and telephone communications will cease. Major cities come to a near standstill during power blackouts; traffic lights go out, elevators and subway trains stop. In fact, life without electricity is unimaginable for most of us.

Generating electricity requires primary energy sources such as coal, oil, natural gas, nuclear fuels, falling water, geothermal sources, sunlight, or wind. Fossil fuels, nuclear fuels, geothermal sources, and solar energy provide heat, which converts water to pressurized steam, and the steam drives turbines, which turn electric generators. Moving water or air (wind) produce electricity by rotating the blades of water turbines or windmills coupled to generators.

World electrical production accounts for 33% (107 quads) of the total energy consumed each year. In the United States, the percentage is slightly higher at 36.5%. Fossil fuels are used to generate 69% of the world's electricity; nuclear and hydropower account for most of the remainder. Coal is the leading source of energy used to generate electricity in the United States (fig. 18.12). How should power be supplied to meet the electricity demands of the future?

In the United States, today's conventional energy resources will continue to generate most of the electric power well into the next decade. The use of renewable energy sources, while projected by the United States Department of Energy to grow at an annual rate of 2.4% will still satisfy only a small fraction of projected total energy demand over the next two

Table 18.4	Renewable energy consumption in the United States (1990)	
	Quads	**Percent**
residential, commercial, industrial		
biofuels	3.12	46.0
hydropower	0.06	0.9
geothermal	0.14	2.0
solar thermal	0.09	1.3
wind	0.03	0.4
transportation		
biofuels (alcohol fuels)	0.07	1.0
electric utilities		
biofuels	0.02	0.3
hydropower	3.04	44.8
geothermal	0.22	3.2
total	6.79	99.9

Source: Energy Information Administration, *Annual Energy Outlook 1990*, Washington, D.C.: U.S. Department of Energy.

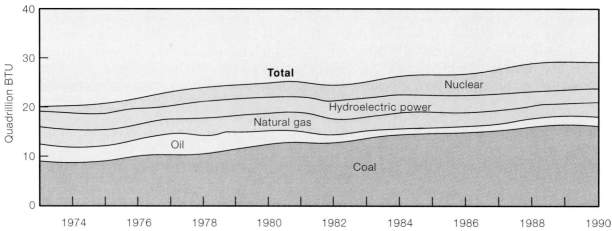

Figure 18.12 Energy resources used to produce electricity in the United States (1973-1990).

Source: *Monthly Energy Review*, December 1990, Energy Information Administration.

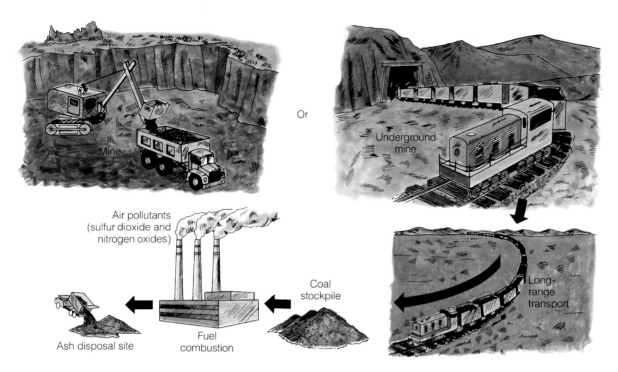

Figure 18.13 The coal cycle for the production of electricity. The coal cycle requires the transport of approximately 100 thousand times more fuel material than the nuclear cycle.

decades. (Projections see hydropower remaining as the largest renewable contributor to the generation of electricity.) Oil is not considered a viable fuel for generating electricity because of its relatively high cost, short supply, and value as a liquid fuel for transportation. The role of natural gas remains uncertain; it may be used more heavily to generate electricity over the short term.

Natural gas has four major advantages over coal for electric power generation. Natural gas (1) burns cleaner, (2) produces less carbon dioxide per BTU of electrical energy, (3) can be converted to electricity more efficiently, and (4) requires simpler and less expensive turbines that can be planned and constructed in a much shorter time frame. Its principal disadvantages are its more limited supply and its cost. These disadvantages will severely limit reliance on natural gas for generating electricity in the United States. Thus, coal and nuclear fuel are likely to be the principal means of meeting future electrical demand in the United States.

Choosing an Electrical Energy Future

Valid comparisons between coal and nuclear energy can be made only by comparing the entire **fuel cycle** of each source. Each fuel cycle has unique characteristics (figs. 18.13 and 18.14). In the coal cycle, familiar to most people, coal is mined, loaded on trains, trucks, barges, or ships, and often hauled long distances to power plants, where it is stockpiled. Before the coal can be converted to electricity, it is pulverized and then burned to produce steam that drives turbines.

The nuclear fuel cycle begins with either subsurface or open-pit uranium mines. Once uranium ore has been mined,

it is processed and placed in fuel-rod assemblies for nuclear reactors. In a reactor, the heat released by nuclear fission is used to convert water to steam, which, in turn, drives a turbine that generates electricity. Special Topic 18.2 on page 407 presents a more detailed explanation of how electricity is generated from nuclear fuels. The advantages and disadvantages of generating electricity with coal and with nuclear fuels are summarized in table 18.5.

Nuclear energy carries a unique risk: the possibility of a catastrophic accident. One serious nuclear reactor accident has occurred in the United States at the Three-Mile Island plant on the Susquehanna River near Middletown, Pennsylvania on 28 March 1979. The plant suffered serious damage to its core and released some radioactive materials into the environment. A special presidential commission report, called the Kemeny Report, found that public exposure to radiation was low and that it would have negligible current and future effects on human health. The accident did not release as much radioactivity as analysts had predicted, but the damage to the facility and the costs of cleanup and repair nearly drove the utility that owned the plant into bankruptcy. (The damaged fuel-rod assemblies were removed in 1990, eleven years after the accident.) Health effects on nearby residents were found to be related to stresses caused by unknown risks at the time of the accident. The undamaged twin reactor at the power plant site has resumed operation.

The most devastating nuclear power accident occurred on 26 April 1986 at the Chernobyl nuclear power plant located 3 kilometers (2 miles) from the Ukrainian city of Prypyat. The accident was caused by human error and exacerbated by safety design problems. A steam explosion ripped open the reactor, causing the graphite core to burn. An estimated 2-8%

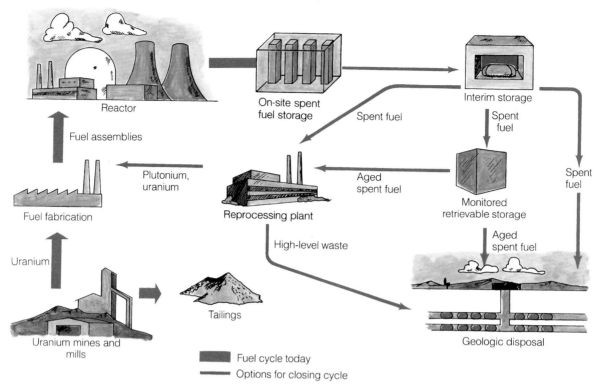

Figure 18.14 The commercial nuclear fuel cycle is used to produce electricity. In 1990, the reprocessing option was not being pursued, and the option of geologic disposal was meeting stiff political opposition from persons who did not want disposal sites near them.
Adapted with permission from *Chemical & Engineering News*, July 18, 1983, 61(29), p. 26. Copyright 1983 American Chemical Society.

Table 18.5	Advantages and disadvantages of generating electricity using coal versus nuclear fuel	
Fuel	**Advantages**	**Disadvantages**
Coal	Inexpensive Ample resources Small capital investment	Costly removal of air pollutants Releases large amounts of carbon dioxide (CO_2) More deaths per BTU of electricity generated Surface mining disturbs landscape Increased rail and road traffic Ash-disposal problem
Nuclear	Fewer deaths per BTU of electricity generated Releases no carbon dioxide Small volume of solid waste	Unresolved high-level radioactive waste problem High operational costs Large capital investment Technology must be updated Costly accidents possible (though risk is slight)

of the radioactivity within the nuclear power plant was ejected into the atmosphere and spread by winds as radioactive dust that settled on parts of the Ukraine, Belorussia, and Europe. Some 35,000 people living within a 30-kilometer (18-mile) radius of the plant were eventually evacuated. In total, an estimated 850,000 people live in areas affected by the radioactivity. Scientists studying the aftermath of the accident project that approximately 17,000 excess cancer deaths will occur over the seventy years following the accident.

Scientists and the public remain divided on the wisdom of using nuclear power. For example, Herbert Inhaber, a nuclear engineer, has analyzed the safety of nuclear power and has concluded that it is ten times safer, including potential catastrophic effects, than all other energy systems that he has investigated, with the exception of electric power gener-

ated by natural gas. His analysis included solar power and hydro-electric dams.

John P. Holdren, professor of energy resources at the University of California at Berkeley, is also concerned about the detrimental effects of coal and nuclear power. He opposes nuclear power for reasons unrelated to technical considerations. His primary concern is that nuclear power will speed the proliferation of nuclear weapons among nations. Nuclear power plants produce (breed) fissionable materials, such as plutonium-239, which can be isolated by sophisticated reprocessing technologies and used in the construction of nuclear weapons. Theft of materials for nuclear weapons is next in Holdren's order of personal concerns, followed by sabotage and accidents in the nuclear fuel cycle. Routine emissions and exposure to radiation are at the bottom on his list of concerns.

How a Nuclear Power Plant Works

*M*uch of the operations of a nuclear power plant is similar to a coal-fired power plant (see diagram below). Generating energy from nuclear fuels, however, is fundamentally different in that a nuclear chain reaction rather than a chemical reaction is sustained within the fuel.

Fuel for conventional nuclear reactors consists of uranium-235, which is the only form (isotope) of uranium that is readily split (fissioned). Once uranium is separated from its ore, the uranium-235 content is only about 0.7%. A complex enrichment process is used to increase the uranium-235 content to

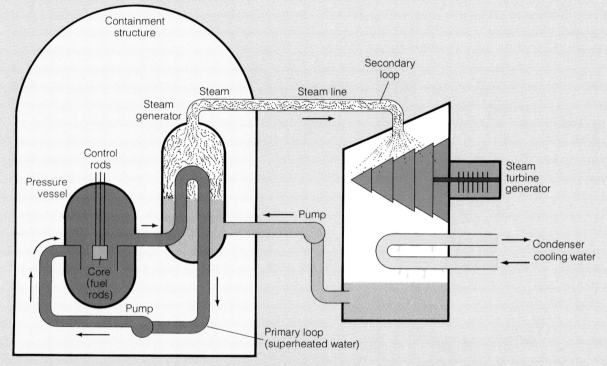

(a)

(b)

(*a*) Schematic diagram of a pressurized nuclear reactor and (*b*) exterior aerial view of a nuclear generating facility. The containment structure is the tall building in the center of the photo.

nearly 3%. After enrichment, the uranium is formed into ceramic pellets, and the pellets are encased in fuel rods, which are placed in fuel-rod assemblies.

The assembled fuel rods are transported to nuclear power plants. A typical power plant (which produces 1,000 megawatts of electricity) requires about 30 metric tons (34 tons) of nuclear fuel per year. On average, fuel rods remain in the reactor core for three years, and during that time uranium-235 undergoes *nuclear fission* (see drawing below). A neutron (an electrically

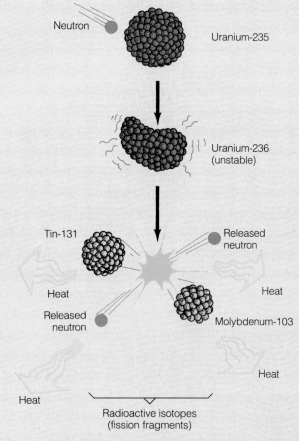

The process of nuclear fission is carried out in the core of a nuclear reactor. In the sequence shown here, uranium-235 is fissioned to form tin-131 and molybdenum-103. Two to three neutrons (an electrically neutral part of the nucleus of an atom) are released per fission event and continue the chain reaction. Fissionable materials yield tremendous amounts of heat per kilogram of material relative to other fuels.

neutral subatomic particle) strikes the nucleus of a uranium-235 atom, causing the unstable nucleus to split apart (i.e., undergo fission). Fission releases heat and produces approximately two hundred different types of radioactive by-products. Two or three neutrons are released per fission event and go on to split other uranium-235 atoms, thus maintaining a *chain reaction.*

Heat released by fission reactions in the core is absorbed by water, which is pumped through a closed, high-pressure primary loop to a steam generator. Pressurized steam is then fed through a secondary loop to steam turbines. Because the primary loop is isolated from the secondary loop, radioactive materials that leak from the fuel rods or that form when water passes through the core are retained within the primary loop. From this point on, a nuclear-powered electric generating plant is similar to a coal-fired plant: Steam turbines turn electrical generators. Heated condenser water is either directly discharged to surface waters or sent on to cooling towers or cooling ponds (chapter 14).

Nuclear power plants release only a very small amount of radioactivity into waterways through the cooling water and into the atmosphere in the form of gases. The Nuclear Regulatory Commission (NRC) requires that these releases be kept at levels that protect people who live near the perimeter of a nuclear power plant.

After the fissionable uranium-235 in the fuel rods has decreased from 3% to about 1%, they are considered to be spent; that is, they can no longer be used to generate electricity efficiently. The spent fuel-rod assemblies, which contain highly radioactive wastes, are removed from the reactor core and stored at nuclear power plants in pools of water that absorb the heat released by the radioactive materials. This method of temporary storage is used because no nuclear-fuel reprocessing centers or permanent storage sites exist for high-level radioactive wastes in the United States. However, many nuclear power plants are approaching the limits of their storage capacity. Addressing the spent-fuel storage problem is a serious concern that many operating nuclear power plants will have to face during the late 1990's. Without additional storage space, many would be forced to shut down. We discuss in more detail the health hazards of radioactive waste and the challenges of permanent radioactive waste disposal in chapter 20.

To reduce these and other risks, a number of nuclear scientists and international policy experts have proposed the establishment of a global commission to ensure that nuclear power plants are designed and operated more safely. Furthermore, they propose to inventory nuclear wastes held at nuclear power plants prior to reprocessing or disposal. The framework for such an organization already exists—the International Atomic Energy Agency. It was established under the Nuclear Non-Proliferation Treaty of 1970, which has been signed by 140 countries. However, in light of the fact that Iraq managed to purchase what it needed to start a nuclear weapons program, future treaties will need more political clout to prevent diversion of nuclear fuel wastes into nuclear weapons programs.

The prospects of global warming provides a powerful argument for reassessing the role of nuclear power, because it produces no greenhouse gases. For example, a worldwide conference (The Changing Atmosphere: Implications for Global Security) held in Toronto in 1988, called for a worldwide reduction in carbon dioxide emissions by 20%, that is, from 5.4 billion metric tons (6 billion tons) per year down to 4.4 billion metric tons (4.8 billion tons) per year. Other scientists have called for carbon dioxide emissions to drop even further to 3.6 billion metric tons (4.0 billion tons) per year by the year 2030. The technological means proposed to accomplish these reductions include a combination of (1) improved conservation, (2) a shift from coal to natural gas, (3) greater reliance on solar power and biomass combustion, and (4) nuclear power. None of these strategies alone are seen as sufficient to achieve the carbon dioxide reduction goals.

Scenarios proposed by scientists who are concerned about the potential for global warming include as many as 2,500 nuclear reactors operating worldwide to meet world energy demands by the year 2030. Clearly, the chance of a major accident increases with so many reactors operating, even though the probability of an accident at any one of them is very small. Furthermore, the public reacts to the consequences of a catastrophic accident and does not feel protected by the low probability of such an event.

There are now ways to reduce even further the low risk of a catastrophic nuclear accident. By incorporating advanced, passive safety features into future light-water reactors, the possibility of a catastrophic accident can be substantially reduced. Passive safety design features take advantage of natural laws to slow the fission process and cool the reactor core in the unlikely event that materials or cooling systems fail.

To further ensure safety, the nuclear industry now generally supports a standardized design for future plants. Thus, when a design problem arises that requires corrective action, the same measures can be prescribed for all plants of that design. Standard designs would also save money.

The operation of nuclear power plants demands a commitment to excellence by a utility. Although some utilities have had an excellent safety and operational record, others have had serious problems with faulty components, fraudulent record keeping, and lax operating procedures. The Kemeny Commission, in its investigation of the Three-Mile Island accident, found many questionable procedures in the operation of nuclear power plants. As a result, the industry has been required to implement new safety and evacuation procedures to improve safety. These procedures have increased the cost of nuclear power.

Thus, the problems that have arisen in the nuclear electrical utility industry have undermined the public's confidence in nuclear power. Unpredictable costs and erosion of public confidence have made utility managers unwilling to plan construction of new nuclear plants. In fact, as a consequence of an economic downturn in the late 1970s and early 1980s, several nuclear power plants under construction were never completed. Despite a projected increase in demand for electricity after the year 2000, not a single new order for a nuclear power plant has been placed since 1978 in the United States. (Canada placed an order in 1990.) As a result, no new nuclear electric generating facilities in the United States could come on line before the year 2010.

What are some of the consequences of abandoning nuclear power? Failure to restore the viability of nuclear power will mean higher electric rates and worsening of some global environmental problems. On the other hand, the problem of long-term disposal of nuclear waste remains unsolved (chapter 20). Some engineers and industrialists are concerned that failure by the United States to develop advanced technologies for nuclear generation of electric power would also mean loss of the opportunity to serve as a world leader in the further development of nuclear power. In contrast, other people believe that a greater emphasis on energy conservation would resolve energy problems. They argue that the more energy we conserve the less reliant we are on the problematic fuels, coal and uranium. However, energy experts confidently state that conservation is far from being the total solution. Therefore, difficult choices remain as we decide the mix of coal and nuclear resources that should be used to generate electricity in the twenty-first century.

Conclusions

Our examination of past consumption and future supplies of oil points out that we live in a unique oil era. Access to oil and its products (and to natural gas) has been relatively easy and therefore inexpensive, but as these resources diminish, prices will rise, and other energy sources will become economically competitive. Continued reliance on oil is likely to produce increasingly severe economic, political, and social disruptions, similar to the Arab oil embargo and the Persian Gulf War. The challenge for society today is to evaluate which means of providing future energy services makes the most sense from economic and environmental viewpoints. We must continue to assess the most efficient means of supplying our society with the safest, cleanest, and most reliable sources of energy possible. A government policy aimed at these goals is long overdue. In addition, we must keep in mind the time frame for switching from one technology to another; such a conversion will require at least one or two decades.

Finally, it has become very clear that meeting the energy needs of the world will require a worldwide cooperative effort. Understanding and tolerance of different cultures must grow as the people of the world work out agreements that achieve greater equity in the distribution of the resources necessary to meet global energy needs. Without such cooperation, nearly all nations will experience a decline in their standard of living, and many less-developed countries will slip even further down the economic ladder. Social and economic disruptions such as embargoes and wars over energy resources may become more common and more severe. In chapter 19, we evaluate ways to conserve energy and alternatives for supplying energy services, as well as the consequences of implementing these technologies.

Key Terms

nonrenewable energy
  resources
giant oil fields
liquefied natural gas (LNG)
eminent domain
synthetic fuels (synfuels)
coal gasification
low-, medium-, and high-BTU
  gas
substitute natural gas

in situ recovery
coal liquefaction
oil shale
kerogen
tar sands
light water reactor
fast breeder reactor
renewable energy resources
biofuels
fuel cycle

Summary Statements

The past century has seen the rapid development of energy-consuming technologies, powered primarily by fossil fuels. Crude oil production in the United States has been decreasing since 1985 and hence, the nation must import more oil. The cost of imported oil is a major reason why the United States continues to have a negative balance of payments.

Natural gas consumption is increasing as more of it is used to generate electric power. This clean-burning fuel releases less carbon dioxide per BTU of energy released than either coal or oil. Like coal, most of the natural gas consumed in the United States is domestic. Unlike coal, natural gas reserves within U.S. boundaries are in relatively short supply. Known reserves will probably last 50-60 years at present consumption rates.

Coal reserves in the United States are extensive and can last several centuries at present consumption rates. More coal is likely to be used to generate electricity in the future, but the cost will probably rise because of air-quality standards mandated under the Clean Air Act amendments of 1990.

Both oil and natural gas supplies can be augmented through techniques for gasification and liquefaction of coal and oil shale. These synfuels can be produced either after the resource is mined or in situ. Only minor amounts of synfuels are produced today because synfuels are still more costly than natural gas and petroleum liquids from conventional sources. World prices for oil and natural gas will have to rise before synfuel production can contribute more significantly to domestic energy supply.

Nuclear energy can be used to augment production of electricity. Studies indicate that the nuclear fuel cycle is a safer means of generating electricity than the coal cycle, but it has other social risks, such as the chance of a catastrophic accident and the diversion of radioactive wastes for the production of nuclear weapons.

Projections indicate that the United States will need additional electric generating capacity after the year 2000. No new nuclear power plants are on order; thus, none can come on line before the year 2010. Loss of public confidence, high construction costs, and existing regulations have halted new plant construction.

Questions for Review

1. Distinguish between renewable and nonrenewable energy resources.
2. Is energy from the sun a renewable resource? Explain your response.
3. What major world political events have had a major impact on the price of oil?
4. Classify the United States as either a net importer or net exporter for each of the fossil fuels.
5. Give four different viewpoints on the role of oil in modern society. What is the major premise of each?
6. From an air-quality perspective, which one of the fossil fuels is the most desirable? Explain your answer.
7. Assess the prospects for U.S. imports of natural gas from the Middle East.
8. Speculate on which one of the fossil fuels will run out first.
9. What are synfuels? What raw materials can be used to manufacture them?
10. What is oil shale? Where are the richest deposits located? What are some of the factors that limit its use as an energy resource?
11. We often hear that a particular fossil fuel will last *X* number of years. Why are such statements oversimplifications?
12. Describe the rate at which you use toothpaste from a tube. Does your rate of use depend on how much is left in the tube? Does an analogous situation apply to fossil fuel reserves?
13. Describe two different ways to produce synthetic natural gas and synthetic liquid petroleum products.
14. Is it possible for the United States to meet its energy demands without experiencing conflicts with other nations? Explain your response.
15. Compare and contrast the hazards of the nuclear and coal fuel cycles for the generation of electric power.
16. What fraction of the energy used in the United States in 1990 was supplied by renewable energy sources? Which renewable resource supplied the greatest amount of energy?
17. How is natural gas transported (two methods) from countries that export it? Where are the world's largest reserves of natural gas?
18. What special significance do giant oil fields have? Where and when was the last giant oil field discovered in the United States?
19. The price of oil is expected to increase toward the end of the 1990s. List factors that might contribute to this price rise.
20. List the environmental impacts of a greater reliance on coal as a source of energy.

Projects

1. Visit a local fossil fuel power plant. Determine how air pollutants, thermal pollution, and solid wastes are handled.
2. Visit a nuclear power generating plant. Determine what pollutants they must monitor and how their spent fuel is handled.
3. Assume that you are in charge of building a new electric power plant. What criteria would you apply in selecting the type of plant and its location?

Selected Readings

Burnett, W. M., and S. D. Ban. "Changing Prospects for Natural Gas in the United States." *Science* 244(1989):305-10. Proposes that natural gas should carry a larger burden in meeting near-term energy demand.

Chandler, W. U., A. A. Makarov, and Z. Dadi. "Energy for the Soviet Union, Eastern Europe, and China." *Scientific American* 263(Sept. 1990):120-27. A look at the special challenge facing countries that have largely ignored the environmental consequences of industrialization and how they might move forward.

Cohen, B. L. *Before It's Too Late: A Scientist's Case for Nuclear Power.* New York: Plenum Press, 1983. A comparision of the risks people face upon implementing various technologies.

Davis, G. R. "Energy for Planet Earth." *Scientific American* 263(Sept. 1990):54-63. A historical look at energy consumption in industrialized and less-developed nations and how we might go forward in an ecologically enlightened manner.

Flavin, C. "The Case Against Reviving Nuclear Power." *Worldwatch* July/August, 1988, 27-35. Presents arguments against nuclear power as a future means of meeting electricity demands.

Fulkerson, W., R. R. Judkins, and M. K. Sanghvi. "Energy from Fossil Fuels." *Scientific American* 263(Sept. 1990):128-35. A review of worldwide fossil fuel resources and an examination of some techniques for increasing the efficiency of fossil fuel use and minimizing their environmental impacts.

Golay, M. W., and N. E. Todreas. "Advanced Light-Water Reactors." *Scientific American* 262(April 1990):82-89. A discussion of the technology envisioned for the next generation of nuclear reactors.

Hafele, W. "Energy from Nuclear Power." *Scientific American* 263(Sept. 1990):136-45. A discussion of international issues that must be addressed if nuclear power is to play an expanded role in meeting future energy demands.

Hirsch, R. L. "Impending United States Energy Crisis." *Science* 235(1987):1467-73. An examination of why the United States will have serious energy problems in the future.

Levi, B. G. "International Team Examines Health in Zones Contaminated by Chernobyl." *Physics Today* 44, no. 8 (1991): 20-22 (Part 1). Review of the Chernobyl accident and projected health effects.

National Academy of Sciences. *Energy: Production, Consumption, and Consequences.* Washington, D.C.: National Academy Press, 1990. Examines past errors in meeting energy demand and how future demand might be met.

National Academy of Sciences. *Fuels to Drive Our Future.* Washington, D.C.: National Academy Press, 1990. A look at how fuels can be manufactured from existing energy reserves.

Pollock, C. *Decommissioning: Nuclear Power's Missing Link.* Washington, D.C.: Worldwatch Institute, 1986. A discussion of what is to be done with nuclear power plants when they are worn out.

Reddy, A. K. N., and J. Goldemberg. "Energy for the Developing World." *Scientific American* 263(Sept. 1990)110-19. A discussion of some of the special considerations in meeting the energy needs of less-developed nations if they are to raise their standard of living.

Taylor, John J. "Improved and Safer Nuclear Power." *Science* 244(1989):318-25. Examines designs for the next generation of nuclear power plants.

Chapter 19

Meeting Future Energy Demands

As an alternative to nuclear and fossil fuel energy sources, this device captures and concentrates solar radiation.
© Greg Vaughn/Tom Stack & Associates

In 1980, Swedish citizens voted to phase out their nuclear power plants by 2010. Some people saw this move as a means for the nuclear power industry to gain time to establish a good track record. If the industry could show that nuclear power was safe and dependable, it would pave the way for rescinding the referendum. However, the Chernobyl accident in April of 1986 further solidified attitudes against nuclear power in Sweden. How a country with little or no coal, natural gas, or oil will find a new way to generate 50% of its electric energy supply remains an open question.

All nations must balance their supply and demand for energy. How to achieve this balance without economic, social, and environmental upheaval is, of course, a major concern. Because most energy technology has a working lifetime of 20-40 years, today's decisions on energy policy will affect the world energy picture and the environment for decades to come.

There are three basic approaches to providing future energy services: technological solutions, social/economic solutions, or both. In the past, technological solutions usually addressed the supply side of the supply/demand equation by developing new technologies to enhance the energy supply. For example, as demand increased for energy in the 1970s and 1980s, and domestic oil supplies in the United States dwindled, scientists and engineers directed major efforts toward research and development of coal-burning technologies, nuclear fission for electric power, production of synthetic fuels, and (to a lesser extent) solar and wind technologies.

Social/economic solutions are usually directed at the demand side of the supply/demand equation. Social solutions reduce demand by changing behavior patterns; for example, encouraging people to use public transportation or carpools, to lower thermostats in winter, and to raise thermostats in the summer reduces energy demand. Usually, however, economic considerations determine the demand for energy and how that demand is met (fig. 19.1).

Technological solutions continue to predominate, probably because policymakers believe that they are easier to achieve. Most people prefer policies that allow them to continue their present life-style; they do not readily accept strategies that might inconvenience them and reduce their access to the good things in life.

The political structure of a country influences the way that country meets its demand for energy services. The open, decentralized, democratic form of the U.S. government permits citizen groups to block the construction of large-scale energy projects such as nuclear power plants, coal-fired power plants, or hydropower facilities. In countries such as France, where the political structure is less democratic, large, centralized energy projects are more readily implemented to meet increased demand. Thus, France has reduced oil imports by relying more heavily on nuclear power, while the United States is increasingly dependent on imported oil, and its nuclear power industry is essentially stagnant.

Every country is faced with major energy policy decisions because fossil fuels (particularly oil) are in tighter supply and

Figure 19.1   In the long run, homeowners can realize a saving in energy cost by investing in sufficient insulation.
© Wendy Neefus/Earth Scenes

because many citizens reject nuclear power. Consequently, other means of furnishing energy services are now being intensively investigated. Alternative energy sources range in scale from solar hot-water systems for homes to hydroelectric facilities large enough to inundate a small state. Some alternatives can be implemented now; others require one or more decades of further development. In this chapter, we examine several strategies to provide future energy services, some of the major obstacles to their implementation, and their environmental consequences.

Conservation of Energy

Although it is not as dramatic as the construction of large power-production projects, energy conservation can be an essential part of a nation's energy policy. In fact, we can view conservation measures as a source of energy whose contribution is every bit as real as new sources of coal, oil, natural gas, or nuclear power. Simply put, conservation reduces the amount of energy that must be produced to meet our demands.

In the late 1970s, energy experts used energy consumption trends to predict that the United States would consume 91 quads of energy in 1991. In fact, the United States consumed 82 quads that year. A nine-quad reduction in demand was largely due to conservation measures adopted in all segments of the economy throughout the 1980s. Because of energy savings between 1980 and 1990, the United States saved an estimated $300-500 billion in imported oil.

Conservation not only saves money, it produces many other benefits. The extraction and combustion of fossil fuels release large quantities of pollutants into the environment (chapters 14, 16, 17, and 20); thus, reduced demand for these fuels translates into less environmental contamination. Conservation also reduces our reliance on foreign oil. Consequently national security improves because we do not have to send our troops to foreign soil to stabilize a region and guarantee the supply of abundant and cheap oil. Furthermore, when fewer dollars are spent on imported oil, more dollars

Table 19.1 Energy consumption and cost for an average U.S. household of four (1990)

Use	BTUs/Year (in millions)	Percent	Estimated Expenditure
gasoline (transportation)	125	47.4	$1,100
space heating (gas)	70	26.6	390
water heating (gas)	27.5	10.4	150
air-conditioner (central)	12.2	4.6	280
refrigerator	7.7	3.0	170
freezer	6.1	2.3	140
other appliances	15	5.7	325
total	263.5	100	$2,555

remain in the United States to help solve our social and environmental problems.

Most conservation measures are low-risk investments: They perform well and have short pay-back periods—some less than a year. In fact, some conservation measures are essentially cost-free because they only involve a change in personal habits. Turning off excess lighting, resetting thermostats, and unplugging a seldom-used refrigerator are just a few simple energy conservation measures that each of us can take in our own homes. Compared to the six or more years required to plan and build large power plants, conservation measures can be implemented quickly—often with a series of small purchases instead of a single huge investment.

Building and Home Measures

In the United States in 1991, residential demands accounted for 36.2% (29.6 quads) of total energy used. To conserve energy in our homes effectively, we must make changes where they count most. Table 19.1 illustrates that heating and cooling account for the greatest amount of energy used in our homes. They are therefore the focal points for major savings in energy consumption.

Properly weatherized buildings greatly reduce demands on both heating and cooling systems. Even today, many homes are inadequately insulated, caulked, and weather-stripped, and many of them are equipped with single-glazed windows (that is, windows that have a single sheet of glass). Such houses not only have high rates of heat loss during the winter, but they also gain heat rapidly during summer. Hence, the cost of adding insulation to poorly weatherized homes is often recovered within one or two years. The performance of insulation methods for structures like doors, windows, walls, and ceilings is measured in terms of **R-ratings.** The higher the R-rating, the better the insulating ability (see Special Topic 19.1).

Installing a more efficient furnace can also save energy. New pulse-combustion furnaces are 90-95% efficient, in marked contrast to the 60-65% efficiency of most conventional natural gas and oil furnaces. In moderate climates, heat pumps can also accomplish space heating more economi-

cally, as compared to electric-resistance systems. Heat pumps can save 25-50% over electrical-resistance heating if outdoor temperatures remain above about -7° C (20° F).

Passive energy-saving features that reduce demand for space heating and cooling can easily be included in new buildings. Builders and architects can incorporate features such as overhangs, large windows that face the south (in the northern hemisphere), and movable, insulated window panels. (We discuss these features again later in this chapter.) Older homes can be retrofitted with some of the same energy-saving features, but the cost can be considerable because this often requires extensive remodeling.

Most electric utilities offer free home inspections and specific suggestions for energy conservation. They answer such questions as how much insulation to add, where to add it, its cost, and its pay-back time.

The heating of water consumes the second-largest amount of energy in our households (table 19.1). As a general rule, we should use as little hot water as possible because heating water requires considerable energy. Short showers, full dishwashers and washing machines, and washing clothes at the lowest water-level setting feasible should be the rule. Furthermore, the thermostat on water heaters is usually set much higher than necessary. Lowering the temperature setting to between 110° and 120° F will save more energy, as will covering the water heater with an insulation blanket.

Large appliances are also major consumers of energy (table 19.2). Improving the operating efficiency of such household appliances as refrigerators and freezers reduces significantly the overall demand for electricity. New models of refrigerators contain more insulation, more efficient motors, and improved cooling systems. According to federal estimates, such models consume 45% less energy than many models now in use. Because energy prices are likely to rise in the future, energy-efficient appliances are good long-term investments, even though their purchase price is typically higher than that of conventional, energy-hungry appliances.

Other major energy-consuming appliances in homes (see table 19.2) include clothes dryers, air conditioners, electric stoves, furnace motors, and dishwashers. Using these appli-

What Are R-ratings?

R–ratings are a means of defining how well a barrier, say a wall, prevents heat losses. (One of the consequences of the second law of thermodynamics is that heat only flows in one direction, i.e., from a warm object to a colder one.) Devices with high R-ratings lose or gain heat slowly.

R-ratings are measured relative to the ability of a single pane of glass to reduce heat flow. A pane of glass is defined to have an R-rating of 1. Double-paned windows, often called *thermopanes,* have an R-rating of 2, which means they transmit heat away from a warm room at one-half the rate of a single pane of glass. Thus, thermopane windows are said to have twice the insulating capacity. A ceiling with an R-rating of 18 would lose heat at 1/18th the rate of that of an equivalent area protected by single-paned glass. These examples show that

energy savings increase as the R-rating of a material increases. There is a point of diminishing return however, where the additional cost of increasing the R-rating can no longer be justified by the small incremental energy savings.

R-ratings are particularly important for windows, which usually have the lowest R-rating of any part of a building. Research on window construction methods, however, over the past decade has increased substantially the R-ratings of windows from 2 for typical thermopanes to as high as 10 or 12 for new, high-tech windows. With these windows, new building designs can incorporate large window areas without costly heat losses. R-ratings, however, fail to predict rates of heat loss through cracks and air leaks caused by poor construction or aging.

Table 19.2 Average annual energy consumption by home appliances	
Appliance	**Kilowatt Hours**
water heater	4,200
refrigerator	2,250
freezer	1,820
clothes dryer (electric)	990
air conditioner (room)	860
kitchen range (with oven)	700
furnace fan motor	650
dishwasher	360
color television	320
microwave oven	190
coffee maker	140
washing machine (electricity only)	100
vacuum cleaner	45
toaster	40
hair dryer	25
clock	17

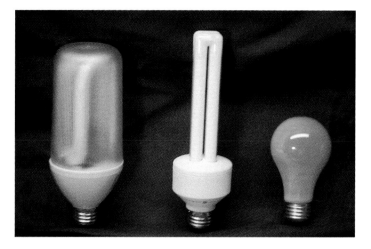

Figure 19.2   The two energy-saving bulbs on the left produce as much light as the conventional incandescent bulb on the right, but they use only one-fourth as much electricity.

ances in an efficient manner can further increase energy conservation. For example, it requires less energy to heat food in a microwave oven than it does in (or on) a conventional stove.

We can save considerable energy by our choice of light bulbs. Ordinary fluorescent lights emit nearly four times as much light per unit of energy as incandescent light bulbs. Recently an even more energy-efficient fluorescent light bulb

that screws into sockets used for ordinary incandescent bulbs has become available (fig. 19.2). This easy retrofit eliminates the need to buy and install the special fixtures required to operate older fluorescent lights. One of these new 40-watt fluorescent lights delivers more light than a 150-watt incandescent bulb but uses one fourth the energy. **Wattage** is a direct measure of the rate of energy consumption by a device. Although these lights cost more, they also last nearly ten times longer. Experts estimate that replacing incandescent bulbs with highly efficient fluorescent bulbs would eliminate the need for twenty-nine power plants across the United States.

High-efficiency modern homes, sometimes called *smart houses,* are well insulated and operated by computer-controlled

Figure 19.3 High-tech homes, called *smart houses*, use a combination of active and passive features to keep energy use to a minimum: (1) Energy-saving appliances; (2) solar greenhouse; (3) well-insulated walls and ceilings; (4) high R (low E) windows; (5) heat pump; (6) central computer to control energy-consuming devices; (7) energy-saving landscape designs.

monitoring systems (fig. 19.3). These systems automatically make adjustments for heating, cooling, and lighting in accordance with optimal energy use. For example, the computer automatically lowers the thermostat at night and turns off lights if no one remains in a room after a certain amount of time. Smart homes also include passive solar features such as south-facing windows. These superinsulated, high-tech homes cost more than ordinary homes, but their lower operating costs allow a buyer to recover the difference in five to seven years. Even so, most people are reluctant to buy them.

How much energy savings can we realize in our homes? A recent study conducted by the Electric Power Research Insti-

tute (EPRI) addressed this question. Table 19.3 shows the range of potential improvement for various uses of electricity in the residential sector of the U.S. economy. These estimates are based on the case where all electrical equipment is instantly replaced in the year 2000 with the most efficient models available today. Overall, EPRI predicts that residential electrical savings of 27-46% are possible with this optimistic scenario. Although no one expects such dramatic savings to be realized, the analysis demonstrates both the potential energy savings available to the individual and the goals that electrical utilities and consumers can strive to attain.

Table 19.3 Potential reductions in residential electricity consumption (1987–2000)

Residential Use	Reduction Possible %	Types of Improvement Required
space heating	40–60	weather-stripping; caulking; ceiling, floor, and wall insulation; increased use of heat pumps and solar heating
water heating	40–70	better-insulated tanks; low water-flow devices on showers
central air-conditioning	46–50	more efficient models
room air-conditioning	35–56	more efficient models
dishwasher	10–30	no-heat drying cycle; reduce volume of hot water used
cooking	10–20	better-insulated ovens; reflective pans
refrigeration	40–60	more efficient, better-insulated models
freezers	55–60	more efficient, better-insulated models
lighting	20–40	use compact fluorescent bulbs
other appliances	14–29	more efficient models

Source: Estimates from "Efficient Electricity Use: Estimates of Maximum Energy Savings," Electric Power Research Institute, March 1990, p. 39.

Today, many electric utilities find that it costs them less to continue to operate at existing production levels than it would to increase their energy-production capacity. Thus, they aggressively pursue energy savings by subsidizing conservation measures. For example, in 1989, the New England Electric System together with the Conservation Law Foundation formulated a plan that encourages energy conservation and still allows the utility to realize a profit. One subsidiary of the utility is spending $65 million in one fiscal year to install high-efficiency lighting, cooling, and heating systems in selected industries, small businesses and new and existing homes. The loss of energy sales by the utility will be compensated for by a rate hike and a 15% profit incentive for the utility.

The electric utility industry in the United States expects to spend $1 billion per year through the end of the century on conservation measures as a means of managing demand. The most generous subsidies for conservation come from utilities whose demand exceeds their productive capacity and who must therefore buy more costly energy from a neighboring utility. In regions where conservation efforts reduce demand below production capacity, utilities are likely to request permission from rate-governing boards to raise their rates to ensure satisfactory returns for their owner-investors.

Many of the energy conservation measures described for homes can also be used in commercial and office buildings. Because lighting is a major percentage of commercial electric bills, particular attention should be paid to energy savings in this sector.

It is important to realize that conserving electricity pays triple dividends. Every BTU of electric energy that is saved at the point of use reduces total energy consumption by approximately 3 BTUs. Why such a major savings? Recall from chapter 3 that the second law of thermodynamics applies to all energy-transforming systems. In accordance with this law, power plants are inefficient and convert only about one-third of the energy that is in fossil or nuclear fuels to electricity. The remaining two-thirds is unavoidably lost in the form of heat at the plant. Much smaller losses also occur along transmission lines. We can also conclude that on an energy-equivalent basis, electricity is the most expensive form of energy we routinely purchase (chapter 18). The inherent inefficiencies in generating electricity provide another financial incentive to conserve electricity as much as possible.

Transportation Measures

In the United States in 1991, transportation accounted for 27.3% (22.2 quads) of the energy used, most of which is attributed to the automobile. The greater individual mobility provided by the automobile has influenced the shape of our cities, increased the average commuting distances between home and work, and molded (to a great extent) the patterns of our individual lives. The automobile is one of our most important symbols of personal freedom; it allows us to go anywhere at any time without being regimented by bus or train schedules. Consequently, in the United States, cars and trucks travel 2.4 trillion kilometers (1.35 trillion miles) each year and, in the process, burn more than 430 billion liters (114 billion gallons) of fuel. Gasoline accounts for over 60% of total transportation energy demand and over 15% of the nation's total energy demand. The amount of gasoline consumed in the United States has increased about 20% since the early 1970s and is the major reason for the increase in oil imports (fig. 19.4).

Five strategies to reduce energy consumption in the transportation sector follow:

1. Improve the efficiency of vehicles
2. Operate vehicles more efficiently
3. Increase the load carried by each vehicle
4. Shift travel from less-efficient to more-efficient modes
5. Reduce the total amount of travel

We now evaluate these five strategies.

Without improvements in the energy efficiency of vehicles during the past twenty years, our demand for petroleum-related fuels would have increased even more than it did. The

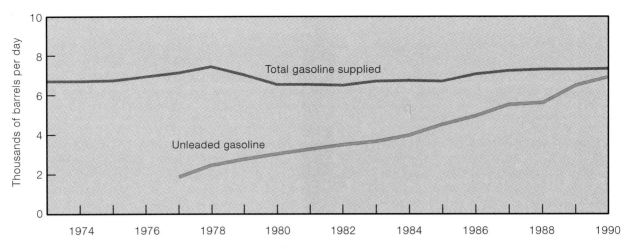

Figure 19.4   Gasoline consumption in the United States between 1973 and 1990. Note the gradual switch to unleaded gasoline.
Source: Energy Information Agency.

fleet of cars on the road today is 50% more fuel efficient than it was two decades ago. Some of these improvements can be traced to the passage of the Energy Policy and Conservation Act, which required automobile manufacturers to produce passenger cars with an average fuel economy of 12 kilometers/liter (27.5 miles/gallon [mpg]) by 1985. Not only was this goal never achieved, but the EPA bowed to political pressure and backed down on mandated fuel efficiency standards. Consequently, during the period from 1989 to 1990, the average fuel efficiency of new cars actually dropped by about 4%. The fuel efficiency of Japanese cars sold in the United States during the same period dropped even more.

Improving the fuel economy of vehicles is a logical goal; it would reduce gasoline consumption and thus cut demand for imported oil and reduce air pollution. Energy experts in Congress's Office of Technology Assessment argue that average automobile mileage could reach 20 kilometers/liter (40 mpg) with existing technology. If consumers are willing to accept smaller cars, fuel efficiencies approaching 80 mpg are possible. However, there are major stumbling blocks.

Automobile manufacturers are reluctant to market more fuel-efficient cars because fuel costs are small compared to the overall operating costs of an automobile. Another stumbling block to increasing mileage is safety. The amount of fuel that a motor vehicle burns depends on its weight. Thus, to meet mandated fuel-economy standards, automobile manufacturers have been downsizing new cars. Unfortunately, smaller cars are not as safe as larger cars. According to safety experts, occupants of smaller cars are 3.4 to 8 times more likely to die in an accident than are occupants of a larger car. This negative factor can be countered to some extent by adding additional safety equipment, such as air bags.

Recent trends in car sales suggest that many consumers prefer cars that give them performance and safety to cars that provide better fuel efficiency. In fact, in 1992 U.S. car buyers ranked fuel efficiency eighth on their list of priorities when choosing a new vehicle. Futhermore, they are more attracted to options that decrease fuel efficiency (e.g., high performance engines, anti-skid brakes, and four-wheel drive). Thus,

some experts suggest that effective conservation can only be achieved by raising the price of gasoline to reflect its true cost. (True cost includes negative externalities such as air-pollution damage to human health and the environment and costs associated with the military protection of oil-producing regions.) By not adding negative externalities to the price of gasoline at the pump, policy makers encourage greater consumption and thwart conservation efforts. Furthermore, these artificially-low prices provide few economic incentives for new domestic drilling ventures or synfuel production (chapter 18). Since 1981, the number of exploratory wells drilled in the United States has plummeted 68%. All of these consequences lead to an ever increasing U.S. dependence on oil imports and further exacerbates the U.S. negative balance of trade. Ultimately, fewer dollars remain in the nation to support domestic programs.

A second conservation strategy is to encourage people to operate their motor vehicles more efficiently. Reducing speed on open highways lowers the rate of fuel consumption. The nationwide highway speed limit of 88 kilometers (55 miles) per hour that was imposed in response to the 1973 Arab oil embargo may have saved 200,000 barrels of oil per day. (It has also been credited with saving nearly 200,000 lives.) Today, however, the speed limit along most stretches of the interstate highway system is 100 kilometers per hour (65 miles/hour). Furthermore, speed limits are minimally enforced in most states, and most people drive at speeds similar to those of the early 1970s. Table 19.4 will quiz your knowledge of other measures that all drivers can use to conserve fuel.

Increasing the load carried by a vehicle is a third way to reduce energy consumption. Almost everyone in the United States drives to work alone, and greater use of car and van pools, buses, and urban rail systems would diminish fuel consumption. In van pooling, an employer makes a van available to perhaps ten employees, who use it exclusively for commuting to and from work. These measures not only conserve energy, they reduce traffic congestion, air pollution, and parking space requirements.

Table 19.4 What can we do to save gasoline?

To assess your understanding of ways to improve the fuel efficiency of automobiles, answer each of the following questions true or false:

1. Automatic transmissions are more efficient than manual transmissions because the car is automatically kept in the correct gear at all times.
2. Steel-belted radial tires can save as much as 3% on gasoline over regular tires.
3. You use more fuel letting your car idle for two minutes than you do by turning off the engine and restarting when you are ready to drive again.
4. From a fuel-efficiency perspective, it is best to warm your engine for at least one minute each morning before driving your car.
5. In order to achieve a quick, low-fuel start each time, it is a good idea to "rev up" the engine just before turning it off.
6. Because of their greater power, it is more efficient to use lower gears while driving rather than higher gears.
7. In late-model cars, you will save fuel during warm weather by closing windows and turning on the air conditioner when driving over 30 MPH.
8. Many drivers report savings of from $300 to $1,000 per year by riding to work or going shopping in a carpool rather than driving alone.
9. You can actually use less gasoline by taking a longer route with fewer stops and hills than a shorter route with more stops and hills.
10. Underinflated tires will cause an auto to consume 5% more fuel than properly inflated tires.

(Answers: 1–F, 2–T, 3–T, 4–F, 5–F, 6–F, 7–T, 8–T, 9–T, 10–T.)

Source: Excerpted from "Auto Energy Efficiency Quiz," *Wisconsin Energy News,* February 1984.

Table 19.5 Energy use by various urban and intercity travel modes

| | Urban Travel | Intercity Travel | |
| | Total energy* BTUs per passenger mile | BTUs per passenger mile | BTUs per seat mile |
Mode			
automobile (compact)	—§	1,900–2,700	1,000–1,400
automobile (average)	14,000‡	2,400–7,600	1,200–2,000
car pool	5,400	—†	—†
van pool	2,400	—†	—†
bus	3,100	1,100–1,800	300–600
heavy rail	6,500	1,800–3,700	400–1,900
rail (commuter)	5,000	1,400–3,200	700–1,300
light rail (streetcar)	5,100	—†	—†
aircraft (wide body)	—†	4,800–6,100	2,000–4,100
aircraft (average)	—†	5,600–9,600	2,600–6,100

Source: *Reducing U.S. Oil Vulnerability,* U.S. Department of Energy, 1980.
*Includes all energy—construction, maintenance, and operation, including terminals over a lifetime.
†Generally not used for stated purposes.
‡The value for average occupancy is 10,200.
§Not available.

People can also switch to more efficient modes of transportation such as buses and computer-controlled commuter trains. In comparing modes of transportation, it is important to consider the amount of fuel that a given system uses to transport one person a specified distance, usually 1 mile. Table 19.5 lists the energy efficiencies for different modes of travel for urban and intercity trips. Surprisingly, modern rail systems, such as San Francisco's BART system and Washington D.C.'s METRO, are more inefficient than old rail systems. However, rail transportation becomes more efficient as more people use it. (The high cost of constructing rail transit systems make them the most heavily subsidized mode of travel per passenger kilometer. Furthermore, engineers and planners are finding that it is more costly to upgrade rail transportation than to expand bus service.) If fuel efficiency of automobiles improves over the long run, intercity travel will be more energy-efficient by car than by rail. However, buses will remain the most energy-efficient mode of passenger transportation.

Finally, we can conserve energy by cutting travel. Fewer trips to the grocery store, walking or using bicycles as often as possible, and finding a residence close to work all reduce travel.

Industrial Measures

In 1991, industry in the United States used 29.8 quads of energy, or 36.5% of total U.S. energy consumption. Improvements in industrial energy efficiency fall into four general categories:

1. Better housekeeping, such as shutting off equipment that is not in use
2. Increasing efficiency of operating equipment, such as installing computerized boiler controls

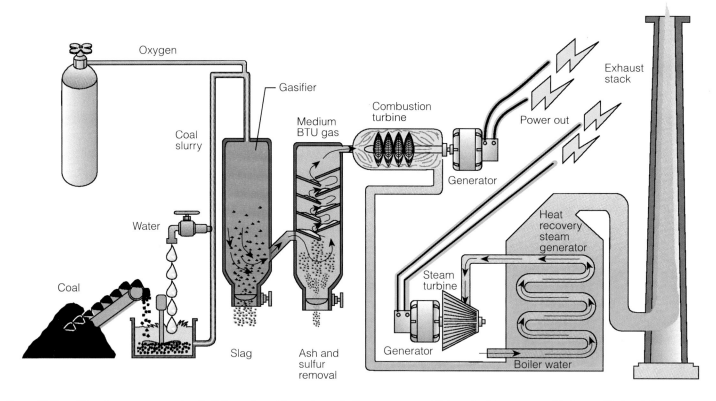

Figure 19.5   Electric power generating facilities such as this are dramatically more energy efficient. In this system, coal is first gasified and then converted to electricity using two separate turbines. Such systems are called *combined cycle* systems. The waste heat from electricity generation is used for an industrial process, an arrangement called *cogeneration*.

3. Shifts to new, more energy-efficient processes, such as using gas turbines to generate electricity
4. Manufacture of less energy-intensive products (products whose manufacture consumes less energy), such as automobile parts made of plastic instead of metal

The first two measures usually have short pay-back periods, and most industries have already implemented them. In contrast, plans to change manufacturing procedures and to manufacture new products that are less energy-intensive often take years to develop and implement. Industrial experience has shown that for each doubling of productivity, improvements in efficiency cut energy costs by approximately 20%. Thus, greater energy efficiency is closely tied to an industry's ability to increase productivity or improve quality. During the past two decades, such changes have allowed industry to increase output substantially, while reducing total energy consumption. Further improvements are possible, but they will not be as easy as those of the past and pay-back periods will be longer than the 2-4 years that industry has typically used as a guide in the past.

Industries and utilities can improve energy efficiency through a cooperative strategy known as **cogeneration**—a process whereby a useful form of energy is produced, and the waste heat from that same process is put to another use. For example, a utility uses its steam to produce electricity and sells its waste heat to a nearby industry for use in its production facilities. The technology for cogeneration of electricity and

process heat is well established. Cogeneration processes can raise the overall efficiency of electricity production from coal from 37% if only electricity is produced to 60-76% if the waste heat is used in an industrial process. The federal government actively encourages cogeneration under the Public Utility Regulatory Act (PURA) of 1978. That law provides a number of incentives to cogenerators, including federal tax breaks.

The electric utility industry is interested in several new technologies that produce electricity more efficiently. One new technology, called *integrated coal gasification combined cycle*, burns syngas in a gas turbine (fig. 19.5). The hot exhaust gases from the turbine are directed into a boiler to generate steam to run a second turbine. Several utilities around the world are testing this system, including Southern California Edison, which is planning two new plants. These systems produce lower levels of nitrogen oxides (a major air pollutant described in chapter 16) and can be quickly built. Such systems can reach a conversion efficiency near 52%. While these gains in efficiency may sound small, the financial rewards to industry are substantial, and consumers may see electricity rates climb more slowly.

Because fossil fuels will undoubtedly carry much of the burden of meeting our near future energy needs, we must not forget that research into their more efficient use could pay significant dividends to consumers and the environment. We now consider the potential of such alternative energy sources as solar, geothermal, and nuclear fusion.

Solar Power

Solar power is a renewable energy source that will last as long as the sun itself—several billion years. Collecting solar energy is not without cost, but the energy itself is free. Despite these significant advantages, it was not until 1974 that Congress provided funds to encourage large-scale research and development of solar-powered systems for heating, cooling, and generating electric power. Today, public funding for solar power remains small relative to government subsidies for the fossil fuel and nuclear industries.

Solar energy occurs in many forms: electromagnetic radiation (chapter 15), organic products of photosynthesis (chapter 3), wind (chapter 15), and running water (chapter 13). In the following sections, we examine the potential of each form of solar energy to reduce our reliance on fossil fuels.

Collecting Sunlight

A variety of devices powered directly by solar energy are now available, including such familiar items as calculators, watches, and even patio walkway lights. When exposed to the sun, these devices store small amounts of solar energy in rechargeable batteries and then release it upon demand. In remote regions where electricity is not available or is too expensive to bring in, direct collection of solar energy is economically competitive. For example, two utilities in Wisconsin donated a solar power system to a state park because bringing in power lines would have doubled the cost. In addition, this meant that the environment was not disrupted by construction of power lines. Many more similar opportunities exist around the world.

The total amount of solar energy that falls on the United States each year is approximately six hundred times greater than the total amount of energy consumed in the United States during the same period of time. If 100% of the energy that falls on the roof of a small house (93 square meters, or 1,000 square feet) in a year could be collected and sold at average rates for electricity, its value would be approximately $9,000.

Collection of solar radiation, however, presents some problems. Unlike fossil and nuclear fuels, sunlight is diffuse (it is spread out over the surface of the planet) and its availability varies with cloud cover and season (chapter 15). For example, in some mid-latitude sites where fair weather predominates, a higher average annual amount of sunlight strikes the earth's surface than reaches the ground in equatorial regions that experience persistent cloudiness. Seasonal differences in intensity, however, are considerably less in tropical regions. Seasonal changes in the length of the day are especially troublesome in middle and high latitudes. In eastern Washington state, for example, which is a relatively sunny location, the average daily solar power varies from a low of 50 watts per square meter in December to a high of 340 watts per square meter in June—a sevenfold difference.

The average daily intensity of solar radiation that reaches the ground across the United States is given in figure 19.6 for December, the month of lowest irradiance, and June, the month of highest irradiance. The region with the greatest potential for solar power is the American Southwest where cloudiness is minimal and sunlight is intense.

Seasonal and daily changes in the sun's position in the sky reduce the efficiency of solar collection devices, but this problem can be partially solved by tilting solar collectors so that they remain perpendicular to the solar beam. More expensive systems use computer-driven machinery to track the sun.

Solar energy can be collected by passive or active systems. **Passive solar systems** use solar energy directly, without concentrating it or converting it into another form of energy. Examples of passive systems include a greenhouse, a solarium, or a large window that faces south. **Active solar systems** convert the sun's energy into another form, such as heat or electricity. Some active systems, including all those that produce electricity, focus (concentrate) the sun's energy before converting it to another form. Active systems are generally attached to or located next to the system that uses the energy.

Solar Low-Temperature Systems

Solar low-temperature systems include both active and passive devices that convert solar radiation to heat that raises the temperature of a conducting fluid up to 100° C (212° F). Low-temperature active solar systems collect sunlight for space and water heating in buildings (e.g., homes, apartments, schools, and businesses) that can make use of the 65°-100° C (150°-212° F) temperatures achieved by flat-plate panels (fig. 19.7). Because these panels do not provide temperatures high enough to produce steam, they are not adequate for most industrial purposes.

Solar panels are framed panels of glass that trap solar energy (fig. 19.7). In most systems, solar radiation passes through two layers of glass separated by an insulating layer of still air before it is absorbed (converted to heat) by a blackened metal plate (fig. 19.8). The heat is transferred from the absorbing plate to a liquid that will not freeze in cold weather (for example, propylene glycol). The heated liquid is subsequently pumped wherever it is needed. Some systems use fans to circulate air through solar panels, eliminating the need for a liquid. Typically, solar panels capture 30-50% of the solar energy that strikes them.

Solar panels often collect more heat during the day than can be used at that time. The excess heat is usually stored in insulated water tanks or in compartments filled with rocks. In middle and high latitudes, conventional heating systems are needed as a backup to solar panels, particularly during cold periods or prolonged periods of cloudiness.

The most formidable obstacle to widespread and immediate use of solar energy for heating is cost. Although the solar radiation is free, the initial cost of the equipment and its installation is high. In 1991, a typical solar system that would provide 40-60% of the space-heating and hot water needs for a house in the upper Midwest would cost $15,000 installed, or $12,000 for do-it-yourselfers. Federal tax credits were once available as an incentive to install solar systems, but they were terminated in 1985. Because a conventional heating unit is also often needed, the capital cost for a total home-heating

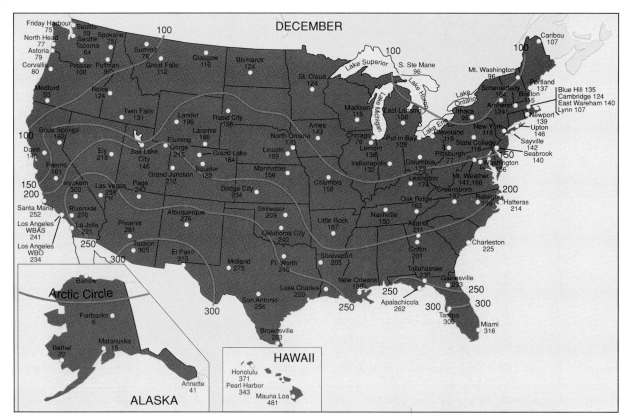

(a)

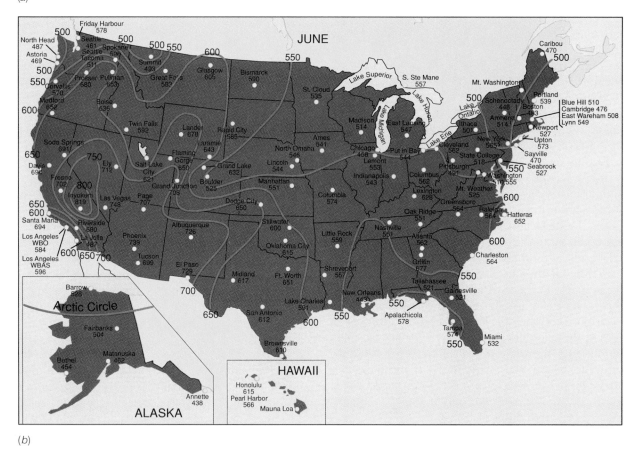

(b)

Figure 19.6 Mean daily solar radiation at the earth's surface in the United States during (a) December—the month with the lowest levels and (b) June—the month with the highest levels. (Units are in calories per square centimeter.)

Source: Federal Energy Administration.

Figure 19.7   Solar panels can provide space heating and hot water for homes. When energy demands exceed the capacity of the panels, conventional furnaces and water heaters are used as supplements.
© Steve Allen/Peter Arnold, Inc.

Figure 19.9   Deciduous trees surround these homes providing shade in the summer and admitting sunlight and slowing the wind in the winter. Landscaping such as this is an example of passive solar design.
© David R. Frazier Photolibrary

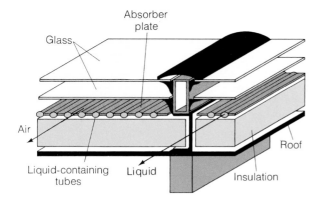

Figure 19.8   A cutaway view of a solar panel. Heat that is absorbed by the blackened absorber plate is moved from the collector in the form of heated air or heated liquid. A double layer of glass reduces heat loss.

system can be several thousand dollars more. Although capital costs for solar heating (and cooling) systems are much higher than for conventional systems, solar heating systems promise to become more economical as the cost of conventional fuels rises. Many variables influence the feasibility of solar heating, however, and people who are considering it should analyze their particular situation carefully.

Using solar energy for cooling is more complex than for heating. Although solar cooling systems are not economically competitive at present, they are particularly attractive in the long term because the time of peak demand for cooling coincides with the time of peak sunlight intensity.

Passive solar systems offer another low-temperature option for meeting part of our heating and cooling needs. A building can be oriented and constructed so that it captures more solar energy during cold months and less solar energy during the summer. An energy-efficient passive solar building incorporates five essential features: (1) good weatherization (including movable window insulation and plugging of air leaks); (2) large, transparent surfaces that face south to admit solar radiation; (3) building mass (brick or concrete) that stores heat during the day; (4) provisions for heat distribution throughout the building; and (5) consideration of wind, shading, and the placement of shrubbery on the building site to minimize heat losses from the wind and maximize gains from the sun in the winter (fig. 19.9).

In a passive solar building, the floor, walls, and furniture absorb the sunlight that enters a room (fig. 19.10*a*). Dark colors absorb sunlight most efficiently, but they are subject to fading. This problem is overcome in some passive solar homes by constructing a rough-surfaced concrete or brick wall (good absorbers of sunlight) inside the home close to a south-facing window (fig. 19.10*b*). These so-called *Trombe walls* store heat during the day and release it to the cooler surroundings at night. An attached sun space, such as a greenhouse or solarium, can also be attractively designed into a building and used for capturing solar energy as well as for growing plants (fig. 19.10*c*). Natural air movements and fans help to circulate heat throughout the building. Planting deciduous trees so that they shade windows in summer and allow sunlight to enter in winter reduces energy consumption. Passive solar designs are gradually gaining public acceptance as the public and builders begin to understand how to modify conventional home and building designs so that they are more energy efficient.

Solar High-Temperature Systems

Over the past decade, scientists and engineers have developed **solar high-temperature systems** that produce temperatures that exceed 100° C (212° F)—hot enough to produce steam. As we might expect, most developmental efforts occur in the desert Southwest. In one system, hundreds of rows of computer-controlled parabolic mirrors track the sun. Each mirror focuses sunlight on a single collection tube (fig. 19.11). The concentrated solar energy heats oil in each tube to

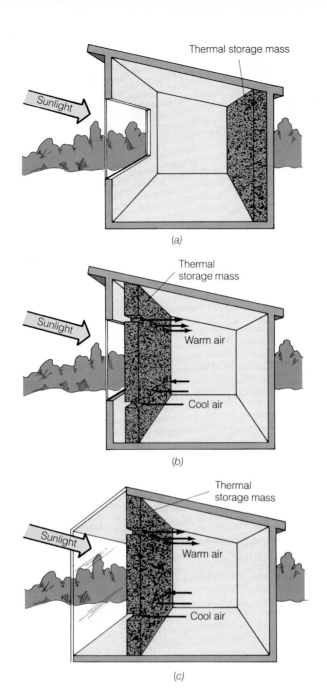

Figure 19.10   Three design strategies for buildings heated with passive solar systems: (*a*) direct gain, (*b*) Trombe wall with natural or mechanical circulation, and (*c*) sunspace with provision for air circulation.
Source: D. R. Schram and D. M. Utzinger, *Passive Solar Heating for the Home*, University of Wisconsin Cooperative Extension Programs, 1983, Madison, Wisconsin.

426° C (735° F). The hot oil is then pumped into a generator and produces steam that turns an electric turbine. The Luz International system at Kramer Junction, California, produces nearly 200 megawatts of power, enough for about 270,000 people in the Los Angeles area. It produces power for 8 cents per kilowatt-hour, which makes it competitive with coal and nuclear power. Because the system has been successful, a billion-dollar expansion that will triple its output is planned for completion in 1994. In addition to the Luz facility, Southern California Edison currently generates 275 megawatts of

Figure 19.11   This trough-style tubular solar collector produces temperatures higher than the boiling point of water. Such temperatures are required for many industrial purposes. The system concentrates the sun's rays on a receiver tube that has a liquid inside that transfers heat.
U.S. Department of Energy

power with solar generating facilities, and plans to increase production by another 300 megawatts in the near future.

Central-receiver systems such the ones just described cause minimal pollution, but they require a significant amount of land. Usually there are no major competitive uses for desert habitats. Even in a sun-drenched location, however, solar-electric systems are not able to produce the same amount of electricity every day. Such operations will become more reliable if and when large-scale technologies become available for storing energy. Other solar collector designs are available for different uses, including the pumping of irrigation water.

Photovoltaic Systems

Solar **photovoltaic systems** produce electricity directly from sunlight and are therefore considered the ultimate solar energy technology. These systems, which have no moving parts, are fundamentally different from all other means of producing electricity and have applications in all segments of society. The first important applications were on space satellites. Today, photovoltaic systems power refrigerators that store vaccines in remote regions of less-developed nations. You may own a watch or a calculator powered by a photovoltaic cell. The energy saved through such uses, however, are inconsequential. Only when photovoltaic systems can be used on a much larger scale will they help to reduce the demand for nonrenewable energy resources.

Photovoltaic cells use sunlight to create a **voltage,** that is, a difference in electrical potential, in an electrical device called a *diode* (see fig. 19.12). When sunlight falls on the diode, an electric current (electricity) flows through the circuit to which it is connected. Only certain materials develop the necessary voltage to produce a direct current when they are illuminated. These materials, known as **semiconductors,** are composed of highly purified silicon to which tiny amounts of specific impurities have been added. Semiconductors are manufactured in the form of thin wafers or sheets that are then placed perpendicular to incoming sunlight. Electricity moves through metal contact wires on the front and back

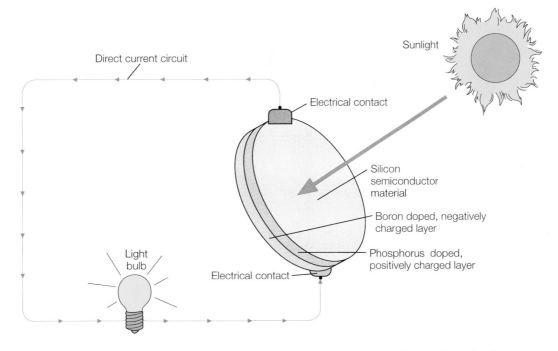

Direct current circuit

Sunlight

Electrical contact

Silicon semiconductor material

Boron doped, negatively charged layer

Phosphorus doped, positively charged layer

Electrical contact

Light bulb

Figure 19.12   Silicon photovoltaic cell, which converts sunlight directly into electricity. Sunlight frees electrons from the silicon atoms, producing an electric current that moves through the circuit. The process has an efficiency of approximately 15%.

sides of the wafer. Groups of wafers are wired together to form photovoltaic modules, and these are interconnected to form a photovoltaic panel.

The amount of electricity produced by photovoltaic panels is a function of their ability to transform sunlight to electricity. The most inexpensive, commercial, flat-plate photovoltaic panels that use silicon-based materials are 10-12% efficient. Using other, more expensive and exotic materials, researchers have developed laboratory models that achieve conversion efficiencies as high as 31%.

Currently the low efficiencies of commercially available photovoltaic panels prohibit their use in the generation of electricity. In 1989, the cost of electricity generated from a photovoltaic panel was 25-35 cents per kilowatt-hour, three to five times greater than the approximately 7 cents per kilowatt-hour that consumers now pay for coal- or nuclear-generated electricity. Thus, even though prices for photovoltaic systems have declined by 90% during the past decade, solar power from these systems is still expensive.

One Japanese firm has developed photovoltaic collection devices in the form of roof tiles and wall panels that can generate electricity and also serve as structural elements of a building (fig. 19.13). How rapidly such systems are accepted in the marketplace depends on how quickly low-cost, large-scale manufacturing processes expand. Photovoltaic experts predict further drops in production costs and gains in efficiency. Thus, they believe that electricity generated by solar cells will eventually be competitive with electricity generated by conventional methods.

Large-scale use of photovoltaic systems would greatly reduce many of the environmental problems associated with generating electricity with nuclear and fossil fuels. A great deal of space would be required, however, for grids of photovoltaic panels. Furthermore, manufacturing these panels produces hazardous wastes that require treatment, and the disposal of worn-out panels will produce a large amount of solid waste. Developers are striving to produce panels with functional lifetimes of approximately thirty years.

Wind Power

Harnessing the energy of the wind is a technology that was well established as early as the twelfth century in portions of the Middle East where water power was not available. In North America, the energy crisis of the mid 1970s spurred renewed interest in this ancient technology. Today, scientists are employing modern aerodynamic principles and space-age materials to design and construct modern wind-driven turbines that convert some of the wind's kinetic energy into electricity. A modern windmill, like those shown in figure 19.14, is capable of generating electricity for about 8 cents per kilowatt-hour. Thus, wind-generated electricity is only slightly more expensive than power from conventional sources.

Recall from chapter 15 that the sun drives the winds. Although only about 2% of the solar energy that ultimately reaches the earth's surface is converted to the kinetic energy of wind, that is still a tremendous quantity of energy. Theoretically, windmill blades can convert a maximum of 60% of the wind's energy that they intercept into electricity. In practice, however, wind generators only convert about 25% of the wind's energy to electricity.

Wind speed is by far the most important consideration in evaluating a region's potential for wind energy. Even small changes in wind speed translate into large changes in energy

Figure 19.13 Photovoltaic panels used as a component of the roof of a building.
© Albert Copley/Visuals Unlimited

Figure 19.14 Wind turbines at Tehachapi Pass, California.
© Wm. C. Brown Communications, Inc./Photo by Doug Sherman

output. Windmill energy output increases with the cube of the wind velocity (V^3); thus, a doubling of wind velocity multiplies the available wind energy by a factor of eight ($2 \times 2 \times 2 = 8$). Both wind speed and direction vary continually with time, and wind speed also varies with exposure of the site, with elevation, and with the season.

As a general rule, the average wind speed must be at least 20 kilometers (12 miles)/hour before most wind-powered generating systems operate economically. Thus, they can only be located in the windiest sites. Most of today's wind-generating power facilities in the United States are sited in the windy, mountainous areas of California (fig. 19.14). Here, 16,000 wind turbines, primarily located on windmill farms, generate about 1% of California's electricity, equivalent to that used by the residents of San Francisco. Other regions where winds are relatively strong and consistent in direction include the western high plains, the Pacific Northwest coast, the eastern Great Lakes, the south coast of Texas, and exposed summits and passes in the Rocky Mountains and the Appalachian Mountains.

Economy of scale suggests that centralized windmill farms that add power to existing power grids are preferable to individual household wind turbines. However, low-power systems have potential in small, isolated communities and on individual farms and ranches.

The impact of wind systems on the environment is minimal compared to coal- and nuclear-powered systems. Their primary drawbacks are that they are somewhat noisy, they detract from the beauty of the landscape, and they kill birds that fly into the blades.

Overall, wind power suffers from the same drawback as active solar power systems: They do not produce power consistently. Thus, as with direct solar systems, the full potential of wind power will not be realized until high-capacity energy-storage systems (e.g., batteries) are developed.

Nevertheless, wind power, like solar power, will undoubtedly make a larger contribution to electrical energy supplies in the future.

Biofuels

During the growing season, green plants more or less continually produce organic materials via photosynthesis that can be burned directly or converted to other usable forms of energy, such as alcohols. Worldwide, some 50 quads of energy are derived from biofuels, mostly in the less-developed nations. Currently, burning of biofuels accounts for approximately 2.2 quads, or about 3% of the total energy consumed in the United States. David Pimentel of Cornell University estimates that approximately 0.6 billion metric tons of the 3.2 billion metric tons of biofuels that are potentially available each year in the United States could be converted to energy. This amount of energy represents 11 quads, or 12% of the energy expected to be consumed in the United States in the year 2000.

Wood has always been a desirable biofuel because it is relatively easy to harvest, store, and handle. Wood furnishes half of the energy used for space heating in some rural regions of the northeastern states. Other states (e.g., Michigan and Oregon) that have large forested areas also use significant quantities of wood. In many small towns in these regions, wood stoves far outnumber oil or gas furnaces. Wood wastes, bark, and sawdust have been used as an energy resource in the wood-products industry (lumber and paper) for a long time. Only a few utilities, such as the 50-megawatt plant in Burlington, Vermont, burn wood to generate electricity. Experts predict that wood and forest residues (bark and branches) will continue to be the largest source of biofuels (table 19.6).

Harvesting large quantities of biofuels has serious implications for soil fertility, water and air quality, and habitats for native plants and animals. Removal of wood, forest residues, and crop residues is likely to intensify soil erosion, which already significantly reduces soil fertility and degrades waterways (chapters 8 and 14). Furthermore, diversion of crop and animal wastes to fuel production would greatly reduce the organic content of soil. Currently, about 90% of animal wastes in the United States are returned to the land. Harvesting crop residues and trees also removes nutrients (table 19.7); for example, one half of the nutrients in corn plants are contained in the residues. Thus, farmers would have to use twice as much fertilizer on their fields to achieve similar yields if the residues were harvested for fuel. Although harvesting aquatic plants would help to alleviate eutrophication problems (chapter 14), ecologists know little about how other aquatic organisms would be affected by this practice.

Replacing natural forests with "energy plantations" would alter habitats for native plants and animals. To obtain maximum energy yield from these plantations, they must be harvested in two- to ten-year cycles. Thus, even the simplified habitat that develops after replanting would quickly be eliminated. Furthermore, these monocultures require

Table 19.6	Projected energy from biofuels in the United States for the year 2000 (quads)	
Resources	**Audubon Society Forecast**	**Office of Technology Assessment Forecast**
wood	5.8	5–10
grasses and legumes	—	0–5
grains	0.3	0–1
crop residues, garbage, animal manure	2.8	0.9–1.5
total	8.9	5.9–17.5

Source: Larry Medsker, *Side Effects of Renewable Energy Sources.* New York: National Audubon Society, December 1982.

Table 19.7	Nutrient removal rates per hectare of area harvested for biofuels			
	Corn*		**Forest (Northeast)**[†]	
Nutrient	*Kilograms*	*Pounds*	*Kilograms*	*Pounds*
nitrogen	224	494	600	1,300
phosphorus	37	82	63	140
potassium	140	309	325	716
calcium	6	13	800	1,760

Source: D. Pimentel et al., "Environmental and Social Costs of Biomass Energy," *Bioscience* 34(1984):89.
*Corn yield: 125 bushels/acre.
[†]Forest yield, including residues: 500 metric tons/hectare.

heavy applications of fertilizers and pesticides to maintain high productivity.

Burning wood has created serious air-pollution problems in many wood-burning regions. Studies by the Department of Energy indicate that wood smoke contains eighteen potential carcinogens plus six irritants of the respiratory system. Some cities in wood-burning regions have passed ordinances that require catalytic converters on wood stoves to reduce air pollution.

Rather than burning plant and animal materials to release energy, they can be converted to other usable forms of energy. Fermentation of starches and sugars in corn produces the alcohols methanol and ethanol, which are used as fuels or fuel additives (e.g., gasohol, a blend of ethanol and gasoline). Processing sunflower seeds yields substantial amounts of oils that can be used as fuels after further processing. Recall from chapter 14 that animal and human wastes can be digested anaerobically to produce methane gas.

Production costs for ethanol in 1990 ranged from $1.25 to $1.30 per gallon compared to about $0.95 per gallon for gasoline. Given its higher price, ethanol accounted for less than 1% of the 22 quads of fuel consumed in the transportation sector that year. The U.S. Department of Energy predicts a fourfold increase in the consumption of ethanol by the year 2010. Thus, conversion of biomass to fuels for transportation will still only make a minor contribution in the United States.

Fill It Up with Methanol, Please!

So far, the United States has dealt with dwindling petroleum supplies by importing more and more oil. One strategy to lessen our dependence on foreign sources of petroleum is to switch from gasoline to methanol as a fuel for motor vehicles. Methanol-fueled cars were first tested because of their potential to cut air-pollutant emissions (chapter 16). Events in the Persian Gulf spurred increased interest in methanol as a home-grown fuel. Methanol is likely to be the first choice among alternate fuels that can totally replace gasoline, but what are the environmental and performance trade-offs?

Fuel Performance: The octane rating for methanol is 100, versus ratings in the low 90s for most gasoline blends currently available. As a consequence, cars fueled by methanol give snappier performances in acceleration tests than gasoline-powered vehicles. Fuel efficiency for methanol-powered engines using pure methanol is projected to be 30% higher than for an equivalent gasoline-powered engine. Methanol's energy density (BTUs per gallon), however, is approximately one-half that of gasoline which means that tanks would have to be nearly twice as large to achieve the same range of a vehicle. Fuel cost per kilometer for a methanol-powered car is projected to be comparable to that of gasoline. So far, in cold climates, test cars require a blend of methanol and gasoline to overcome starting problems. There are likely to be other problems that require refinements in the design of the fuel-injection and ignition systems of methanol-powered motor vehicles.

Air-Pollution Concerns: Vehicles fueled by pure methanol have the potential of reducing hydrocarbon emissions overall by 90% but exhausts will release more aldehydes, substances that contribute to smog and ozone production. Some of the aldehydes may be carcinogens, but the gasoline in use today is also far from carcinogen-free. The first large-scale commercial testing of methanol-powered vehicles will take place in California's smoggiest cities.

Availability of Vehicles: The Chevrolet Lumina model was first available in California in 1992; Ford, Chrysler, and foreign manufacturers are also developing competitive models in all sizes.

Availability of Fuel: Methanol will be marketed in California, but its availability will be limited. Thus, the first vehicles will be designed to burn a wide range of mixtures of gasoline and methanol. Once enough service stations that sell methanol are established, engines that run only on pure methanol will be marketed; only then will the dual benefits (lower air-pollutant emissions and nonpetroleum fueled vehicles) of methanol-powered vehicles be realized.

Brazil, on the other hand, uses ethanol extensively for transportation. Brazil consumes more than 12 billion liters (3 billion gallons) of ethanol annually. About 30% of that nation's 13 million vehicles run on 100% ethanol, and the remainder burn a blend of ethanol and gasoline. Brazil ran into trouble in 1990, however, when higher world prices for sugar caused sugar cane growers to sell more of their product to world sugar markets rather than to local fermenting industries. As a result, Brazil had to import 400 million liters (100 million gallons) of ethanol from the United States, one-ninth of the total U.S. production that year.

Interest in methanol and ethanol is increasing dramatically in the United States. These fuels are sometimes called **oxygenates** because they have higher oxygen content than gasoline. This higher oxygen content affords several significant advantages: (1) engines that burn oxygenates produce less carbon monoxide and nitrogen oxides; thus, they are being used in regions with severe air-pollution problems (chapter 16); (2) oxygenates increase the octane rating of gasoline, which raises engine efficiency; and (3) oxygenates can be blended with gasoline and thus extend petroleum resources.

Methanol is likely to be the favored oxygenated fuel because it is also produced from syngas (manufactured from coal or oil shale) or natural gas (chapter 18). Mobil Oil has developed a method to convert methane (natural gas) to methanol. While a worldwide methanol fuel market has yet to develop, most methanol consumed in the United States would probably be imported from nations that have large natural gas reserves, but a limited market. Special Topic 19.2 discusses the initial efforts in the United States to develop methanol as fuel for internal combustion engines.

Hydropower

The kinetic energy of falling water is a renewable source of energy that produces about 25% of the world's electricity. Hydroelectric generating facilities range in size from waterwheel generators in mill pond outlets to Brazil's giant Itaipu Dam, a 12,600-megawatt facility on the Parana River along the Brazil-Paraguay border (fig. 19.15). The average efficiency of converting the energy in falling water to electricity is 75-80%; in some newer installations, efficiencies run as high as 90%. Nations with mountainous regions or rugged topography have the greatest hydropower potential because they have a large number of rivers with **high-head sites**—those with a relatively large vertical drop between the head waters and the dam site. The greater this drop, the greater the amount of hydropower that can be produced. Norway, for example, obtains 99% of

Figure 19.15   Itaipu Dam in Parana, Brazil.
© Luiz Claudio Marigo/Peter Arnold, Inc.

its electricity from high-head hydropower dams. Other nations (e.g., Canada and the United States) use many **low-head sites** which have a relatively small vertical drop between the headwaters and the dam site. Low-head sites usually inundate much larger land areas than high-head sites. Canada meets 67% of its demand for electricity by hydropower and also exports electricity to the United States. In the United States, hydropower accounts for about 10% of electrical power.

Even though the United States has been building hydroelectric dams for decades, it has tapped only 42% of its hydroelectric potential. Worldwide, the total amount of developed potential is just 17%. However, the United States is unlikely to expand its hydropower generation significantly.

Construction of hydroelectric dams on many U.S. rivers is prohibited by the federal Scenic and Wild Rivers Act (chapter 10) or by legislation enacted by individual states. Dams are expensive to construct. Furthermore, good hydroelectric dam sites are usually located in areas with rugged topography. Because these regions are often far from cities that need electricity, long-distance transmission lines are required. Such long distances increase both the cost of electrical transmission equipment and the amount of electricity that is lost in transmission. Moreover, construction of transmission lines often runs into stiff opposition because they cut unsightly paths across private and public lands. Additional environmental shortcomings of dams are described in chapters 13 and 21. For these reasons, few large hydroelectric facilities will be built in the United States—except perhaps, in Alaska, which has tremendous hydroelectric potential.

Worldwide, many large, expensive facilities are being planned, but they too are running into opposition. For example, the world's largest proposed hydroelectric plant, called the "Three Gorges," on the Yangtze River in China, may not be built. Three factors hamper this development: (1) public acceptance—up to one million people would be displaced by the reservoir; (2) cost—$10 billion at 1990 prices; and (3) environmental factors—it would inundate some of China's most scenic areas, which also serve as habitat for rare and endangered species.

Geothermal Energy

Geothermal energy, a nonrenewable resource, is heat energy generated deep in the earth's interior. Much of that heat, a by-product of the decay of certain radioactive elements contained within rock, is slowly conducted to the surface of the earth. In most areas, geothermal heat flow is so diffuse that it cannot be tapped as an energy source. In certain regions of the earth's crust, however, hot molten material (*magma*) invades the bedrock, producing local hot zones (chapter 17). Heat is conducted from magma to bedrock and then to water contained within aquifers. In this way, groundwater may be heated to above 100° C (212° F) and may be tapped as a significant energy source.

The most promising areas for tapping geothermal energy have the following characteristics: (1) a history of recent volcanic activity, (2) a highly permeable reservoir rock (aquifer), and (3) an overlying impermeable *caprock* that essentially traps the heated water in the aquifer. Within the aquifer, water is under great pressure, which causes it to remain mostly in liquid form, even though the temperature rises above the boiling point (100° C). That is, the water is **superheated**. When a well taps such a geothermal aquifer, water rises to the surface and flashes into a mixture of hot water and steam. The water is separated out, and the pressurized steam is piped to a turbine that generates electricity.

One major problem with geothermal wells is disposal of wastewater, which often contains at least trace amounts of sulfur dioxide (SO_2), hydrogen sulfide (H_2S), ammonia (NH_3) and boron (B) that can be toxic to plants and animals. The wastewater is also often extremely saline. The yield of some geothermal wells in the Imperial Valley of California is only 20% steam; the rest is brine that is up to thirty times more saline than ocean water. Saline water must either be disposed of at sea or pumped back underground.

Problems with contaminants in geothermal steam can be avoided by passing all the superheated water through a heat exchanger, where a hydrocarbon mixture is gasified and used to turn turbines. These systems are called **binary cycle systems.** This method allows all geothermal brines to be pumped back into the ground. Binary cycle systems are more efficient (17%) than **flash cycle systems** (12%), in which superheated water is depressurized.

Another serious limitation of geothermal power is that many of the best potential geothermal energy sites are in scenic areas. Sites within national parks, such as Yellowstone National Park and Hawaii Volcanoes National Park, may not legally be developed without an act of Congress. Furthermore, some geologists fear that developing geothermal resources adjacent to these parks may siphon off heat that maintains various geothermal features (e.g., geysers, boiling mud pots, and hot springs), thereby destroying part of the United States' natural heritage.

Geothermal energy sources are considered to be nonrenewable because the rate of heat removal is greater than the natural rate at which heat is conducted to the reservoir rock (aquifer). Thus, after a period of time, a geothermal field, like an oil well, plays out.

In spite of limitations, geothermal energy competes economically with other sources of electric power in certain areas. Plant construction costs are generally 66-75% of comparable fossil fuel power plants, and geothermal operations require less maintenance.

The first geothermal energy facility in the United States was The Geysers (fig. 19.16), located about 145 kilometers (90 miles) north of San Francisco. Power production at The Geysers began in 1960 and in 1990 accounted for 69% of all U.S. geothermal electricity. Other regions of the United States where geothermal energy is tapped include Nevada, Utah, and California's Imperial Valley. In 1990, U.S. geothermal plants produced 2,719 megawatts of electricity (about 30% of total world geothermal energy production). Planned world capacity and plants under construction are expected to increase production by about another 35% in the 1990s.

Figure 19.16 The geothermal electric-power generating facility at The Geysers, California, north of San Francisco.
© Science VU/Visuals Unlimited

Although geothermal sources only contribute a small amount of the total energy generated in the United States, the portion can be significant locally. In Iceland and New Zealand, geothermal sources are used extensively to produce electric power and to heat living space and commercial buildings such as greenhouses. In these applications, hot water is pumped directly from the ground and circulated where it is needed.

Nuclear Fusion

In contrast to nuclear fission, in which atoms are split to release enormous amounts of energy (chapter 18), **nuclear fusion** releases similar amounts of energy via the fusion of atoms. In this process, the nuclei of elements, mostly isotopes of hydrogen, are fused together to form heavier elements. The sun and other stars are nuclear-fusion reactors.

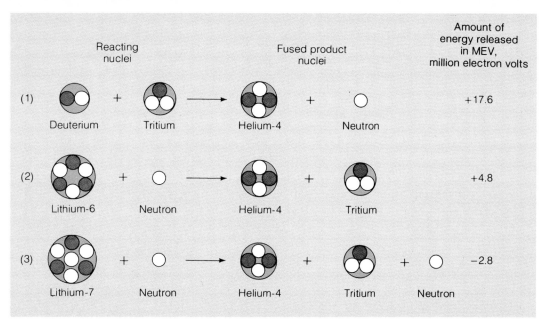

Figure 19.17   The fusing of deuterium and tritium, a promising, energy-releasing nuclear fusion reaction (reaction 1). Deuterium is abundant in sea water. The tritium that is required for reaction 1 would be produced from two common isotopes of lithium (lithium-6 and lithium-7) by means of reactions 2 and 3. A positive value for release of energy means that the reaction produces energy. Reaction 3 absorbs energy rather than releasing it, but its product, tritium, releases much energy in a subsequent step (reaction 1). Thus the net energy balance for reactions 1 and 3 combined is +14.8 MeV.

Physicists are trying to duplicate on a small scale the processes that occur in the sun to learn how to achieve controlled fusion. Nuclear fusion, however, is proving to be quite complex and difficult to control. A fusion reactor must maintain temperatures of 50-100 million degrees Celsius if the hydrogen isotopes are to collide with enough force to overcome the natural repulsive forces that keep them apart. (The fusion reaction is shown as reaction 1 in figure 19.17.) Since no materials can withstand these super-hot temperatures, scientists must contain the fuel gases (known as plasmas) in doughnut-shaped magnetic bottles or use other very sophisticated technologies to achieve these temperatures.

If—and it is currently a big *if*—the technology of nuclear fusion is perfected, fusion reactors promise three significant advantages over both nuclear fission and fossil fuel systems. First, the ocean contains vast amounts of the required hydrogen isotope, deuterium (D). The fusion of the deuterium in just one cubic kilometer of sea water would yield an amount of energy nearly equivalent to that contained in the total oil reserves originally on earth. Second, the economic and environmental costs of extracting deuterium from sea water would be minimal. Third, fusion does not pose the hazards that nuclear fission does, because fusion-reaction products are inert. However, some of the equipment associated with fusion reactors would become radioactive. What is the status of nuclear fusion today?

In July 1991, four international participants, the former Soviet Union, the European Communities, Japan, and the United States began a joint six-year project to design a new nuclear fusion reactor that is to be one thousand times more powerful than any of the previous experimental fusion reactors. Each participant has pledged $40 million per year just for the design phase of the so-called International Thermonuclear Experimental Reactor. Plans are for construction to be completed in 2005. A demonstration power plant based on the results of the experimental reactor might be tested within three decades. The next two decades of research will give an indication of the potential viability of nuclear fusion technology.

Choosing an Energy Future

Many alternatives are available to meet our future energy needs. As we have described in chapter 18 and this chapter, however, all methods of providing energy services have both benefits and drawbacks. Recall, for example, that (1) although oil is a versatile and relatively clean-burning fuel, world reserves are declining rapidly; (2) although world reserves of coal will last for centuries, its combustion has numerous environmental impacts; (3) although renewable energy sources rely upon energy from the sun, implementing these technologies has some environmental and safety impacts; (4) although synfuels offer great promise, they will not be marketed until oil prices climb to the point that synfuels are economically more attractive.

The U.S. Energy Information Administration (EIA) periodically predicts future energy supply and demand (fig. 19.18). By the year 2010, the EIA expects U.S. energy demand to increase by about 25 quads, or 1% per year. The EIA's projection of how these energy needs are to be met is mirrored in President Bush's energy policy, announced just after the close of hostilities in the Persian Gulf. Both the EIA and President Bush expect future energy needs to be met by increasing the supply of fossil fuels and revitalizing the

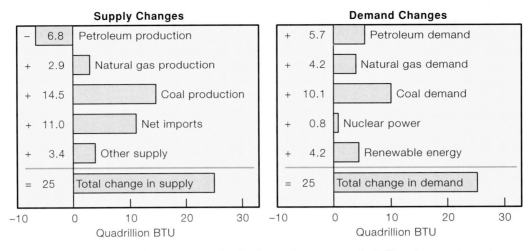

Figure 19.18   The U.S. Department of Energy base case projection for changes in energy supply (left), and energy demand (right) between 1988 and 2010.

Source: *Annual Energy Outlook 1990,* U.S. Government Energy Information Agency.

nuclear power industry. Both downplay the importance of conservation and alternative fuels.

Many people take issue with Bush's energy policy, which fundamentally seeks to maintain the status quo and relies primarily on market forces. To continue with business as usual is to fail to recognize that (1) domestic reserves of oil are being rapidly consumed, so that we depend increasingly on foreign sources of oil, and (2) we have no viable alternatives to oil in place. These people argue that government subsidies should be shifted away from conventional energy to renewable energy sources and conservation strategies. Government subsidies for fossil fuels and nuclear power are ten times what is spent each year on alternative fuels and conservation.

Conclusions

Nobody really knows the exact mix of renewable and nonrenewable fuels that we will use in the future. Certainly oil will play a lesser role. Use of natural gas may increase until its supply declines to the point that its price rises enough to make synfuels economical and they begin to serve as a major source of energy. Certainly coal and renewable resources will see their roles increase. Which renewable energy sources will offset some of our dependence on fossil fuels will depend on which alternatives make the most economic and environmental sense for a specific geographic region.

A change to a new mix of energy sources will not occur overnight. It took decades to build existing energy-delivery systems, and many of these systems will operate for many more decades. They will be scrapped only when the cost of operation exceeds the operating cost of replacement technology.

The United States must develop a sound energy policy, a policy that extends the lifetime of conventional fuels and moves toward renewable energy sources and conservation. In the past, we have too often approached energy issues as though some quick technological fix would provide unlimited energy

services at low cost without worrying much about the environmental consequences.

Key Terms

R-rating	semiconductor
wattage	oxygenates
cogeneration	high-head sites
passive solar system	low-head sites
active solar system	geothermal energy
solar low-temperature systems	binary cycle system
solar high-temperature systems	flash cycle system
solar panels	superheated water
photovoltaic system	nuclear fusion
voltage	

Summary Statements

Existing fossil fuel reserves can be extended through conservation. Energy conservation requires us to evaluate all our energy-consuming practices and determine how we can accomplish each by using less energy.

Energy conservation in the residential sector means better weatherization of buildings, reducing heat loss from hot-water heaters, and greater use of energy-efficient appliances.

Energy conservation in the transportation sector involves relying upon smaller, fuel-efficient vehicles, using car or van pools, operating vehicles efficiently, and planning trips carefully.

Some industries can conserve more energy by using cogeneration, a strategy that makes use of waste heat from an electricity-generating plant.

Solar energy is abundant, but it is also diffuse and highly variable. Both passive and active systems can be used to cut fossil fuel use.

Photovoltaic cells convert the sun's energy directly into electricity. Mass production has lowered their costs substantially, but they are still relatively expensive.

Expanded use of biofuels as an energy source is possible, but large-scale expansion would require tremendous amounts of land, which, in turn, would aggravate environmental problems.

Wind generation of electricity by moderate-sized windmills is economically competitive with conventional electric generation methods but is unreliable because winds are highly variable.

The United States has used approximately 42% of its favorable hydroelectric dam construction sites. Construction of new dams is highly controversial because dams cause social and environmental disruptions and are expensive to build.

In some regions it is possible to extract energy from geothermal resources, particularly for generating electricity. Some favorable sites cannot or should not be used because extraction of energy might jeopardize geothermal features in public parks. Geothermal sites produce pollutants that must be dealt with.

Nuclear fusion yields enormous quantities of energy. Development of nuclear fusion technology that will permit sustained generation of electricity is technologically challenging, and thus, extremely expensive. Its ultimate development is still highly questionable.

The mix of fossil fuels, nuclear fuels, and renewable resources will change somewhat by the year 2010, with coal and renewable sources picking up most of the increase (25 quads) in demand.

Questions for Review

1. List the top three energy-consuming activities in the average household.
2. What measures are some utilities taking to reduce the demand for electricity in their service areas?
3. What passive features can be incorporated into a new home to cut its energy consumption? Which of these features could be incorporated into existing buildings?
4. How much more efficient are the new screw-in-fluorescent bulbs than ordinary incandescent bulbs?
5. List six ways to conserve energy in your home or apartment. Rank them in order of pay-back time and the likelihood of implementation.
6. Why is fuel efficiency in automobiles of primary concern? What are some of the factors that keep fuel efficiency from reaching its full potential?
7. How can a community increase the efficiency of its transportation network? How could you cut the amount of energy you use for personal transportation?
8. Cite reasons why car and van pools are not used more widely. Suggest some ways to overcome some of the barriers to their use.
9. Give two examples of passive and active solar collection systems.
10. Use figure 19.6 to determine the seasonal variation in the amount of solar radiation in your region. What are some of the more common methods used in your region to take advantage of solar energy?
11. What are photovoltaic cells? Why are they considered to be an ideal solar energy collection device. What are some of their limitations?

12. How are biofuels used to supplement traditional energy sources? Can the use of biofuels be expanded?
13. Identify the advantages and disadvantages of harvesting the sun's energy using the following methods: solar collectors, solar photovoltaic cells, windmills, hydroelectric dams, and biofuels. Which systems work best for homes? Which systems work best for large-scale applications?
14. Summarize the arguments of those who favor and those who oppose expansion of hydroelectric power in the United States.
15. Describe why geothermal energy should be considered a nonrenewable resource.
16. Peak electrical demand in most communities occurs at about 4 P.M. each day. On an annual basis, peak demand occurs during the summer. What are the advantages of reducing peak demand? Suggest some policies that would lower peak demand.
17. More efficient energy use during the last decade is the result of substantially higher energy prices. If energy costs continue to rise, will it be possible to conserve as much energy as was conserved in the 1980s?
18. Distinguish between nuclear fission and nuclear fusion.
19. What efforts are underway to make nuclear fusion a viable technology?
20. What changes in the energy delivery system are possible for the United States over the next two decades? Are these changes likely to be controversial?

Projects

1. Locate a solar home in your community, and obtain some data on its performance. Report on changes that might make it even more energy-efficient.
2. Do you think mandatory performance standards imposed by federal or local governments for vehicles, appliances, and new buildings are a good idea? Explain why you agree or disagree with them.
3. Contact your local electric utility and determine which energy conservation measures they are attempting to implement in your community. What types of incentives are they using to reach their energy conservation goals?

Selected Readings

Abelson, P. H. "Energy Futures." *American Scientist* 75 (1987):584-93. A discussion focusing on why energy policies must change if the United States is to avoid severe shortages of liquid fuels.

Conn, R. W., V. A. Chuyanov, N. Inoue, and D. R. Sweetman. "The International Thermonuclear Experimental Reactor." *Scientific American* 266 (April 1992):102-10. A discussion of efforts underway to build a single large experimental nuclear fusion reactor.

Dostrovsky, I. "Chemical Fuels from the Sun" *Scientific American* 265 (Dec. 1991):102-7. A description of fuels that could be produced using sunlight and then transported to the point of use.

Echeverria, J. et al. *Rivers at Risk: The Concerned Citizen's Guide to Hydropower.* Covelo, Calif.: Island Press, 1989. Discusses how to become involved when a hydroelectric generating dam is proposed for a scenic area.

Fickett, A. P., C. W. Gellings, and A. B. Lovins. "Efficient Use of Electricity." *Scientific American* 263 (Sept. 1990): 64-74. An interesting look at how more-efficient uses of electricity can be achieved.

Flavin, C., and A. B. Durning. *Building on Success: The Age of Energy Efficiency.* Washington, D.C.: Worldwatch Institute, 1988. A discussion of how today's society can be more energy-efficient using proven methods.

Gibbons, J. H., and P. D. Blair. "U.S. Energy Transition: On Getting from Here to There" *Physics Today* 44, no. 7 (1991):22-30. A critical discussion of current energy policies and energy needs for the United States over the next two decades.

Holdren, J. P. "Energy in Transition." *Scientific American* 263 (Sept. 1990): 156-63. An examination of how to manage the transition from cheap, convenient sources of energy to more expensive but less polluting sources.

Hubbard, H. M. "Photovoltaics Today and Tomorrow." *Science* 244 (1989):297-304. A discussion of the advances made during the 1980s that have brought photovoltaic systems closer to being economically competitive.

Hubbard, H. M. "The Real Cost of Energy." *Scientific American* 264 (April 1991):36-42. A discussion of the various subsidies provided to the energy sector of the U.S. economy.

Kerr, R. A. "Extracting Geothermal Energy Can Be Hard." *Science* 218 (1982):668. A discussion of technical problems encountered in developing geothermal sources.

MacKenzie, J. J., and M. P. Walsh. *Driving Forces*. Washington, D.C.: World Resources Institute, 1990. Examines the impact of motor vehicles on global warming.

Ogden, J. M., and R. H. Williams. *Solar Hydrogen: Moving Beyond Fossil Fuels.* Washington, D.C.: World Resources Institute, 1989. Discusses how sunlight could be used to manufacture hydrogen as a potential fuel.

Pimentel, D., et al. "Environmental and Social Cost of Biomass Energy." *Bioscience* 34 (1984):89-94. A still-relevant discussion of the potential of biofuels as an energy source and the negative consequences of this technology.

Rocky Mountain Institute. *Resource-Efficient Housing Guide.* Old Snowmass, Colo.: Rocky Mountain Institute, 1989. A compilation of practical steps for keeping energy use to a minimum.

Rocky Mountain Institute. *An Energy Security Reader.* 2d ed. Old Snowmass, Colo.: Rocky Mountain Institute, 1988. Explains how to conserve sizeable amounts of energy, thus reducing the need to expand energy technology.

Starr, C., M. F. Searl, and S. Alpert. "Energy Sources: A Realistic Outlook." *Science* 256 (1992):981-87. A presentation of the major factors influencing future energy resources and technology options.

Weinberg, C. J., and R. H. Williams. "Energy from the Sun." *Scientific American* 263 (Sept. 1990) 263:146-55. An examination of the future of solar-derived technologies in response to mounting environmental concerns.

Waste Management and Resource Recovery

Chapter 20

Much of the trash thrown out by residents of cities ends up permanently entombed in a sanitary landfill. Ironically, a significant fraction of the trash could be reused.
© Thomas Kitchin/Tom Stack & Associates

In the summer of 1979, bulldozers leveled 228 bungalows and small ranch houses in a peaceful subdivision near Niagara Falls, New York, once home to about one thousand people (fig. 20.1). A year earlier, laboratory analyses showed that soil and sediment samples from that subdivision contained more than two hundred industrial chemicals, including dioxin and benzene. Some of these chemicals are known contributors to serious health problems, including leukemia, nervous system disorders, and respiratory distress. Testing of the soil and sediment and subsequent demolition of the subdivision were an unprecedented response of state and federal agencies to residents' concern over an alarming above-average incidence of birth defects among their children, including mental retardation, cleft palate, and deformed ears and teeth. Whether a cause-effect relationship actually existed between the chemical waste and the birth defects and other health problems in the area is not known, but government agencies judged the weight of circumstantial evidence to be more than enough to warrant drastic action.

Public records reveal that the subdivision was built near the site of a former canal that was later used for chemical waste disposal. Between 1942 and 1953, Hooker Electrochemical Corporation dumped 19,800 metric tons (22,200 tons) of waste into Love Canal. In 1953, the company sold the site to the Niagara Falls Board of Education for $1, and one year later an elementary school was built on the land. Hooker had informed the Board of Education in writing about its original use of the site. A neighborhood quickly developed around the school, but the new residents were totally unaware of the toxic legacy adjacent to their homes.

By September of 1988, much of the remedial work on Love Canal was complete. The 6.5-hectare (16-acre) dump is now contained by clay barriers both underground and on top, and monitoring wells and a water treatment facility have been installed. The state health commissioner even declared part of the area north and west of the canal to be safe again for human habitation. In August 1990, the first nine houses in that area went up for sale.

Events at Love Canal focused worldwide attention on the potential dangers of abandoned hazardous-waste disposal sites. Love Canal also spurred the federal government to give much higher priority to the identification and cleanup of inactive or abandoned hazardous-waste dumps. (To date, more than 31,000 such sites have been found nationwide.) Furthermore, Love Canal demonstrates how ordinary citizens can become actively involved in correcting environmental problems. The voices of concerned citizens of Love Canal were eventually heard by state and federal agencies.

We no longer put up with smoldering open dumps, noxious incinerator smoke, and industries that disgorge toxic chemicals onto the land, into waterways, or into the air. Today, strict federal legislation aims to guard against further deterioration of land, water, and air, and to spur upgrading of environmental quality. Nonetheless, we continue to generate massive amounts of waste, some of which is toxic and hazardous. This has given renewed impetus to efforts to reduce waste generation and manage waste disposal in ways

Figure 20.1   Public reaction to the discovery of an abandoned hazardous waste dump in the Love Canal neighborhood of Niagara Falls, New York, eventually spurred passage of the Superfund Act.
© Andy Levin/Photo Researchers, Inc.

that help to preserve environmental quality. We must also solve the serious problems created by abandoned and poorly-designed disposal sites.

In this chapter, we examine requirements for safe disposal of municipal wastes, hazardous materials, and radioactive substances, and consider some of the factors that cause certain disposal problems to remain unresolved. We also consider the potential for recovering valuable materials and energy from those wastes.

Historical Perspective

Throughout much of our nation's history, people have freely tossed garbage and other refuse into backyards, streets, fields, wetlands, and rivers. In colonial days, hogs wandered freely through city streets, rooting through garbage and leaving their

Figure 20.2 Heaps of refuse clogging the streets of New York City in the 1880s.
The Bettmann Archive, Inc.

excrement and stench behind. After the Civil War, cities swelled with new industries and workers, and the problem of waste disposal became acute. Chicago's population soared from only 5,000 in 1840 to almost 1 million by 1890. At the time, sanitation efforts were meager and were often overwhelmed by huge piles of garbage that clogged city streets. In the late nineteenth century, heaps of rotting refuse actually impeded pedestrian and vehicular traffic in New York City (fig. 20.2).

Public concern over the growing health hazard eventually forced some reforms; for example, hogs were banned from city streets, and garbage was hauled to hog farms, to open dumps adjacent to cities, or to nearby rivers, or it was barged out to sea. Still, the waste problem was not solved, merely displaced. Not until recent decades has the United States had health and environmental quality regulations that require the disposal of waste in an environmentally responsible manner.

Although hauling wastes out of congested cities improved conditions in urban areas, it often worsened environmental conditions in areas surrounding disposal sites. Uncompacted wastes left open to the atmosphere made ideal breeding grounds for disease-carrying pests, such as flies and rats. Over the years, rainwater and melting snow seeped into dump sites, producing contaminated leachate, which percolated into groundwater reservoirs (chapter 14). **Leachate** is the liquid produced as water seeps through and dissolves materials. Many dumps received hazardous chemical wastes and seriously contaminated leachates entered aquifers and slowly migrated some distance from the dump sites. Also, smoldering fires in open dumps polluted the air, a persistent source of irritation for nearby residents. In some cases, hazardous wastes caught fire, sending toxic fumes downwind, forcing people from their homes.

Correcting problems caused by irresponsible dumping in the past is very difficult. In many states, dump sites were not regulated; hence, their exact locations and contents are sim-

ply unknown. Compounding the problem is the fact that a small (1 or 2 hectare) hazardous-waste dump can pose just as much threat to the health of nearby residents as a much larger dump that is free of hazardous wastes. Furthermore, once a dump is located, it is expensive to determine whether it is leaking and what specific chemicals are present. Then the challenge is to obtain financing to clean it up.

Until recently, our affluent society has chosen not to acknowledge that much of what we throw away can be recovered and reused. In fact, *waste* is a concept unique to humankind; as we have seen in this book, nature reuses and recycles. In recent decades, recycling has waxed and waned in popularity. Recycling has fluctuated with the general economic climate, even though it clearly saves energy, cuts the volume of waste, and reduces the environmental impact associated with mining and processing raw materials. Now, with fewer environmentally acceptable disposal sites available, rising disposal costs are forcing communities and industries to pursue recycling more aggressively.

Legislative Response

The rising tide of environmental consciousness that began in the late 1960s eventually spurred a response by the federal government in the form of new, strict legislation directed at waste management. A few states, notably Massachusetts and California, have had standards for the collection and disposal of solid waste for many decades. Typically, however, until recently most states exercised little control over land disposal sites. The first federal legislation, the Solid Waste Disposal Act of 1965, called for research on solid-waste problems and provided funds to states and municipalities for planning and developing waste-disposal programs. The 1965 law was amended by the 1970 Resource Recovery Act, which provided funds to states for constructing waste-disposal facilities. Also, the 1970 law was the first federal legislation to encourage the recycling of solid waste.

Strict regulation of solid-waste disposal was a major objective of the Resource Conservation and Recovery Act of 1976. This law required the states to close all open dumps by 1983 and provided help for developing waste-reduction programs. The 1976 act also called on the Environmental Protection Agency (EPA) to draw up guidelines for the siting of landfills to minimize the environmental impact of waste disposal and provided for strict regulation of hazardous wastes. The law required the states to identify hazardous wastes and set national standards for their generation, transport, treatment, storage, and disposal.

Hazardous materials were further regulated in 1976 with passage of the Toxic Substances Control Act, which requires premarket screening of new chemicals (excluding pesticides) for potential toxicity. The EPA can regulate chemicals that pose an unreasonable risk in their manufacture, distribution, use, or disposal. Although the intent of this law was laudable, its provisions have generated considerable controversy, thereby delaying full compliance. Disagreements developed over the

Table 20.1	Composition of compacted refuse in a typical municipal landfill	
Waste Type		**Percent by Volume**
paper (e.g., newspapers, packaging)		50%
organic (e.g., yard waste, food waste)		13
plastic		10
metal		6
glass		1
miscellaneous (e.g., construction debris, diapers, textiles, rubber)		20

Source: Data from W. L. Rathje, "Once and Future Landfills," *National Geographic* 179:116–35, No. 5, 1991.

degree of risk associated with exposure to different chemicals. Also, chemical companies still fear that the law will force public disclosure of trade secrets and thereby aid their competitors. These problems are compounded by the great number of chemicals that are currently in commercial production (approximately 86,000), which makes the screening process both expensive and time-consuming.

Passage of the Superfund Act* in 1980 was an important milestone in hazardous-waste management. That law initially established a $1.6 billion fund for the cleanup of abandoned hazardous-waste sites. In addition, the law requires the discharger to compensate victims for illness and injury as well as for property damage. The Superfund is financed primarily (over 80%) by taxes on the chemical and petroleum industries, with the remainder coming from other federal revenues. In 1986, Congress reauthorized the program for another five years and appropriated an additional $8.6 billion. Congress took similar action in 1990, this time adding $5.1 billion to the fund. Nonetheless, many hazardous-waste managers consider the Superfund to be seriously underfunded.

Sources of Waste

Every human activity generates some waste. We create waste when we extract and process rock, minerals, and fuels, and when we grow and process food and fiber. Basically, air and water pollution are problems of waste disposal. When products wear out, become outmoded, or otherwise outlive their novelty or utility, we usually throw them out. People in the United States directly and indirectly dump an estimated 10 billion metric tons (11.2 billion tons) of all forms of solid waste each year. Although a vast variety of materials constitute waste, for convenience of study, we classify waste by origin as municipal, mining, industrial, and agricultural.

Components of municipal solid waste are as diverse as those of our surroundings: bricks, tree branches, aluminum cans, newspaper, paints, sewage sludge, cinders, food, and so on. Table 20.1 shows the typical composition of refuse in a municipal landfill. In 1990, each person in the United States discarded an average of 1.9 kilograms (4.3 pounds) of gar-

*More formally known as the Comprehensive Environmental Response, Compensation, and Liability Act.

Figure 20.3 Coal-burning electric power plants produce massive amounts of fly ash that must be disposed of in an environmentally sound manner.
EPRI

bage and other refuse each day. This is enough to fill about 65,000 garbage trucks daily and, over the course of a year, totals over 173 million metric tons (195 million tons).

The amount of municipal waste pales in comparison to the massive quantities of waste produced by mining, industrial, and agricultural activities. Mining and mineral wastes include mill tailings, slag, and various mine wastes exclusive of overburden (chapter 17). Such wastes are generated at a rate approximately eleven times greater than the rate at which municipal solid wastes are produced. For example, for every metric ton of copper produced by open-pit mining (including extraction, separating, smelting, and refining), more than 500 metric tons of solid waste are generated. Air pollutants emitted in the process are not included in this total. However, mining and mineral processing usually take place in remote areas, so that these wastes are not as visible as municipal wastes.

The industrial sector generates a wide variety of waste depending on the kinds of raw materials used, energy source, and products. For example, coal-burning electric power plants yield tremendous quantities of waste. Typically, about 10% of the mass of coal is noncombustible and accumulates as a residue known as **fly ash.** A 1,000-megawatt plant (about the size of most plants being built today) must dispose of 250,000 metric tons (280,000 tons) of fly ash each year. If fly ash is dumped in a landfill (fig. 20.3), then during a plant's forty-year life span, disposal of fly ash requires about 2.5 square kilometers (1 square mile) of land.

Agricultural wastes are generated at a rate approximately fifteen times that of municipal wastes. About 75% of the 2.1 billion metric tons (2.4 billion tons) of agricultural wastes produced each year consist of animal manure, crop residue, and various by-products of food production (fig. 20.4). However, most animal waste and crop residue is worked into the soil, enhancing its fertility and capacity to retain moisture.

Figure 20.4   Agricultural wastes include livestock manure and crop residue. These biodegradable materials are usually worked back into the soil to improve fertility.

Courtesy of Steve Bennett

Of the total amount of waste generated in the United States each year, approximately 260 million metric tons (approximately 1 metric ton per person) are considered to be potentially hazardous. Hazardous wastes are produced in the manufacture of inorganic and organic chemicals (including plastics), in the smelting and refining of ores, in electroplating, and in petroleum refining. Nearly two-thirds of hazardous wastes are generated by the chemical industry and its associated industries. Paper- and glass-product industries and metal production each account for between 3% and 10% of the nation's hazardous-waste output.

For more than forty years, U.S. nuclear power plants and weapons manufacturers have been producing radioactive wastes that are stored *temporarily,* pending development and public acceptance of a permanent repository. Such wastes emit high-energy radiation that is extremely hazardous to humans as well as other organisms.

Properties of Waste

Proper management of waste hinges on an initial evaluation of certain key properties of the waste—specifically whether the waste is hazardous or not and its potential persistence in the environment.

The degree of hazard posed by waste depends upon its flammability, corrosivity, reactivity, or toxicity. These criteria are specified by the Resource Conservation and Recovery Act. **Hazardous waste,** for example, includes acids, explosives, combustible solvents, and toxic chemicals. Radioactive wastes present a unique hazard, and are therefore managed separately by the Nuclear Regulatory Commission.

Wastes persist in the environment if natural physical, chemical, and biological processes do not reduce them to harmless by-products. Heavy metals such as lead, mercury, and cadmium persist because they occur as elements and cannot be further broken down chemically. Our greatest concern is with persistent hazardous waste that enters food webs.

In passing from one trophic level to the next higher trophic level, these materials bioaccumulate and cause health problems in species that occupy upper trophic levels, such as birds of prey (chapter 9) and humans (chapter 12).

Common practice describes the potential environmental persistence of a substance in terms of its biodegradability. **Biodegradable waste** is subject to natural biological decay through the action of aerobic or anaerobic decomposers. **Nonbiodegradable waste** resists such decomposition and thus persists in the environment unless degraded by physical or chemical processes. Most biodegradable waste is organic and includes paper, lawn clippings, and discarded wood. Agricultural wastes (animal manure, crop residue) are mostly biodegradable, whereas mining wastes are primarily nonbiodegradable (chapter 17). Urban refuse and industrial waste are complex mixtures of biodegradable and nonbiodegradable components.

In recent years, scientists have developed two new plastics that partially break down in the environment. In time, both biodegradable and photodegradable plastics fragment into small pieces. *Biodegradable* plastics contain embedded starch, which is decomposed by soil microorganisms. *Photodegradable* plastics lose their strength upon exposure to the ultraviolet (UV) portion of solar radiation; the UV breaks the polymer chains composing such plastics.

While degradable plastics would appear to be an important innovation in view of today's widespread use of plastics, there are serious drawbacks to the two types developed so far:

1. While both biodegradable and photodegradable plastics fragment, the individual pieces do not further degrade.
2. The sunlight, oxygen, and moisture necessary to break down such plastics are absent in landfills.
3. Both plastics have very limited use because they lack some of the important properties of other plastics (e.g., they lose strength upon contact with oil and solvents).

At this point, prospects for a completely degradable plastic with all of plastic's desirable properties are not promising. Hence, many solid-waste experts argue that management efforts should focus on recycling plastics.

The remainder of this chapter describes the management of waste by category: nonhazardous, hazardous, and radioactive.

Management of Nonhazardous Waste

Most of the waste collected from households and commercial establishments, along with nonhazardous industrial waste, ends up in a sanitary landfill. The remainder is either incinerated or recycled. In 1990, about 67% of the United States' municipal solid waste was disposed in about 5,500 active landfills. Some coastal communities have traditionally dumped sewage sludge and dredge spoils at sea, although new legislation is ending ocean dumping.

The 1976 Resource Conservation and Recovery Act recognizes two ways of managing municipal waste besides disposal: (1) reducing the waste potential of products before they reach the consumer, and (2) recovering resources from solid waste, that is, recycling. Both strategies ultimately reduce the number of facilities needed for disposal. Waste reduction lowers collection and disposal costs and lessens demands on raw materials, energy, and land for disposal. Recycling converts a liability into an asset by adding to the supply of available materials and energy.

In this section, we examine the disposal alternatives for nonhazardous waste: sanitary landfilling and incineration. We also address the status of ocean disposal and examine strategies to recover resources from the waste stream and reduce the generation of waste.

Land Disposal Alternatives

Today, a potential landfill site must meet rigorous engineering and environmental standards set by the EPA or an appropriate state regulatory agency, in sharp contrast to the casual operation of the open dumps of earlier decades. A **sanitary landfill** eliminates most of the problems associated with an open dump because wastes are sealed between successive layers of clean dirt each day. This practice prevents infestation by insects and rodents and greatly reduces odors. In addition, no open burning is allowed at sanitary landfills. Furthermore, landfills can be designed to restore landscapes scarred by mining. Indeed, many communities are creating new parks and other recreational sites by revegetating completed landfills.

The foremost concern in designing a sanitary landfill is to safeguard groundwater quality. Where there is danger of groundwater contamination, the bottom of the landfill site is lined with a thick layer (2 meters or so) of impermeable clay (and in some cases, plastic sheets). In some areas, surface water that seeps through wastes (leachate) must be pumped out and then either piped or hauled to a sewage treatment plant. As an added precaution, special wells monitor groundwater around the perimeter of the landfill to determine whether it is being contaminated by unexpected leaks. Of the nation's municipal landfills, currently only about 15% are lined, 5% collect leachate, and 25% monitor groundwater. These statistics are expected to change soon because new standards set by EPA require monitoring of groundwater and installation of clay and plastic liners at all new sanitary landfills.

Municipalities practice three basic types of sanitary landfilling—trench, area, and mounded—depending upon the terrain and depth of the water table.

A **trench landfill** (fig. 20.5) is most appropriate where the water table is deep and the subsurface soil and sediment are thick and relatively impermeable. Initially, a broad trench is excavated. Trucks dump their loads of refuse into the trench, and then special heavy equipment spreads and compacts the trash. Each day, the compacted waste is covered with a layer of clean dirt (obtained from the original trench excavation). This process is repeated until the trench is full. Meanwhile, another trench is dug and readied for new shipments of garbage.

Figure 20.5 A trench-type sanitary landfill is appropriate in areas where the water table is relatively deep.
© Frank M. Hanna/Visuals Unlimited

Figure 20.6 Mounded sanitary landfills are constructed in areas where the water table is close to the surface.

An **area landfill** is developed in natural valleys or canyons as well as abandoned pits and quarries. Refuse is dumped on the bottom liner, compacted, and covered with soil each day until the site is filled. Area landfills are often more expensive to operate because of the need to haul in additional fill, but the cost of transporting waste is the key economic consideration. Hence, area landfilling may be chosen over trench landfilling if a suitable site is closer to the source of refuse.

In regions where the water table is close to the surface, a **mounded landfill** is often necessary to prevent groundwater contamination. Alternating layers of compacted refuse and dirt are mounded over a clay base on the surface to form a hill (fig. 20.6). A completed mounded landfill is contoured and vegetated to control erosion.

Besides the potential for groundwater contamination, another drawback of sanitary landfills is the possible migration of hazardous gases formed during the anaerobic decomposition of waste. Recall from chapter 14 that anaerobic decomposition produces methane (CH_4) and hydrogen sulfide (H_2S). These gases can seep through the sides of a landfill and follow layers of permeable soil and sediment into the basements of neighboring buildings. Since methane is explosive in confined spaces, such migration must be prevented. In addition, hydrogen sulfide can act as an asphyxiating agent (chapter 16). Therefore, if gas migration from a landfill is a potential problem, vent pipes and trenches must be installed to collect the gases so that they can be flared (burned), allowed to escape slowly into the open air, or used as an energy source.

Recall from chapter 18 that methane is the principal component of natural gas. The first methane-recovery landfill went into operation in 1975 at Rolling Hills Estates in California. Since then, many other large landfills have begun to recover methane by using a relatively simple technology. Perforated collecting pipes are inserted into vertical holes drilled into the landfill. A slight pressure reduction in the pipes draws methane from the landfill. Water vapor and other gases (such as hydrogen sulfide) are removed before the methane is piped to a nearby facility that uses it as a fuel. The Fresh Kills landfill on Staten Island in New York City (the nation's largest) produces enough methane to heat ten thousand homes through a New York winter. That landfill receives about half of the city's garbage, about 15,400 metric tons (17,000 tons) per day, six days a week. In 1992, methane was recovered at 114 landfills in the United States, supplying about 344 megawatts of power.

One popular myth about sanitary landfills is that the organic portion of the waste decomposes rapidly. Compacted garbage, however, is effectively sealed from oxygen and moisture, which are necessary for rapid decay. In fact, William Rathje, a University of Arizona archeologist, and his students probed the contents of fourteen sanitary landfills in North America and found that only 20-50% of food and yard waste biodegrade within the first 15 years of burial. They also found that newspapers were still readable after more than a decade of burial (fig. 20.7).

In the future, sanitary landfills will be designed for one of two purposes. Some will entomb mostly unrecyclable wastes while others, primarily near large metropolitan areas, will produce methane.

Finally, a significant nuisance problem can occur during daily operation of a sanitary landfill when the wind scatters litter and raises dust. Maintenance personnel can minimize those problems by installing litter fences around the site, hand collecting litter, and regularly applying water from sprinkler trucks.

Once a sanitary landfill is complete, the site must be inspected periodically for uneven settling of the ground, which can expose wastes and increase gully erosion. Settling always occurs, primarily because of the anaerobic decay of the wastes, which causes further compaction of the refuse. About 90% of total settling usually occurs within about five years. After a substantial amount of settling, more filling and grading may be necessary to control erosion.

Figure 20.7 Garbage archeologists have found that newspapers show little sign of decay after being buried in a sanitary landfill for more than a decade.
© James A. Sugar/Black Star

Completed landfills cannot be used as building sites, but they are suitable for recreational areas (fig. 20.8), botanical gardens, or wildlife preserves. Even though landfills offer many advantages over open dumps for solid-waste disposal, they often encounter strong public opposition. Most people equate a sanitary landfill with a dump, and are also concerned about increased traffic and the noise of large garbage trucks. They simply refuse to accept a sanitary landfill in or near their neighborhood, even though it may eventually become a park that would benefit the entire community. This negative reaction is sometimes referred to as the NIMBY (Not In My Backyard) response. A proposed landfill often provides a classic example of the sometimes intractable problem of reconciling individual preferences with the public good.

An alternative to landfilling is incineration. A **municipal incinerator** burns combustible solid waste and melts certain noncombustible materials. Incineration is better than landfilling from a health perspective, since the high

Figure 20.8 After completion, a sanitary landfill can be converted into a community asset. Pictured is a completed mound landfill in Riverview, Michigan, now used as a ski hill in winter and a park in summer.
Courtesy of the City of Riverview, Michigan

temperatures destroy pathogens and their vectors. Installation, maintenance, and operational expenses are higher, however, so the average cost of incinerating solid waste is generally greater than for a sanitary landfill. There are exceptions. In regions where appropriate nearby sites for sanitary landfills are unavailable and where land prices are high, incinerators are often less expensive than landfills. Ironically, those circumstances usually prevail in and around large urban areas, which produce the most household, commercial, and industrial wastes. The solution in such situations may be a combination of incineration and sanitary landfilling. Incineration reduces the volume of solid waste by about 80% to 90%. Hence, disposing of the residue after incineration requires considerably less landfill space. On the other hand, incinerator residue may contain relatively high concentrations of heavy metals that are hazardous and require special disposal measures.

Incineration has additional advantages over landfills of any kind because it does not directly endanger groundwater quality. If equipped with adequate air-pollution control devices and attractively housed and landscaped, incinerators are somewhat more likely to gain public acceptance than are landfills. However, most of the incinerators operating in the United States in the late 1960s were constructed before air-quality control devices were required. When the Clean Air Act amendments were enacted in 1970, incinerators were forced to comply with air-quality standards, and the cost of installing the pollution-control equipment forced most of them to shut down. Conventional incinerators are also unable to handle certain types of wastes, for example, potentially smokey discards such as tires.

Considerable progress was made in the 1970s and 1980s in incineration technology and in air-pollution control devices for incinerators. Hence, in some large metropolitan areas, new high-tech incinerators play a major role in waste management. In the early 1990s, the United States

incinerated about 16% of all municipal solid waste; the EPA expects this percentage to grow substantially over the next five years as more sanitary landfills are completed or closed (because of environmental problems).

Ocean Disposal

Until recently, **ocean disposal** was a traditional alternative to landfilling and incineration (fig. 20.9) in some coastal areas. More than 90% of the waste barged out to sea and dumped consisted of spoils generated by dredging ship channels in harbors and rivers. Typically, one-third or more of these spoils are contaminated by industrial and municipal effluents and runoff from agricultural lands. Most of the remaining waste dumped at sea consisted of sewage sludge. (Dumping industrial waste in the ocean was banned in 1988.) Sludge from sewage treatment facilities poses a special hazard to ocean life because it usually contains heavy metals and pathogenic microorganisms.

Concern about the dumping of such wastes into the ocean or onto the seabed focuses not only on toxicity to organisms, but also on the possibility of bioaccumulation in food webs. Also, because mixing processes in the oceans are poorly understood, ocean dumping that seems safe enough in the present may cause unanticipated problems in the future.

At this point, the principal question about ocean dumping has to do with the ocean's ultimate assimilative capacity, that is, its ability to receive waste without long-term disruption of marine ecosystems. Most marine scientists agree that the shallow marine environment over the continental shelf, the usual site of ocean dumping, has a very limited assimilative capacity. Greater disagreement surrounds the assimilative capacity of the open ocean beyond the continental shelf. Some scientists point to the vastness of the ocean and argue that the open ocean is a biological desert whose productivity can only benefit from inputs of nutrient-rich waste such as sewage sludge. Others argue that the assimilative capacity of the entire ocean is limited and press for a ban on all ocean dumping. The EPA has sided with the latter argument.

The Marine Protection Research and Sanctuaries Act of 1972—commonly referred to as the Ocean Dumping Act—required municipalities to obtain an EPA permit to transport wastes and dump them in oceans. That law also empowered the EPA (and the U.S. Army Corps of Engineers, in the case of dredge spoils) to locate proper ocean disposal sites (beyond the continental shelf, where possible) and to regulate the types and amounts of substances that could be dumped. In the past, most ocean dumping sites were within 40 kilometers (25 miles) of shore. From those sites, ocean currents could transport waste back to shore, where it could adversely affect commercial shellfish beds and force the closing of beaches. The long-term objective of the 1972 law was the eventual phasing out of all ocean disposal.

In 1977, the EPA called for an end to ocean disposal of wastes that threaten to "unreasonably degrade" the marine environment. Amendments to the 1972 Ocean Dumping Act,

Figure 20.9   In the past, coastal cities commonly barged their waste out to sea. Under recent federal legislation, all ocean disposal is being phased out.
© David R. Frazier Photolibrary

enacted in 1977 and 1980, prohibited disposal of all sewage sludge in the Atlantic Ocean (the site of 90% of the nation's ocean dumping) after 31 December 1981. However, federal court action taken by New York City against the EPA in November of 1981 delayed compliance with these regulations and forced the EPA to examine the environmental impacts of sludge disposal on land and to reevaluate its ocean dumping policy. New York City and eight surrounding municipalities have long relied on the ocean as a dump for their sewage sludge. However, the more than 8 million wet metric tons of sludge dumped each year has high levels of PCBs and heavy metals, including cadmium, lead, and mercury. Thus, in 1987, the EPA ordered the city to close its ocean dump site (the New York Bight) located in 40 meters (130 feet) of water 19 kilometers (12 miles) outside of New York harbor. That site had been operating since 1924. A new dump site was authorized 171 kilometers (106 miles) out to sea on the edge of the continental shelf and in 2,000 meters (6,500 feet) of water.

Events of the summer of 1988, however, precipitated an end to ocean disposal of sewage sludge once and for all. Many East Coast residents seeking relief from oppressive summer heat found their recreational beaches closed because of polluted water and fouled sands. Congress responded by banning ocean sludge disposal by 1 January 1992. (New York City and its neighbors were granted an extension until June 1992.) The alternatives to ocean disposal are landfilling, incineration, or recycling.

While New York City struggles to find an alternative to ocean disposal of sewage sludge, metropolitan Boston has opted for sludge recycling. In December 1991, one of the nation's largest sludge recycling operations opened in Quincy, Massachusetts. The $87-million facility serves the forty-three cities and towns of the metropolitan Boston sewage district. Previously, sewage sludge had been dumped into Boston Harbor. Sludge is barged to the recycling facility, where presses squeeze out most of the water—sludge is more than 90% water. Drying is completed in rotary heat drums that also kill bacteria and remove odors. The solid residue is pelletized and sold for use in agriculture and landscaping.

Resource Recovery

To reduce the amount of waste that must be landfilled or incinerated, municipalities and industries are increasingly opting for **resource recovery,** the retrieval of energy or reusable materials from the waste stream. In 1991, the EPA estimated that about 17% of municipal solid waste was recycled, but in the future, this figure could reach 45%. In this section, we describe various methods of resource recovery.

Reclamation or **recycling** involves separating materials, such as rubber, glass, paper, oils, and scrap metal, from refuse and reprocessing them for reuse. In the United States, waste reclamation is a multibillion-dollar-a-year business, based primarily on the recycling of scrap metal from the nine million motor vehicles junked each year. Also, a significant fraction of the aluminum, lead, copper, and zinc used by industry is derived from recycling those metals. For example, industries that machine metal parts keep metal turnings and scraps separate from other plant refuse and recycle them. Automobile repair shops send worn-out batteries to recyclers, who recover the lead they contain.

Today, many people recycle aluminum cans in return for cash at recycling centers. More than half of the aluminum cans purchased since 1981 have been reclaimed, and recycled aluminum offers a 90% energy savings over aluminum derived from bauxite ore.

What can be done with the 285 million rubber tires discarded each year in the United States? One solution is to grind them up and mix them with crushed stone to form an aggregate for asphalt road surfaces. In fact, federal law now requires that asphalt road-building projects funded by federal sources utilize asphalt containing at least 5% scrap rubber.

Paper makes up a rapidly growing portion of landfills, and at 18% of the total volume of compacted trash, newspapers are the largest single component of a landfill. Hence, public pressure is on for paper recycling. Nationwide, thousands of companies and government facilities routinely sell waste paper (about 50% of which is high-grade) to paper recyclers. One-third of the nation's six hundred paper mills process waste paper exclusively, and about three hundred others utilize 10-30% waste paper that comes directly from collectors and some fifteen hundred waste-paper processing plants in the United States.

Contrary to popular opinion, the amount of plastic entering landfills is not growing and has stabilized at about 10% of the total volume. Although we use more plastic items now than we did two decades ago, each item uses less plastic. For example, the weight of a plastic milk jug has declined from 98 grams (3.5 ounces) to 60 grams (2.1 ounces). Nonetheless, great interest in recyling plastic persists. Of the several types of plastic, the ones most commonly recycled from post-consumer sources are (1) PET (polyethylene terephthalate), used in soft-drink containers and microwavable food pouches, and (2) HDPE (high-density polyethylene) used for milk jugs, detergent bottles, and trash bags. Research is currently underway to develop methods of recycling products composed of a mixture of different plastics.

The popularity of community (public sector) recycling surged in the 1990s for both environmental and economic reasons. Landfill space is shrinking, especially in congested urban areas, forcing a sharp increase in tipping fees (the charge for dumping garbage at a landfill). Compounding the problem, an estimated 20% of existing landfills will close by 1993 because they threaten environmental quality. In many areas, recycling now costs less than disposal. Hence, by mandating recycling, communities with high disposal fees can cut the overall cost of waste management. By 1991, more than half the states had laws that either provided incentives for or required recycling of residential refuse. At least eight states have set a goal to recycle half their garbage within at least the next fifteen years.

Today's public-sector involvement in recycling contrasts with the mostly private-sector recycling efforts of the early 1970s. Most of the latter were small-scale ventures that had difficulty selling recyclables in a traditionally volatile market and ultimately failed during the 1974-75 recession. Today, centralized processing facilities help coordinate the marketing of recyclables and thus aid the economic viability of community recycling. Consider an example.

Because of environmental problems, Wilton, New Hampshire, was forced to close its old landfill. Wilton and six neighboring towns then chose to emphasize recycling and established a recycling dropoff facility. About 65% of the area's population participates; the others pay a private hauler to take away their trash. Nearly half the trash is recycled (aluminum and steel cans, glass, newspaper, textiles, large appliances, used motor oil, scrap metal, auto batteries, and yard waste), 35% is incinerated, and the remaining 15% is landfilled. Wilton has realized a 33% savings in waste-disposal costs.

An important key to a successful community recycling program is citizen motivation coupled with convenience of participation (fig. 20.10). Curbside collection (recyclables are put out in separate bins or bags alongside other trash for pickup by sanitation trucks) is generally more convenient and energy efficient than a dropoff center (which requires people to transport their recyclables to a central facility). In 1990, about 2,700 U.S. communities had curbside pickup programs (up 120% from 1988). A serious drawback to community recycling is the variable quality of recycled materials; some people are more willing than others to clean glass containers, separate clear glass from colored glass, crush aluminum cans, and separate clean paper from paper covered with food. Industries that use recycled materials are especially concerned that they be clean and of uniform quality.

Composting is an age-old process that simulates nature's methods of recycling (chapter 3). Organic refuse is heaped into piles and allowed to decompose biologically (fig. 20.11). After about six months, a humuslike substance remains that makes an excellent soil conditioner. In the composting of municipal wastes, the organic portion (e.g., paper, food wastes, leaves, and lawn clippings) must first be separated. It is then piled into windrows that are aerated approximately once a month by mechanically turning them over. The final humus product is then marketed.

In 1990, there were over 1,400 yard-waste composting programs in the United States, primarily in the northeast and upper midwest. Direct land application of yard wastes, without formal composting, is also becoming more common. Finally, large-scale composting of municipal solid wastes is growing in popularity as technologies used for decades in Europe are adapted to U.S. needs. In 1990, thirteen such facilities were operating. Potentially, 15-20% of municipal solid waste could be composted.

Reclamation is not limited to the recycling of materials for reuse as new products. Some refuse is reclaimed directly as a fuel, called **refuse-derived fuel (RDF).** When municipal waste is shredded or milled, combustibles are separated from noncombustibles, and the combustible portion is then burned with coal, using a mixture that contains 5-25% refuse, to generate steam. Currently, only 130 small-scale facilities in the

(a)   (b)

Figure 20.10   Curbside pickup of recyclables (*a*) is more convenient than centralized drop-off facilities (*b*). In either case, the consumer must clean and separate recyclable paper, glass, and metal.
© Steve Elmore/Tom Stack & Associates

Figure 20.11   A small backyard compost heap consisting of grass clippings, yard and garden debris, and food wastes.
© Ronald Specker/Earth Scenes

United States employ refuse-derived fuels. Together, they generate about 2,000 megawatts of electricity, enough to power about 1.1 million homes. As part of his national energy plan in 1991, President Bush set a goal of a sevenfold increase in refuse-derived energy within two decades. Because of high capital costs for recovery facilities and limited markets, energy recovered from wastes will probably never meet more than 1% of the total energy demand in the United States. Although it will not take the place of other fuels for homes and transportation, RDF can be an important local source of energy for industrial boilers and some power plants.

Waste Reduction

A complement to landfilling, incineration, and resource recovery is **waste reduction,** that is, strategies that decrease the waste potential of consumer goods *before* they enter the waste stream. Waste-reduction strategies include bottle laws, cutting the amount of raw materials and packaging used for consumer goods, and designing products for durability.

Following Oregon's lead in 1972, eight other states (Vermont, Maine, Michigan, Connecticut, Iowa, New York, Massachusetts, and Delaware) have passed **bottle laws** (more formally referred to as *container deposit legislation*) that impose a deposit on plastic, aluminum, and glass soft-drink containers ($0.05 to $0.10 per container). Deposits are refunded when the containers are returned for subsequent reuse or recycling (fig. 20.10 *b*). Bottle laws are primarily litter-reduction strategies. In states without bottle laws, beverage containers constitute up to 60% of all litter. In bottle-law states, all but about 20% of soft-drink containers are returned. The majority of returned containers are not reused. Most are recycled, but some are simply landfilled. In fact, until recently, most plastic containers returned in Iowa were landfilled because that was less expensive than recycling.

Another waste-reduction strategy is the redesign of products so that each unit requires less raw material. That approach has a double advantage for automobiles. Cars that require less metal are smaller and lighter, and hence more fuel efficient. Spurred by foreign competition and by federal requirements to improve fuel economy, U.S. automobile manufacturers have trimmed the weight and size of most models now on the market. In fact, since 1975, the mass of a typical auto has been reduced by about 400 kilograms (880 pounds); about one-quarter of this was the result of substituting aluminum and plastic for steel. The trade-off is that plastic is more difficult to recycle than steel. (On today's cars, 10-15% of the weight is plastic, and it ends up in a landfill). In the future, U.S. auto manufacturers will be designing cars whose components, including plastics, will be easier to recycle. Already certain European auto manufacturers have adopted this design-for-disassembly approach, which is more cost-effective for recycling.

Packaging consumes large amounts of raw materials and creates tremendous amounts of waste. Packaging of consumer goods uses about 75% of the glass, 40% of the paper, 29% of the plastic, 14% of the aluminum, and 8% of the steel consumed each year in the United States. Typically, 30-40% of a city's solid waste consists of discarded packaging materials. One way to reduce such waste is to develop packaging methods that conserve raw materials.

Pending changes in the packaging of compact discs (CDs) show what can be done to reduce packaging. Traditionally, an 11.5-centimeter (4.5-inch) CD was packaged in a plastic "jewel box" inside a cardboard box about 30.5 centimeters (12 inches) long. This long box was chosen for security reasons and so that retailers could fit CD containers in existing record album racks. In response to consumer concern about excessive packaging, the Recording Industry Association of America in early 1992 agreed to abandon long-box CD packaging after April 1993. As of this writing, record companies and manufacturers are testing several alternative packaging plans, including recyclables.

Planned obsolescence, an operating principle of many important industries, especially since World War II, contributes significantly to our waste management problems. Advertisers encourage consumers to purchase new products (and dispose of old ones) by changing styles at regular intervals. We could counter the effects of that practice to some extent by developing more durable products that are more easily and economically repaired, but the most fundamental waste-reduction strategy is for each of us to avoid wasteful consumption and strive to use products efficiently.

During the 1990s, many communities are finding their waste-management problems compounded by declining landfill capacity and rising costs of potential landfill sites. One way to alleviate these problems for most communities may be **integrated waste management,** that is, some combination of strategies that reduces waste generation, encourages recycling, and utilizes incineration along with sanitary landfilling. Some of the things that all of us can do to reduce waste are listed in table 20.2.

In some large metropolitan areas, the cost of managing solid wastes is so high locally that they opt to export their refuse out of the region. This issue is explored on page 447.

Management of Hazardous Waste

The EPA considers hazardous-waste management as one of the principal environmental concerns of the 1990s. In this section, we describe how hazardous waste is classified, meth-

Table 20.2 What we can do to reduce the volume of refuse
1. Participate in community recycling programs.
2. If possible, use a mulching lawn mower, which returns grass clippings to the soil and reduces the need for fertilizer.
3. Compost yard and food waste.
4. As much as possible, reuse items such as paper and plastic bags.
5. Instead of disposing of old but still functional items such as clothing and appliances, donate them to charity.
6. Given the option, select durable rather than disposable goods (cloth rather than paper napkins, metal instead of plastic cutlery, ceramic coffee mugs instead of styrofoam cups).
7. Avoid overpackaged products (nonbiodegradable material, single-serving units).
8. Reject disposable diapers, razors, flashlights, and so on.
9. Remove your name from junk mailing lists.
10. Support by your patronage companies that minimize waste and practice recycling.

ods of predisposal treatment to reduce or eliminate the hazard, advantages of resource recovery, and disposal alternatives.

Waste Classification

Prior to disposal, an industry must first classify its waste as either hazardous or nonhazardous. Under provisions of the Resource Conservation and Recovery Act (RCRA) of 1976 and subsequent amendments, the EPA is responsible for defining hazardous wastes and protecting the public's health and the environment from them. The law directs the EPA to "develop and promulgate criteria for identifying the characteristics of hazardous (chemical) wastes, . . . taking into account toxicity (including carcinogenicity), persistence, and degradability in nature, potential for accumulation in tissues, and other related factors, such as flammability, corrosiveness, and other hazardous characteristics." In an attempt to meet those mandates, the EPA first had to develop policies that would review for possible hazardous properties the estimated 86,000 different chemical formulations produced by some 67,000 companies. Today, the EPA regulates several hundred commercial chemical products and the residues of certain manufacturing processes. Although an estimated one thousand chemicals are considered to be hazardous, the EPA, at present, does not have specific regulations for many of them.

If a waste is classified as hazardous, the particular hazardous substance(s) that it contains must be identified, so that the proper method of disposal can be selected. Heavy metals, for example, are not destroyed by incineration, so they are disposed of in a landfill designed to accomodate hazardous materials. On the other hand, organic wastes (e.g., pesticides and solvents) can be incinerated. Other methods of hazardous-waste disposal include chemical and biological detoxification, and deep-well injection.

Industries that generate more than 1,000 kilograms (2,200 pounds) of hazardous waste each month at any one site must

Exporting Garbage

Residents of Center Point, Indiana (population 250) saw their town landfill expand from 7.3 hectares (18 acres) to 25 hectares (62 acres) after local owners sold it to a Camden, New Jersey partnership in 1989. Each day 20-30 semitrailer rigs haul in garbage from New York and New Jersey.

In 1991, the Rosebud Sioux Tribal Council of South Dakota agreed to allow a Connecticut-based waste-management firm to construct a landfill on Sioux land. The proposed 2,000-hectare (5,000-acre) site is on what is thought to be an ancient Indian burial ground. The landfill will receive garbage from Denver, Minneapolis, and other cities. In return the tribe will receive about $1 per metric ton of waste.

Toronto, faced with imminent closure of its two landfills, plans to send 3,600 metric tons (4,000 tons) of compressed garbage each day by train to an abandoned open-pit mine at Kirkland Lake in northern Ontario. A recycling plant will be constructed at the mine, and the community will receive $388 million in benefits. All told, over the next twenty years, an estimated 27 million metric tons (30 million tons) of garbage will be dumped into the abandoned mine.

These are three of a growing number of examples of densely populated metropolitan areas that choose to export their solid waste for disposal. Typically, these cities are facing (1) closure of existing landfills either because they are nearing capacity or are environmentally unsound, or (2) excessive costs for developing new landfills or incineration facilities locally.

In the United States, transport of garbage is mostly from the Northeast to the Midwest, where landfill space is greater and tipping fees are usually lower (fig. 20.12). More than half the garbage traveling between states originates in New Jersey and New York. The number of operating landfills in New Jersey plummeted from about three hundred in the 1970s down to only eleven in 1991. Consequently, New Jersey now exports about 50% of its garbage. New York's garbage exports climbed from 5% in 1988 to 11% in 1990. An estimated 20-30% of garbage landfilled in Indiana is imported; in Ohio, garbage imports tripled between 1988 and 1990.

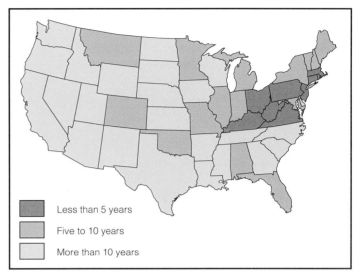

Figure 20.12   The life expectancy of landfills. As of 1991, the northeastern states had only ten years or less before landfills were at capacity.
Source: National Solid Wastes Management Association.

A backlash against this trend is gaining momentum. Some Midwestern states have tried to outlaw garbage imports entirely. To date, however, such efforts have been thwarted by federal court rulings that such prohibition unreasonably interferes with interstate commerce.* Some states have responded by boosting tipping fees for out-of-state garbage haulers or by requiring importers to adhere to strict local recycling regulations. Conservationists also warn that exporting garbage fosters an out-of-sight, out-of-mind attitude that does little to encourage recycling or waste reduction.

The EPA expects that about 75% of the nation's existing landfill space will be exhausted or closed by the turn of the century. Thus, in the coming years, more rural communities, especially those with economic problems, are likley to debate the issue of importing garbage for cash as large urban areas seek sites to unload their growing heaps of refuse.

*Only Rhode Island, which declared its single, large landfill a *state natural resource,* has been able to mandate limits on garbage imports.

report the nature of the material and the quantity that is generated to the EPA or to their state regulatory agency. In the case of a few extremely toxic substances, generation of as little as 1 kilogram (2.2 pounds) of waste per month must be reported. A manifest (cargo-invoice) system tracks the movement of hazardous wastes from their point of origin to their disposal site. All groups or organizations en route are responsible for that waste while it is in their possession. Copies of the manifest go to the appropriate regulatory agency to ensure that the waste was delivered to a designated disposal site and that it was disposed of properly.

Currently, thirteen states are net importers of hazardous waste; for example, 90% of the hazardous waste treated in Alabama comes from outside the state. There is also a flourishing international traffic in hazardous waste. Faced with high disposal costs at home, some U.S. industries have shipped their hazardous waste abroad. Typically, the waste ends up in less-developed nations, where environmental regulations are lax and dollars for waste disposal are welcome. To date, the United States does not have a law banning hazardous-waste exports. Many less-developed nations, however, are closing the door on such imports. For example, in January 1991, all African states except South Africa, signed the Bamako Convention prohibiting trade in hazardous waste. Of course, the individual nations must pass and enforce laws in accord with this international agreement in order for the ban to be effective. As of this writing, many South and Central American countries also have banned hazardous waste imports.

Predisposal Treatment

Many procedures are available to reduce or eliminate the hazards or toxicity of hazardous wastes so they can be managed in traditional ways. Such procedures constitute **predisposal treatment.** Disposal costs for hazardous waste are considerably higher than for nonhazardous waste, hence, an economic incentive exists for predisposal treatment.

The toxicity of some hazardous wastes is virtually eliminated by mixing them with various solidifying agents, such as cement, bitumen, and glass. Also, these materials effectively seal wastes from the leaching action of water. Another alternative is to encapsulate wastes in plastic containers that can be disposed of in sanitary landfills.

Some industries, especially the petroleum industry, have found that soil microorganisms can detoxify certain types of waste. Refineries produce small amounts of unusable toxic wastes that can be incorporated into the soil where bacteria decompose and thereby detoxify them. Tilling, which aerates the soil, accelerates the decomposition rate. The EPA estimates that more than half of the 275 refineries in the United States are using this so-called **land-farming** technique to dispose of toxic wastes.

Many chemical industries recognize that the best way to manage hazardous wastes is to avoid generating them in the first place. New chemical plants are incorporating technologies that minimize the generation of hazardous wastes. Unfortunately, it is often too costly to retrofit older plants.

Resource Recovery

In some cases, the hazardous-waste stream can be reduced by recovering and marketing discarded chemicals. For example, at one time the Monsanto Company landfilled a toxic acid byproduct of a herbicide that it manufactured. Because the chemical is useful for removing sulfur emissions from coal-fired electric power plants, the company now sells that acid to Cities Utility of Springfield, Missouri, for use in its scrubbers. Both companies benefit; waste-disposal costs are eliminated, and the cost for chemicals to remove sulfur from stack gases is reduced.

In some areas, a **waste exchange** facilitates marketing of waste chemicals that are useful to another industry. Waste exchanges act as brokerage houses; industries submit a list of available waste chemicals to the waste exchange, and prospective customers examine the list and purchase what they need. About a dozen waste exchanges currently operate in the United States.

Some industries have found that they can recover valuable products from their own wastes. Silver is recovered from photographic wastes, and industrial solvents from the liquid wastes in electroplating facilities. When chlorinated hazardous wastes (certain solvents and some plastics) are destroyed by incineration, hydrochloric acid can be recovered. In fact, enough is recovered each year to provide approximately 90% of the hydrochloric acid used in the United States. Usually, these facilities also recover much of the heat energy that is liberated by incineration.

For industries faced with the staggering costs of hazardous-waste disposal, resource recovery can offer significant economic incentives. Ciba-Geigy, for example, spent about $300,000 between 1985 and 1989 on new recycling equipment and process modification to minimize waste generation at its Toms River, New Jersey, plant. This investment allowed the company to save more than $1.8 million in reduced disposal costs.

Disposal Alternatives

The three principal methods of disposal of hazardous wastes are secure landfills, incineration, and deep-well injection. Hazardous wastes may be disposed of at special landfill sites regulated by the EPA or state agencies that are designated as **secure landfills** (fig. 20.13). Secure landfills must meet strict requirements for operation; for example, they must be higher than the hundred-year floodplain* and away from fault zones. They must have impermeable liners along with a network of pipes (beneath the landfill and just above the liner) to collect leachate, which must be pumped out and treated. Monitoring wells are also required to check the quality of groundwater in the area. The Resource Conservation and Recovery Act also mandates specific procedures for operating secure landfills and for monitoring them after they are filled.

Incineration is used to dispose of combustible liquid and solid hazardous wastes, such as solvents, pesticides, and petroleum-refinery wastes. This method greatly reduces the volume of these wastes, but more importantly, incineration destroys organic chemicals by converting them to carbon dioxide, water, and other gases that are removed by scrubbers (chapter 16). Many incinerators also recover some of the energy released during combustion. At present, only 1-3% of the hazardous waste generated in the United States is incinerated.

Before a particular type of hazardous waste can be incinerated, the EPA must authorize burning that type of waste. That requirement is especially important when the chemical substances to be destroyed are very stable at high temperatures (PCBs, for example). Low-temperature incinerators ($650°$ C to $850°$ C) do not completely destroy those

*The hundred-year floodplain is an area that might be expected to flood once in one hundred years.

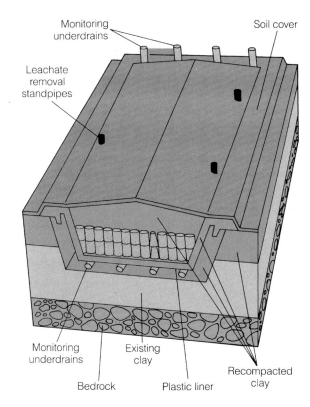

Monitoring underdrains

Leachate removal standpipes

Soil cover

Monitoring underdrains

Existing clay

Recompacted clay

Bedrock

Plastic liner

Figure 20.13   A cross section of a secure landfill.

substances and thus release them to the atmosphere, so high-temperature incinerators (around 1100° C) must be used. For EPA authorization, incinerators must demonstrate that they can destroy 99.99% of the combustible toxic wastes, and must be equipped with air-pollution control devices that remove acidic and noncombustible by-products.

Deep-well injection is a waste-disposal technique that the petroleum industry developed in the late 1800s. Oil exploration and extraction often bring brines to the surface, and the harmful effects of these solutions can be avoided by pumping them back into the subsurface rock formation from which they were drawn. Today, this same technology is used to dispose of other types of hazardous liquid wastes.

Only certain kinds of subsurface rock formations are suited for deep-well injection. The formations must be deep, porous enough to provide adequate storage space, and sandwiched between impermeable layers of rock. In addition, the formation should not contain any valuable mineral or fuel resources. Installing a deep injection well requires drilling and encasing the shaft to a depth of 1,000 meters (3,250 feet), and such wells can cost $1 million or more; however, their operating cost is relatively low (except when the casings corrode and have to be replaced). Some authorities oppose deep-well injection primarily because of potential blowouts, leaks, and ground tremors. And, for these reasons, some states have banned deep-well injection of wastes.

Two of the more perplexing problems related to hazardous waste management are abandoned waste sites and illegal dumping. These topics are described in the issue on page 450.

Management of Radioactive Waste

Since scientists first learned how to manipulate the nucleus of the atom a half-century ago, society has had to contend with radioactive wastes. Nuclear weapons production and nuclear power plants generate most of today's radioactive wastes. These wastes differ from other hazardous wastes in several significant ways. Hence, we describe the unique properties of radioactive materials, so that we can understand the extraordinary precautions required to safely handle and store them. In addition, we consider prospects for long-term storage of radioactive wastes.

Unique Characteristics

Some nuclei of certain elements are unstable and are called **radioactive isotopes.*** They become more stable by spontaneously emitting high-energy rays and/or particles, collectively called **ionizing radiation.** When this radiation strikes a molecule, the molecule is converted to an **ion** (that is, a molecule or molecule fragment that carries either a positive or negative electrical charge) or other reactive fragments.

Radioactive isotopes emit three types of ionizing radiation: alpha particles, beta particles, and gamma rays. (Recall from chapter 15 that gamma rays are a form of electromagnetic radiation.) All three types of radiation are sufficiently energetic that they break chemical bonds in any medium through which they travel, and that is the reason for their potentially adverse effects on human health.

When ionizing radiation penetrates living cells, it collides with and breaks apart molecules in the cells. The health effects of this molecular damage can be categorized as either somatic or genetic. **Somatic effects** are direct impacts on the human body and often include the conversion of normal cells to cancer cells and other changes that can shorten life expectancy. Large doses of ionizing radiation can lead to radiation sickness (vomiting, diarrhea, and nausea) and possible death. **Genetic effects** are changes in the genetic makeup of sex cells (egg and sperm) that are transferred to offspring. Scientists have only a limited understanding of the genetic effects caused by exposure to radioactive materials, and currently such effects can only be detected as an increased (above-average) incidence of specific health problems in the children of people who have been exposed to ionizing radiation (chapter 12).

Radioactive isotopes differ in the types of high-energy radiation they emit. Some are primarily alpha emitters; others are primarily beta or gamma emitters. Some emit more than one kind of radiation; uranium-238 emits both alpha particles and gamma rays, for example. This difference is important because all three types of radiation differ in their ability to penetrate materials (fig. 20.14). For example, it takes a block of concrete that is 1-2 meters (3-7 feet) thick to stop gamma rays, which are the most energetic form of ionizing radiation. Thus, gamma rays readily penetrate an entire organism, resulting in what is known as **whole-body exposure.**

*The nuclei of isotopes of an element contain the same number of protons, but the number of neutrons varies.

Abandoned Hazardous-Waste Dumps and Illegal Dumpers

*I*n the United States, more than 31,000 abandoned or inactive hazardous-waste dumps pose significant threats to human health and the environment. In addition, an estimated 375,000 leaking underground storage tanks threaten groundwater quality (chapter 14).

The majority of chemical dump sites received wastes before any regulatory guidelines were established. Many of the owners of these sites are no longer in business or cannot afford to clean up their leaky dumps. Most of the problem sites are located in heavily industrialized parts of the East and Midwest and are associated with chemical and petrochemical industries.

When legal responsibility can be established, the owners of a hazardous-waste site must pay the cleanup costs. However, records of who dumped what, when, where, and in what quantity are usually nonexistent, which makes the establishment of financial liability difficult if not impossible. For example, at one site investigators determined that more than one hundred companies had dumped many different kinds and amounts of wastes. To deal with such problems Congress passed the Comprehensive Environmental Response and Compensation Act of 1980. That legislation, more commonly referred to as the Superfund Act, provided funding for cleaning up sites where legal responsibility cannot be established or where the owners cannot afford to clean up their wastes.

Environmentalists and the Congressional Office of Technological Assessment are concerned about the relatively slow pace of the EPA's administration of the Superfund. By 1992, 1,134 sites were on the EPA's National Priorities List (NPL); that is, they were identified as hazardous-waste dumps in need of immediate and major cleanup. Of those NPL sites, about 70% were under study, 20% had cleanup underway, and only 4% had cleanup completed. The time from site identification to the onset of cleanup activities has averaged 7-9 years, and actual cleanup has typically taken 2-3 years.

The slow pace of dealing with hazardous dump sites is attributed in part to insufficient funding. Estimated cleanup costs average $21 million to $30 million per site, so that the total bill for cleaning up all 1,224 NPL sites will probably exceed $30 billion, about twice the amount authorized by Congress for the Superfund since 1980. Furthermore, this is only a small fraction of the total cost of cleaning up the almost 30,000 abandoned or inactive hazardous-waste sites not on the National Priorities List. The total bill for remediation of all hazardous-waste dumps could top $1 trillion.

Unfortunately, the relatively high costs for disposal of hazardous waste have provided an incentive for so-called midnight dumpers who dispose of hazardous wastes illegally. A few companies that generate hazardous wastes subcontract disposal of those wastes to firms offering such services, paying the going rate for disposal. Then, rather than pay the high fees charged by a secure landfill or incinerator, unscrupulous hazardous-waste disposal firms collect the wastes and simply dump them into a storm sewer, along an isolated roadway, or into a nonapproved dump site, and pocket the profit.

Illegal disposal of hazardous waste continues, and no one knows the full scope of the problem. A 1985 report prepared by the General Accounting Office (GAO) indicated that the EPA received more than 240 allegations of illegal disposal between 1982 and 1984—more than it can check. Some so-called midnight dumpers have been linked to organized crime, and the FBI and state police have become involved in efforts to stop them. In fact, it has been proposed that EPA officials be given the power to make arrests and carry weapons. The EPA, for the first time in its history, has criminal investigators on its staff. Of the 36 disposal cases involving enforcement proceedings (included in the 1985 GAO report) charges were brought against 27 generators of hazardous waste and 9 transporters. Of the 36 cases, 34 were found as the result of tips from private citizens, employees, or local government employees. EPA inspectors found the other 2 cases.

Beta particles (high-energy electrons) have less penetrating power than gamma rays. To protect oneself from beta particles from an external source, a person need only place several centimeters of solid material (this book, for example) between herself and the source. Even body tissue impedes the penetration of beta particles. Thus, although the skin and tissue directly beneath it can be damaged by an external source of beta particles, internal organs are generally protected. If a material that emits beta particles becomes concentrated within an organ of the human body, that organ will receive a higher dose, perhaps resulting in significant local radiation damage.

Strontium-90 is a beta emitter with chemical properties similar to those of calcium, so that strontium concentrates in bone. If a person ingests food (e.g., milk) that is contaminated with strontium-90, the strontium-90 would be taken up by the blood and concentrate in the bones. Since bone marrow produces red blood cells, one possible consequence of such an accumulation is leukemia.

Alpha particles (each composed of two protons and two neutrons) are the least penetrating of the three types of ionizing radiation. Even the most energetic alpha particles cannot penetrate beyond the outer layer of skin. Thus, exposure to sources of alpha particles can be prevented by a thin

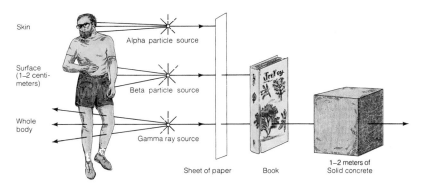

Exposure to external radiation source

Skin — Alpha particle source

Surface (1–2 centimeters) — Beta particle source

Whole body — Gamma ray source

Sheet of paper   Book   1–2 meters of Solid concrete

Figure 20.14   The relative penetrating power of particles and rays emitted by radioactive materials.

protective covering, such as a sheet of aluminum foil, placed over the source. However, alpha emitters that enter the body through breathing or ingestion can cause serious health problems. As we saw in chapter 16, breathing air that contains a relatively high concentration of radon gas can cause lung damage, because alpha particles are emitted by the decay products of radon. Protection from gaseous alpha emitters requires that they be stored in closed gas containers or diluted through better ventilation.

Ionizing radiation is measured in rems (Roentgen equivalent man). One millirem is 0.001 rem. Average background radiation from cosmic rays and naturally occurring radioactive isotopes in soil and bedrock expose people to 100 to 250 millirems per year. Medical and dental X rays add another 80 millirems. Other sources of background radiation include nuclear weapons-testing fallout, airline travel, and television sets; together these account for an additional 10 millirems. Thus, the average total exposure is 190–340 millirems per year. For comparison, continuous exposure to more than 1 million millirems over a period of two weeks can result in death.

As a radioactive isotope emits ionizing radiation, it gradually transforms to a stable form that is no longer radioactive. That shift toward stability is called **radioactive decay,** and scientists specify the rate of decay of a particular material by measuring its **half-life,** the time it takes for half the nuclei of a radioactive material to decay to a more stable form. Figure 20.15 illustrates the radioactive decay of phosphorus-32 (^{32}P), which has a half-life of fourteen days. Phosphorus-32 decays to sulfur-32 (^{32}S), which is a stable, nonradioactive isotope. If we start with 1 gram of phosphorus-32, after fourteen days, 0.5 gram of phosphorus-32 remains. The other 0.5 gram has converted to sulfur-32. After another 14 days have passed, 0.25 gram of phosphorus-32 remains, and so on.

Half-lives of radioactive wastes produced in nuclear reactors vary widely—from 0.96 seconds for xenon-143 to as long as 160 million years for iodine-129 (table 20.3). Usually at least ten half-lives must elapse before a radioactive isotope decays to the point that it is no longer considered to be a

serious hazard. After ten half-lives have elapsed, only one one-thousandth of the original radioactivity remains. Thus, the half-life of a radioactive isotope is important in determining how long it must be kept isolated from exposure to humans (and all other organisms).

The long half-lives of many radioactive wastes mean that they persist in the environment. Thus, once radioactive wastes enter ecosystems, they may accumulate in food webs. This is one of the lessons learned from above-ground (atmospheric) testing of nuclear weapons. Fallout from bomb testing in the 1950s spread strontium-90 over cattle grazing land. When the cows ate the grasses, they also ingested the strontium-90. As noted earlier, strontium has properties similar to calcium, and thus some of it found its way into cows' milk. When people drank the contaminated milk, they also accumulated small amounts of strontium-90 in their bones. This transfer of strontium-90 through food webs to humans was one of the reasons that atmospheric testing of nuclear weapons was halted.

Sources of Radioactive Waste

Nuclear power plants and nuclear weapons production facilities produce high-level radioactive waste, that is, isotopes that emit intense levels of ionizing radiation and have long half-lives. Low-level wastes pose much less of a health threat and are more easily managed. The estimated 16,000 sources of low-level wastes include hospitals, research laboratories, universities, and some industries.

Radioactive wastes, including low-level wastes, are generated at about one-sixtieth the rate of other hazardous chemical wastes. Although their volume is relatively small, the danger per unit mass is extremely high, especially for high-level and **transuranic wastes**—radioactive, alpha-emitting elements that are heavier than uranium. Those wastes are isolated during the reprocessing of nuclear fuel and nuclear weapons production.

Recall our description of the operation of a nuclear power plant in chapter 18 (see Special Topic 18.2). High-level

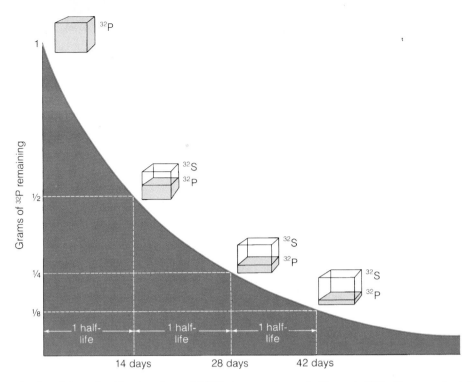

Figure 20.15 The amount of radioactivity emitted by phosphorus-32 (^{32}P) decreases by one-half every half-life (fourteen days). The decay of ^{32}P results in the formation of the stable isotope, sulfur-32 (^{32}S).

Table 20.3	Half-lives of typical radioactive waste components
Radioisotope	**Half-Life (years)**
americium-241	433
americium-242	150
cesium-135	3,000,000
cesium-137	30
curium-242	0.45
curium-243	32
curium-244	18
iodine-129	17,000,000
iodine-131	0.022
neptunium-237	2,100,000
plutonium-239	24,000
plutonium-241	13
radon-226	1,600
strontium-90	28
technetium-99	200,000
thorium-230	76,000
tritium	13

Figure 20.16 Radiation emanating from spent nuclear fuel-rod assemblies under water.
© Science VU/Visuals Unlimited

radioactive wastes accumulate in the spent **fuel-rod assemblies** (fig. 20.16) that are stored in water-filled basins at nuclear power plants. During annual refueling operations, a typical nuclear power plant (1,000 megawatts) adds approximately 30 metric tons (33 tons) of spent fuel rods to its storage facility. In late 1989, a Congressional panel warned that

without a permanent repository, we can expect a great surge in the amount of nuclear waste stored at power plants, especially as existing reactors complete their life cycles between 2009 and 2030. They project that the cumulative amount of spent fuel will reach 87,000 metric tons (97,000 tons) by 2045.

The annual addition of spent-fuel assemblies to *temporary* storage facilities is already causing a storage problem at many nuclear power plants. Over the next several years, storage problems probably will be dealt with by expanding the size of

(a)

(b)

Figure 20.17   Steel storage tanks for high-level liquid radioactive wastes at the U.S. Department of Energy's Hanford, Washington, facility. These are views of the tanks (*a*) before and (*b*) after burial.
U.S. Department of Energy

temporary storage facilities and by storing older spent fuel-rod assemblies closer together.

High-level military wastes come from nuclear-powered submarines and nuclear reactors designed primarily to produce plutonium for weapons. Spent fuel rods from these sources are reprocessed to extract unused uranium (which is recycled) and plutonium (for weapons). Reprocessing generates high-level radioactive liquid wastes. Currently, those

reprocessing wastes are stored in steel tanks at federal facilities at Hanford, Washington, and Savannah River, South Carolina (fig. 20.17). The long-term plan is to convert the liquid waste to a solid (most likely a kind of glass) and deposit it in a permanent nuclear waste repository.

Because much military waste is in liquid rather than solid form, it occupies considerably greater volume than the spent fuel of commercial reactors. On the other hand, because

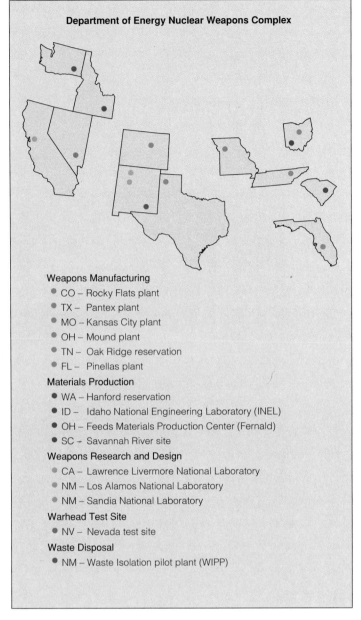

Department of Energy Nuclear Weapons Complex

Weapons Manufacturing
- CO – Rocky Flats plant
- TX – Pantex plant
- MO – Kansas City plant
- OH – Mound plant
- TN – Oak Ridge reservation
- FL – Pinellas plant

Materials Production
- WA – Hanford reservation
- ID – Idaho National Engineering Laboratory (INEL)
- OH – Feeds Materials Production Center (Fernald)
- SC – Savannah River site

Weapons Research and Design
- CA – Lawrence Livermore National Laboratory
- NM – Los Alamos National Laboratory
- NM – Sandia National Laboratory

Warhead Test Site
- NV – Nevada test site

Waste Disposal
- NM – Waste Isolation pilot plant (WIPP)

Figure 20.18 Facilities in the Department of Energy's nuclear weapons complex.

Source: Data from the Office of Technology Assessment, 1991.

uranium and plutonium are removed prior to disposal, military waste is only about 10% as radioactive as the waste from nuclear power plants.

Hanford and Savannah River are two of fifteen facilities comprising the nation's Nuclear Weapons Complex (fig. 20.18). More than four decades of improper disposal practices at many of these facilities have left an extensive legacy of contaminated soil, sediment, groundwater, and surface water. Nearly a million liters (260,000 gallons) of liquid wastes have leaked from storage tanks at Hanford and Savannah River. The worst spill occurred in 1973, when 450,000 liters (120,000 gallons) leaked into the ground from a tank at the Hanford site. Fortunately, no human casualties (as far as we know) have resulted from any of those spills. Most of those wastes

were retained in the soil at the site, although at Hanford, a small amount seeped through the soil and into the Columbia River, only 16 kilometers (10 miles) away.

The Department of Energy (DOE), which administers the Nuclear Weapons Complex, has drawn up a five-year plan for cleaning up contaminated sites. The cost of environmental restoration and waste management activities is expected to top $20 billion by 1996, and this is only the initial phase. DOE's ultimate goal is to clean up all sites within thirty years, an enormous project that is likely to cost hundreds of billions of dollars. However, in 1991, the Congressional Office of Technology Assessment reported that the DOE plan may be either too costly or technically impossible to achieve. In any event, as of this writing, very little cleanup is taking place, and waste is likely to remain stored at those sites until a permanent repository is operational.

Permanent Storage

Decay of radioactive isotopes is spontaneous and cannot be prevented. Therefore, if we are to protect life from exposure to these sources of hazardous radiation, we must isolate radioactive wastes, so that they have virtually no chance of reaching the earth's surface and cycling through the environment. Because ten half-lives must elapse before radioactive waste is considered safe, and because many radioisotopes have half-lives of decades or longer, how long must those wastes remain isolated?

The EPA considers ten thousand years to be a sufficient time span for storage of spent fuel rods because it takes that long for the total hazard (effective radiation dose) to diminish to the natural radioactive decay rate of uranium ore. In 1983, a National Academy of Sciences panel evaluated the technology for containment of radioactive wastes and reconfirmed the ten thousand-year storage criterion set by the EPA. They ruled that *effective disposal* would limit the release of radioactive materials to levels no more than 10% higher than the radiation dose we normally receive from natural background radiation. The panel concluded that only a small fraction of the radioactivity in properly designed storage facilities could ultimately reach the environment during the ten thousand-year storage period proposed by the EPA. Furthermore, they concluded that a storage system does not have to be effective in safeguarding people and the environment beyond 10,000 years because only small amounts of radiation would be released slowly for hundreds of thousands of years.

Several methods have been proposed for permanent storage of nuclear waste. Today, scientists consider burial in deep underground repositories to be the most feasible. Such a facility must be (1) dry, (2) at least 600 meters (2,000 feet) below the surface, and (3) in a stable geological formation that is chemically nonreactive. Potential sites must be evaluated with the same criteria used for storage facilities for hazardous chemical wastes, that is, groundwater movement, rock type and strength, potential for earthquake activity, possibility of human intrusion, surface characteristics, economic

Figure 20.19 The nation's first permanent repository for high-level radioactive waste is proposed for Yucca Mountain, a barren volcanic ridge 150 kilometers (93 miles) northwest of Las Vegas, Nevada.
U.S. Department of Energy

impact on nearby communities, and impact on the environment. Suitable sites require approximately 160 hectares (400 acres) of land, with an additional 1,200 hectares (3,000 acres) serving as a protective buffer. Other drilling and mining activities in the area would be strictly forbidden.

Isolation of radioactive wastes in an underground repository will be achieved through a system of natural and engineered barriers. First, the wastes will be solidified into a durable, leach-resistant material, such as vitrified glass, which is similar to a ceramic glaze. Next, solidified wastes will be sealed in corrosion-resistant canisters, which will be placed in holes lined with corrosion-resistant sleeves in the floor of the repository. Each sleeved hole will then be surrounded by materials that have high sorptive characteristics, so the migration of chemicals will be prevented if water should somehow leak into the repository. (*Sorption* is the property of a material to remove soluble chemicals from water.) Initially, waste canisters will be monitored and remain retrievable if problems arise. However, if monitoring indicates that future problems are unlikely, the holes will be sealed, and the entire excavation will be filled with sorptive materials. With such precautions, even if a repository were to experience a problem, only very small releases would be likely.

To hasten the disposal schedule, President Reagan signed the Nuclear Waste Policy Act in early 1983, which authorized the Department of Energy to locate two sites for high-level nuclear waste repositories. Finally, in the summer of 1989, Yucca Mountain, Nevada, was selected as the nation's first permanent repository for high-level radioactive waste. Identification of the second site, expected to be in the eastern portion of the country, has been delayed indefinitely.

Yucca Mountain is located about 150 kilometers (93 miles) northwest of Las Vegas (fig. 20.19). The plan is to excavate the repository out of layers of compacted volcanic ash, known as *tuff.* Among the advantages of the Yucca Mountain site are (1) relatively deep water table (more than 615 meters [2,000 feet] below the surface), (2) isolation from population centers, (3) ownership by the federal government, and (4) prior conduction of nuclear tests nearby. Still, five years of detailed on-site evaluation are needed. This will involve excavation of tunnels and emplacement and monitoring of a small number of containers. The cost of site evaluation alone is likely to exceed $1 billion.

Not surprisingly, opposition to the site selection has been intense. By law, the state governor or legislature, or an Indian tribe affected by the site selection has sixty days to submit a notice of disapproval to Congress. Once that happens, the site remains unapproved unless Congress, within ninety days, overrides the state or tribal disapproval. Soon after final selection, the State of Nevada refused to grant permits for preliminary surface clearing and drilling. Numerous suits have been filed to block the project but, as of this writing, all suits have been dismissed. Nonetheless, opposition is likely to delay the opening of the repository until at least 2010.

The nation's first permanent repository for low-level nuclear waste has been constructed at a site about 40 kilometers (25 miles) east of Carlsbad, New Mexico. At the Waste Isolation Pilot Plant (WIPP), the Department of Energy plans to store encapsulated mixtures of transuranic and hazardous wastes in chambers carved out of salt beds 650 meters (2,100 feet) below the surface. The plan has drawn considerable fire from many people living near the WIPP site as well as along likely highway routes for trucks transporting waste to the site. Questions raised concern the security of containers, the potential for accidental leakage while containers are en route, and whether the DOE has sufficiently informed citizens of the composition and potential hazards of the waste. While the facility is now ready to receive waste, as of this writing, environmental and safety concerns have delayed opening. Initially, the WIPP site is scheduled to receive up to 8,500 drums containing 200 liters (55 gallons) each of waste during a five-year test phase. This volume represents only 1% of the facility's total capacity.

Conclusions

Public concern about human health risks and environmental quality have resulted in laws that require municipalities and industries to adopt better methods of solid waste management. Strict federal legislation regulates where and how municipal wastes, hazardous chemical wastes, and radioactive wastes can be disposed of, stored, or both. The key to safe disposal is proper isolation. In most cases, eliminating or minimizing migration of water through disposal sites is the most critical factor. Although new disposal technology should lessen future problems, the enormous task of cleaning up more than 31,000 old, leaky, environmentally unsound disposal sites still remains for the 1990s and beyond.

Reclamation of materials and energy from wastes is an environmentally sound alternative to waste disposal. Rising landfill costs are making waste-reduction strategies and recycling more attractive economically. In the future, pressure will no doubt increase for alternatives to landfilling, reduction of hazardous-waste generation, and permanent repositories for radioactive waste.

Key Terms

leachate
fly ash
hazardous waste
biodegradable waste
nonbiodegradable waste
sanitary landfill
trench landfill
area landfill
mounded landfill
municipal incinerator
ocean disposal
resource recovery
reclamation
recycling
composting
refuse-derived fuel (RDF)
waste reduction
bottle laws

planned obsolescence
integrated waste management
predisposal treatment
land-farming
waste exchange
secure landfill
deep-well injection
radioactive isotopes
ionizing radiation
ion
somatic effects
genetic effects
whole-body exposure
radioactive decay
half-life
transuranic wastes
fuel-rod assemblies

Summary Statements

Waste is a concept unique to humankind. Nature reuses and recycles.

Our activities generate a wide variety of municipal, mining, industrial, and agricultural wastes. Proper management requires evaluation of the basic properties of waste. We distinguish between hazardous and nonhazardous, biodegradable and nonbiodegradable, and persistent and nonpersistent wastes.

Disposal alternatives for municipal wastes include sanitary landfilling and incineration. Unless properly regulated, those measures may adversely affect human health and environmental quality.

Trench landfills are appropriate where the soil is thick and the water table is deep. Area landfills are constructed in canyons or surface-mined areas. Mounded landfills are necessary where the water table is near the surface. With landfills, protection of groundwater quality is the primary environmental concern and all new ones require impermeable liners.

Incineration offers many advantages over sanitary landfilling, especially where land prices are soaring and where heat from the incineration process is recoverable. Incineration greatly reduces the volume of waste that must be landfilled; however, incinerators must be equipped with effective air-quality control devices, and the solid residue may be hazardous.

Ocean disposal is used primarily for nonpolluted dredge spoils and is regulated by the U.S. Army Corps of Engineers and the EPA.

Recycling and waste-reduction strategies lessen the demand for landfill space and add to the supply of energy and material resources. The organic portion of nonhazardous waste may be composted or used as a fuel.

The EPA sets criteria for classifying waste as hazardous. Hazardous waste may be flammable, corrosive, reactive, or toxic.

Predisposal treatment and resource recovery reduce the amount of hazardous waste that must be disposed. Some hazardous wastes can be treated or encapsulated so that they can be managed as nonhazardous materials.

Hazardous-waste disposal methods include secure landfills, incineration, and deep-well injection. Disposal costs for hazardous wastes are much higher than those for municipal wastes. Illegal dumping of hazardous wastes enables unscrupulous disposal firms to realize large profits and is an environmental nightmare.

Alpha particles, beta particles, and gamma rays are products of radioactive decay. Gamma rays are the most penetrating of these three forms of ionizing radiation, and can cause whole-body exposure.

High-level radioactive wastes must be isolated to protect organisms from exposure to ionizing radiation and to prevent radioactive substances from cycling through the environment. The disposal technology for radioactive wastes has been developed and is being tested. At present, deep underground storage is the preferred method for managing high-level radioactive wastes.

Questions for Review

1. Explain how waste is a concept unique to humanity.
2. How is solid waste classified by origin? How do these categories of waste compare in magnitude?
3. Distinguish between wastes that bioaccumulate and those that are biodegradable. Is it possible for a substance to have both properties?
4. Explain why heavy metals such as lead and cadmium persist in the environment. Do all hazardous wastes have the same potential to bioaccumulate in food webs?
5. What are the major advantages of sanitary landfills over open dumps? Are there disadvantages of sanitary landfills?
6. Distinguish among trench, area, and mounded landfills. What is the principal environmental concern in designing a sanitary landfill?
7. Is the ocean's assimilative capacity for waste limited? Explain your response.
8. Identify and describe several methods of resource recovery. What are the environmental advantages of resource recovery?
9. Speculate on why so few states have adopted so-called bottle laws.
10. Contrast a sanitary landfill with a secure landfill.
11. Summarize the various methods for disposal of hazardous wastes. Arrange the alternatives in order, from the least costly methods to the most costly ones.
12. Have conflicts occurred in your community regarding the siting of landfills or resource-recovery centers? If so, what factors contributed to those conflicts?
13. Should metals be given priority over glass and paper in recycling efforts? Support your opinion.
14. Suggest several reuses in the home for each of the following common household items: (a) newspapers, (b) metal cans, and (c) nonreturnable glass bottles.
15. How does the proper management of solid waste reduce air-pollution problems? Identify some incentives that would reduce the per-capita generation of wastes.
16. What precautions must be taken during deep-well injection of hazardous waste to protect the quality of groundwater?
17. Distinguish among the penetrating power of alpha, beta, and gamma rays. How is penetrating power linked to potential health effects?
18. What fundamental property of radioactive substances determines how long they persist in the environment?
19. What happens to radioactive wastes generated at nuclear power plants? How are they to be ultimately disposed?
20. Compare and contrast radioactive wastes produced by a nuclear power plant with those produced by nuclear weapons development.

Projects

1. How would you assess your community's strategies for solid-waste management? Is progress evident or are conflicts interfering with progress? What can be done to help resolve the difficulties?
2. What types of waste-reclamation efforts are underway in your community? What other materials might be recovered? Describe your role in this effort.
3. Arrange a field trip to a sanitary landfill, a public recycling center, and a local industry that recycles.

Selected Readings

Frosch, R. A., and N. E. Gallopoulos. "Strategies for Manufacturing." *Scientific American* 261 (Sept. 1989):144-52. Argues for a closed industrial ecosystem that emphasizes resource recovery and recycling.

Goldoftas, B. "Recycling: Coming of Age." *Technology Review* 90 (Nov. 1987):28-35. A detailed account of the rising popularity of community recycling.

Kovacic, D. A., R. A. Cahill, and T. J. Bicki. "Compost, Brown Gold or Toxic Trouble?" *Environmental Science and Technology* 26 (1992): 38-41. Raises questions concerning potentially hazardous components of composted yard waste.

Lenssen, N. "Confronting Nuclear Waste" in *State of the World 1992,* The Worldwatch Institute, New York: W. W. Norton & Company, 1992. Includes a discussion of the political aspects of nuclear waste management both in the United States and abroad.

Lewis, J. "Superfund, RCRA, and LUST: The Clean-Up Threesome." *EPA Journal* 17 (July/Aug. 1991): 7-14. Describes the steps involved in remediation of hazardous-waste sites.

Moore, J. N., and S. N. Luoma. "Hazardous Waste from Large-Scale Metal Extraction." *Environmental Science and Technology* 24, no. 9 (1990): 1278-85. Describes the production of hazardous waste as a by-product of mining.

O'Leary, P. R., P. W. Walsh, and R. K. Ham. "Managing Solid Waste." *Scientific American* 259 (Dec. 1988):36-42. Discusses the advantages of a system of integrated solid-waste management.

Rathje, W. L. "Once and Future Landfills." *National Geographic* 179 (May 1991):116-34. A University of Arizona archaeologist describes the excavation of landfills.

Slovic, P., J. H. Flynn, and M. Layman. "Perceived Risk, Trust, and the Politics of Nuclear Waste." *Science* 254 (1991): 1603-07. Traces the basis for public mistrust of the nation's management of nuclear waste.

Suflita, J. M., C. P. Gerba, R. K. Ham, A. C. Palmisano, W. L. Rathje, and J. A. Robinson. "The World's Largest Landfill." *Environmental Science and Technology* 26 (1992): 1486-95. A multidisciplinary study of the content and properties of the Fresh Kills landfill.

Chapter 21

Land-Use Conflicts

Our growing demands on finite lands for a
variety of purposes, such as housing, can stress
the environment and conflict with other uses,
such as farming, recreation, and habitat
preservation.

© Greg Vaughn/Tom Stack & Associates

458

*C*edar Island is a narrow strip of sand some 9.5 kilometers (6 miles) long situated just off Virginia's eastern shore. Since the mid-1980s, the island has been the focus of controversy between prodevelopment interests and groups that want to protect the island from further development. Currently, the privately-owned island has several dozen cottages and no roads. Residents must use private generators for electricity, and access is by boat only. The island is home to nesting colonies of several species of terns and other shorebirds, including the piping plover, a threatened species. The U.S. Fish and Wildlife Service has recommended wildlife refuge status for the island.

Because of its exposed location facing the open Atlantic, Cedar Island is vulnerable to powerful storm waves and flooding. Erosion washes away an average of 4.5 meters (15 feet) of beach each year, and in the past decade several shoreline cottages have been lost to the sea. In such an area, law prohibits use of federal funds for development activities (roads and bridges, for example) or flood insurance. Despite this financial obstacle and over objections by conservation groups and government agencies, developers have gone ahead and subdivided the island into lots, and many have already been sold.

Cedar Island, Virginia, is just one example of how conflicts over land use can pit human activities against the forces of nature. In the end, of course, nature prevails, but along the way the environment may be seriously disrupted. Today, more people realize that working with rather than against nature is the best way to resolve conflicts over land use. For example, we no longer view wetlands as worthless; today, we recognize their value for flood control and water filtration (chapter 10).

In this chapter, we examine how human activity affects the coastal zone, the dynamic interface between land and sea. We also examine how land use is limited by various natural hazards (floods, earthquakes, and volcanic eruptions). We close with a discussion of the environmental stress caused by community growth and the strategies employed by communities to control that growth.

Land-Use Conflicts: The Coastal Zone

The **coastal zone** broadly encompasses the relatively narrow region that is transitional between the continent and the open ocean. Hence, the coastal zone includes, for example, estuaries, coastal forests, beaches, and barrier islands, as well as the ocean margin.

Rapid population growth is stressing the coastal zone. About 53% of the U.S. population lives within 80 kilometers (50 miles) of the 160,000 kilometers (100,000 miles) of coastline that borders the Atlantic and Pacific Oceans and the Great Lakes, only 10% of the nation's total land area. Moreover, the coastal population is growing at about three times the national rate. Seventy percent of the world's population lives on coastal plains. The sheer density of population alone stresses the coastal environment in numerous ways: waste disposal; recreation; the siting of industry, homes, and power plants; and the exploitation of fisheries, minerals, oil, and natural gas.

Although perhaps 95% of U.S. sewage system facilities now meet federal standards for wastewater treatment, many of the 130 that do not meet these standards serve major coastal cities such as San Diego and Los Angeles. The problem can become serious during rainy periods in localities served by combined sewer systems (chapter 14). In early 1992, the Center for Marine Conservation in Washington, D.C., estimated that along the nation's coast, 15,000 to 20,000 pipes discharge raw sewage into the ocean during rainy periods. On any given day, polluted water closes roughly one-third of shellfish beds to fishing. Furthermore, as we saw in chapter 13, the heavy use of groundwater in low-lying coastal plains enhances the risk of salt-water intrusion.

Estuaries

An **estuary** is an ecosystem created by the mixing of fresh and salt water at the mouths of rivers and in tidal marshes and bays. In an estuary, water undergoes daily tidal oscillations (that is, the regular rise and fall of sea level due to the gravitational attraction of the moon and sun). Hence, organisms that reside in an estuary are adapted to frequent fluctuations in currents, temperature, salinity, and concentrations of suspended sediment. Among the nation's best-known estuaries are Chesapeake Bay (Maryland), San Francisco Bay, Puget Sound (Washington), and Long Island Sound (New York).

Because of a special combination of biological and physical characteristics, estuaries are among the most productive ecosystems on earth. Many estuaries are located at the endpoints of rivers and consequently receive water that is rich in nutrients and organic matter. More important, estuaries receive a daily influx of nutrients from ocean tides. Sea water averages sixty times the dissolved mineral content of fresh water, and it is also higher in organic constituents including phytoplankton, zooplankton, and detritus.

The unique circulation of water within estuaries traps and recirculates nutrients and detritus (refer to fig. 3.27). These conditions support luxuriant plant growth (both phytoplankton and sea and marsh grasses) as well as large populations of detritus feeders. Animals that occupy the lower trophic levels include zooplankton, oysters, mussels, clams, lobsters, blue crabs, abalone, bay scallops, shrimp, and snails. Those consumers, in turn, are an abundant food source for consumers at higher trophic levels, particularly fish and birds. The unique environmental conditions of an estuary favor development and protection of juvenile fish: Food is abundant, and the low salinity and shallowness of the water serve as deterrents to many ocean predators. Estuaries also provide habitat for juvenile anadromous fish (fish that swim from the sea upriver to spawn), such as striped bass and salmon.

Figure 21.1   Coastal erosion at Newport Beach, Oregon.

Ironically, the very characteristics that make estuaries highly productive also make them vulnerable to pollution. Many estuaries are downstream from large cities and receive a significant input of waterborne industrial waste (including heavy metals and synthetic organic chemicals), and pathogens from inadequately treated sewage and agricultural waste. Circulation patterns within estuaries responsible for high productivity also inhibit flushing of those pollutants out to sea. As an extreme example, the once commercially important harbor at New Bedford, Massachusetts, is now closed to all fishing because of PCBs accumulated in the harbor's bottom sediments.

Beaches and Barrier Islands

A particularly troublesome problem in many coastal areas is accelerated shoreline erosion, caused by shortsighted human disturbances of the natural processes of beach formation. A **beach** is an accumulation of wave-washed sediment along a coast. An estimated 90% of the nation's sandy recreational beaches are undergoing some erosion (fig. 21.1).

The combined action of sea waves, rivers, and coastal currents supplies sand (and other sediment) to beaches. Sea

waves breaking against coastal headlands loosen and break down rock into sediment. Also, as rivers and streams empty into the sea, they slow and deposit enormous quantities of suspended sediment. Longshore currents then transport sediment along the shoreline. A **longshore current** is produced by sea waves that approach the shore at some angle, directing a component of water motion parallel to the shore (fig. 21.2). The current transports sand along the shore and thereby provides a continual supply of sand for coastal beaches.

Artificial structures such as dams and breakwaters, however, can interrupt the longshore flow of sand. When a river is dammed upstream, sediments settle out in the quiet waters of the reservoir behind the dam, thus reducing the supply of sand for downstream beaches. In that case, instead of supplying sand to beaches, longshore currents transport sand away from beaches until all that remains behind is rocky rubble. In some coastal areas, breakwaters are constructed to provide boaters calm waters for docking (fig. 21.3); these structures reduce the force of sea waves and thereby weaken longshore sand transport. Reduced wave energy causes deposition of sediments behind a breakwater which may eventually choke the harbor and require costly dredging. Also, because most of the sediments end up in the harbor, they are unavailable for

(a)

(b)

Figure 21.2 (a) Longshore currents transport a river of sand along a beach on Long Island, New York. A series of groins (structures perpendicular to the beach) trap sand and reduce coastal erosion by storm waves. Wave movement and sand transport are shown diagrammatically in (b).

© Frank Hanna/Visuals Unlimited

deposition on down-current beaches. Just such a situation developed along the coast of Santa Barbara, California, where it became necessary to pump sand from the upcurrent to the downcurrent side of the harbor (fig. 21.4). Energy-consuming pumping must be used to transport the sand, which was formerly accomplished by the energy of natural longshore currents.

Powerful ocean storms can make life difficult if not dangerous for coastal residents, because the driving force of waves and water can undermine and destroy seaside roads, homes, and businesses. However, the greatest danger exists for residents of the 295 **barrier islands,** long and narrow strips of

sand that fringe portions of the U.S. Atlantic and Gulf coasts. Cedar Island, Virginia, described at the opening of this chapter, is such an island. Barrier islands form the first line of defense against winter storms and hurricanes for eighteen states from Maine to Texas. They protect coastal beaches, wetlands, and estuaries by absorbing much of the energy of approaching storm waves. Barrier islands and associated wetlands, beaches, and sand dunes are also important habitats for a variety of wildlife and fish. In fact, they are home to a greater variety of bird species than any other ecosystem in the continental United States.

Barrier islands are dynamic systems. Under natural conditions, their sands gradually migrate landward and in the direction of longshore currents. Sea waves breaking on the shores of barrier islands dissipate their energy by shifting sands and, in the process, change the shape of the islands. However, when barrier islands are developed for human habitation, sands are stabilized and less able to absorb (or dissipate) the energy of storm waves. For decades, barrier islands have undergone rapid commercial development and urbanization; and some coastal cities and resorts (including Atlantic City, Virginia Beach, Miami Beach, and Galveston Island) are built entirely on barrier islands (fig. 21.5).

Population growth on barrier islands worries meteorologists and others, who note that such exposed locations are particularly vulnerable to the ravages of tropical storms and

Figure 21.3   A breakwater such as this one in the harbor of Honolulu, Hawaii, absorbs much of the energy of ocean waves and thereby provides relatively calm waters for boats.
© Brian Parker/Tom Stack & Associates

Figure 21.4   The breakwater outside the harbor at Santa Barbara, California, disrupts the longshore transport of sand so that sand washes into the harbor, and costly dredging is required to keep the harbor navigable.
© Frank Hanna/Visuals Unlimited

Figure 21.5   Resort hotels built along a barrier island, Miami Beach, Florida.
© Southern Living/Photo Researchers, Inc.

hurricanes (fig. 21.6). Much of this population growth oc-curred during a period of relatively low storm activity. The concern now is that residents might not receive adequate warning to evacuate the islands and avert disaster if a hurri-cane strikes. For more on the hurricane threat to the Atlantic and Gulf Coasts, see Special Topic 21.1.

Ocean Margin

Biological productivity differs significantly from one portion of the ocean to another. The differences stem primarily from contrasts in two limiting factors: sunlight penetration and nutrient supply.

Figure 21.6   Timely evacuation of coastal communities no doubt saved many lives as Hurricane Hugo slammed ashore near Charleston, South Carolina, around midnight on 21 September 1989.
© Dr. Fred Espenak/Science Photo Library/Photo Researchers, Inc.

Hurricane Threat to the Southeastern United States

Hugo and Andrew did much to heighten public awareness of hurricane threat to the southeastern United States. Hurricane Hugo slammed ashore near Charleston, South Carolina, about midnight on 21 September 1989 (see fig. 21.6). The powerful storm, packing winds of up to 218 kilometers (135 miles) per hour, took 21 lives and caused property damage estimated at $7 billion. In the three hours it took Hurricane Andrew to cross extreme southern Florida on the morning of 24 August 1992, its winds, gusting in excess of 265 kilometers (164 miles) per hour, contributed to the deaths of 25 people and left 180,000 homeless. With property damage estimated at $30 billion, Hurricane Andrew was the most costly natural disaster in U.S. history. To make matters worse, Andrew continued to push northwestward into the Gulf of Mexico and two days later struck the Louisiana coast where it claimed 4 more lives and caused $400 million in property damage.

A *hurricane* is an intense storm that develops over warm tropical seas, usually in late summer or fall. By convention, a hurricane has a maximum wind speed over 119 kilometers (74 miles) per hour. A hurricane derives its energy from the ocean through heat released when water vapor (from the ocean surface) condenses. Hence, the storm weakens and its winds rapidly diminish once it makes landfall. Most hurricane wind damage is confined to a band that stretches about 200 kilometers (125 miles) inland of the coastline.

Hurricanes also produce torrential rains with 24-hour totals of 13 to 25 centimeters (5 to 10 inches) not unusual. Even if the storm tracks well inland, heavy rains often persist and may trigger flash flooding. The most deadly and devastating aspect of a hurricane, however, is the *storm surge*, a wall of ocean water driven by strong winds over barrier islands, mangrove swamps, and coastal plains (see the diagram below). The height of a storm surge may range from 1 to 2 meters (3 to 7 feet) for a relatively weak hurricane to more than 6 meters (20 feet) for the most powerful hurricanes.

Atmospheric scientists classify a hurricane from 1 to 5 (weak to catastrophic) on the basis of wind speed, height of storm surge, and the potential for property damage

Normal high tide — Mean sea level

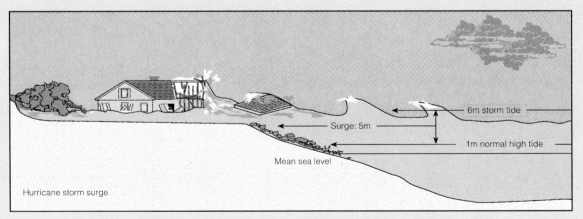

Hurricane storm surge — Mean sea level — Surge: 5m — 6m storm tide — 1m normal high tide

Perhaps the most destructive aspect of a hurricane along a low-lying coastal region is the storm surge. Hurricane winds drive a wall of water ashore.

(see the table below). Hurricanes Hugo and Andrew both rated 4 on the Saffir-Simpson Scale. The last category 5 hurricane to strike the coast of the United States was Hurricane Camille in August 1969.

On average, two hurricanes strike the United States each year. However, infrequent hurricane activity during the 1970s and much of the 1980s lulled many residents of the coastal southeast into a false sense of security and encouraged population growth. (In fact, fewer hurricanes affected the United States in the 1970s than in any prior decade of the century.) More and more resort hotels, condominiums, and homes were built perilously close to the shoreline, among coastal sand dunes, and on barrier islands. Population along the Gulf Coast from Florida west to Texas climbed from 5.2 million in 1960 to 10.1 million in 1990. Along the Atlantic Coast from Florida north to Virginia, the population more than doubled from 4.4 million in 1960 to 9.2 million in 1990. Today, perhaps 75% of Atlantic and Gulf Coast residents have never experienced a major hurricane.

Public safety officials have responded to the hurricane hazard by drawing up evacuation plans for residents of low-lying coastal areas and barrier islands. For example, 1 million people in Florida and 1.7 million in Louisiana and Mississippi were either asked or ordered to evacuate their homes in advance of Hurricane Andrew. No doubt, the death toll in both Hugo and Andrew would have been considerably higher without evacuation. Successful evacuation, however, hinges on sufficient advance notice of a hurricane's approach. Unfortunately, hurricanes often follow erratic paths and are notorious for sudden changes in direction and forward speed. This is especially troublesome for certain congested cities and isolated localities where the required evacuation time may be lengthy. For example, estimated evacuation time is thirty-seven hours for the Florida Keys and fifty hours for New Orleans.

Other strategies adopted or advocated by public safety and other government officials to minimize loss of life and property to hurricanes are (1) stringent building codes, (2) preservation of mangrove swamps, and (3) elimination of federal flood insurance. Building codes may require new homes to be equipped with bolts that anchor the floor to the foundation and steel brackets that attach the roof to the walls. In addition, in many areas, all new buildings must be elevated above the once-in-a-century flood level. Unfortunately, many structures were built prior to these tough codes and, as the aftermath of Hurricane Andrew demonstrated, those codes were not always enforced. Mangrove swamps, the final line of natural defense against hurricanes, help dissipate the energy of a storm surge. In too many cases, however, shoreline development has destroyed these swamps—often because the trees obscured the ocean view for seaside dwellings. A controversial strategy is to eliminate federal flood insurance for flood-prone coastal areas as has been done on undeveloped portions of barrier islands. Some people argue that by providing policies at low cost, federal flood insurance programs encourage development in flood-prone areas. Furthermore, this insurance enables homeowners to rebuild structures destroyed by a storm surge in the same hazardous location.

Saffir-Simpson Hurricane Intensity Scale			
Category	Wind Speed km/hr (mi/hr)	Storm Surge meters (feet)	Damage
1	119–154 (74–95)	1–2 (4–5)	minimal
2	155–178 (96–110)	2–3 (6–8)	moderate
3	179–210 (111–130)	3–4 (9–12)	extensive
4	211–250 (131–155)	4–6 (13–20)	extreme
5	>250 (>155)	>6 (>20)	catastrophic

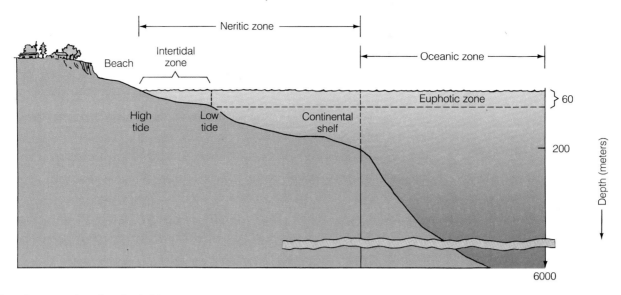

Figure 21.7   A cross section of marine habitat zones.

Seaward from the high-tide mark, the marine environment can be divided into the neritic zone and the oceanic zone (fig. 21.7). The **neritic zone** is the ocean margin; by definition, it extends seaward to the edge of the continental shelf, where water depth averages about 200 meters (650 feet). Nutrients and sunlight for photosynthesis are relatively abundant. While the neritic zone represents only about 10% of the ocean's surface and less than 0.5% of sea-water volume, it accounts for about 30% of all ocean productivity and is the principal site of commercial fish harvests (chapter 8). Sea life is abundant and very diverse throughout the depths of this zone. The **oceanic zone** includes the deep water that stretches beyond the edge of the continental shelf. Although enormous in area and volume, the open ocean is quite unproductive because it is relatively nutrient-poor, and sunlight penetration is limited to surface waters.

One major environmental concern in the neritic zone is excessive algal blooms. In temperate latitudes under natural conditions, small blooms occur in both spring and autumn, but where the input of nutrients is excessive, algae populations soar. As we learned in chapter 14, fertilizer runoff, septic tank leakage, sewage effluents, and runoff from farms and feedlots contribute nutrients (particularly nitrogen and phosphorous compounds) to rivers and streams. The final destination of some of these nutrients is coastal waters, where they can trigger dense, huge algal blooms that reduce dissolved oxygen concentrations and cut light penetration for photosynthesis—effects that stress marine ecosystems.

Excessive nutrient input sometimes leads to toxic algal blooms, the best known of which is the **red tide.** The name comes from the reddish-brown discoloration of sea water caused by a population explosion of certain species of dinoflagellates, a group of phytoplankton. During a bloom, the number of organisms per liter of sea water soars from a normal level of about 1,000 to as many as 60 million! These microscopic organisms release substances that are toxic to some fish, whales, and dolphins. These toxins can also harm humans who eat contaminated shellfish.

In 1987 and 1988, a red tide developed along the North Carolina coast, forced the closing of shellfish beds (at a loss of $25 million), and was implicated in a massive kill of dolphins. In that case, dinoflagellates released a neurotoxin, a poison that attacks the nervous system. More recent examples of the effects of algae-produced toxins come from Canada where an outbreak of diarrhetic shellfish poisoning was reported in 1990 and from Alaska in 1992 where a toxin was discovered in the guts of the Dungeness crab.

In the scientific community there is considerable debate as to whether toxic algal blooms are becoming more common. Most agree, however, that more research is needed to determine what factors contribute to a bloom and why certain algae produce toxins.

Government Regulations

The need to balance the use and preservation of the coastal zone has spurred states with shorelines to develop guidelines for coastal management. Delaware's regulations are probably the most stringent; they prohibit the siting of all new industry within 3 kilometers (2 miles) of the shoreline. The federal government is helping coastal states to manage and protect their coastal resources through provisions of the 1972 Coastal Zone Management Act. As a consequence of that law, most of the eligible states now have federally approved and funded plans. Those plans aim to inventory coastal resources and set priorities for the orderly use of coastal land and water.

The 1982 Coastal Barriers Resources Act discourages development of undeveloped barrier islands by prohibiting use of federal funds for flood insurance and construction of bridges, roads, and sewers. Initially, the law applied to 182,000 hectares (450,000 acres), and in 1990 it was expanded to include an additional 320,000 hectares (788,000 acres).

One provision of the 1987 Water Quality Act established the National Estuary Program (NEP), which calls for identification of nationally significant estuaries threatened by pollution or development. The objective of NEP is to design inno-

Global Warming, Rising Sea Level, and Coastal Erosion

*L*ow-lying coastal communities are likely to face major problems from accelerated beach erosion should global warming actually materialize. Recall from chapters 3 and 15 that steadily rising concentrations of carbon dioxide and other greenhouse gases may trigger a significant rise in global temperatures by the middle of the next century. Higher global temperatures, in turn, would cause sea water to expand and might accelerate melting of the polar ice caps. According to numerical models of the earth-atmosphere system, the predicted warming that would accompany a doubling of atmospheric carbon dioxide concentration might cause sea level to rise between 0.2 and 1.5 meters (1-5 feet) within the next sixty years. By EPA estimates, an increase of only 0.3 meters would cause beaches to recede by 15 to 60 meters along the East Coast, 60 to 120 meters along the California coast, and up to 300 meters along coastal Florida.

What can be done? The traditional approach to stem shoreline erosion is to erect *hard structures* to hold back the sea and stabilize the coast. This may involve some combination of seawalls, breakwaters, jetties, or revetments. Today, an increasingly popular alternative is *soft engineering*, whereby sand is imported (perhaps dredged or pumped from offshore) to replace beach sand lost to erosion. In fact, Maine and North Carolina have banned seawalls and other hard erosion-control structures from their beaches, and South Carolina requires existing seawalls to be torn down within forty years. In the late 1970s, sand pumped from offshore was used to form a new beach 100 meters (325 feet) wide along a 16-kilometer (10-mile) stretch of Miami Beach. Beach nourishment, however, is only a temporary solution that must be repeated every 2-10 years depending on frequency and severity of storms. Futhermore, beach nourishment is expensive, typically costing more than $1 millon for every 2 kilometers that are restored.

In some areas, strategic retreat is the only feasible alternative in the face of rising sea level. Some coastal communities are considering requiring all new homes and other buildings to be portable so that they can be moved inland should the sea rise. Some areas require minimum setbacks for new construction to buffer buildings from storm damage. For example, North Carolina law now requires all new single-family homes and other large buildings to be set back from the shoreline a distance equal to sixty times the annual erosion rate.

vative management approaches to deal with those threats. Currently, twelve estuaries are in the program, including Puget Sound, Long Island Sound, Galveston Bay, and Narragansett Bay (Rhode Island).

Another major effort to preserve the natural resources of coastal zones is the designation of **marine sanctuaries.** Point Lobos, California (near Monterey), was founded in 1960 as the nation's first underwater sanctuary. Since then, California has set aside eleven other sites as sanctuaries and has been studying several more potential sites. Although the federal government has moved more slowly than California, federal legislation holds much promise for this endeavor. Provisions of the 1972 Marine Protection, Research, and Sanctuaries Act authorize the President to designate marine sanctuaries in coastal waters of the continental shelf and in the Great Lakes. Any of several characteristics may qualify a region as a marine sanctuary, but basically, potential sites must have special biological, esthetic, or historical significance. The objective of federal legislation is to preserve and protect those areas by managing the multiple demands placed on them. Hence, those sanctuaries are not, in the strictest sense, places of refuge and protection for marine life. Indeed, some harvesting of aquatic resources is permitted. An example of an existing federal marine sanctuary is Key Largo in Florida, which consists of coral reefs that are home to more than five hundred species of fish.

Another contemporary concern is prospects for accelerated coastal erosion should global warming occur. For more on this topic, refer to the issue discussed above.

Land Use and Natural Hazards

A primary objective of federal, state, and local land-use strategies is to provide for our resource needs now and in the future, while minimizing environmental disturbances. That goal requires assessment of the land in terms of its capabilities and limitations. Simply put, some land is suitable for some purposes but not for others. As we saw in chapter 10, some land can be managed for multiple purposes as, for example, when a national forest is used for recreation, grazing of livestock, and lumbering. However, natural hazards, such as floods or earthquakes, severely limit the utility of some lands. Thus, comprehensive land-use policies must also regulate the habitation of those areas to minimize the risk of life, limb, and property. Such policies must reflect an understanding of geological characteristics and the potential hazards created by those characteristics.

Flood Hazard

Floods have caused more damage in the United States than any other natural hazard, and they account for the majority of events that are officially declared disasters by U.S. Presidents (whereby residents become eligible for federal assistance through low-interest loans). Floodwater drowns people and livestock, topples trees, flattens buildings, erodes valuable topsoil, disrupts municipal water and sewage systems, and interferes with communications, transportation, and commerce. Furthermore, as floodwaters recede, they leave behind thick layers of silt and mounds of debris. The National Weather Service estimates that in an average year, floods claim the lives of two hundred Americans and cause up to $3 billion in property damage.

Nationwide, fatalities and property damage attributable to flooding are increasing, because more people live on floodplains. Fertile soils, plentiful water for irrigation, and the potential for inexpensive transportation have always attracted people to floodplains, and the shift from an agrarian to an industrial-based society has made the attraction even stronger. The flat terrain of floodplains is well suited for highway, railroad, industrial plant, and home construction. With the steady growth of urban areas, the population of floodplains has soared. Today, many cities in the United States (such as New Orleans) are built almost entirely on floodplains. Approximately 15% of the nation's total urban land and almost 10% of all agricultural land are floodplains and are therefore vulnerable to flooding.

Another factor that has contributed to increased flood fatalities in recent decades is that remote areas (where flood warnings are not readily communicated) have become more accessible to campers and other visitors; more people are driving their cars, campers, and mobile homes into flash-flood-prone mountainous areas. Vacationers unwittingly set up camps and build cabins in narrow canyons, which are subject to unexpected and rapid rises in stream levels. For example, on 31 July 1976, more than 130 people (mostly campers) lost their lives in the Big Thompson Canyon in Colorado, when heavy rains caused a flash flood.

To determine whether floods can be prevented or whether flood damage can at least be reduced, we need to understand the causes of flooding. A **flood** occurs whenever runoff exceeds the discharge capacity of a river channel, flows over the river's banks, and pours over the floodplain (chapter 13). Runoff that is sufficient to cause flooding is due to natural events, human activities, or some combination of the two. Hurricanes and intense thunderstorms produce heavy rainfall, which often exceeds the infiltration capacity of soil and consequently causes flooding. In the summer of 1972, torrential rains from Hurricane Agnes triggered flooding that claimed 122 lives and caused property damage that exceeded $2.1 billion in the Northeastern states. That same year, 237 lives were lost in Rapid City, South Dakota, when exceptionally heavy rains caused flash flooding. Rapid spring snow melts also cause a sharp rise in river levels, which is sometimes compounded by the break-up of river ice. Frequently, huge slabs of ice pile up at bridges or narrow stretches of a river channel, causing the water to back up and flood.

Human modification of the land can also increase the flood potential. A major contributor to flooding is removal of the land's protective vegetative cover. Vegetation slows the flow of runoff, allowing more water to infiltrate the soil and thereby reducing the flood threat. Deforestation, overgrazing, and mining are notorious for removing vegetative cover and contributing to the hazard of floods. As we noted in chapter 10, the filling in and draining of wetlands eliminates

Figure 21.8 Automobiles trapped by a flash flood under a viaduct. Urban sewer systems may not be able to carry off all the runoff from a heavy rainfall.
© Miro Vintoniv/Stock Boston

a natural flood-control mechanism. In addition, failure of dams, levees, and other structures that are designed, ironically, to prevent flooding, is sometimes a major factor in calamities caused by floods.

As cities grow and expand, new roads and buildings render increasing areas of land impervious to water, and sewer systems in many cities are unable to accommodate the enormous volumes of water that can accompany, for example, a summer downpour. As a result, basements, viaducts, and other low-lying areas are subject to rapid inundation (fig. 21.8).

Until the mid-1970s, U.S. flood-control efforts consisted almost entirely of engineering projects designed to keep floodwaters out of populated areas. That approach matched structural flood-control measures to the specific characteristics of a flood-prone area. Thus, in areas where it was possible, the land along river valleys was shaped into terraces, benches, or levees to confine floodwaters. As we noted in chapter 10, stream channelization helps curb local flooding by enlarging the discharge capacity of a river. In other cases, earthen or concrete dams, dikes, or floodwalls have been built to detain or divert floodwaters (fig. 21.9). Those structural flood-protection measures can be enhanced by erosion control, which protects vegetative cover and thus the soil of the watershed.

In spite of more than $15 billion spent over the years on structural flood control, flood-related deaths, injuries, and property damage have continued to rise. The presence of flood-protection devices often engenders a false sense of security among local residents and encourages further land development and home construction. However, since most flood-control structures are actually designed to accommodate only moderate flooding, the stage is set for tragedy when a great flood occurs.

Congress responded to catastrophic floods of 1972 by enacting the Federal Flood Disaster Protection Act of 1973, which reflects a new emphasis on **floodplain management** as opposed to structural flood-control measures. That law re-

Figure 21.9 A concrete wall at Cape Girardeau, Missouri, designed to protect the community from floodwaters of the Mississippi River.

quires local governments to adopt floodplain development regulations before they can be eligible for federal flood insurance. Also, any construction projects planned within regions identified as flood-hazard areas (which do not qualify for flood insurance) are denied federal funding altogether. Floodplain development regulations require that buildings be elevated or flood-proofed and that provisions be made for a floodway that will allow floodwaters to pass through a community without causing severe damage.

The goal of the 1973 law is to foster the wise use of floodplains and other flood-prone areas and thereby reduce the toll of floods. It encourages communities to use floodplains in ways that are compatible with periodic flooding (e.g., agriculture, forestry, recreational areas, parking lots, and wildlife refuges) rather than as construction sites for homes and industries. For example, in the 1960s, the U.S. Army Corps of Engineers proposed construction of a concrete drainage channel through the center of Scottsdale, Arizona, which would carry off floodwaters from Indian Bend Wash. However, local civic leaders, fearing that the channel would divide the community both physically and psychologically, searched for an alternative. The result was the Scottsdale greenbelt, which was completed in 1983 (fig. 21.10). The Scottsdale greenbelt is an 11-kilometer (7-mile) grass-lined drainage-way that consists of

Figure 21.10 The Scottsdale (Arizona) greenbelt, completed in 1983, was built as an alternative to more traditional flood-control structures.
© Tom Johnson/Azscene

Land-Use Conflicts 469

Figure 21.11 On 27 March 1964, a powerful earthquake triggered a tsunami that crashed into Resurrection Bay at Seward, Alaska. Property damage in Resurrection Bay topped $14 million.

National Geophysical Data Center

Table 21.1	The world's ten most disasterous earthquakes in terms of lives lost	
Location	**Year**	**Estimated Deaths**
Shaanxi, China	1556	830,000
Calcutta, India	1737	300,000
T'ang-shan, China	1976	240,000
Gansu, China	1920	180,000
Messina, Italy	1908	160,000
Tokyo and Yokohama, Japan	1923	143,000
Chihli, China	1290	100,000
Beijing, China	1731	100,000
Naples, Italy	1693	93,000
Shemaka, U.S.S.R.	1667	80,000

five parks featuring tennis courts, golf courses, bicycle paths, and other recreational amenities. Although the main purpose of the greenbelt is to carry off floods waters, it also serves other community needs.

Earthquake Hazard

In late July, 1976, one of the greatest natural disasters of recorded history struck T'ang-shan in northeast China. Perhaps 240,000 persons perished when severe earthquakes devastated the homes and businesses in that congested city. T'ang-shan's buildings had not been constructed to withstand earthquake tremors, and the city was situated on unstable river sediments that shifted during the violent quakes. More recently, the September, 1985, Mexico City earthquake killed more than 10,000 people and left 250,000 people homeless. On 7 December 1988, a severe earthquake leveled Spitak, Armenia, killing at least 25,000 people.

An **earthquake** is the vibration or shaking of the ground caused by an abrupt release of energy accumulated in bedrock. Severe earthquakes can cause buildings, bridges, and elevated highways to collapse, and they can break sewer lines and water and gas pipes. However, violent shaking of the ground is not the only cause of earthquake damage. Earthquakes can trigger landslides and avalanches of rock, mud, and snow; they can disrupt the flow of rivers and groundwater, causing rivers to flood and wells to run dry. In some coastal areas, an earthquake can generate a destructive tsunami (sometimes called a *seismic sea wave*).

A **tsunami** is a huge ocean wave that is triggered by an abrupt flexing of the seafloor, often as the consequence of a major submarine earthquake. Underwater landslides or the collapse of a coastal volcano can also cause a tsunami. At sea, the energy of a tsunami is dispersed throughout a huge volume of water, so the wave poses no threat to ships and, in fact, is not even noticeable. The wave has an inconspicuous height of less than 2 meters (6.5 feet) but travels along at a tremendous speed (perhaps in excess of 800 kilometers, or 500 miles, per hour). However, when a tsunami approaches certain shorelines (particularly around the Pacific), its energy gradually

becomes focused into a wall of water that can be taller than a three-story building. During the Alaskan earthquake of 27 March 1964, a tsunami that was more than 10 meters (33 feet) above high-tide level crashed into Resurrection Bay at Seward, Alaska, causing extensive damage to ships and harbor buildings (fig. 21.11). The loss of life that accompanies a tsunami can be staggering. In August 1976, a strong earthquake centered off the island of Mindanao in the Philippines created a series of tsunamis that claimed 5,000 lives—even though a tsunami warning system has operated in the Pacific since 1948.

Millions upon millions of earthquakes have occurred through the course of human history, some catastrophic (table 21.1), but most so slight as to be inconsequential. More than 1 million tremors occur each year (about one every half-minute), and nearly 80% of those are barely strong enough to rattle dishes.

Scientists commonly use the Richter scale as an index of the total energy released during an earthquake, that is, *earthquake magnitude*. The **Richter scale** (named for Charles Richter of the California Institute of Technology, who devised it in 1935) is open-ended and based on the amplitude of earthquake vibrational waves as recorded by sensitive instruments known as *seismographs*. Like the pH scale (chapter 16), the Richter scale is logarithmic; that is, each increment of 1 on the scale corresponds to a ten-fold increase in wave amplitude. Also, each increment of 1 on the Richter scale roughly corresponds to a thirty-fold increase in energy release. Thus, a magnitude 6 earthquake releases 27,000 times more energy than a magnitude 3 earthquake ($30 \times 30 \times 30 = 27,000$).

The largest earthquake on record rated about 9.5 on the Richter scale. The most violent North American earthquake, the 1964 Alaskan quake, had a magnitude of 8.4–8.6. Any earthquake of 8.0 or above has the potential of causing catastrophic damage, but such tremors (on average worldwide) occur only once every 5–10 years. As table 21.2 shows, frequencies rise sharply for weaker earthquakes. Without instruments, humans cannot detect earthquake vibrations below a threshold of about 3.4.

Table 21.2	Worldwide average annual earthquake frequency by magnitude	
Richter Magnitude	Number per Year	Potential Damage in Populated Areas
< 3.4	800,000	detected by seismographs only
3.5–4.2	30,000	felt by some people
4.3–4.8	4,800	felt by many people
4.9–5.4	1,400	felt by all
5.5–6.1	500	slight damage to buildings
6.2–6.9	100	much damage to buildings
7.0–7.9	20	considerable damage to structures
> 8.0	1 every 5–10 years (average)	catastrophic damage

In addition to energy release (rank on the Richter scale), the destructiveness of an earthquake depends on (1) depth of the earthquake's origin (focal depth), (2) proximity of an earthquake epicenter to a populated region (the **epicenter** is the location on the earth's surface directly above the point of earthquake origin), (3) duration of vibrations, (4) nature of subsurface soil, sediment, and bedrock, and (5) structural design of buildings. As a rule, for a given earthquake magnitude, damage is greatest close to the epicenter of a shallow earthquake that persists for several minutes (most earthquakes last less than a minute). Rigid buildings constructed of unreinforced concrete or masonry are most vulnerable to collapse. Also, soft, unconsolidated sediments (e.g., where land has been filled, or in floodplains) tend to amplify vibrations and are notoriously unstable during an earthquake. In fact, earthquake waves may transform wet sediments into a slurry that provides absolutely no support for buildings, and they collapse.

Almost all major earthquakes occur along boundaries of the gigantic plates described in Special Topic 17.1. Recall that these boundaries are fault zones. Adjacent plates grind against one another and occasionally lock together, which causes strain energy to accumulate within the rock. Then, abruptly, when the plate breaks free, the stored energy is released in the form of vibrations (seismic waves), which travel through the earth or along the earth's surface. In other instances, adjacent plates creep slowly past one another producing more frequent low-magnitude earthquakes.

The West Coast is the principal site of earthquake activity in the United States. Seven major fault zones are potential sites of violent earthquakes that could devastate major population centers in California. Foremost among these is the San Andreas fault produced by the slow (about 5 centimeters per year) sliding of the Pacific plate northwestward past the North American plate (fig. 21.12). It is a 430-kilometer (270-mile) branching fracture zone that has been the locus of many destructive earthquakes. Eight or more major earthquakes have occurred along the San Andreas fault near Los Angeles over the past 1,200 years, the most recent one in 1857.

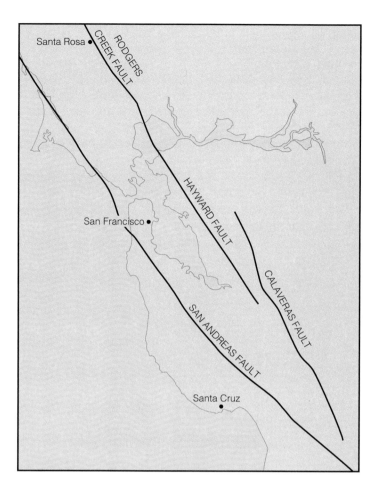

Figure 21.12   Location of faults in the San Francisco Bay area.

Scientists are concerned about a recent upswing in earthquake activity along the southernmost segment (Coachella Valley) of the San Andreas fault. Between 1948 and 1986, that segment experienced no earthquakes ranking higher than 5 on the Richter scale. Between 1986 and 1992, however, six earthquakes ranking above 6 on the Richter scale have occurred along that segment. The principal concern is that this increased activity may portend a much more powerful quake that could spread northward into the more populous San Bernardino Mountain segment and then into the Los Angeles basin and San Fernando Valley.

On 17 October 1989, a magnitude 7.1 earthquake occurred along the 3-kilometer (24-mile) Loma Prieta segment of the San Andreas fault, 87 kilometers (54 miles) south of San Francisco and east of Santa Cruz. This segment had remained locked and inactive since the 1906 earthquake that destroyed much of San Francisco. Fortunately, the strongest tremors during the Loma Prieta quake lasted only about 6-10 seconds; otherwise, damage to structures built on sandy, loose fill would have been considerably greater. Nonetheless, a 2-kilometer stretch of the double-deck Nimitz freeway, part of a major bridge, and numerous buildings collapsed (fig. 21.13). Most damage was to buildings constructed prior to modern earthquake building codes and located on soft soils and sediments that were used for fill. The death toll was sixty-seven, but probably would have been much higher if people had not been home watching a World Series baseball game. Injuries totaled 3,757, and property damage estimates topped $7 billion.

On a positive note, the Loma Prieta earthquake triggered renewed emphasis on earthquake preparedness in California. By law, all of the state's homeowners now must have earth-

Figure 21.13 The upper deck of Interstate 880 (the Nimitz Freeway) in Oakland, California, collapsed onto the lower deck, killing more than 40 people during the Loma Prieta earthquake on 17 October 1989.
© Francois Gohier/Photo Researchers, Inc.

quake insurance. San Francisco has installed an emergency communications system and a crisis command center. The city has also upgraded its firefighting capabilities and organized neighborhood response teams. Statewide, engineers have begun the costly task of shoring up bridges and roadways, and retrofitting unreinforced masonry and concrete buildings.

Although earthquake activity is most likely in the geologically active Pacific coast states, earthquakes have been reported far from plate boundaries along deeply buried faults, and some of these were severe. The 1886 earthquake that destroyed much of Charleston, South Carolina, and claimed sixty lives rated at least 7 on the Richter scale and was probably the most destructive U.S. earthquake of the nineteenth century. Also, three of the most energetic earthquakes in the nation's history struck New Madrid, Missouri, during the winter of 1811-1812, changing the course of the Mississippi River. Figure 21.14 shows the variation in earthquake risk across the lower United States.

While earthquakes are less frequent east of the Rocky Mountains, they are potentially more dangerous than West Coast quakes for a variety of reasons: (1) in the east, because of differences in bedrock geology, damage from major earthquakes (above 7.5 on the Richter scale) tends to affect larger areas; (2) the public is generally unaware of the earthquake risk, and there is no adequate emergency response plan; (3) building codes in the East generally do not consider earthquake hazards. The biggest potential risk is in old congested cities made up largely of masonry and concrete buildings on filled lands. Considering that a strong earthquake is likely only once every century or two, the prevailing attitude seems to be one of accepting the risk, especially in view of the

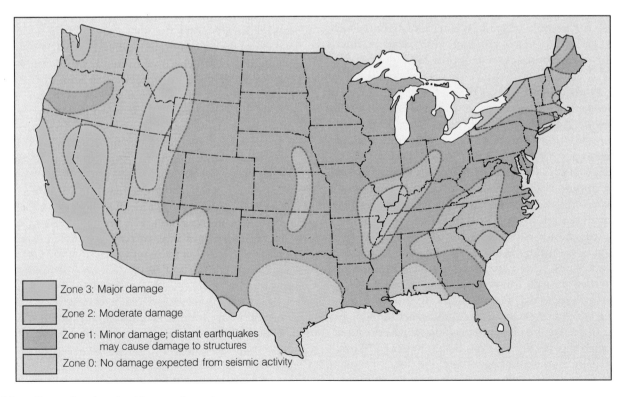

Zone 3: Major damage

Zone 2: Moderate damage

Zone 1: Minor damage; distant earthquakes may cause damage to structures

Zone 0: No damage expected from seismic activity

Figure 21.14 Zones of earthquake risk across the nation.
Source: U.S. Coast and Geodetic Survey, *ESSA Rel. ES-1*, January 14, 1969.

Earthquake Prediction

*I*n recent years, with the growth of population in seismically active regions, predicting earthquakes has become a high-priority objective for earth scientists. Several long-term and short-term forecasting schemes are currently being studied.

One long-term technique relies on the computation of recurrence rates, that is, determination of earthquake frequency in a specific area from historical records and geological reconstructions. That method is not generally satisfactory, however, since earthquakes are largely random events and several violent earthquakes have occurred in locations with no prior record of significant earthquake activity. An interesting and rare exception comes from central Chile, where major earthquakes have occurred since 1575 with remarkable regularity every eighty-three years, plus or minus nine years. This was the basis for forecasting a major quake in the 1980s, and one did occur on 3 March 1985.

Another approach to earthquake forecasting involves the identification of *seismic gaps,* —regions that are relatively inactive within a seismically active belt. An example is the Loma Prieta segment of the San Andreas fault prior to 17 October 1989. Although seismic gaps may indicate that energy is gradually accumulating in bedrock in advance of a major earthquake, it is not possible to predict the occurrence of an earthquake by this method alone. Seismic gaps may only pinpoint areas that are not appropriate for further urban development.

Long-term earthquake prediction typically specifies the probability of an earthquake of a certain magnitude occurring during some time interval. For example, in the summer of 1988, the U.S. Geological Survey proposed a 30% probability of a major earthquake in the following thirty years along the Loma Prieta segment of the San Andreas fault. In July of 1990, a panel of earthquake experts predicted a 67% chance of a magnitude 7 or greater earthquake in the San Francisco Bay area by the year 2020.

Short-term earthquake prediction focuses primarily on identifying *premonitory seismic events,* that is, signals that immediately precede potentially destructive earthquakes. Scientists use extremely sensitive instruments to monitor those events by measuring the buildup of strain (deformation) in rocks along fault zones. Other instruments detect slight changes in land elevation or relatively weak tremors (foreshocks) that may occur before a major earthquake. Changes in the behavior of animals prior to an earthquake are also being studied; some animals apparently can sense premonitory events.

Overall, scientists have had better luck with long-term than short-term earthquake prediction. The 1989 Loma Prieta earthquake is a good example. Although scientists anticipated significant earthquake activity in the San Francisco Bay area for at least a decade prior to the quake, the lack of detectable premonitory phenomena meant that they were unable to more precisely pinpoint the timing of the earthquake.

Even if our ability to predict earthquakes greatly improves, we are unable to prevent them from happening. Hence, precautions must be taken to minimize the death, injuries, and property damage caused by major earthquakes. For example, building codes and zoning regulations can limit the height and structural composition of buildings and otherwise minimize earthquake damage, and we can restrain construction on terrain prone to sliding during earthquakes or in coastal areas where tsunamis are a threat.

---

additional cost (1-10%) of incorporating earthquake resistant features in a new building.

In view of the potential death and destruction caused by an earthquake, Special Topic 21.2 considers the prediction of earthquakes.

Volcano Hazard

On 18 May 1980, a massive explosion blew the top off Mount St. Helens, a Washington State volcano, reducing its summit from 2,950 meters (9,588 feet) to 2,560 meters (8,320 feet) above sea level (fig. 21.15). The eruption triggered huge avalanches of rock and mud, and a cloud of hot gases (over 600° C) and pumice swept down the mountain's north slope, incinerating everything in its path. The eruption devastated 518 square kilometers (200 square miles) of forest and leveled trees some 24 kilometers (15 miles) away. Total damage was estimated at $1 billion to $3 billion.

Until the catastrophic eruption of Mount St. Helens, little concern was expressed over the hazard of volcanoes in the United States. Most active Alaskan volcanoes are isolated and too remote to threaten population centers, and the Hawaiian eruptions generally consist of relatively quiescent flows of lava (see Appendix D). But the Mount St. Helens eruption occurred near major population centers of the Pacific Northwest (only 80 kilometers or 50 miles north of Portland, Oregon). Some scientists fear that the eruption of Mount St. Helens may herald a reawakening of volcanic activity in the Cascade Mountain Range and in other portions of the western United States.

In the ninety years prior to the Mount St. Helens eruption, only two relatively minor eruptions occurred among Cascade volcanoes: an ash eruption at Mount Hood in 1906, and a series of small eruptions of Lassen Peak, between 1914 and 1917. Mount St. Helens itself had been dormant since 1857. In the mid-nineteenth century, however, northern

Figure 21.15 The eruption of Mount St. Helens on 18 May 1980 is thought by some scientists to signal an awakening of volcanic activity in the western United States.

U.S. Geological Survey

Figure 21.16 The volcanic hazard in the western United States. Volcanoes that have erupted on an average of every 200 years or less, during the past 300 years, or both include the following: (1) Mount St. Helens, (2) Mono-Inyo Craters, (3) Lassen Peak, (4) Mt. Shasta, (5) Mt. Rainier, (6) Mt. Baker, and (7) Mt. Hood. Volcanoes that erupt less frequently than every 1,000 years and have not erupted in more than 1,000 years include the following: (A) Three Sisters, (B) Newberry Volcano, (C) Medicine Lake Volcano, (D) Crater Lake, (E) Glacier Peak, (F) Mt. Adams, (G) Mt. Jefferson, and (H) Mt. McLoughlin. Volcanoes that last erupted more than 10,000 years ago include: (AA) Yellowstone, (BB) Long Valley, (CC) Clear Lake Volcanoes, (DD) Coso Volcanoes, (EE) San Francisco Peak, and (FF) Socorro.

From R. A. Kerr, "Volcanoes to Keep an Eye On." *Science* 221:634-635, 1983. Copyright 1983 by the AAAS based on data from U.S. Geological Survey *Open-File Report* 83-400. Used by permission.

Worldwide, about six hundred volcanoes are considered active, and each year an average of fifty erupt. An estimated 360 million people live on or near potentially dangerous volcanoes.

A **volcano** is defined as a landform composed of eruptive material, which accumulates around a vent in the earth's crust. The material that is extruded depends on the type of activity the volcano exhibits. Although some eruptions are relatively peaceful, others are explosive and violent. Furthermore, some volcanoes have an eruptive history that begins explosively and then becomes relatively quiet.

Volcanoes whose ejecta consist mainly of lava flows are generally the most peaceful and the most massive. Their gently sloping flanks are built up of numerous flows of viscous lava that form layers. The Hawaiian Islands are composed of many such overlapping volcanoes. Nevertheless, lava flows can cause considerable damage, setting forests and other vegeta-

Cascade volcanoes exhibited considerable activity. Thus, the eruption of Mount St. Helens may mark a renewal of that activity. Such speculation prompted the U.S. Geological Survey to assess the likelihood of future volcanic activity throughout the western states. Based on historical and reconstructed records of past volcanic activity, many volcanoes were singled out as being potentially hazardous (fig. 21.16).

Figure 21.17 Lava flows, such as this one in Hawaii, can ignite forest fires and destroy buildings.
© Francois Gohier/Photo Researchers, Inc.

tion ablaze, destroying dwellings, and covering the land with a blanket of rock rubble (fig. 21.17). As a rule, however, lava flows so slowly that the threat to human life is minimal. In some cases, lava flows can even be diverted from populated areas by structures such as diversion walls. Nonetheless, extreme caution should be taken in the vicinity of any lava flow.

The most hazardous volcanoes are those that are prone to explosive eruptions. Apparently, eruptive intensity depends on the temperature and composition of the magma (molten rock) that feeds the volcano. Magmas with the greatest potential for explosive activity have relatively low temperatures and are rich in silica (SiO_2). Such magmas are very viscous and retain large quantities of gases. If those gases cannot escape readily (e.g., if the vent through which the magma is moving becomes blocked) pressures may build to explosive proportions.

An explosive volcanic eruption can send fine ash particles as well as steam and other gases into the stratosphere and catapult huge rock slabs and partially congealed masses of lava down the volcano's flanks. Sometimes an incandescent cloud of ash mixed with steam and other gases rolls down the flanks of a volcano at tremendous speeds, incinerating all vegetation and animals in its path. If the volcanic summit was originally covered with snow and ice, the heat of the eruption can cause rapid melting, triggering avalanches of mud and rock. All this happened during the Mount St. Helens eruption.

Earthquake activity often accompanies a volcanic eruption whether that eruption consists of a violent explosion or quiet lava flows. In fact, on 29 November 1975, one of the most powerful earthquakes of the century in the United States originated under the southern flanks of the Kilauea volcano on the big island of Hawaii. That earthquake ranked 7.2 on the Richter scale.

Some volcanic eruptions may affect global climate and even stratospheric ozone. Apparently, only those that are relatively rich in sulfur oxide gases and sufficiently explosive to send ejecta well into the stratosphere can disturb the climate or the ozone shield.

If sulfur oxide reaches the stratosphere, the gas reacts with moisture to form tiny droplets of sulfuric acid. The acid droplets are so small (less than one micrometer in diameter) that they may remain suspended in the stratosphere for months to several years. While in the stratosphere, the acid droplets absorb solar radiation, causing warming of the stratosphere and reducing the amount of solar radiation that reaches the lower atmosphere. Furthermore, some solar radiation is scattered by acid droplets back to space. The net result is cooling at the earth's surface. Based on studies of past sulfur-rich violent volcanic eruptions, the drop in global mean temperature is likely to be less than 1 C° (1.8 F°). Recall, however, from our discussion of climate in chapter 15 that a small change in global mean temperature may translate into greater temperature changes in specific localities.

The aftermath of the June 1991 eruption of Mt. Pinatubo in the Philippines brought new concerns regarding the potential environmental impact of violent volcanic eruptions. After being dormant for almost six hundred years, Mt. Pinatubo produced what was perhaps the most violent eruption of this century. Millions of metric tons of sulfur dioxide were blasted into the stratosphere, and the resulting veil of sulfuric acid droplets spread around the globe. The acid droplets are not only likely to affect climate (perhaps temporarily offsetting greenhouse warming), they also may be speeding the destruction of stratospheric ozone by causing chlorine (from CFCs) to be more reactive (chapter 15).

Occasionally, a volcanic eruption can be predicted, but at present such predictions are more an art than a science. In making such predictions, geologists rely primarily on premonitory events. For example, as magma flows in the subsurface, it apparently sets off a swarm of distinctive minor earth tremors. Also, prior to an eruption, intruding magma may cause a volcano to swell, thereby causing slight changes (a few centimeters) in the tilt and elevation of the land. Sensitive instruments on land and on satellites can detect those events, but they are not always indicators of a pending volcanic eruption. Nonetheless, only two years prior to the Mount St. Helens eruption, two U.S. Geological Survey scientists, D. R. Crandell and D. R. Mullineaux, after analyzing premonitory events in the volcano's area, warned that an eruption was likely sometime before the end of this century.

Land-Use Conflicts: Community Growth

Another group of land-use problems stems from community growth and urban expansion. Many metropolitan areas want both the economic benefits of continued growth and a high quality of life and healthy environment for their citizens. Too often, however, community growth is at odds with quality of life and the environment. This problem is becoming particularly acute in the nation's densely populated Northeast and mid-Atlantic regions. A poll published in the September 1990 issue of *Money* magazine cited various factors that were important to readers in selecting an ideal place to live. Of forty-four factors, clean water rated number one and clean air rated number three.

One negative consequence of community growth is the loss of prime agricultural land to development for industry, housing, highways, reservoirs, or other urban-related land use. Of the half million hectares of cropland that are removed from production each year, more than half goes to urban and suburban development—a trend that conflicts with the future food needs of a growing population (chapter 8).

During the 1970s, for perhaps the first time since the founding of the United States, human population in rural areas grew about as fast as the population of urban areas. This situation changed again in the 1980s, largely for economic reasons; new jobs opened up in and near cities, while job prospects declined in rural areas. The U.S. Bureau of the Census reports that 90% of the nation's population growth during the 1980s took place in supercities. A **supercity** is a coalescence of adjacent cities and towns having a combined population of at least 1 million. In 1950, the nation's fourteen supercities were home to less than 30% of the population. The number of supercities reached thirty-five in 1980 and thirty-nine in 1990. Now, for the first time, more than half (50.2%) of the U.S. population resides in supercities.

Expressways that link suburbs and the inner city have contributed significantly to urban expansion. They allow industries to locate on the fringes of metropolitan areas, and they have spurred the growth of suburbs. Expressways or freeways also crisscross cities, breaking up old, established neighborhoods, threatening the economic vitality of central business districts, encouraging the relocation of industries and populations to formerly rural areas, and forcing greater reliance on motor vehicles. Highway construction has also had adverse effects on the landscape, destroying natural habitats, disrupting drainage patterns, and distorting scenic panoramas.

The original goal of the National System of Interstate and Defense Highways (the I-system), begun by President Eisenhower in 1956, was to construct a network of highways that would link 90% of cities having a population greater than 50,000. Today, with the system nearly complete, the 70,000 kilometers (43,000 miles) of interstate highways cover 0.73 million hectares (1.8 million acres) of land. Although states and cities enthusiastically participated in the extensive interstate highway construction during the 1950s and 1960s, today many communities are reevaluating the merits of more new freeways. Further growth will probably involve upgrading existing highways rather than new construction.

In large metropolitan areas, freeways are plagued by increasing traffic congestion (fig. 21.18), which increases air pollution and causes rapid deterioration of the roadway. In the three decades since the interstate system began, highway travel has tripled. In some large cities (e.g., New York City), gridlock is becoming the norm rather than just a rush-hour phenomenon. In Los Angeles, the average freeway speed is expected to be cut in half in twenty years. Heavy use has left many segments of the I-system badly in need of repair. The Federal Highway Administration predicts that by the year 2000, 90% of the I-system will require capital improvement.

What can cities do to relieve highway congestion? Chapter 16 raised this question in the context of strategies to curb

Figure 21.18 In most larger cities, heavy traffic congestion and even gridlock last many hours longer than the usual rush hours.
© Steve Elmore/Tom Stack & Associates

urban air pollution: car pools, special lanes for bus (and car pool) use only, more flexible work hours, so that everyone does not head for work and home at the same time, and mass transit. In New York City, more than one thousand companies currently provide their employees with fare subsidies to encourage them to take buses and subways. In recent years, many cities have shown renewed interest in building subways and light-rail (streetcar) systems to alleviate traffic congestion. Light-rail service returned to Los Angeles in July 1990, after a twenty-nine-year hiatus (fig. 21.19). A 35-kilometer (22-mile) line linking Los Angeles and Long Beach is the first segment of a planned 240-kilometer (150-mile) light-rail network. In addition, a $3.9 billion subway is currently under construction in downtown Los Angeles that will link the city to the San Fernando Valley.

Critics of the new mass transport movement cite the problems, particularly financial, that plague existing urban commuter lines; many of which (e.g., in New York City and Boston) run deeply into debt each year and are burdens for taxpayers. New equipment generally has been less than satis-

Figure 21.19   Many cities are considering the return of the streetcar to relieve traffic congestion. This is the new light rail line in Santa Clara County, California.

© Hans Halberstadt/Photo Researchers, Inc.

factory, with costly breakdowns and delays. In some cities, subway cars are targets for vandals and thieves, and those threats usually necessitate a costly security system.

Growth-Control Strategies

Individual states are becoming more involved in resolving conflicts in land use and in preserving land from development. Dozens of states now have laws that regulate the siting of power plants and control mining. Many have passed laws to protect wetlands and to identify and protect *critical areas*, which are either particularly sensitive to disturbances or have historic significance. Some states even provide tax incentives that encourage landowners to keep their land in agriculture or to preserve it for its esthetic or recreational value.

Many towns and suburbs have long employed zoning ordinances to control residential growth. Usually, those ordinances specify minimum lot size for homes, control the size of structures, and specify how far structures must be set back from the road lot lines. Also some communities have growth-control ordinances that limit the number of new dwellings. In recent years, some communities have even declared a one- to two-year moratorium on development, so that they can properly assess the impact of continued growth on such services as water supply, sewage treatment, highway maintenance, and police and fire protection.

Zoning that specifies minimum lot size for new homes can actually spur urban and suburban sprawl. Consequently, developers build on land beyond the edge of an urban area, thus accelerating urbanization of rural areas. Ironically, the inner portion of most large cities has a considerable amount of vacant or underutilized space that could be developed for housing, light manufacturing, and commercial enterprises. M. D. Lowe of the Worldwatch Institute* argues that urban sprawl (and inner-city decay) can be checked by land-use planning that encourages greater housing density and integration of commercial and housing structures within the existing boundaries of a city. Portland, Oregon, for example, has curbed its urban sprawl by encircling the city with an

*For more on this topic, refer to the article by Lowe cited at the end of this chapter.

urban growth boundary beyond which no new development is allowed. This strategy forces development of available space within the city. Consequently, more people live in the city, stores and work are more conveniently located, and there is less reliance on automobile transportation.

In response to a general perception that the federal government was not doing enough during the 1980s to curb suburban sprawl, some state legislatures appropriated funds to set aside lands from development. A land-stewardship law in Wisconsin earmarks $250 million for land conservation during the 1990s. In 1990, Florida Governor Robert Martinez proposed that the state spend $3.2 billion over the next decade to help prevent development of threatened beaches, mangrove islands, and other natural habitats. In New Jersey, a voter-approved $300 million bond issue is intended to halt development of open lands.

An increasingly popular conservation and growth-control strategy is the **land trust,** in which citizens donate undeveloped land to a nonprofit, private conservation group (such as The Nature Conservancy or the National Audubon Society) for a specified period of time, with the stipulation that the land will not be developed. In return, the donor may earn an income tax deduction. Alternatively, a conservation group raises funds to buy up undeveloped land, which it either manages itself or sells or donates to a state agency for preservation. A third approach is enactment of a conservation easement whereby the development rights to a parcel of land are granted to a conservation group while the owner is permitted to continue living on the land.

Several communities on Cape Cod, Massachusetts, have adopted land trusts as a way of curtailing the region's rapid rate of development. Cape Cod is a 113-kilometer (70-mile) long arm of sand that reaches into the Atlantic Ocean, and is the fastest growing section of New England. Between 1970 and 1989, the number of new housing units on the Cape nearly doubled. Each summer, the population swells by hundreds of thousands of tourists. As a result, more shops, fast-food restaurants, motels, cottages, and condominiums are springing up on the Cape and threatening the environmental assets that draw tourists in the first place.

Nationwide, the number of land trusts increased from 429 in 1980 to almost 900 by 1991. As of this writing, land trusts protect about 1.1 million hectares (2.7 million acres) of forests, farmland, wetlands, and historic sites.

Growth-control strategies, however, have been criticized by people who argue that the regulations create economic barriers to open housing and unjustly restrict free enterprise. Some people view zoning ordinances as unwarranted intrusions on property rights and personal freedom. Others argue that community growth-control measures are a legitimate means of exercising wise stewardship over the land and its use. No doubt, this issue will continue to be controversial.

Urban Planning and New Towns

Virtually every city has a planning commission that typically hears challenges to zoning regulations and decides on the location of highways and other public service facilities. Unfortunately, policy is often determined by political pressures applied by special interests rather than by carefully reasoned and visionary planning. As a result, in most cities, costly demolition of houses and other structures becomes necessary to make way for expressways, skyscrapers, and water and sewer lines.

Many city planning agencies have come under severe criticism for urban renewal projects that favor the well-to-do over the low-income segments of the population. Critics point out that too often slums are cleared only to be replaced by luxury high-rise apartment complexes. Meanwhile, housing opportunities for low-income persons decline, pushing up demand and rents for existing low-income housing. Critics also complain that urban renewal projects have produced sterile, bland, and uninviting buildings, walkways, and public parks.

Although they are generally politically independent, a city and its suburbs are highly interdependent for such public services as police protection, water supply and treatment, transportation, and garbage collection. Many suburbs, in fact, are nothing more than bedroom communities, whose residents depend on city services but provide the city with little or no tax support. Hence, the most logical approach to planning is for the various component communities of a metropolitan area (or supercity) to coordinate planning through a single authority. Yet very little regional-scale planning occurs in the United States.

Theoretically at least, some of the problems of existing cities can be avoided or relieved by planning and constructing new communities virtually from scratch. So-called **new towns** would operate efficiently and protect the environment. New towns would relieve population pressures on urban areas and permit a more orderly and deliberate approach to the problems of rising housing costs and social integration. Actually, the concept dates back to the turn of the century, although popular interest in new towns was spurred by the housing needs of post-World War II years.

Among the best-known and most successful of the new town projects is the one authorized by the Greater London Plan of 1944. New towns were intended to curb population growth and reduce congestion in London. The 1944 plan set limits on population density within London (250 persons per hectare) and established a permanent greenbelt (a parkway) surrounding the city. Citizens and industry were encouraged to locate in new planned communities that were situated well beyond the greenbelt (at least 40 kilometers, or 25 miles). Although the new town approach reduced congestion in London to some extent, population density increased dramatically in areas just beyond the greenbelt. The original plan was then revised, so that now all new towns must be located at least 100 kilometers (62 miles) beyond London.

New towns in the United States are quite different from those in Europe in that they typically feature lower population densities, greater emphasis on leisure facilities and automobile transport, and primarily white-collar employment opportunities. Also, median incomes tend to be relatively high.

Some new towns were developed near research or educational facilities (e.g., Irvine, California), or regional shopping centers (e.g., Columbia, Maryland), while other new towns have very specialized purposes (e.g., retirement communities).

Until the early 1960s, new town development in the United States seldom adhered to a master regional plan. That practice changed with the planning and construction of Reston, Virginia, and Columbia, Maryland. Reston, which is 23 kilometers (15 miles) west of Washington, D.C., is a leisure-oriented community that features much open space and green areas; employment is limited principally to offices and research facilities. Columbia, located midway between Baltimore and Washington, D.C., was originally developed around a large shopping center and then attracted substantial industrial employment opportunities.

The late 1980s saw renewed interest in new town projects largely in response to traffic congestion in rapidly growing suburbs. In contrast to earlier designs, which were primarily residential, these proposed new towns were self-contained villages in which people live, work, and shop, thus reducing dependency on automobiles. Montgomery Village, New Jersey, is one such town; when completed in the mid-1990s, it will be 50% residential and 50% offices and stores.

Conclusions

As human population grows, competition for finite lands also increases, and conflicts over land-use are bound to become more frequent. In addition, as we step up our efforts to exploit new sources of fossil fuels, minerals, fresh water, and other resources, land-use conflicts will grow more intense. As our cities expand, more people may live in regions that are prone to floods, earthquakes, and other natural hazards. Assessment of land capability is a logical first step in wise land-use planning. Ideally, that process will enable us to balance multiple uses of land and permit both use and preservation.

Key Terms

coastal zone	floodplain management
estuary	earthquake
beach	tsunami
longshore current	Richter scale
barrier islands	epicenter
neritic zone	volcano
oceanic zone	supercity
red tide	land trust
marine sanctuaries	new towns
flood	

Summary Statements

The coastal zone encompasses estuaries, coastal forests, beaches, barrier islands, and the ocean margin. Rapid population growth and consequent development are stressing the coastal zone.

Estuaries are the most productive ecosystems. They are transitional between land and sea and feature a circulation pattern that retains and recirculates nutrients. Unfortunately, that same circulation pattern also retains pollutants.

Disturbance of beach-forming processes (e.g., by damming rivers or building coastal breakwaters) has accelerated shoreline erosion.

Barrier islands along the Atlantic and Gulf coasts are the first line of defense against powerful storm waves. They are dynamic systems that under natural conditions absorb wave energy and thereby continually change shape. Homes and roads stabilize barrier islands and alter their basic protective function.

The ocean margin (neritic zone) is the principal site of commercial seafood harvests. It is also vulnerable to excessive inputs of nutrients that can trigger massive algae blooms.

Land-use control strategies must take into account the vulnerability of certain lands to natural hazards such as flooding, earthquakes, and volcanic eruptions.

Floods may be caused by natural events such as excessive rainfall or by human activities that alter the ability of the ground to soak up heavy rains or snowmelt (e.g., urban development).

Traditionally, U.S. flood-control efforts focused almost exclusively on structural alternatives such as dams and levees. Today, however, the emphasis is on floodplain management.

An earthquake is a violent shaking of the ground that can trigger landslides; avalanches of snow, mud, and rock; floods; and tsunamis. While most earthquakes occur along plate boundaries, major earthquakes have occurred in the mid-section of the North American continent.

The Mount St. Helens eruption in 1980 may signal a resurgence of volcanic activity in the western United States. Prediction of volcanic eruptions relies on premonitory phenomena and is currently more an art than a science.

During the 1980s, urban and suburban areas grew, reversing the trend away from metropolitan areas that had characterized the 1970s.

Traditionally, communities control growth through zoning ordinances, which are sometimes criticized for creating economic barriers to open housing. Land trusts are a relatively new strategy in preserving ecologically or historically important areas from urban and suburban development.

Increasing traffic congestion, especially in suburban areas, has brought renewed interest in mass transit systems and planned communities (new towns).

The future is likely to bring increasing competition for finite lands as our resource demands continue to grow.

Questions for Review

1. Identify and describe some of the biological and geological features of the coastal zone. List some of the human activities that currently stress the coastal zone.
2. Speculate on why people are drawn to coastal areas.
3. What is an estuary? Explain why estuaries are such productive ecosystems.
4. Why are estuaries ideal habitats for juvenile fish?
5. Why are estuaries particularly vulnerable to pollution?
6. Explain how a longshore current forms. How does it supply sand to a beach? What is the ultimate source of that sand?
7. Describe how damming a river might affect a downstream estuary or beach.

8. In what sense are barrier islands dynamic systems? How do they protect coastal regions from storm waves? What is the impact of housing and commercial development on barrier islands?

9. Why is the marine neritic zone so productive biologically?

10. Identify the sources of nutrients that cause an excessive algae bloom in coastal waters.

11. What causes a red tide? Why are they a concern?

12. How and why is economic development of the coastal zone being regulated?

13. List some human activities that increase the potential for flooding. What factors contribute to urban flooding?

14. Identify activities that are appropriate along a river floodplain. What criteria do you use?

15. Explain the shift away from structural control of floods and toward floodplain management.

16. Explain how the Richter scale is an index of energy released by an earthquake.

17. List some of the factors that affect the amount of property damage caused by an earthquake.

18. What measures can be taken in earthquake-prone areas to reduce the toll of deaths, injuries, and property damage?

19. Why is the West Coast considered to be the most earthquake-prone region of the United States? Why might an earthquake be much more destructive in eastern North America?

20. How do cities and towns limit land development? Some critics argue that community growth-control ordinances are discriminatory. What is your view? Support your opinion.

Projects

1. Find out how land-use planning is done in your community. What local and state agencies are involved, and how can ordinary citizens participate in decision making?

2. Is your locality vulnerable to any natural hazards such as flooding or earthquakes? What is being done to protect lives and property?

Selected Readings

Dolan, R., and H. Lins. "Beaches and Barrier Islands." *Scientific American* 257 (July 1987):68-77. Describes the natural processes that produce barrier islands and how those islands are threatened by development.

Dvorak, J. J., C. Johnson, and R. I. Tilling. "Dynamics of Kilauea Volcano." *Scientific American* 267 (Aug. 1992): 46-53. Describes what is understood about subsurface magma movements that control the eruptions of Kilauea, Hawaii.

Environmental Protection Agency. "Can Our Coasts Survive More Growth?" *EPA Journal* 15, no. 5 (1989). Collection of articles that address the environmental problems caused by development pressures in the coastal zone.

Francis, P., and S. Self. "Collapsing Volcanoes." *Scientific American* 256 (June 1987):91-97. Describes the effects of the catastrophic collapse of volcanoes.

Johnston, A. C., and L. R. Kanter. "Earthquakes in Stable Continental Crust." *Scientific American* 262 (March 1990):68-75. Reviews the occurrence of powerful earthquakes well away from plate boundaries.

Lowe, M. D. "Shaping Cities." *State of the World, 1992*. Worldwatch Institute. New York: W. W. Norton, pp. 119-37. Advocates an urban-planning ethic that is sensitive to both environmental and societal needs.

Monastersky, R. "Perils of Prediction." *Science News* 139 (1991):376-79. Describes efforts by the U.S. Geological Survey to develop a public warning system for seismic and volcanic events.

Pilkey, O. H., and W. J. Neal. "Save Beaches, Not Buildings." *Issues in Science and Technology* 8 (Spring 1992): 36-41. Argues for a more aggressive government policy to discourage development of beachfront property.

Plummer, C. C., and D. McGeary. *Physical Geology*. 5th ed. Dubuque, Iowa: Wm. C. Brown, 1991. Includes well-illustrated chapters on volcanic activity and earthquakes.

Soil Conservation Society of America. "Managing Floodplains to Reduce the Flood Hazard." *Journal of Soil and Water Conservation* 31 (1976):44-62. A collection of articles on floodplain dynamics, the nation's increasing vulnerability to floods, and approaches to floodplain management.

Stein, R. S., and R. S. Yeats. "Hidden Earthquakes." *Scientific American* 260 (June 1989):48-57. Discusses the occurrence of large earthquakes on faults that do not cut the earth's surface.

Epilogue
Toward a Sustainable Environment

The Promise of Technology
Reducing Resource Demand
Sustainable Development
What Each of Us Can Do
Conclusions

Our industrial systems are best advised to follow
nature's example of sustainable resource
utilization.
© Doug Sherman

E1

Since the turn of the century, certain global trends have increased the stress on the quality of the environment. Human population has tripled, the global economy has grown twentyfold, fossil fuel consumption is up by a factor of thirty, and industrial productivity has soared fiftyfold (mostly since 1950). Among the many environmental consequences of these trends are contaminated waterways, eroded soil, shrinking tropical forests, a decline in species diversity, and modifications of the atmosphere that may change the global climate. All these adverse consequences have prompted more and more scientists to question how much future growth the planet can sustain.

History attests to the strong link between changes in economic systems and the integrity of ecological systems. Industrialization spurred economic growth, and with industrialization came waste and pollution. Affluent societies are wasteful. Most of the environmental problems described in this book arise from our acquisition, use, and disposal of natural resources to support an affluent life-style. For example, soil salination, soil erosion, and waterlogging via agriculture degrades the land; toxic gases enter the atmosphere as by-products of manufacturing; and disposal of industrial wastes threatens groundwater quality. At the same time, supplies of many important resources, such as oil, are declining. Furthermore, as our demands for water, food, housing, energy, transportation, commercial goods, and recreation continue to grow, competing interests increasingly come into conflict over the use of finite lands.

Unforeseen negative consequences often overshadow the benefits of technological advances. CFCs, DDT, and PCBs are prime examples of innovations that in time had to be scrapped because of their dire environmental impacts. Hence, it is not surprising that some people view economic development and environmental protection as incompatible goals.

If we accept those goals as incompatible, then the U.N. projection that world population will top eleven billion within the century is particularly troubling. Today, the difference between the haves (more-developed nations) and have-nots (less-developed nations) is staggering. The wealthiest 23% of the global population consumes more than half the world's energy resources and more than one-third of the fertilizer supply. At any time of year, perhaps 25% of the world's population goes hungry. Most future population growth will take place among poor, less-developed nations. If we assume that they will approach the same standard of living as today's more-developed nations, then the demands on the planet's finite resources will grow, and the generation of waste will accelerate, further degrading the environment. Compounding the problem, impoverished people in the more-developed nations strive to improve their lot in life, and the most affluent citizens of more-developed nations tend to want a bigger house, another auto, or the latest electronic gadget.

What, if anything, can we do to meet the demands of future generations and at the same time preserve environmental quality? In this epilogue, we attempt to answer this question by building on topics already discussed in this book.

The Promise of Technology

Perhaps the most important lesson of our exploration of space is that earth is our only home. We will not establish major colonies on other planets in the foreseeable future; nor will we import resources from other worlds. The resources on this planet are all we have. For some of these resources, such as metals, minerals, and fuels, the regeneration rate by geological processes is so slow that the supply is essentially fixed and finite; that is, those resources are nonrenewable. Renewable resources such as crops require fertile soils and adequate fresh water to ensure a continuous supply. However, soil erosion and air and water pollution threaten the future production of food and timber.

Some scientists, economists, and government officials believe that the concept of limited resources is misleading. They point out that the planet's crust and oceans contain enormous quantities of minerals, albeit in very dilute concentrations. They argue, too, that abundant energy sources are available to us. For example, trillions of barrels of oil are locked up in western oil shales. Those experts believe that as economic conditions become more favorable and exploration and mining technologies improve, copious supplies of many valuable, nonrenewable resources will become available—enough to meet the needs of a growing economy far into the future.

It is true that vast, untapped reserves of minerals and energy sources exist that are currently too expensive to exploit. However, for very low-grade mineral deposits, recovery may never be feasible unless inexpensive fuel becomes available—an unlikely prospect. In the long run, all sources of energy are expected to become more expensive. Furthermore, as we saw in chapter 18, the requirement of huge inputs of fresh water plus the prospect of generating enormous amounts of waste are likely to delay significant development of the energy potential of oil shale well into the next century.

Many Americans have an almost blind faith that technological innovations will eventually come to our rescue. Indeed, history encourages this belief. Technology has helped us to control many diseases, expand our resource base, and raise our standard of living. Although past technological success may make us optimistic, we must temper our optimism by recalling the inherent limitations of technological research and development. Scientists and engineers may sometimes lack sufficient understanding to focus research efforts appropriately. For example, research on alternative energy sources is proceeding in many directions simultaneously because researchers do not yet know which of their efforts (or combination of efforts) will be most fruitful. Furthermore, some technological problems are more difficult to solve than others, assuming that solutions exist at all. Developing a commercially available nuclear fusion reactor, for instance, poses much greater scientific challenges than does travel to the moon.

Advances in technology usually involve complex efforts expended over long periods of time and require the skills of many highly trained and creative people. In addition, new technologies often employ rare and therefore expensive materials. Typically, a new technology must undergo lengthy testing and gradual scaling up from a pilot, or prototype, stage to full-scale operation. Some critics of nuclear power argue that many of the problems that the industry faces today can be traced to a deployment that was much too rapid. In retrospect, they see the nation as having rushed the application of nuclear fission technology without taking enough time to develop adequate safeguards, especially regarding the permanent disposal of radioactive waste.

Experience also tells us that just about every technological advance creates side effects that detract from its benefits. For example, any new technique that we might develop for exploiting low-grade deposits of minerals is certain to require substantial amounts of fuel and other resources (such as water) and to generate huge quantities of waste. More waste, in turn, is likely to degrade other valuable resources (e.g., surface or groundwater, or natural habitats).

The human mind is a fertile source of ideas, and human ingenuity will no doubt continue to devise technological solutions to our resource supply problems. However, the promises of technology are limited. As we have noted, new technology is expensive; it typically requires abundant and inexpensive fuel for implementation; it takes time for development; and it may produce unacceptable side effects. Hence, it is unwise to base our future solely on the belief that the planet is a vast storehouse of resources that our technology will inevitably unlock. A more fruitful course of action would be to rely more on technology that is energy-efficient, suited to the real needs of society, and sensitive to limitations set by nature.

Reducing Resource Demand

If prospects for technological innovations that will increase our resource supply are limited, what about the other side of the supply/demand equation? Can we somehow curb demand and thus ensure adequate resources and environmental quality for future generations?

We must consider demand from two perspectives: cumulative demand by the entire global population and per capita demand. The population growth rate has subsided greatly in the United States (as well as in the other affluent, more-developed nations), and will probably stop growing within about forty years (disregarding migration). This projection is a good sign for the future of more-developed nations; it means that Americans, for example, appear to have a better opportunity to cope with the problems of resource supply and demand than do people in less-developed nations.

On the other hand, per capita consumption in more-developed nations such as the United States is quite another matter. Despite increasing costs, each succeeding generation of Americans has so far enjoyed an ever-rising level of affluence. Many American families now have air-conditioned homes equipped with at least two bathroooms, two television sets, two cars, a stereo, and a personal computer. Many people own recreational vehicles, such as powerboats and campers. Furthermore, recycling has yet to supplant planned obsolescence as an intrinsic operating principle of our economy.

We continue to be infatuated with disposable items; too often we use an appliance for a short time and then replace it with a newer model that is more stylish or offers one or two new features. How many of us, for example, have made the transition from records to audiotapes to compact discs (CDs)?

In recent years, some people have voluntarily reduced their energy consumption. More people are riding bicycles and small motorcycles for transportation and emphasizing self-propelled sports, such as jogging, backpacking, canoeing, and cross-country skiing. Given a choice and/or an economic incentive, more people opt for energy-efficient appliances and automobiles. Faced with increasing disposal costs, more communities are mandating recycling as part of their waste-management programs. These are encouraging signs because such activities consume fewer resources, generate less waste, and reduce environmental damage. This is just the beginning, however, and much more could be done to curb per capita demand.

A relative handful of Americans have chosen to dramatically change their life-styles and consumption habits by leaving the comforts of home and taking up a pioneer existence in such remote regions as Alaska or northern Maine. However, few of us really have any desire to take on the hardships of the so-called good old days. We value the advantages that our society provides in making our lives easier and, presumably, more enjoyable. Furthermore, few people have the knowledge or skills needed to grow most of their own food, to build and maintain their own houses, or to sew and mend their clothing. Surviving without modern conveniences would be extremely difficult if not impossible for most of us.

In less-developed nations, the problem is not so much per capita demand but the cumulative demands of a burgeoning population. Understandably, the less-developed nations wish to raise their standard of living and to do so as rapidly as possible. The People's Republic of China, for example, refused to agree to the Montreal Protocol (on ozone depletion) and the international ban on the use of CFCs. Chinese officials argued that their people desperately needed refrigeration to preserve food and could not wait for the development of environmentally safe alternatives to CFCs for their refrigerators.

The likelihood of holding in check the demands of 77% of the world's population (people of the less-developed nations) is slim. In fact, there is the ethical question of whether the affluent 23% of the world's population is justified in witholding from less-developed nations the beneficial aspects of our technology, even though that technology is also resource-intensive and wasteful. Perhaps a wiser course of action is development of low-input, low-impact technology that is sustainable in the long run.

Sustainable Development

With growing world demand for finite resources, will we eventually exceed the planet's carrying capacity? Can we avoid global ecological catastrophe? A growing number of scientists and economists argue that we can if we strive for sustainable development. Fundamentally, *sustainable development* refers to economic growth that takes place within the limits set by nature; it occurs in a way that maintains environmental quality and allows a reasonable level of global prosperity in both more-developed and less-developed nations. The concept of sustainable development is a direct outgrowth of the growing environmental awareness of the past two decades and our increasing understanding of how ecosystems function and interact and how human activities affect the orderly operations of nature.

One of the principal messages of our discussion of environmental problems in this book is that the modern industrial world has undergone *unsustainable development;* that is, economic growth has taken place by using up resources and despoiling the environment at a rate faster than the natural rate of restoration. It is as if humankind were somehow apart from nature and free to control and manipulate nature for personal gain. This is a short-sighted view that discounts or ignores the consequences for future generations. Consider, for example, the mounting legacy of nuclear waste that we are leaving to our descendants. Also, with unsustainable development, we tend to react to environmental problems by applying some technological solution such as scrubbers or water filtration equipment. Such an approach addresses merely the symptoms, not the root cause of the problem.

Living within limits set by nature seems to be a logical alternative to the way we have been living. We have pointed out numerous strategies that work with and not against the natural functioning of the environment. We described, for example, the advantages of recycling, ways to increase energy efficiency, the value of protecting cropland from erosion, and the use of wetlands to upgrade water quality (fig. E.1). Americans certainly value environmental quality—public opinion polls consistently show this. Our grassroots concern for the environment spurred passage of the most stringent environmental regulations of any nation, and continues to do so. Yet, we are the most wasteful society on earth. Although we may have started on the road to sustainable development, we have not progressed very far. What's missing?

In the opinion of William D. Ruckelshaus, former head of the EPA, Americans require greater personal motivation and social institutions to translate it to action. Furthermore, the movement toward sustainable development must encompass all nations, regardless of their level of development. This is a formidable task that Ruckelshaus warns may be comparable in scale to the agricultural or industrial revolutions.

Sustainable development requires (1) significant reduction in the rate of global human population growth, (2) inclusion of the environmental cost of production, use, and disposal in all economic activity, and (3) operation of industries to more closely approximate ecosystems. In this section we explore these requirements in greater detail.

The foremost challenge in achieving sustainable development on a global scale is to slow human population growth until birth rates balance death rates. This has already happened in a few more-developed nations. Although population

Figure E.1 Degraded water is filtered free of most contaminants as it flows through this marsh.
© C. C. Lockwood/Earth Scenes

growth rates are slowing in many less-developed nations, there is still a pressing need for further reductions. As we saw in chapter 7, this requires (1) better educational and employment opportunities for women, (2) reduced rates of infant mortality (one of the prime reasons why people in poor nations choose to have large families), and (3) increased availability of safe, effective, and appropriate fertility-control information and methods.

A second requirement for sustainable development is that all economic activity must include the full environmental cost of production, use, and disposal. That is, all industries and consumers must assume the cost of protecting the environment from pollution or other disturbances resulting from their actions.

Energy is a prime example of a commodity whose market price falls far short of its real cost, which includes environmental and other impacts. The hidden costs of fossil fuels include, for example, health care for people suffering from respiratory illnesses, effects of toxic gas emissions, destruction

Figure E.2 This cartoon illustrates how the market price of fuel fails to account for its true cost.

© ED STEIN reprinted by permission of NEA, Inc.

of natural habitats, and loss of species diversity. Other hidden costs of fossil fuels include military expenditures to protect Persian Gulf oil and government subsidies to energy suppliers in the form of tax credits and research funding (fig. E.2). The magnitude of the gap between market price and actual cost of fossil fuels is difficult to assess, but estimates range from $100 billion to $300 billion per year for the United States.

The difference between the market price and the actual cost of a product illustrates the economic concept of *negative externality*. Recall from chapter 2 that a negative externality is the cost of undesirable side effects of some activity (such as generating electricity in a coal-fired power plant) that must be borne by people who are not directly involved in that activity. This type of negative externality leads to what biologist Garrett J. Hardin calls the "tragedy of the commons."

The commons are the resources that we all share, such as the atmosphere, the oceans, and the rivers. Each of these resources contributes directly or indirectly to our personal and collective well-being. When an individual (or industry) despoils the commons for personal (or collective) gain, that gain accrues exclusively to the individual (or industry) while the degradation affects *all* users of the commons. For example, it may be economically advantageous for an industry to dump its waste in a river, while those who depend on the river as a food source, for recreation, or as a source of drinking water must suffer the consequences of reduced water quality. Traditionally, when human activity degrades the commons, the offender has no responsibility to pay for the damage.

Thus, sustainable development requires a government-mandated pricing policy that accounts for the environmental costs of goods and services. This may, for example, take the form of a surcharge on gas-guzzling automobiles (encouraging the purchase of more fuel-efficient vehicles), a greater tax on gasoline (to spur development of renewable energy sources), or a climate-protection tax on coal-fired electric generating plants (to cut emissions of the greenhouse gas, carbon dioxide). The state of New York, for example, assesses a fee per kilowatt-hour of electricity used based on the associ-

ated environmental degradation. The Congressional Budget Office estimates that a carbon tax on fossil fuels would reduce projected carbon dioxide emissions for the year 2000 by 37%. Similar taxes may be imposed on hazardous waste, pesticides, and groundwater mining.

Obviously, environmental ("green") taxes are passed on to the consumer in the form of higher prices for goods and services. Such a pricing policy provides the consumer with economic incentives to conserve, recycle, or select a less expensive alternative that is also less offensive to the environment.

Sustainable growth also requires industry to model its operations after the natural functioning of ecosystems (chapter 3). That is, industry must operate as a more closed system (an *industrial ecosystem*) with a considerably greater commitment to waste reduction, recycling, and energy recovery than is now the case. This requires, for example, more waste exchanges and product designs that facilitate recycling (chapter 20). Such strategies cut energy demand and the raw material content of products and services. The consequence for the environment is less mining, reduced pollution of air and water, and reduced disturbance of natural habitats (chapter 17).

If sustainable development is to succeed, less-developed nations must curb their inclination for rapid development at all costs. They must avoid the short-sighted and ecologically unsound technologies that now plague more-developed nations and opt instead for more closed industrial systems that optimize the use of resources and environmental protection. This is a tall order that calls for international cooperation and much more financial aid to less-developed nations. The World Bank reported in 1988 that the seventeen poorest countries paid out $31.1 billion more (to debt holders) than they received in aid from more-developed nations. Clearly, this negative cash flow must reverse if less-developed nations are to adopt technologies that protect the environment at the expense of slower economic growth. Already, we are seeing increasing interest in debt-for-nature swaps, in which a debt is forgiven in return for the setting aside of some threatened ecosystem (chapter 11).

Even as we embark on the road toward sustainable development, we must watch for new environmental problems. This means continued support for scientific research and monitoring of the global environment, and the development and application of new technologies and strategies to conserve resources, reduce pollution, and reclaim scarred lands. Many environmental problems have international or global dimensions (e.g., global warming, thinning of the ozone shield), and solving them requires the cooperation of nations representing a wide spectrum of political, social, and economic viewpoints. Hence, the need is great for international coordination of research and free exchange of information in spite of political barriers.

In June 1992, the United Nation's Conference on Environment and Development (UNCED) convened in Rio de Janeiro, Brazil. UNCED was a follow-up to the UN's Stockholm Conference on the Human Environment held twenty years ago. Some 130 heads of state plus delegates from more than 160 nations grappled with the major challenges of protecting the global environment while spurring economic development in less-developed nations. Numerous statements and recommendations were produced dealing with global issues such as loss of biodiversity, prospects for global climate change, management of biotechnology, protection of oceans and freshwater resources , and international cooperation in technology development. Some representatives of the United States government were widely criticized for not taking a more pro-environment stance at Rio de Janeiro. However, UNCED marks merely the beginning of international debate and negotiations on global environmental problems that is likely to continue for many decades to come.

What Each of Us Can Do

What, if anything, can one person do about environmental issues? Actually, by reading this book, you have taken an important first step. You are now better informed about the roots and implications of environmental problems. We cannot anticipate all of tomorrow's environmental issues, nor can we be sure which of today's problems will fade in importance. By studying this book, however, you have gained a clearer idea of how the various segments of our society contribute to and mitigate environmental problems.

For the individual, perhaps the best operating principle in dealing with problems of the environment is to think globally and act locally. We can all modify our life-styles and thereby contribute to solving environmental problems. For example, we could decide to drive a smaller car that gets better mileage and emits fewer pollutants, or we could join a car pool to save gasoline and wear on our vehicle. We could improve energy conservation in our home by turning down the thermostat in winter, adding insulation, and using air conditioning only when really needed. These are only a few examples; perhaps you have already decided on taking these steps or others.

Personal efforts to modify our life-style in order to conserve resources and reduce pollution are essential and praiseworthy, but there are other effective ways to become personally involved in environmental decision making. One way is to participate in the shaping of public policy by electing environmentally aware and responsive public officials at the local, state, and national levels. As we have seen (chapter 2), environmental issues are matters of public policy as well as matters of science and technology; such issues raise important and often controversial questions involving personal values, competing interests and goals, and the use of cost-benefit or risk-benefit analysis.

Progress in improving air and water quality has come about largely as a result of federal pollution-control legislation. Efforts to conserve scarce resources and clean up the environment are, at least in part, responses to government-imposed financial incentives or disincentives. For example, nine states have bottle laws that encourage recycling and discourage littering by requiring purchasers of canned or bottled beverages to pay a deposit added to the price of beverages.

How can we know which public officials are environmentally responsible? The League of Conservation Voters, located in Washington, D.C., can provide some help. The League is a nonpartisan, national campaign organization that promotes the election of environmentally aware public officials at the national level. Prior to each presidential election, the League publishes a thoroughly researched background study of each candidate's views on such issues as air and water pollution, endangered species, pesticides, and nuclear power. The report is entitled *The Presidential Candidates: What They Say, What They Do on Energy and the Environment.* The League also publishes, once a year, the voting records of incumbent United States senators and representatives on environmental issues, and it assigns an environmental score to each member of Congress on the basis of his or her voting record.

For nonincumbents at the national level and all candidates at state and local levels, it is more difficult to determine who stands where on the issues, but it can be done. For example, one way is to check with the League of Women Voters, which prepares information on the views of candidates for state and federal office. Another approach is to write or telephone a candidate's campaign office and find out the candidate's views on environmental matters.

How can we influence government policymakers after election day (as well as our neighbors and fellow citizens)? We can start by joining an environmental organization, such as the National Wildlife Federation or the Sierra Club, that works to influence public policy by lobbying elected and appointed public officials. Such groups work to pass or defeat legislative bills and sometimes propose environmental legislation, and they monitor the implementation of environmental laws. Their representatives meet with government policymakers to discuss environmental issues as well as pending environmental legislation. To locate such environmental groups and to learn more about them, consult the *Conservation Directory* (published annually by the National Wildlife Federation), which lists organizations, government agencies, and government officals concerned with natural resource use and management in both the United States and Canada. Appendix C also lists the names and addresses of major environmental organizations.

Knowledgeable citizens can also exert a more direct influence on government policymakers. Perhaps the most straightforward way is to write a well-informed letter to appropriate policymakers on the environmental matter that concerns us. Such letters—even just a handful of them—can, and often do, make a difference. Elected officials are looking for cues: Very few of them have a good understanding of environmental science, and they do not necessarily have a preformed opinion about every issue that comes before them. Moreover, astute politicians want to know both sides of an issue. When they get information from a lobbyist, they know that they are getting only one side of the story. Consequently, even a small number of well-thought-out letters from their constituents can make a difference in how they see the issue in question, and how they vote on it.

Figure E.3 One way that each of us can contribute to environmental quality is to volunteer our time and energy for community cleanup projects.
© C. C. Lockwood/Earth Scenes

Consider the following, however, when you write to elected government officials: (1) Avoid form letters prepared for mass mailings.(2) Be sure that you write to the ones who represent *you*. If you are from Iowa, your letter will not have much influence on a senator or representative from New Mexico or Delaware.

Telephone calls as well as letters to the local offices of state legislators and to local officials (elected or appointed) can also be effective. Calls to congressional representatives and bureaucrats in Washington are generally less effective than letters, because the message may not get past the office staff. For example, a representative might have twenty or more persons working for him or her. One of those staff members will take your call, but not all of them have regular direct contact with their boss. In contrast, a state legislator is likely to have only one or two assistants. Those aides see the legislator regularly, and they might put you through directly to him or her.

Another way to influence government decision makers is to attend and participate in public hearings on environmental matters. The federal EPA and state environmental agencies, as well as their respective regional offices, can tell us when and where hearings will be held. They can also provide background information for the hearings. Be assured that any hearing about changes in rules or regulations that could adversely affect an industry will be attended by representatives of that industry. If such a hearing is not well publicized—and many are not—the public officials who preside over it may hear only the industry's side of the issue unless people or groups that are proenvironment have been alerted.

Your class or campus student organization can invite public officials to make presentations. Most officials are usually willing to do this. The best format is a brief talk by the official about what he or she does or (if an elected official) what his or her views are, followed by a question-and-answer session. Such a program allows you to ascertain the policymaker's views and also to inform him or her of your perspective. As noted earlier, few policymakers have a clear understanding of environmental science and environmental problems; they are looking for cues. Remember, though, that public officials rarely consider a problem solely in terms of its scientific aspects. The problem's economic and political dimensions also weigh heavily (chapter 2).

Still another means of influencing public officials is to take carefully prepared, well-timed legal action. For example, it is posssible to file civil law suits against the EPA for failure to perform many of its nondiscretionary duties. Nearly all the major federal environmental laws, except the Superfund Act, allow anyone to file suit to compel the EPA to fulfill its legal mandates. Several national environmental organizations have used those so-called citizen-suit provisions to force the EPA to issue standards for hazardous air pollutants. Legal action can, of course, be expensive. However, some environmental laws (such as the Clean Air Act Amendments) allow the courts to impose the plaintiff's litigation costs on the government if it loses the case.

You can influence other people's views on the environment and contribute to environmental problem solving by your choice of a career. For example, you may enter business and help your company incorporate environmental sensitivity as a high priority in strategic planning. You may enter the political arena, where reforms and changes in priorities are clearly needed. You may pursue scientific research and seek new knowledge or work on more efficient technologies. Or you may become a teacher, and teach your students the value of a healthy environment.

We can also influence the quality of the environment as a consumer. What we choose to buy or not to buy has a great influence on manufacturers and distributors of durable goods. For example, purchase goods that are recyclable or made from recycled materials. In this way, we can help create a market for such goods and contribute to sustainable development.

Another option for you, your class, or civic group is to initiate or get involved in a project that benefits the environment. Some communities operate "clean-sweep" programs for collecting toxic and hazardous household waste. Student volunteers, especially those with a chemistry background, often help to coordinate the effort. Some communities call for volunteers to clean up litter from local beaches, recreational areas, or highways (fig. E.3). In fact, the adopt-a-highway program is now well accepted nationwide.

The more we know, the more we can do. Above all, then, you can become—and remain—as informed as possible about environmental matters. Every day, the electronic and print media bombard us with details about the latest environmental problems. We must interpret that information in the context of basic ecological principles, and if we are serious about keeping environmentally informed, we will seek out more information. The National Wildlife Federation's *Conservation Directory* also lists periodicals of interest, directories, and sources of information in audiovisual format. (See also Appendix C.)

One of the best sources of information about contemporary environmental issues is the annual *State of the World* reports of the Worldwatch Institute and the *Worldwatch Paper* series on environmental topics. Publications from federal agencies are another source of useful information on the environment. The Government Documents section of college and university libraries probably has (or can quickly obtain) most reports of agencies such as the EPA, the Department of Energy, the Departmentof the Interior, and the Department of Agriculture. If you have not done so already, acquaint yourself with the Government Documents section of your library. Librarians in such departments will usually provide special instruction on researching government publications. Note that government documents often are not included in the card or on-line catalog. If you do research without the benefit of government documents, you will miss much relevant and current information.

Finally, if we are patient and persistent, we can use the Freedom of Information Act to obtain information that a federal agency has not published. For example, suppose that you are researching air pollution in your town or county and you wish to know the emission levels of a nearby industrial plant. The EPA occasionally publishes information concerning emissions for air-quality control regions, but it does not publish data for specific installations. By invoking the Freedom of Information Act, however, you can obtain the emission data for that plant if the EPA has it. Be sure to invoke the act when you write: "Under the provisions of the Freedom of Information Act, I request the following information. . . ." If you do not use this wording, the recipient agency may not feel obliged to provide the specific information you seek. Federal agencies are supposed to answer requests that are made under this act within ten working days. If you do not receive an answer within a reasonable period of time, you can appeal through the same agency. If your appeal fails, as a last resort, you can take the agency to court.

Do not overlook state agencies as sources of unpublished information; every state but Mississippi now has a law that is comparable to the federal Freedom of Information Act.

Conclusions

Prospects for the future are unclear. Throughout this book, we have seen that in recent decades significant progress has been made to enhance the resource base and to improve environmental quality: (1) the world is feeding more people than ever before; (2) alternatives to environmentally harmful pesticides are being developed; (3) some endangered species have been rescued from impending extinction; (4) water and air quality in some regions have improved significantly; and (perhaps most important) (5) public interest in seeking solutions to environmental problems is at an all-time high.

Yet we are well aware that many serious problems remain: (1) erosion of the stratospheric ozone shield; (2) the potential for global warming; (3) toxic materials in the land, air, and water; (4) shrinking global supplies of oil and natural gas; (5) a loss of perhaps 25% of the planet's biodiversity; (6) degradation of cropland; and (7) a continued and unprecedented rate of human population growth. What are the

solutions? The limitations of technology make it unlikely that innovations will provide sufficient resources to sustain continued population growth. In more-developed nations, populations are approaching zero growth. Although this would appear to ease demands on raw materials and reduce environmental degradation, the effect is offset by a wasteful and often excessive per capita demand. In less-developed nations, the population has already exceeded the carrying capacity.

A growing number of scientists, however, argue that we can avoid future ecological disaster by adopting sustainable development, that is, economic growth that occurs within limits set by nature. Among the requirements for sustainable development are a slowing of global population growth, a pricing policy for goods and services that includes environmental costs, and industrial systems modeled after ecosystems. Furthermore, we should not underestimate the role of the individual. Each of us can contribute significantly to conserving resources and improving environmental quality.

Selected Readings

Brown, L. R., C. Flavin, and S. Postel. "Picturing a Sustainable Society." in *State of the World, 1990.* Worldwatch Institute. New York: W. W. Norton, 1990, pp. 173-90. Projects what the world would be like in 2030 if we adopt sustainable development.

Clark, W. C. "Managing Planet Earth." *Scientific American* 261 (Sept. 1989):47-54. An introduction to an issue of *Scientific American* that is devoted to the topic of sustainable development.

Frosch, R. A., and N. E. Gallopoulos. "Strategies for Manufacturing." *Scientific American* 261 (Sept. 1989):144-52. Argues that industrial systems should emulate natural ecosystems in resource recovery.

Getis, J. *You Can Make a Difference: Help Protect the Earth.* Dubuque, Iowa: Wm. C. Brown, 1991. Lists specific steps that an individual can take to help conserve resources and reduce environmental degradation.

Grove, R. H. "Origins of Western Environmentalism." *Scientific American* 267 (July 1992): 42-47. Discusses the historical roots of the modern environmental movement.

Hardin, G. "The Tragedy of the Commons," in *Managing the Commons,* edited by G. Hardin and J. Baden. New York: W. H. Freeman and Company, 1977, pp. 16-30. A classic work on a concept that is central to an understanding of environmental issues.

Hubbard, H. M. "The Real Cost of Energy." *Scientific American* 264 (April 1991):36-42. Presents the case for including environmental and social costs in the market price of energy sources.

MacNeill, J. "Strategies for Sustainable Economic Development." *Scientific American* 261 (Sept. 1989): 155-65. Discusses the relationship between ecological systems and economic systems.

Renner, M. "Creating Sustainable Jobs in Industrial Countries," in *State of the World 1992,* Worldwatch Institute. New York: W. W. Norton & Company, 1992, pp. 138-54. Describes strategies whereby environmental goals are compatable with economic goals.

Rosenberg, N., and L. E. Birdzell, Jr. "Science, Technology, and the Western Miracle." *Scientific American* 263 (Nov. 1990):42-54. Discusses the historical links between scientific understanding and technological development.

Ruckelshaus, W. D. "Toward a Sustainable World." *Scientific American* 261 (Sept. 1989):166-74. Describes how to implement sustainable development.

Appendix A
Conversion Factors

	Multiply	By	To Obtain
Length	inches	2.540	centimeters
	feet	0.3048	meters
	statute miles	1.6093	kilometers
	nautical miles	1.853	kilometers
	centimeters	0.3937	inches
	meters	3.281	feet
	kilometers	0.6214	statute miles
	kilometers	0.5397	nautical miles
Mass and weight	ounces (avdp)	28.350	grams
	pounds	0.4536	kilograms
	tons	0.907	metric tons
	grams	0.03527	ounces (avdp)
	kilograms	2.205	pounds
	metric tons	1.120	tons
Volume	fluid ounces	0.02957	liters
	gallons	3.785	liters
	liters	0.2642	gallons (U.S.)
	liters	33.82	fluid ounces
Area	square yards	0.8361	square meters
	acres	0.4047	hectares
	square miles	2.590	square kilometers
	square meters	1.196	square yards
	hectares	2.471	acres
	square kilometers	0.3861	square miles
	hectares	0.0100	square kilometers
	square kilometers	100.0	hectares
Pressure	bars	0.9869	atmospheres
	inches of mercury	25.40	millimeters mercury
	bars	1000.0	millibars
Energy	joules	0.2389	calories
	kilocalories	1000.0	calories
	joules	1.000	watt-seconds
	calories	0.003968	BTUs
	BTUs	252.0	calories
	kilowatt-hours	859,850.0	calories
	kilowatt-hours	3413.0	BTUs
	therms	2.52×10^7	calories
	therms	100,000.0	BTUs
	quads	1.0×10^{15}	BTUs
	barrels of crude oil	5.80×10^6	BTUs
	gallons of gasoline	1.25×10^5	BTUs
	tons (2,000 lb) of coal	2.35×10^7	BTUs
	cubic feet of natural gas	1020.0	BTUs
Power	horsepower	745.7	watts
	BTUs per minute	0.0176	kilowatts
	watts	0.0569	BTUs per minute
	watts	0.00134	horsepower
	watts	0.0143	kilocalories per minute
	joules per second	1.00	watts

Temperature
Conversion Equations
$^\circ F = 9/5 \, ^\circ C + 32$
$^\circ C = 5/9 \, (^\circ F - 32)$

Appendix B
Expressing Concentrations

Scientists express concentrations in a number of different ways because gases, liquids, and solids have different properties. We are perhaps most accustomed to expressing concentration as a percentage. We breathe air that is approximately 78% nitrogen, 21% oxygen, and 1% other gases. This means that 78 parts (molecules) out of a base of 100 total parts of air (molecules) are nitrogen. Of the remaining 22 parts, 21 are oxygen, and 1 part is composed of other gases. Visualize a box filled with 100 colored ping pong balls: 78 are green, 21 are blue, and 1 is yellow. The concentration of green balls in the box is 78%, or 78 parts per hundred (pph), and so forth. Hence, percent concentration is the same as parts per hundred. Concentrations that are only a tiny fraction of a percent are often important; for example, only 0.035% of air is carbon dioxide (CO_2), a gas essential for photosynthesis. For such small concentrations, scientists increase the size of the base. Thus, the concentration of carbon dioxide in the atmosphere can be expressed as 350 parts per million (ppm) rather than 0.035%.

The table below shows how we can express the same concentration using different bases. Concentrations listed for the sodium chloride (NaCl; table salt) content of sea water use different bases, but the salt content is identical for each expression of concentration.

Concentrations of substances dissolved in water are often stated as the mass (or weight) of substance per volume of water, for example, milligrams per liter (mg/L), because water is more frequently measured by volume than by weight. Concentrations in water stated as mg/L or parts per million (ppm) have the same value, because 1 liter of water weighs 1 million milligrams. Thus, the concentration of sodium chloride in sea water (29,600 ppm) can also be stated equivalently as 29,600 milligrams per liter (mg/L).

For solid materials such as soil or food, which are easier to weigh, concentrations are expressed as mass (or weight) per kilogram of material, for example, milligrams per kilogram (mg/kg). For solids, ppm and mg/kg have the same value, because 1 kilogram equals 1 million milligrams, that is, 10 ppm = 10 mg/kg. For air, concentrations of contaminants are expressed as milligrams per cubic meter (mg/m^3), or parts per million (ppm). Stating concentrations as mg/m^3 is only useful when atmospheric pressure and temperature are measured at the time a sample is analyzed. Because air becomes thinner with altitude (less dense at lower atmospheric pressure), it is more convenient to express such concentrations in terms of ppm. This eliminates the problem of dealing with the changes in volume that an air sample undergoes with changes in temperature and/or pressure.

Sodium chloride content of sea water expressed in different units			
Parts per Hundred	**Parts per Thousand**	**Parts per Million**	**Parts per Billion**
2.96 pph	29.6 ppt	29,600 ppm	29,600,000 ppb

Appendix C
Resources on the Environment

Publications

Ambio: A Journal of the Human Environment. Royal Swedish Academy of Sciences, Box 50005, S-104 05, Stockholm, Sweden.

American Forests. American Forestry Association, 1516 P St. NW, Washington, DC 20005.

Amicus Journal. Natural Resources Defense Council, 122 E. 42nd St., New York, NY 10168.

Annual Review of Energy. Department of Energy, Forrestal Building, 1000 Independence Ave. SW, Washington, DC 20585.

Audubon. National Audubon Society, 950 Third Ave., New York, NY 10022.

BioScience. American Institute of Biological Sciences, 730 11th St. NW, Washington, DC 20001.

Conservation Biology. Blackwell Scientific Publications, Inc., 52 Beacon St., Boston, MA 02108.

Demographic Yearbook. Department of International Economic and Social Affairs, Statistical Office, United Nations Publishing Service, United Nations, New York, NY 10017.

The Ecologist. MIT Press Journals, 55 Hayward St., Cambridge, MA 02142.

Environment. Heldref Publications, 4000 Albemarle St., NW, Washington, DC 20016.

Environmental Science & Technology. American Chemical Society, 1155 16th St., NW, Washington, DC 20036.

EPA Journal. Environmental Protection Agency. Order from Government Printing Office, Washington, DC 20402.

Issues in Science and Technology. National Academy of Sciences, 2101 Constitution Ave. NW, Washington, DC 20077.

Journal of Environmental Education. Heldref Publications, 4000 Albemarle St. NW, Washington, DC 20016.

Journal of Environmental Health. National Environmental Health Association, 720 S. Colorado Blvd., Suite 970, Denver, CO 80222.

National Geographic. National Geographic Society, P.O. Box 2895, Washington, DC 20077.

National Parks and Conservation Magazine. National Parks and Conservation Association, 1015 31st St. NW, Washington, DC 20007.

National Wildlife. National Wildlife Federation, 1400 16th St. NW, Washington, DC 20036.

Nature. 711 National Press Building, Washington, DC 20045.

New Scientist. 128 Long Acre, London, WC 2, England.

Not Man Apart. Friends of the Earth, 530 Seventh St. SE, Washington, DC 20003.

Population Bulletin. Population Reference Bureau, 1875 Connecticut Ave. NW, Washington, DC 20009.

Science. American Association for the Advancement of Science, 1333 H St. NW, Washington, DC 20005.

Science News. 1719 N St. NW, Washington, DC 20036.

Scientific American. 415 Madison Ave., New York, NY 10017.

Sierra. 730 Polk St., San Francisco, CA 94108.

State of the World. Worldwatch Institute, 1776 Massachusetts Ave. NW, Washington, DC 20036.

Weatherwise. Heldref Publications, 4000 Albemarle St., NW, Washington, DC 20016.

Wilderness. The Wilderness Society, 1400 I St. NW, 10th Floor, Washington, DC 20005.

World Rainforest Report. Rainforest Action Network, 300 Broadway, Suite 298, San Francisco, CA 94133.

World Resources. World Resources Institute, 1735 New York Ave. NW, Washington, DC 20008.

Worldwatch Papers. Worldwatch Institute, 1776 Massachusetts Ave. NW, Washington, DC 20036.

Yearbook of World Energy Statistics. Department of International Economic and Social Affairs, Statistical Office, United Nations Publishing Service, United Nations, New York, NY 10017.

Organizations

Alan Guttmacher Institute, 2010 Massachusetts Ave. NW, 5th Floor, Washington, DC 20036.

American Forestry Association, 1516 P St. NW, Washington, DC 20005.

American Meteorological Society, 45 Beacon St., Boston, MA 02108.

Center for Marine Conservation, 1725 DeSales St. NW, Suite 500, Washington, DC 20036.

Center for Science in the Public Interest, 1501 16th St. NW, Washington, DC 20036.

Clean Water Action, 317 Pennsylvania Ave. SE, Washington, DC 20003.

Conservation Foundation, 1250 24th St. NW, Suite 500, Washington, DC 20037.

Cousteau Society, 930 W. 21st St., Norfolk, VA 23517.

Defenders of Wildlife, 1244 19th St. NW, Washington, DC 20038.

Environmental Defense Fund, Inc., 257 Park Ave. South, New York, NY 10010.

Friends of the Earth, 218 D St. SE, Washington, DC 20003.

League of Women Voters of the United States, 1730 M St. NW, Washington, DC 20036.

National Audubon Society, 950 Third Ave., New York, NY 10022.

National Clean Air Coalition, 801 Pennsylvania Ave. SE, Washington, DC 20003.

National Environmental Health Association, 720 S. Colorado Blvd., Suite 970, Denver, CO 80222.

National Geographic Society, 17th and M Sts. NW, Washington, DC 20036.

National Parks and Conservation Association, 1015 31st St. NW, 4th Floor, Washington, DC 20007.

National Wildlife Federation, 1400 16th St. NW, Washington, DC 20036.

Natural Resources Defense Council, 40 W. 20th St., New York, NY 10011, and 1350 New York Ave. NW, Suite 300, Washington, DC 20005.

The Nature Conservancy, 1814 N. Lynn St., Arlington, VA 22209.

The Oceanic Society, 218 D St. SE, Washington, DC 20003.

Population Reference Bureau, 1875 Connecticut Ave. NW, Washington, DC 20009.

Rainforest Action Network, 300 Broadway, Suite 298, San Francisco, CA 94133.

Resources for the Future, 1616 P St. NW, Washington, DC 20036.

Scientists' Institute for Public Information, 355 Lexington Ave., New York, NY 10017.

Sierra Club, 730 Polk St., San Francisco, CA 94109, and 408 O St. NE, Washington, DC 20002.

Smithsonian Institution, 1000 Jefferson Dr. SW, Washington, DC 20560.

Society of American Foresters, 5400 Grosvenor Lane, Bethesda, MD 20814.

Soil and Water Conservation Society, 7515 N.E. Ankeny Rd., Ankeny, IA 50021.

Student Conservation Association, Inc., P.O. Box 550, Charlestown, NH 03603.

Union of Concerned Scientists, 26 Church St., Cambridge, MA 02238.

The Wilderness Society, 900 17th St. NW, Washington, DC 20006.

World Resources Institute, 1735 New York Ave. NW, Washington, DC 20006.

Worldwatch Institute, 1776 Massachusetts Ave. NW, Washington, DC 20036.

World Wildlife Fund, 1250 24th St. NW, Suite 500, Washington, DC 20037.

Zero Population Growth, 1400 16th St. NW, 3rd Floor, Washington, DC 20036.

Federal and International Agencies

Agency for International Development, State Building, 320 21st St. NW, Washington, DC 20523.

Bureau of Land Management, U.S. Department of Interior, 18th and C Sts., Room 3619, Washington, DC 20240.

Bureau of Mines, 2401 E St. NW, Washington, DC 20241.

Bureau of Reclamation, Washington, DC 20240.

Congressional Research Service, 101 Independence Ave. SW, Washington, DC 20540.

Conservation and Renewable Energy Inquiry and Referral Service, P.O. Box 8900, Silver Spring, MD 20907.

Consumer Product Safety Commission, Washington, DC 20207.

Council on Environmental Quality, 722 Jackson Pl. NW, Washington, DC 20006.

Department of Agriculture, 14th St. and Jefferson Dr. SW, Washington, DC 20250.

Department of Commerce, 14th St. between Constitution Ave. and E St. NW, Washington, DC 20230.

Department of Energy, Forrestal Building, 1000 Independence Ave. SW, Washington, DC 20585.

Department of Health and Human Services, 200 Independence Ave. SW, Washington, DC 20585.

Department of Housing and Urban Development, 451 Seventh St. SW, Washington, DC 20410.

Department of the Interior, 18th and C Sts. NW, Washington, DC 20240.

Department of Transportation, 400 Seventh St. SW, Washington, DC 20590.

Environmental Protection Agency, 401 M St. SW, Washington, DC 20480.

Federal Energy Regulatory Commission, 825 N. Capitol St. NE, Washington, DC 20426.

Fish and Wildlife Service, Department of the Interior, 18th and C Sts. NW, Washington, DC 20240.

Food and Agriculture Organization (FAO) of the United Nations, 101 22nd St. NW, Site 300, Washington, DC 20437.

Food and Drug Administration, Department of Health and Human Services, 5600 Fishers Lane, Rockville, MD 20852.

Forest Service, P.O. Box 98090, Washington, DC 20013.

Geological Survey, 12201 Sunrise Valley Dr., Reston, VA 22092.

Government Printing Office, Washington, DC 20402.

Inter-American Development Bank, 1300 New York Ave. NW, Washington, DC 20577.

International Whaling Commission, The Red House, 135 Station Rd., Histon, Cambridge CB4 4NP England.

Marine Mammal Commission, 1625 I St. NW, Washington, DC 20006.

National Academy of Sciences, Washington, DC 20660.

National Aeronautics and Space Administration, 400 Maryland Ave. SW, Washington, DC 20548.

National Cancer Institute, 9000 Rockville Pike, Bethesda, MD 20892.

National Center for Appropriate Technology, 3040 Continental Dr., Butte, MT 59701.

National Center for Atmospheric Research, P.O. Box 3000, Boulder, CO 80307.

National Marine Fisheries Service, U.S. Dept. of Commerce, NOAA, 1336 East-West Highway, Silver Spring, MD 20910.

National Oceanic and Atmospheric Administration, Rockville, MD 20852.

National Park Service, Department of the Interior, P.O. Box 37127, Washington, DC 20013.

National Science Foundation, 1800 G St. NW, Washington, DC 20550.

National Technical Information Service, U.S. Department of Commerce, 5285 Port Royal Rd., Springfield, VA 22161.

Nuclear Regulatory Commission, 1717 H St. NW, Washington, DC 20555.

Occupational Safety and Health Administration, Department of Labor, 200 Constitution Ave. NW, Washington, DC 20210.

Office of Ocean and Coastal Resource Management, 1825 Connecticut Ave., Suite 700, Washington, DC 20235.

Office of Surface Mining Reclamation and Enforcement, 1951 Constitution Ave. NW, Washington, DC 20240.

Office of Technology Assessment, U.S. Congress, 600 Pennsylvania Ave. SW, Washington, DC 20510.

Organization for Economic Cooperation and Development (U.S. Office), 2001 L St. NW, Suite 700, Washington, DC 20036.

Soil Conservation Service, P.O. Box 2890, Washington, DC 20013.

Solar Energy Research Institute, 1617 Cole Blvd., Golden, CO 80401.

United Nations, 1 United Nations Plaza, New York, NY 10017.

United Nations Environment Programme, Regional North American Office, United Nations, Room DC2-0803, New York, NY 10017, and 1889 F St. NW, Washington, DC 20006.

World Bank, 1818 H St. NW, Washington, DC 20433.

Appendix D

Environmental Concerns of Hawaii, Mexico, and Canada

Introduction

Throughout this text we present examples of environmental issues from numerous nations around the world, but our primary focus is the continental United States. In this section we summarize issues of environmental quality and resource management in three regions: Canada, Mexico, and the state of Hawaii. Canada is a more-developed nation; hence, its environmental problems are similar in many respects to those of the United States, but they differ in several significant ways. Although Mexico has undergone a period of rapid industrial growth, it remains a less-developed nation and thus faces issues of explosive population growth, seriously degraded air and water quality, and resource exploitation that differ fundamentally from those facing the United States and Canada. Hawaii is a collection of small, volcanic islands in the middle of the Pacific Ocean. The isolated location of Hawaii means that its people must cope with environmental issues that differ significantly from those facing their mainland counterparts.

Focus on Hawaii

Hawaii is unique among the fifty states. It consists of eight main islands in the middle of the tropical Pacific Ocean (fig. D.1). Its isolation and volcanic history have important implications for both its people and its diverse ecosystems. Isolation from distant genetic stock favored the evolution of many plants and animal species found nowhere else on the planet. Isolation plus a limited resource base have forced Hawaiians to rely upon expensive imports, including energy. Furthermore, continuing volcanic activity on the big island of Hawaii poses a long-term hazard to everyday life on parts of that island.

Geothermal Power

The state of Hawaii faces a variety of problems that are similar to those of other small, geographically isolated regions with rapidly growing populations. Between 1980 and 1990, Hawaii's population grew nearly 15%, or 50% faster than the national average. Furthermore, the U.S. Bureau of the Census projects that by 2010 the population of Hawaii will grow by another 45%, well above the national average of about 13%. Meanwhile the state is struggling to

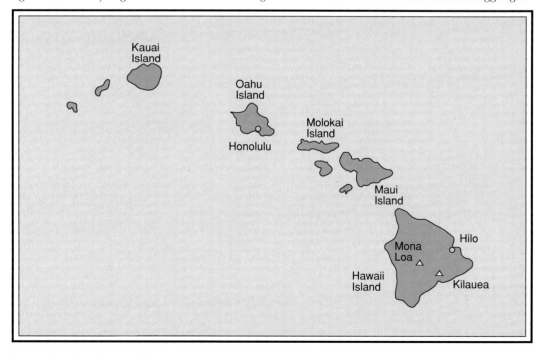

Figure D.1 The islands of Hawaii.

obtain sufficient supplies of certain resources, particularly inexpensive fuel to generate electricity. Currently Hawaii produces much of its electricity by burning expensive, imported oil. Furthermore, the state has a limited reserve capacity to generate electricity, so that equipment breakdowns frequently result in blackouts.

To resolve the energy-supply problem, planners began over ten years ago to consider development of Hawaii's seemingly abundant geothermal energy resources (chapter 19). The largest source of geothermal energy lies beneath the East Rift Zone on the big island of Hawaii, a ridge that extends eastward from the summit of Kilauea (the most active volcano in the world) far out into the Pacific Ocean. In recent years drilling has begun on sites near the eastern boundary of Volcanoes National Park. Plans call for an initial production of 25 megawatts of electric power from geothermal energy. Eventually the proposed Hawaii Geothermal/ Deep Water Cable project may utilize numerous geothermal wells to supply energy to run a 500-megawatt plant. Much of this electricity will be transmitted to the island of Oahu (where most Hawaiians live) by three undersea cables (a distance of over 160 kilometers or 100 miles). Since its inception the plan has generated considerable controversy.

Proponents of geothermal energy point to the economic advantages of reducing the state's dependence on costly imported petroleum. Furthermore, they argue that an in-state source of energy would be less subject to disruption and therefore would help to ensure continued economic growth. They also contend that extraction of geothermal energy has certain environmental advantages over oil combustion (chapter 19).

Opponents of geothermal energy raise fundamental questions of public safety. They are concerned about the potential release of toxic concentrations of such substances as hydrogen sulfide, lead, chromium, and mercury from geothermal well-heads. Uncontrolled releases of steam carrying minute amounts of these substances have already occurred at two drilling sites. Subsequently, drilling was halted until the summer of 1992 when more stringent safety measures were implemented. Proponents argue that the new measures will ensure public safety, while opponents counter that too little is known about the intricate maze of volcanic conduits that underlie the drilling sites to be confident that accidents will not occur.

Opponents also strongly object to the construction of geothermal facilities so near to a 12-kilometer-long (7.5-mile-long) fissure from which lava has been pouring more or less continuously since 1983. These lava flows have covered nearly 100 square kilometers (39 square miles), an area roughly equal in size to the island of Manhattan. Flows have crushed and burned more than 180 homes. In fact, the site proposed for initial geothermal drilling had to be abandoned in the mid-1980s because it was gradually overrun by Kilauea's continual eruptions. A new drilling site was acquired in 1985 when the state government and private developers agreed to a land swap. The state obtained 10,000 hectares (25,000 acres) of private land (which included the first proposed drilling site) adjacent to Volcanoes National Park in exchange for the state's 7,200 hectare (16,000 acre) Wao Kele o Puna Natural Area Preserve.

Now opponents of geothermal drilling are raising questions over the vulnerability of the new drilling sites at Wao Kele o Puna to future lava flows. They also object to drilling in Wao Kele o Puna because this tract contains the last sizeable tropical lowland rain forest in the United States. Opponents contend that it is hypocrisy for the United States to expect less-developed nations to stop clearing their rain forests when we refuse to protect our own. Hawaii has already cut 80% of its rain forests over the last two centuries. Proponents of drilling counter that portions of Wao Kele o Puna have

Figure D.2 Lava flows from Kilauea.
USGS/Photo by J. D. Griggs

already been highly disturbed by wild pigs, whose destructive activities have permitted the invasion of many exotic weedy plants.

At the time of this writing, the future of geothermal power in Hawaii is in doubt. Large-scale projects such as the proposed 500-megawatt plant remain far from becoming reality. Meanwhile Hawaii continues to pay high prices for imported oil and to experience electrical blackouts.

The Volcanic Hazard

Volcanic activity on the big island of Hawaii is potentially beneficial in supplying geothermal energy, but in some parts of the island it is hazardous to life. Here we examine some additional environmental issues surrounding the impact of volcanoes on life in Hawaii.

The only active volcanoes in the state of Hawaii are Kilauea and Mauna Loa (see fig. D.1), both on the island of Hawaii. However, this island is also home to two other volcanoes, Mauna Kea and Haulalai, that have been dormant for 4,000 and 190 years respectively. How much of a threat do these volcanoes pose to the 120,000 inhabitants of the island and the hundreds of thousands of tourists who visit each year?

In contrast to explosive volcanic eruptions that produce large, fragmented debris, most eruptions of Mauna Loa and Kilauea are relatively quiescent outpourings of fluid lavas (chapter 21). Yet most of us would consider the sight of fountains of lava bursting hundreds of meters into the atmosphere to be both spectacular and dangerous. These fountains feed rivers of liquid rock that flow down the gentle slopes of a volcano (fig. D.2). Herein lies the major threat from Hawaii's volcanoes. Since written records began nearly 200 years ago, the lavas of Kilauea have blanketed some 15,000 hectares (38,000 acres), while those of Mauna Loa have covered over 64,000 hectares (160,000 acres). Although Kilauea is the world's most active volcano, Mauna Loa has also erupted numerous times in recent centuries. Furthermore, geologists expect both volcanoes to remain active for centuries to come.

Recent eruptions of Kilauea demonstrate the dangers of volcanic eruptions. Since early 1983, Kilauea has had numerous periods of lava flows punctuated by periods of inactivity. Downhill from the erupting fountains are the southeastern corner of Volcanoes National Park and the adjoining small communities of Royal Gardens, Kapa'ahu, and Kalapana, which lie near the coastline.

On 3 January 1983, a line of 60-meter (200-foot) fountains appeared in Kilauea's East Rift Zone. Throughout the following

Figure D.3 Lava overrunning a house near Volcanoes National Park.
Hawaiian Service Slides

months, lava repeatedly rolled down the slopes from the rift. Eventually the flows began to encroach on sparsely-settled Royal Gardens. On 1 March, a new flow pushing toward the settlement forced the evacuation of those in its immediate path. The next day the molten lava ignited the first home (fig. D.3). Then the fountaining stopped, the lava slowed and cooled, and the advance ceased. Soon afterwards, the evacuated residents returned to their homes. But their return to normal life did not last long. A month later, on 8 April, fountaining began anew. This time the entire community was evacuated as the lava moved quickly through the area, destroying several more homes. Again the flow stopped, and the residents returned. This process was repeated several times throughout the year. By the end of the first year of eruption, fifteen homes had been destroyed by lava, but remarkably no one had been injured.

The years 1984 and 1985 were relatively quiet. Lava flowed into Royal Gardens only twice in 1984 and once in 1985, and occasional evacuations became a way of life. Even so, homes were rebuilt, and new residents moved in amid considerable controversy over the appropriate use of land along a lava path.

On 18 July 1986, a series of fissures along a 12-kilometer (7.5-mile) segment of the East Rift opened and spewed fountains of lava that began to flow down the mountainside. Lava flows continued throughout the following months, and by late November, the advancing front had reached Kapa'ahu, a rural residential area between Royal Gardens and Kalapana. Blocked by the Kalapana highway, the lava flowed laterally and ignited and buried nine homes. As the flow continued down the mountainside, it fed a thickening slab of lava along the highway until it was higher than the highway. Then the lava began to creep across the road and flowed to the nearby ocean, eventually adding over 120 hectares (300 acres) of new land to the island of Hawaii.

The blocked highway cut off Royal Gardens from its neighboring communities to the east. The children of Royal Gardens now had to commute to school an additional 64 kilometers (40 miles). Furthermore, the dangers of the intervening active lava field forced people from Royal Gardens to drive 96 kilometers (60 miles) to see friends and relatives who lived only 2 kilometers (1.2 miles) away. In January 1987, crews began to rebuild the Kalapana highway to reunite the communities and to reopen the loop route between Volcanoes National Park and Hilo.

Meanwhile lava continued to flow from the fissures, and new flows further east began to invade Kalapana Gardens (a subdivision near the old coastal village of Kalapana), igniting house after house. All told, seventeen homes were destroyed. To illustrate the seemingly random movements of lava flows, the Royal Gardens subdivision was untouched during those years.

In the following two years, the volcano continued to send lava flows toward the ocean. Homes were occasionally lost, and the repaired highway was again overrun by lava. Life was hardly normal for the residents along the Kalapana coastline. Some were fed up with the constant threat of the volcano and moved out. Refugees from destroyed homes in Royal Gardens and Kapa'ahu had rented some homes in Kalapana, but although land prices had been slashed dramatically, no one was buying. Economic depression deepened. Instead of being part of a scenic loop of forests and volcanic lands between Volcanoes National Park and Hilo, Kalapana was a dead end.

By January of 1990 Kapa'ahu was entirely gone and Royal Gardens was mostly destroyed. The following months were the final days for Kalapana. In April the flows resumed with full vigor, and lava first reinvaded Kalapana Gardens. As homes were burned and crushed, utility workers moved through the village, removing propane tanks and taking down transformers from power poles. Some residents brought in large equipment and hauled away their homes. One family with the aid of friends and neighbors totally dismantled and carted away a home that was only a few months old. As the dismantling continued, only a few services remained. In early April a UPS delivery van arrived at a driveway only to find that the house was gone. As the lava began to spread into Kalapana Village and homes continued to burn one by one, controversy raged when some insurance companies announced that they would no longer insure homes in the area. The state insurance commissioner threatened severe action against companies that canceled policies with people whose homes lay in the path of the lava flows.

Throughout the following months the destruction continued. By the end of June, only sixteen homes remained standing in Kalapana Gardens and Kalapana Village. In seven and a half years the volcano had destroyed more than 180 homes, a church, and a store, and had caused nearly $70 million in property damage. Furthermore, the lives of hundreds of former residents had been changed forever.

We should note that the potential for volcanic destruction on the island of Hawaii is much greater than that suffered during the recent Kilauea eruptions. On 25 March 1984, Mauna Loa also erupted with lava volumes that were many times greater than those spewing from Kilauea. The lava crept downslope toward Hilo, a city of 38,000 some 50 kilometers (30 miles) away. Fortunately, by the first week of April the volume of lava from Mauna Loa was declining. The flow slowed and finally stopped. On 15 April, geologists declared the eruption over, and the citizens of Hilo and its environs rested more easily. Yet, if the eruption of Mauna Loa had continued, the consequences could have been devastating.

The Kilauea eruptions illustrate how volcanic disasters differ from other natural disasters such as floods, tornadoes, and hurricanes. First, volcanic eruptions that consist mainly of lava flows can go on for a decade or longer. Second, property that lava flows claim is permanently inundated and thus cannot be recovered.

The Kilauea incidents raise many questions about how people deal with volcanoes. Should land that may be subject to lava flows be strictly zoned? If yes, what land uses can be permitted in such regions? Should construction be permitted? If yes, should structures be built so that they can be easily transported out of the path of an advancing lava flow? Should structures be insurable? If yes, should the premiums be subsidized by government agencies or by lower-risk policy holders? Ultimately the questions come down to how much risk individuals and society are willing to accept and who will pay if disaster strikes.

Preserving Biodiversity

A third set of environmental issues in the Hawaiian Islands concerns the preservation of biodiversity. Recall from chapter 11 that island species are particularly vulnerable to extinction. Species on the Hawaiian Islands are no exception. Ecologists are particularly concerned about the loss of biodiversity in the state of Hawaii because a high percentage of its species are endemic; that is, they are found nowhere else on the planet. For example, 98% of native flowering plant species are found only in Hawaii. The flora and fauna are unique because of the islands' location; they are the most geographically isolated islands on earth. The nearest continents are over 3,800 kilometers (2,400 miles) away. Recall also from chapter 11 that small, isolated islands experience few colonizations. Thus, the few species that came across these small islands were effectively isolated from distant genetic stock. Over millions of years, these colonizers diverged and evolved into many species that are unique to the islands.

Evolution of numerous species in Hawaii was aided by the fantastic diversity of habitats on the islands. Each island is the consequence of volcanic activity. As eruption after eruption built the volcanic islands ever higher above sea level, a gradient of environments from coastal dunes to alpine slopes developed. Geological forces have contributed to habitat diversity in yet another way. Although the islands are all of volcanic origin, they have formed at different times. Kauai, the oldest of the eight main islands, formed about six million years ago through the activities of a single volcano. With time and heavy rainfall, erosion of the volcanic mountain has created a landscape of deeply dissected canyons and steep ridges. In contrast, the island of Hawaii is the youngest of the main islands. It has formed within the past million years and, as we know, is still growing. Its landscape is composed of five overlapping volcanoes whose smooth, gentle slopes remain relatively untouched by erosional forces.

The diversity of habitats is further enhanced by the location of the islands in the path of the northeast trade winds. As the winds flow up the windward slopes, rainfall dramatically increases (chapter 13). However, as air descends on the leeward slopes, rainfall declines just as dramatically. On Mount Waialeale on the island of Kauai, the contrast in precipitation is spectacular. Annual rainfall varies from 11,700 millimeters (460 inches) on the windward side to less than 460 millimeters (19 inches) on the leeward side. Low rainfall in combination with highly permeable volcanic substrates produces a remarkably arid environment. Hence, the windward side of the islands support rainforest communities, while the leeward sides are dominated by sparsely vegetated cinder deserts.

The combination of habitat diversity and isolation created a rich diversity of species that were poorly prepared to cope with the arrival of humans and their fellow travelers—goats, cows, pigs, dogs, cats, and rats. Since the arrival of Polynesians around the year 400 A.D. and Europeans in 1778, at least fifty-six species of Hawaiian birds and unknown numbers of Hawaiian plants have been driven to extinction. Today, the state of Hawaii can claim the dubious honor of being the extinction capital of the United States. Twenty-five percent of the federally listed threatened and endangered species of plants and birds, or those proposed as candidates for listing, reside in Hawaii. In early 1992 the federal list included 63 species of Hawaiian plants (with another 126 species slated for listing) and 28 species of Hawaiian birds. Vanishing insects such as butterflies remain uncounted.

Threats to biodiversity in Hawaii differ little from those that endanger species elsewhere. Much of the original habitat below 950 meters (3,800 feet) elevation has been lost or seriously degraded by human activities. Native vegetation has been cleared for agriculture, ranching, housing, and commercial and tourist development. Activities on Hawaii's numerous military bases have also contributed to habitat degradation.

Feral animals have massively altered native ecosystems. For example, feral goats and cattle consume native vegetation and trample roots and seedlings. Feral pigs uproot precious ground cover. These destructive activities create favorable conditions for invasion of alien weeds that outcompete native vegetation for light, water, and soil nutrients. Furthermore, some of these weedy aliens can form such a thick vegetative mat that they prevent the re-establishment of native plant seedlings. The destructive activities of feral animals prevents regeneration of the native forests and can also lead to severe soil erosion that, in turn, produces silt-clogged streams and silt-covered coral reefs.

Predation by alien animals poses a major threat to numerous species of birds. Introduced black rats inhabit all eight main islands and have already contributed to the extinction of several species of birds, including the Hawaiian rail and the Laysan rail. These rats prey upon eggs or hatchlings in nests and also compete with some fruit-eaters for food. Feral cats have been common on the islands since the 1860s. They raise havoc with ground-nesting birds and species of the understory.

Efforts to save some of Hawaii's endangered species have brought some success. For example, cattle and goats once grazed in what is now the Haleakala National Park on the island of Maui. Some years ago the Park Service began an aggressive feral animal control program which included the construction of exclosures and systematic hunting. As a result of this program, the number of Haleakala's famous silversword plants (fig. D.4) has increased from a low of between 100 and 1,500 plants to an estimated 50,000 today.

The Hawaiian duck was once found on Kauai, Oahu, Maui, Molokai, and the island of Hawaii. It is now extinct on Maui and Molokai, and the populations on Oahu and Hawaii have been reintroduced following extinction. The species has been threatened by wetland loss and predation by feral cats and mongooses for some time. Since 1972 five wetlands have been set aside as national waterfowl refuges. With protection from poaching and predation, managers believe that the duck populations in these five refuges will become self-sustaining.

The Hawaiian state bird is the nene, or Hawaiian goose (fig. D.5). Previously found on all of the larger islands, it is now restricted to the islands of Hawaii and Maui. The Maui population is contained within Haleakala National Park, and portions of the Hawaii population are protected within Volcanoes National Park and four other sanctuaries. A successful captive breeding program was begun on the island of Hawaii in 1918, followed by the establishment of another program in England in the early 1950s. Recovery programs also include placing exclosures around nests, controlling predators, providing water and food in key areas, and replacing exotic vegetation with native food plants and cover. Although populations are currently holding steady at an estimated 390 to 425

Figure D.4 Haleakala silversword plant, an endangered Hawaiian plant.
© Kevin Schafer/Peter Arnold, Inc.

Figure D.5 Nene (Hawaiian goose), an endangered bird.
© S. J. Krasemann/Peter Arnold, Inc.

birds, they are only maintained with the introduction of captive-bred recruits.

Efforts to save Hawaii's legacy of endangered species have expanded in recent years. Additional lands have been set aside as preserves, efforts by plant nurseries and botanic gardens to propagate endangered species have increased, and captive-breeding programs have expanded. Yet the outlook for many species remains grim. Little is known about the majority of rare species, and minimal financial resources are available to develop and implement recovery plans for most endangered species. Assaults by humans and feral animals continue to fragment or degrade critical habitats, while predation and competition by alien species still take their toll.

Many rare species are particularly imperiled by their small numbers and minute refugia. For example, the alani, a tree in the citrus family, is known from two populations, one on Molokai and the other on Maui. The total population numbers only five individuals. The only known population of a shrubby member of a group of plants known as haha totals fewer than thirty individuals and occupy an area on Maui of only 9 square meters (100 square feet).

Similar situations are found among species of endangered birds. The only wild flock of the Hawaiian crow has dwindled to eleven birds. (Furthermore, only ten birds live in captivity.) In 1890, the large Kauai thrush was the most common forest bird on the island of Kauai. As of 1981, less than twenty-five birds remained, and they had retreated to a remote portion of an isolated swamp. Similarly, the Molokai thrush was once widespread and abundant throughout the island of Molokai. Surveys taken in the early 1980s found that less than twenty birds were left.

Unfortunately, many other examples can be given of Hawaiian species that are severely depleted and restricted to small patches of often marginal habitat. Such conditions greatly enhance their vulnerability to extinction from catastrophic natural events such as hurricanes (such as Iniki, which struck Kauai in September of 1992), floods, fire, or lava flows. Furthermore, such reduced gene pools favor the likelihood of inbreeding and a consequential loss of population vigor.

Focus on Mexico

Mexico (fig. D.6), with a 1992 population estimated at 88 million, ranks in the upper third of world economies. For Mexico, however, rapid economic growth over the past five decades has taken a terrible toll on environmental quality. Only in very recent years has the government taken steps to stem the tide of environmental degradation. In this section, we briefly review some of Mexico's most pressing environmental problems, including air pollution in Mexico City,

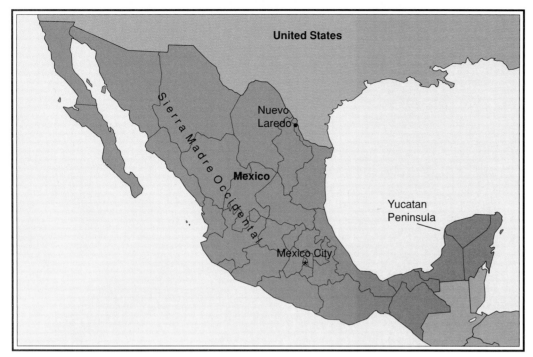

Figure D.6 Mexico.

excessive reliance on synthetic pesticides, industrial pollution in the maquiladoras, water quality, deforestation, and human population growth.

Economics and the Environment

In 1946, Mexico was primarily an agrarian society. It was then that the government embarked on a massive program of industrial growth based on exploitation of the nation's abundant natural resources. Until the 1970s, the program met with considerable economic success. The nation's GNP (gross national product) grew by more than 6% annually, which was faster than the population growth rate. A combination of severe inflation and a drop in world oil prices caused Mexico's economic growth to sputter during the 1970s and 1980s, however, and GNP began to lag population growth.

Economic growth favored the cities over the rural areas. People moved to the cities seeking employment in new industries, so that today, 70% of the nation's population lives in urban areas. Mexico City, the world's second largest city (after Tokyo) at nearly 20 million, is home to 35,000 industries and accounts for about 35% of the nation's GNP. The city's population is up from 10 million in 1970, and continues to grow by at least 300,000 per year (by birth and immigration), so that it could hit 25 million by the turn of the century.

Mexico's economic roller coaster has had considerable impact on its environment. Today, Mexico City is arguably the smoggiest city in the world. About three-quarters of the nation's river basins are seriously polluted. Forests are being harvested at rates that greatly exceed sustainability, and habitat destruction is leading to a considerable loss of biodiversity.

In the early 1970s, the same environmental awareness that swept the United States also reached the people of Mexico. But from an economic standpoint, environmental concern came at the wrong time. Environmental laws were passed setting goals for clean-up, but very little was accomplished because the nation's economy was too weak. Hence, environmental quality continued its downward spiral. This situation began to change in the late 1980s with a rebirth of grass roots concern over environmental quality and human health. In response, Carlos Salinas de Gortari, elected president in 1989, began to deal aggressively with environmental issues. For example, in May 1990, Salinas banned the commercial sales of marine turtles and announced plans to join the Convention on the International Trade in Endangered Species [CITES (chapter 11)].

Recently, government revenue has increased, and more funds are now available for environmental clean-up. The Secretariat of Urban Development and Ecology (SEDUE), the government agency responsible for environmental matters, was given a larger budget and greater enforcement powers. In fact, between 1989 and 1991, SEDUE's budget climbed from $5 million to $39 million.

What has motivated the Salinas administration to do something about the nation's environment? Some experts argue that the new emphasis on environmental quality has much to do with current negotiations with the United States and Canada over the proposed North American Free Trade Agreement (NAFTA). NAFTA would remove trade barriers and is designed to stimulate the economies of all three nations. Some opponents to NAFTA argue that complete removal of trade barriers will encourage U.S. industries to relocate to Mexico to avoid more stringent environmental regulations. This migration, in turn, will create economic pressures to weaken environmental standards in the U.S. (Actually, such relocations have been going on for the past two decades.) On the other hand, some proponents of the plan point out that Mexico requires the economic growth promised by NAFTA in order to adequately deal with its environmental problems. Although negotiations were completed in the summer of 1992, there is no doubt that the process of ratifying NAFTA will stir up even more controversy. Assuming ratification, full implementation will likely require another fifteen years.

Air Quality in Mexico City

So far, the most visible aspect of Salinas' environmental clean-up campaign is directed at improving Mexico City's badly deteriorated air quality (fig. D.7). As in Los Angeles, a combination of natural and human factors is responsible for the high air-pollution potential of Mexico City. The city is located in a mountain basin at an elevation of 2,215 meters (7,200 feet). This makes the city prone to a high frequency of persistent radiational temperature inversions, especially during the winter. As noted in chapter 16, a radiational temperature inversion forms near the earth's surface in response to extreme radiational cooling. In topographic basins such as the setting of Mexico City, drainage of cold, dense air down the slopes of surrounding mountains strengthens the inversion. Furthermore, those mountain ranges block horizontal winds that could disperse air pollutants.

Figure D.7 Air pollution in Mexico City, one of the smoggiest cities in the world.
© Gary and Elizabeth L. Hovinen

During an air-pollution episode, the city's factories and 3 million motor vehicles emit pollutants that are trapped within the low-level inversion layer. Motor vehicles are estimated to be responsible for 80% of pollutant emissions. In addition, SEDUE lists 1,500 of the city's factories as "major" polluters. Industry is responsible for 78% of sulfur dioxide (SO_2), 68% of particulates, and 24% of nitrogen oxide emissions. The most serious problems are suspended particulates and ozone (O_3), a component of photochemical smog. As noted in chapter 16, photochemical smog is an unhealthy mixture of gases and aerosols that forms when sunlight acts on motor vehicle exhaust and some industrial emissions.

In 1990, Mexico City exceeded World Health Organization (WHO) air-quality standards on 310 days. In 1991, this figure climbed to 354 days. In fact, on some days, air-pollution levels were three times WHO standards! In March, 1991, air-pollution levels were so hazardous that an emergency was declared and President Salinas ordered the closing of one of the nation's largest petroleum refineries. Located just outside Mexico City, the government-owned Azcapotzalco refinery was responsible for an estimated 4% of the region's air pollutant emissions. The shutdown reduced the nation's refinery capacity by about 7% at a cost of about $500 million. The refinery was allowed to reopen only after submitting a plan for reducing emissions and posting a bond with the government to ensure compliance. Again, in March 1992, dangerously high smog levels triggered another emergency declaration. This time the government temporarily ordered 40% of all private autos off the road and forced factories to cut production in half.

In response to Mexico City's serious air-pollution problems, Salinas has adopted many strategies and allocated several billion dollars to implement them. He has continued the mandatory vehicle emissions testing program begun by his predecessor in 1988. In 1989, he started the no-driving-day program, whereby motorists were prohibited from driving their autos on one workday per week, based on their license plate numbers. (Some people have tried to circumvent this regulation by purchasing a second car.) Unleaded gasoline has been available only since September 1990, and through Salinas' efforts, its price is coming down, and it is gradually replacing leaded gasoline. (Recall that unleaded gasoline is required for vehicles equipped with catalytic converters.) All 3,600 city buses have been either replaced with less-polluting vehicles or fitted with catalytic converters. (About 85% of the population uses public transportation.) All pre-1984 taxis are being phased out and the entire fleet (about 40,000) must eventually be equipped with catalytic converters. Those that have already made the switch are painted green, clearly distinguishing them from the more polluting yellow taxis. Furthermore, industry is encouraged to shift from high-sulfur fuel oil to cleaner-burning energy sources, including natural gas and low-sulfur diesel fuel.

In March 1992, President Salinas ordered the city's worst 220 polluters to reduce particulate emissions by 90% within eighteen months and to cut in half nitrogen oxide emissions within one year. If an industry fails to meet emissions standards by 1994, it will be closed.

While President Salinas' plan to rid Mexico City of its reputation as the world's smoggiest city is promising, the challenge is awesome. A surprise inspection of 1,150 factories in 1991 found that only 5% met environmental standards. Furthermore, Mexico City is plagued by many other environmental problems that demand attention. About 30% of the population has no access to a sewer system, municipal waste management is badly needs upgrading, and until recently, hazardous waste was sent to local landfills or dumped into the sewers.

Issues in Agriculture

While the Salinas government has started to tackle the air-quality and other environmental problems facing Mexico City, many chronic environmental problems persist in other parts of the nation, especially in rural areas. One major problem involves heavy application of synthetic pesticides, some of which are so toxic that their use is banned in other nations. Unlike the United States, Mexico is struggling to feed its people; thus, the Mexican government traditionally has taken a much more lenient approach toward the use of highly toxic synthetic pesticides.

Mexico's reliance on synthetic pesticides is linked to the nation's economic development program, started in the 1940s. As part of that program, the government requested and received the assistance of the Rockefeller Foundation to increase crop yields. New higher yielding hybrids were made available. At first, large commercial agricultural estates in northern Mexico resisted change, preferring instead to boost production by increasing the area of land under cultivation. Eventually, however, a combination of government subsidies and impressive demonstration plots convinced the northern farmers to switch to the new hybrids. This was the origin of Mexico's Green Revolution (chapter 8). As a consequence, between 1950 and 1970, the food crop yield per hectare quadrupled, and the land planted to crops doubled.

While the new hybrids delivered higher per hectare yields, they required much more intensive management and, as in other parts of the world, the new hybrids were more sensitive to weather extremes. This meant a need for more irrigation water in a part of the country where water is in short supply, and heavier applications of both fertilizers and synthetic pesticides. Alternative practices such as biological control and integrated pest management still are not widely used.

Heavy reliance on synthetic pesticides has led to a host of adverse environmental effects, including contaminated drinking water and killing of non-target species. Inadequate regulation of synthetic pesticides has also caused the direct poisoning of people. The actual number of cases is unknown, but poisoning is thought to be primarily a consequence of misuse of pesticides. In some cases, pesticide containers have instruction and warning labels written in English rather than Spanish. Futhermore, field workers may not be given proper safety instructions, and the hot weather prompts them to discard protective clothing.

Interestingly, the poor subsistence farmers in the south never experienced the Green Revolution and appear to be better off without it. Traditionally, those farmers have practiced a type of farming that succeeds without reliance on commercial pesticides. They select crops that are best suited to the climate and relatively resistant to pests. Furthermore, they rely on intercropping, whereby several different crop species are grown together. This practice significantly reduces the loss of any single crop to pests.

An estimated 22% of Mexicans suffer from malnutrition. And since the mid-1960s there has been a troubling trend in Mexican agriculture. More and more farmers are switching from food crops, including rice, beans, corn, and wheat, which are the staples of the poor, to sorghum. Sorghum is fed to chickens and pigs, which are primarily marketed to better-off urban consumers. More than 30% of Mexico's grain now goes to livestock. Given the realities of food-

web inefficiencies (chapter 3), it is evident that this trend reduces the total amount of food energy available.

Excessive soil erosion and soil salinization also adversely affect Mexico's ability to feed its growing population. For example, it is estimated that the annual crop loss due to soil salinization on irrigated land is enough to feed 5 million people.

Maquiladoras

Beginning in the mid-1960s, lower labor costs, access to Mexican markets, weaker pollution standards, and lax enforcement of environmental regulations prompted some companies based in the United States and other more-developed nations to open facilities in towns just south of the U.S.-Mexico border. Known as maquiladoras, these facilities manufacture or assemble imported raw materials and export finished products. Maquiladoras consist primarily of chemical, electronics, automotive, and textile industries. About 1,900 foreign-owned maquiladoras employ 500,000 people and earn Mexico about $3.5 billion in foreign exchange, second in importance only to petroleum exports.

The concentration of maquiladoras in border towns has seriously degraded the regional environment. Border towns are growing rapidly and are plagued by industrial pollution, overcrowding, and neglect by government officials. As of this writing, there is not a single municipal sewage treatment plant along the Mexican side of the border. Nuevo Laredo, for example, dumps an estimated 100 million liters (25 million gallons) of raw sewage into the Rio Grande each day. (The Rio Grande serves as a source of water for many communities on both sides of the border.) This is a classic case of a less-developed nation in desperate need of jobs for its burgeoning population overlooking the attendant health risks. Furthermore, industrialists seeking lower production costs appear to have little concern about the impact of their pollution on the living conditions of employees.

Environmental problems continue in spite of a 1983 agreement between the U.S. and Mexico to protect public health and environmental quality within 100 kilometers (62 miles) on either side of the border. Under this law, for example, shipments of hazardous materials across the border are to be closely monitored. Any hazardous by-product of raw materials imported from the United States and processed in Mexico is to be sent back to the United States for disposal. In many instances, however, hazardous materials, including mutagenic and carcinogenic wastes, have been dumped into surface waterways.

In early 1992, Mexico and the United States agreed to a three-year, $800 million project that would reduce pollution in the border towns. Mexico's SEDUE will spend a total of $460 million for new wastewater treatment facilities, solid waste management programs, road construction and public transportation, and acquisition of land for territorial reserves. Some experts argue that further improvement in environmental conditions in the border towns will come with passage of NAFTA, since it is likely to shift industrial growth to Mexico's interior, thus relieving pressure on the border towns.

Water Supply and Quality

Mexico's freshwater resource is not distributed equitably. In the north, demand exceeds supply; in the south, supply exceeds demand; and in the urban-industrial area in between, water shortages are common. About one-third of the nation's total water consumption is used for irrigation, mostly in the north. About two-thirds of the population has access to potable water, and one-half are served by wastewater treatment facilities. Most of these beneficiaries live in urban areas.

In Mexico City, groundwater supplies 70% of the drinking water. However, the demand greatly outstrips the natural recharge, so that the source aquifers are shrinking. In response to the worsening problem of groundwater mining (chapter 13), over the past three years, the city has seized 4,100 hectares (10,131 acres) of land to set aside as part of the recharge area. Additional land acquisitions for this purpose are planned. Furthermore, in an effort to reduce consumption and encourage conservation, water rates were elevated 400% in only two years. Surface waters are not a healthy alternative. The Lerma River, the nation's largest, flows near Mexico City but is heavily polluted, and diversions for urban and agricultural uses have reduced its flow to a trickle.

As noted in chapter 13, groundwater is particularly susceptible to contamination where soils are thin and bedrock is permeable. Such conditions prevail in the Yucatan peninsula of Mexico. There the bedrock consists of highly permeable limestone. Pollutants from cesspools and animal waste heaps seep into crevasses in the bedrock, which act as conduits into the drinking-water wells. Contaminated drinking water has caused infectious hepatitis to reach epidemic proportions in the Yucatan, and dysentery is a major cause of death.

Deforestation

A logging and forest management project in the Sierra Madre Occidental Mountains along Mexico's west coast has been soundly criticized by conservationists for not adequately protecting the environment. The project is funded by a $45.5 million loan by the World Bank and is intended to help improve Mexico's trade deficit by reducing its dependence on foreign sources of wood pulp. Three percent of these funds were earmarked for conservation measures.

This project well illustrates the often competing interests of economic development and environmental preservation. Today, with heightened concerns over the possibility of an enhanced greenhouse effect (and potential global warming), many people are calling for a commitment to reverse the trend of deforestation. Added to this is the pressing need to stem the continuing loss of biodiversity. Recognizing the need for both environmental preservation and economic development, many experts are advocating sustainable development. But in southern Mexico's tropical forests, deforestation continues unabated.

Less than 10% of the original rain forests of southern Mexico remain. Today, the deforestation rate in Mexico is estimated at 7,000 square kilometers (2,700 square miles) per year. By comparison, Brazil has the world's greatest deforestation rate at 13,800 square kilometers (5,300 square miles) per year.

Over the past fifty years, burning and clear-cutting have reduced Lacandona forest, the nation's largest remaining virgin tropical forest, to only 30% of its original area. In just the past twenty years, the population of the region (Peten) has soared from 20,000 to 300,000. Conservationists have tried to convince petroleum interests, loggers, and peasants that the forest is more valuable standing than cut down. But without alternative employment opportunities, the people continue to seek their livelihoods by exploiting the forest at an unsustainable pace. Recently, some conservationists from Mexico and neighboring countries of Belize and Guatemala that share the forest have been successful in setting aside some parts of the forest as reserves, buffer zones, and multiple-use areas. In the long run, however, the strategy that

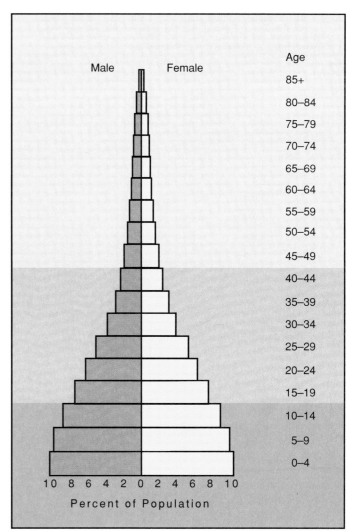

Figure D.8 Age-structure diagram for Mexico (1990). The dark-blue portion of the diagram represents the prereproductive age group, the medium-blue portion represents the reproductive age group, and the light-blue portion represents the postreproductive age group.

Source: Data from the United Nations and Population Reference Bureau.

appears to offer the best potential for slowing deforestation is debt-for-nature swaps involving the United States and other more-developed nations (chapter 11).

Population Growth

Although there are hopeful signs that Mexico is beginning to deal with its environmental problems, rapid population growth may stifle those efforts. A burgeoning human population makes increasing demands on finite resources including fresh water, food, energy, and living space, and increases the amount of waste, land degradation, and habitat destruction.

As in all less-developed nations, Mexico's population growth is the central concern for its environmental future. The age structure of Mexico features a broad base; that is, the number of individuals in the prereproductive age group is relatively large (fig. D.8). (In 1992, 38% of the population was under age 15.) As described in chapter 6, such an age-structure imparts to the population a momentum that virtually guarantees continued explosive growth. Mexico's annual rate of natural population change is 2.3%, which means that its population doubles every thirty years.

On an encouraging note, since the late 1970s, the proportion of Mexican women of child-bearing age using contraceptives has increased from 30% to 53%. But even if Mexico were to achieve replacement-level fertility immediately, the population would double in fifty years. Indeed, unless the death rate climbs, it will take more than a century for Mexico to achieve a stationary population.

Focus on Canada

Canada is home to 27.4 million people who inhabit a vast land of nearly 10 million square kilometers (4 million square miles) (fig. D.9). These lands, comprising 7% of the world's total land mass, touch three oceans and have the world's longest coastline. Canada's abundant resources provide its citizens with their livelihood, food, water supply, and recreation. These resources comprise 10% of the earth's forest, 9% of the fresh water, and important reserves of oil, natural gas, coal, and minerals. Despite this abundance, Canadians are experiencing problems of pollution and overexploitation of resources.

Canada's relatively low population density (2.7 people per square kilometer, or 1.0 per square mile) might lead one to expect that its environmental problems would be minor. But most people reside within 100 kilometers (62 miles) of the U.S.-Canadian border. In fact, over half of Canadians live in southern Ontario and Quebec. This narrow strip of land runs along the north shores of lakes Erie and Ontario. Furthermore, 77% of Canadians live in urban areas—a fact that is in sharp contrast with the popular perception of Canada.

Canada faces many of the same environmental problems as other more-developed nations. However, the Canadian economy is strongly dependent on its timber and fisheries. Consequently, Canadians are particularly concerned about maintaining the sustainability of these resources. Recent polls show that Canadians rank environmental quality high on the list of the issues that federal and provincial governments should address.

As in the United States, debate in Canada over environmental problems has undergone continual evolution since the early 1970s. The Canadian approach to environmental problems has evolved from inefficient piecemeal schemes to recognition of the need for holistic approaches that integrate ecology, public policy, and economics. Implementation of this enlightened approach is progressing slowly, however, and much remains to be accomplished. In this section we describe three major environmental problem areas: forest management, fisheries management, and persistent toxic pollutants.

Forest Management

Spanning the nation from the Atlantic to the Pacific coast, a diversity of forests covers nearly half of the Canadian landscape. Few nations rely more heavily on their forest resources. Forestry contributes more to Canada's balance of trade than agriculture,

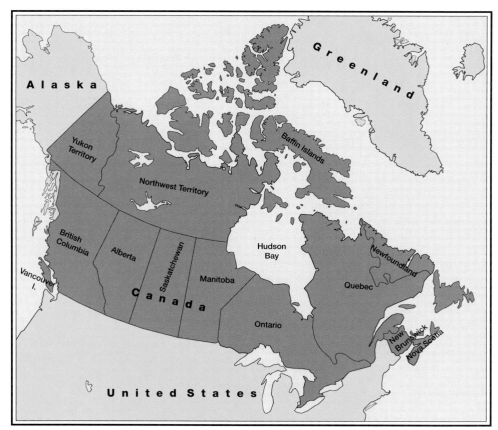

Figure D.9   Provinces of Canada.

fisheries, mining, and energy combined. All told nearly 900,000 Canadians (one in every fifteen employees) work either directly in the forest industry or for companies that support it. Together, the United States, the former Soviet Union, and Canada supply more than half of the world's industrial timber.

Recognizing the importance of forests for Canada's future, Parliament in 1989 passed the Department of Forestry Act. This legislation designated Forestry Canada a federal department and charged it with the responsibility to ensure that Canada's forests meet current needs without jeopardizing future productivity, ecological diversity, and capacity for regeneration. Under the so-called Green Plan undertaken by the federal government, the nation's forestry managers will endeavor to practice sustainable development while managing for wildlife, fisheries, recreation, and other ecological values. Although these objectives are laudable, they will not be easily achieved. Like their U.S. counterparts, Canadian foresters face major challenges in changing policies that emphasize timber production from relatively few woody species to policies that promote sustainable development of the many resources that comprise forest ecosystems (chapter 10).

Nowhere are conflicts between these approaches more evident than in the temperate rain forests of British Columbia. Although the plight of the tropical rain forests is well-publicized, the public remains generally unaware of the continuing loss of these highly productive, diverse forests. Yet in British Columbia, logging of ancient (old-growth) rain forests proceeds at the rate of 270,000 hectares (670,000 acres) per year. Furthermore, timber companies hold licenses to harvest essentially all of the forests that remain. Although it is difficult to calculate the rate of current forest production required for sustained yields, some experts estimate that current logging rates exceed sustained yields by at least 30%. There is no doubt that by logging these large trees and hugh volumes of wood per hectare, the timber industry is harvesting a resource capital that has amassed over several centuries. They will never again have such a bountiful harvest.

What is the legacy of this logging? Recall from chapters 10 and 11 the role of forests in stabilizing soil and water resources, maderating climate, and safeguarding biodiversity. The first of these values is particularly significant in British Columbia, where a large salmon industry once thrived in its rivers and coastal waterways. Today the natural reproduction of chinook salmon in western Canada is seriously damaged, due in part to the clear cutting of old-growth forests on steep slopes that accelerates soil erosion. The eroded sediments subsequently cover gravel stream beds, the spawning grounds for salmon. Only six of 89 large watersheds on Vancouver Island are unlogged.

Calls for a halt to the logging of old-growth forests usually are met by vocal concerns over the loss of jobs. To resolve this conflict, some resource economists suggest that two points must be recognized. First, many jobs have already been lost in the timber industry because of increased mechanization and export of raw logs. Second, even if the cutting of old-growth forests were to continue, more jobs would soon be lost anyway as the industry runs out of these ancient forests. Some economists, therefore, emphasize that shifting to sustainable development of natural resources will require retraining those workers whose jobs have been eliminated.

What is the fate of old-growth forests in British Columbia? At the time of this writing the question remains unanswered. Many other issues concerning sustainable development of Canada's forest resources also are unresolved. How will the establishment of ecological reserves affect the existing level of timber harvesting? Will the managed forests that replace today's virgin forests produce wood of the same high quality? Can genetic diversity be adequately preserved? What is the best array of environmentally-sound management technologies and strategies to ensure sustainable development? How will forest ecosystems be affected by continuing air pollution and possible climate change? As answers to these questions become clearer, Canadians will have many opportunities to determine the course of forestry development in their nation.

Fisheries Management

When one thinks of Canada, images of commercial fishing boats plying the North Atlantic or perhaps sports fishermen casting lures on one of Canada's hundreds of thousands of freshwater lakes may quickly come to mind. The main commercial fisheries in Canada are located off the Atlantic and Pacific coasts, in the St. Lawrence Basin, and the larger lakes located on and near its southern border. A portion of the harvest is sold in nearby urban centers. Another portion may be iced (but not frozen), shipped by air freight, and served as the entree in a restaurant in Europe two or three days later. Other fisheries freeze their catch for distribution to worldwide markets. The economic importance of fisheries, both freshwater and marine, to Canadians cannot be overstated. Commercial fishing is a $3.1 billion industry employing some 130,000 persons (fig. D.10). In 1,500 villages, fishing is the only real opportunity for employment. Sport fishing offers recreation for more than 5 million Canadians and an additional 1 million visitors each year. These pursuits add $1.8 billion to the Canadian economy. Yet, Canadian fisheries are facing major problems. In this section we examine the key issues that must be resolved to place Canadian fisheries on a sustained-development basis.

Fisheries are affected by stresses on the aquatic ecosystems that sustain them. These stresses include overharvesting, introduction

Figure D.10 Fisherman in Nova Scotia working with commercial nets on a family-owned fishing boat.
© Bob Krist/Leo de Wys, Inc., NY

of toxic and infectious effluents, physical transformation (e.g., building dams), siltation, excessive nutrient cycling, acid precipitation, and introduction of exotic species. The response of aquatic ecosystems depends on the amount of stress placed on them. Unfortunately, the relative importance of these stresses is difficult to evaluate, and it is a topic of continual debate in fisheries management circles.

The oceans off Canada's shores provide rich, diverse natural resources. However, oceans have long been viewed as a global commons--available for exploitation by all, but the responsibility of no one. As a result, Canadian marine ecosystems are experiencing considerable stress. The two major stresses on fish stocks are overfishing and habitat destruction. The shellfish industry suffers from infectious agents contaminating their harvests, particularly in nearshore areas.

Sharp declines in Atlantic commercial catches testify to the overharvesting of fisheries in those coastal waters. For example, harvests of cod, the mainstay of the Atlantic commercial fishery, were 50% less in the mid-1980s than their peak 1968 levels. Landings of herring declined by 80% over essentially the same period. In an attempt to reduce overfishing, Canada extended its jurisdiction to 320 kilometers (200 miles) offshore. This zone is designated for the exclusive use of Canadian fishermen. However, in order to conserve fish resources and to rebuild certain fish stocks, catch and gear (e.g., types of nets) restrictions were imposed on Canadian fishermen. These programs saw some successes, but the all-important cod stocks have continued to decline in recent years. Furthermore, the groundfish stocks, such as cod, continue to be overharvested beyond the 320-kilometer (200-mile) limit by fishing vessels of the European Community nations and other nations.

More recently, in 1990, Canada implemented the more stringent Atlantic Fisheries Adjustment Program. The problem is to set harvesting quotas and to control the type of fishing gear used without future damage to the fishery. Thousands of commercial fishermen have invested their savings to purchase equipment to support their families; under the provisions of this management program, many feel they can not survive economically. Yet, the Atlantic fishery will only support a limited harvest, a sustainable harvest that is difficult to define precisely.

On the Pacific coast, concern has focused on certain salmon species that are the mainstay of this commercial fishery. Herring and halibut rank second and third respectively in relative commercial importance. In recent years stocks of salmon originating from major Canadian rivers have seen significant declines. Chinook salmon have declined dramatically in the Frazer River and the Straits of Georgia; sockeye salmon have dropped along the central Pacific coast and the Rivers-Smith Inlet; and chum salmon numbers are reduced in most areas. Overharvesting, siltation of spawning grounds, and hydroelectric dams are thought to have negatively impacted the spawning success of wild salmon.

Commercial fisheries in the Great Lakes, both Canadian and U.S., have experienced major declines in some fish species because of a combination of factors. Competition from new and existing species, changes in habitat (shoreline development), toxic chemicals, and changes in nutrient status all have affected these ecosystems. Introduction of the sea lamprey all but eliminated lake trout populations in Lake Superior and Lake Huron. Current commercial harvests of lake trout are at less than 10% of traditional levels. One bright spot is Lake Erie, where walleye stocks have rebounded tremendously since being depleted in the 1960s.

In Canada the issue of who gets to use the production from rivers and inland lakes continues to be a point of contention. Regulatory struggles between high-kill (commercial fishing) and low-kill (sports fishing) are pervasive across much of Canada. How to allocate production between these two competing groups and still maintain a sustainable fishery remains a serious issue. Fishery allocation takes on yet another dimension in the Great Lakes. Canadians argue that they have contributed more to water pollution abatement on the Great Lakes than their American counterparts. Thus they believe that they are entitled to a greater share of the fish stocks.

For Canada's hundreds of thousands of inland lakes, fishery managers are setting more restrictive limits for sports anglers, so that the condition of this fishery is not severely damaged by overfishing. Many of these lakes contain large fish which are aggressively pursued by sports fishermen. Because most of these lakes are not very productive, these fisheries are easily overharvested. Increased use of these resources will require more attention in the future.

Shellfish on both coasts suffer from a different threat: contamination from untreated sewage and agricultural runoff. Recall from chapter 14 that fecal coliform bacteria are used as indicator species for the potential presence of pathogens in water samples. If the coliform count is too high, shellfish beds are closed to harvesting. Such closures have a significant impact on local economies. In recent decades, the total area closed to shellfishing on both the Atlantic and the Pacific coasts has increased. In 1989, on the Atlantic Coast, about 37% of the area identified for harvesting of shellfish, some 165,000 hectares (410,000 acres), was closed. In British Columbia, 500,000 hectares (1,250,000 acres) of shell-fishing grounds were closed. Nationally, poorly treated municipal discharge waters are implicated in approximately half of the closures. Such occurrences indicate the need for improved sewage treatment facilities.

Acid precipitation continues to threaten Canadian lakes, primarily those east of the Manitoba-Ontario border (chapter 16). These lakes have a low capacity to neutralize acids in precipitation. Consequently, exposure to acid rain and snowmelt lowers the pH (creates a higher level of acidity) of lake water. Fish, particularly lake trout, are sensitive to lower pH. When the pH of a lake drops below 5.2 , lake trout stocks are vulnerable to extinction. Some 2,200 lakes, or about 1% of all lakes in Ontario, have lake trout populations. Lake trout have been eliminated in 3% of these lakes and another 16% are considered acidified or are extremely sensitive to acidification. Canadian scientists suggest that most of the extinctions in Ontario are due to acid pollutants from sulfide ore smelting operations at Sudbury, Ontario. Many other eastern Canadian lakes are also experiencing acidification from acid precipitation, much of which is imported from the United States. Canadian scientists report that one in seven of their eastern lakes have been damaged by acid precipitation. Although other species such as perch and common shiners have been eliminated from lakes, the loss of lake trout is of particular concern because each population appears to be specifically adapted to its native lake. Thus successful restocking after extirpation is difficult to achieve.

In recent years Canada has initiated schemes to restore degraded aquatic ecosystems. In British Columbia, an extensive Salmon Enhancement Program has been underway since 1977. Efforts range from large salmon-rearing projects that produce 250,000 mature fish per facility to small habitat restoration projects. The Strategic Plan for Ontario Fisheries has undertaken the restoration of fisheries in degraded lakes in southern Ontario. Ecosystems of the Great Lakes are being given serious attention through the Great Lakes Water Quality Agreements as overseen by the International Joint Commission. Some formerly degraded parts of the Great Lakes, especially Lake Erie, have been restored to a relatively healthy condition.

Persistent Toxic Chemicals

The chemical revolution following World War II produced a myriad of new formulations and products that found worldwide use. Some of these chemicals resist physical, chemical, and biological degradation. The fact that these persistent substances bioaccumulate has left Canadians and much of the rest of the world with a unique legacy: the problem of protecting humans and wildlife from the harmful effects of these substances. In Canada, the freshwater ecosystems of the Great Lakes and the marine ecosystems of the Arctic region are contaminated with PCBs, DDT, hexachlorohexane (HCH), hexachlorobenzene (HCB), mercury, and lead.

As described in chapter 12, toxic substances adversely affect the health of humans and other organisms and thereby impair the functioning of ecosystems. In the past, scientists were most concerned about acute exposure to toxins. Today, attention has shifted to the more subtle effects of chronic exposure to low levels of toxins. Chronic exposure to persistent toxins leads to bioaccumulation which, in turn, may threaten reproductive success. Bioaccumulation is a particularly serious problem for animals that feed at the upper trophic levels of aquatic ecosystems. In some cases, their offspring die. For example, studies that have tracked healthy bald eagles that immigrated to the Great Lakes found that after feeding on contaminated fish for a few years, the eagles often produced hatchlings with defects that eventually proved fatal. PCBs and DDE (both chlorinated organic compounds) in the birds' diet are strongly implicated. These substances are persistent and are known to readily bioaccumulate in Great Lakes fish—a major component of eagles' diets. In other instances, offspring that do survive develop various abnormalities. For example, male birds may develop ovarian tissue, female birds grow excessive oviduct tissue, and male fish sometimes fail to reach sexual maturity.

Levels of toxic chemicals in drinking water drawn from the Great Lakes do not pose an immediate threat to human health. However, long-term effects may occur through consumption of fish harvested from these waters. Sports fish such as salmon and trout have the highest levels of PCBs and other contaminants because they occupy upper trophic levels.

Biological indicators are used to assess trends in levels of persistent toxins. Consider two examples. Lake trout are a popular sport fish in the Great Lakes. They feed at the top of the food web and are relatively long-lived, typically living for a half dozen years or more. Hence bioaccumulation of toxins in lake trout may be substantial. (Salmon are not as good an indicator because their life span is shorter.) Each year scientists capture, grind up, and analyze lake trout for the presence of PCBs, DDT, and other toxins. Since 1977 the levels of PCBs in Lake Ontario lake trout have remained above 2 milligrams per kilogram (ppm), well above the Great Lakes Water Quality Agreement objective of 0.1 milligrams per kilogram. (This agreement between Canada and the United States was signed in 1987 to meet certain water-quality objectives in the Great Lakes.) In addition, since 1985, DDT levels in lake trout have remained essentially constant at near 1 milligram per kilogram, a level considered acceptable by the Great Lakes Water Quality Agreement.

While contaminant levels in lake trout are used primarily to assess the potential for human exposure, other biological indicators provide a more general measure of persistent toxins in the environment. For example, herring gull eggs are used as an indicator for the Canadian Great Lakes. Herring gulls eat primarily fish and remain on the lakes all year. The eggs from one colony located on Snake Island, Lake Ontario, has been monitored since 1974 for PCBs, DDE (a metabolic by-product of DDT), dieldrin, mirex, dioxins (all chlorinated organics), and mercury. From 1974 to 1986, levels of PCBs declined from near 150 milligrams per kilogram to near 20 milligrams per kilogram—a level that is still of concern. Herring gull eggs from other colonies in the Great Lakes exhibited smaller declines in contamination during the same period. Although these trends were encouraging, levels of PCBs among all colonies have remained essentially the same since 1986.

In view of the fact that production of PCBs was halted in 1977, what explains the continued presence of PCBs in Canadian ecosystems? As noted in chapter 14, PCBs are extremely stable compounds that are continually cycled in the environment. For example, dredging and lake currents disturb PCB-laden sediments, releasing PCBs into the water column and making them available to organisms that live in the sediments. Also, atmospheric deposition delivers significant amounts of PCBs to the Great Lakes. That process accounts for perhaps 90% of the PCB loading of Lake Superior and 30% for Lake Ontario.

In addition to the Great Lakes, persistent toxic chemicals may be posing problems for Arctic-based ecosystems. Although far less thoroughly investigated, studies of the remote ecosystems of the Canadian Arctic indicate that they are not immune from the effects of industrial societies located thousands of kilometers away. Over the past two decades, scientists have detected a variety of pesticides and industrial chemicals in various species, including humans, that reside in these northern ecosystems.

Transport of chemicals to northern ecosystems by air currents, north-draining rivers (particularly the Yenisey, Lean, and Ob rivers located in the former Soviet Union), and ocean currents is particularly worrisome. Reduced sunlight and low temperatures slow the rate of breakdown of persistent organic chemicals. Even more troublesome is the diet of the native peoples. Organochlorine compounds are known to bioaccumulate in whale and seal blubber and in caribou fat, food sources considered as delicacies by these people.

Scientists are investigating potential connections between contaminant levels in Arctic ecosystems and the health of humans and wildlife. Limited studies to date have not found any direct health effects, but concerns remain, especially for the young. For example, polar bear cubs typically nurse for nearly two years. They may be receiving significant doses of persistent toxins through their mother's milk because she eats contaminated prey (mostly seals). Similar bioaccumulation can occur in humans. In fact, human milk fat from Inuit women contains levels of PCBs that are five times greater than samples from Caucasian women residing further south. A diet rich in marine mammals is thought to be the source of contamination.

See the Geographic Index for further citings on Hawaii, Mexico, and Canada.

Selected Readings

Hawaii

Dvorak, J. J., C. Johnson, and R. I. Tilling. "Dynamics of Kilauea Volcano." Scientific American 267 (Aug. 1992):46-53.

Ehrlich, P. R., D. S., Dobkin, and D. Wheye. Birds in Jeopardy: The Imperiled and Extinct Birds of the United States and Canada, Including Hawaii and Puerto Rico. Stanford, CA: Stanford Univ. Press, 1992.

Kaye, G. Hawaii Volcanoes: The Story Behind the Scenery. Las Vegas: KC Publications, 1990.

Kepler, A. K. Hawaiian Heritage Plants. Honolulu: Oriental Publishing, 1983.

Reeves, L. N., ed. Hawaii's Kilauea Volcano: Documenting Destruction from 1983 into the 90's. Sun City West, AZ: C. F. Boone Publishing, 1990.

Mexico

Mumme, S. P. "Cleaning the Air, Environmental Reform in Mexico." Environment 33, no. 10 (1991):7-30.

Selcraig, B. "Up Against the Wall in Mexico." International Wildlife, (March/April 1992):12-16.

Canada

Anonymous. 1991. "Toxic Chemicals in the Great Lakes and Associated Effects (Synopsis)." Environment Canada, Department of Fisheries and Oceans, and Health and Welfare Canada. Minister of Supply Services Canada.

Healey, H. C., and R. R. Wallace. 1987. Canadian Aquatic Resources. Department of Fisheries and Oceans Ottawa.

Twitchell, K. 1991. "The Not-So-Pristine Arctic." Canadian Geographic (Feb/Mar 1991):53-60.

Credits

Line Art

Chapter 1

1.4: Reprinted with the permission of Macmillan Publishing Company from *Meteorology*, by Moran and Morgan. Copyright 1991 by Macmillan Publishing Company.

Chapter 3

Pages 54–55: Adapted from P. R. Ehrlich, A. H. Ehrlich, and J. R. Holden, *Ecoscience: Population, Resources and Environment.* Copyright © 1977 W.H. Freeman and Company. Reprinted by permission.

3.30: Graphic from R. A. Kerr, "Global Temperature Hits Record Again," *Science* 251:274, January 18, 1991, copyright 1991 by the AAAS, based on data from University of East Anglia/Hadley Center. Used by permission.

Chapter 4

4.8: From R. L. Smith, *1990 Ecology and Field Biology*, 4th ed. Copyright © Robert Leo Smith. Used by permission.

4.11 and 4.12: From F. H. Bormann and G. E. Likens, *Pattern and Process in a Forested Ecosystem.* Copyright © 1979 Springer-Verlag, New York. Used by permission.

Chapter 5

5.5: © Royal Society of South Australia Inc. Used by permission.

5.9: From V. Stern, et al., "The Integrated Control Concept" in *Hilgardia* 29:81–101. Copyright © *Hilgardia*, University of California. Used by permission.

5.15: Adapted from G. F. Gause, "Experimental Analysis of Vito Volterra's Mathematical Theory of the Struggle for Existence," in *Science* 79:16–17. Copyright © 1934 by the AAAS. Used by permission.

5.16: From "Population Ecology of Some Warblers in Northeastern Coniferous Forests," by R. MacArthur, *Ecology* 1958, 36: 533–536. Copyright © 1958 by the Ecological Society of America. Reprinted by permission.

Chapter 6

6.5: From Carl Haub, Mary Mederios Kent, and Machiko Yanagishita, *1989 World Population Data Sheet* (Washington, D.C.: Population Reference Bureau, Inc., 1989). Used by permission.

6.9: From G. Tyler Miller, *Living in the Environment,* 7th ed. Copyright © 1992 Wadsworth Publishing Company. Used by permission.

6.10: From Carl Haub, Mary Mederios Kent, and Machiko Yanagishita, *1990 World Population Data Sheet,* (Washington, D.C.: Population Reference Bureau, Inc., 1990). Used by permission.

Chapter 7

Pages 128, 129 (bottom), 131: From John W. Hole, Jr., *Human Anatomy and Physiology,* 5th ed. Copyright © 1990 Wm. C. Brown Communications, Inc., Dubuque, Iowa. All Rights Reserved. Reprinted by permission.

Pages 129, 130 (right): From Leland G. Johnson, *Biology,* 2d ed. Copyright © 1987 Wm. C. Brown Communications, Inc., Dubuque, Iowa. All Rights Reserved. Reprinted by permission.

7.4: Adapted with the permission of The Alan Guttmacher Institute from Elise F. Jones and Jacqueline Darroch Forrest, "Contraceptive Failure in the United States: Revised Estimates from the 1982 National Survey of Family Growth," *Family Planning Perspectives,* Vol. 21, No. 3, May/June 1989.

7.10: From the Family Life Education Programme, Ministry of Health, Singapore. Used by permission.

7.12: From Peter J. Donaldson and Amy Ong Tsui, "The International Family Planning Movement," *Population Bulletin*, Vol. 45, No. 3 (Washington, D.C.: Population Reference Bureau, Inc., 1990), p. 20. Used by permission.

7.13: From Thomas W. Merrick, "World Population in Transition," *Population Bulletin,* Vol. 41, No. 2 (Washington, D.C.: Population Reference Bureau, Inc., 1991 reprint), p. 4. Used by permission.

7.14: From "Population by Age and Sex for Developing and Developed Countries: 1989," *Population Today,* Vol. 18, No. 1 (January 1990), p. 2. Used by permission.

Chapter 8

8.1: Reprinted with the permission of Macmillan Publishing Company from *Meterology*, by Moran and Morgan. Copyright 1991 by Macmillan Publishing Company.

8.2: From P. J. Lamb, "On the Development of Regional Climatic Scenarios for Policy-Oriented Climatic-Impact Assessment," *Bulletin of the American Meteorological Society,* Vol. 68, No. 9, September 1987. Used by permission of the American Meteorological Society.

Glossary

Abiotic Pertaining to nonliving components of the environment.

Abortion Voluntary termination of an established pregnancy before the fetus has attained the ability to survive outside the uterus.

Absorption (of radiation) Process whereby a portion of the radiation that strikes an object is converted to heat.

Acclimatization Adjustment of an organism to environmental change.

Acid deposition The combination of acid precipitation and acidic particles that settles to the earth's surface.

Acid mine drainage (AMD) Runoff that becomes excessively acidic as water seeps through mine wastes.

Acid rain Rain that has become more acidic than usual by intercepting and dissolving tiny droplets of sulfuric (H_2SO_4) and nitric (HNO_3) acid. By convention, acid rain has a pH less than 5.6.

Active solar system Device that captures or focuses solar radiation and converts it to another form of energy, such as electricity.

Acute exposure A single and potentially damaging exposure to radiation or toxic substances that lasts seconds, minutes, or hours.

Adaptation A genetically controlled characteristic that enhances an organism's chances of survival and reproduction in its environment.

Aerobic decomposers Microorganisms that require oxygen to break down organic materials into carbon dioxide (CO_2) and water (H_2O).

Aerobic environment Where molecular oxygen (O_2) is present.

Aerosols Tiny solid and liquid particles suspended in the atmosphere.

Age structure The distribution of individuals by age level in a population.

Agroclimatic compensation The situation where good growing weather and relatively high crop yields in one region make up for poor growing weather and relatively low crop yields in another region.

Air mass A huge volume of air covering thousands of square kilometers that is relatively uniform in water vapor concentration and temperature.

Air pollutant An airborne substance that adversely affects the well-being of organisms or the life-support systems on which they depend.

Air-pollution episode A period in which air-pollutant concentrations reach or exceed levels that are considered hazardous to human health.

Air pressure The weight of the atmosphere over a unit area of the earth's surface.

Air sacs (alveoli) Tiny pouches at the terminus of bronchioles within the lungs; sites of gas exchange between the bloodstream and atmosphere.

Albedo The fraction of radiation striking a surface that is reflected by that surface.

Anaerobic decomposers Microorganisms that can live without molecular oxygen (O_2). They break down organic molecules into methane (CH_4), ammonia (NH_3), hydrogen sulfide (H_2S), and water (H_2O).

Anaerobic environment Where molecular oxygen (O_2) is absent.

Antagonism An interaction in which the total effect is less than the sum of two or more effects acting independently.

Antifeedant A chemical that prevents certain insects from feeding by destroying their sense of taste.

Aquaculture The growing and harvesting of fish and shellfish in confinement.

Aquifer A permeable layer of rock, sediment, or soil that holds and transmits water; groundwater reservoir.

Area landfill A sanitary landfill located in a natural topographic depression (e.g., canyon) or an abandoned surface mine.

Area strip mining A method of surface mining carried out in relatively flat terrain by means of a trench from which overburden has been removed; usually used for recovery of near-surface coal deposits.

Areas of concern Specifically designated areas within the Great Lakes' drainage basin that are the principal focus of efforts to reduce water pollution.

Arithmetic growth Increase at a constant rate over a specific period of time.

Artesian well A free-flowing well tapped into a confined aquifer in which groundwater is under pressure.

Asphyxiating agents Air pollutants that deprive the blood of its oxygen supply when inhaled; carbon monoxide (CO) is an example.

Assimilated food energy The portion of ingested food absorbed by an animal's digestive system.

Assimilative capacity The ability of a system such as the ocean to receive waste without long-term adverse effects.

Atmosphere The thin envelope of gases (mostly nitrogen [N_2] and oxygen [O_2]) and suspended aerosols that surrounds the planet.

Atmospheric fixation The process by which the high temperatures associated with lightning cause nitrogen gas (N_2) to combine with oxygen gas (O_2) to form nitrate (NO_3^-).

Atmospheric windows Infrared wavelength bands for which there is little or no absorption by the principal greenhouse gases, water vapor (H_2O) and carbon dioxide (CO_2).

Barrier islands Long and narrow strips of sand that fringe portions of the U.S. Atlantic and Gulf Coasts.

Base discharge Basis for determining the reliable water supply from a river; derived from discharge when runoff is zero.

Beach An accumulation of wave-washed sand or other sediment along a coast.

Binary cycle system A technology for converting geothermal energy to electrical energy.

Bioaccumulation Process whereby persistent chemicals become more concentrated as they move from one trophic level to another.

Biochemical oxygen demand (BOD) The amount of oxygen required by microorganisms to decompose the organic wastes in a given volume of water.

Biodegradable waste Materials that can be decomposed by microorganisms.

Biodiversity The total number of species and the genetic variability within each species.

Biofuels Biological energy sources, including wood, organic waste, methanol (CH_3OH), and ethanol (CH_3CH_2OH).

Biological control Regulation of a pest population by natural predators, parasites, or disease-causing bacteria or viruses.

Biological fixation Conversion of atmospheric nitrogen gas (N_2) to ammonia (NH_3) by specialized bacteria.

Biomass The total weight, or mass, of organisms in a given area.

Biotic Pertaining to the living components (organisms) of the environment.

Blue-baby disease (methemoglobinemia) A disease that causes babies to appear blue because of the excessive intake of nitrate, which reduces the oxygen carrying capacity of red blood cells.

Bottle laws Laws that require payment of a deposit on beverage containers to reduce litter and encourage recycling and reuse.

Broad-spectrum pesticide A chemical used to kill a wide variety of pests.

Bronchioles Small tubes that connect the bronchi and alveoli (air sacs) within the lungs.

Calorie A measure of the energy content of food.

Captive propagation Breeding of animals in zoos and plants in botanical gardens for subsequent release back into the wild.

Carbamate A type of synthetic insecticide.

Carbon cycle Circulation of carbon among various reservoirs of the environment; major carbon reservoirs include the atmosphere, the oceans, and vegetation.

Carcinogen A chemical substance or physical agent (such as radiation) capable of causing cancer.

Carnivore An animal that feeds only on other animals.

Carrying capacity The maximum population that the total resources of a habitat can sustain.

Catalytic converter A motor vehicle exhaust control device that causes nitric oxide (NO) to oxidize carbon monoxide (CO) and hydrocarbons, thereby releasing nitrogen (N_2), carbon dioxide (CO_2), and water vapor.

Cell The basic structural and functional unit of all organisms.

Cellular respiration The process of breaking down and releasing energy from sugar.

Chlorinated hydrocarbons A class of chlorine-containing chemicals, some of which are used as insecticides or solvents.

Chlorosis A yellowing of plant leaves due to chlorophyll loss.

Chronic exposure Continuous or recurring exposure to radiation or toxic substances that lasts for days, months, or even years.

Cilia Tiny undulating hairs lining the respiratory tract that help protect the respiratory system from inhaled pollutants.

Circumpolar vortex A band of strong winds that encircles the margins of the Antarctic continent.

Clear-cutting Removal of all the trees in a plot in a single cutting.

Climate The weather conditions of an area averaged over a period of time, plus extremes in weather behavior that occurred during the same period.

Climatic anomaly Departure of temperature, precipitation, or other weather elements from long-term average values.

Climax species The assemblage of plants and animals at the endpoint of ecological succession.

Climax stage The relatively stable stage of ecological succession characterized by an assemblage of species that is able to maintain itself on a site.

Cloud seeding The introduction of nucleating agents (such as silver iodide) into clouds in an attempt to stimulate natural precipitation-forming processes.

Clouds Visible suspensions of tiny water droplets and/or ice crystals in the atmosphere.

Coal gasification One of several processes whereby coal is heated in the absence of oxygen and in the presence of steam to produce methane (CH_4), carbon monoxide (CO), and hydrogen (H_2).

Coal liquefaction One of several processes for manufacturing liquid fuels from coal.

Coastal zone Encompasses the relatively narrow region that is transitional between the continent and the open ocean.

Coevolution The process by which two interacting populations (e.g., a parasite and a host) adapt to each other, to their mutual benefit.

Cogeneration A process by which a useful form of energy (e.g., electricity) is produced and the waste heat from that process is put to another use.

Cold front A narrow zone of transition between relatively cold and warm air masses where the cold air replaces the warm air.

Coliform bacteria A type of bacteria that resides in the human intestine and whose presence in water is used as an indicator of polluted water.

Combined sewer system A single-pipe network that transports both runoff and sanitary wastes to a sewage treatment plant.

Competition Interactions among organisms to secure a resource in short supply.

Competitive exclusion The generalization that two species having the same resource requirements cannot coexist indefinitely in the same habitat.

Composting The production of a soil conditioner through the aerobic breakdown of organic wastes, such as grass clippings, leaves, and paper.

Compressional warming The temperature rise that occurs in a unit volume of air that is subjected to increasing pressure. For example, air undergoes compressional warming when it descends within the atmosphere, and the pressure on the air consequently increases.

Conceptual model A framework for organizing information on some physical or biological system (e.g., a food web).

Condensation The process whereby water changes from a vapor to a liquid; accompanied by a release of heat to the environment.

Condensation nuclei Tiny solid or liquid particles upon which water vapor condenses into droplets.

Condom A sheath that fits over the penis and prevents entry of sperm into the vagina.

Cone of depression A cone-shaped lowering of the water table formed when groundwater is withdrawn.

Confined aquifer A subsurface layer of rock or sediment containing groundwater and sandwiched between impermeable earth materials.

Consumers Organisms that use other organisms as a food source.

Contour farming A soil-conservation practice in which fields are cultivated parallel to contours of elevation.

Contour strip mining A surface mining method used to extract near-surface mineral deposits (usually coal) in hilly or mountainous terrain. Earth-moving equipment removes the overburden and the deposit from a series of benches cut along contours of elevation.

Contraceptive pill Contains a combination of female hormones that suppresses the release of eggs from the ovaries.

Control system The unperturbed system that is used during an experiment as a standard for comparison.

Controlled experiment Involves comparison of a test system that is perturbed with a control system that is not perturbed.

Convection current A circulation pattern in which a relatively light fluid (such as warm air) rises and a denser fluid (such as cold air) sinks.

Cost-benefit analysis A technique employed by economists in which all the financial gains or benefits of a project are compared with all the financial losses or costs of that project.

Cost-effectiveness analysis A technique employed by economists to determine how a particular goal can be achieved for the lowest possible cost.

Criteria air pollutants Those air pollutants for which the U.S Environmental Protection Agency has established national standards for ambient (surrounding) air. Standards are based on the threshold concentration for adverse effects on human health and environmental quality.

Critical-link species Species that play an essential role in ecosystem functions, regardless of the magnitude of their biomass or the trophic level at which they function.

Crop rotation A farming practice in which a field is planted to a different crop every 1–3 years.

Crossing-over During the formation of gametes, the exchange of genetic material between paired chromosomes.

Crown fire A fire that burns through the forest canopy.

Crude birth rate The annual number of births per one thousand people in a population.

Crude death rate The annual number of deaths per one thousand people in a population.

Crust The solid, relatively thin, outermost shell of the planet; source of rock, mineral, and fuel resources.

Cultural eutrophication Acceleration of the natural nutrient enrichment process of surface waters by human activities that discharge nutrient-rich wastes.

Cycling rate Amount of material that moves from one environmental reservoir to another within a specified period of time.

Cyclone collector A device that removes particulates from industrial air emissions by inducing gravitational settling.

Debt-for-nature swap A program in which a more-developed nation forgives the debt of a less-developed nation in return for the preservation of some natural resource within the debtor nation.

Decomposers Microorganisms, primarily bacteria and fungi, that utilize detritus (the remains of plants and animals) as a source of energy and nutrients.

Deep-well injection Disposal of liquid wastes by pumping them under pressure into deep subsurface cavities and pore spaces in bedrock.

Demographic transition A pattern of change in birth and death rates in a population from high birth and death rates to low birth and death rates.

Denitrification Conversion of nitrate (NO_3^-) to nitrogen gas (N_2) by a specialized group of bacteria.

Density-dependent factor Interactions such as predation, parasitism, and competition, whose influence on population size varies with population density.

Density-independent factor Environmental factors, such as weather, whose influence on population size is little affected by population density.

Deposition Process whereby water vapor changes directly to the solid (ice) phase without first becoming a liquid.

Desalination Process whereby the salt content of water is greatly reduced so that the water can be used for human consumption or irrigation.

Detoxification mechanisms Means by which a toxic substance in the body is eliminated or converted to a less toxic substance.

Detritivore An organism that consumes detritus.

Detritus Freshly dead or partially decomposed remains of plants and animals.

Detritus food web A food web based on detritus.

Diaphragm A dome that covers a woman's cervix and prevents entry of sperm into the uterus.

Dissolved oxygen (DO) Oxygen (O_2) dissolved in water; the concentration is usually expressed in milligrams per liter or parts per million (ppm).

Distillation Process whereby water is purified by evaporation and subsequent condensation.

DNA The cell component that contains the genetic code.

Doubling time The time needed for a population to double in size when it grows at a constant rate.

Drainage basin The geographical region drained by a river and all its tributaries.

Dredging Process for excavating streambed or harbor sand and gravel deposits, using chain buckets and drag lines.

Earthquake Shaking of the earth's crust often caused by an abrupt release of energy when large rock masses fracture.

Ecological succession The replacement of one ecosystem by another through a sequence of colonization and replacement of species until an assemblage of species is established that is able to maintain itself on a site.

Economic injury level The population size of a pest below which no further reduction in population size is profitable.

Ecosystem A community of organisms that interact with one another and with the chemical substances and physical conditions that characterize their environment.

Ecosystem stability A measure of the ability of a community of plants and animals either to resist or to recover quickly from a disturbance.

Efficient economy An economic system in which resources are used in the best possible way to satisfy the largest number of consumer demands.

El Niño Large-scale changes in wind systems and ocean currents in the Pacific equatorial zone. Such changes influence regional fisheries as well as global weather patterns.

Electromagnetic radiation Energy that can travel through a vacuum in the form of oscillating waves (e.g., visible light).

Electromagnetic spectrum The various types of radiational energy arranged by wavelength, frequency, and energy level.

Electrostatic precipitator A device that removes particulates from industrial air emissions by inducing an electrical charge on the particulates. The changed particulates then collect on plates having the opposite electrical charge.

Emigration Migration out of a population.

Eminent domain The right of a government body to purchase land for the public good.

Endangered species A species that is in immediate danger of extinction.

Energy The ability to do work or to produce change.

Energy efficiency The fraction of the total energy input of a system that is transformed into work or some other usable form of energy.

Environment The entire assemblage of external factors or conditions that influence living organisms in any way.

Environmental heterogeneity A measure of the degree of diversity of habitats within a region.

Environmental impact statement An assessment of the potential environmental disturbance accompanying a federally-sponsored project.

Environmental patchiness A measure of the degree of diversity of habitats within a region.

Environmental science Study of the fundamental interactions between living and nonliving things with the goals of understanding how the environment works, finding solutions to environmental problems, or both.

Enzymes Protein molecules that accelerate specific chemical reactions.

Epicenter Geographical location above the point of release of earthquake energy.

Erosion The transport of soil and sediment by wind, running water, or glacial ice.

Essential amino acids Building blocks of protein that must be present in an organism's diet if it is to synthesize all of its required proteins.

Estuary An exceptionally productive coastal ecosystem where fresh water and salt water meet and mix to some extent.

Eutrophic ecosystems Ecosystems that are relatively high in fertility and biological productivity.

Eutrophication A natural process of nutrient enrichment that gradually makes lakes and reservoirs more productive.

Evaporation Change in phase of water from liquid to vapor.

Evapotranspiration Vaporization of water through a combination of transpiration by plants and direct evaporation from wet surfaces.

Evolution Changes in the genetic makeup of a population with time.

Expansional cooling The drop in temperature that accompanies the expansion of a gas. For example, air undergoes expansional cooling when it ascends within the atmosphere and the pressure on the air consequently decreases.

Experiment Procedure designed to study some phenomenon under known conditions with the goal of discovering or illustrating some scientific principle or identifying the cause of some effect or problem.

Exponential growth An increase by a constant percentage of the cumulative whole over a specific period of time.

Family planning A voluntary program of fertility control used by a couple to achieve the family size of their choice.

Fast breeder reactor A nuclear fission reactor capable of producing fissionable materials from substances (e.g., uranium-238) that normally do not undergo fission.

Filter A device composed of material (e.g., fiberglass) that is porous enough to permit flue gases to pass through while trapping particulates; a common industrial air-pollution control method.

First law of thermodynamics Energy can neither be created nor destroyed, but can change from one form to another; also known as the law of conservation of energy.

Flash cycle system A process in which superheated water is depressurized and thereby transformed into steam instantly.

Flood Flow of water over the banks of a river when the discharge exceeds the capacity of the river channel.

Floodplain management Use of floodplains in ways that minimize the hazards to human safety and property.

Fly ash Solid residue from burning coal.

Food chain A sequence of organisms, such as green plants, herbivores, and carnivores, through which energy and materials move within an ecosystem.

Food web A network of interconnected food chains.

Fossil fuels The remains of ancient plant and animal life that have been gradually transformed into coal, oil, and natural gas.

Free markets Markets in which resources are allocated and prices established on the basis of individual, voluntary exchange among producers and consumers.

Fuel-rod assemblies Containment devices for radioactive fuel within a nuclear fission power reactor.

Fungicide A chemical meant to kill fungi.

Gamete A sexual reproductive cell; an egg or a sperm.

Gene A portion of a chromosome that codes for a particular characteristic.

Gene pool All the genes of all the individual members of a given population.

Gene transfer Process whereby a beneficial gene from a species is introduced into the cells of another species.

Genetic drift A change in gene frequency determined by chance rather than by evolutionary advantage.

Genetic effects Refers to changes in the genetic makeup of sex cells due to exposure to mutagens (e.g., ionizing radiation).

Geometric growth An increase by a constant percentage of the cumulative whole over a specific period of time.

Geothermal energy Heat energy conducted from the earth's interior.

Giant oil fields Deposits of petroleum that exceed 100 million barrels.

Graphic model Approximate representation of a real system in graphic form (e.g., a map).

Gravitational settling Gravity-induced downward motion of particles suspended in air or water.

Grazing food web A food web based on living plants.

Green Revolution Popular term for the dramatic increases in crop yields per hectare that sometimes accompany the introduction of new crop varieties and the improvement of soil, water, and pest management.

Greenhouse effect The absorption and reradiation of infrared radiation by atmospheric water vapor (H_2O), carbon dioxide (CO_2), and ozone (O_3); substantially elevates the earth's surface temperature.

Greenhouse gases Substances (e.g., H_2O, CO_2, and O_3) that absorb infrared radiation within the atmosphere.

Ground fire A fire that creeps along the ground, burning only litter, herbaceous plants, shrubs, and small trees.

Ground subsidence The sinking of the earth's surface that occurs when subsurface mineral deposits or fluids are withdrawn without replacement.

Groundwater Water that fills the pore spaces or crevices within soil, sediment, or rock and can be withdrawn via wells.

Groundwater mining Withdrawal of groundwater at a rate that exceeds the natural recharge rate.

Haber process The industrial synthesis of ammonia (NH_3) from natural gas (methane, CH_4) and nitrogen gas (N_2).

Habitat island A habitat fragment that is surrounded by dissimilar habitats.

Half-life The amount of time required for the radioactivity emanating from a particular radioactive substance to decrease by one-half.

Hard water Water with high levels of calcium (Ca^{2+}) and magnesium (Mg^{2+}).

Hazardous air pollutants Pollutants that pose a serious hazard to human health and well-being.

Hazardous waste Substances that pose a serious hazard to human health and well-being.

Heavy metals A group of elements that are potentially toxic to organisms; examples are cadmium, mercury, copper, nickel, chromium, lead, zinc, and arsenic.

Herbicide A chemical compound that kills vegetation, used for weed control.

Herbivore An animal that feeds only on plants.

Hibernation A state of inactivity exhibited by some animals during winter; characterized by reduced heart and breathing rates, and a lower body temperature.

High-head sites A location characterized by a relatively great vertical drop between the headwaters and dam.

Holistic perspective The view that considers a system as a whole rather than focusing only on the individual parts.

Humus Organic material in soil that decays slowly.

Hygroscopic nuclei Condensation nuclei having a special affinity (attraction) for water molecules.

Hypothesis A conclusion, based upon scientific observations, that is tested and used to stimulate further research.

Igneous rocks Rocks formed by the crystallization and solidification of magma (or lava), either within the earth's crust or on the earth's surface.

Immigration Migration into a population from elsewhere.

Impaction Removal of particulates from the air when they strike and adhere to the surfaces of buildings and other structures.

In situ recovery Processing of a resource, such as coal or oil shale, within the natural deposit.

Inbreeding Mating between close relatives.

Incremental decision making A process used to formulate public policy; it involves small, step-by-step changes in existing laws.

Individual risk Risks, known or unknown, that a person assumes when involved in an activity (e.g., driving a car).

Inefficient economy An economic system in which resources are not used to their greatest advantage to satisfy consumer demand.

Infiltration Seepage of surface water through soil and rock layers.

Initiator A substance that causes the initial changes in DNA that may lead to cancer.

Inorganic chemicals Relatively simple substances, usually composed of elements other than carbon, including water (H_2O), oxygen (O_2), and nutrient minerals such as calcium (Ca) and phosphorus (P).

Inorganic fertilizer Plant nutrients derived from mineral deposits (phosphate or potash) or from natural gas (ammonia).

Insecticide A chemical meant to kill insects.

Integrated pest management A combination of procedures that are environmentally compatible, intended to keep a pest population below the economic injury level.

Integrated waste management A combination of waste-disposal and waste-reduction techniques appropriate for local or regional economic and environmental conditions.

Interdisciplinary A subject that crosses the usual boundaries of traditional areas of study.

Internal geological processes Volcanic activity, earthquakes, and the deformation of rock; process maintained by energy in the earth's interior.

Interspecific competition Competition for a resource among populations of two or more species.

Intraspecific competition Competition for a resource among members of the same species.

Ion An electrically charged atom or group of atoms.

Ionizing radiation Highly energetic emissions of alpha and beta particles and gamma radiation by radioactive materials.

IUD (intrauterine device) A small metallic or plastic device that is inserted into the uterus to prevent conception.

Juvenile hormone A chemical produced by an insect that initiates various changes in its life cycle.

Kerogen Organic material contained within oil shale that is a source of fuel.

Keystone species A species that plays a central role in controlling the relative abundance of other species in an ecosystem.

***K*-strategists** Species that maximize their probability of surviving to reproduce in the future, but therefore have a relatively low current rate of reproduction.

Kwashiorkor A form of protein malnutrition characterized by swollen abdomen, hair loss, failure to grow, and depressed mental ability.

Land-farming A technique for disposing of industrial toxic wastes whereby bacteria in soil break down toxins into innocuous by-products.

Land reclamation Measures that foster ecological succession on land that has been disturbed (e.g., by mining or construction).

Land trust A community growth-control strategy in which private citizens donate undeveloped land to a nonprofit conservation agency for a specified time period with the stipulation that the land not be developed.

Lava Melted rock that reaches the earth's surface either through fractures in bedrock or the vents of volcanoes.

Law of conservation of matter Matter can neither be created nor destroyed, but can change in form.

Law of energy conservation Energy can neither be created nor destroyed, but can change from one form to another; also known as the first law of thermodynamics.

Law of the minimum The growth and well-being of an organism ultimately is limited by the essential resource that is in lowest supply relative to what the organism requires for survival.

Law of tolerance For each physical condition and chemical substance in an environment, minimum and maximum levels exist beyond which no member of a particular species can survive.

LD$_{50}$ The quantity of a substance that will kill 50% of a test population when administered as a single dose.

Leachate Water that has seeped through soil, sediment, rock, or waste and contains dissolved substances.

Less-developed nations Countries that are low to moderately industrialized, with low to moderate GNP per capita and moderate to high rates of natural population increase, including most of the nations of Africa, Asia, and Central and South America.

Limiting factor Any component of the environment that limits the ability of an organism to grow or reproduce.

Linear growth Increase at a constant amount over a specific period of time.

Liquefied natural gas (LNG) Natural gas that has been liquefied by cooling to -162° C (-259° F), reducing its volume to 1/600 of normal.

Longshore current A component of water motion parallel to the shoreline, the result of coastal wave action, that supplies sand to beaches.

Lower tolerance limit The minimum level of an environmental factor, below which no member of a species can survive.

Low, medium, and high BTU gas Terms used to describe the relative energy content of synthetic natural gas.

Low-head sites A location characterized by a relatively small vertical drop between the headwaters and dam.

Macrophages Specialized, free-living cells in the air sacs (alveoli) of the lung that can engulf and digest foreign particles.

Magma Melted rock that occurs beneath the earth's surface; at the earth's surface, it is called lava.

Male sterilization A pest-control method in which the release of sterilized males greatly reduces the pest population's rate of reproduction.

Malnutrition Inadequate nutrition resulting from a diet deficient in amino acids, vitamins, or minerals or some combination of these nutrients.

Manganese nodules Deposits of manganese, copper, cobalt, and nickel found on the deep ocean bottom.

Marginal costs The extra cost of removing one less unit of pollutants.

Marine sanctuaries Areas of the coastal zone set aside to preserve certain natural resources or historical or esthetic features.

Metabolites Products that result from the metabolism of food or toxins.

Metamorphic rocks Rocks formed when a preexisting rock is modified by high temperatures, confining pressures, and chemically active fluids deep within the earth's crust.

Mimicry Superficial close resemblance in form, color, or behavior of certain organisms (mimics) to other organisms or objects, resulting in concealment from predators or some other advantage for the mimic.

Mineral (geological) A solid characterized by an orderly internal arrangement of atoms, fixed chemical composition, and definite physical properties.

Minimum viable population The smallest number of individuals that is needed to ensure the survival of an isolated population.

Mixed-market economies Economic systems, such as that in the United States, that combine private competitive enterprise with some government involvement.

Mixing layer Lower layer of the atmosphere in which air is thoroughly mixed by convection.

Monoculture Cultivation of a single crop in a particular area.

More-developed nations Highly industrialized countries with high GNP per capita and a low rate of natural population increase, including nations of Europe and North America, Russia, Japan, Australia, and New Zealand.

Mortality The death rate.

Mounded landfill A sanitary landfill formed by dumping solid waste on the ground, compacting the waste, and then sealing it with dirt to eventually produce a large hill. This technique is necessary where the water table is close to the earth's surface.

Municipal incinerator A device that burns combustible solid waste and melts certain noncombustible materials.

Mutagen A chemical substance or physical agent (e.g., radiation) capable of producing a change in an individual's chromosomes.

Mutation A random, inheritable change in the DNA sequence of a chromosome.

Mutualism A relationship between two species whereby the survival of each depends on the presence of the other.

Natality The rate of production of new individuals by birth, hatching, or germination.

Natural family planning methods A variety of procedures for planning abstinence during the transient period of a woman's greatest fertility.

Natural pesticides Chemicals produced by organisms that are injurious to their enemies.

Natural selection Differences in the rates of survival and reproduction of individuals in nature that lead to an increase in the frequency of some characteristics and a decline in the frequency of others.

Negative externalities Undesirable side effects (such as industrial pollutants, noise, and litter) of activities that do not involve us directly.

Neritic zone The region of the ocean that extends from the high-tide line to the outer edge of the continental shelf.

Net primary production The rate at which plants accumulate energy by means of growth and reproduction during a specific period of time.

New towns Planned communities designed to operate efficiently, protect the environment, and be as self-sufficient as possible.

Nitrification Conversion of ammonia (NH_3) to nitrate (NO_3^-) by specialized bacteria.

Nitrogen cycle Circulation of nitrogen among various reservoirs of the environment; major reservoirs include the atmosphere and ocean.

Nonattainment areas Regions that have failed to meet federal ambient air-quality standards.

Nonbiodegradable waste Materials that resist degradation by microorganisms.

Nonpoint sources Sources of pollutants in the landscape; for example, agricultural runoff.

Nonrenewable resources Resources that are not regenerated (at least within the time frame of humanity).

Nonreserve resources Rock and mineral deposits that are subeconomic, hypothetical, or speculative.

No-observed-effect level (NOEL) Highest dose level of a toxin that produces no undesirable effects in a test population.

Normal climate Average weather plus extremes in weather for some location over some period of time.

No-till cultivation Agricultural practice whereby loosening of the surface soil, planting, and weed control are combined into one operation. This technique reduces soil erosion.

Nuclear fusion The fusing together of atomic nuclei of light elements to form heavier elements, thereby releasing energy.

Numerical model Approximate representation of a system as one or more mathematical expressions.

Ocean disposal Dumping of waste such as dredge spoils or sewage sludge at sea.

Oceanic zone The region of open ocean that lies beyond the continental shelf.

Oil shale Deposits containing organic materials that yield oil-like products when heated to high temperatures.

Old-field succession Ecological succession following the abandonment of farmland.

Oligotrophic systems Ecosystems that are low in fertility and biological productivity.

Omnivore An animal that feeds on plants and other animals.

Oncogene Gene in DNA that if activated can cause cancer.

Open pit mines Surface mines where the overburden is removed from a large area so that rock and mineral deposits can be excavated to considerable depth; usually used to extract metals such as copper and iron.

Optimum concentration The level of an environmental factor that sustains the maximum population of a species.

Ore An economically important mineral deposit that occurs naturally within the earth's crust.

Organic chemicals Materials that have carbon as a primary component; many are produced by or derived from organisms.

Organophosphate A type of synthetic insecticide.

Orographic lifting Upward motion of an air mass as it encounters the windward slopes of hills or mountains.

Overburden Vegetation, soil, sediment, rock or anything else that overlies a mineral or fuel deposit.

Overfishing Reduction of fish stocks by harvesting more fish than are produced in a given time.

Overgrazing Degradation of grasslands or rangelands by stocking more livestock than the available forage can sustain.

Oxygen cycle Circulation of oxygen among various reservoirs of the environment; major reservoirs include the atmosphere and oceans.

Oxygen sag curve Decline and subsequent recovery of dissolved oxygen concentrations downstream from point sources of organic waste.

Oxygenates Vehicle fuels with a relatively high oxygen content (e. g., methanol) that burn more cleanly than traditional blends of gasoline.

Ozone hole Area of reduced ozone concentration within the stratosphere; best-known occurrence is over Antarctica in spring.

Ozone shield A layer of ozone in the stratosphere that absorbs potentially harmful intensities of ultraviolet radiation.

Parasitism An interaction in which one organism (the parasite) obtains energy and nutrients by living within or upon another organism (the host).

Partial cutting A continuum of activities that range from selectively cutting only a few trees to harvesting most, but not all, of the trees on a site.

Passive solar system Features designed into a building to reduce the amount of energy required for heating and cooling (e.g., locating windows to minimize solar capture in summer and maximize solar capture in winter).

Pathogen An organism that causes disease.

Permeability The capability of soil, sediment, or rock to transmit water or other fluid.

Pest Species that compete with humans for food or fiber, transmit pathogens, cause disease, or otherwise threaten human health or well-being.

Pesticide A chemical used to kill pests.

Pesticide resistance The evolution of pest populations to the point that they are no longer harmed by a particular pesticide.

pH scale A system to specify the degree of acidity and alkalinity; a pH of 7 is neutral, below 7 is acidic, and above 7 is alkaline.

Phosphorus cycle Circulation of phosphorus among various reservoirs of the environment. Major reservoirs include soil, sediments, and phosphate-bearing rock.

Photosynthesis The process whereby green plants transform light energy into food (chemical) energy.

Photovoltaic system Device that uses semiconducting materials to convert sunlight directly to electricity.

Physical model An approximate representation of a real system, usually on a reduced scale.

Phytoplankton Free-floating, microscopic aquatic plants.

Pioneer species Organisms present in the initial stage of ecological succession.

Pioneer stage The initial stage in ecological succession.

Placer deposits Heavy-mineral deposits found mixed with sand and gravel in streambeds or coastal areas.

Planetary albedo The percentage of solar radiation reflected and scattered by the entire earth-atmosphere system back to space.

Planned obsolescence The deliberate design of consumer goods with a limited life span to encourage consumption.

Point sources Discernible conduits, such as pipes, chimneys, ditches, channels, sewers, tunnels, or vessels, that discharge pollutants.

Poleward heat transport The circulation of heat from tropical latitudes into middle and high latitudes, primarily by the exchange of air masses.

Pollutant A type of material or a form of energy that adversely affects the well-being of organisms or the natural processes upon which they depend.

Pollution An environmental disturbance that adversely affects the well-being of organisms or the natural processes upon which they depend.

Population A group of individuals of the same species occupying the same geographic region at the same time.

Population momentum The tendency of a growing population to continue to grow for some time.

Porosity The volume of open spaces within soil, sediment, or rock.

Positive crankcase ventilation (PCV) A device that reduces hydrocarbon emissions from motor vehicles by channeling crankcase blow-by gases back through the engine.

Positive externalities Benefits that accrue to many people as a consequence of the actions of others.

Precipitation The return of water from clouds to the earth's surface in the form of rain, snow, ice pellets, or hail.

Predation An interaction in which one organism (the predator) kills and eats another organism (the prey).

Predisposal treatment Alteration of the properties of a hazardous waste prior to disposal so that the waste no longer poses a hazard.

Prescribed burns The setting of ground fires to reduce fuel buildup or to prevent invasion of successional species or both.

Primary air pollutants Substances introduced into the atmosphere in sufficient concentrations to pose serious hazards to environmental quality.

Primary air quality standards Maximum concentrations of criteria pollutants in ambient air set by the U.S. Environmental Protection Agency and based on potential health risk to humans.

Primary succession Ecological succession that begins where no soil exists (bare rock, for example).

Private goods Goods owned and/or controlled by individuals or companies.

Producers Organisms, mainly green plants, that manufacture their food by utilizing raw materials from air and soil.

Promoter Chemical or physical agent that activates genes in DNA to cause cancer.

Protective coloration The physical appearance of an organism that allows it to blend unnoticed with its background.

Public goods Things and services (such as police protection) that benefit a large number of people.

Public lands Lands (e.g., forests, parks, and wildlife refuges) owned and managed by local, state, or federal governments.

Public policies Governmental statements of what should or should not be done.

Quarries Small-scale surface mines used for the recovery of rock.

Radiational cooling Temperature drop of a substance that accompanies the emission of radiation.

Radiational heating Temperature rise of a substance that accompanies the absorption of radiation.

Radiational temperature inversion A temperature inversion formed by cooling of a surface air layer by emission of infrared radiation so that the coldest air is at the earth's surface and air temperature increases with altitude.

Radioactive decay The shift of an unstable nucleus to greater stability; accompanied by emission of highly energetic particles and/or radiation.

Radioactive isotopes Forms of elements that emit highly energetic particles and/or radiation.

Rangelands Grazing lands that are essentially unsuited for rain-fed crop cultivation or forestry.

Ranks of coal Stages in the sequence of change from peat to lignite to bituminous coal to anthracite coal; in general, the

energy yield per unit weight increases with rank.

Rate of natural population change The rate at which a population is increasing or decreasing in a given unit of time due to a surplus or deficit of births over deaths.

Reclamation Separation of useable resources (including fuels) from a waste stream.

Recovery plan Management plan to protect and increase the population size of endangered or threatened species.

Recycling The recovery and reuse of resources from discarded products.

Red tide Reddish-brown discoloration of sea water caused by a population explosion of certain phytoplankton that may produce toxins.

Reflection The throwing back of radiation by some surface.

Refuse-derived fuel (RDF) Combustible fraction of solid waste that is used as a source of heat.

Relative humidity The ratio of the concentration of water vapor in air to the concentration of water vapor in saturated air at the same temperature; usually expressed as a percentage.

Remedial action plan Plan that establishes priorities for water pollution cleanup and prevention efforts in a designated area.

Renewable energy resources Resources that are regenerated within a reasonable length of time.

Replacement level fertility The number of births per woman that must occur in a population so that it neither increases or decreases.

Reproductive strategies A continuum of strategies employed by species to balance the use of resources required for current reproduction against the probability of future survival.

Reserve That portion of a rock, mineral, or fuel deposit that can be extracted economically.

Reserve base That portion of a resource that could be mined given current mining and production technology.

Reservoir rock Porous and permeable rock that contains water and/or petroleum.

Resource conservation Measures taken to reduce the rate at which resources are depleted.

Resource recovery The recovery of materials, energy, or both from a waste stream.

Respiratory system Pathway by which air and air pollutants enter the human body, and waste gases such as carbon dioxide (CO_2) leave the body.

Reverse osmosis A desalination process whereby water under pressure is forced through a thin plastic membrane, leaving behind dissolved substances.

Richter scale An open-ended measure of energy released in an earthquake.

Riparian areas Lands adjacent to streams and rivers where vegetation is strongly influenced by the supply of water.

Risk The chance of injury, disease, or loss.

Risk assessment Critical examination of activities that lead to injury or loss with the objective of reducing exposure to harmful activities or substances.

Risk-benefit analysis An analytic approach employed by economists in which the risks of a product or activity are weighed against its benefits to determine if the product or activity should be allowed.

Rock cycle The transformation of one rock type into another as a consequence of changes in the rock's geological environment.

R-rating A measure of the insulating ability of some material, such as window glass or wall board; the higher the R-rating, the better the insulation.

r-strategists Species that maximize their current rate of reproduction and therefore do not survive for long periods of time.

Runoff Water that flows over the earth's surface as rivers or streams.

Saltwater intrusion Seepage of saline groundwater into a reservoir of fresh groundwater, usually along a low-lying coastal plain.

Sand and gravel pits Small surface excavations used for the recovery of sand and gravel.

Sanitary landfill A landfill consisting of compacted solid waste sealed between layers of clean dirt.

Sanitary sewer A system that transports raw sewage to a centralized treatment facility.

Saturation The condition in which a medium such as water or air contains its maximum possible concentration of some dissolved substance.

Scattering The dispersal of radiation in random or almost random directions when the radiation is intercepted by particles.

Scavenging (of pollutants) Removal of pollutants from air through washout by rainfall and snowfall.

Schistosomiasis A debilitating waterborne disease, common in many less-developed nations, caused by parasitic worms that live in calm water, especially irrigation canals.

Scientific method A systematic form of inquiry that involves observation, speculation, reasoning, and the development and testing of hypotheses.

Scientific model An approximation or simulation of a real system: omits all but the most essential variables of the system.

Scrubbing An industrial air-pollution control strategy that removes soluble gases from effluents, usually by channeling the effluent through some chemical exchange medium.

Second law of thermodynamics In every energy transformation, some energy is converted to heat and is thereafter unavailable to do further useful work.

Secondary air pollutants Products of reactions among primary air pollutants; photochemical smog and acid deposition are examples.

Secondary air-quality standards Maximum concentrations of criteria pollutants in ambient air set by the U.S. Environmental Protection Agency, designed primarily to minimize damage to property and crops.

Secondary recovery Measures that substantially increase the yield of petroleum from subsurface reservoir rock.

Secondary succession Ecological succession that begins in a disturbed area where soil still exists.

Secure landfill A landfill designed according to rigid specifications for long-term storage of hazardous wastes.

Sedimentary rocks Rocks formed by the compaction and cementation of particles (sediments) of abiotic or biotic origin.

Sediments Fragments of rocks, minerals, or organic matter.

Seed bank Storage of seeds to save genes for future uses.

Self-thinning Mortality associated with severe intraspecific competition among plants.

Semiconductor Special material that produces a direct electrical current when exposed to sunlight; used in photovoltaic cells.

Separated sewer system A network of two pipes, one to transport surface runoff to a waterway and the other to transport sewage to a treatment facility.

Sex attractant A chemical released by female insects that makes it easier for males of the same species to locate them.

Shelterbelts Rows of trees or shrubs that reduce wind erosion of soil; also known as windbreaks.

Sigmoid growth curve An S-shaped curve that illustrates one model of population growth when some environmental factor begins to limit that growth.

Sink population A population that would become extinct without immigration to compensate for the excess of deaths over births.

Slash-and-burn agriculture A farming method used mostly in the tropics; involves clearing overgrowth on a several-hectare plot, burning residue, planting crops for several years, and abandoning the site to allow the forest to reinvade.

Social hierarchy A pattern of dominant-subordinate relations among members of a population.

Societal risk Involuntary risks to society that governments often attempt to regulate.

Soil salinization An accumulation of salts at the soil surface due to excessive evaporation of water.

Solar high-temperature system A system that collects sunlight to produce temperatures that exceed 100° C (212° F).

Solar low-temperature system A system that collects sunlight to produce temperatures in the 65° C to 100° C (150° to 212° F) range.

Solution mining Removal of deep deposits of soluble minerals by pumping water down an injection well to dissolve the minerals and retrieving the solution via extraction wells.

Somatic effects The direct impact of exposure to hazardous materials or high levels of radiation on cells; often includes the conversion of normal cells to cancer cells and other changes that reduce life expectancy.

Source population A population that produces excess individuals.

State implementation plans (SIPs) Measures defined by individual states in order to meet federally mandated air-quality goals.

Sterilization A procedure, usually surgical, that alters the reproductive system, thereby preventing pregnancy. Exposure to radiation and certain chemicals also may produce sterility.

Storm sewer Pipe system designed to carry runoff from rain and snow melt.

Stratosphere The subdivision of the atmosphere that lies between the troposphere and an altitude of about 50 kilometers (30 miles); site of the ozone shield.

Stream channelization Straightening and ditching of meandering rivers for the purpose of transporting water more rapidly downstream, thereby reducing the threat of upstream flooding.

Stream discharge Volume of water that passes a given point in a stream in some period of time (e.g., cubic meters per second).

Strip cropping Planting technique used on sloped surfaces that reduces soil erosion by alternating a crop of low soil-anchoring ability with a crop of high soil-anchoring ability.

Strip mining Surface mining method usually used for recovery of near-surface coal deposits. Depending on the terrain, it may be of the area or contour type.

Sublimation Process whereby ice or snow changes directly to water vapor without first becoming a liquid.

Subsidence temperature inversion An elevated temperature inversion formed as air sinks over a wide area and is warmed by compression.

Substitute natural gas High-BTU gas that is produced via coal gasification.

Subsurface mining Recovery of deposits by excavation of subsurface tunnels or by drilling.

Supercity A coalescence of adjacent cities and towns with a combined human population of at least one million.

Superheated water Water heated under pressure that remains liquid even though its temperature tops its normal boiling point.

Suppressor gene Segment in DNA that prevents expression of cancer by cells.

Surface geological processes Weathering and erosion by wind, water, and glaciers that help shape the earth's surface; ultimately maintained by solar energy.

Surface mining Extraction of near-surface deposits of rock, minerals, or fuels following removal of the overburden.

Sustainable agriculture Farming practices that promote long-term food sufficiency without destroying the natural resource base.

Synergism An interaction in which the total effect is greater than the sum of two or more effects acting independently.

Synthetic fuels (synfuels) Gaseous and liquid fuels produced primarily from coal and oil shale.

Synthetic pesticide Chemical formulated in a laboratory and designed to kill pests.

System Assembly of components that behave (function and interact with one another) in some regular and predictable manner.

Tailings The waste residue of ore processing.

Tar sands Natural deposits of sand coated with petroleum tar.

Temperature inversion A temperature profile in the atmosphere characterized by an increase of temperature with altitude.

Teratogen A chemical substance or physical agent (such as radiation) capable of causing a birth defect.

Territorial behavior Any activity whereby an individual defends an area against intruders of the same species.

Test system That which is perturbed in a controlled experiment.

Theory of island biogeography Biodiversity on an island remains more or less constant because the extinction rate on the island is balanced by new species migrating to the island.

Thermal plume Area of heated discharge water usually associated with electric power plants.

Thermal pollution Emission of heated water into a body of water, such as a lake or river, that stresses aquatic organisms.

Threatened species A species that is still abundant in some parts of its range, but whose continued existence is in question because its numbers have declined significantly in other areas.

Tolerance limits The range of an environmental factor beyond which no members of a particular species can survive.

Total fertility rate (TFR) The average number of children that women in a population bear in their lifetime.

Toxicity A measure of the poisonousness of a substance.

Toxicology The scientific study of poisons.

Toxins Chemicals (poisons) that cause disease or dehabilitation; can be natural or synthetic in origin.

Transgenic plants and animals Organisms with genes that were transferred to them via biotechnology.

Transpiration The loss of water vapor from plants through leaf pores.

Transuranic wastes Radioactive alpha-emitting elements that are heavier than uranium.

Trench landfill A sanitary landfill formed by digging wide trenches for waste disposal, then daily covering the compacted waste with dirt that was excavated from the trench.

Trophic level The feeding position occupied by a given organism in a food chain, measured by the number of steps it is removed from producers.

Troposphere The lowest subdivision of the atmosphere, and the site of most weather events.

Tsunami A sea wave often triggered by submarine earthquakes that builds to destructive heights in certain coastal areas.

Tubal ligation Surgical alteration of female reproductive system to prevent pregnancy.

Unconfined aquifer A groundwater-bearing rock or sediment layer overlain by permeable earth materials and recharged locally.

Undernutrition Inadequate nutrition resulting from a diet deficient in calories.

Upper tolerance limit The maximum level of an environmental factor, above which no member of a species can survive.

UVB Portion of ultraviolet radiation that contributes to the development of skin cancer.

Vasectomy Surgical alteration of the male reproductive system to prevent pregnancy.

Vector An organism (e.g., tick, mosquito) that transmits a pathogen from one organism to another.

Volcano A mountain or hill composed of the eruptive products of igneous activity.

Voltage Difference in electrical potential.

Warm front Narrow zone of transition between warm and cold air masses, in which the warm air displaces the cold air.

Waste exchange An organization that inventories waste chemicals and facilitates their reuse by other industries.

Waste reduction Measures such as recycling that reduce the amount of waste that must be disposed of.

Water conservation Measures that reduce the per capita use of fresh water.

Water cycle Circulation of water among oceanic, atmospheric, and terrestrial reservoirs.

Water table The upper surface of the groundwater reservoir.

Watershed The geographical region drained by a river or stream and its tributaries.

Watershed transfer Movement of fresh water from one watershed to another, usually via canals.

Wattage Measure of the rate of energy consumption of an electrical device.

Weather The state of the atmosphere at some place and time described in terms of such variables as temperature, cloudiness, and precipitation.

Weathering The chemical decomposition and mechanical disintegration of solid rock into sediments.

Westerlies Bands of winds encircling middle-latitudes and blowing in a wavy pattern from west to east.

Wetlands Transitional areas between terrestrial and aquatic ecosystems that are flooded for part or all of the year, but not necessarily every year.

Whole-body exposure The result of exposure to high-energy radiation (gamma rays) that readily penetrates an entire organism.

Zero population growth (ZPG) A state in which a population ceases to grow.

Zone of aeration A layer of soil, sediment, or rock in which pore spaces are filled with a mixture of air and water.

Zone of saturation A layer of soil, sediment, or rock in which the pore spaces are completely filled with water; the groundwater reservoir.

Zooplankton Microscopic, free-floating aquatic animals.

Index

Export of waste, 447–48
Extinction. *See* Species extinction
Exxon Valdex oil spill, 300, *301,* 302

Facilitation model of species replacement, 73
Family planning, **135**
Famine, 149–51
Fast breeder nuclear reactor, **403**
Federal Flood Disaster Protection Act, 468
Federal government, United States. *See also*
     Environmental legislation
  agencies with environmental responsibilities,
     19, A4
  coastal land use regulations, 466–67
  land ownership by, *194* (*see also* Public lands)
  policies of (*see* Public policy)
Federal Insecticide, Fungicide, and Rodenticide
     Act, 18
Female reproductive system, *129, 130, 131*
Fertility control, 125–34
  current status and future of, 133–34
  effectiveness of, 126–32
  estimated use of select methods of, 127
    (table)
  failure rates of select methods of, 127 (table),
    132
  human female reproductive system and,
    129–31
  human male reproductive system and, 128–29
  methods of, 127 (table)
  moral and ethical considerations in, 133
  safety of, 132–33
  use of, in select world regions, *140*
Fertilizers, 161–64
  animal manure as, 161–62
  inorganic, 161
  water pollution caused by, 56, 283
Fetal development, effect of toxins on, 248–49
Fires
  ecological succession and, 71, *72,* 74
  policies on managing, in national parks,
    205–8
  in Yellowstone ecosystem, *72,* 205–8
First law of thermodynamics, **37**
Fish
  acid rain effects on, 349
  bioaccumulation of PCBs in, *240,* 241
  contaminated, 238
  salmon as endangered species, 234, A15
  standards for contaminants found in, 241
    (table)
Fish and Wildlife Service, U.S., 193
Fish food resources, 167–70
  aquaculture and mariculture of, 169–70
  in Canada, A15–A16
  distribution of world's fisheries, *169*
  global catch, *167*
  overfishing, 168
  pollution and habitat destruction, 169
Flash cycle systems, **430**
Fleming, Alexander, 10
Floc, 292
Flood hazards, 467–70
Floodplain management, **468**–69
Flue-gas desulfurization (FGD), **359**–60
Fly ash, **438**

Food
  climate change and procurement of, in Norse
    Greenland, 332
  energy content of, 151
  minerals in, 153 (table)
  toxins in, 239
  types of molecules in, 151
  vitamins in, 152 (table)
  waste produced from metabolism of, 6, 8
Food, Drug, and Cosmetic Act, 238
Food and Drug Administration (FDA), 238, 241
Food chain, **34**
Food production. *See* Agriculture; Crop
    production; Livestock
Food resources, 148–72
  cultivating new land to increase, 153–58
  famine and lack of, 149–51
  harvesting oceans for, 167–70
  hunger and lack of, 151–53
  increased, as factor in human population
    growth, 114
  increasing production of, on cultivated land,
    158–67
Food web, 34–41
  agricultural, 40–41
  bioaccumulation of toxins in, *240*–41
  as conceptual model, 10
  defined, **34**
  efficiency of, 37–41
  energy flow into and within, 37
  model of, *34*
  trophic levels of, 35
  types of, 35–37
Foreign workers, 137, *139*
Forest and Rangeland Resources Planning Act of
    1974 (RPA), 194
Forests. *See also* Tropical rain forests
  Canadian, A14–A15
  as carbon reservoirs, 57, *58*
  clear-cutting, 195, *196*
  decline and death of, 69–70
  fire in, 71, *72,* 74
  habitat destruction in, 222
  national, 194–96
  partial cutting, 196
  potential effects of elevated carbon dioxide
    on, 59–60
  size reduction of U.S., 40
  succession to climax, 70–71
  vertical stratification of vegetation in
    deciduous, *104*
Forest Service, U.S., 193, 194–96
Formaldehyde, 246
Fossil fuels, **369**–71. *See also* Coal; Natural gas; Oil
  air pollution caused by combustion of, 343
  carbon cycle and production of, **43,** *44*
  consumption of, 393, 395
  greenhouse gases and, 333, 334
  supplies of, *396*
Freedom of Information Act (FOI), E8
Free markets, **21**–22
Freshwater, 262. *See also* Water cycle; Water
    resources
  enhancing supplies of, 275–79
  groundwater (*see* Groundwater)
  human activities affecting quality of, 283, *285,*
    286
  lakes, 274–75

laws on, 274
  runoff in rivers and streams, 268, 271–74
Fuel cycle, coal vs. nuclear, *405, 406*
Fuel-rod assemblies in nuclear power plants, **452**
Fuels. *See also* Coal; Fossile fuels; Oil; Natural gas
  animal manure as, *162*
  biofuels, 427–28
  ethanol, 427–28
  nuclear, 371, 403
  oxygenated, 357, 428
  refuse-derived, 444–45
  synthetic, 399, **401**–3
Fungicides, **177**

Gametes, **79,** *80*
Gamma rays, 318, 319, 449, *451*
Gasoline
  conserving, 419 (table)
  consumption patterns, *418*
  as groundwater pollutant, *306,* 307 (table)
Gastrointestinal tract, human, *250*
Gender and tolerance limits, 67
Gene, **81**–82
  oncogene, **246,** *247*
  suppressor, **246,** *247*
Gene pool, **218**
  inbreeding in, 96, 228
General Mining Law of 1872, 386
Genetic drift, **228**
Genetic effects of ionizing radiation, **449**
Genetic variability
  evolution and, 78, 79–80
  excessive inbreeding and, 96, 228
  in food crops, 167
Gene transfer, 165, **183**
Geology, 9
Geometric growth of populations, **90.** *See also*
    Exponential growth of populations
Geothermal energy, **429**–30
  in Hawaii, A5–A6
Giant oil fields, **397**
Glaciation, North American, *221,* 330, *331, 368*
Glacier National Park, *199*
Global warming, 10
  agriculture and, 159
  coastal zones and effects of, 61, 467
  electricity generation and, 409
  greenhouse effect and, 323–24, 333–35
  mean temperature trends, *61*
  models of, *12*
  potential effects of elevated carbon dioxide
    and, 59–61
  as public policy subject, 16
Global water budget, 267–68
Governmental policy. *See* Public policy
Government regulation, environment-related,
    22–23. *See also* Environmental legislation
Grand Canyon National Park, 203, *204*
Graphic model, **10,** *11*
Gravitational settling as air cleansing process, **355**
Grazing food webs, **35**
Great Barrier Reef of Australia, 230, *231*
Great Lakes
  introduction of sea lamprey into, 227
  PCB bioaccumulation in, *240,* 241
  toxin in fish of, 238
  water quality and activities near, 283, *284,* A16

Geographical Index